# GROUND CHEMISTRY – IMPLICATIONS FOR CONSTRUCTION

PROCEEDINGS OF THE INTERNATIONAL CONFERENCE ON THE
IMPLICATIONS OF GROUND CHEMISTRY AND MICROBIOLOGY FOR
CONSTRUCTION UNIVERSITY OF BRISTOL/UK/1992

Ground Chemistry Implications for Construction

Edited by
A. BRIAN HAWKINS

A.A.BALKEMA/ROTTERDAM/BROOKFIELD/1997

Authorization to photocopy items for internal or personal use, or the internal or personal use of specific clients, is granted by A.A. Balkema, Rotterdam, provided that the base fee of US$1.50 per copy, plus US$0.10 per page is paid directly to Copyright Clearance Center, 222 Rosewood Drive, Danvers, MA 01923, USA. For those organizations that have been granted a photocopy license by CCC, a separate system of payment has been arranged. The fee code for users of the Transactional Reporting Service is: 90 5410 866 5/97 US$1.50 + US$0.10.

Published by
A.A. Balkema, P.O. Box 1675, 3000 BR Rotterdam, Netherlands
Fax: +31.10.4135947; E-mail: balkema@balkema.nl; Internet site: http://www.balkema.nl

A.A. Balkema Publishers, Old Post Road, Brookfield, VT 05036-9704, USA
Fax: 802.276.3837; E-mail: info@ashgate.com

ISBN 90 5410 866 5

© 1997 A.A. Balkema, Rotterdam
Printed in the Netherlands

FOREWORD

During the 1970s and 80s, there was an increasing awareness of the importance of ground chemistry in construction. The disturbance of the status quo as a consequence of engineering works may give rise to problematic situations, create new problems or exacerbate existing ones. For example, the change in conditions may lead to chemical reactions inducing aggressivity or the creation of environments particularly conducive to the proliferation of bacteria.

Of concern also is the disposal of contaminated materials and the use of contaminated land. Again, not only the immediate situation but the chain of reactions which may take place in the short or long term needs to be considered. Current work is addressing not only how to control unwanted consequences of ground chemistry but the way in which chemical and bacterial activity can be positively used, in prevention and remediation.

The subject is one of immense importance to both engineers and environmentalists. In bringing together representatives of the various disciplines closely involved in different aspects of the **Implications** of changes in **Ground Chemistry** for engineering **Construction**, it is hoped this conference will be the first of many such occasions when practising engineers and scientists can come together, not only to present their own case histories and their current research topics but for a lively exchange of views.

The Editor wishes to express his grateful thanks to David Greenwood, Fynn Jardine and Ken Tiller who joined him to act as an informal "steering committee" and gave so freely of their advice, enthusiasm and support.

Thanks are also offered to Marian Trott who undertook the administration and helped with the editing of the scripts and to the many postgraduate students who gave unsparingly of their time to assist in the preparation of the pre-conference proceedings and during the meeting itself.

Particular thanks are expressed to Marcus Hawkins and Robbie Narbett who worked so hard with both scripts and diagrams to produce the final proceedings.

CONTENTS

SESSION 1

1-1 Compressibility of clay-water systems 3
R.H. Ottewill, U.K.

1-2 Inside mudrocks - a review of the petrology of siliciclastic mudrocks 17
H.F. Shaw, U.K.

1-3 Microbiology of soils 35
R. Campbell, U.K.

SESSION 2 - FOUNDATIONS

2-1 Understanding sulphate generated heave resulting from pyrite degradation 51
A.B. Hawkins & G.M. Pinches, U.K.

2-2 Some geotechnical problems associated with pyrite bearing mudrocks 77
J.C. Cripps & R.L. Edwards, U.K.

2-3 Investigation of the effects of bacterial action on the chemistry and mineralogy of pyritic shale 89
S.D. Jackson & J.C. Cripps, U.K.

2-4 The generation of sulphates in the proximity of cast in situ piles 101
A.B. Hawkins & M.D. Higgins, U.K.

2-5 Roadford Dam: geochemical aspects of construction of a low grade rockfill embankment 111
S.E. Davies & J.M. Reid, U.K.

2-6 The assessment of volumetric loss for clay/sand and clay/clay mixes 133
S.H. Al-Jassar & A. B. Hawkins, U.K.

2-7 Effect of some chemical additives on the strength development of soil-cement 147
R. Angelova, Bulgaria

2-8 Methane generation from void formers used in foundations on clay sites 161
R.W. Johnson, U.K.

2-9	Gaseous pollutants and their implications for site safety J.S. Butterworth, D.P. Creedy & J.S. Edwards, U.K.	167
2-10	Hydrogen sulphide generation beneath a deep basement in Gault Clay D.F.T. Nash, M.L. Lings & S.A. Jefferis, U.K.	187

SESSION 3 - TRANSPORT

3-1	Influence of groundwater chemistry and motion on highway construction materials W.J. French, U.K.	199
3-2	Physico-chemical aspects of soluble salt damage to thin bituminous road surfacing B. Obika, R.J. Freer-Hewish & D. Newill, U.K.	211
3-3	Generation of acid groundwater beneath City Road, London N.S. Robins, D.G. Kinniburgh & M.J. Bird, U.K.	225
3-4	Biological stability of geosynthetics in critical earth structures D.G. Bright, U.S.A.	233
3-5	The physical and chemical characteristics of geotextiles and their effects on in situ degradation P.R. Rankilor, U.K.	253
3-6	The physical environments of geotextiles in civil engineering structures P.R. Rankilor, U.K.	277
3-7	Evaluation of soil chemistry and its role in highway embankment erosion - a case study R.K. Srivastava, A.V. Jalota, R.P. Tiwari, T. Nath & A.K. Sahu, India	283
3-8	The influence of calcite on the index properties of Triassic mudrocks J. Collins & A.B. Hawkins, U.K.	293
3-9	Contribution of sulphides and carbonates to the weathering of sedimentary rocks T. Oyama, M. Chigira & T. Sidahara, Japan	309
3-10	Strength characteristics of root systems in mudstone J-W. Chen & D-H. Lee, Taiwan	321
3-11	Engineering properties of mudstone-lime-slag mixtures D-H. Lee & Y-M. Tien, Taiwan	329

SESSION 4 - MARINE & ESTUARINE

4-1 An overview of the factors responsible for the degradation of materials exposed to the marine/estuarine environment 337
A.K. Tiller, U.K.

4-2 Biofouling and microbial corrosion in estuarine waters 355
R.G.J. Edyvean & H.A. Videla, U.K./Argentina

4-3 Susceptibility of steel piles to corrosion in the marginal lands of the Niger Delta 369
S.C. Teme, Nigeria

4-4 The influence of microbial assemblages on sediment erodibility: an engineering perspective 383
D.M. Paterson, U.K.

4-5 Organic residues as a factor in the plasticity of clay soil from the SERC soft clay research site, Bothkennar 401
M.A. Paul, U.K.

4-6 Chemical stabilisation of estuarine alluvium 413
A.B. Hawkins, J.C. Lonsdale & R.W. Narbett, U.K.

4-7 Laboratory study of soil-fluid interaction behaviour 431
R.K. Srivastava, A.V. Jalota & B. Singh, India

SESSION 5 - ENVIRONMENTAL

5-1 Performance of inactive barrier clays in contact with municipal solid waste leachate 439
R.M. Quigley, Canada

5-2 Assessment of chemical buffering capability of a micaceous soil 451
R.N. Yong, B.K. Tan & A.M.O. Mohamed, Canada/Malaysia

5-3 A European perspective on landfill engineering 463
A. Street, U.K.

5-4 Classification of excavated materials for disposal purposes - a case history on the analysis of ground chemistry 473
C. Mirza, Canada

5-5 Physical and chemical reactions between seepage ground water and fault gouge within a dam foundation rock mass 481
Z. Huyuan & H. Wenfeng, China

5-6	Factors controlling the solubility of substances in natural waters A. Sahinci & A. Turkman, Turkey	489
5-7	Microbiological aspects of the generation of impermeability in high sand content greens infested with black plug layering D.R. Cullimore & S. Nilson, Canada	499
5-8	Biogenic sulphuric acid attack on concrete in sewer environments T. van Mechelen & R. Polder, Netherlands	511
5-9	Determination of the chemical resistance of cements using small scale accelerated tests V. Paul & S. Uberoi, U.K.	525
5-10	Calcium aluminate mortars and concretes: an application to sewer pipes in harsh environments T. Dumas, France	541
5-11	The influence of geoenvironmental dissolved sulphates on the swelling process in concrete W. Prince & R. Perami, France	553
5-12	Damage to concrete caused by alkaline metal hydroxides W. Prince & R. Perami, France	559
5-13	Standard hydrogeochemical models and their possible application to engineering construction problems involving ground water J.H. Tellam & J.W. Lloyd, U.K.	569
5-14	Development of a computer assisted program to project the well plugging risk index (WPRI) for existing or planned water wells D.R. Cullimore & G. Alford, Canada/U.S.A.	591
5-15	Chemical and microbiological effects on construction dewatering systems W. Powrie, T.O.L. Roberts & S.A. Jefferis, U.K.	607
5-16	Chemical erosion in lateritic soil: a case study M.R. de Ruijter, Netherlands	617
5-17	Ultrasonic monitoring of the acid rain impact on building stones J. Pininska & P. Lukaszewski, Poland	623
5-18	Environmental biochemical decay control on Jeronimos Monastery cloister, Lisbon, Portugal L. Aires-Barros & A.M. Mauricio, Portugal	633

Concluding remarks
D.A. Greenwood, U.K. 647

Reference index 651

SESSION 1

1-1	Compressibility of clay-water systems *R.H. Ottewill*	3
1-2	Inside mudrocks - A review of the petrology of siliciclastic mudrocks *H.F. Shaw*	17
1-3	Microbiology of soils *R. Campbell*	35

1-1 COMPRESSIBILITY OF CLAY-WATER SYSTEMS

R.H. OTTEWILL
School of Chemistry, University of Bristol, Cantock's Close, Bristol BS8 1TS

ABSTRACT

An experimental approach is described which allows the properties of clay beds to be examined under laboratory conditions. The effects of salt concentration on homoionic montmorillonites are described and related to the surface properties of the clay plates. The results show specificity in terms of the counter-ions used, magnesium having a significant effect. An attempt is made to relate the data obtained to practical and environmental conditions.

INTRODUCTION

Isomorphous substitution in the tetrahedral or octahedral layers which is a feature of many naturally occurring clays leads to defects in the lattice and a deficiency of positive charge (van Olphen, 1963). As a result, there is an electrostatic potential difference between the surface of the clay mineral and the surrounding solution phase; this constitutes the so-called electrical double layer (Verwey and Overbeek, 1948). Consequently, at distances of the order of a few tens of nanometres, electrostatic interaction occurs between the clay plates and thus the electrostatic forces generated play a significant role in determining the properties of clay-water dispersions.

It has been found that the most direct method for the examination of electrostatic forces is to use macroscopic surfaces with an intervening solution of electrolyte between them (Lubetkin, Middleton and Ottewill, 1984). This involves the use of cleaved mica in a specially constructed apparatus designed for the direct measurement of surface forces. However, from a practical point of view, measurements on concentrated clay-water dispersions under various conditions are important and investigations on these in a compression cell are described in this communication. In particular, measurements have been made on plate-like particles of montmorillonite, in some cases dispersed as single plates with a thickness of about 1nm so that the surfaces were anticipated to be smooth within the limits of atomic dimensions. A number of the experiments were carried out using as a starting material lithium montmorillonite as this could be dispersed essentially as single clay plates. The range of the measurements essentially conform to those anticipated in clay beds under dry atmospheric conditions and under wet conditions. The results are compared with those obtained on macroscopic mica surfaces and with theoretical expectations.

SYNOPSIS OF THEORY

As mentioned above, isomorphous substitution in the lattice of clay particles results in the generation of an electrostatic charge at the clay surface. Correspondingly, an electrostatic potential develops at the planar surface of clay particles such as montmorillonite, beidellite etc. For a single isolated plate this potential decays in approximately an exponential fashion with distance (x) from the surface. Simplistically this can be represented as:

$$\psi_x = \psi_s \exp(-\kappa x) \qquad [1]$$

where ψ_s can be taken as the surface potential and ψ_x the potential at a distance x from the surface. The parameter κ is very important as it moderates the rate at which the potential decreases with distance. It is given by:

$$\kappa^2 = 2 n_o v^2 e^2 / \varepsilon_r \varepsilon_o kT \qquad [2]$$

where, for a symmetrical electrolyte,
- v = the valency of the ions, for a symmetrical electrolyte
- n_o = the number of ions per unit volume of solution,
- e = the fundamental unit of charge,
- ε_r = the relative permittivity of the medium and
- ε_o = that of free space.
- k = the Boltzmann constant
- T = absolute temperature

When two plates having the same surface potential approach in an aqueous environment, electrostatic repulsion occurs as a consequence of the overlap of the potential profiles (see Figure 1). This can be represented as an excess osmotic pressure given by (Verwey and Overbeek, 1948; Lubetkin, Middleton and Ottewill, 1984),

$$P_{el} = 2 n_o kT [\cosh u - 1] \qquad [3]$$

where u = ve ψ_d/kT and ψ_d is taken as the electrostatic potential mid-way between the plates, at a distance d from each plate. The distance between the surfaces of the plates can be taken as h. For long-range interaction, defined by the condition that h > 2/κ, equation [3] can be simplified to:

$$P_{el} = 64 n_o kT \tanh^2 (ve\psi_s/4 kT) - \kappa h \qquad [4a]$$

or in natural logarithmic form,

$$\operatorname{Ln} P_{el} = \operatorname{Ln} [64 n_o kT \tanh^2 (ve\psi_s/4 kT)] - \kappa h \qquad [4b]$$

Thus, experimental results plotted in the form Ln P_{el} against h should give a linear result with a slope of -κ. Moreover, the intercept at h → 0 gives the term in square brackets and as n_o, k, T, v and e are known, an estimate of ψ_s

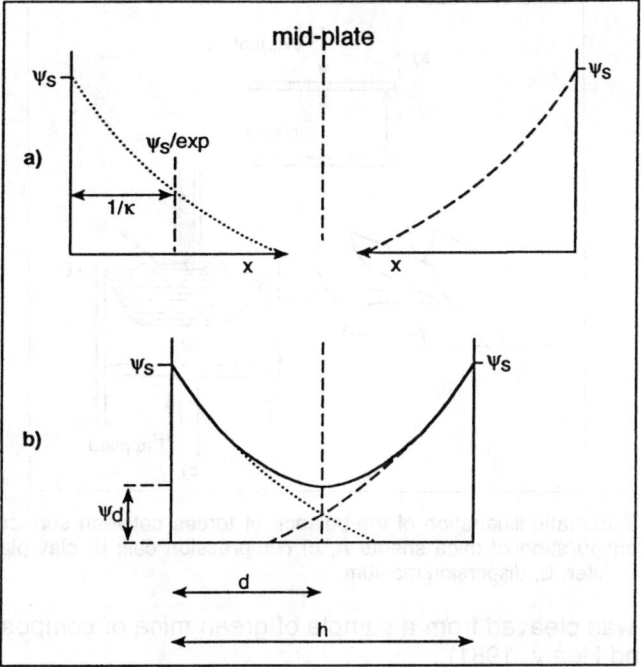

Figure 1: Interaction between electrostatically charged plates: a) isolated plates, b) overlap of the electric fields from the plates.

can be obtained. Detailed arguments about the electrical double layer are beyond the scope of this paper, hence as discussed earlier, ψ_s will be taken as the diffuse layer potential.

INTERACTION BETWEEN MACROSCOPIC SURFACES

The most direct experiments to test the validity of this hypothesis have been carried out using mica, which can be obtained in very thin sheets of about 1-2 μm and with a surface which is molecularly smooth. However, the experiments have been carried out not with plates but with the mica attached to hemicylindrical formers as illustrated in Figure 2. For this situation equation [4b] must be re-written as:

$$\text{Ln}\left[\frac{F_{el}}{2\pi R}\right] = \text{Ln}\left[\frac{64\, n_0\, kT}{\kappa} \tanh^2 \frac{ve\, \psi_s}{4\, kT}\right] - \kappa h \quad [5]$$

where F_{el} = the measured force,
h = the measured inter-surface separation and
R = the radius of the hemicylinder.

The force was measured using a linear transducer and the distance using a multiple beam interferometer (Lubetkin et al, 1984; Tolansky, 1948).

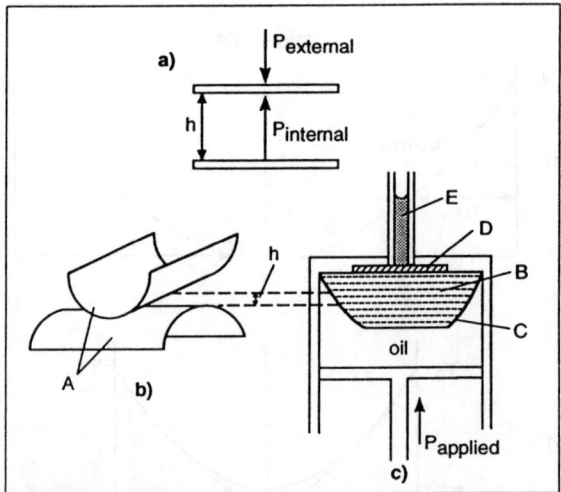

Figure 2: Schematic illustration of the balance of forces between surfaces: a) between plates, b) configuration of mica sheets A, c) compression cell; B, clay plates; C, rubber membrane; D, filter; E, dispersion medium.

The mica was cleaved from a sample of green mica of composition (Lyons, Furlong and Healy, 1981)

$$\left| Si^{4+}_{3.12} Al^{3+}_{0.88} \right| \left| Al^{3+}_{1.72} Fe^{3+}_{0.27} Mg^{2+}_{0.09} \right| (OH)_2 O_{10} \; K^+_{0.93} \; N^+_{0.08}$$

Figure 3 shows the results obtained in experiments of this type with the mica immersed in aqueous solutions of potassium chloride at concentrations of 10^{-3} and 10^{-2} mol dm^{-3}. As can be seen from the figure, the range of the electrostatic force in 10^{-3} mol dm^{-3} electrolyte extends out to distances of the order of 50 nm indicating the long range of these forces on a molecular scale. It is also clear that the range is greater in 10^{-3} than in 10^{-2} mol dm^{-3} electrolyte. In the former ψ_s was estimated to be 84 mV and the value of $1/\kappa$ obtained from the slope of the linear curve was 9.6 nm; this compares favourably with the theoretical expectation of 9.62 nm calculated from equation [2]. In 10^{-2} mol dm^{-3} potassium chloride solution the values obtained were 79 mV for ψ_s and 3.5 nm for $1/\kappa$; the calculated value for the latter was 3.04 nm. The points corresponding to $h = 2/\kappa$ are marked on Figure 3 and indicate that good linearity is obtained for $h > 2/\kappa$. From these and similar experiments with macroscopic surfaces it can be concluded that the "long distance" interaction between charged plates is satisfactorily described by equation [5].

COMPRESSION OF CLAY BEDS

The link between measurements on macroscopic surfaces and those carried out on dispersions of clay particles is illustrated in Figure 2.

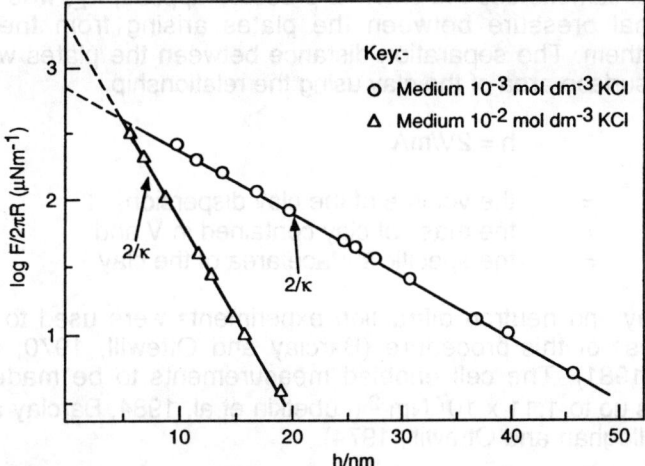

Figure 3: Results obtained for the interaction between mica surfaces.

A practical cell for compression measurements is shown schematically as a cross-section in Figure 4. The clay dispersion was contained in a compartment composed of a rubber membrane which took up a hemispherical shape and a filter membrane through which the medium but not the particles could pass. Pressure was applied using a hydraulic system, which was also coupled to gauges for direct pressure measurement. The attainment of equilibrium in the cell was ascertained by measurement of the level of expelled electrolyte in the capillary using a cathetometer (Barclay and Ottewill, 1970; Callaghan and Ottewill, 1974).

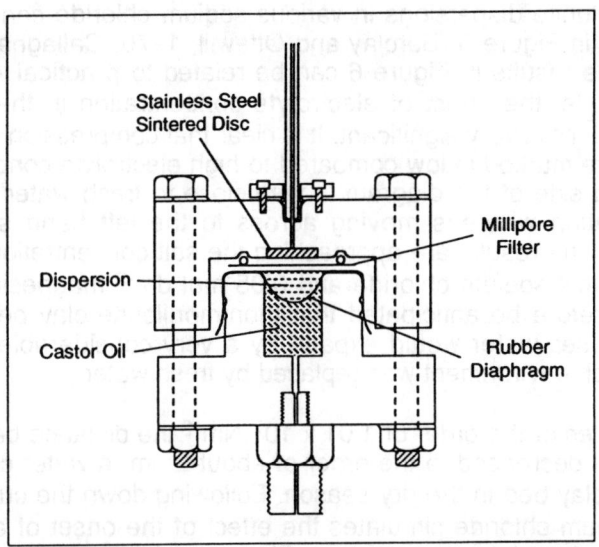

Figure 4: Cross-sectional diagram of compression cell used for studies on the compression of clay beds.

In these circumstances the external pressure applied, P_E, was balanced by the internal pressure between the plates arising from the interaction between them. The separation distance between the plates was obtained from the surface area of the clay using the relationship:

$$h = 2V/mA$$

with V = the volume of the clay dispersion
      m = the mass of clay contained in V and
      A = the specific surface area of the clay

Both X-ray and neutron diffraction experiments were used to confirm the correctness of this procedure (Barclay and Ottewill, 1970; Cebula and Ottewill, 1981). The cell enabled measurements to be made at applied pressures up to 1.11×10^7 Nm^{-2} (Lubetkin et al, 1984; Barclay and Ottewill, 1970; Callaghan and Ottewill, 1974).

Curves of measured pressure against the distance of plate separation for homoionic sodium montmorillonite in 10^{-4} mol dm^{-3} sodium chloride are shown in Figure 5. It was found that the first compression on each sample gave a larger pressure at a given distance up to a pressure of about 2.02×10^6 Nm^{-2}. After this, however, subsequent decompression and recompression cycles gave concordant results and this was taken as the equilibrium pressure. This effect was interpreted as the reorientation of clay platelets into ordered domains under the influence of the applied pressure. Diffraction studies on samples oriented by the second compression gave h values in close agreement with those calculated from the surface area (Cebula and Ottewill, 1981).

Results for equilibrium pressure against distance obtained on sodium montmorillonite dispersions in various sodium chloride concentrations are illustrated in Figure 6 (Barclay and Ottewill, 1970; Callaghan and Ottewill, 1974). The results in Figure 6 can be related to practical circumstances. For example, the effect of electrolyte concentration in the range 10^{-4} to 10^{-1} mol dm^{-3} is very significant. It is clear that compression or expansion is much more marked in low compared to high electrolyte conditions; thus the right hand side of the diagram is very close to fresh water with a low salt concentration whereas moving across to the left hand side to shorter distances, the results are approaching the salt concentration of sea water, 0.5 mol dm^{-3} sodium chloride and 0.05 mol dm^{-3} magnesium chloride. It could therefore be anticipated that montmorillonite clay beds which were formed in sea water would expand by a very considerable amount when the salt-rich environment was replaced by fresh water.

At pressures of the order of 1.01×10^7 Nm^{-2} the distance between the clay plates has decreased to the order of about 2 nm, a water content close to that of a clay bed in the dry season. Following down the curve in 10^{-4} mol dm^{-3} sodium chloride simulates the effect of the onset of a rainy season when the clay plates move apart. The inter-plate water content changes from 0.38 g per g of clay at 1 nm to say 3.8 g water per g of clay at 10 nm. Between a pair of plates this is a small motion but in a bed of clay of 10^n

COMPRESSIBILITY OF CLAY-WATER SYSTEMS

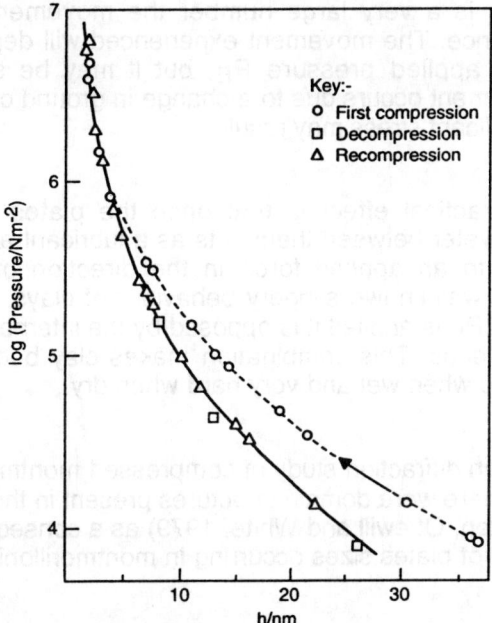

Figure 5: Measured pressure against estimated distance between plate surfaces for sodium montmorillonite in 10^{-4} mol dm^{-3} sodium chloride solution.

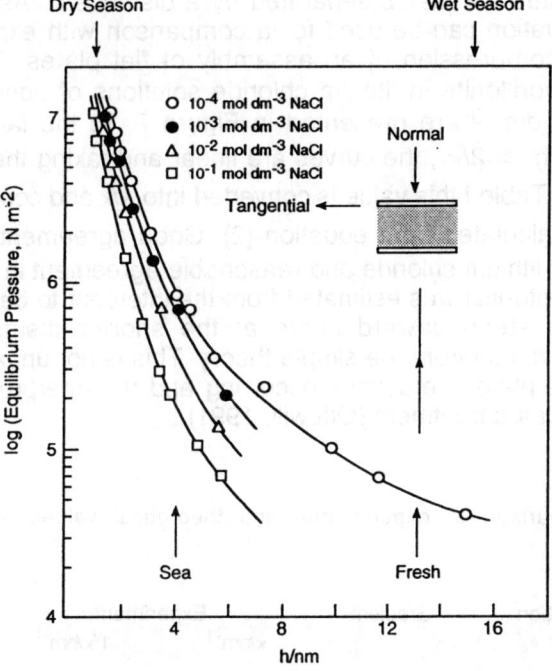

Figure 6: Equilibrium pressure against the distance between the clay plate surfaces for sodium montmorillonite dispersions in various concentrations of sodium chloride.

plates, where n is a very large number the movement $\Delta h \times 10^n$ is a substantial distance. The movement experienced will depend on the load, ie the external applied pressure P_E, but it may be substantial and if differential movement occurs due to a change in ground conditions beneath a structure, significant stress may result.

An additional practical effect is that once the plates are in a parallel orientation, the water between them acts as a lubricant and hence there is little resistance to an applied force in the direction of the plates; this explains the the well-known slippery behaviour of clays. However when a normal pressure P_E is applied it is opposed by the internal pressure normal to the plate surfaces. This combination makes clay beds difficult to deal with, visco-elastic when wet and very hard when dry.

A detailed neutron diffraction study of compressed montmorillonite particles suggested that there were domain structures present in the system (Cebula, Thomas, Middleton, Ottewill and White, 1979) as a consequence of the very large distribution of plates sizes occurring in montmorillonite clays.

COMPARISON WITH MODEL

In equation [4b] an expression is given for interaction between a pair of charged isolated flat plates separated by a distance h. As in the case of mica, this equation can be used for a comparison with experimental data obtained by compression of an assembly of flat plates. The results for lithium montmorillonite in lithium chloride solutions of concentration 10^{-3} and 10^{-2} mol dm^{-3} are presented in Figure 7. At the largest distances measured, ie $h > 2/\kappa$, the curves are linear and taking the slope gives a value of $-\kappa$. In Table I this value is converted into $1/\kappa$ and compared with the value of $1/\kappa$ calculated from equation [2]. Good agreement is obtained in 10^{-3} mol dm^{-3} lithium chloride and reasonable agreement in 10^{-2} mol dm^{-3}. The surface potential was estimated from the intercept to be of the order of 150 mV. The steep upward curve at the shorter distances shows a substantial deviation from the simple theory. This is not unexpected in view of the multiple plate interactions occurring and the interpretation needs a more sophisticated treatment (Ottewill, 1991).

Table I: Comparison of experimental and theoretical values of $1/\kappa$ for lithium montmorillonite

LiCl Concentration (mol dm^{-3})	Gradient	Experiment κ/nm^{-1}	$1/\kappa$/nm^{-1}	Theory $1/\kappa$/nm^{-1}
10^{-3}	0.045	0.104	9.62	9.62
10^{-2}	0.134	0.309	3.24	3.04

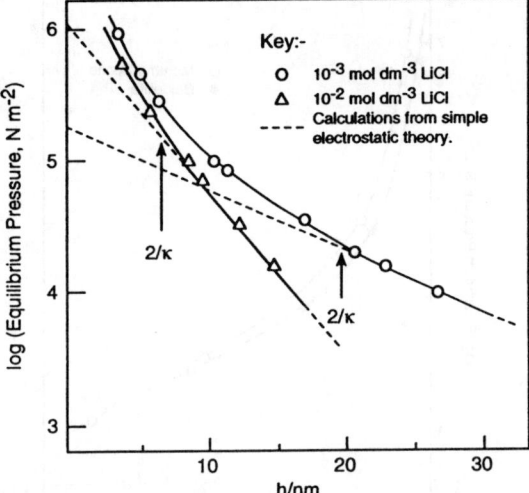

Figure 7: Comparison between theory and experimental results for a dispersion of lithium montmorillonite in lithium chloride solutions.

ISOMORPHOUS SUBSTITUTION - BENTONITE AND BEIDELLITE

Bentonite and beidellite have very similar structures as 2-layer clays. The essential difference between them is that in the case of the bentonite the isomorphous substitution occurs in the octahedral layer and in the case of beidellite it occurs in the tetrahedral layer.

Figure 8 presents some results comparing the compressibility behaviour of the two types of clay in 10^{-4} mol dm^{-3} lithium chloride solution (Verwey and Overbeek, 1948). The results for the beidellite suggest that the plates approach more closely at the higher pressures but expand more than bentonite as the clay dispersion is diluted. These results suggest in principle that there is a lower intercept at h = 0 and hence there is a lower surface potential at the beidellite-water interface than at the bentonite-water interface. The inference from this is that the deficit of charge in the tetrahedral layer causes a much stronger binding of Li$^+$ ions to the surface than occurs with the bentonite. Consequently, this allows the beidellite plates to approach more closely under pressure.

THE INFLUENCE OF COUNTER-IONS

Figure 9 shows the effects of a series of counter-ions on the compressibility of a number of homoionic montmorillonites in 10^{-4} mol dm^{-3} solutions of the alkali metal chlorides of Li$^+$, Na$^+$, K$^+$ and Cs$^+$. It was noted that there were changes in the optical appearance of the samples in the presence of K$^+$ and Cs$^+$ These samples appeared much whiter when compared with the pale yellow translucent Li$^+$ clay samples (Lubetkin et al, 1984). Evidence

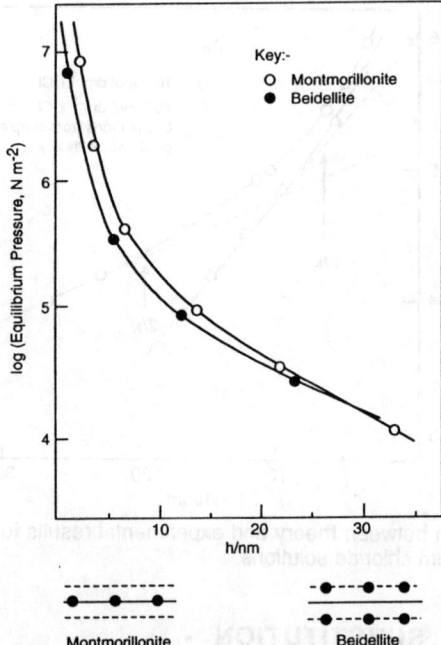

Figure 8: Comparison of experimental results obtained for lithium montmorillonite and lithium beidellite in 10^{-4} mol dm^{-3} lithium chloride solutions. Isomorphous substitution also shown schematically for montmorillonite and beidellite.

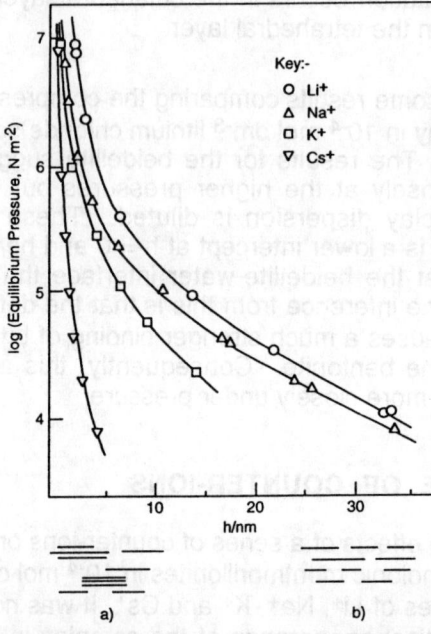

Figure 9: Results for the compression of homoionic montmorillonites in 10^{-4} mol dm^{-3} salt solutions of various counter-ions. Schematic illustrations of clay beds shown for: a) tactoids of caesium montmorillonite, b) single plates of lithium montmorillonite.

from scattering experiments and from electron microscopy suggests that in the case of Cs^+, in addition to the greater counter-ion binding there was also some association of the plates into face-face stacks containing two or three plates with layers of water between. This is illustrated schematically in Figure 9.

The restricted range of compression, or conversely the restricted swelling of the clay bed is also evident when Mg^{2+} is used as the counter-ion. This can be seen from the results of experiments carried out on a magnesium montmorillonite in 10^{-4} and 10^{-3} mol dm^{-3} magnesium sulphate (Figure 10). This implies that the effect of adding Mg^{2+} ions to the montmorillonitic clays can be beneficial if restricted swelling is required.

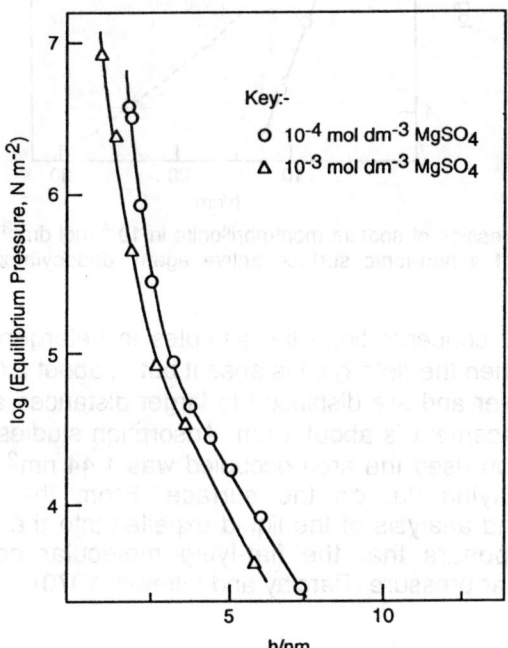

Figure 10: Results for the compression of magnesium montmorillonite in magnesium sulphate solutions.

SURFACE ACTIVE AGENTS

Amongst the many materials which influence clay swelling processes are surface active agents. Figure 11 shows some compression results obtained on sodium montmorillonite in the presence of the non-ionic surface active agent, dodecylhexaoxyethylene glycol monoether. $C_{12}H_{25}(OCH_2CH_2)_6OH$; this is similar in structure to some non-ionic detergents. The results were obtained using a concentration of 7×10^{-5} mol dm^{-3} of the detergent, just below the critical micelle concentration and 10^{-4} mol dm^{-3} sodium chloride. In the figure they are compared with results obtained with sodium montmorillonite in 10^{-4} mol dm^{-3} sodium chloride

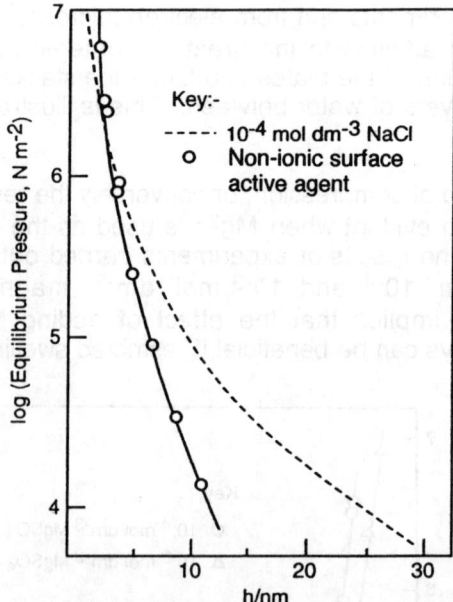

Figure 11: Compression of sodium montmorillonite in 10^{-4} mol dm^{-3} sodium chloride and in the presence of a non-ionic surface active agent, dodecylhexaoxyethylene glycol monoether.

solution. At low concentrations the samples in detergent are more compressed than when the detergent is absent but at about 2.02×10^6 Nm^{-2} the curves cross over and are displaced to larger distances and at 1.01×10^7 Nm^{-2} the displacement is about 1 nm. Adsorption studies indicated that at the concentration used the area occupied was 1.44 nm^2 corresponding to the molecule lying flat on the surface. From the displacement in compression and analysis of the liquid expelled into the external capillary (Figure 4) it appears that the flat-lying molecular conformation was maintained under pressure (Barclay and Ottewill, 1970).

DISCUSSION AND CONCLUSIONS

The results presented in this study were obtained in order to obtain a fundamental physico-chemical understanding of the behaviour of concentrated dispersions of clay particles, ie clay beds. However, the results obtained are also of interest in a practical sense to soil scientists and engineers as they attempt to correlate the properties of individual clay plates and their surfaces with the bulk properties of a clay bed and to relate these properties to environmental conditions.

A feature of the results is the presence of interlamellar water between the plates, considerable under low salt conditions and low pressure but compacted to a thin film under compressive conditions. In the present work the highest pressure applied was c 10^{-7} Nm^{-2} (100 atmospheres) and this gave an interlamellar spacing of 1.8 nm; this corresponds to a thickness of about 6 to 8 molecules of water. Indications are that even pressures of up to 2.5×10^{-8} Nm^2 do not completely remove the water from between the plates. Current evidence indicates that the organisation of the water molecules in these very thin layers is sufficient to create a force which prevents the van der Waals forces bringing the thin plates into intimate contact (Baclay and Ottewill, 1970; Low, 1987). It has been termed a hydration force (Ninham, 1985). A consequence of this conclusion is that even apparently atmospherically "very dry" clay beds should be able to rehydrate although the rate of rehydration may be slow as the diffusion path for the water molecules to enter the interlammelar spaces is tortuous. Rehydration, however, will accelerate once the osmotic swelling described in this work comes into effect.

In terms of binding clay plates together it can be seen from the results obtained that cations play a significant role. The results illustrated in Figure 10 for magnesium montmorillonite suggest that the swelling of this material is significantly reduced compared to sodium montmorillonite. Although the effect of calcium ions was not studied, Ca^{2+} would be anticipated to behave in the same manner as Mg^{2+} and indeed there is practical evidence that this is so, eg adding lime to clay soils in agriculture. However, continued washing with fresh water can displace Ca^{2+} ions from between the plates and inundation with sea water (0.5 mol dn^{-3} Na$^+$) can ion-exchange Ca^{2+} by Na.

Although the work reported has been directed primarily towards the behaviour of montmorillonites, studies have been carried out using both kaolinites and illites (Callaghan, 1975; Middleton, 1978). Many similar features were observed although, as anticipated, the swelling effects were less pronounced, in particular because of the greater thickness of the particles.

ACKNOWLEDGEMENTS

The author gratefully acknowledges the excellent help of Drs Callaghan and Middleton in studies on the compressibility of clays.

REFERENCES

BARCLAY, L.M. & OTTEWILL, R.H. (1970). Measurement of Forces between Colloidal Particles. *Special Discussions of the Faraday Society*, **1**, 138-147.

CALLAGHAN, I.C. & OTTEWILL, R.H. (1974). Interparticle Forces in Montmorillonite. *Faraday Discussions of the Chemistry Society*, **57**, 110-118.

CEBULA, D.J. & OTTEWILL, R.H. (1981). Neutron Diffraction Studies on Lithium Montmorillonite-Water Dispersions. *Clays and Clay Minerals*, **29**, 73-75.

CEBULA, D.J., THOMAS, R.K., MIDDLETON, S.R., OTTEWILL R.H. & WHITE, J.W. (1979). Neutron Diffraction from Clay-Water Systems. *Clays and Clay Minerals*, **27**, 39-52.

ISRAELACHVILI, J.N. & ADAMS, G.E. (1978). Measurement of Forces between Two Mica Surfaces in Aqueous Electrolute Solutions in the Range 0-100 nm. *Journal of the Chemistry Society Faraday Transactions*, **74**, 975-1001.

LOW, P.F. (1987). Structural Component of the Swelling Pressure of Clays. *Langemuir*, **3**, 18.

LUBETKIN, S.D., MIDDLETON, S.R. & OTTEWILL, R.H. (1984). Some Properties of Clay-Water Dispersion. *Philosophical Transactions of the Royal Society of London*, **311**, 353-368.

LYONS, J.S., FURLONG, D.N. & HEALY, T.W. (1981). The Electrical Double-layer Properties of the Mica (muscovite)-Aqueous Electrolyte Interface. *Australian Journal of Chemistry*, **34**, 1177-1187.

NINHAM, B.W. (1985). The Background to Hydration Forces. *Chemica Soripha*, **25**, 3.

OTTEWILL, R.H. (1991). *Interactions between the Surfaces of Clay Particles. Clay swelling and expansive soils, NATO Advanced Research Workship*, Cornell.

TABOR, D. & WINTERTON, R.H.S. (1969). The Direct Measurement of Normal and Retarded van der Waals Forces. *Proceedings of the Royal Society, London*, **A312**, 435-450.

TOLANSKY, S. (1948). *Multiple Beam Interferometry of Surfaces and Films*. Oxford University Press.

VAN OLPHEN, H. (1963). *An Introduction to Clay Colloid Chemistry*. Interscience, London.

VERWEY, E.J.W. & OVERBEEK, J.TH G. (1948). *Theory of Stability of Lyophic Colloids*. Elsevier, Amsterdam.

1-2 INSIDE MUDROCKS - A REVIEW OF THE PETROLOGY OF SILICICLASTIC MUDROCKS

H.F. SHAW
Department of Geology, Imperial College, Prince Consort Road, London

ABSTRACT

Mudrocks are the most abundant but least studied of the sedimentary rocks and much of the petrological knowledge of them is based on bulk whole rock mineralogical and geochemical analyses. However, modern electron microscopy techniques, especially backscatter electron imaging, now allow detailed micro-petrographic analyses to be made of mudrocks. This paper provides a general review of our current knowledge on the petrology of siliciclastic mudrocks. It illustrates how techniques are now readily available to provide detailed information on the macro- and micro- petrofabrics of mudrocks and thus help in gaining a better understanding of the role of petrofabrics in controlling geotechnical properties.

INTRODUCTION

Mudrocks are the most abundant of the sedimentary rocks occupying about 65% of the stratigraphic column. Surprisingly they are also the least studied even though their importance has long been recognised as shown by this observation by Sorby in the last century:

> "Possibly many may think that the deposition and consolidation of fine grained mud must be a very simple matter, and the results of very little interest. However, when carefully studied experimentally, it is soon found to be so complex a question and the results dependent on so many variable conditions, that one might feel inclined to abandon the enquiry were it not that so much of the history of our rocks appears to be written in that language".

Despite their importance the number of published papers specifically on mudrocks shows an inverse relationship to their abundance and very few of these are concerned with their petrology.

The reason for this apparent lack of interest in the mudrocks has been the difficulty in studying them using standard techniques because of their fine grain size. As a consequence, most studies have been based on interpretations of variations in their whole-rock mineralogy and geochemistry. However, the use of electron microscopy techniques, especially backscatter imaging now allows petrographic studies of mudrocks to be made on a routine basis.

This paper reviews the current state of knowledge of mudrock petrology and highlights those features of most significance for their engineering properties.

CLASSIFICATION

There is widespread agreement that mudrocks may be considered as fine grained rocks whose dominant grain size is less than 0.06 mm. In the geological classifications a mudrock is defined as having more than 50% mud grade material. However, some classifications proposed for engineering applications have used the criterion of more than 35% mud grade material (Grainger, 1984). This percentage boundary is the one included in the British Standard Classification of Soils (BS 5930:1981) to define fine grained soils as above this value the mechanical properties of the soils are predominantly controlled by the mud grade material. It is often assumed that mudrocks must be siliciclastic but fine grained carbonates and the lithified equivalents of biogenic oozes (e.g. chalks and cherts) can also be regarded as mudrocks (Figure 1). In this paper only those materials with more than 50% silicate material are discussed, although it should be noted that elsewhere other definitions of siliciclastic have been used e.g. more than 65% silicate minerals of which 50% are quartz and clay minerals (Grainger, 1984).

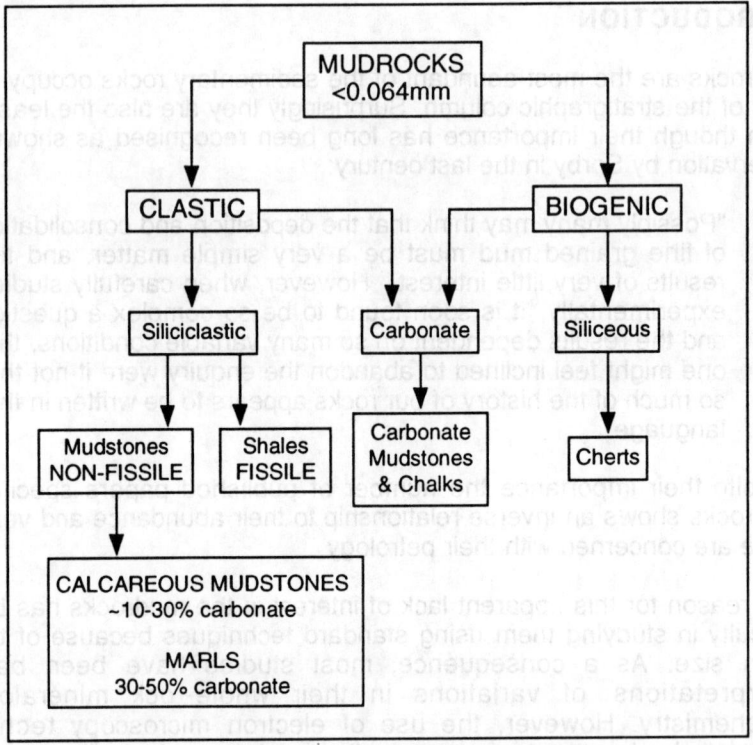

Figure 1: A classification of mudrocks.

The most widely used geological/sedimentological classification for siliciclastic mudrocks is that of Blatt, Middleton and Murray (1972), see Table I. However, for engineering studies, modified classifications have been proposed such as those by Hawkins and Pinches (1992) which, being based on grain size, correlate better with the observed engineering properties of the mudrocks (Table II). When describing mudrocks, therefore, it is essential to state clearly which classification system has been used.

Table I: Classification of siliciclastic mudrocks (after Blatt et al., 1972).

Grain Size	Fissile (laminae <10 mm)	Non-fissile (beds >10 mm)
>2/3 silt size	silt-shale	siltstone
silt and clay size	mud-shale	mudstone
>2/3 clay	clay-shale	claystone

Mud - effective grain diameter = <0.063 mm
Silt - effective grain diameter = 0.063-0.002 mm (or 0.063-0.004 mm)
Clay - effective grain diameter = <0.002 mm (or 0.004 mm)

Table II: Classification of mudrocks for engineering purposes (after Hawkins and Pinches, 1992).

Grain Size	Fissile (laminae <20 mm)	Non-fissile (laminae >20mm)
<25% clay size	shaley siltstone	siltstone
25-40% clay size	shaley mudstone	mudstone
>40% clay size	shaley claystone	claystone

Mud - effective grain diameter = <0.06 mm
Silt - effective grain diameter = 0.06-0.002 mm
Clay - effective grain diameter = <0.002 mm

MINERALOGY

Based largely on X-ray diffraction analyses the 'average' composition of siliciclastic mudrocks is:

Clay minerals	60%
Quartz	30%
Feldspar	5%
Carbonates	4%
Iron oxides	<1%
Organic matter	<1%

Although the mineralogy is usually dominated by the clay minerals, especially in the claystones and clay shales, this is less true of the siltstones where there are increasing amounts of quartz and other non-clay minerals. Mineralogical studies of mudrocks have tended to over-emphasise the clay minerals which play an important but not always dominant role in controlling rock properties.

Most of the quartz and feldspar are detrital but there is new evidence from electron microscopy of the diagenetic formation of quartz cements and, to a lesser extent, of feldspar in mudrocks. The carbonates originate often as biogenic debris but a significant proportion of the carbonate can be diagenetic in origin forming intraclast cements, concretions or interbeds.

CLAY MINERALS

The clay minerals are a group of hydrous aluminosilicates, the majority of which are sheet silicates. The sheet silicate clay minerals can be subdivided into two main types (Figure 2):

(a) the two layer or 1:1 type
(b) the three layer or 2:1 type

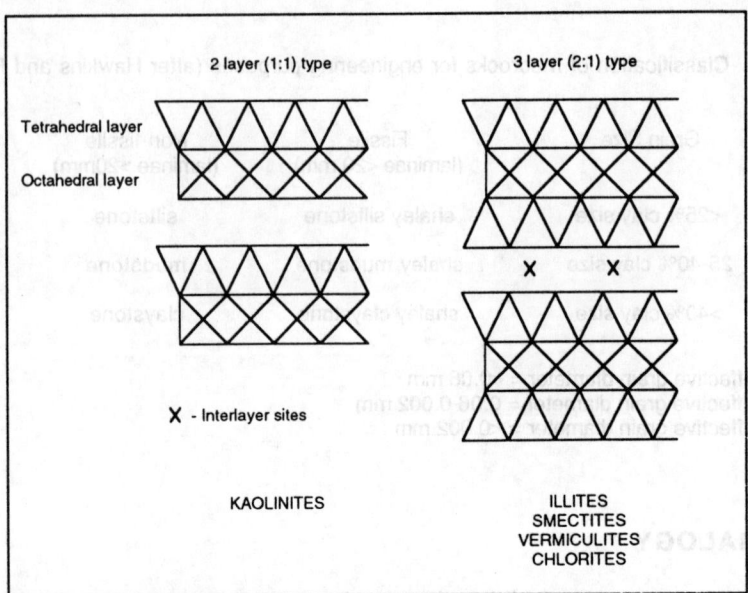

Figure 2: The structural classification of clay minerals.

In addition, there are mixed layer clay minerals, usually composed of coherent structural intergrowths of two of the 2:1 clay minerals e.g. illite-smectites or chlorite-smectites (Figure 3), although kaolinite smectites have also been reported. These intergrowths may be random or ordered in nature. There has currently been much debate about the true nature of these mixed layer phases. Nadeau, Wilson, McHardy and Tait (1984) proposed

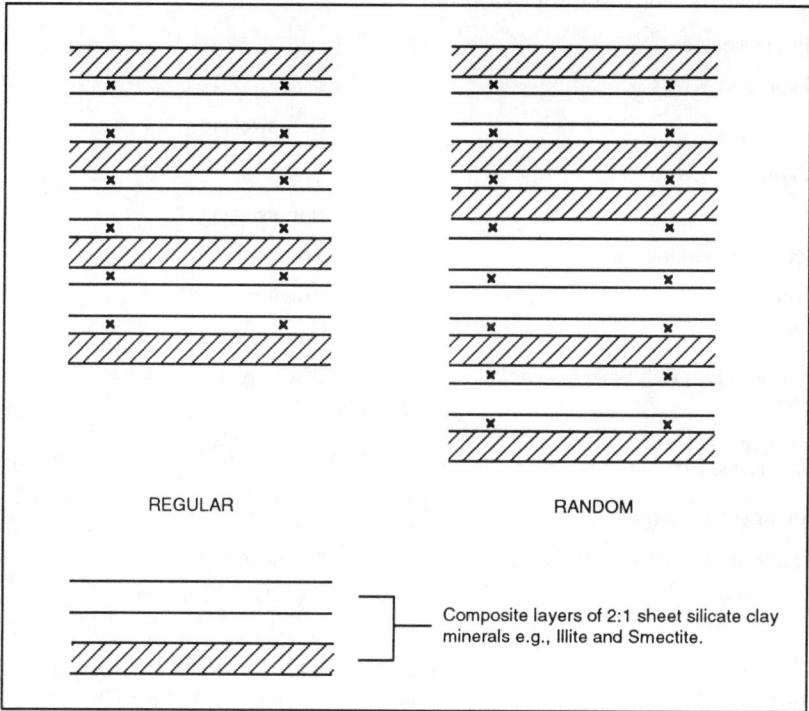

Figure 3: Mixed layer clay minerals.

that interparticle diffraction effects could mean that it would be difficult to differentiate very thin illite particles (< 20Å thick) from smectite based only on X-ray diffraction analyses. The X-ray diffraction characteristics of ordered mixed layer illite-smectites could be produced by aggregates of very thin illite particles and of random mixed layer illite-smectites by aggregates of thin illite and smectite particles as well as true interstratification.

In addition to the sheet silicate clay minerals, palygorskite and sepiolite also have essentially amphibole-like band silicate structures.

In addition to being classified according to their crystal structures the clay minerals can also be differentiated into swelling or non-swelling types based on whether or not they show intraparticle crystalline swelling in the presence of water and other fluids (Table III).

ORIGIN OF CLAY MINERALS

The major source of clay minerals is pedogenically from the weathering of pre-existing silicates, although they can form authigenically in sediments (e.g. smectites from the submarine alteration of volcaniclastics and glass; palygorskite in evaporitic sediments). The type of clay minerals formed is a

Table III: Classification of clay minerals.

A) Sheet silicates

1:1 Type

Kaolinite	Non-swelling

2:1 Type

Illite	Non-swelling
Smectite (Montmorillonite)	Swelling
Vermiculite	Swelling
Chlorite	Non-swelling
Mixed layer clays with Smectite and/or Vermiculite	Swelling
Mixed layer clays without Smectite and/or Vermiculite	Non-swelling

B) Non-sheet silicates

Palygorskite - Sepiolite (= Attapulgite)	Non-swelling

function of climate, original rock type, drainage, topography, vegetation and weathering time (Singer, 1980). Chlorite and illite are usually formed and preserved in weathering regimes associated with cold temperate climates whilst more intense chemical weathering in humid tropical climates favours the formation of kaolinite. Smectites are usually generated in weathering regimes with contrasting seasons and especially a pronounced dry season. Significant amounts of organic matter can affect the behaviour of the inorganic chemical components in the weathering environment via the generation of organo-metallic complexes. These can fundamentally influence the products of the weathering process compared to what might be expected from thermodynamic stability relationships.

CLAY MINERAL ASSEMBLAGES IN RECENT SILICICLASTIC MUD SEDIMENTS

Clay mineral assemblages in Recent marine sediments are largely inherited and detrital in origin and can be broadly related to the soil clays of the adjacent land masses. This general picture can be complicated by other factors, i.e. the clays may be hydrothermal rather than pedogenic in origin; the clays may originate from palaeo rather than modern soils or jet-stream aeolian transport may carry the clays large distances from their original sources.

In the marine environment smectites are the principal neoformed clay minerals. They are usually associated with volcanic or hydrothermal sources (Figure 4) which explains the abundance of smectites in the Pacific and Indian Oceans (Griffin, Windom and Goldberg, 1965). In modern deltaic environments also variations in the relative abundance the clay mineral assemblages can frequently be related to differential settling of the clay particles of different compositions (Figure 5). Thus more kaolinite and illite occur close to shore while the amount of smectite increases away from the delta shoreline (Gibbs, 1977).

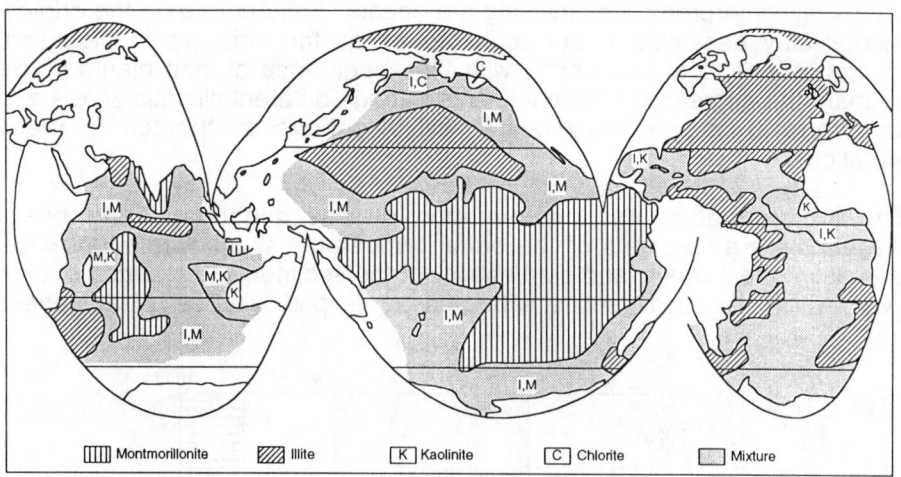

Figure 4: Distribution of the principal clay minerals in Recent marine sediments (after Griffin et al., 1965).

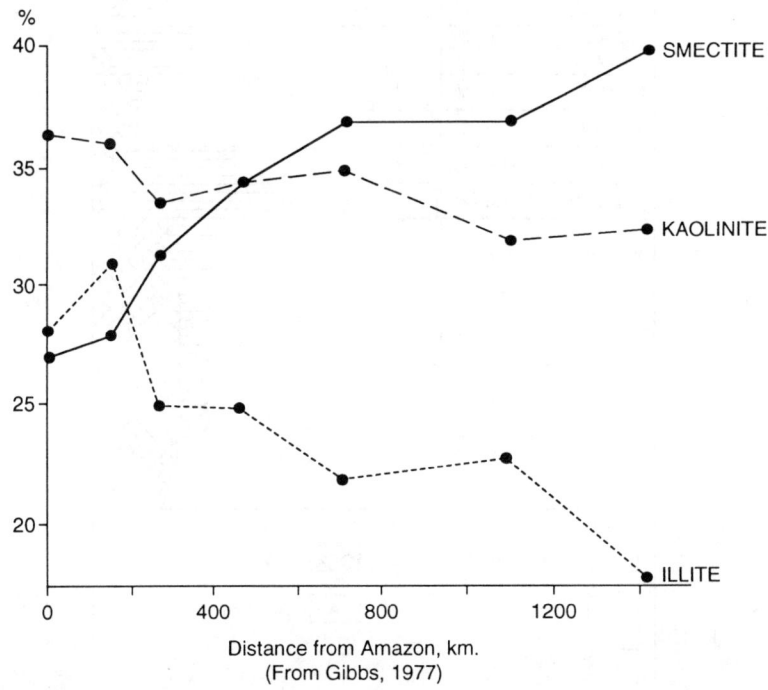

Figure 5: Variations in the relative abundance of clay minerals with distance from the mouth of the Amazon River (after Gibbs, 1977).

CLAY MINERAL ASSEMBLAGES IN ANCIENT SILICICLASTIC MUDROCKS

In ancient siliciclastic mudrocks there is generally an increase in illite and chlorite with geological time, at the expense of kaolinite and smectite and the mixed layer smectite minerals (Shaw, 1981). These trends (Figure 6) are usually interpreted as reflecting the effects of diagenesis on the original detrital clay assemblage but such factors as the changes in biological control on chemical weathering with the appearance of land plants in the Silurian, the migration of the continents through different climatic zones and variations in source provenance, will also have contributed to these variations.

The transformation and neoformation of clay minerals during burial diagenesis are functions of the burial temperatures, the formation water chemistry, the porosity and permeability of the sediments, the period of time over which the sediment is subjected to a particular set of diagenetic

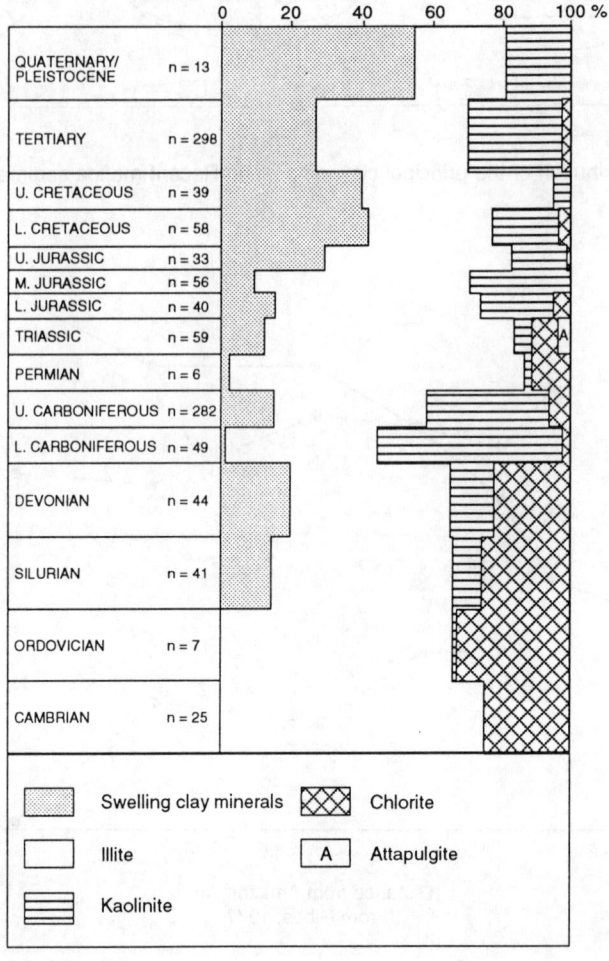

Figure 6: Variations in the clay mineralogy of siliciclastic mudrocks from the U.K.

conditions, the original composition of the sediments and their effective degree of compaction. The general effects of burial diagenesis on clay minerals based on the original model of de Segonzac (1970) are summarised in Figure 7. One of the principal and most discussed diagenetic reactions is the illitisation of smectites whereby smectite is converted via mixed layer illite-smectites to illite. These changes are usually considered to occur in the temperature range of 80 to 200°C and in an environment where an available source of K and Al allows the reaction to occur:

$$\text{Smectite} + K + Al = \text{Illite} + Si + H_2O$$

The nature and course of this illitisation reaction varies within different sedimentary basins indicating that it is a function of a wide variety of different factors and not an "automatic" event which always occurs over a particular depth/temperature range during burial. One consequence of the illitisation reaction is that the interlayer water in the smectite structure will be released, which may have important implications regarding fluid migration, overpressuring/undercompaction and related effects in mudrocks.

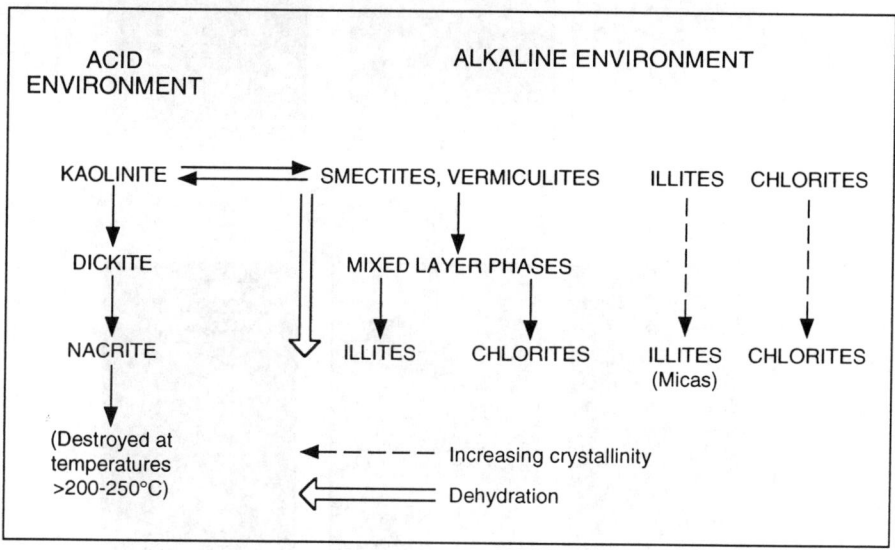

Figure 7: Model for clay mineral diagenesis (after de Segonzac, 1970).

PETROGRAPHY

LAMINATIONS AND FISSILITY

Laminations are thin (geologically <10mm, geotechnically <20mm) bands of various origins which may or may not be planes of fissility. Fissility is the property whereby mudrocks split along laminae and is used to differentiate shaley types of mudrocks. Although the term implies lateral continuity, laminations have also been described as wavy or lenticular (O'Brien and

Slatt, 1990). Laminations can frequently be identified in hand specimens, especially if they are coloured, but sometimes mudrocks with no laminations visible to the naked eye are found to be strongly laminated when examined using X-radiography (Plate 1). In this type of analysis thin (2-3mm) slabs of the rock are X-rayed and differential absorption of the X-rays allows lithological variations in the rock to be observed on the X-ray film placed beneath the specimen.

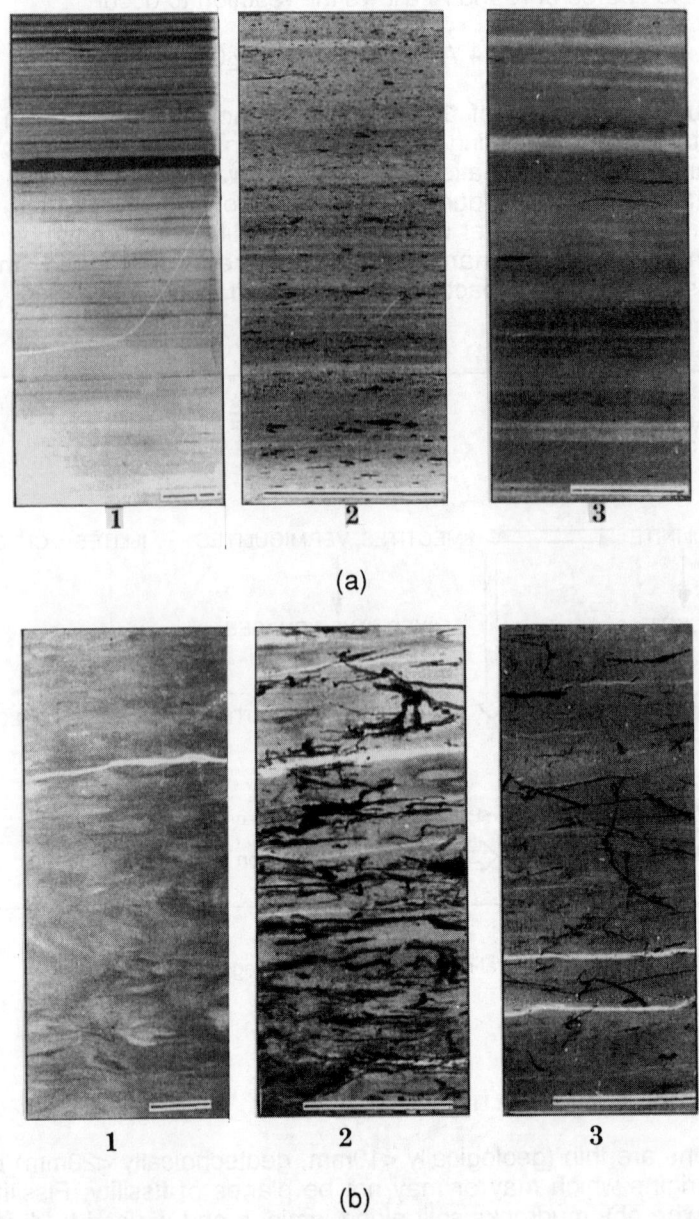

Plate 1: X-radiographs showing (a) well developed laminations (b) bioturbated mudrocks (from O'Brien and Slatt, 1990 - reproduced with the permission of the authors).

The development and preservation of laminations is largely a function of the depositional environment thus they indicate variations in sedimentary processes and sediment supply, being organic-rich / organic-poor, silt- or clay-rich for example. Colour changes will also be informative, with for example red and green representing changes in the redox state of iron in the sediment.

Well laminated sediments are often fissile but not necessarily so. Fissility arises from the orientation of platy minerals, usually parallel to bedding (Plate 2). The development of this orientation is obviously a function of the abundance of the platy, usually clay mineral, particles while the presence of increasing amounts of silt size quartz will disrupt the development of fissility (Plate 3). However, even mudrocks with similar amounts of platy clay minerals will not necessarily develop the same degree of fissility. The platy clay particles will not separate out as individual particles but will show varying degrees of inter-particle adhesion or flocculation as a function of the surface charge on the particles and the depositional environment. Greater degrees of flocculation will occur in saline waters than in fresh waters, the process of flocculation is also encouraged by the presence of organic matter. Although flocculated clays initially produce sediments with relatively unoriented fabrics, this effect is usually lost on burial and compaction of the sediment.

Plate 2: Secondary electron image of a laminated mudrock showing well developed parallelism of the platy clay minerals (from O'Brien and Slatt, 1990 - reproduced with the permission of the authors).

However, exceptions to this are found in mudrocks deposited in highly saline evaporitic environments (see Plate 4 and O'Brien and Slatt, 1990). Bioturbation of muddy sediments by bottom dwelling organisms will prevent the initial development of fissility and this effect is often preserved in the

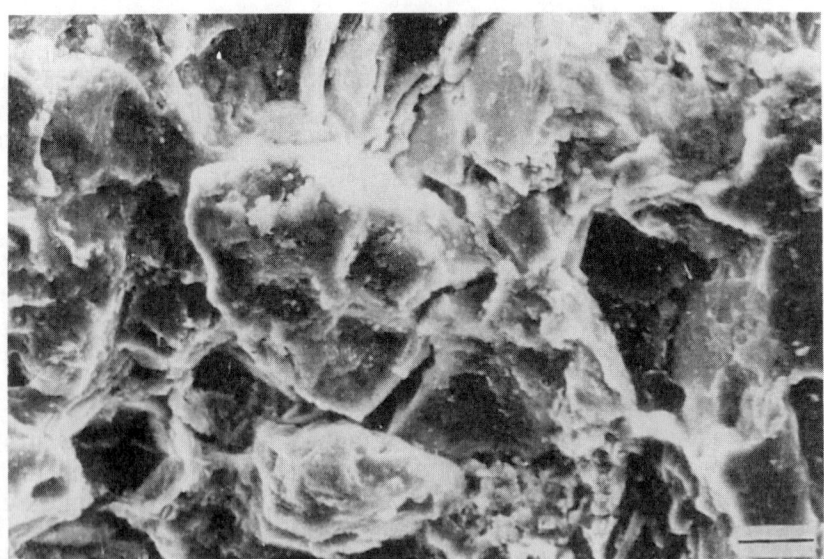

Plate 3: Secondary electron image showing a random fabric within a silt-rich mudrock (from O'Brien and Slatt, 1990 - reproduced with the permission of the authors).

Plate 4: Secondary electron image showing random orientation of platy clay minerals in a mudrock deposited in an evaporitic environment (from O'Brien and Slatt, 1990 - reproduced with the permission of the authors).

lithified sediments as mottling. Conversely the presence of well developed fissility, particularly in black shales, is often interpreted as evidence of anoxic bottom water conditions at the time of deposition; again, however this is not always the case. Whilst diagenetic effects involving the growth of clay and non-clay cements can cause disruption of pre-existing fissility (Plate 5) more usually it is burial and compaction which enhance fissility.

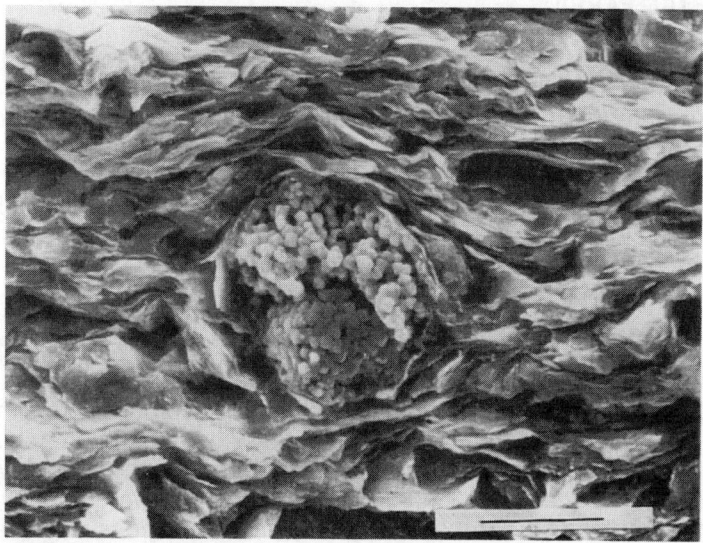

Plate 5: Secondary electron image showing authigenic framboidal pyrite disrupting the orientation of the platy clay minerals (from O'Brien and Slatt, 1990 - reproduced with the permission of the authors).

COLOUR

The colour of siliciclastic mudrocks is principally a function of their organic matter content and Fe^{2+}/Fe^{3+} ratios (see Figure 8; and Potter, Maynard and Pryor, 1980). These factors reflect both the rate of production and preservation of organic matter and the sedimentation rate, which in turn determine whether the depositional conditions were reducing or oxidising.

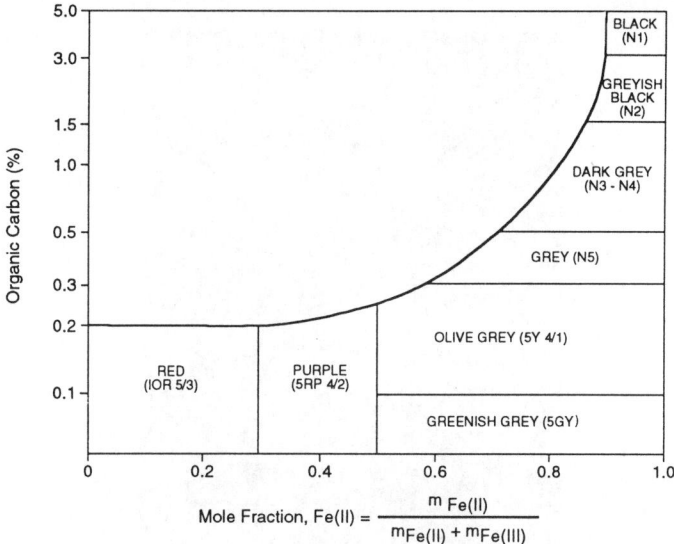

Figure 8: The colour of wet shales as a function of organic matter and Fe^{2+}/Fe^{3+} contents (after Potter et al, 1980).

DIAGENETIC FABRICS

Prior to the use of backscatter and secondary electron microscopy techniques for the study of mudrock petrography, diagenetic processes in mudrocks were interpreted on the basis of X-ray diffraction and geochemical analyses of whole rock and clay fraction samples. These studies led to the development of models which gave the impression that during diagenesis mudrocks behaved as homogeneous materials. Generalised reactions were written to summarise these processes, such as that produced for the Tertiary Gulf Coast (Hower, Eslinger, Hower and Perry, 1976):

$$\text{K-feldspar} + \text{Smectite} = \text{Illite} + \text{Chlorite} + \text{Quartz}$$

Electron microscopy observations indicate the heterogeneity of mudrock fabrics that reflect both depositional and diagenetic processes (Primmer and Shaw, 1987; Shaw and Primmer, 1991). They illustrate the complexity of the diagenetic reactions taking place involving both clay and non-clay minerals and that at best equations such as the one written above can only represent broad summaries. Examples of diagenetic processes revealed by backscatter electron microscopy are shown in the micrographs taken of samples of the Kimmeridge Clay Formation (Plates 6 to 8) and the London Clay (Plates 9 and 10). Plate 6 illustrates how the use of electron microscopy imaging enables detrital and diagenetic minerals to be differentiated and their distributions within the rocks to be determined, which can be of critical importance in understanding variations in their physical properties. These micrographs demonstrate how the diagenetic precipitation (eg., pyrite in Plate 6; calcite in Plate 7) and dissolution (eg., calcite in Plate 7; K-feldspar in Plate 8) of minerals can fundamentally affect the petrology of the mudrocks and thus their physical properties. They also indicate the variability of the mudrocks taken from the same formations (eg., the Kimmeridge Clay Formation and the London Clay), even within one location.

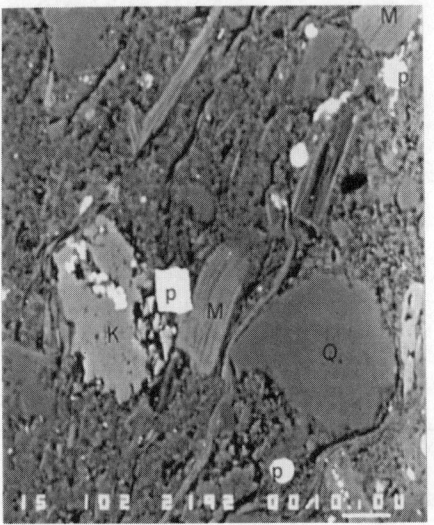

Plate 6: Backscatter electron image of Kimmeridge Clay Formation clay-rich mudrock showing dissolution at some mineral boundaries.

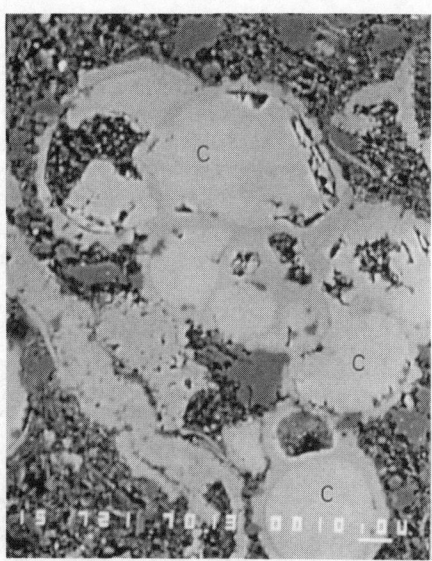

Plate 7: Backscatter electron image of Kimmeridge Clay Formation calcareous mudrock showing dissolution around calcite-infilled microfossils.

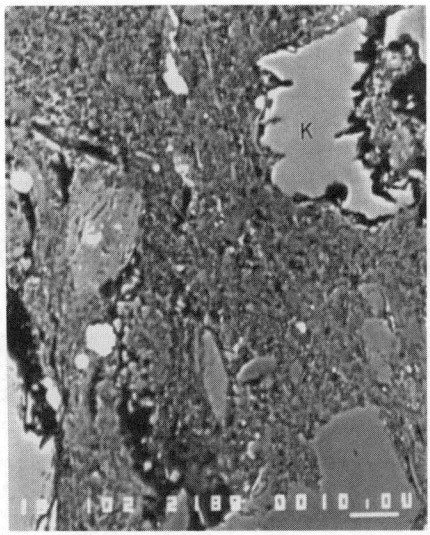

Plate 8: Backscatter electron image of Kimmeridge Clay Formation silty mudrock showing authigenic minerals and some mineral dissolution.

CONCLUSIONS

The geotechnical properties of the siliciclastic mudrocks are controlled by their mineralogies, petrofabrics and the physico-chemical properties of the surface active clay minerals and other colloidal phases which occur in

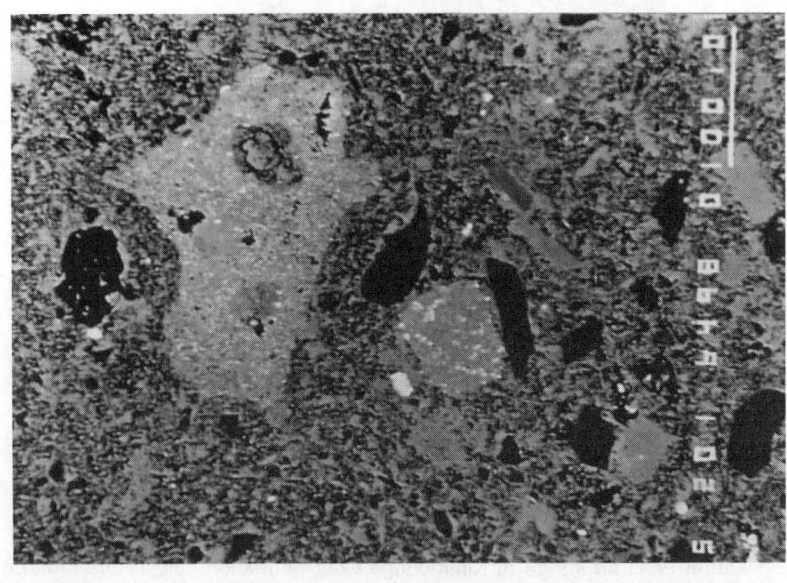

Plate 10: Backscatter electron image of London Clay (South Ockenden) showing the presence of carbonate cemented concretions and lenses.

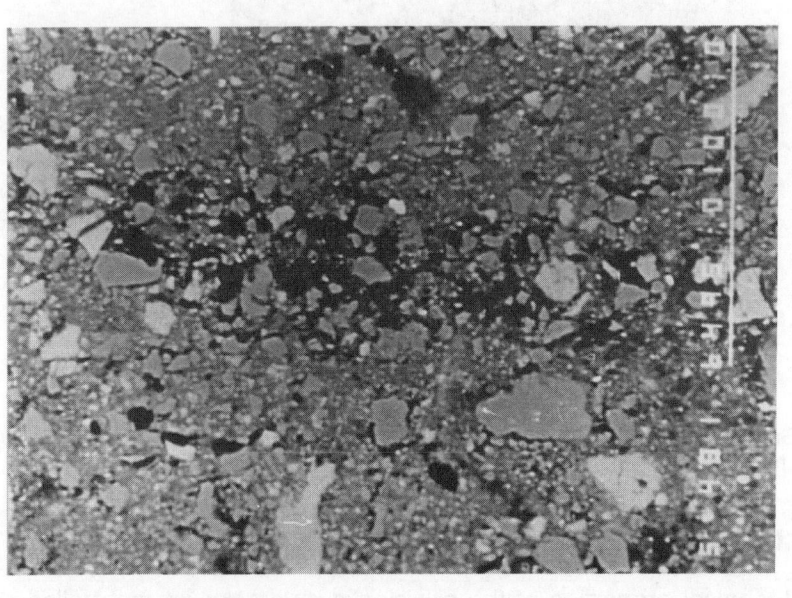

Plate 9: Backscatter electron image of London Clay (South Ockenden) showing a silt laminae.

varying proportions in such rocks. However, in the past undue emphasis has been placed on the role of clay minerals and insufficient attention given to the influence of the petrofabrics on the physical behaviour of mudrocks. Now that techniques are readily available for the characterisation of mudrock petrofabrics there is an opportunity to study in detail the role of petrofabrics in determining engineering properties. It is important that geotechnical engineers acquire information on petrofabrics and not rely on mineralogical and chemical analyses to characterise the mudrocks.

REFERENCES

BLATT, H., MIDDLETON, G. & MURRAY, R. (1972). *Origin of sedimentary rocks.* Prentice-Hall, New Jersey, pp 634.

DE SEGONZAC, G.D. (1970). The transformation of clay minerals during diagenesis and low grade metamorphism: a review. *Sedimentology,* **15**, 281-346.

GIBBS, R.J. (1977). Clay mineral segregation in the marine environment. *Journal of Sedimentary Petrology,* **47**, 237-243.

GRAINGER, P. (1984). The classification of mudrocks for engineering purposes. *Quarterly Journal of Engineering Geology,* **17**, 381-387.

GRIFFIN, J.J., WINDOM, H. & GOLDBERG, E.D. (1965). Distribution of clay minerals in the world's oceans. *Deep Sea Research,* **15**, 433-459.

HAWKINS, A.B. & PINCHES, G.M. (1992). Engineering descriptions of mudrocks. *Quarterly Journal of Engineering Geology,* **25**, 17-30.

HOWER, J., ESLINGER, E.V., HOWER, M.E. & PERRY, E.A. (1976). Mechanisms of burial metamorphism of argillaceous sediments: 1 - Mineralogical and geochemical evidence. *Bulletin of the Geological Society of America,* **87**, 725-737.

NADEAU, P.H., WILSON, M.J., MCHARDY, W.J. & TAIT, J.M. (1984). Interparticle diffraction: a new concept for interstratified clays. *Clay Minerals,* **19**, 757-769.

O'BRIEN, N.R. & SLATT, R.M. (1990). *Argillaceous rock atlas.* Springer-Verlag, New York, pp 141.

POTTER, P.E., MAYNARD, J.B. & PRYOR, W.A. (1980). *Sedimentology of shale.* Springer-Verlag, New York, pp 306.

PRIMMER, T.J. & SHAW, H.F. (1987). Diagenesis in shales: evidence from backscattered electron microscopy and electron probe analyses. In: *Proceedings of the International Clay Conference, Denver,* 1985 (Eds. L.G. Schultz, H. van Olphen & F.A. Mumpton) Clay Minerals Society, 135-143.

SHAW, H.F. (1981). Mineralogy and petrology of the argillaceous rocks of the U.K. *Quarterly Journal of Engineering Geology,* **14**, 277-290.

SHAW, H.F. & PRIMMER, T.J. (1991). Diagenesis of mudrocks from the Kimmeridge Clay Formation of the Brae Area, U.K. North Sea. *Marine Petrology Geology,* **8**, 270-277.

SINGER, A. (1980). The palaeoclimatic interpretation of clay minerals in soils and weathering profiles. *Earth Sciences Review,* **15**, 303-326.

1-3 MICROBIOLOGY OF SOILS

R. CAMPBELL
Department of Botany, University of Bristol, Bristol BS8 IUG

ABSTRACT

The great diversity of micro-organisms is considered in terms of morphology and metabolism. Micro-organisms in soils and sediments live in microhabitats which vary greatly in their chemical and physical properties and are discontinuous in space and time, although the metabolic versatility of micro-organisms means that they can also live in extreme environments on earth and elsewhere. They affect nutrient availability, aggregate structure and the levels of key chemicals, especially sulphur compounds and oxygen. We can try and control and manipulate micro-organisms but we cannot ignore them.

INTRODUCTION

All soils contain micro-organisms, sometimes in very large numbers. They are major factors in the control of fertility, plant nutrient status, general chemistry, redox potential, crumb structure and the water retention characteristics of the soil. Micro-organisms are also important in determining the amount of biodeterioration of materials in the soil, whether this be the breakdown of xenobiotics or the destruction of construction materials and services buried in the soil.

The microbiology of soils is the subject of many texts and reviews. Among more recent works Jensen, Kjoller and Sorensen (1986) cover most aspects of soil and Campbell (1990) provides a brief review while Grivorova and Norris (1990) cover methodology and Howsam (1991) some aspects of importance to engineers.

MICROBIAL DIVERSITY

The concept of micro-organisms is a very wide one but in general terms can be taken to encompass all organisms which cannot be seen with the naked eye; the protozoa and small algae, the fungi and the bacteria (including the cyanobacteria or blue-green algae). There are basic differences in structure between these groups; the bacteria being more simple without defined nuclei in their cells while the fungi and protozoa have nuclei and other complex membrane bound organelles. These micro-organisms are generally single-celled or have groups of similar cells without extensive differentiation into complex organ systems such as are found in higher plants and animals. Viruses are also included in the term "micro-organism" but as they are all pathogens of other micro-organisms, plants and animals rather than free-living in the environment they will not be considered further here.

This heterogeneity of micro-organisms is reflected in their shape and size (Atlas and Bartha, 1987). The smallest bacteria are less than 1 μm in diameter but others may be up to 200 μm long and of almost every shape. The protozoa and algae have a very diverse morphology, ranging in size from the flagellate organisms of <1 μm diameter to large ciliates and amoebae visible with the naked eye (Patterson and Hedley, 1992). The fungi are usually composed of filaments which elongate by tip growth (allowing them to penetrate substrates rather than just grow on surfaces like most micro-organisms) and they reproduce by spores which may be carried on elaborate fruiting structures such as the mushrooms and toadstools. They range from microscopic filaments (< 1 μm across) to fruit bodies of up to 1 m in diameter.

Bacteria have a density of about 1 g/cm^3 and a mass measured in picograms (10^{-12} g). They live in water and being almost the same density they sink only very slowly. If they move they have almost no momentum - a bacterium cannot stop swimming and glide up to the surface. Much more important are molecular and atomic forces causing attraction between particles and electrostatic repulsion forces due to the surface charge on bacteria and clay platelets (Berkeley, Lynch, Melling, Rutter and Vincent, 1980; Campbell, 1983). Adsorption phenomena due to charge are also important in concentrating nutrients, toxins and other ions; for example the acidity of a surface (the concentration of H^+) can be up to 100 times greater than the bulk fluid. Clay and other soil colloids have a great effect on micro-organisms.

Micro-organisms may exist as vegetative structures (single cells, filaments) or in a variety of dormant states which are resistant to desiccation and high or low temperatures (spores, cysts). These dormant structures may have special dispersal mechanisms but more usually, being so small, micro-organisms are spread accidentally in dust, soil moving water, rain splashes and other biotic and abiotic forces which move materials around the environment.

METABOLIC STRATEGIES OF MICRO-ORGANISMS

There are very few places on earth where micro-organisms are not found - indeed they have been transported to the moon and survived for months or even years. Whilst they flourish in moderate temperatures, with water, sun and oxygen in the air, they are also found in extreme tropical and arctic regions (Edwards, 1990; Kushner, 1978); in dry valleys in Antarctica or deep freezers; in volcanic hot springs on land (80°C) and hydrothermal vents in the mid-ocean volcanic ridges with temperatures of 200 to 300°C, in complete darkness in deep oceans with pressures of 1000 atmospheres and temperatures of less than 4°C, in highly saline lakes (pH >10) and acid mine drainage waters (pH <2).

All organisms require a source of energy in order to live, build their bodies and move but there are a variety of ways in which it may be obtained, stored and released when needed (Atlas, 1984; Brock and Madigan, 1991).

Some bacteria and the algae use light energy from the sun to reduce carbon dioxide to sugars (the photo-autotrophs). The stored sugars can later be oxidised to produce energy for chemical syntheses and movement. Such organisms are independent of external supplies of organic compounds but must live in the light. Other bacteria (the chemo-autotrophs) obtain their energy by the oxidation of inorganic chemicals (such as hydrogen sulphide, ferrous iron (iron II), reduced forms of manganese etc) and use this to 'fix' carbon dioxide to sugars. Many of these chemo-autotrophs are also independent of external sources of organic compounds and can live and grow in the dark. Autotrophs are responsible for many of the changes in oxidation states of inorganic compounds in environments and are the ultimate source of organic material in the biosphere.

Not only the animals (including humans) but also all the fungi and some bacteria and protozoa depend on reduced forms of organic material. These heterotrophs must acquire reduced carbon and oxidise it to obtain their energy.

The difference between the photo-autotrophs, chemo-autotrophs and heterotrophs is their original energy source: they all produce energy for metabolism, growth and movement by the oxidation of organic compounds. Such oxidation to yield energy involves reducing some other compound, the terminal electron acceptor. The nature of the terminal electron acceptor is used to classify further the autotrophs and heterotrophs. If it is oxygen then the micro-organisms are aerobic (like humans) and can only survive in the presence of oxygen, but by using that oxygen they may deplete it to very low values in soils and sediments. However bacteria and fungi in particular can use other terminal electron acceptors; these organisms are the anaerobes. For fungi and some bacteria these other acceptors can be organic compounds and the process is known as fermentation and produces alcohol and lactic acid. Inorganic terminal electron acceptors, such as sulphate, nitrate and carbon dioxide (or carbonate) are also used by anaerobes which produce reduced inorganic compounds (N_2, NH_4^+, S, H_2S, CO, CH_4) in the process. Some bacteria, for example, obtain energy from oxidation of ferrous to ferric iron (iron II to iron III) and/or by the oxidation of sulphide to sulphate. They can do this in the absence of oxygen by using nitrate as the terminal electron acceptor; thus *Thiobacillus denitrificans* can live and grow in the dark, with sulphide, carbon dioxide and nitrate for 'food' and producing sulphate, sugar and reduced nitrogen compounds. The sulphate forms sulphuric acid so the optimum pH for the bacterium is 1 or 2 and only low temperatures ($< 10°C$) are required.

In short, many different compounds are oxidised by micro-organisms to release energy and the same or other micro-organisms reduce compounds when using them as terminal electron acceptors under anaerobic conditions. The balance of the micro-organisms in the soil or sediment and the types of substrates available for oxidation (to produce energy) or for reduction (as terminal electron acceptors) will to a large extent determine the chemistry of the soil or sediment.

METHOD OF STUDYING MICRO-ORGANISMS IN SOILS AND SEDIMENTS

This has been reviewed in various texts, most recently by Grigorova and Norris (1990). There are three main ways of estimating micro-organisms; growing them, measuring some consequence of their activity, or simply looking for them in situ. Each of these methods has its advantages and disadvantages and often several different methods are used in an attempt to obtain different views of the microbial abundance and importance in the environment of interest.

In order to grow micro-organisms they must be supplied with a medium suited to their varied nutritional requirements, thus several media and different environmental conditions may be necessary. It may be possible to determine how many bacteria per gram of soil were able to grow in the given conditions but this is by no means the same as how many bacteria are present in that soil. However, if only 1 to 10% are grown, this at least provides a culture to be studied and identified; many organisms seen in sediments simply will not grow in any system yet devised.

If all that is required is to know whether an organism is present or not, the conditions and media which are particularly favourable for that organism may be provided by inoculating the medium with soil and the organism thus encouraged. This technique is known as enrichment culture and is useful for isolating particular physiological groups of organisms or those which are rare in the environment under study.

To measure activity, some product of metabolism is monitored, eg carbon dioxide, or radioactive labelled material is introduced and the rate of incorporation into the microbes determined. This may provide a great deal of information about the environment but not about what lives there. It will also give average effects between, say, the production and release of carbon dioxide, and its use by autotrophs and as a terminal electron acceptor.

Finally, it may be possible to observe the micro-organisms. As they generally live in opaque environments, incidence light microscopes, fluorescence microscopy and scanning electron microscopes will be required. Even then it may not be possible to identify the organism but some information on where it lives may be acquired.

NUMBERS, ACTIVITY AND DISTRIBUTION OF MICRO-ORGANISMS IN SOIL

Most of the methods described above give a very general picture of the micro-organisms in soil but numbers and especially activity varies greatly on different scales. There are different populations in different soil types or under different types of vegetation. In general fertile soils with high levels of organic matter have the most micro-organisms, especially heterotrophs which are the group most usually measured (Campbell, 1983).

In any natural soil there are usually layers or horizons resulting from an accumulation of organic matter on the surface and the downward leaching of minerals and colloids by rain water. There may be a much greater variation between the different horizons of a particular soil than between equivalent horizons in different soils (Campbell, 1983). Activity is concentrated in the surface horizons where energy is available from organic matter derived from the leaves, roots etc of plants growing on or in the soil. As there is most microbial activity here, there is the greatest use of oxygen hence most anaerobic bacteria occur near the soil surface rather than at depth. Anaerobic pockets and microhabitats can develop wherever there is organic matter (eg in tree root runs or where surface soil is buried during construction work).

Even within an horizon there is a sub-structure affecting micro-organisms. Soils are formed into crumbs by the action of micro-organisms physically binding the soil particles together and also by producing mucilage to stick particles together. The soil crumbs and the individual particles are usually surrounded by water films. As water is a poor carrier of oxygen, the water filled pores or particles with water films have a very limited rate of oxygen supply and anaerobic centres develop in soil crumbs. Even in a basically aerated soil with a good crumb structure and therefore good drainage, there can be many anaerobic microhabitats (Campbell, 1983; Smith and Arah, 1986) and the bacteria flourishing inside a crumb will differ from those found on its surface.

Microhabitats are very important to micro-organisms; where it is matters to within a few micrometers as there are very steep gradients of nutrients and oxygen which may determine whether it can live or at least whether it can be active. Microhabitats are transitory and are discontinuous in space and time as environmental conditions and the micro-organisms change and vary.

In general bacterial numbers are in the order of 10^7 or 10^8 per gram dry weight of soil with 10^5 or 10^6 fungi, 10^4 protozoa and lesser numbers of the autotrophic algae concentrated in the light on the surface. Most of these microbes will be limited by the amount of available carbon, oxygen or water. One of the few places where there is great activity is around plant roots which exude available carbon. There is also seasonal activity in the soil responding to leaf fall (available nutrients), wet seasons and favourable temperatures. The pH of the soil rarely limits microbial activity except in special saline soils or those affected by mine drainage, but in neutral to slightly alkaline soils (eg most limed agricultural soils) bacteria tend to be the dominant component whilst fungi become relatively more important in acid soils (heathland soils, coniferous forest soils).

IMPORTANCE OF MICRO-ORGANISMS IN SOILS

Micro-organisms are essential for the normal fertility of soils. They are responsible for the breakdown of all organic matter and the recycling of nutrients to higher plants and, through food chains, to animals. Nutrient cycles (Campbell, 1983; Krumbein, 1978) are interlocked so that the release of minerals is dependent on the breakdown rates of organic matter and the chemistry of that material. Breakdown is often limited by environmental factors such as temperature and the availability of nitrogen and sometimes phosphorus. Organic matter of plant origin is low in nitrogen and that available is sequestered by the micro-organisms and only released again when most of the carbon has been respired and there is excess nitrogen. During the process of release of nutrients therefore, the micro-organisms can induce a deficiency of the element for plants if it is in short supply.

Apart from the decay of organic matter, micro-organisms change the oxidation states of various essential elements such as nitrogen, sulphur, iron and manganese which may greatly affect their solubility, toxicity and availability to plants and animals. The metabolic reasons for these changes have been outlined above and are summarized in Figure 1 (Campbell, 1990). Chemo-autotrophs are responsible for the deposition of some insoluble ores (Krumbein, 1978).

Some bacteria, usually in association with the roots of plants (especially legumes) can 'fix' atmospheric nitrogen gas into ammonium which is available to plants. Before artificial fertilizers this was the only way of using nitrogen gas and it is still the only way in natural ecosystems and for farming in the many countries which cannot afford fertilizers. Nitrogen is often a limiting factor in crop production and supplements may be required to replace loss by leaching, burning and by microbial denitrification.

The soil also contains many micro-organisms harmful to plants and to a lesser extent to animals. Fungi and viruses in particular cause plant diseases and spend part of their life cycle in the soil or survive unfavourable environmental conditions or the absence of the host by lying dormant in the soil.

MICRO-ORGANISMS IN AQUIFERS AND GROUND WATER

The general microbiology of subsurface soil, rocks and aquifers is poorly studied, apart from a few public health aspects of aquifers used for public water supply. In recent years there has been a growing interest in ground water contamination by micro-organisms which are genetically engineered and subject to special control and release regulations. Another area of concern is the contamination of aquifers from waste dumps and their possible bioremediation. It was once thought that percolation through soil would allow micro-organisms to 'purify' leachates from dumps. Although this does occur to a great extent, micro-organisms are not 100% efficient and some chemicals are not degraded or only partly degraded. This is now the subject of active research and several reviews of the general problems of the microbiology of aquifers have been published, eg Ghiorse and Wilson, 1988.

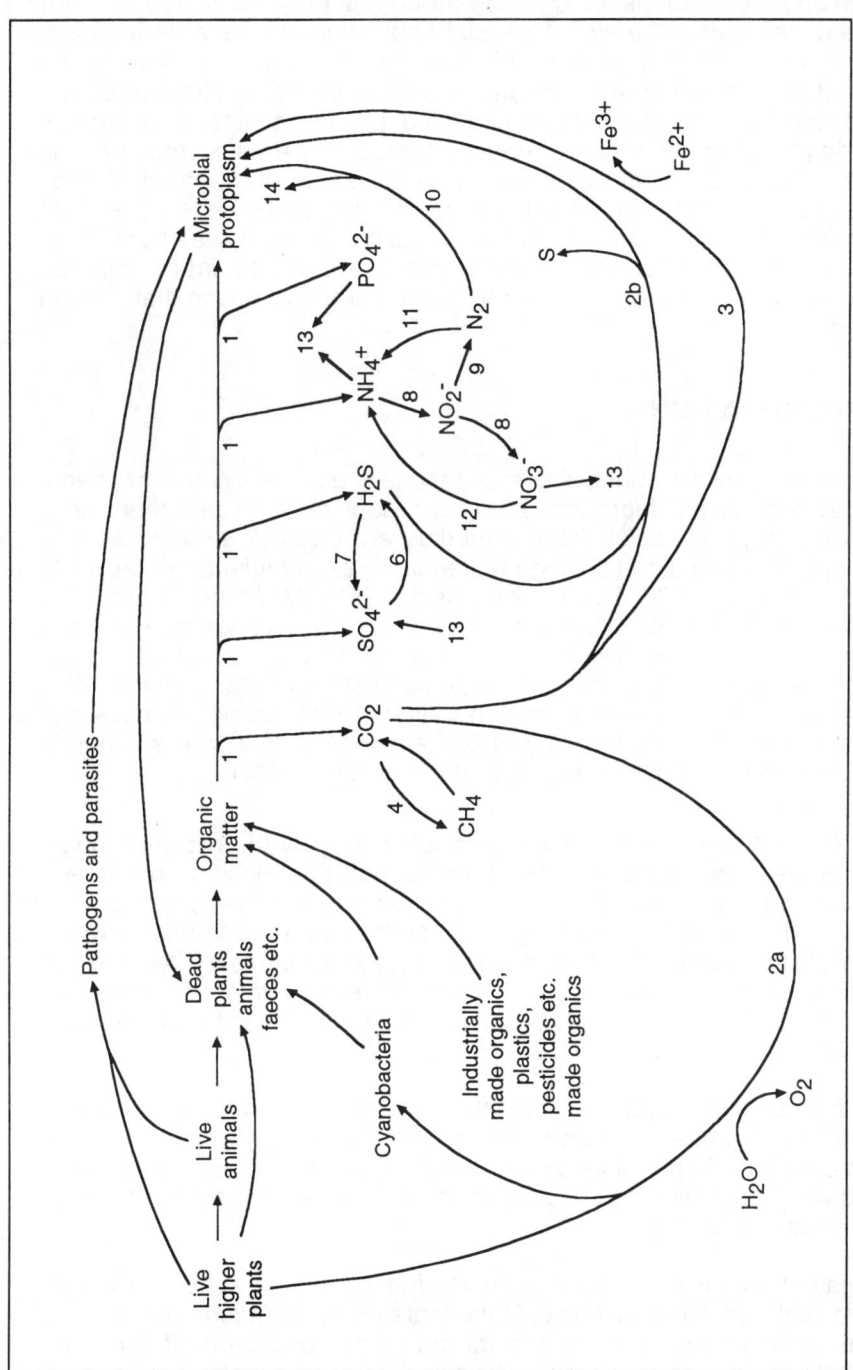

Figure 1: Metabolic reasons for the change in oxidation state of various essential elements by micro-organisms (after Campbell 1990). 1 = decay, 2 = photoautotrophic C fix, 4 = methane formation, 5 = methane oxidation, 6 = sulphate reduction, 7 = sulphide oxidation, 8 = nitrification, 9 = denitrification, 10 = nitrogen fixation, 11 = fertilizers, 12 = nitrate reduction, 13 = plant uptake, 14 = plant uptake via microbes.

Most aquifers are in porous rocks with low nutrient levels for heterotrophs (<1 mg carbon per litre) and being dark, photo-autotrophs are not active. Oxygen levels, albeit low, are probably not limiting in uncontaminated aquifers and populations of chemo-autotrophs may exist if there are supplies of reduced minerals. Microbial populations in general decrease with depth and fungi and protozoa (if recorded) are rare. Bacteria may be present at 10^4 or 10^6/ml even in unpolluted, so-called 'pristine' aquifers. Ghiorse and Wilson (1988) have compiled extensive lists of pollutants whose degradation rates have been monitored in samples from various aquifers. Some of the rates are very slow but as the throughput of most aquifers is also small there is a long residence time for the water. Terminal electron acceptors in these systems are oxygen in pristine sites but nitrate, carbonate and occasionally sulphate are important in more polluted aquifers where breakdown rates of introduced carbon are such that oxygen is limiting.

BIODETERIORATION

Almost all organic materials, especially those made by living organisms, are capable of being degraded by micro-organisms who use them as a source of energy. Many of these materials are used in industry as raw materials, building and construction materials or as manufactured products (Allsopp and Seal, 1986; Rose, 1981; Seal, 1991). Micro-organisms can also cause biodegradation of man-made materials such as plastics and metals and of materials which may not be used *per se* but which are damaged by the release of metabolic acids and other products which affect chemical processes such as corrosion (Miller, 1971). Stone surfaces can be damaged by *Thiobacillus* in polluted environments where sulphur is available for conversion to sulphuric acid (Strzelczyk, 1981).

The destruction of wood and wood products is largely the result of fungal attack but the chemistry of the decay varies with the species concerned; some may use cellulose and leave the lignin and others the opposite. The type of rot greatly influences the strength properties and the use to which the timber can be put. The best method of preservation is by keeping the wood dry, or conversely anaerobic by immersion in deep water or waterlogged soil. Wood products such as paper and board products also decay quite easily.

The deterioration of rubber and plastics affects electrical insulation and seals on buried pipes, especially if they contain sewage which acts as a supplementary carbon source and a ready supply of microbial inoculants. Breakdown of polysulphide sealants is known and may be associated with sulphur oxidising bacteria.

The decay of oil-based products including fuel of various sorts, lubricants, hydraulic fluids etc can cause loss of the material itself but more usually the problem is secondary such as metal corrosion, blockage of fuel and hydraulic lines and the breakdown of cutting emulsion stability. In the past the main problem with fuel has been with that for aircraft based on kerosene, but now that the excellent biocide, lead, is being removed from car petrol, this could also cause problems; although the petrol fractions are intrinsically less degradable than kerosene.

Paints are a good example of where deterioration has become worse because of environmental concerns and modern requirements for ease of application. Paints used to be oil based, with no water, hence they did not deteriorate in the can. Furthermore, many of the pigments were lead based, but lead is now banned in most paints and water/oil emulsions are used. The easy-application, thickened, non-drip paints have a further advantage as far as the micro-organisms are concerned, in providing a readily available carbon source such as methyl cellulose. The activity of anaerobes within cans may be particularly hazardous when it results in the production of gases such as CO_2 and CH_4. Deterioration in use is a separate problem and paints now routinely contain considerable quantities of biocides.

Metal corrosion can and clearly does take place without microbes, but they can exacerbate the effects in a number of ways (Miller, 1985 in Rose; Miller, 1971; Tiller, 1990 in Howsam). The basic process is one of metal ionisation. Often a protective layer of corrosion products forms over the surface, such as rust or hydrogen on the cathodic part of the metal. Some anaerobes (*Desulfovibrio*) can use hydrogen as an energy substrate and oxidise it to water if oxygen is present or more usually to hydrogen sulphide if sulphate is the terminal electron acceptor. This causes cathodic depolarisation and allows ionisation to proceed. Elimination of electron acceptors and the carbon sources for the heterotrophic sulphate reducers is therefore a useful method of control and for this reason pipe trenches should be backfilled with inert gravel rather than nutrient rich top-soil. Sulphur oxidisers (*Thiobacillus*) can produce sulphuric acid and so increase corrosion. The *Cladosporium* fungus causes corrosion of aircraft fuel tanks by producing metabolic acids. Another mechanism by which micro-organisms such as *Cladosporium* can assist corrosion is by creating relatively anaerobic patches when a colony grows on the metal surface. There is an anodic area in the middle of the colony and a cathodic area surrounding the colony. Pit corrosion occurs as the metal ionises in the centre and electrons move to the outer edges of the colony where they are removed by the conversion of O_2 to the hydroxide which in the case of iron or aluminium is insoluble.

BIOREMEDIATION OF POLLUTION AND XENOBIOTICS

Xenobiotics are man-made, 'unnatural' substances introduced into the environment, either by design as in the case of pesticides or landfill dumping, or by accident as in the case of spillages. It is surprising how little affected microbes are by even very toxic substances in the soil; indeed micro-organisms can and do degrade many xenobiotics rendering them less harmful to the environment. This has led to the conscious use of micro-organisms for *in situ* bioremediation.

Typical compounds are the many organic pesticides (of which there are now several hundred active ingredients formulated in more than a thousand products in the UK alone). Numbers of active compounds have decreased recently as products have had to be re-tested to meet EC guidelines and in some cases the chemical has been withdrawn when it was considered that the market was not worth the expense. However about ten new active ingredients are marketed each year in the UK. Most of these are designed to be biodegradable to meet the requirements of the registration and testing agencies; Cork and Krueger (1991) provide extensive information on the breakdown of such products in the soil.

Wastes from industry may result in ground or water contamination on or near old industrial sites. A wide range of chemicals may be involved including toxic phenolics and heavy metals. The chemical and allied industries account for about 62% of the hazardous substances produced and the metals industry a further 15% (Hicks, Stotzky and Von Voris, 1990). Once in the soil the compound may be physically changed in terms of volatilisation and adsorption on clays and other colloids or may be changed chemically by the effects of micro-organisms. Microbial abundance is sometimes drastically reduced by very toxic spills but more usually the microbial numbers increase or are unaffected. The increase may represent the renewed activity of micro-organisms if they can use the substance as a carbon and/or nitrogen source in the nutrient limited soil environment. Alternatively the substance may kill the more sensitive higher animals and plants (insecticides and herbicides for example). These provide a food base for the soil micro-organisms, although the effect is secondary rather than directly on the microbes themselves.

In general, micro-organisms of some sort will survive much higher concentrations of more toxic compounds than will higher plants and animals, albeit with reduced metabolic rates. Different groups of micro-organisms respond in different ways to the pollutants. Some pathways, such as nitrification (the conversion of ammonium to nitrate), are particularly sensitive and with the present state of knowledge it is difficult to predict the effect of a particular chemical.

There are special micro-organisms selected and marketed to remove or reduce specific forms of pollutants, particularly the petroleum products and some of the phenols Such organisms have been used to clean up oil spills and make old industrial sites less toxic (Ellis and Bewley, 1990). However, as there are so many micro-organisms in the soil there are often some which can be effective without the specific addition of a prepared inoculum, especially if the spill is quite old and there has been time for an 'enrichment culture' to select organisms which can handle the particular pollutant. For example, there are reports that natural soil with no previous history of petroleum contamination may have 10^5 or 10^6 micro-organisms which can degrade petroleum products per gram of soil but soils which have been subjected to such contamination in the past may have 10^8 to 10^9 per gram (Bossert and Bartha, 1985). In this case all that is necessary is to supply any missing nutrients (such as nitrogen and phosphorous or sometimes a carbon source if the xenobiotic is not likely to have sufficient available carbon to support growth) and to improve aeration by moving or spreading the soil or injecting air into it. Improved aeration is invariably needed to sustain the bioremediation (Bossert and Bartha, 1985). After the first degradation of the available fraction in the oil there is frequently a much slower degradation of more recalcitrant fractions and material adsorbed onto clays or otherwise closely bound in the soil.

When micro-organisms destroy petroleum products in lubricants or fuel it is seen as a problem of biodeterioration yet if they destroy the same products accidentally spilled it becomes the much valued process of bioremediation.

Plants and soil animals are usually more sensitive to oil spills than the micro-organisms. Vegetation may take many years to recover. Even with lesser spills, the microbial demand for oxygen and nitrogen can induce mineral deficiency and make the soil anaerobic, even if the plant is not directly damaged by the oil. Another aspect of pollution control is the use of microbes to prevent its occurrence. The systems now used to treat sewage are dependent on micro-organisms to break down the excess organic matter and prevent the imposition of a large biological oxygen demand on the environment. Activated sludge tanks can be used to reduce the biological oxygen demand of industrial effluents, for example, and also to reduce the levels of toxic components in the effluent before discharge direct or via the sewage system. There are also special fermentation systems which reduce the biological oxygen demand of food processing waste etc and at the same time produce sufficient microbial biomass to be used as animal feed or even for human consumption.

As some bacteria, fungi and algae adsorb or even concentrate metals against concentration gradients, fermentation and other microbial systems may be used to remove heavy metals from effluents. For example, a fungus in a continuous fermenter system reduced the levels of cadmium from 5 ppm to less than 1 ppm while at the same time removing more than 90% of the metal from the liquid flowing through and accumulating nearly 3% metal by weight in the fungal biomass (Campbell and Martin, 1990).

Micro-organisms, especially the bacteria in the genus *Thiobacillus*, are used to extract minerals from low grade ores by leaching large dumps. The bacteria convert the sulphide ores into soluble sulphates and increase the acidity to release the copper or iron which is then collected and recovered by chemical precipitation (Brierley , Brierley and Davidson, 1989). The same activity leads to deleterious effects in old mine workings which become flooded when they are abandoned; the drainage water is very acid and usually contains toxic levels of metals which are environmentally damaging.

CONCLUSIONS

For the engineer, micro-organisms have advantages and disadvantages. Indeed the same micro-organism in some situations may be regarded as "good" and in others as "bad". As they can exist in almost any kind of environment it is essential that their presence is appreciated and understood so that where possible they may be manipulated to man's advantage. Having existed for several billion years, it is likely micro-organisms will continue to inhabit the earth long after man has become extinct.

REFERENCES

ALLSOPP, D. & SEAL, K.J. (1986). *Introduction to Biodeterioration*. Edward Arnold, London. pp. 136.

ATLAS, R. M. (1984). *Microbiology, Fundamentals and Applications*. MacMillan Publishing Co., New York. pp. 987.

ATLAS, R.M. & BARTHA, R. (1987). *Microbial Ecology, Fundamentals and Applications.* 2nd Edition. Benjamin/Cumming Publishing Co., Menlo Park, California. pp.533.

BERKELEY, R.C.W., LYNCH, J.M., MELLING, J., RUTTER, P.R. & VINCENT, B. (1980). *Microbial Adhesion to Surfaces.* Ellis Horwood Publishers, Chichester, England. pp. 559

BOSSERT, I. AND BARTHA, R. (1985). The fate of petroleum in soil ecosystems. In: *Petroleum Microbiology*, Atlas, R. (ed), MacMillan Publishing Co., New York. pp. 435-473.

BRIERLEY, C.L., BRIERLEY J.A. & DAVIDSON, M.S. (1989). Applied microbial processes for metal recovery and removal from wastewater. In: *Metal Ions and Bacteria*, Beveridge, T.J. and Doyle, R. J. (eds), John Wiley and Sons, New York. pp. 359-382.

BROCK, T.M. & MADIGAN, M.T. (1991). *Biology of Micro-organisms*. 6th Ed. Prentice Hall International Inc., Englewood Cliffs, California. pp. 874.

CAMPBELL, R. (1983). Microbial Ecology. In: *Basic Microbiology*, volume 5. 2nd Edition. Blackwell Scientific Publications, Oxford. pp. 191.

CAMPBELL, R. (1990). The ecology of soil bacteria. In: *Topley and Wilson's Principles of Bacteriology, Virology and Immunity, Volume 1, General Bacteriology and Immunity, Eighth Edition*, Linton A.H. and Dick, H.M. (eds), Edward Arnold, London. pp. 213-223.

CAMPBELL, R. & MARTIN, M.H. (1990). Continuous flow fermentation to purify waste water by the removal of cadmium. *Water, Air and Soil Pollution*, **50**, 397-408.

CORK, D.J. & KRUEGER, J.P. (1991). Microbial transformations of herbicides and pesticides. *Advances in Applied Microbiology*, **36**, 1-66.

EDWARDS, C. (1990). *Microbiology of Extreme Environments*. Open University Press, Milton Keynes, UK. pp.218.

GHIORSE, W.T. & WILSON, J.T. (1988). Microbial Ecology of the Terrestrial Subsurface. *Advances in Applied Microbiology*, **33**, 107-172.

GRIGOROVA, R. & NORRIS, J.R. (1990). *Methods in Microbial Ecology, Volume 22, Techniques in Microbial Ecology.* Academic Press, London. pp. 627.

HICKS, R.J., STOTZKY, G. & VAN VORIS, P. (1990). Review and evaluation of the effects of xenobiotic chemicals on micro-organisms in soil. *Advances in Applied Microbiology*, **36**, 195-253.

HOWSAM, P. (1991). *Microbiology in Civil Engineering.* FEMS Symposium 17. E & F N Spon (Chapman and Hall), London. pp. 382.

JENSEN, V., KJØLLER, A. & SØRENSEN, L.H. (1986). *Microbial Communities in Soil.* Elsevier Applied Science Publishers, London. pp. 447.

KRUMBEIN, W.E. (1978). *Environmental Biogeochemistry and Geomicrobiology.* 3 vols. Ann Arbor Science Publishers, Ann Arbor, Michigan. pp. 1055.

KUSHNER, D.J. (1978). *Microbial Life in Extreme Environments.* Academic Press, London. pp. 465.

MILLER, J.D.A. (1971). *Microbial Aspects of Metallurgy.* Medical and Technical Publishing Co., Aylesbury, England. pp. 202.

PATTERSON, D.J. & HEDLEY, S. (1992). *Free-living Freshwater Protozoa, a Colour Guide.* Wolfe Publishing Ltd., London. pp. 223.

ROSE, A.H. (1981). *Microbial Biodeterioration.* Academic Press, London. pp. 516.

SMITH, K.A. & ARAH, J.R.M. (1985). Anaerobic Micro-environments in soil and the occurrence of anaerobic bacteria. In: *Microbial Communities in Soil.* JENSEN, V., KJØLLER, A. & SØRENSEN, L.H. (eds), Elsevier Applied Science Publishers, London, 247-261.

Session 2 - Foundations

2-1 Understanding sulphate generated heave resulting from pyrite degradation
A.B. Hawkins & G.M. Pinches — 51

2-2 Some geotechnical problems associated with pyrite bearing mudrocks
J.C. Cripps & R.L. Edwards — 77

2-3 Investigation of the effects of bacterial action on the chemistry and mineralogy of pyritic shale
S.D. Jackson & J.C. Cripps — 89

2-4 The generation of sulphates in the proximity of cast in situ piles
A.B. Hawkins & M.D. Higgins — 101

2-5 Roadford Dam: geochemical aspects of construction of a low grade rockfill embankment
S.E. Davies & J.M. Reid — 111

2-6 The assessment of volumetric loss for clay/sand and clay/clay mixes
S.H. Al-Jassar & A. B. Hawkins — 133

2-7 Effect of some chemical additives on the strength development of soil-cement
R. Angelova — 147

2-8 Methane generation from void formers used in foundations on clay sites
R.W. Johnson — 161

2-9 Gaseous pollutants and their implications for site safety
J.S. Butterworth, D.P. Creedy & J.S. Edwards — 167

2-10 Hydrogen sulphide generation beneath a deep basement in Gault Clay
D.F.T. Nash, M.L. Lings & S.A. Jefferis — 187

2-1 UNDERSTANDING SULPHATE GENERATED HEAVE RESULTING FROM PYRITE DEGRADATION

A.B. HAWKINS
Engineering Geology Research Group, University of Bristol, Wills Memorial Building, Queens Road, Bristol BS8 1RJ.

G.M. PINCHES
Scott Wilson Kirkpatrick & Partners, Bayheath House, Rose Hill West, Chesterfield S40 IJF.

ABSTRACT

None of the international engineering codes of practice adequately addresses the potential for ground heave due to sulphate generation. This paper describes the natural or man-induced chemical/microbial reactions which can produce such heave and illustrates the phenomenon with a case history from Llandough Hospital, Cardiff. It then considers methods of ground investigation to assess whether a potential problem exists and how the effects can be minimised by appropriate engineering design and construction.

INTRODUCTION

Pyrite (FeS_2) occurs in many sedimentary rocks but rarely forms more than 5% by mass in ancient sediments (Taylor and Cripps, 1984). In mudrocks it is found most frequently as framboids with diameters varying between 2 and 40 μm, sometimes as fine disseminated crystals and rarely as visible cubes. Generally the iron sulphides are formed in the anoxic conditions of deep marine environments although they also occur in shallow coastal water and occasionally in intertidal marshes. Often the formation of pyrite is associated with the presence of organic matter, although even in the very dark grey mudrocks evidence of the previous existence of such matter may be difficult to detect. In addition, some sedimentary sulphides are formed as a result of the microbial reduction of dissolved sulphates.

The deleterious effects of sulphate on both concrete and any contained reinforcement has long been appreciated, eg Fookes and Collis (1975). This has been recognised in various international standards, eg British Standard BS 1377:1975 and 1990, while in BRE Digest 363 (previously 250) guidance is given on how the ground chemistry should be measured in order to assess its potential aggressivity towards concrete. However, despite the known occurrence of heave related to both colliery shale fill (Nixon, 1978) and dark shale bedrock (Hawkins and Pinches, 1987a), for the engineer there is still little guidance in the literature as to the processes involved in sulphate development and what action can be taken to minimise the risk of problems associated with this.

In the last forty years a number of cases have provided evidence of heave being created by sulphate growth. Initially work in Norway by Moum and

Rosenqvist (1959) highlighted the capacity of aerated dark shale fill to oxidize, facilitating other chemical reactions which in turn result in significant ground heave. Further, it is known that mine floors may experience heave related not only to stress release but also to an increase in volume of the floor rocks due to the development of ground sulphates (Coveney & Parizek, 1977). In Britain Nixon (1978) described floor heave in houses at Teesside where compacted colliery shale had been used to support the ground bearing floor slabs. He concluded that mudrocks should be regarded as "troublesome" if the sulphur content exceeds the acid soluble sulphate content and if more than 0.5% $CaCO_3$ is present.

Earlier in the 1970s, Canadians working in the Ottawa region were the first to describe significant sulphate-related heave problems occurring in undisturbed ground beneath the Bell Canada building and the Rideau Health Centre (Quigley and Vogan, 1970; Penner, Eden and Gillott, 1973). These buildings were founded respectively on the Billings Formation and Lorraine Formation, both of which are Lower Palaeozoic fissile mudrocks. In each case ground levelling prior to construction had exposed less weathered material at formation level and in the case of the Rideau Health Centre the situation was exacerbated by the design of the building.

The first occurrence of sulphate induced ground heave developing in natural undisturbed strata to be reported in Britain was at Llandough Hospital, Cardiff (Hawkins and Pinches, 1987a); the relevant part of this structure being of almost identical design to that of the Rideau Health Centre in Ottawa (Figure 1). Again almost fresh shale was exposed when up to 3 m of ground was excavated at the eastern end of the hospital in order to maintain the 200 m long main corridor at a constant elevation. As at the Rideau Health Centre, side heating ducts were incorporated in the design and it is likely the resultant increase in temperature accelerated the development of the ground sulphates.

Studies have shown that the degradation of pyrite involves both chemical and microbial processes and that the natural rate of degradation may be considerably enhanced as a consequence of engineering construction. This paper proposes possible mechanisms for the accelerated geochemical and microbial degradation of pyrite and the way in which consequential heave may lead to structural failure. A case history is provided to illustrate the phenomenon and guidance is offered as to how such structural failure could be predicted and/or recognised. Engineering solutions to prevent the deleterious effects of pyrite degradation are discussed.

CHEMISTRY AND GEOLOGY OF PYRITE AND SULPHUR

Being composed of reduced iron and reduced sulphur, pyrite is metastable in atmospheric environments. Both parts of the pyrite molecule are characterised by multivalent behaviour, the iron ions having an oxidation state of either +2 or +3. As illustrated in Table I, sulphur demonstrates an extreme multivalent behaviour. Sulphur can form anions with oxidation states -2 or -1 but may also form a whole range of compound ions with oxygen where it demonstrates oxidation states between +2 and +6. With such a variation in oxidation states, it is possible for numerous

SULPHATE GENERATED HEAVE FROM PYRITE DEGRADATION

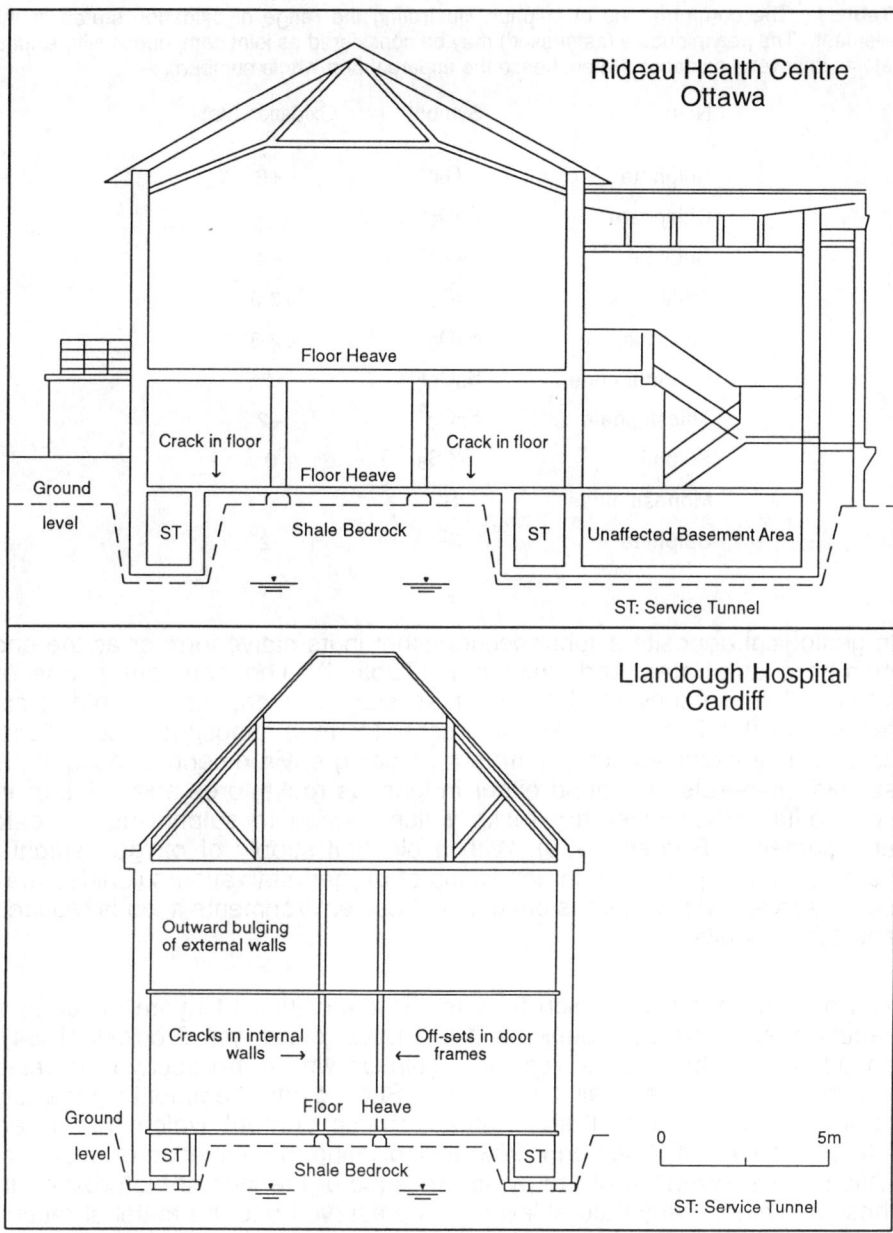

Figure 1: Cross sections showing style of structure and evidence of distress at the first well-documented cases of ground heave due to sulphate generation in Canada (above) and the UK (below).

thermodynamic pathways to be created by means of a large series of redox reactions. These can provide the sulphur required by some microbial organisms for the derivation of their metabolic energy while other microbes are able to substitute sulphur for oxygen as the terminal electron acceptor.

Table I: The common ions of sulphur, illustrating the range of oxidation states of this element. The polythionates (asterisked) may be considered as joint compounds with sulphur atoms in several oxidation states, hence the apparent non-whole numbers.

Name	Symbol	Oxidation State
Sulphate	SO_4^{2-}	+6
Dithionate	$S_2O_6^{2-}$	+5
Sulphite	SO_3^{2-}	+4
Trithionate	$S_3O_6^{2-}$	+3.3*
Tetrathionate	$S_4O_6^{2-}$	+2.5*
Pentathionate	$S_5O_6^{2-}$	+2*
Thiosulphate	$S_2O_3^{2-}$	+2
Sulphur	S^o, S_8	0
Monosulphide	S^-	-1
Sulphide	S^{2-}	-2

In geological deposits sulphur occurs either in its native form or as the end members sulphates and sulphides (Table I). The compound ions of intermediate valency tend to occur as stages in chemical or biological reactions, often in solution. As pyrite cannot form in atmospheric conditions, its presence indicates survival from a reducing environment. Consequently sulphide minerals are found either in igneous rocks, ores and veins or in soft sediments/sedimentary rocks which formed in sulphur-rich anoxic environments (Berner, 1970). With a plentiful supply of oxygen sulphur develops as sulphate but in conditions of oxygen starvation sulphides may be preserved. Where iron is present in these environments it too is reduced and pyrite results.

The anoxic conditions which lead to the deposition of pyrite occur in a variety of environments (Jenkyns, 1980; Love, Al-Kaisy & Brockley, 1984), notably below the zone of regular circulation where the decay of organic material uses all the available oxygen. Such pyritic sediments generally retain a proportion of their original organic content which has been interpreted by van Breemon (1988) as indicating the influence of microbial action in the formation of the pyrite. Because of the lack of bioturbation in these sediments, any natural layering is preserved and the material retains its fissility (the shales described by Hawkins and Pinches, 1992). Although the classic conception of pyrite is the visible cubes of "fool's gold", in sedimentary formations it often occurs as disseminated grains of 0.1 to 0.5 μm diameter or as the 2 to 40 μm size pyrite clusters known as framboids. The large specific surface of these crystals and fine aggregates enhances their susceptibility to chemical or bacterial attack. In addition, the high density of the sulphide minerals is important in explaining how their degradation leads to heave. When pyrite, which has a specific gravity of 4.8-5.1, combines with calcite (specific gravity 2.7) for example, the resultant mineral is gypsum ($CaSO_4.2H_2O$) which has a specific gravity of only 2.3.

THE DEGRADATION OF PYRITE

CHEMICAL ATTACK

As many museum curators would confirm, pyrite specimens frequently degrade, albeit over several decades. Initially the oxidation of sulphide leads to the generation of sulphuric acid

$$2FeS_2 + 2H_2O + 7O_2 \rightarrow 2FeSO_4 + 2H_2SO_4 \quad [1]$$

Further oxidation is then inhibited as Fe(II) is stable in acid conditions. In order for the reaction to continue by purely chemical processes, the acid must be either removed or neutralised. The oxidation reactions require both oxygen and moisture. In a closed system (without acid loss) the oxidation of both parts of the pyrite molecule involves many stages; in some of which the action of bacteria has been recognised (Hawkins and Pinches, 1987a). One of the products of the degradation process is ferric sulphate [$Fe_2(SO_4)_3$], which is itself an oxidation agent capable of attacking pyrite according to the equation:

$$7Fe_2(SO_4)_3 + FeS_2 + 8H_2O \rightarrow 15FeSO_4 + 8H_2SO_4 \quad [2]$$

Alternatively, a two stage process could be involved, with free sulphur produced at an intermediate stage:

$$Fe_2(SO_4) + FeS_2 \rightarrow 3FeSO_4 + 2S \quad [3]$$

$$6Fe_2(SO_4)_3 + 2S + 8H_2O \rightarrow 12FeSO_4 + 8H_2SO_4 \quad [4]$$

Ferric sulphate can also be hydrolysed:

$$Fe_2(SO_4)_3 + 2H_2O \rightarrow 2Fe(OH)SO_4 + H_2SO_4 \quad [5]$$

Although in neutral conditions the solubility of Fe(III) is too low to have much effect, in acid conditions (pH <3) it can become operative as an oxidizing agent which acts far more rapidly than pure oxygen (van Breemon, 1988). As a result of the chemical degradation of pyrite iron sulphates and sulphuric acid are produced which can then attack other components present in the ground, leading to the creation of new minerals.

THE THIOPHILIC BACTERIA

A number of genera of chemotrophic organisms derive their metabolic energy from reactions involving inorganic sulphur to fix carbon dioxide. Some of these are capable of respiring using oxidized sulphur rather than oxygen as the terminal electron acceptor. Most of the thiophiles tend to live in special environments such as solfataras but bacteria of the genus *Thiobacillus* also occur in both rocks and soils. The *Thiobacilli* are generally aerobes, oxidizing sulphides to sulphates. A number of different species have been recognised; their carbon sources, oxygen conditions and temperatures are given in Table II. Only *Thiobacillus ferrooxidans* is capable of converting Fe(II) to Fe(III), which it does in an acidic environment. This characteristic is a key component of pyrite degradation. Colmer, Temple and Hinkle (1950) first identified *Thiobacillus ferrooxidans* from the acid drainage from bituminous coal mines and were able to show that the purely chemical reaction previously considered to be significant in sulphate generation was in fact accelerated by the activity of the species.

Table II: Species of the genus *Thiobacillus* and some of their growth conditions (from Vishniac, 1974). Note that pure autotrophs are obligate chemolithotrophs and autotrophic and heterotrophic species are facultative chemolithotrophs.

SPECIES	CARBON SOURCE		OXYGEN CONDITIONS		TEMPERATURE	
	Auto-troph	Hetero-troph	Aerobic	Anaerobic	Optimum (range)	Optimum (range)
T thioparus	✓	x	✓	possible on nitrate	28°C	pH 6.6-7.2 (4.5-7.8), max pH 10.0
T neapolitanus	✓	x	✓	x	28°C	pH 6.2-7.0 (3.0-8.5)
T thiooxidans	✓	x	✓	x	28-30°C (10-37)	pH 2.0-3.5 (0.5-6.0)
T denitrificans	✓	x	x	✓		pH (6.0-8.0)
T ferrooxidans	✓	possible on some strains	✓	x	15-20°C (max 25)	pH 2.5-5.8 (1.4-6.0)
T novellus	✓	x	✓	x	30°C	pH 7.8-9.0 (5.0-9.2)
T intermedius	✓	✓	✓	x	30°C	pH 6.0-7.0 (1.9-7.0)
T perometabolis	x	✓	✓	x	30°C	pH (2.8-6.8)
T acidophilus	✓	✓	✓	x		pH 3.0 (2.0-4.0)
T A2	✓	✓	✓	✓		pH (7.0-9.0)
T organoparus	✓	✓*	✓	x		pH 3.0 (2.0-5.0)

* may be chemolithotrophic on sulphur compounds or organolithotrophic on glucose

Fliermans and Brock (1972), studying solfataras from the Yellowstone National Park, isolated two bacteria - the spherical *Sulfolobus* which thrives between 40 and 90°C and the rod-shaped *Thiobacillus* which decreased rapidly in numbers above 50°C, with none identified at 60°C. These authors suggested that *Thiobacillus thiooxidans* could grow in temperatures of up to 55°C.

Most *Thiobacilli* obtain the carbon they require from the CO_2 in the atmosphere. When sulphuric acid reacts with calcite (Equation 9), CO_2 is released which would also provide a carbon source and hence assist the proliferation of the bacteria.

MICROBIAL ATTACK

Proving that the *Thiobacilli* directly attack pyrite rather than simply act as a catalyst is not easy. However, Kelly, Norris and Brierly (1979) have described the presence of etch marks along the edge of pyrite crystals which match the 0.4 to 1 μm size and rod-shaped habit of these acidophilic microbes and were able to show the physical attachment of *Thiobacillus* and *Sulfolobus* to the mineral surface. In the acid conditions created by the reactions shown in Equation 1, the ferrous sulphate generated would remain around the grain. However *T. ferrooxidans* is able to oxidize Fe(II) to form the oxidizing agent ferric sulphate as follows:

$$4FeSO_4 + O_2 + 2H_2SO_4 \rightarrow 2Fe_2(SO_4)_3 + 2H_2O \qquad [6]$$

Once the bacteria are established, therefore, a system is set up in which pyrite is quite rapidly degraded by a combination of chemical and microbial processes. The ferric sulphate generated by the *T. ferrooxidans* chemically converts more pyrite into sulphuric acid and ferrous sulphate thus providing further substrate for the bacteria.

Although *T. ferrooxidans* is the most important of the *Thiobacilli* in pyrite oxidation, most species of the genus could play a part if others of the numerous alternative pathways are followed. For instance, if free sulphur is generated by the action of ferric sulphate on pyrite (Equation 3), *T. thiooxidans* can convert the resultant sulphur to sulphuric acid:

$$2S + 3O_2 + 2H_2O \rightarrow 2H_2SO_4 \qquad [7]$$

Some authors believe that the reactions given in Equation 1 can also be assisted by autotrophic bacteria and Kelly et al (1979) report an alternative pathway directly affected by bacteria to be:

$$4FeS_2 + 15O_2 + 2H_2O \rightarrow 2Fe_2(SO_4)_3 + 2H_2SO_4 \qquad [8]$$

Although both *T. thiooxidans* and *T. ferrooxidans* are acidophiles, species of *Thiobacilli* favouring neutral conditions have also been recorded (Table II).

GYPSUM

In most case studies where structural failure has been ascribed to pyrite degradation the sulphate mineral gypsum has been identified. In the first described case histories from both Canada and the UK, gypsum in the crystal form of selenite (Plate 1) was found along bedding plane laminations in the shaley mudrocks below the distressed structure. In each example it was considered that the gypsum resulted from the reaction between sulphuric acid and the calcium minerals present in the bedrock:

$$H_2SO_4 + CaCO_3 + 2H_2O \rightarrow CaSO_4.2H_2O + H_2O + CO_2 \qquad [9]$$

Plate 1: Bladed gypsum crystals (selenite) found along bedding planes of black shale recovered from interior trial pit, Llandough Hospital.

In addition to its density, the very low solubility of gypsum (120 mg/l) is also a significant factor in its role in ground heave. While gypsiferous waters may be mobile within the ground, once gypsum crystallises it is not easily re-dissolved or removed by ground water flushing.

OTHER MINERALS

Although gypsum is the most important mineral associated with pyrite degradation, a range of other minerals may be formed including the brown iron oxide and hydroxide minerals (eg goethite and limonite). It is likely that the bright yellow mineral jarosite is the most important non-gypsum mineral causing heave. It can be created as a result of the chemical attack of the common clay mineral illite by the reaction product ferrous sulphate:

$$12FeSO_4 + 4(KAl_2Si_3O_8(OH)_2) + 48H_2O + O_2 \rightarrow$$
$$4(KFe_3(OH)_6(SO_4)_2) + 8Al(OH)_3 + 12Si(OH)_4 + 4H_2SO_4 \quad [10]$$

Jarosite is a member of the rhombohedral alunite group of basic sulphates with complete solid solution between jarosite, natrojarosite and hydronium jarosite. Jarosites usually occur as massive aggregates of well developed rhombohedral crystals about 1 μm in diameter (van Breemon, 1988) and give a characteristic X-ray diffraction pattern. Gillott, Penner and Eden (1974) reproduce photomicrographs showing 10-30 μm size spheroids of jarosite composed of 2 μm tabular octahedral and pseudotetrahedral crystals. However Quigley, Zajic, McKyes and Yong (1973) show jarosite carpeting a discontinuity as an amorphous mat rather than a series of discrete crystals.

Jarosite is relatively more stable than the Fe(III) oxides with its solubility field limited to extremely acid conditions. As a consequence jarosite is usually observed only in soils with a pH below 4 and even under these conditions it is slowly hydrolysed to goethite. At pH 1-2 and in the presence of *Thiobacillus ferrooxidans*, it is known that jarosite may be formed within weeks. It will also persist for several decades even after the addition of lime and it has been observed in Tertiary sediments.

FACTORS AFFECTING PYRITE DEGRADATION AND RELATED HEAVE

The following conditions are typical and in most cases essential for the production of heave as a result of pyrite degradation:

1. The presence of pyrite in saturated ground.

2. A change in ground water regime and partial pressure which allows oxygen and carbon dioxide to enter the soil and come into contact with the pyrite.

3. Oxidation of pyrite. Chemically this is a relatively slow process with oxygen as the oxidizing agent but when the resultant ferrous sulphate is utilised by *Thiobacillus ferrooxidans*, ferric sulphate is formed. Ferric sulphate is itself an oxidizing agent whose action on pyrite is much more rapid than that of pure oxygen, provided the pH is below 3.

4. *Thiobacillus ferrooxidans* must already be present in the sediment or must colonize the oxidizing environment. A continuous supply of moisture, a low pH and the availability of atmospheric oxygen and carbon dioxide will increase the activity and encourage the proliferation of this bacteria.

5. The sulphuric acid generated will naturally migrate within the ground until other substances are encountered with which it can combine. The proportion of calcite present in the host material is critical. To permit the acidic conditions appropriate to the chemical and microbiological activity, only small quantities are required; excess calcite would neutralise the environment and gypsum would not form. Although the reaction is preferentially with calcite, when this is not present but the host rock contains the clay mineral illite, a chemical reaction will take place with this mineral to form jarosite.

6. Sulphate crystals develop in the capillary zone and as they grow preferentially in areas of least stress, they tend to be localised along discontinuities. If a near-horizontal laminated bedding fissility is present (eg in shales) vertical heave is greatest while in a massive mudrock with random fissuring, the expansive effects will be dissipated.

7. Sulphate-induced heave is a direct consequence of the density differences resulting from the chemical reactions. However, as the blade-shaped selenite crystals occupy only a limited proportion of the bedding surfaces, the effect is proportionately much larger than if the whole aperture of each discontinuity had been completely infilled.

8. If a structure spans more than one geological condition, differential ground heave may result in significant distress to the building.

9. Raised temperatures associated with boilers etc may contribute to an acceleration of pyrite degradation and the associated heave processes. Thus particular features in the building design may lead to a locally enhanced growth rate, again resulting in differential stress.

CASE HISTORY

RECOGNITION

Quigley and Vogan (1970) reported that in the Rideau Health Centre which was constructed in 1950, floors had heaved by 76 mm and cracked. In the Bell Canada Building, constructed in 1961, the deep basements were out of level by 1966 and the heavy equipment installed had ceased to be efficient (Penner, Eden and Gillott, 1973). The first recognised British case was at Llandough Hospital, Cardiff where floors built in the 1930s had risen by at least 60 mm after approximately fifty years.

LLANDOUGH HOSPITAL

The full case history of heave at Llandough Hospital is given elsewhere (Hawkins and Pinches, 1987a) but the salient points are reported and expanded here to illustrate the postulated processes in sulphate related heave and to draw attention to the way in which the occurrence of this may be recognised or predicted.

The area of concern was the eastern ward which had experienced a history of problems; the doors had been planed several times, the floor re-screeded and the internal walls had cracked. In 1982 it became obvious that the outer walls were bulging so badly that the roof could become unsafe and as a result the ward was closed to allow investigation and remedial works.

The distress was confined to a small part of the hospital - that constructed over an outcrop of the Westbury Formation of the Penarth Group of Rhaetic (Triassic) age (Figure 2). The Westbury Formation is characteristically a weak, dark grey to black, fissile, thinly laminated mudrock (shale) with occasional limestone horizons. It is highly pyritic (1 to 5% FeS_2) and organic (maximum 10%) with a calcite content of 2 to 6%. No distress was recorded in the other hospital buildings to the south of the Penarth Fault which were constructed on the more massive, non-fissile, non-pyritic blue-grey mudstones of the Blue Anchor Formation (formerly Tea Green Marl) of

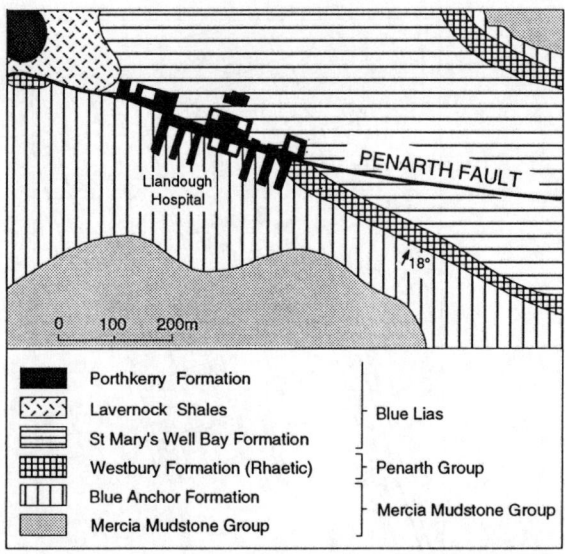

Figure 2: Geological map of the Llandough Hospital area.

the Mercia Mudstone Group or to the north where the bedrock was the interbedded limestones and mudstones of the Blue Lias Formation.

In August 1982 trial pits were excavated adjacent to the east and west walls of the ward block and beneath the interior floor, where the distress appeared to be centred (Figure 3). They confirmed the geology shown on the 1:10,560 map (Sheet ST 17 SE) and by means of two limestone bands correlation across the ward was possible (Figure 4). On visual inspection the interior pit proved strikingly different to the exterior ones. The shale in the interior pit contained significantly more oxidation products while between 0.85 m depth and the top of the upper limestone band at 1.75 m, a large amount of crystalline gypsum in the form of selenite was present along the laminations and in the joints. Laboratory tests on samples taken at 0.25 m intervals confirmed the visual differences observed and the initial results provided sufficient evidence for remedial works to be planned. These included underpinning of the internal walls by means of needle piles taken to a limestone band at about 4 m which was considered to be well below the zone of enhanced gypsum growth.

When the pile holes were opened in February/March 1983 further samples were taken, again at 0.25 m intervals but to a depth of about 4 m. Whole rock chemistry was analysed by XRF and wet chemical methods and the mineralogy determined by XRD. The major oxides identified are given in Table III.

The analyses indicate that the fresher black shales typically have a calcium content of 2 to 4% CaO and a carbonate content typically of 1.5 to 2% CO_2, although as would be anticipated the figures are different adjacent to the limestone bands. The data suggest that at depth the fresh mudrock material

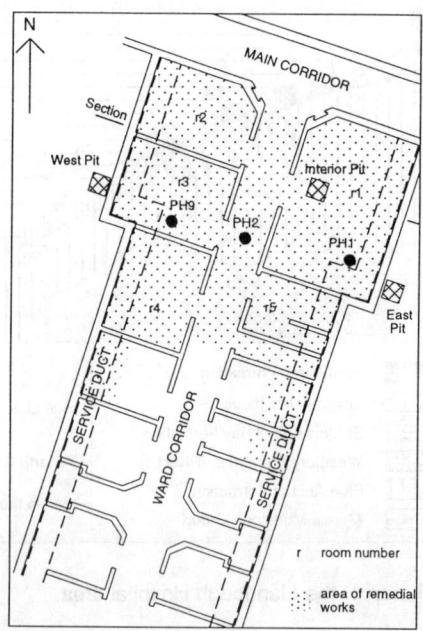

Figure 3: Plan of affected ward at Llandough Hospital showing location of trial pits and sampled pile holes.

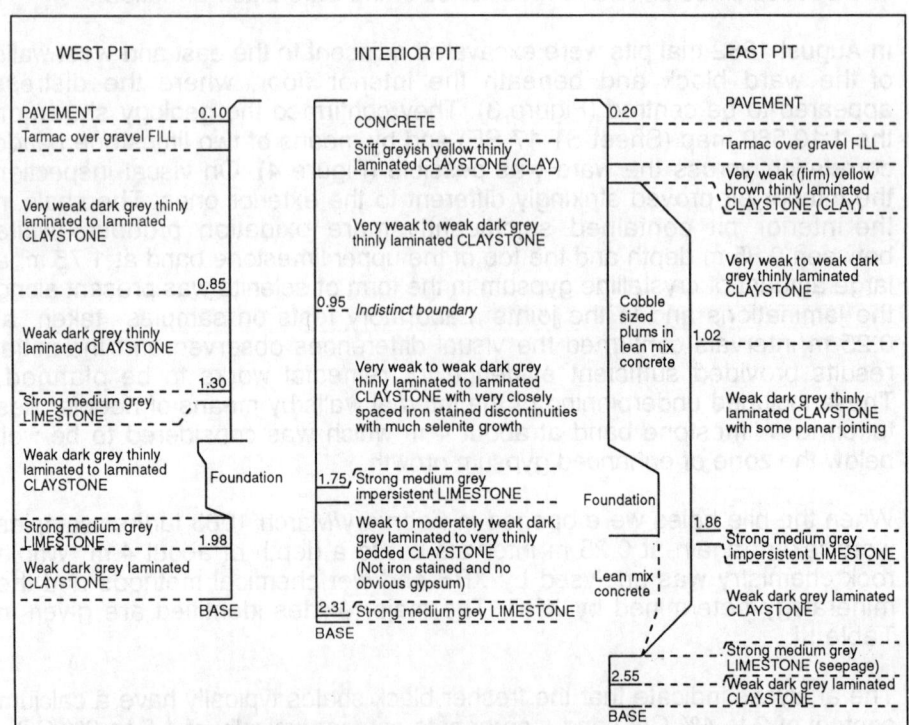

Figure 4: Trial pit logs from the 1982 site investigation at Llandough Hospital.

Table III: Analyses for selected major oxides (%) for some samples from Llandough Hospital: SiO_2, Al_2O_3 total iron oxide, MgO and CaO determined by XRF; FeO determined by the method of Riley (1958); Fe_2O_3 by difference; CO_2 by the method of Grimaldi, Shapiro & Schnepfe (1966) and SO_3 determination according to BS 1377 (1975).

Depth (m)	SiO_2 %	Al_2O_3 %	Fe_2O_3 %	FeO %	MgO %	CaO %	CO_2 %	SO_3 %
INSIDE PIT								
0.25	49.05	16.26	5.53	0.92	2.68	6.39	4.23	0.07
0.50	53.10	18.72	5.91	1.20	3.01	0.71	trace	0.04
0.75	55.11	19.20	6.67	1.10	2.81	0.44	0.00	0.10
1.00	54.17	18.67	6.23	1.20	3.62	1.15	0.00	1.13
1.25	51.92	18.23	5.77	0.98	2.99	0.49	0.00	3.44
1..50	53.25	18.59	5.62	1.30	3.13	0.67	0.00	3.09
1.75	49.90	17.93	4.75	2.55	3.37	3.04	1.35	1.87
2.00	50.03	17.89	5.48	1.90	3.23	4.22	2.77	1.75
2.25	53.36	18.87	5.32	2.10	4.43	2.99	2.88	0.96
2.50	51.56	18.05	5.14	2.09	4.03	1.81	-	1.63
EAST PIT								
0.50	56.41	19.74	6.47	1.52	3.16	1.06	0.26	0.29
0.75	54.49	19.28	5.66	2.02	3.38	0.63	0.44	0.75
1.00	54.31	18.52	5.37	1.92	3.44	0.99	0.80	0.94
1.25	54.29	18.91	5.44	1.85	3.38	0.69	0.54	0.85
1.50	54.34	9.15	4.82	2.32	3.50	1.41	7.26	0.36
1.75	52.87	18.51	6.07	1.73	3.69	2.91	2.13	0.77
2.00	47.92	17.47	5.15	1.93	4.27	4.37	2.94	1.03
2.25	50.29	15.85	4.46	1.73	6.38	4.71	0.96	0.96
PILE HOLE								
0.50	51.07	17.92	6.50	1.50	3.11	1.91	0.00	4.48
0.75	50.35	18.30	5.31	1.33	3.38	1.67	0.00	3.68
1.00	51.92	17.93	5.68	1.25	3.09	1.12	0.00	3.72
1.25	54.04	19.31	5.43	1.40	3.15	0.93	0.00	3.11
1.50	52.23	18.14	6.04	1.44	2.94	0.97	0.00	2.92
1.75	50.07	17.70	5.67	1.95	3.43	2.91	1.38	1.69
2.00	9.23	3.13	0.73	0.65	0.75	42.60	32.42	0.65
2.25	50.33	17.85	4.77	2.09	4.05	2.74	2.33	0.76
2.50	43.84	15.27	4.02	1.90	3.13	11.00	7.17	1.10
2.60	13.36	4.50	1.45	0.60	1.13	40.11	30.29	0.66
2.75	41.18	13.80	3.79	1.84	3.28	13.53	9.65	0.87
3.00	49.75	16.84	5.02	1.78	3.23	4.03	2.18	0.74
3.25	53.60	18.46	5.18	1.81	3.50	2.30	1.57	0.62
3.50	51.34	17.64	4.79	1.65	3.38	2.50	1.84	0.64
3.75	53.73	18.15	4.35	2.17	3.29	1.48	1.23	0.79
4.00	52.56	17.08	4.56	1.60	3.07	1.55	1.35	0.68

has a calcite content of 4-5% but that near the surface it is slightly lower, probably as a consequence of past leaching. In the internal pile holes there was an almost complete depletion of CO_2 and a corresponding increase in total acid soluble sulphates. Calculations on the CaO and CO_2 percentages indicate an excess CaO, probably largely in the form of calcium sulphate. An average of the analyses for the four interior profiles shows an increase of CaO relative to CO_2 closer to the surface while the average for the outside pits shows no distinct relationship (Figure 5).

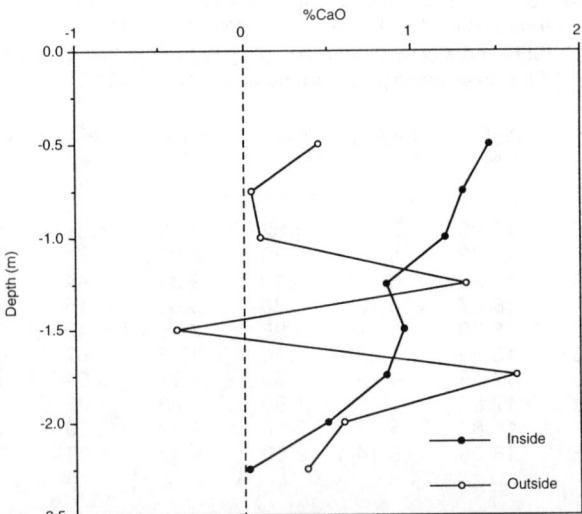

Figure 5: Llandough Hospital: comparison between inside and outside profiles of mean surplus calcium (as %CaO) after allocation of calcium due to calcium carbonate.

Laboratory testing also confirmed enhanced sulphate levels beneath the internal floor slab compared with those outside (Table IV), presumably due to gypsum generation. At depths of below 2 m, however, the total acid soluble sulphate values were similar, both beneath the building and outside, ie 0.6-1.1%. This proportion is maintained in the upper levels of the outside pits but is much higher above 2 m for the interior pile hole, with a good depth/SO_3 correlation. At intermediate depths this reflected the trend in the interior trial pit but near the surface increased sulphates were noted in the pile hole. In the trial pits there was a marked reduction in total acid soluble sulphate towards the surface, which was interpreted as being due to leaching.

Table IV records the amount of sulphur in the form of sulphate (ie oxidized) for two internal profiles. Again the very low levels for the interior trial pit suggest leaching. It is assumed the remainder of the sulphur is reduced, probably in the form of sulphide from unreacted pyrite. The results indicate that although reduced sulphur is present at depth, almost all of this has been oxidized near the ground surface. This was confirmed using X-ray diffraction; the diffractogram peaks indicating a significant sulphate presence between 0.75 and 1.75 m depth.

As it is known that "abundant" pyrite exists within the shales of the Westbury Formation it was considered realistic to assume that the FeO measured in the analysis was related to this mineral. It must be appreciated that there are difficulties in analysing for pyrite iron as its low solubility in the reagent acids may lead to very low FeO values and consequently raised Fe_2O_3 percentages. The ratio between the different irons is given in Table V. Although an increasing Fe_2O_3/FeO ratio towards the surface was noted in the outside pits, the marked contrast is between the average values for the

Table IV: Distribution of total sulphur and acid soluble sulphate for the inside trial hole and pile hole 1. Calculations to show percent sulphur in the sulphate minerals are included; the difference between this and total sulphur is expressed as excess (unoxidized) sulphur. The final column indicates the proportion of sulphur which has been oxidized to sulphate.

Depth (m)	Total sulphur %S	BS 1377 sulphate % SO_3	Sulphate sulphur %S	Excess sulphur %S	% Ratio of sulphate sulphur to total sulphur
PILE HOLE PH 1					
0.50	1.80	4.48	1.79	0.01	99
0.75	1.08	3.68	1.47	-0.39	136
1.09	1.17	3.72	1.49	-0.32	127
1.25	1.52	3.11	1.25	0.27	82
1.59	1.99	2.92	1.17	0.82	59
1.75	1.80	1.69	0.68	1.12	38
2.00	0.26	0.65	0.26	0	100
2.25	0.53	0.76	0.30	0.23	57
2.50	1.21	1.10	0.44	0.77	36
3.00	1.74	0.75	0.30	1.44	17
3.50	1.66	0.64	0.26	1.40	16
INSIDE TRIAL PIT					
0.25	0.25	0.07	0.03	0.22	8.3
0.50	0.16	0.04	0.02	0.14	12.5
0.75	0.12	0.10	0.04	0.08	33.3
1.00	0.63	1.13	0.45	0.18	71.4
1.25	1.10	3.44	1.8	1.28	125.5
1.50	1.47	3.09	1.24	0.23	84.4
1.75	1.63	1.87	0.75	0.88	46.0
2.00	0.86	1.44	0.58	0.28	67.4
2.25	0.64	0.96	0.38	0.26	59.4
2.50	1.12	1.50	0.60	0.56	53.6

internal and external samples. This is consistent with the more advanced deterioration of pyrite below the floor slab, notably at the levels at which the selenite crystals were observed, Figure 6.

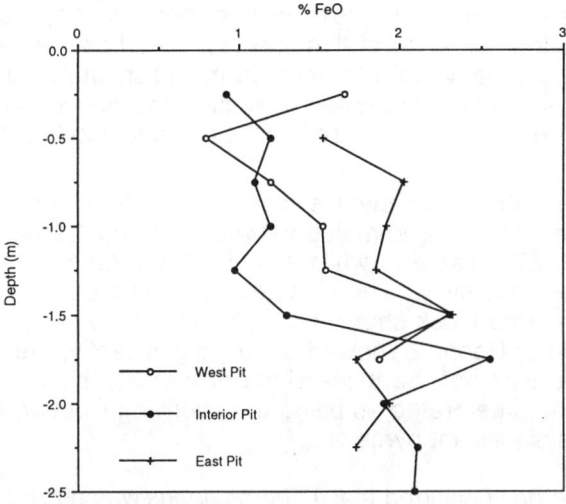

Figure 6: Vertical distribution of reduced iron oxides (%FeO) for the three trial pits at Llandough Hospital.

Table V: Fe_2O_3/FeO ratios on Llandough samples

Depth (m)	West Pit	PH 9	PH 2	Inside Pit	PH 1	East Pit	Average profiles Inside	Outside
0.25	3.3			6.0			6.6	6.5
0.50	8.7	7.6	7.7	4.9	6.2	4.3	6.7	4.2
0.75	5.6	9.7	7.2	6.1	4.0	2.8	6.3	3.4
1.00	4.0	9.1	6.2	5.2	4.5	2.8	5.8	3.0
1.25	3.1	7.6	5.7	5.9	3.9	2.9	4.9	2.0
1.50	1.8	7.5	3.5	4.3	4.2	2.1	2.9	3.1
1.75	2.7		4.1	1.9	2.9	3.5	3.3	2.7
2.00			3.6	2.9		2.7	2.5	2.6
2.25		3.5	2.6	2.5	2.3	2.6		
2.50		3.0		2.5	2.1			
2.75		3.1	3.1		2.1			
3.00		2.9	3.8		2.8			
3.25		2.5	3.3		2.9			
3.50		2.9	3.0		2.9			
3.75		3.6	2.7		2.0			
4.00					2.9			

Measurements of pH also supported the hypothesis that pyrite degradation was taking place below the interior floor slab. Acidic conditions (pH 3 - 5) were recorded in the interior trial pit between 1 and 2 m depth whereas neutral conditions (pH 6 - 7) were measured in the deeper samples and in those from the exterior pits.

The work undertaken in Canada in the 1970s had implied a possible involvement of bacteria in the pyrite degradation processes. To investigate this further, microbiological analyses were undertaken using the method of Kuenen and Tuovinen (1981). Flasks of acidic ferrous sulphate ($FeSO_4$) media were inoculated with samples from pile hole 9 and two other short profiles from trenches below the floor slab. As *Thiobacillus ferrooxidans* is the only bacteria which can grow on this medium, any bacterial activity would indicate the presence of this species. The flasks were inspected at intervals and a steady discoloration from bluish green through yellow, orange, red and brown was noted indicating the conversion of $FeSO_4$ to $Fe_2(SO_4)_3$ as the Fe(II) was oxidized by the bacterium to Fe(III).

From pile hole 9 there was clear evidence of bacterial activity in a sample from 2.35 m and in all the samples between 0.5 and 1.95 m (see Hawkins and Pinches 1987a, Table 7) while samples from 2.2 m and below 2.35 m showed little or no activity. The samples from the profile of black shales in Room 5 of the ward block showed a high bacterial activity whereas those from the profile in Room 2 showed none. Significantly, the geology of the Room 2 profile was not the black shale of the Westbury Formation but a green mudstone, interpreted as being the overlying Cotham Member which rarely contains significant pyrite.

It can therefore be concluded that *T. ferrooxidans* was absent below a depth of about 2.5 m and where the natural material does not contain pyrite. Chemical analyses also showed pyrite to be in decline and gypsum at elevated levels here, ie a zone of ongoing pyrite degradation. Although the full role of the bacterium has still not been satisfactorily established, the correlation of bacterial proliferation and SO_3 content strongly suggests a relationship between the activity of the bacterium and the degradation process.

CONTRIBUTORY FACTORS

It is of note that both the Rideau Health Centre and Llandough Hospital buildings are two storey constructions incorporating heating pipes in ducts adjacent to the exterior walls (Plate 2). The presence of the service ducts effectively isolated an upstanding plug of shale while the adjacent backfill permitted access of air into the mudrock laminations. At the same time, the ducts formed barriers to the lateral movement of fluids resulting from the oxidizing reaction, which in other situations may have been transported out of the system.

Plate 2: Service duct under the eastern ward of Llandough Hospital.

Being used for medical purposes, both buildings were substantially heated. Raised temperatures contribute to an enhanced rate of pyrite degradation and the heating gradient would undoubtedly have encouraged the rise of moisture. This would not only have replenished the water involved in the chemical reactions but would also have carried the oxidation products upwards, encouraging crystallisation in zones of least pressure. Localised heating of the plug was further enhanced by the hot water pipes carried within the service ducts.

While chemical reactions have a direct relationship with temperature, biological processes usually operate around an optimum. For *Thiobacillus ferrooxidans* this is variously reported as 15 to 20°C (Vishniac, 1974) or 30 to 35°C (Nixon, 1978) while Hawkins and Pinches (1987b) suggest that activity is still increasing at 40°C (Figure 7). These authors report that *Thiobacillus* is probably most active in the 30 to 40°C temperature range

(Table VI). Recent work by Hawkins and Higgins (1992) has shown that in London Clay samples kept at 40°C in cabinets with a retained 100% relative humidity the SO_3 content increased from 0.2 to 0.96% after a period of only six weeks. Figure 7 shows the rise in SO_3 content and fall in pH values noted in specimens of dark mudrock tested after 1, 2, 4 and 15 weeks.

Table VI: Variations in percentage SO_3 and pH over time in temperature controlled conditions for a sample of black shale of the Westbury Formation (Penarth Group).

	Storage period	Refrigerator 7.5°C	Cupboard ~18.5°C	Incubator 29.5°C	Oven 41.5°C
SO_3 (%)	Start	0.60	0.60	0.60	0.60
	1 week	0.59	0.55	0.63	1.03
	2 weeks	0.60	0.62	0.86	1.21
	4 weeks	0.75	0.88	1.08	1.35
	15 weeks	0.78	1.26	1.50	1.80
pH	Start	3.8	3.8	3.8	3.8
	1 week	3.8	3.5	3.6	2.7
	2 weeks	3.8	3.2	2.8	2.5
	4 weeks	3.8	3.3	2.8	2.5
	15 weeks	2.95	2.6	2.7	

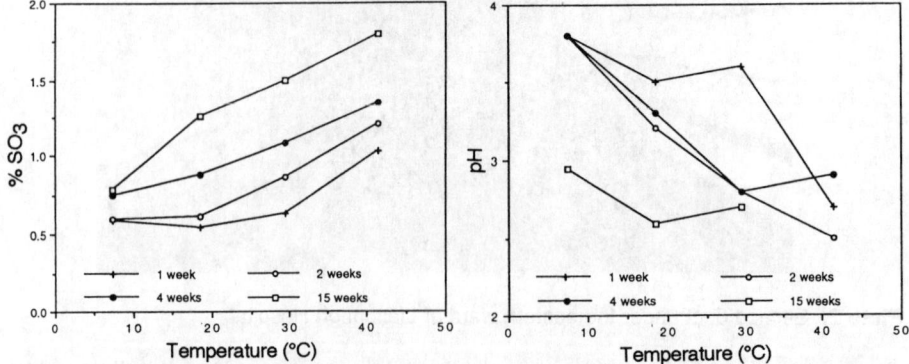

Figure 7: The influence of time and temperature on sulphate generation and change in acidity of laboratory stored specimens of a sample of Westbury Formation black shale from Chipping Sodbury, north east of Bristol.

In all three documented cases of floor heave referred to in this paper, the floor slab was ground bearing, thus any movements of the underlying strata would be transferred directly to the floor slab. If the floor slab is integrally connected to the walls, as at Llandough Hospital, any heave of the slab will result in a rotational movement of the walls and bulging of the external walls. As the pressure beneath the floor slab is substantially less than under the load bearing walls, the pressure gradient would encourage lateral migration of pore fluids before crystallisation of further gypsum. This is consistent with the observations of Coveney and Parizek (1977) who reported heave in a mine floor in the Hushpuckney pyritic shales at Kansas City. These authors note that the maximum heave took place towards the centre of the mine rooms where there was minimum stress while there was no apparent growth of selenite beneath the load bearing mine pillars.

It should not be assumed that higher stresses will prevent crystallisation however. In laboratory investigations on alum Becker and Day (1905) and Correns (1949) showed that crystals may grow against an exerted pressure and calculations on the principles of growing crystals show that gypsum can crystallise under significant loads (Winkler and Singer, 1972). Observations on spalled building stone and concrete (Fookes, 1978) indicate that the growth pressure of the gypsum crystals must be greater than 20 N/mm^2, ie the inherent strength of the host material.

Such data suggest that although gypsum growth from pyrite degradation is opportunistic in that it tends to occur in zones of lower stress, it may also take place beneath a substantial load. Work is being undertaken to establish the effect of varying loads such as would occur beneath buildings and the ultimate limits under which gypsum growth can occur; no conclusive results are yet available.

RECOMMENDATIONS

IDENTIFYING THE POTENTIAL FOR HEAVE IN DARK MUDROCKS

From the foregoing it is clear that heave due to pyrite degradation may be predicted or avoided if at the conceptual stage of a project consideration is given to the nature of ground and the long term effects of the proposed construction. Attention is drawn to the following points:

1. PRESENCE OF PYRITE

    With typical diameters of approximately 30 μm, the common framboidal pyrite is not visible with the naked eye and photomicrographs are beyond the scope of most normal site investigations. Occasionally, cubes of pyrite are distinguishable in dark mudrocks.

2. COLOUR AND ORIGIN

    In a fresh rock, a dark grey colour may indicate contained sulphides and/or organic material. Pyrite is rarely found in sediments which accumulated in a terrestrial environment.

3. ASSESSMENT OF WEATHERING

    The potential for pyrite degradation at a construction site depends on the pyrite content of the engineering soil/rock. Where the foundation material is extremely weathered and the oxidation products have leached, sulphate-induced heave is unlikely. The presence of secondary minerals within the soil profile, such as gypsum, jarosite or iron oxides, would indicate that reaction processes have already commenced. The presence of sucrose selenite is currently believed to be of particular importance in indicating a recent growth of gypsum.

4. ROCK STRUCTURE

Heave tends to be more pronounced in a rock with a marked fissility (shaley structure). The presence of discontinuities not only facilitates movement of the reaction components but also provides a convenient lower stress zone for the commencement of crystallisation and, with bedding partings, its unidirectional upward effects.

5. SULPHUR ANALYSIS

Used alone, the standard analysis for total acid soluble sulphate in accordance with BRE Digest 363:1991 provides little assistance in assessing the likelihood of ground heave due to sulphate generation. In areas of potential ground heave the standard test should be accompanied by an analysis of the total sulphur; the difference between total sulphur and sulphur in the form of sulphate indicates the potential for sulphide oxidation. If possible it is prudent to repeat sulphate analyses at monthly intervals for a period of four months or for two months at temperatures of 40°C (Hawkins and Pinches, 1986, 1987b). Laboratory stored samples have shown significant increases in sulphate within short periods when oxygen and carbon dioxide are available (Table VII). It must be assumed that a similar situation could occur in the field, albeit more slowly.

Table VII: Changes in sulphate contents and pH with time in samples at 1.5 m depth from Llandough Hospital.

		West Pit	Int. Pit	East Pit
SO_3%	Nov 1982	0.29	2.20	0.36
	Mar 1984	1.20	4.15	1.60
Water soluble SO_3 g/l in	Nov 1982	1.25	4.70	1.55
2:1 water soil extract	Mar 1984	3.20	7.80	2.40
pH	Nov 1982	7.10	3.75	7.10
	Mar 1984	6.25	2.65	5.30

6. IRON OXIDE ANALYSIS

Total iron analysis should be undertaken in the first instance and if iron is present the proportions of Fe_2O_3 and FeO should be established as this will indicate the potential for the oxidation of reduced iron. Using the method of Riley (1958) the data from Llandough Hospital suggests that ratios of 2:1 to 4:1 (Fe_2O_3:FeO) indicate material with a significant reduced iron content [Fe (II)] whereas ratios of 5:1 to 10:1 were found where much pyrite degradation had already taken place.

7. CALCIUM ANALYSIS

For gypsum to grow from the oxidation products a source of calcium is required, commonly calcite ($CaCO_3$). If there is no calcium with which the sulphuric acid from pyrite degradation can react, other minerals may be involved; eg the clay mineral illite to form jarosite. It is likely that the optimum conditions for selenite growth include the presence of only a small amount of calcium. This would ensure that the acid conditions are not completely neutralised and a conducive environment for the microbes is maintained. At Llandough Hospital 0.6 to 1.5%.CaO was found at the level from which the crystals were recovered.

8. X-RAY DIFFRACTION

Mineralogical assessment by XRD would indicate the presence of pyrite and calcite in the fresh rock and the oxidation products gypsum, jarosite and iron oxide in a weathered material. It must be remembered that XRD is differentially sensitive so that although small percentages of gypsum and calcite may produce significant peaks, similar quantities of pyrite may not produce a diffractogram signature.

9. ACIDITY

A low pH could imply the presence of sulphuric acid from ongoing oxidation of pyrite, although in itself this is not a sufficient criterion to conclude sulphate generation is taking place (Hawkins and Pinches, 1988).

10. MICROBIOLOGY

The method of proving the presence of the organism *Thiobacillus ferrooxidans* described by Kuenen and Tuovinen (1981) involves the inoculation of a sterile acidic solution of ferrous sulphate with soil samples. If oxidation takes place in the inoculated solutions this would indicate the presence of bacteria in the soil and presumably therefore, the oxidation of pyrite or its derivatives.

11. GROUND AERATION

Stripping of the upper weathered layers of the formation or lowering of the ground water table could permit fresh pyrite-bearing ground to come into contact with atmospheric oxygen and carbon dioxide leading to the onset of oxidation processes.

12. PROPOSED STRUCTURE

Consideration should be given to an appropriate design including the location of heating equipment which could raise the ground temperatures beneath the building and accelerate the chemical reactions. In addition temperature gradients could create a capillary zone along which oxidation products in solution could be transferred to areas of least stress. Such areas occur mainly beneath ground bearing floor slabs, which would deflect in direct response to heave stresses in the underlying formation.

Due cognisance should be taken of not only the proposed areal extent of the development but of the dangers of constructing a building across a significant geological change where the potential for sulphate heave is different.

Although there is much circumstantial evidence pointing to the role of gypsum in creating heave, it cannot be assumed that other minerals, such as jarosite, may not have a similar effect. However the aggregate crystals or amorphous form of jarosite commonly observed suggest it is a void filling mineral which does not grow against pressure.

AMELIORATION OF THE EFFECTS OF SULPHATE GENERATED HEAVE

If conceptual studies indicate an engineering structure may suffer heave from pyrite degradation or the phenomenon has been found to occur, its effects can be countered using a variety of approaches.

1. Avoid construction on pyritic material. If this is not possible, ensure the whole building is founded on the same stratum. If only a thin horizon of potentially hazardous material is present it may be economically viable to remove it prior to construction.

2. Adopt a scheme which does not encourage the reactions:

    a) maintain or raise the current ground water level;
    b) construct on the existing mantle of weathered or leached material;
    c) in zones of least pressure reduce exposure of fresh material to a minimum and as far as practical keep drain runs at right angles to the structure;
    d) attempt to seal the ground from atmospheric oxygen by laying impermeable membranes or coating fresh excavations with gunite or bitumen;
    e) avoid sub-floor heating and/or insulate the ground floor;
    f) keep ground water levels high by prohibiting the establishment of water-loving vegetation near the structure.

3. Adopt a solution which can absorb the heave:

    a) extend the foundations below the material in which sulphate growth is likely or below the long term ground water level;

b) use suspended floor slabs or cast the slabs onto intervening collapsible (clay board) units;
   c) install flexible slabs or joints;
   d) use rafts for the total unit construction.

4. Actively counter the effects of oxidation:

   a) encourage lateral drainage to remove any oxidation products;
   b) raise the pH by alkali treatment;
   c) consider the use of bactericides;
   d) flood the sub-floor zones with water to reduce the entry of air into the natural ground.

CONCLUSIONS

1. The material most susceptible to ground heave from pyrite degradation is fresh dark grey to black fissile mudrock (shale) with a small calcium content (1-2% CaO) and a pyrite content exceeding 1%. Care should be taken if chemical analysis indicates the difference between the total sulphur content and the percentage sulphur in sulphate (SO_3) is above 0.5% or the Fe_2O_3:FeO ratio is 2:1 to 4:1. Less effectively, pyrite may be detected by XRD or electron photomicroscopy.

2. Reactions are initiated when relatively fresh susceptible material comes into contact with readily available oxygen and moisture. The breakdown of pyrite and formation of ultimate oxidation products can be by a number of pathways and involve both chemical and microbial agents.

3. The crystallisation of oxidation products in zones of lower stress leads to upward movement of the ground and consequently distress in any overlying structure.

4. These processes are accelerated by heating of the ground.

5. Typically, cases of heave are first observed within 5-30 years of construction but the processes can be simulated and accelerated in the laboratory.

6. The deleterious effects can be accentuated by individual features of the structure, which may facilitate localised crystal growth.

7. To reduce or eliminate the effects of the phenomenon either construction on susceptible material must be avoided or methods adopted which as far as practical will preclude the reactions taking place. Where this is not possible, it is essential that the structures should be designed appropriately with deep load bearing walls and suspended floor slabs.

ACKNOWLEDGEMENTS

The authors thank Marian Trott for her help in preparing this paper and the MRM Partnership on behalf of whom the initial work was undertaken.

REFERENCES

BECKER, G.F. & DAY, A.L. (1905). The linear force of growing crystals. *Proceedings Washington Academy of Science*, **7**, 283-288.

BERNER, R.A. (1970). Sedimentary pyrite formation. *American Journal of Science*, **268**, 1-23.

BRITISH STANDARDS INSTITUTION, (1975 & 1990). Methods of test for soils for civil engineering purposes BS 1377.

BUILDING RESEARCH ESTABLISHMENT, (1991). Digest No 363. Concrete in sulphate-bearing soils and ground water.

COLMER, A.R., TEMPLE, K.L. & HINKLE, M.E. (1950). An iron oxidizing bacterium from the acid drainage of some bituminous coal mines. *Journal of Bacteriology*, **59**, 317-328.

CORRENS, C.W. (1949). Growth and dissolution of crystals under linear pressure. *Discussion of the Faraday Society*, **5**, 267-271.

COVENEY, R.M. & PARIZEK, E.J. (1977). Deformation of mine floors by sulfide alteration. *Bulletin of the Association of Engineering Geologists*, **XIV**, 131-156.

FLIERMANS C.N. & BROCK, T.D. (1972). Ecology of sulfur oxidizing bacteria in hot acid soils. *Journal of Bacteriology*, **111**, 343-350.

FOOKES, P.G. (1978). Middle East - inherent ground problems. *Quarterly Journal of Engineering Geology*, **11**, 33-49.

FOOKES, P.G. & COLLIS, L. (1975). Problems in the Middle East. *Concrete*, **9**, 12-17.

GRIMALDI, F.S., SHAPIRO, L. & SCHNEPFE, M. (1966). Determination of carbon dioxide in limestone and dolomite by acid-base titration. *US Geological Survey Professional Paper*, **550 B**, B186-188.

HAWKINS, A.B. & HIGGINS, M. (1992). The generation of sulphates in the proximity of cast in situ piles. *Proceedings of the International Conference on the Implications of Ground Chemistry and Microbiology for Construction, Bristol*, 251-259.

HAWKINS, A.B. & PINCHES, G.M. (1986). Timing and correct chemical testing of soils/weak rocks. *Engineering Geology Special Publication No. 2, Geological Society*, 273-277.

HAWKINS, A.B. & PINCHES, G.M. (1987a). Cause and significance of heave at Llandough Hospital, Cardiff - a case history of ground floor heave due to gypsum growth. *Quarterly Journal of Engineering Geology*, **20**, 41-57.

HAWKINS, A.B. & PINCHES, G.M. (1987b). Sulphate analysis on black mudstones. *Geotechnique*, **37**, 191-196.

HAWKINS, A.B. & PINCHES, G.M. (1988). Discussion on Sulphate analysis on black mudstone. *Geotechnique*, **38**, 322-323.

HAWKINS, A.B. & PINCHES, G.M. (1992). Engineering description of mudrocks. *Quarterly Journal of Engineering Geology*, **25**, 17-30.

JENKYNS, H.C. (1980). Cretaceous anoxic events from continents to oceans. *Journal of the Geological Society*, **137**, 157-170.

KELLY, D.P., NORRIS P.R. & BRIERLY, C.L. (1979). Microbiological methods of the extraction and recovery of metals. In: *Micro Technology: Current*

State, Future Propsects. Society of General Microbiology Symposium 29, Bull, Edwood & Rattledge (eds), Cambridge University Press.

KUENEN, J.G. & TUOVINEN, O.H. (1981). The genera Thiobacillus and Thiomicrospira. In: *The Prokaryotes*, Starr, M P., Stolp, H.; Truper, H G; Balows, A & Schlegel, H G (eds), I, 1023-1032, Springer-Verlag, New York.

LOVE, L.G., AL-KAISY, A.T.H. & BROCKLEY, H. (1984). Mineral and organic material in matrices and coatings of framboidal pyrite from Pennsylvanian sediments, England. *Journal of Sedimentary Petrology*, **54**, 869-876.

MOUM, J. & ROSENQVIST, I. (1959). Sulphate attack on concrete in the Oslo region. *Journal of the American Concrete Institution*, **56**, 257-264.

NIXON, P.J. (1978). Floor heave in buildings due to the use of pyritic shales as fill material. *Chemistry and Industry*, March, 160-168.

PENNER, E., EDEN, W.J. & GILLOTT, J.E. (1973). Floor heave due to biochemical weathering of shale. *Proceedings 8th International Conference on Soil Mechanics and Foundation Engineering, Moscow*, II. 151-158.

QUIGLEY, R.M. & VOGAN, R.W. (1970). Black shale heaving at Ottawa, Canada. *Canadian Geotechnical Journal*, **7**, 106.

QUIGLEY, R.M., ZAJIC, J.E., MCKYES, E. & YONG, M. (1973). Chemical alteration and heave of black shale. Detailed observations and interpretations. *Canadian Journal of Earth Science*, **10**, 1005-1015.

RILEY, J.P. (1958). The rapid analysis of silicate rocks and minerals. *Analytica Chimica Acta*, **19**, 413-428.

TAYLOR, R.K. & CRIPPS, J.C. (1984). Mineralogical controls on volume change. In: *Ground movements and their effects on structures*, Attewell, P B & Taylor, R K (eds), Surrey University Press, 268-302.

VAN BREEMON, N. (1988). Redox processes of iron and sulfur involved in the formation of acid sulfate soils. In: *Iron in Soils and Clay Minerals*, Stucki, J W., Goodman, B A & Schwertmann, U (eds), NATO ASI Series Vol 217, 85-841.

VISHNIAC, W.V. (1974). Thiobacillus In *Bergey's Manual of Determinative Bacteriology*, Buchanan & Gibbons (eds), Williams & Wilkins, Baltimore, 456-461.

WINKLER, E.M. & SINGER, P.C. (1972). Crystallisation pressure of salts in stone and concrete. *Bulletin of the Geological Society of America*, **83**, 3509-3514.

2-2 SOME GEOTECHNICAL PROBLEMS ASSOCIATED WITH PYRITE BEARING MUDROCKS

J.C. CRIPPS
Department of Earth Sciences, University of Sheffield, Dainton Building, Brook Hill, Sheffield S3 7H

R.L. EDWARDS
AMEC Civil Engineering Ltd, Chapel Street, Adlington, Lancs PR7 4JP

ABSTRACT

Problems associated with the presence of pyrite in bedrock and constructional materials include the generation of acid groundwater conditions, rapid deterioration of the mechanical behaviour of materials and precipitation of minerals leading to ground or foundation heave. Although oxidation of this mineral proceeds slowly by a chemical pathway, much more rapid pyrite consumption and acid aqueous solution production can arise due to the activities of various autotrophic bacteria. In view of the widespread occurrence of pyrite-bearing formations in the UK, the potential for such problems is great.

The paper includes consideration of particular foundation design features and constructional practices that can favour the microbial alteration of pyrite and lead to problems for construction. The examination of a case history in which potential heave was linked to pyrite alteration and the precipitation of gypsum provides some guidance as to the circumstances that would be likely to exacerbate or ameliorate the potentially harmful effects of the presence of pyrite in the bedrock materials.

INTRODUCTION

Foundation heave is one of the main constructional problems resulting from reactions between the oxidation products of pyrite and other rock, soil, or construction materials. Uplifts of the order of 100mm are reported (Penner, Eden and Gillot, 1973), although much greater movements are possible under certain conditions; for instance Penner et al (1973) make reference to differential heaves as great as 300mm. Care must be taken to thoroughly investigate foundation movements as problems due to heave have been misinterpreted as settlement of adjacent areas (Anon, 1960). As can be seen in Table I, constructional problems involving heave due to pyrite oxidation have been encountered in a number of countries.

Pyrite commonly occurs in dark coloured marine shales. Bacterial alteration of organic matter under low redox conditions may lead to the formation of sulphides of iron and manganese, of which the most common are pyrite and marcasite (FeS_2) and pyrrhotite (FeS). According to Potter, Maynard and Pryor (1980) these minerals are most likely to form in the upper levels of the sediment where, due to restricted circulation or a high input of organic matter, the conditions become anoxic. Microbial action then leads to the reduction of sulphate derived by diffusion from the overlying sea water (see also Raiswell and Berner, 1986).

As shown in Table II, in the UK mudstones and shales of various geological ages contain significant amounts of pyrite. Pyrite is also found in certain metamorphic rocks, such as slates, which were derived from pyrite-bearing mudrocks. Some pyrite occurs as mineralised veins emplaced by hydrothermal action and, although not very stable in weathering environments, it may also be present as a detrital mineral in sandy and silty sediments.

The oxidation of pyrite during weathering proceeds chemically in moist air to produce sulphuric acid and ferrous sulphate, the latter usually occurring as melanterite ($FeSO_4.7H_2O$) which, due to high solubility, may be removed from the reaction site by groundwater. As explained by Jackson and Cripps (this volume) and others (Hawkins and Pinches 1987; Pye and Miller 1990) who outline the relevant formulae, this ferrous sulphate may be converted by the autotrophic bacteria, *Thiobacillus ferrooxidans*, to ferric sulphate. This then reacts with fresh pyrite to produce more ferrous sulphate and sulphuric acid. The intervention of *Thiobacillus ferrooxidans* greatly increases the rate of pyrite consumption and acid production. The bacteria require an acidic environment for optimum growth and they are most active at an ambient temperature of about 35°C. The process results in the production of acid which may give rise to the appropriate pH conditions while the exothermic nature of pyrite oxidation can raise the temperature sufficiently to favour microbial activity.

Quigley and Vogan (1970)	Heave of a floor slab on 0.46m of granular fill and first floor auditorium in a two storey building
Penner et a (1973)	Heave of 350mm thick floor slab on 150mm limestone fill in a three storey building with pad foundations
Nixon (1978)	Heave of domestic housing on shale fill, Teesside, England
Hawkins and Pinches (1987)	Heave of floors and disruption to internal and external load bearing walls in a two storey hospital building, Cardiff, Wales

Table I: Some cases of foundation heave due to pyrite oxidation

The oxidation potential is increased for impure pyrite and also if is amorphous, fine grained or framboidal in form. Marcasite and pyrrhotite are more reactive than pyrite and may act as a catalyst for pyrite oxidation.

POTENTIAL FOR HEAVE

The effects of this biochemical and chemical activity on construction depend on reactions between the reaction products, particularly ferrous sulphate and sulphuric acid and other soil, rock, groundwater and construction materials. If carbonates are present (eg calcite) then gypsum will be produced. The carbon dioxide evolved as a consequence of this reaction may itself prove to be a problem during construction, its presence being suspected during the construction of the Carsington Reservoir. Other minerals, including clay minerals, may also be involved to produce a number of sulphate minerals, such as jarosite and alunite. Various iron minerals may also be produced, eg limonite, coquimbite and haematite.

The removal of pyrite and other minerals such as calcite from a rock and their direct replacement with gypsum entails an increase in volume (see Table III). It should be appreciated however that, as indicated later, the process of dissolution and precipitation need not occur in the same location, so both expansion and void creation can be involved and expansion may take place at locations not containing pyritic rocks.

Table II: Pyrite in UK mudrock formations; from Taylor and Cripps (1984) except for H - Hawkins and Pinches (1987) and A - Anderson and Cripps 1992; N - not known

Geological age/ Formation	Location	Pyrite (%)	
Oligocene			
Bembridge Beds	Isle of Wight	N	
Barton Clay	Hampshire	N	
Palaeogene			
London Clay	London Basin & Hants	3.3	
	Kent	0-4	
Cretaceous			
Gault Clay	Buckinghamshire	1.0	
Speeton Clay	Yorkshire	N	
Fuller's Earth	Surrey	0.5	
Weald Clay	Sussex	N	
Jurassic			
Kimmeridge Clay	Dorset	4.0	
Oxford Clay	Oxon and Cambs	3-5	
	E and S England	5-15	
Lr Oxford Clay	E and S England	0-17	
Fuller's Earth	Somerset	3.0	
Whitby Shale	Teesside	3-9	
Upper Lias Clay	Northants	3-5	
Lwr and Mid Lias Clay	Lincs and Glos	N	
Stonesfield Slate	Gloucestershire	N	
Triassic and Permian			
Westbury Formation	South Glamorgan	4-6	H
Marl Slate	Durham	<4	
Carboniferous			
Coal Measures Shale	England	0.7-1.4	
Namurian Shale	Derbyshire	0-6	A
Calciferous Sandstone	Midlothian	1.1	
Carboniferous Lst.	Yorks and Derbyshire	5-10	
Colliery Spoil	UK	0-12	
Tansley Shale	Derbyshire	0-5	

The oxidation of pyrite gives rise to the availability of soluble minerals, in particular gypsum and jarosite, which may remain in solution until precipitation occurs, as a result of either a reduction in temperature or an increase in solution concentration. In practice the latter is the most likely mechanism and would be most probable within the capillary fringe where evaporation leads to enhanced solution concentrations. Precipitated minerals may either serve a void-filling role or precipitation may generate space by exerting crystallisation pressures.

Most of the cases of heave listed in Table I have arisen due to the interaction of pyrite from in situ shales or mudstones and the precipitation of gypsum. Nixon (1978) attributed heave in shale fill to these processes and Fasiska, Wagenblast and Dougherty (1974) report the precipitation of sulphates of aluminium, titanium, iron and chromium where pyrite reaction products had been carried into flyash. Where detailed studies have been carried out it has been noted that pyrite and calcite are present in the fresh mudrock below a zone of alteration. Closer to the ground surface these minerals decrease in quantity and gypsum becomes more prevalent.

Table III: Theoretical volume changes due to mineralogical changes (volume of dissolved components not included)

Original minerals	Product minerals	Expansion (%)
Calcite	Gypsum	103
Pyrite	Gypsum	518
Pyrite + Calcite	Gypsum	53
Pyrite	Jarosite	125
Illite	Alunite	160
Illite	Jarosite	38

From the cases of foundation heave reported in Table I, it is clear that crystallisation pressures can be sufficient to open joints and bedding planes against overburden and loading pressures. Gillot et al. (1974) show prismatic gypsum crystals in open bedding planes. The c-axes lie perpendicular to the bedding planes and the crystals appear to be propping or pushing them open.

Measurements of heave pressures reported in the literature give widely divergent values, from 14 kN/m^2 in the laboratory tests reported by Anon, (1960) to 500 kN/m^2 as given by Sherrell (1979). Winkler and Singer (1972) indicate that the pressure (P - kN/m^2) produced by crystal growth can be determined theoretically, being dependent on the degree of saturation, temperature ($T°K$), molecular volume (V litres/mol) and gas constant (R=8.3143 J/Kmol) as follows:

$$P = \frac{RT}{V} \ln \frac{c}{c_s}$$

where c is the concentration of the solution (g/litre) and c_s is the concentration of a saturated solution (g/litre).

Thus, as shown in Figure 1, depending on the concentration and the temperature, gypsum crystallisation can result in uplift pressures of 20 MN/m² at low levels of supersaturation (c = 1.5cs).

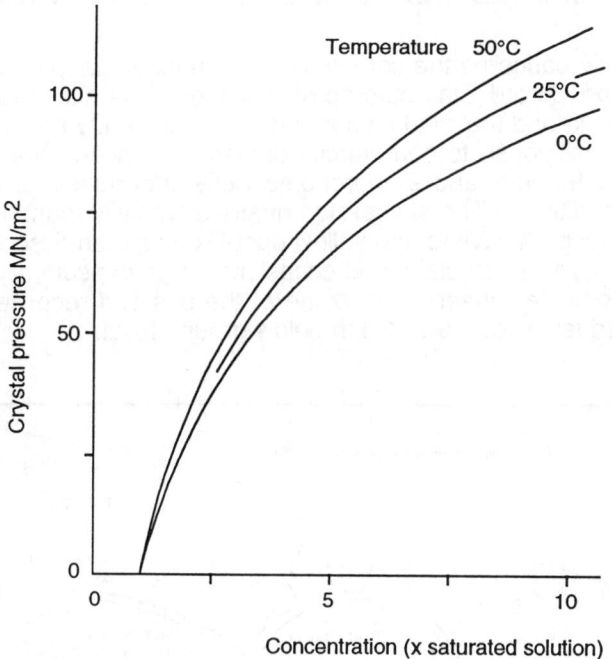

Figure 1: Variation of theoretical gypsum growth pressures for different conditions, after data in Winkler and Singer (1972).

Wellman and Wilson (1965) point out that to minimise the chemical free energy of crystallisation, large crystals will grow in preference to small ones, hence large crystals may grow against a pressure even if pore space is available for the unrestricted growth of smaller crystals. In a material consisting of large and small pores, once the large pores have been filled the crystals in them will continue to grow and therefore to exert pressure, provided this entails less effort than that required to enlarge the crystals. As the infilling of capillary pore space would result in a relatively large increase in the surface area of crystals for a small increase in their volume, less energy may be expended by increasing the size of the large crystals.

Using the pyrite content of soil or rock samples, attempts can be made to estimate the total heave that could occur if all the pyrite were converted into gypsum. The measurement of the pyrite available for oxidation is by no means simple although the recently revised British Standard (BS1377:1990) does suggest a method; see also Hawkins and Wilson (1990). However, it is important to appreciate that a heave calculation based on the pyrite content assumes a closed system for the reactions, a situation unlikely to arise in practice. Depending on the particular pattern of groundwater movement, heave much greater than this calculated figure could occur if pyrite oxidation was present over a wide zone and/or gypsum precipitation was concentrated within a small one. Naturally, the reverse could also apply.

POTENTIAL GROUND HEAVE IN SOUTH WEST ENGLAND

This case study concerns the construction of an industrial park on a green field site on the Oxford Clay outcrop of Wiltshire. The site (Figure 2) has an area of 26 Ha and the road layout was prepared ready for development unit by unit in response to commercial demand. Between May 1987 and June 1989, 71 trial pits and 21 boreholes were undertaken to investigate the ground conditions. The succession of strata typically found is shown in Figure 3 for trial pit A. White and yellow dust-like paste on fissure surfaces, small 'sugary' gypsum crystals, and occasional large gypsum crystals were noted. Groundwater observed in some of the pits and recorded in stand pipes indicated levels of 1.5 to 1.8 m below ground level.

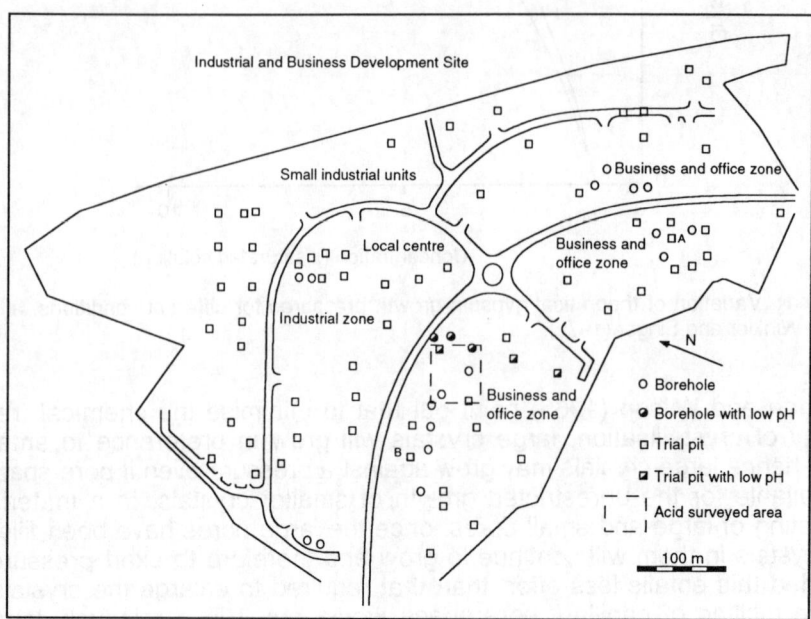

Figure 2: Site plan for southwest England site.

X-ray diffraction analysis of samples from various depths confirmed the presence of calcite, gypsum and pyrite as well as quartz and several clay minerals. Plaster of Paris ($CaSO_4 \cdot 0.5H_2O$) was also detected and was attributed to the dehydration of gypsum during the preparation of the samples which were heated to 105°C overnight. It was concluded that the small white nodules present in the samples were composed of calcite. These

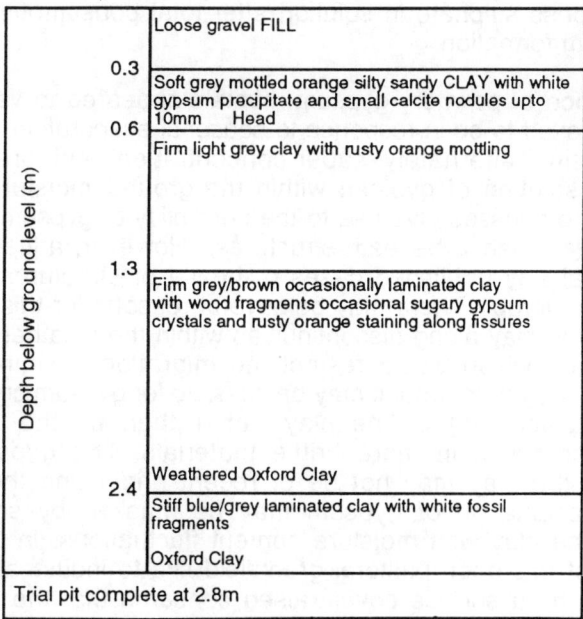

Figure 3: Ground conditions at the south west England site.

Table IV Composition of waters at south west England site

Depth below ground level (m)	Composition of ground waters (ppm)								
	Na	K	Mg	Ca	Al	Fe	Cl	SO_4^{2-}	pH
1.5	22	4	7	179	7	18	9	117	7.2
3.0	167	25	37	474	11	38	87	508	7.2
3.6a	105	18	74	513	0	1	114	1028	1.2*
3.6b	104	16	73	507	2	1	114	1023	1.2*
3.6c	103	14	74	504	3	1	117	1023	1.2*

* Acidified before analysis

analyses were consistent with the progressive weathering of Oxford Clay (see Russel and Parker 1979). The deepest and least weathered samples contain pyrite and calcite which become successively replaced by gypsum in the upper levels. Near the ground surface, the gypsum is probably removed from the weathered clay due to infiltration by fresh water while calcite becomes concentrated into nodules.

Water samples from trial pit B were analysed for major ions. As shown in Table IV they contain a high percentage of sulphate and calcium. The compositions can be explained by the oxidation of pyrite and subsequent reactions with other components of the clay. Calculations of the amount of gypsum available for precipitation give values of between 1.1 and 4.9 g/l of solution where in the case of the samples from 3.0 and 3.6m depth there would be excess sulphate in solution after total consumption of calcium during gypsum formation.

The occurrence of powdery gypsum crystals appeared to vary with time. This was believed to be in response to seasonal evaporation at the ground surface causing alternately super-concentration and precipitation or dilution and solution of gypsum within the ground moisture at shallow depth. Such processes give rise to the possibility of gypsum precipitation with consequent heave beneath structures. However, a literature search did not reveal any published cases of heave of structures built on the Oxford Clay outcrop. There may be several reasons for this; for instance softening of the clay along discontinuities within the weathered zone may lead to closure which would restrict the migration of mineral-carrying solutions. On the other hand, it may be possible for gypsum crystals to grow by local displacement of the clay rather than by the widening of discontinuities noted in more brittle materials. The gypsum crystals observed in the clay may not be of recent origin and the effect of a seasonal precipitation of gypsum may be masked by shrinkage and swelling of the clay with moisture content fluctuations. In this case the conversion of this open pastoral greenfield site to industrial use and the creation of a hard surface cover raised concerns that the reductions in water absorption into the ground surface coupled with heat ingress from the buildings could result in renewed gypsum precipitation. For this reason the clay exposed during floor construction was immediately protected by an impermeable membrane to prevent evaporation which in turn was covered with a granular layer to which rainfall was diverted to maintain a saturated condition in the underlying ground.

As five of the trial pits and two of the boreholes on one part of the site encountered acidic ground water conditions (pH 4.7 to 3.6) the possibility of continuing pyrite oxidation within the relatively fresh clay was a serious consideration. In this case the foundations would require protection from acid and sulphate attack. The need for an additional ground survey was advised to establish the lateral extent of the acidic area. However, before this was undertaken, site preparation involving the removal of topsoil and subsoil and the placement of a layer of limestone gravel was carried out. Surface water trapped in this layer was found to have a pH of 8.7 . To provide a check on the pH of the underlying clay, samples were taken at 0.5m intervals down to a depth of 2m on a grid of 25 number probes at 10m centres. As the pH of this material was in the range 5.2 to 7.2, it was concluded that the previously acid conditions had been neutralised by the alkaline water from the limestone fill and thus the potential for pyrite oxidation by bacterial action was greatly reduced.

CONCLUSIONS

Although the presence of pyrite in various geological formations is reasonably well documented in the geological literature and the formations liable to contain this mineral can be reasonably predicted from geological information, the implications for construction have come to light relatively recently. Published Canadian experience dates from the early 1970s but the first case of heave in the UK was not published until 1978. Even today few case histories exist from which it might be possible to derive guidance for remedial or preventative measures.

The avoidance of ground heave problems requires geotechnical/ foundation engineers to be aware of the geological situations in which the problem is liable to occur. In addition, to avoid the use of unnecessary preventative measures, engineers need to have some knowledge of the factors controlling the underlying processes and to have access to the necessary geochemical data to identify the areas liable to be affected. Pyrite oxidation requires the presence of oxygenated water and if the process is to be assisted by microbial action, warm, acidic conditions.

Features of design and construction which would result in the passage of oxygenated water through pyrite bearing strata beneath or near to structures should be avoided. This requires an assessment to be made of the local pre-construction hydrogeology and a prediction for the post-construction period. It is unlikely that a problem will arise if the water table is sufficiently low beneath the foundation for evaporation not to occur from the capillary fringe. However, it should be remembered that the height of capillary rise can be as great as 3-4 m in silty and clayey soils (see Lane and Washburn 1946). Where chemical precipitation is likely to occur, changes to the ground level profile and the water table beneath structures should be avoided.

It would be beneficial to position the foundations of structures within the zone of permanent saturation. In the case of a floor slab supported directly on the ground, the floor should be supported independently from the walls and be free to move. It is also important that steps or changes in level of the floor slab should be avoided as these allow greater penetration of warmth into the ground. Suspended ground floors with a void beneath may be considered, but as in practice it is impossible to predict the size of the void required, some monitoring of the situation would need to be undertaken.

As indicated in this paper, given the composition of pore waters and geological materials it is possible to determine the potential volume changes and hence the heave magnitude using estimates of the quantities of gypsum and other minerals that might be formed. However, such calculations are subject to large errors. The amount of heave liable to occur is not necessarily limited by the amount of pyrite and calcite in the immediate vicinity. In fact it is highly unlikely that oxidation and precipitation will occur concurrently in the same place as they require different conditions. In the most difficult scenario pyrite oxidation would occur over a wide area and supply gypsum in solution to a small zone where conditions favoured precipitation. The avoidance of problems requires a detailed knowledge of the groundwater chemistry and movements in the post construction period; such features are highly likely to be modified by the presence of the structure.

The pressures generated by gypsum precipitation may also be predicted theoretically. Again such calculations are subject to much uncertainty as they are heavily dependent on the concentration of solutions, the presence of minor impurities and the pore size distribution of the host material.

It is hoped that by considering the case history described in this paper and the other instances mentioned in the literature, an increased awareness of the potential causes of, and controls on, gypsum precipitation following pyrite oxidation will be forthcoming. In practice, when dealing with pyrite bearing strata, thorough investigations of the chemistry of the ground and the groundwaters should precede construction. Careful consideration should also be given to the changes in the ground conditions, particularly the movement of groundwater, which will be brought about both by the process of construction and the structure itself and due cognisance taken of them in the design of the proposed structure and the construction process. It would be advantageous to monitor the success of the precautions and the modifications to the geochemical conditions to enhance our understanding of the actual rates of reactions, although this is not usually possible.

ACKNOWLEDGEMENTS

The authors wish to acknowledge and thank the management and directors of AMEC Building Ltd, AMEC Design and Management Ltd and AMEC Civil Engineering Ltd for their assistance during the investigation and analysis of the data from this site.

REFERENCES

ANDERSON W.F. & CRIPPS., J.C. (1992). The effects of acid leaching on the shear strength of Namurian shale. *Engineering Geology of Weak Rocks, Engineering Geology Special Publication*, **8**, Balkema, Rotterdam, in press.

ANON. (1960). Structures do not settle in this shale, but watch out for heave. *Engineering News Record*, **164**, 46-48.

BRITISH STANDARD BS1377:1990. *Methods of test for soils for civil engineering purposes.* British Standards Institution, London.

FASISKA, E., WAGENBLAST, N. & DOUGHERTY, M.T. (1974). The oxidation mechanisms of sulphide minerals. *Bulletin of the Association of Engineering Geology*, **11**, 75-82.

GILLOT, J.E., PENNER, E. & EDEN, J.W. (1974). Microstructure of Billings Shale and biochemical alteration products, Ottawa, Canada. *Canadian Geotechechnical Journal*, **11**, 482-489.

HAWKINS, A.B. & PINCHES, G.M. (1987). Cause and significance of heave at Llandough Hospital, Cardiff - a case history of ground floor heave due to gypsum growth. *Quarterly Journal of Engineering Geology*, **20**, 41-57.

HAWKINS, A.B. & WILSON, S.L.S. (1990). Technical Note: Sulphate increase in laboratory prepared samples. *Quarterly Journal of Engineering Geology*, **23**, 383-385.

JACKSON, S.D. & CRIPPS, J.C. (1992). Investigation of the effects of bacterial action on the chemistry and mineralogy of pyritic shale. *Proceedings*

of the International Conference on the Implications of Ground Chemistry and Microbiology for Construction, Bristol.
LANE, K.S. & WASHBURN, D.E. (1946). Capillary tests by capillarimeter and soil filled tubes. *Proceedings of the Highways Research Board*, **26**, 460-473.
NIXON, P.J. (1978). Floor heave in buildings due to the use of pyritic shale as fill material. *Chemistry and Industry*, March, 160-164.
PENNER, E., EDEN, W.J. & GILLOT, J.E. (1973). Floor heave due to biochemical weathering of shale. *Proceedings of the 8th International Conference on Soil Mechanics and Foundation Engineering, Moscow*, **2**, 151-158.
POTTER, P.E., MAYNARD, J.B. & PRYOR, W.A. (1980). *Sedimentology of Shale*, Springer-Verlag, New York.
PYE, K. & MILLER, J.A. (1990). Chemical and biochemical weathering of pyritic mudrocks in a shale embankment. *Quarterly Journal of Engineering Geology*, **23**, 365-382.
QUIGLEY, R.M. & VOGAN, R.W. (1970). Black shale heaving at Ottawa, Canada. *Canadian Geotechnical Journal*, **7**, 106-115
RAISWELL, R. & BERNER, R.A. (1986). Pyrite and organic matter in Phranerozoic normal marine shales. *Geochemica and Cosmochimica Acta*, **50**, 1967-1976
RUSSEL, D.J. & PARKER, A. (1979). Geotechnical, mineralogical and chemical interrelationships in weathering profiles of an over-consolidated clay. *Quarterly Journal of Engineering Geology*, **12**, 107-116.
SHERRELL, F.W. (1979). Engineering properties and performance of clay fills. In:*Clay Fills*, Institution of Civil Engineers, London, 241.
TAYLOR, R.K. & CRIPPS, J.C. (1984). Mineralogical controls on volume change. In: *Ground Movements and their Effects on Structures*, Attewell, P.B. and Taylor, R.K. (eds), Surrey University Press, 268-302.
WELLMAN, H.W. & WILSON, A.T. (1965). Salt weathering: a neglected erosive agent in coastal and arid environments. *Nature*, **205**, 1097-1098.
WINKLER, E.M. & SINGER, P.C. (1972). Crystallisation pressures of salts in stone and concrete. *Bulletin of the Geological Society of America*, **83**, 3509-3513.

2-3 INVESTIGATION OF THE EFFECTS OF BACTERIAL ACTION ON THE CHEMISTRY AND MINERALOGY OF PYRITIC SHALE

S.D. JACKSON & J.C. CRIPPS
Department of Earth Sciences, University of Sheffield, Dainton Building, Brook Hill, Sheffield S3 7HF

ABSTRACT

Iron oxidising bacteria now recognised as *Thiobacillus ferrooxidans* were first isolated from acid mine drainage waters in the late 1940s. By deriving energy from the oxidation of thiosulphate and ferrous iron they participate in reactions involving the consumption of sulphides (including pyrite) and the production of acid.

The present study has been directed at investigating the effects of the bacterial removal of pyrite from the pyritic Carboniferous (Namurian) shale being used in the reconstruction of the Carsington Dam in Derbyshire. Exposure of the shale to weathering processes has resulted in the formation of iron hydroxides, sulphates and other minerals, in addition to the production of acid pore waters. The effects of leaching samples of the shale with *Thiobacillus ferrooxidans* in the laboratory are examined in terms of changes in mineralogy and geochemistry. Comparisons are drawn between the microbially leached and control samples. The processes and controls on alteration of the shale are discussed.

INTRODUCTION

Colmer, Temple and Hinkle (1950) report that a bacterium similar in description to *Thiobacillus ferrooxidans* was probably first isolated in the late 1940s. It had been noted that waters associated with bituminous coal mines ranged from slightly alkaline to strongly acidic in composition. Initially it was thought that acid was generated by the chemical oxidation of pyrite in the coal but when it became apparent that the reaction could be arrested or slowed by sterilization, a microbial agent was sought. It was concluded that the organism concerned is a short, rod-shaped, motile, gram-negative bacterium, 0.4 and 1.0 μm in size, which may occur singly or, more rarely, in pairs. As a result of subsequent work, Temple and Colmer (1951) assigned the name *Thiobacillus ferrooxidans* to this microbe.

Thiobacillus ferrooxidans is an aerobic, acidophilic, chemoautotroph. As carbon dioxide, which it normally derives from the atmosphere, is its sole source of carbon, the supply of carbon dioxide is the main control on the growth rate of the organism. The energy required is derived from oxidation reactions involving either ferrous iron and reduced sulphur (Hutchins, Davidson, Brierley and Brierley, 1986) or monovalent copper (Lundgren and Silver 1980); the iron and sulphur reactions possibly assisting in the oxidation of pyrite as follows:

$$4FeS_2 + 15O_2 + 2H_2O \longrightarrow 2Fe_2(SO_4)_3 + 2H_2SO_4 \quad [1]$$

$$4FeSO_4 + O_2 + 2H_2SO_4 \longrightarrow bacteria \longrightarrow 2Fe_2(SO_4)_3 + 2H_2O \quad [2]$$

$$Fe_2(SO_4)_3 + 6H_2O \longrightarrow 2Fe(OH)_3 + 3H_2SO_4 \quad [3]$$

Lundgren and Silver (1980) indicate that the oxidation of pyrite is most rapid if it is fine grained. As shown by equations [1] and [2], as well as carbon dixoide, oxygen is required for this to occur and various nutrients including ammonia, sulphate, potassium, phosphate, magnesium and calcium are also needed. For optimum growth of the bacteria the temperature should be in the range 25 to 45°C. As a result of the reactions, sulphuric acid is produced which provides the low pH conditions required for rapid proliferation (pH approx 1.0-2.5). The resulting ferric hydroxide is susceptible to reactions with carbonates and clay minerals to produce various sulphates including gypsum, jarosite and other secondary minerals.

Recent interest in *Thiobacillus ferrooxidans* has centred on its potential for removing sulphur compounds from coal and in the recovery of metals from sulphide ores (Gaudy and Gaudy 1980). The extraction of sulphide from coal prior to combustion is seen as an attractive alternative method to the scrubbing of flue gases in efforts to reduce acid rain generation. However the presence of pyrite is also an important factor in the weathering of rocks bearing this mineral, particularly in the marine, organic-rich mudrocks. As a consequence of the anaerobic conditions within the bottom waters or the upper layers of sediment during deposition or diagenesis, pyrite forms due to the activities of sulphate reducing bacteria. The subsequent introduction of aerated waters during weathering brings about oxidation of the pyrite and the associated generation of acid.

Work reported by Steward and Cripps (1984) indicated that changes in pore water composition caused by the oxidation of pyrite can have an effect on the residual shear strength of shales containing pyrite. The process contributes to the degradation of the material through the removal of diagenetic cements (particularly carbonates) and the precipitation of gypsum and other minerals. The extensive generation of acid by the oxidation of pyrite in the mudstone fill during the construction of the Carsington Dam (Pye and Miller 1990) had a significant influence on the design and construction of the dam.

The aim of the present research is to investigate the effects of the growth of *Thiobacillus ferrooxidans* on the chemical composition and mineralogy of the shale at Carsington. Various volumes of shale were leached in sterile acid water and in the presence of bacteria for periods of up to 4 weeks. Several attempts were made to culture the bacteria from water or shale samples taken from the Carsington Dam site but as these were not successful the experiments were conducted using commercially available bacteria obtained from Philip Harris Biological Ltd, Avon. These were cultured in a modified 9K medium (Silverman and Lungren, 1959; Barron and Leuking, 1990) using ferrous sulphate as the iron source. The medium was brought to a pH of between 1.7 and 1.9 with 6 M sulphuric acid and sterilised by autoclaving at a pressure of 1 atmosphere before the addition of the ferrous sulphate solution which had previously been sterilised by filtration. A 10% inoculum was used and the medium was maintained at a constant temperature of 25°C and aerated by continuous shaking at 90 rpm. After an incubation period of three to four days, the flasks contained a rusty brown precipitate in contrast to the uninoculated flasks which appeared yellow (Plate 1). Aliquots of the inoculated flasks were plated onto Agar 9K plates, incubated at 25°C and examined for the presence of bacteria.

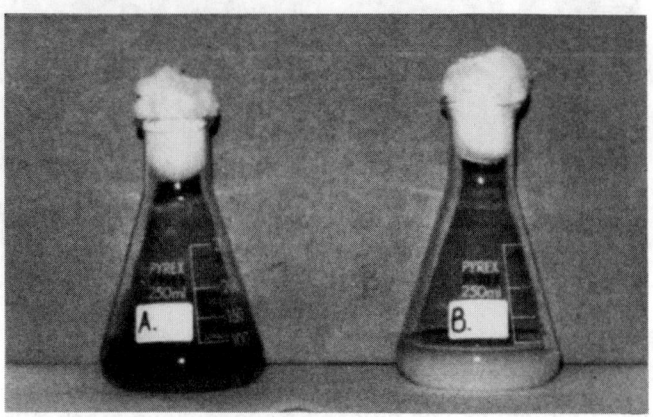

Plate 1: Flasks containing (a) inoculated and (b) uninoculated 9K medium after 4-5 days incubation at 25°C at 90 rpm.

LEACHING EXPERIMENTS

The shale used in the tests was obtained from the deeper parts of borrow pits at the Carsington Reservoir site where Namurian shale of Carboniferous age lying between the *Cravenoceratoides nititoides* and *Hudsonoceras proteus* goniatite marker horizons was being dug for the construction of the new dam. The appearance of this material at outcrop is shown in Plate 2. It consisted of faintly and slightly weathered, laminated and thinly bedded, dark grey, moderately weak and moderately strong, carbonaceous mudstone. The sequence also contained beds and partings of silty mudstone and siltstone. Frequent very thin partings and laminae of coal and some carbonate-rich partings were also present. The closely spaced, sub-horizontal, generally smooth bedding plane discontinuities and the intersecting, generally closely spaced, sub-vertical, slightly open joints with partial clay smears implied the rock would easily disintegrate into angular blocks of gravel and cobble sized fragments. Occasional small crystals and clusters of pyrite were visible within the rock and accicular gypsum crystals occurred on some bedding plane and joint surfaces.

Plate 2: Appearance of shale in the borrow pit.

Prior to leaching the shale was crushed to pass sieve mesh 425 μm and thoroughly homogenised. It was then sterilised by autoclaving at a pressure of 1.0 atmosphere for 20 minutes. The 9K solution without $FeSO_4$ used for the leaching was made acid (pH between 1.5 and 2) by the addition of a predetermined quantity of 25% H_2SO_4 and then sterilised by the same means. Various leaching experiments were then carried out as follows:

1. Small quantities of shale in 250 ml flasks for periods of 2 and 4 weeks respectively. Samples weighing 3, 6, 9 and 12 g were then leached using 90 ml of modified 9K solution containing no $FeSO_4$ and 10 ml of *Thiobacillus ferrooxidans* inoculum. A 6 g uninoculated control sample and some repeat samples were similarly treated.

2. Large quantities of shale in 250 ml flasks. In this case two 150 g lots of powdered shale were leached in either 100 ml of sterile water made acid to pH below 2 by the addition of 25% H_2SO_4 or 90 ml of distilled water to which 10 ml of *Thiobacillus ferrooxidans* inoculum had been added.

To safeguard against the possibility of contamination, the flasks were stoppered throughout the leaching period using gauze and cotton wool bungs. All the samples were subjected to constant agitation by shaking at 90 rpm at a temperature of 25°C. At the end of the leaching period the shale was allowed to settle before the supernatant liquid was poured off. The remaining material was centrifuged, washed with distilled water, centrifuged and then dried at a temperature of 75°C.

After leaching the samples were analysed for major elements and Fe^{2+}. The major element analyses were carried out by X-ray fluorescence (see Table I) and Fe^{2+} was determined by the method of Begheijn (1979). Mineralogical determinations were performed by X-ray diffraction in which smear mounts were prepared by mixing a small sample of the shale with distilled water. This was then air dried and scanned between a 2θ angle of 4 and 60°.

RESULTS AND DISCUSSION

CHEMICAL ANALYSES

The results of the chemical and mineralogical determinations are given in Table I. Begheijn's (1979) method for the determination of iron provides a value for Fe^{2+} and also total iron ($Fe^{2+} + Fe^{3+}$). The amount of Fe^{3+} present is determined by subtraction of these two results. As the iron is taken into solution with non-oxidizing acid, the Fe^{2+} does not include pyrite. However assuming there are no other insoluble iron compounds the quantity of Fe^{2+} present in this form can be calculated by subtracting the Begheijn $Fe^{2+} + Fe^{3+}$ from the total iron given by the XRF analysis. This is achieved by calculating the equivalent quantity of Fe corresponding with the Fe_2O_3 XRF result, having factored the value to take account of the loss on ignition and to bring the total for the analyses to 100%. In turn, the quantity of Fe^{2+} thus determined is expressed as equivalent FeS_2.

A convenient way of considering the progress of chemical weathering processes in rocks is provided by Al_2O_3 v SiO_2 plots. Although some Al_2O_3 often appears in the pore fluids of rocks undergoing weathering, in many environments the changes are minor. Under normal circumstances, due to the removal of lower stability components during weathering, both SiO_2 and Al_2O_3 show relative increases.

Such a trend can be seen in the present results where, as shown in Figure 1, the data cluster in three areas and, compared with the original shale, a trend of increasing Al_2O_3 and SiO_2 with increased leaching occurs. All leaching causes change, including that in acid water. It is noticeable that the larger samples undergo less change than the small ones, however, and in the case of the 6g samples the change is less in the samples leached in acid water than those similarly leached in the presence of bacteria. During leaching there is a reduction in the amount of pyrite present from an approximate average value of 6% (Table I). This is shown in Figure 2, where the apparent increase in Al_2O_3 is probably due to

Table 1: Chemical changes due to leaching

Sample	SiO$_2$	TiO$_2$	Al$_2$O$_3$	Fe$_2$O$_3$	MnO	MgO	CaO	Na$_2$O	K$_2$O	P$_2$O$_5$	SO$_3$	Total	Igl	Fe^B	Fe^T	Fe^X	Fe^{3+}	FeS$_2$
Original Shale	56.21	0.92	20.94	10.04	0.18	2.31	3.65	0.37	3.11	0.20	1.38	99.30	17.73	1.85	3.20	5.81	1.35	5.62

Leached in sterile acid water 6g 2 and 4 weeks and 150g (2 samples) 3 weeks

Sample	SiO$_2$	TiO$_2$	Al$_2$O$_3$	Fe$_2$O$_3$	MnO	MgO	CaO	Na$_2$O	K$_2$O	P$_2$O$_5$	SO$_3$	Total	Igl	Fe^B	Fe^T	Fe^X	Fe^{3+}	FeS$_2$
6g 2 wk	60.22	1.11	22.36	9.05	0.07	1.70	0.34	0.42	3.39	0.46	0.36	99.48	16.51	1.65	3.60	5.31	1.95	3.68
6g 4 wk	60.59	1.12	22.66	8.83	0.06	1.70	0.18	0.40	3.43	0.54	0.27	99.78	15.66	1.90	4.05	5.22	2.15	2.51
150g 3 wk A	57.35	0.93	21.19	10.05	0.16	2.26	3.63	0.36	3.18	0.21	0.39	99.72	20.31	1.70	3.20	5.61	1.50	5.19
150g 3 wk B	57.41	0.95	21.32	10.06	0.16	2.17	3.69	0.39	3.18	0.21	0.38	99.91	20.20	1.85	3.05	5.62	1.20	5.52

Leached with Thiobaccillus 3 (x2), 6, 9 and 12g (2 samples) 2 and 4 weeks and 150g (2 samples) 3 weeks

Sample	SiO$_2$	TiO$_2$	Al$_2$O$_3$	Fe$_2$O$_3$	MnO	MgO	CaO	Na$_2$O	K$_2$O	P$_2$O$_5$	SO$_3$	Total	Igl	Fe^B	Fe^T	Fe^X	Fe^{3+}	FeS$_2$
3g 2 wk A	61.87	1.15	22.79	6.86	0.03	1.70	0.03	0.46	3.59	0.26	0.06	98.80	16.96	1.50	4.0	4.03	2.50	0.07
3g 2 wk B	61.17	1.15	22.79	6.86	0.03	1.70	0.02	0.43	3.69	0.29	0.03	98.74	17.23	1.60	4.30	4.37	2.70	0.14
3g 4 wk	61.00	1.14	22.47	8.31	0.03	1.65	0.03	0.45	3.87	0.33	0.06	99.34	16.61	1.45	4.90	4.88	3.45	0
6g 2 wk	61.57	1.17	22.82	7.15	0.03	1.74	0.17	0.43	3.63	0.26	0.09	99.06	16.90	1.50	4.10	4.19	2.60	0.20
6g 4 wk	60.91	1.12	22.53	7.77	0.03	1.71	0.49	0.40	3.78	0.32	0.15	99.21	17.03	1.40	4.60	4.56	3.20	0
9g 2 wk	60.45	1.09	22.68	8.28	0.03	1.83	0.36	0.46	3.74	0.27	0.09	99.28	18.22	1.45	4.90	4.77	3.45	0
9g 4 wk	60.61	1.14	22.95	8.08	0.03	1.73	0.33	0.45	3.78	0.28	0.14	99.52	16.82	1.60	4.55	4.72	2.95	0.37
12g 2 wk A	60.73	1.14	22.68	7.81	0.04	1.77	0.29	0.44	3.67	0.24	0.13	99.04	17.27	1.60	4.55	4.62	2.95	0.15
12g 2 wk B	60.63	1.10	22.89	7.85	0.04	1.79	0.24	0.41	3.38	0.27	0.12	98.99	17.45	1.75	4.80	4.58	3.05	0
12g 4 wk	60.43	1.13	22.99	8.39	0.03	1.77	0.37	0.41	3.89	0.27	0.11	99.59	17.21	1.50	4.60	4.88	3.10	0.59
150g 3 wk A	56.97	0.95	21.23	10.06	0.16	2.19	3.73	0.38	3.16	0.21	0.39	99.43	21.51	1.85	3.50	5.55	1.65	4.41
150g 3 wk B	56.99	0.93	21.26	10.06	0.16	2.21	3.74	0.34	3.14	0.21	0.40	99.46	21.68	1.80	3.60	5.54	1.80	4.17

SiO$_2$, TiO$_2$, Al$_2$O$_3$, Fe$_2$O$_3$, MnO, MgO, CaO, Na$_2$O, K$_2$O, P$_2$O$_5$, SO$_3$ determined by major elemental XRF analysis.
Igl Ignition loss on heating to 850°C.
Fe^B Ferrous iron determined by method of Begheijn (1979).
Fe^T Total iron determined by the method of Begheijn (1979).
Fe^X Total iron determined from XRF total iron factored to bring total to 100% and with allowance for loss on ignition.
Fe^{3+} Ferric iron calculated from Begheijn results (Fe^T minus Fe^B).
FeS$_2$ Pyrite calculated as equivalent to Fe^X minus Fe^T.

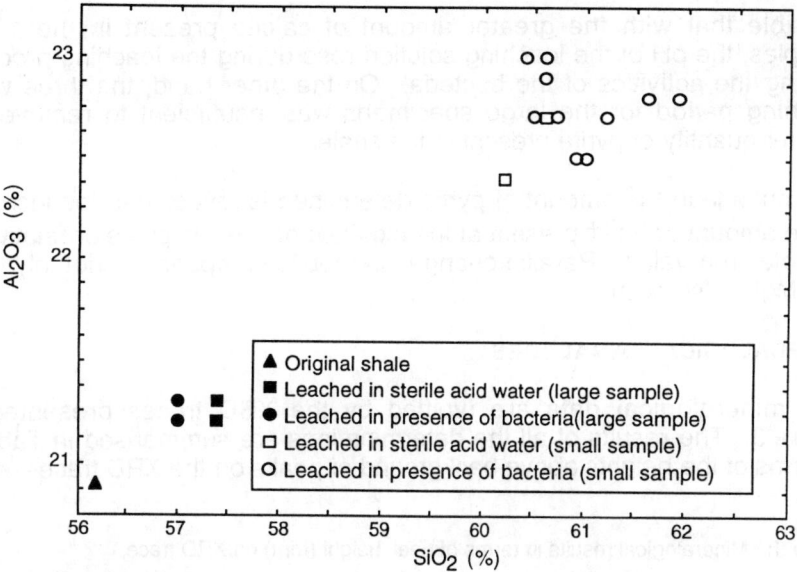

Figure 1: Effects of leaching on the presence of silica and alumina in the shale.

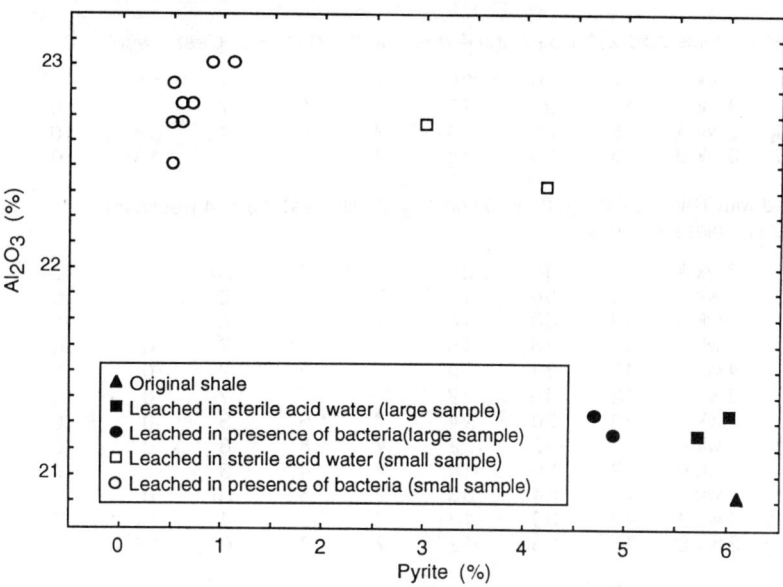

Figure 2: Effects of leaching on the presence of alumina and pyrite in the shale.

the removal of more mobile components but pyrite shows a decreasing trend. The control samples leached under sterile conditions also show a decrease in pyrite which is greater after four weeks of leaching than after two and is presumably due to chemical oxidation. The trend is the same for the small samples as for the large ones, although in the latter case, due to the larger quantities involved, the changes are relatively small. It is possible that with the greater amount of calcite present in the 150 g samples, the pH of the leaching solution rose during the leaching process, limiting the activities of the bacteria. On the other hand, the three week leaching period for the large specimens was insufficient to remove the greater quantity of pyrite present in the shale.

This change in the amount of pyrite determined is reflected in the increase in the amount of Fe^{3+} present at the expense of Fe^{2+} in pyrite or relative to the total iron value. Parallel changes in mobile components, notably CaO and MgO, also occur.

MINERALOGICAL ANALYSES

The mineralogical data are typified by the XRD traces presented in Figure 3. The results of all the determinations are summarised in Table II in terms of the heights above background of peaks on the XRD traces.

Table II: Mineralogical results in terms of peak height (mm) on XRD trace.

Sample		Quartz	Kaolin	Illite	MLSC	Chlor	Felds	Pyrite	Calcite	Gyps
Original Shale		42	17	14	?	8	5	20	12	10

Leached in sterile acid water 6g 2 and 4 weeks and 150g (2 samples) 3 weeks

Sample		Quartz	Kaolin	Illite	MLSC	Chlor	Felds	Pyrite	Calcite	Gyps
6g	2 wk	37	30	20	?	8	7	15	0	0
6g	4 wk	43	33	20	?	?	7	0	0	0
150g	3 wk A	45	30	14	?	8	4	14	0	45
150g	3 wk B	45	30	18	?	?	5	16	0	45

Leached with Thiobacillus 3 (x2), 6, 9 and 12g (2 samples) 2 and 4 weeks and 150g (2 samples) 3 weeks

Sample		Quartz	Kaolin	Illite	MLSC	Chlor	Felds	Pyrite	Calcite	Gyps
3g	2 wk A	43	30	25	?	8	10	0	0	20
3g	2 wk B	40	36	17	?	6	8		0	10
3g	4 wk	40	30	18	?	?	8	0	?	14
6g	2 wk	44	36	18	?	10	7	0	0	10
6g	4 wk	43	33	20	?	?	7	0	0	12
9g	2 wk	42	35	22	?	7	?	0	0	10
9g	4 wk	36	30	24	?	5	5	0	0	8
12g	2 wk A	42	33	22	?	8	8	0	0	8
12g	2 wk B	39	30	18	?	4	8		0	10
12g	4 wk	44	34	33	?	5	10	0	0	55
150g	3 wk A	42	30	14	?	8	4	14	0	45
150g	3 wk B	73	25	12	?	7	0	17	0	48

Kaolin - kaolinite, MLSC - mixed layer swelling clay, Chlor - chlorite, Felds - feldspar, Gyps - gypsum.

Although this does not indicate the proportions of the mineralogical components with respect to each other, as the samples were analysed by the same procedure a change in height of a particular peak does reflect a change in the amount of that mineral from one sample to the next. As shown, the data for small and large samples can be divided into those for the original shale, shale leached in acid water and shale leached in acid

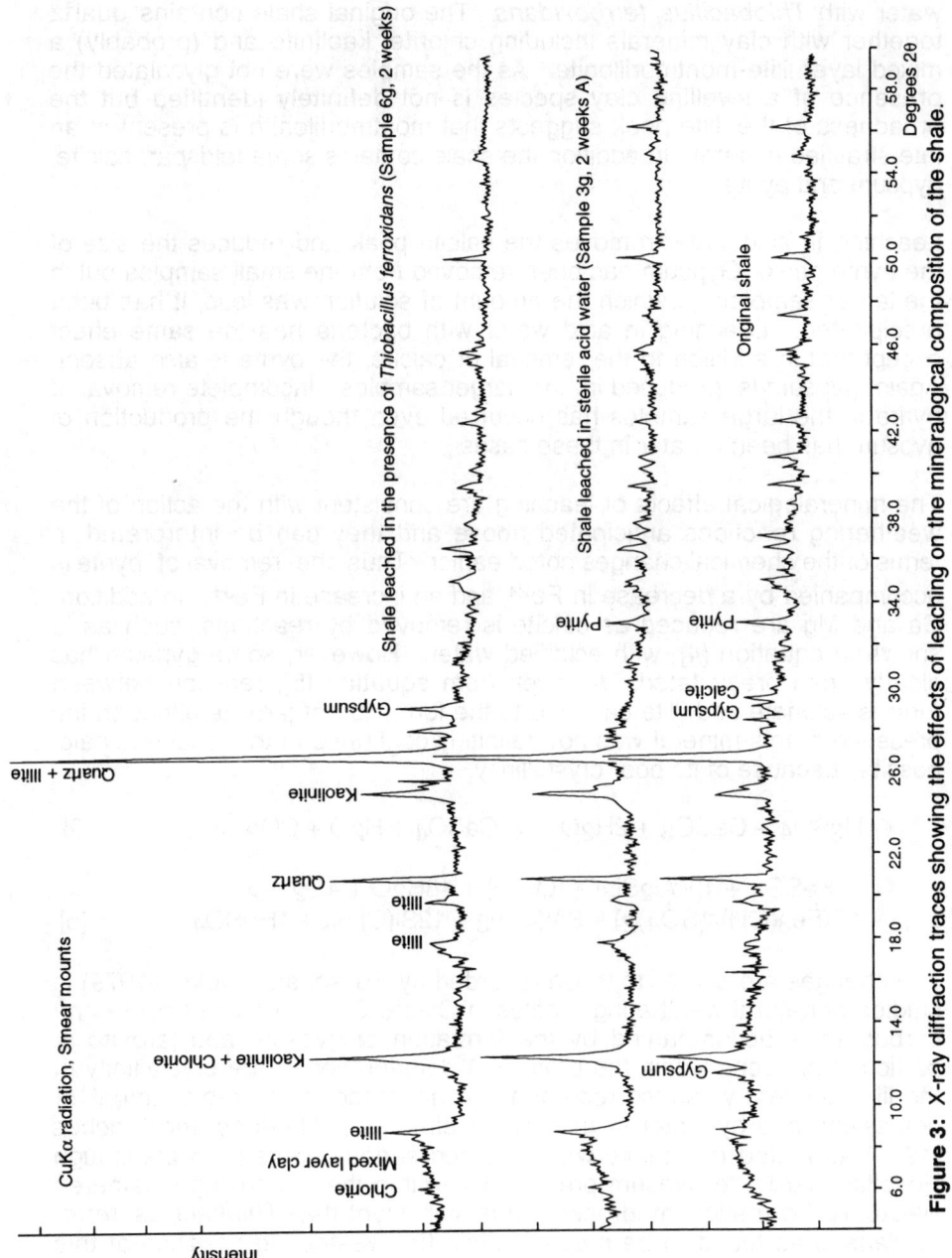

Figure 3: X-ray diffraction traces showing the effects of leaching on the mineralogical composition of the shale.

water with *Thiobacillus ferrooxidans*. The original shale contains quartz together with clay minerals including chlorite, kaolinite and (probably) a mixed layer illite-montmorillonite. As the samples were not glycolated the presence of a swelling clay species is not definitely identified but the broadness of the illite peak suggests that montmorillonite is present in an interstratified mineral. In addition the shale contains some feldspar, calcite, gypsum and pyrite.

Leaching in acid water removes the calcite peak and reduces the size of the pyrite peak. Gypsum has been removed from the small samples but in the larger samples, in which the amount of solution was less, it has been precipitated. Leaching in acid water with bacteria has the same effect except that in addition to the removal of calcite, the pyrite is also absent. Again gypsum is produced in the larger samples. Incomplete removal of pyrite in the large samples has occurred even though the production of gypsum has been greater in these cases.

The mineralogical effects of leaching are consistent with the action of the weathering reactions anticipated above and they can be interpreted in terms of the chemical changes noted earlier. Thus the removal of pyrite is accompanied by a decrease in Fe^{2+} and an increase in Fe^{3+}. In addition, Ca and Mg are reduced as calcite is removed by reactions, such as is shown in equation [4], with acidified water. However, some gypsum has clearly been precipitated. As seen from equation [5], reaction between ferrous sulphate and illite can lead to the formation of jarosite although the presence of this mineral was not definitely confirmed in the leached shale, possibly because of its poor crystallinity.

$$H_2SO_4 + CaCO_3 + 2H_2O \rightarrow CaSO_4 + H_2O + CO_2 \quad [4]$$

$$12FeSO_4 + 4(KAl_2Si_3O_8(OH)_2) + 48H_2O + 4O_2 \rightarrow$$
$$4(KFe_3(OH)_6(SO_4)_2) + 8Al(OH)_3 + 12Si(OH)_4 + 4H_2SO_4 \quad [5]$$

The changes are similar to those recorded by Russel and Parker (1979) in studies of natural weathering profiles in Oxford Clay. A loss of pyrite and carbonate is accompanied by the formation of gypsum and jarosite at particular horizons within the profile. A deterioration in the crystallinity of the illite present was also recorded and attributed to the replacement of potassium ions by water in the crystal structure. Hawkins and Pinches (1987) attributed the heave which caused serious damage to Llandough Hospital, Cardiff to gypsum precipitation within the underlying weathered Westbury Formation mudrocks. It is significant that *Thiobacillus ferrooxidans* was found to be present within the weathering horizons at this site.

Pye and Miller (1990) report the results of chemical and biochemical leaching tests on samples of Carsington Shale. The tests included free draining leaching of shale with a culture medium containing ferrous sulphate and a parallel experiment in which the leaching solution was inoculated with *Thiobacillus ferrooxidans*. Although most chemical and mineralogical changes were similar to those in the present investigation they did not find that pyrite was removed by this biochemical leaching. The fact that in the present investigation a proportion of the pyrite also remained in the larger (150g) samples after leaching in the presence of bacteria may indicate that there was insufficient time to remove the amount of pyrite in Pye and Miller's 50 g samples which were subjected to only temporary wetting.

CONCLUSIONS

Thiobacillus ferrooxidans has been identified in weathering environments and has been implicated in the oxidation of pyrite in marine shales. It is liable to accelerate the degradation of these materials through the removal of diagenetic cements, particularly carbonates, and the precipitation of gypsum and other compounds.

For growth and propagation it requires acidic conditions in the range pH 1.5 to 5.0, optimum conditions for the oxidation of pyrite being in the range 1.0 to 2.5. The optimal oxidation of ferrous sulphide occurs in the temperature range 25 to 45°C. The growth medium requires various nutrients and carbon dioxide which is the only source of carbon. The bacterium is aerobic and oxygen is needed for oxidation. More rapid oxidation of pyrite is liable to occur if it is fine grained.

Tests involving the leaching of shales indicate that laboratory cultures can be maintained but that the conditions, particularly the acidity, must be carefully monitored and appropriately adjusted if required. It would appear that with fresh shale there is a tendency for the bacterial growth to be limited by a rise in pH as the acid produced reacts with available calcite. The reduced amount of calcite present in partially weathered shales can make the pyrite in these materials more susceptible to bacterial action than is the case with the fresh shales.

The mineralogical and chemical effects of the laboratory leaching process appear to follow those recorded in the natural weathering of shales. Leaching in sterile acid removes calcite and other mobile components leading to a relative increase in Al_2O_3 and SiO_2. A small amount of pyrite is also oxidized with a consequent increase in the ferric iron present. In addition to the removal of calcite, leaching in the presence of bacteria has led to the consumption of pyrite. Within the variation in dilution in the small samples (3 to 12g shale in 250 ml of solution) and time for leaching (one to four weeks), the effects of leaching are very similar. Possibly due to the rise in the pH of the solution during the leaching experiments, increasing the amount of shale (150 g) resulted in a reduced effect in terms of dissolution of mobile components and the removal of pyrite. However, the removal of pyrite was greater in the case of leaching with bacteria than in acid alone.

It is anticipated that further research in which the engineering properties of the leached and artificially weathered shale are determined will provide a means of predicting the effects of these processes on the mechanical behaviour of mudrocks containing significant pyrite.

ACKNOWLEDGEMENTS

The work described in this paper was carried out under a Natural Environmental Research Council CASE award. The authors are grateful for the logistical and financial support provided by Babtie Shaw and Morton who were the industrial sponsors for the project.

REFERENCES

BEGHEIJN, L.TH. (1979). Determination of iron(II) in rock, soil and clay. *Analyst*, **104**, 1055-61.

BARRON, J.LAC. & LEUKING, D.R. (1990). Growth and maintenance of Thiobacillus ferrooxidans cells. *Applied Environmental Microbiology*, **56**, 2801-6.

COLMER, A.R., TEMPLE, K.L. & HINKLE, M.E. (1950). An iron oxidizing bacterium from the acid mine drainage of some bituminous coal mines. *Journal of Bacteriology*, **59**, 317-28.

HAWKINS, A.B. & PINCHES, G.M. (1987). Cause and significance of heave at Llandough Hospital, Cardiff - a case history of ground floor heave due to gypsum growth. *Quarterly Journal of Engineering Geology*, **20**, 41-57.

HUTCHINS, S.R., DAVIDSON, M.S., BRIERLEY, J.A. & BRIERLEY, C.L. (1986). Microorganisms in reclamation of metals. *Annual Review of Microbiology*, **40**, 311-336.

GAUDY, A. & GAUDY, E. (1980). *Microbiology for environmental scientists and engineers.* McGraw Hill.

LUNDGREN, D.G. & SILVER, M. (1980). Ore leaching by bacteria. *Annual Review of Microbiology*, **34**, 263-83.

PYE, K. & MILLER, J.A. (1990). Chemical and biochemical weathering of pyritic mudrocks in a shale embankment. *Quarterly Journal of Engineering Geology*, **23**, 365-382.

RUSSEL,D.J. & PARKER, A. (1979). Geotechnical, mineralogical and chemical interrelationships in weathering profiles in a overconsolidated clay. *Quarterly Journal of Engineering Geology*, **12**, 107-16.

SILVERMAN, M.P. & LUNGREN, D.G. (1959). Studies on the chemotrophic iron bacterium Ferrobacillus ferrooxidans. *Journal of Bacteriology*, **77**, 642-47.

STEWARD, H.E. & CRIPPS, J.C. (1983). Some engineering implications of chemical weathering in pyritic shales. *Quarterly Journal of Engineering Geology*, **16**, 281-9.

TEMPLE, K.L. & COLMER, A.R. (1951). The autotrophic oxidation of iron by a new bacterium: Thiobacillus ferrooxidans. *Journal of Bacteriology*, **62**, 605-11.

2-4 THE GENERATION OF SULPHATES IN THE PROXIMITY OF CAST IN SITU PILES

A.B. HAWKINS & M.D. HIGGINS
Engineering Geology Research Group, University of Bristol, Wills Memorial Building, Queens Road, Bristol BS8 1RJ

ABSTRACT

This paper reports a change in the ground chemistry adjacent to large concrete piles. Tests were carried out on samples between contiguous cast in situ piles in the London Clay. In the brown London Clay acid soluble sulphate peaks in the order of 2%, approximately 30 mm from the edge of the pile, were observed. No such peaks were recorded in similar situations in the blue London Clay. It is considered this is related to the difference in ground chemistry and the ground water level but is induced by the change in temperature equilibrium associated with the heat of hydration created within the pile during its curing. Tentative explanations are presented.

INTRODUCTION

The potential degradation of concrete resulting from the growth of sulphate crystals has been known for a long time. Consequently before concrete is emplaced in the ground it is standard engineering practice to measure the sulphates present in ground waters or soils (BS 5930: 1981; BRE Digest 363: 1991). In addition, in the 1970s it was appreciated that ground heave could occur related to the growth of sulphates (Quigley and Vogan, 1970; Penner Eden and Gillott, 1973). A similar induced heave at Llandough Hospital (Cardiff) was described by Hawkins and Pinches who emphasised both the generation of sulphates as a consequence of engineering works and the importance of temperature in the production of ground sulphates. Previously Vishniac (1974) had suggested that the optimum temperature for bacteria known to be important in the oxidation process was 15 to 20°C while Nixon (1978) suggested a range of 30 to 35°C. However Hawkins and Pinches (1992) in their work on the dark shales of the Westbury Formation have shown that the generation of sulphate continues to, and probably past, 40°C. In their Table III they show that in samples from the black shales of the Westbury Beds the SO_3% can increase from 0.6 to 1.8% after 15 weeks in an oven at 41.5°C. Subsequent work on London Clay has shown that the SO_3 content in samples kept at 40°C in cabinets with a retained 100% relative humidity increased from 0.2 to 0.96% after a period of only six weeks.

Following the realisation that temperature was important in enhancing sulphate growth, the possibility that selenite may crystallise within the ground as a consequence of the heat of hydration from large masses of concrete was considered. When the opportunity became available it was decided to examine whether there was an increase in sulphates related to the construction of large diameter piles where the heat evolved from the hydration reactions would cause a rise in temperature of the surrounding clay. This paper describes the work carried out during this initial study and reports the findings.

BACKGROUND GEOLOGY AND INITIAL SITE INVESTIGATION

In 1989 a site investigation was carried out at the location of an underground car park in College Road, Harrow, north London (TQ 153 882). The initial investigation included field testing and the recovery of disturbed and undisturbed samples at 1.5 m intervals. In order to accommodate the large number of cars specified in the restricted area available, ten parking levels were planned, hence the total excavation extended to 18 m. Initially the investigation involved five boreholes drilled to depths of between 34 and 43 m. The depth of the boreholes was controlled by the necessity to obtain sufficient information for the design of the pile foundations within the structure, the design of a contiguous pile wall surrounding the excavation for the car park and an assessment of the potential ground heave following the excavation.

The investigation proved the presence of London Clay beneath a thickness of approximately 1 to 2 m of made ground. This confirmed the evidence on the geological map which implies that the site, at 65 m AOD, is underlain by London Clay. At Harrow on the Hill, 0.5 km to the south of the site, the London Clay is overlain by the Claygate Beds, the junction being at approximately 95 m AOD. The investigation proved 30 m of London Clay beneath the site before the underlying Woolwich and Reading Beds were encountered.

As anticipated the investigation proved the London Clay to consist of a fissured upper brown weathered zone underlain by an unweathered grey clay with less fissures but an increasing amount of silt, generally forming discrete laminae. The boundary between the weathered and unweathered material was at about 8 m, approximately 57 m AOD. The investigation recorded selenite crystals and "pockets or partings of orange brown fine sand" throughout the weathered zone. The grey London Clay was divided into three sub-zones. The upper 8 m or so generally consists of very stiff fissured clay with occasional thin fine sand partings. Below this is a very stiff to hard grey silty to very silty clay with partings of light grey fine sand. In this 4 m zone the clay is only slightly fissured, which may explain the higher undrained shear strengths obtained. The basal zone generally comprises hard grey very silty sandy clay interbedded with clayey sand. Beneath this at a depth of 30 m are the Woolwich and Reading Beds.

Seepages were generally encountered close to the junction of the London Clay and underlying Woolwich and Reading Beds. Undoubtedly in the cable percussive boreholes these seepages were in part a consequence of the more silty sandy nature of the basal London Clay permitting a faster ingress of water as the hole was advanced. Standpipes were installed in two of the boreholes and after three weeks gave water depths of 14.1 and 15.1 m. As is well known, the response time in clay rich horizons is often

SULPHATE GENERATION IN THE PROXIMITY OF CAST IN SITU PILES

Table I: Geological summary compiled from continuous U_{100} samples.

Depth (m)	Description
2.00-2.50	Soft medium brown slightly silty CLAY with occasional organic specks.
2.50-2.75	Firm medium brown slightly silty CLAY with gravel sized pockets of sucrose and micro crystalline selenite.
2.75-7.50	Stiff to very stiff medium brown fissured slightly silty CLAY with some grey gleying along discontinuities and remnant laminations. Occasional isolated pockets of selenite crystals up to 10mm long.
7.50-8.50	Stiff to very stiff medium grey weathered brown fissured slightly silty CLAY with remnant laminations. Lenses of light orange brown friable clayey SILT up to 20mm thick and occasional isolated selenite crystals up to 6mm long.
8.50-16.00	Very stiff medium grey slightly silty CLAY with thin layers (<3mm) of light grey SILT.
16.00-19.83	Very stiff medium grey slightly silty CLAY with very closely spaced discontinuities, remnant laminations and thin (<2mm) layers of grey silt.

long and it is probable that the long term ground water level is between 8 and 9 m depth. The extent of any underdrainage as a consequence of water abstraction from the underlying chalk is not known and this area is outside of the main region being studied to determine the effects of the rising ground water beneath London since abstraction ceased.

Samples were taken from the five boreholes for chemical testing. Samples of both the soil and the ground water indicate values such that from the classification in BRE 250 (current at the time), Aggressivity Classes 3 and 4 would be considered appropriate for all concrete at or below ground level.

As selenite was proved to a depth of 8 m in the initial boreholes it was considered prudent to undertake a further borehole in order to obtain a detailed geological profile. A cable percussive rig was used to take continuous U_{100} samples. Each U_{100} was individually logged and from the descriptions a borehole profile recorded (Table I). The detailed log confirms that the boundary between the brown and blue London Clay was at about 8 m depth.

GROUND CHEMISTRY

As the initial ground investigation proved the site to have a ground water containing significant amounts of sulphates (3.7 gm/litre) and abundant selenite crystals, it was considered an appropriate site on which to test the effect of the heat of hydration from cast in situ piles on the ground chemistry of the adjacent London Clay. Unfortunately at the time of the first visit the work in the "construct and excavate" system had advanced too far to allow a full range of samples to be obtained. In view of the cramped working space and the contractor's "fast-track" work, it was clearly not convenient to clear the excavation floor and sample horizontally away from the large diameter load bearing piles as was originally envisaged.

From an examination of the surrounding contiguous piled wall structure, however, it was appreciated that in several places there were gaps between the piles which continued consistently with depth. It was considered that by detailed sampling of the clays between the piles it would be possible to determine whether there was any relationship between the sulphate content and the distance from the piles. Carrying out such sampling for the full depth of the excavation would also allow a geochemical profile to be established. The results of this are given in Figure 1.

The total iron present in the profile varies between 6 and 8% with one slightly higher figure recorded at 8.5 m depth. Above 8 m there is a ferrous iron content of about 1% indicating the oxidised nature of the brown London Clay. Below 9 m the Fe (II) varies between 2 and 3%, although a single value at 8.5 m reaches 5.5%, corresponding with the higher value obtained in the total iron analysis. The calcite content shows a number of changes with depth. Above 6 m it is less than 1%, probably as a consequence of dissolution during natural ground weathering. Between 6 and 7.5 m values of 2 to 3% were obtained while below 7.5 m there is no consistent level, the values varying between 3 and 8%.

Acid soluble sulphate values of 0.4 to 1.2% were recorded between 5 and 7 m below the ground surface, reaching a peak of 3.2% at 7.5 m. Below 8 m the ground sulphates are less than 0.5%, suggesting this is the equilibrium value in the relatively reducing conditions of the blue London Clay below the ground water table. The acid soluble sulphate chemistry confirmed the visual impression of the profile between the piles. Pyrite contents of 0.5% are seen in the brown London Clay whereas in the less oxidized blue London Clay between 0.75 and 1.75% pyrite is present.

X-ray diffraction analysis confirmed that the clay minerals in the London Clay were mainly illite with some kaolinite and a small proportion of smectites and interlayered illite/smectite minerals. There is a difference between the whole rock mineralogy of the blue London Clay and that of the weathered brown upper material. The former show some pyrite but while gypsum was present on the traces for the brown London Clay no pyrite was indicated.

At a number of the depths sampled, a continuous section of vertical slices of the clay present between the piles was taken from which SO_3 values could be obtained. The analyses are shown diagrammaticaly in Figure 2. It is noted that within the blue London Clay there is little if any difference between the percentage SO_3 across the relatively unweathered clay between the two piles; the minor variations simply reflecting natural changes within the ground. Above 8 m, however, coincident with material above the brown/blue boundary and likely ground water level, there are distinct peaks in the SO_3 percentage some 30 mm away from the pile edges. As this has been recorded at several levels it is clearly a

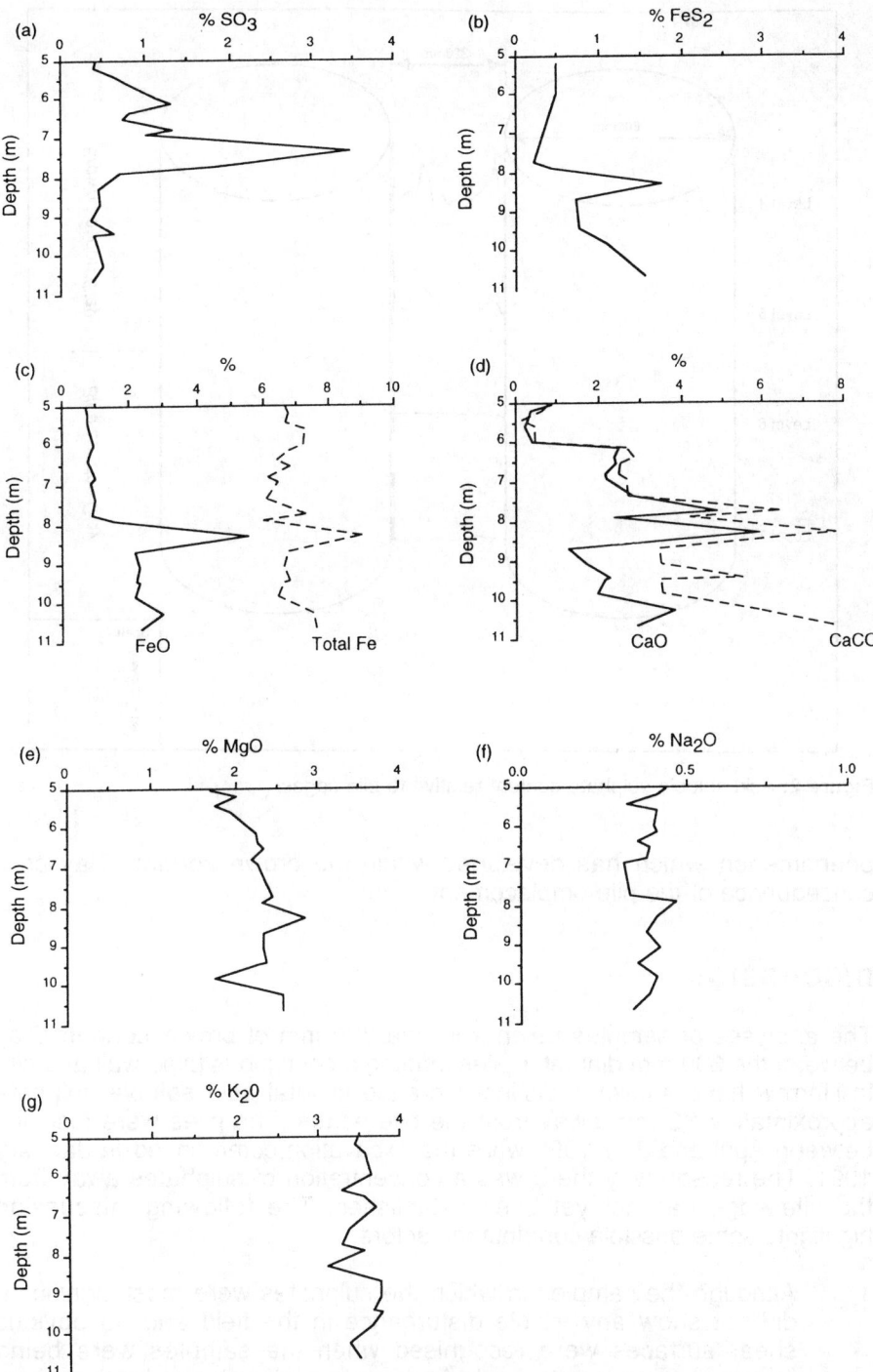

Figure 1: Geochemical profiles through brown and blue London Clay.

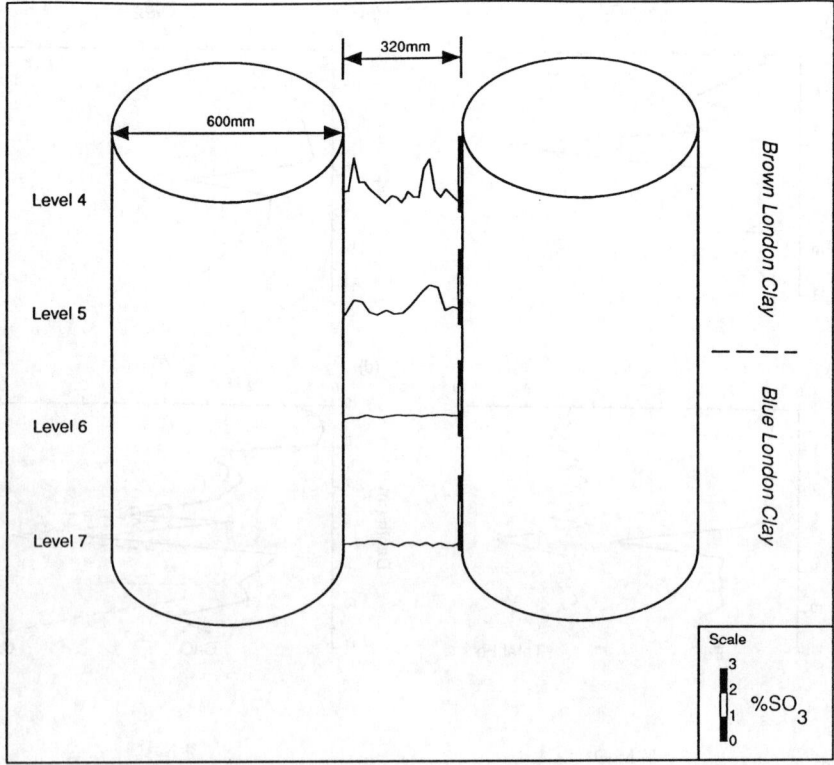

Figure 2: Acid soluble sulphate content relative to pile edges.

phenomenon which has developed within the brown London Clay as a consequence of the pile emplacement.

DISCUSSION

The analyses of samples taken from the 320 mm of brown London Clay between the 600 mm diameter piles forming a contiguous piled wall at a site in Harrow have shown a distinct increase in total acid soluble sulphate approximately 30 mm away from the pile edges. The piles were installed between April and July 1990 while the excavation commenced in January 1991. The reason why there was a concentration of sulphates away from the pile edge has not yet been established. The following discussion highlights some possible contributory factors.

1. Although the samples in which the sulphates were most significant did not show any visible disturbance in the field and no obvious shear surfaces were recognised when the samples were being prepared for sulphate analysis, the stresses induced by the auger cutting the 600 mm diameter piles may well have caused some disturbance in the surrounding soils. It is likely that any development of sulphates would preferentially occur in zones where the soil had

been dilated as a result of shear beyond the circumference of the cut cylinder. The presence of such a zone might be indicated by areas of higher moisture content adjacent to the piles (Figure 3).

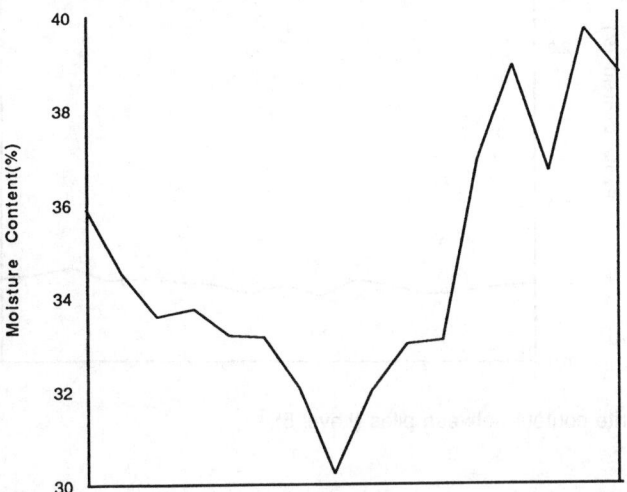

Figure 3: Moisture content between piles (Level 5).

2. At the time of curing, the temperature created by the heat of hydration could induce a movement of existing sulphate-rich ground waters towards the pile. Although the sulphate content of the ground water may have been in equilibrium with the temperature conditions before the pile was constructed, the raised temperatures around the pile would have allowed an increased concentration of sulphates commensurate with the modified temperature conditions at the time of curing. This may have resulted in supersaturation of the liquid and on cooling crystallisation could occur. Such crystals would naturally form in the area of least stress, ie the postulated disturbed zone created by shear dilation some distance from the cutting edge of the pile auger.

3. It is known that during the curing temperatures in the order of 55°C may be produced within the concrete. As a consequence, the natural ground temperatures of 15 to 20°C would be raised to levels at which oxidation processes are enhanced. The occurence of such chemical reactions over a period of several months would be sufficient to cause an increase in SO_3 percentages from 0.5 to 2%. With 0.5% pyrite and 2 to 3% calcite remaining in the brown London Clay, it can be seen that gypsum could develop. However there is no reduction in the pyrite content coincident with the peak SO_3 values (Figure 4). If a shear zone exists within the clays, this too would have facilitated the concentration as crystallisation of sulphates formed throughout the soil between the piles would preferentially take place in areas of lower pressure - having been transported by the temperature driven capillary moisture flow.

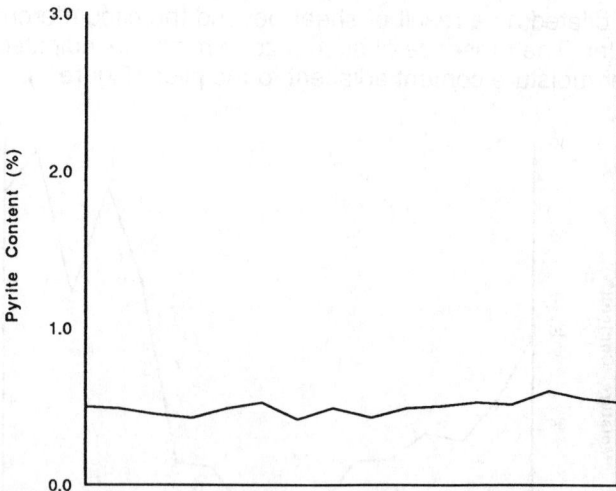

Figure 4: Pyrite content between piles (Level 5).

To date, it has not been possible to determine the cause of the zone of increased sulphate development in the brown London Clay. Unfortunately on an active engineering construction site it was not practical to return and take further samples but this is clearly a topic which requires further research.

CONCLUSIONS

At the location of an underground car park construction in the London Clay at Harrow, an enrichment of ground soil sulphates has been recorded approximately 30 mm from 600 mm diameter piles forming a continguous piled wall. The piles were installed between April and July 1990 while the excavation had reached a depth of 8 m by March 1991.

It is known that significant changes in ground temperature result due to the heat of hydration during the curing of piles. In addition, it is likely that the process of installation creates a zone of disturbed soil adjacent to the pile. It is considered that the combination of these two conditions - a disturbed zone coincident with raised temperature - produces a favourable environment for the production of sulphate.

The concentration of acid soluble sulphates approximately 30 mm from the piles may be purely a response to a localised increase in temperature leading to enhanced sulphate formation. Sulphates formed due to this process or those already present could be transported by capillary water whose movement was induced by the elevated temperature of the piles. Which of these two factors is dominant or whether both play an equal part is not known.

This paper reports a change in the ground chemistry adjacent to large concrete piles. In view of the engineering significance of this, further research should be undertaken into the heat transfer from the pile into the surrounding ground, the movement of water rich in soluble salts as a consequence of the rise in temperature of piles during the curing process, the development of low stress zones external to the pile excavation and the mechanism by which the sulphates increase.

REFERENCES

BRITISH STANDARDS INSTITUTION, (1981). *Code of Practice for Site Investigations, BS 5930.*

BUILDING RESEARCH ESTABLISHMENT, (1991). *No 363. Concrete in sulphate-bearing soils and ground water.*

HAWKINS, A.B. & PINCHES, G.M. (1987a). Cause and significance of heave at Llandough Hospital, Cardiff - a case history of ground floor heave due to gypsum growth. *Quarterly Journal of Engineering Geology*, **20**, 41-57.

HAWKINS, A.B. & PINCHES, G.M. (1992). Understanding sulphate generated ground heave due to gypsum growth. *Proceedings International Conference on the Implications of Ground Chemistry and Biology for Construction, Bristol.*

NIXON, P.J. (1978). Floor heave in buildings due to the use of pyritic shales as fill material. *Chemistry and Industry*, March, 160-168.

PENNER, E., EDEN, W.J. & GILLOTT, J.E. (1973). Floor heave due to biochemical weathering of shale. *Proceedings 8th International Conference on Soil Mechanics and Foundation Engineering, Moscow*, **II**, 151-158.

QUIGLEY, R.M. & VOGAN, R.W. (1970). Black shale heaving at Ottawa, Canada. *Canadian Geotechnical Journal*, **7**, 106.

VISHNIAC, W.V. (1974). Thiobacillus. In: *Bergey's Manual of Determinative Bacteriology*, Buchanan & Gibbons (eds), Williams & Wilkins, Baltimore, 456-461.

2-5 ROADFORD DAM: GEOCHEMICAL ASPECTS OF CONSTRUCTION OF A LOW GRADE ROCKFILL EMBANKMENT

S.E. DAVIES
Babtie Environmental Sciences, Renslade House, Bonhay Road, Exeter, EX4 3A

J.M. REID
Babtie Geotechnical, 95 Bothwell Street, Glasgow, G2 7HX

ABSTRACT

Roadford Dam in Devon has been constructed from locally available mudstone, siltstone and sandstone of Carboniferous age. The mudstone contains significant quantities of pyrite and it was realised that some of this might oxidise during and after construction. The chemistry and mineralogy of the fill was monitored during construction and the chemistry of the drainage waters has been monitored since completion of the embankment in 1988. The results have confirmed that pyrite oxidation is occurring but the quantities are very small and will not significantly affect the geotechnical properties of the fill. The embankment drainage has been satisfactorily treated by woodland irrigation. The chemistry of drainage waters has been used to investigate the source of flows from membrane drains within the embankment.

INTRODUCTION

The construction of major civil engineering projects such as reservoirs can have a significant impact on the environment in both the short and long term. Many of these effects are related to the geochemical properties of the materials used in the construction. This is particularly relevant where major earthworks are involved, as in the construction of embankment dams.

Study of the geochemistry of the proposed construction materials can give an insight into the nature and extent of reactions which may occur during and after construction (Macdonald and Reid, 1991). These can then be taken into account in the design of the works. Monitoring of the chemistry of drainage waters and gases allows an assessment of the scale of the reactions and enables predictions to be made regarding future trends. Remedial works can be undertaken where necessary. These points can be illustrated by the recently completed Roadford Reservoir scheme in south west England.

THE ROADFORD RESERVOIR PROJECT

The Roadford Reservoir project has been described by Gilkes, Millmore and Bell (1991). South West Water wished to construct an embankment dam on the River Wolf at Roadford in west Devon. The water impounded by the dam is used primarily for water supply purposes during the summer months when demand is high due to the tourist industry. A plan of the dam is given in Figure 1.

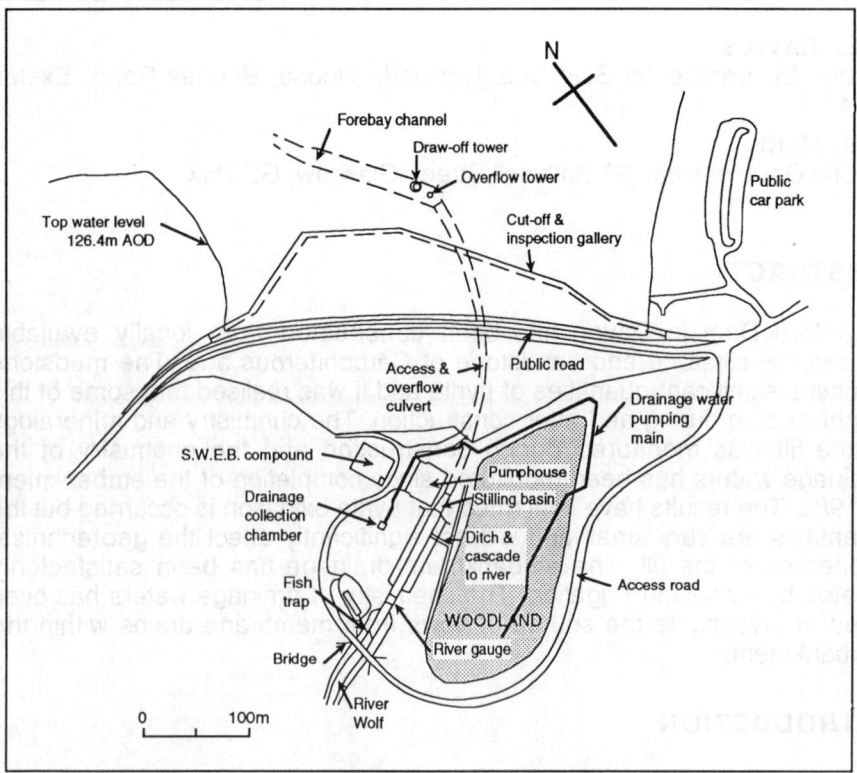

Figure 1: Plan of dam

Babtie Shaw and Morton were commissioned by South West Water to design and supervise construction of the embankment and ancillary works. Construction of the embankment has been described by Wilson and Evans (1991). The chosen design was an embankment of low grade rockfill with an impermeable upstream membrane of asphaltic concrete. The rockfill was obtained from a quarry opened for the purpose immediately upstream of the embankment in the reservoir solum. A drainage blanket of imported material was provided at the base of the embankment and sandwaste material from China Clay workings was incorporated at the upstream toe and beneath the upstream membrane. A minor road runs along the crest of the embankment. A section through the embankment is shown in Figure 2.

The contract for construction of the works was awarded to Alfred McAlpine Construction Ltd in February 1987. The embankment was built in 1988 and the upstream membrane constructed in 1989. The completed embankment

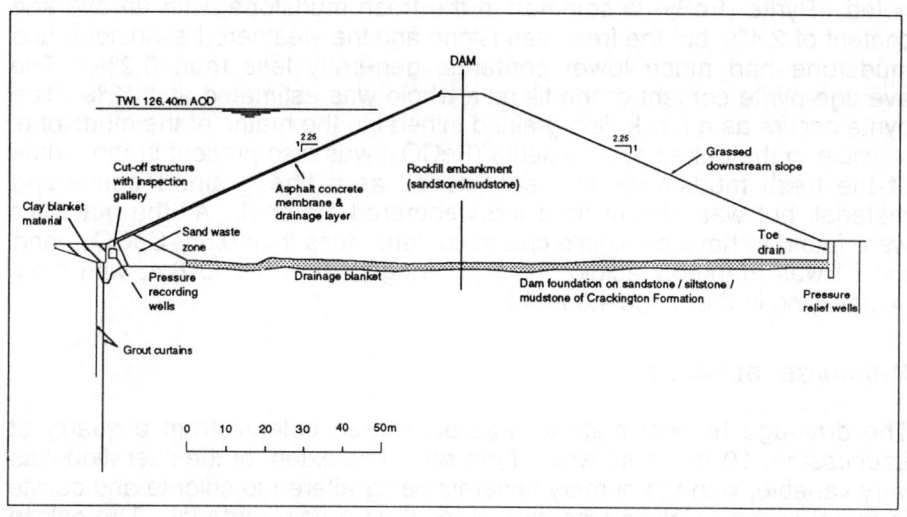

Figure 2: Cross section of dam.

is 430 m long with a maximum height of 41 m. It contains 970,000 m³ of locally won low grade rockfill, 62,000 m³ of drainage blanket material and 40,000 m³ of sandwaste.

Impounding of the reservoir commenced in October 1989. The reservoir was 60% full by Spring 1990 and had almost reached top water level by Spring 1991. The reservoir has a capacity of 37,000 Ml.

NATURE OF MATERIALS

ROCKFILL

The bulk of the embankment is composed of locally available low grade rockfill. This material consists of fresh to moderately weathered thinly interbedded mudstones, siltstones and sandstones belonging to the Crackington Formation of the Culm Measures of Carboniferous age. Highly weathered material and superficial deposits were not used in the embankment. The natural variability of the strata and the method of working the quarry combined to produce a relatively homogeneous embankment of well compacted, reasonably free-draining granular fill, with an even distribution of sandstone and mudstone, fresh and weathered material.

Considerable attention had been given to the mineralogy and geochemistry of the strata during the site investigations for the project and further testing of the fill materials was carried out during construction.

The mineralogy of the mudstones is dominated by illite, with lesser kaolinite and quartz. Traces of chlorite and swelling clay minerals were occasionally noted. Pyrite (FeS_2) is common in the fresh mudstone, with an average content of 2.4%, but the fresh sandstone and the weathered sandstone and mudstone had much lower contents, generally less than 0.2%. The average pyrite content of the fill as a whole was estimated at 1.12%. The pyrite occurs as a black, fine grained mineral in the matrix of the mudstone, invisible to the naked eye. Siderite ($FeCO_3$) was also present in the matrix of the fresh mudstones and sandstones as a fine grained cementing material, but was absent from the weathered material. All the materials were found to have negligible calcite content - less than 0.5% $CaCO_3$ - and the fill was markedly acidic. The pH ranged from 3.1 to 6.4, with most values lying in the range 4.0 to 4.5.

DRAINAGE BLANKET

The drainage blanket material was an altered dolerite from a quarry at Launceston, 10 km south west of the site. The extent of the alteration was very variable, with the primary minerals being altered to chlorite and calcite in many cases. Calcite was also present as a vein material. The calcite content of individual samples varied widely, from 0.7% to 24%, with an average of 11.6%.

SANDWASTE

The sandwaste material was obtained from China Clay workings in Cornwall and consisted of sand sized quartz and feldspar particles. This material had previously been used successfully in the construction of Colliford Dam in Cornwall (Johnston and Evans, 1985). It contains no pyrite or calcite and was used to form a fill of low compressibility at the upstream toe of the embankment to minimise stress on the asphalt membrane.

GROUNDWATER

The natural groundwater in the Crackington Formation at Roadford is generally fairly dilute. The typical analysis given in Table I shows it to be a calcium - magnesium - bicarbonate water with fairly low sulphate but high iron content and a pH near neutral.

GEOCHEMICAL REACTIONS WITHIN THE EMBANKMENT

PRE-CONSTRUCTION ESTIMATES

It was recognised at an early stage in the design process that the presence of pyrite in the strata could lead to problems. When the fill was excavated and placed in a free-draining embankment, the pyrite could oxidise according to the general equation:

$$4FeS_2 + 15O_2 + 14H_2O \rightarrow 4Fe(OH)_3 + 16H^+ + 8SO_4^{2-} \qquad [1]$$

Table I: Chemistry of embankment drainage

Parameter	Reservoir Water (10/1/91)	Natural Groundwater (10/1/91)	Seepage Below Embankment (7/1/91)	Embankment Drainage Summer (23/7/91)	Embankment Drainage Winter (17/1/91)	Trial Embankment of Fresh Material 3/10/86	Trial Embankment of Fresh Material 3/12/85
pH	7.10	6.90	6.60	7.10	7.40	4.5	3.6
SO_4^{2-}	10.1	31.6	23.5	65.1	372	158	6254
HCO_3^-	21.0	62.2	45.0	135.4	127.5	0	0
Ca^{2+}	10.8	19.9	10.1	68.9	105	14.6	310
Mg^{2+}	3.3	16.4	9.2	23.2	76.9	22.9	1370
K^+	2.5	1.0	2.9	2.0	3.5	2.0	4.5
Fe (total)	0.49	2.43	4.44	0.89	0.44	7.6	24
Mn (total)	0.04	0.34	0.54	1.64	2.29	4.6	170
Al (total)	0.20	0.01	0.04	0.01	0.01	0.90	>6.0
SiO_2	3.5	27.0	10.4	12.9	13.8	15.0	26.6
Dissolved O_2	10.8	0.7	1.0	1.0	1.8	n.d.	n.d.
% Saturation	87	6	5	9	17	n.d.	n.d.
Dissolved CO_2	3.2	25.0	22.0	34.0	16.0	n.d.	n.d.
Conductivity (µS/cm)	142	284	187	522	969	340	6050
Saturation Index for Gypsum SI gyp	-2.95	-2.19	-2.61	-1.33	-0.40	-1.62	+1.30

All concentrations in mg/l except pH and Conductivity and $\%O_2$ Saturation

Where $SIgyp = \log \dfrac{[Ca^{2+}][SO_4^{2-}]}{K_{gyp}}$ $K_{gyp} = 10^{-4.6} = 2.51 \times 10^{-5} M$

The acid generated by this reaction could then attack any carbonates or clay minerals in the fill, causing softening and loss of strength to individual particles as well as loss of material by leaching.

The question which had to be addressed was: how far will these reactions proceed in the embankment? This problem was approached from various angles. Laboratory tests were carried out to establish the "worst case" scenario, drainage from trial embankments was analysed, and a literature search was carried out to seek guidance from similar cases.

Leaching tests were carried out on samples of the fill with peaty water and a hydrochloric/nitric acid mixture. The results of these tests suggested that in the worst case of total weathering there could be a loss of up to 5% of the mass of the fill. Leaching with peaty water produced a reduction of about 1% in the mass of the fill.

A number of trial embankments had been constructed during the site investigations for the dam. Drainage from embankments of fresh material was found to be acidic, with high sulphate and iron concentrations; two examples are given in Table I. Orange deposits of ferric hydroxide (ochre) were noted around the edges of the trial embankment. However, the quantities of seepage were small. The experience from the trial embankments indicated that some oxidation of pyrite was occurring within the fresh material. Because of the low calcite content of the fill the drainage waters were acidic and contained no bicarbonate. Some neutralisation had occurred by ion exchange reactions with clay minerals such as chlorite and illite, releasing magnesium, calcium and potassium into solution. The relatively low concentrations of silica and aluminium suggested that there had been relatively little attack on the structure of the clay minerals.

Experience from elsewhere in the U.K. suggested that the extent of chemical degradation would be limited in an embankment such as Roadford. A number of studies have been carried out on old colliery spoil tips by various workers (e.g. Taylor, 1984; Spears, Taylor and Till, 1971). These indicate that significant chemical degradation is largely limited to a surface zone about 1 m deep. Inspection of the downstream shoulder of Burnhope Dam in County Durham 25 years after construction revealed no visual indications of chemical or physical weathering of the shale fill (Kennard, Knill and Vaughan, 1967). At Carsington Dam in Derbyshire, which is constructed with pyrite-rich mudstone, weathering of the exposed surface had produced a zone of degraded clayey material about 0.3 m thick between 1984 and 1987.

In the light of all the available evidence, it was concluded that while some pyrite oxidation and associated reactions would occur, this would not be likely to cause significant degradation of the fill. An allowance for long term degradation was made in the design shear strength parameters taken for the in-service condition (Wilson and Evans, 1991). It was assumed that the angle of friction would decrease from 40° to 35° at low normal stresses and from 32° to 29.5° at high normal stresses. It was considered that most of the chemical reactions were likely to occur during construction and in the following year. After that, the embankment would be sealed by an impermeable membrane on the upstream side and a road on the crest. Herringbone drains were to be laid on the downstream shoulder, which would then be topsoiled and grassed. The amount of air and water which could percolate through the fill would thus be greatly reduced and the rate of the reactions decrease.

It was decided to monitor the chemistry of the embankment drainage water, both for water quality reasons and to assess the extent of chemical reactions occurring within the embankment.

CONSTRUCTION EXPERIENCE

The specification for the drainage blanket included a stipulation that the material should be non-calcareous. However in January 1988 it was discovered that the proposed material had a significant calcite content up to 24%, averaging 11.6%. As the water draining from the embankment fill was anticipated to be acidic, it could react with the calcite to generate carbon dioxide and possibly lead to the precipitation of gypsum.

The implications of this discovery had to be assessed rapidly. If an alternative material had to be found for the drainage blanket, there could be a significant increase in the cost of the project and delay in what was a tight programme for the embankment construction. Preliminary calculations indicated the blanket would contain sufficient calcite to neutralise all the acid which was likely to be generated from the embankment fill. An experiment was carried out on site, in which a sample of the drainage blanket material was leached with dilute sulphuric acid in a constant head permeameter cell for 46 hours. This produced a slight decrease in permeability, from 5.3×10^{-4} m/s to 3.1×10^{-4} m/s, and a loss of about 1% in weight of the material. This was considered acceptable with regard to the design parameters. It was decided to continue using the altered dolerite for the drainage blanket, and to monitor the chemistry of the embankment drainage water closely.

A typical section through the embankment is shown on Figure 2. The embankment is designed so that the water table will be contained within the drainage blanket, with the rockfill almost entirely above the water table. Readings from piezometers in the embankment indicate that this has been successfully achieved to date (Evans and Wilson, 1992). Water emerging from the embankment drainage system is thus a mixture of groundwater seeping below the embankment and water percolating through the rockfill into the drainage blanket, with the former thought to be dominant.

As indicated earlier, the embankment consists of a relatively homogeneous mixture of fresh and weathered mudstone, siltstone and sandstone. Drainage from the embankment commenced in August 1988 before construction was complete and was found to have high sulphate concentrations and neutral pH from the outset. The pattern of sulphate concentration versus time for the embankment drainage is shown on Figure 3. The chemistry of typical samples of embankment drainage, groundwater, reservoir water and underseepage are given in Table I.

Compared to groundwater and underseepage, the embankment drainage water has much higher concentrations of sulphate, bicarbonate, calcium, magnesium and manganese. It also has higher pH, but lower concentrations of iron. These differences can be attributed to oxidation of pyrite in the rockfill followed by subsequent ion exchange reactions with clay minerals in the fill and neutralisation of the acid by reaction with calcite in the drainage blanket. The low concentrations of aluminium and silica suggest that little or no attack on the structure of the clay minerals is occurring, and the low iron concentration suggests that most of the iron released by oxidation of pyrite is being precipitated within the rockfill as ferric hydroxide. Manganese, which is more soluble than iron, remains in solution.

A clear seasonal trend in the embankment drainage can be seen from Figure 3 and Table I. Concentrations of sulphate and other weathering products are much higher in winter than in summer, suggesting greater percolation of rainwater through the rockfill during these periods. The total quantity of seepage is higher in winter than summer, but the seasonal difference is not so pronounced as for the sulphate concentration; the drainage flow is about 750,000 l/day in summer and 950,000 l/day in winter.

The extent to which the chemistry of the drainage water is affected by reactions within the embankment is shown by a comparison of the sulphate concentrations of the embankment drainage (Figure 3) and the underseepage (Figure 4). Samples of the underseepage were obtained from the pressure recording wells at the upstream toe of the embankment (Figure 2). There is an initial flush of weathering products out of the foundation, due to exposure during construction operations, followed by a slow decrease in concentration due to dilution of groundwater with reservoir water. No seasonal changes are evident. As the underseepage makes up the bulk of the embankment drainage water, the water percolating from the rockfill must have very high concentrations of weathering products to produce the composition shown on Figure 3 and Table I.

The extent to which gypsum may be precipitated in the drainage blanket can be assessed in a qualitative fashion by calculating saturation indices for the various waters. Values are shown on Table I, using an equilibrium constant of $2.51 \times 10^{-5}M$ for gypsum (Krauskopf, 1967). Negative values indicate undersaturation with respect to gypsum, and positive values supersaturation. The figures should be used for comparative purposes only, as the equation does not take account of other ions in solution or variations in temperature and pressure. Reservoir water, groundwater and underseepage are all clearly undersaturated with respect to gypsum. The embankment drainage is undersaturated in summer but approaches saturation in winter. The drainage from the trial embankment of fresh

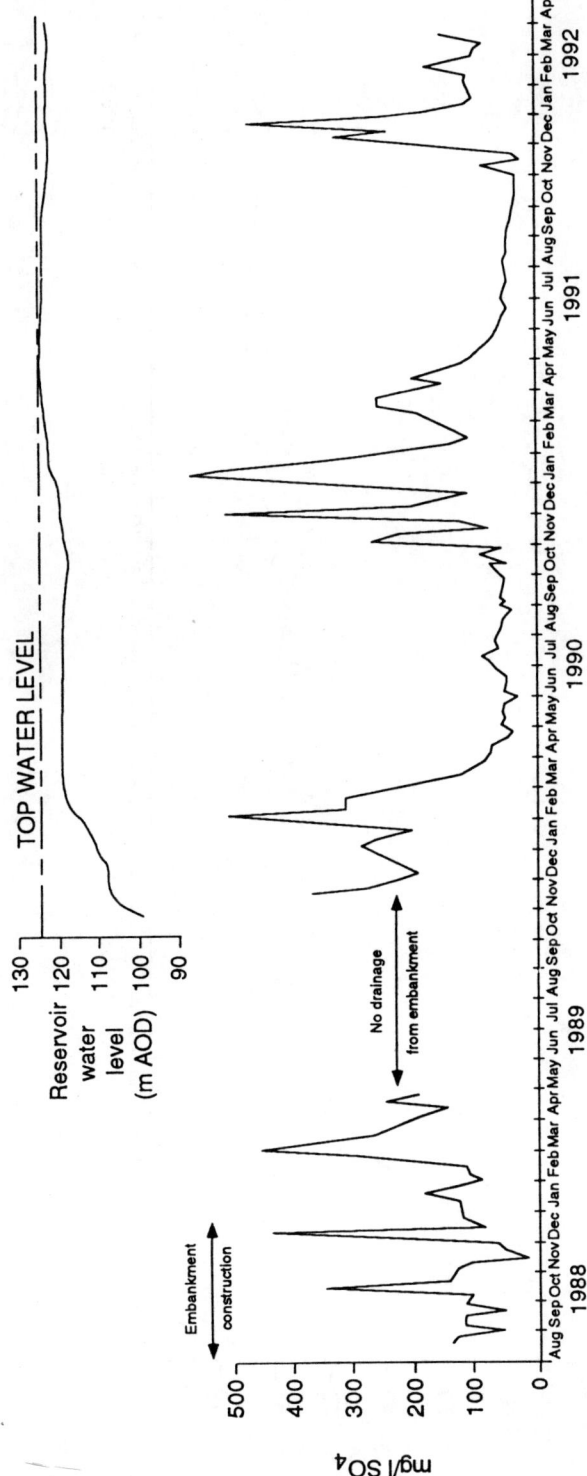

Figure 3: Sulphate content of embankment drainage.

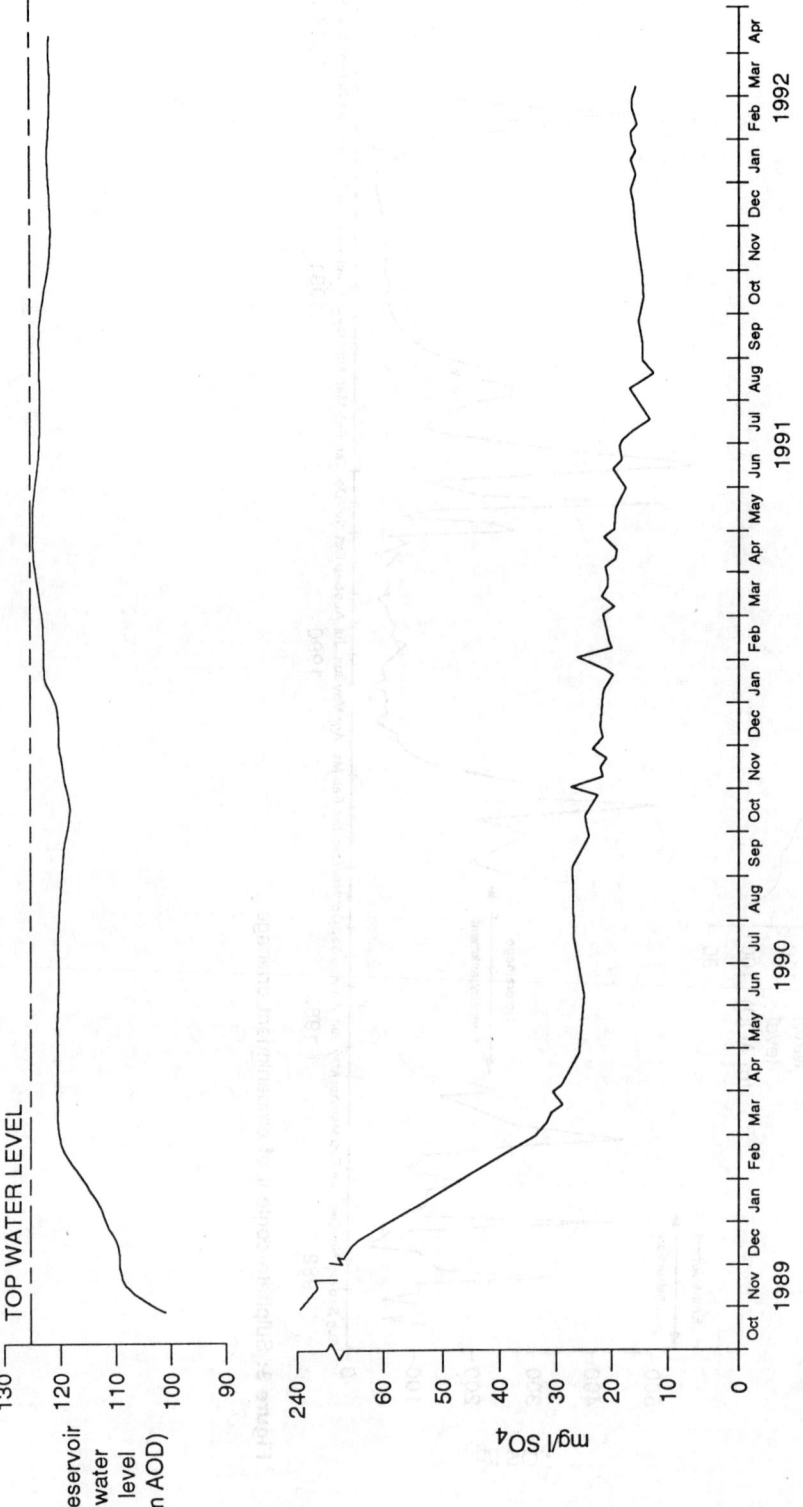

Figure 4: Sulphate content of water seepage.

material was supersaturated on one occasion. However the concentration of calcium was so low that only minor amounts of gypsum could be precipitated.

The figures suggest that some precipitation of gypsum may take place within the drainage blanket in winter, although the steady flow of water through the blanket will diminish the likelihood of this occurring. Gypsum precipitation may occur in local areas where particularly high concentrations of calcium and sulphate are present, or any areas that are above the water table. A limited amount of gypsum may also be precipitated within the rockfill.

QUANTITIES INVOLVED IN THE REACTIONS

The chemical reactions in the embankment have a significant effect on the drainage water chemistry. The quantities of material involved in the reactions were estimated in order to assess the effect they were having on the geotechnical properties of the fill materials.

The sulphate and calcium concentrations of the embankment drainage were used for the calculations, as they can be related directly to oxidation of pyrite and solution of calcite respectively. Calculations were made on a quarterly basis to allow for seasonal variations. The average concentrations, minus the concentrations in groundwater, were multiplied by the average drainage flow to give the total weight loss of sulphur and calcium. These were then converted to equivalent loads of pyrite and calcite and summed to give annual figures. The process is illustrated for the year July 1990 - June 1991 on Table II. It is estimated that 6.95 m^3 of pyrite and 18.2 m^3 of calcite were lost during this period.

It can be seen from Table II that the annual losses represent a very small proportion of the total amounts of pyrite (0.15%) and calcite (0.33%) present in the fill materials. The reactions are thus unlikely to have any significant effect on the geotechnical properties of the fill. The allowances made for chemical degradation in the parameters chosen for the in-service condition are thus adequate to cope with the observed reactions (Wilson and Evans, 1991).

It had been anticipated that the rate of the chemical reactions would decrease fairly rapidly once the embankment was covered. By early 1992 there was no sign of a decrease in the quantities of reaction products in the embankment drainage (Figure 3). This may occur over the next few years. However, because the quantities of materials involved are so small it is possible that the reactions could continue for many years at much the same rate as at present. It is thus likely that the chemistry of the embankment drainage will not change significantly in the foreseeable future.

Table II: Degradation of fill, July 1990-June 1991

Period	Drainage Flow	Chemistry of Drainage*		Sulphur Lost	Pyrite Lost	Calcite Lost
		Ca^{2+}	SO_4^{2-}			
	(l/day)	(mg/l)	(mg/l)	(kg)	(kg)	(kg)
July-Sept 1990	760,000	56	53	1,224	2,289	9,682
Oct-Dec 1990	750,000	73	244	5,562	10,401	12,456
Jan-Mar 1991	950,000	75	286	8,258	15,442	16,209
April-Jun 1991	956,000	50	96	2,784	5,216	10,875
TOTAL				17,833	33,348	49,222

Volume Pyrite Lost (by weight)	6.95 m^3
Percentage Pyrite Lost	0.15%
Percentage Fill Lost (by weight)	0.0017%
Percentage Fill Lost (by volume)	0.00072%
Volume Calcite Lost	18.2 m^3
Percentage Calcite Lost (by weight)	0.33%
Percentage Drainage Blanket Lost (weight)	0.038%
Percentage Drainage Blanket Lost (volume)	0.03%

Assumptions: * Corrected for ground water
SG Pyrite = 4.8
SG Calcite = 2.71
Volume Rockfill = 970,000 m^3
Dry Density Rockfill = 2,070 kg/m^3
Volume Drainage Blanket = 62,000 m^3
Dry Density Drainage Blanket = 2,070 kg/m^3
% Pyrite in Rockfill (by weight) = 1.12%
% Calcite in Drainage Blanket (by weight) = 11.6%

TREATMENT OF EMBANKMENT DAMAGE

MONITORING OBJECTIVES

The quality of the embankment drainage waters was monitored for the following reasons:

a) to assist in the assessment of the likely future quality of the water in the reservoir;

b) to indicate the present quality of the River Wolf and to forecast future changes as a result of the construction of the dam;

c) to help confirm that the embankment is behaving in accordance with the design assumptions and give early warning of any changes which may occur;

d)   to assist in tracing the sources of seepages from the reservoir at the dam or in the valley bottom.

ROUTINE MONITORING

The implemented programme for routine water quality monitoring incorporates the sampling of drainage water component flows and the composite dam drainage. An extensive analytical suite has been used as it was envisaged that many parameters could ultimately be important (Gilkes et al, 1991). Certain 'key' parameters were considered for particular attention as they satisfied at least one of two criteria, ie:

a)   They were included in a list of guidelines formulated in 1989 by South West Water Rivers Unit - now NRA South West - for river water quality leaving Roadford Dam site. These are shown in Table III.

b)   They were considered to be especially indicative of potential embankment rockfill and drainage degradation, as discussed above.

EMBANKMENT DRAINAGE COMPONENTS

Embankment drainages with varying origins are channelled to a common collection chamber. Embankment underdrainage, natural groundwater and membrane drainage are monitored at source as well as the composite dam drainage. The chemistry of the embankment drainage waters is shown in Table I.

After completion of the embankment in the Spring of 1989 an assessment of drainage water quality was made. The assessment indicated that the concentrations of manganese and iron were relatively high, within the ranges 2.0 to 8.8 mg/l and 0.4 to 6.3 mg/l respectively. The drainage was also de-oxygenated, slightly acidic and contained high levels of soluble sulphates. These chemical characteristics are consistent with the expected mineralisation of pyritic shales and sandstones within the embankment.

The quantity of the underdrainage was observed to vary according to rainfall. In February 1989 it was about 0.36 Ml/day but was negligible during a dry spell in May the same year. It had been estimated that the maximum drainage flows that could be expected after impoundment would be around 3.72 Ml/day, although the average would be much less.

Overall drainage water quality would be determined by the source and concentration of the major components of the flow. It was expected that the principal flow would be through bedrock and the quality would therefore be better than pre-impoundment monitoring in 1989 when the flow was mainly through the embankment.

Pre-impoundment embankment drainage water quality compared with guideline values is shown in Table III. These parameters show that apart from iron, manganese, sulphate and dissolved oxygen the drainage water quality was suitable for direct discharge to the River Wolf. The Quality

Regulating Officer (now with the NRA) regarded iron and manganese as being of greatest importance. Treatment of the drainage waters would focus largely on manganese, iron and dissolved oxygen content. Manganese is only regarded as an aesthetic problem while iron and dissolved oxygen have direct implications to the aquatic biology of receiving watercourses.

Table III: Discharge guidelines and drainage water quality before and after irrigation treatment

	Guidelines	Woodland Irrigation			Embankment Drainage		
From		10/89	06/90	10/90	10/89	06/90	10/90
To		06/90	09/90	06/91	06/90	09/90	06/91
pH	6.0-9.0	8.0	7.3	7.4	7.1	7.2	7.3
Soluble sulphate	18.84	179	71.8	232.1	187	60.1	206.5
Iron (total)	1.00	0.47	0.17	0.08	0.59	0.86	1.37
Iron (soluble)	1.00	0.14	0.08	0.05	0.43	0.58	0.42
Manganese (total)	0.14	0.6	0.1	0.07	1.79	1.2	1.53
Manganese (soluble)	0.14	0.36	0.16	0.06	1.57	1.15	1.38
Suspended solids	25.0	24.8	19.7	7.2	10.6	4.8	6.88
Dissolved oxygen	<80%	91	85	91	54	44	77.3
Ammonia	1.0	0.06	0.02	0.02	0.05	0.04	0.03
Phosphate	0.065	0.04	0.02	0.01	0.02	0.01	0.01
Nitrate	2.8	0.33	0.36	0.33	0.37	0.15	0.19

Note: All parameters mg/l except pH

Clark and Crawshaw (1979) document the effects of a ferruginous discharge upon the River Calder in Lancashire. They observed a severely restricted invertebrate fauna and a poor fish population and concluded that a concentration of 1.0 mg/l iron produced an acceptable level of pollution in a stream. Scullion and Edwards (1980) studying brown trout populations in the Taff Bargoed in South Wales found that in a stream of neutral pH there was a decrease in fish density from 0.18 per m^3 above a ferruginous discharge to 0.03 per m^3 below it, the relevant total iron concentrations being 0.71 mg/l and 2.29 mg/l respectively.

In contrast to iron very little information is available for manganese but the aesthetic problem caused by the black staining of the stream bed through manganic deposits was likely to be of greater concern than of direct biological significance.

OPTIONS FOR DRAINAGE WATER TREATMENT

Three treatment methods for manganese and iron removal were considered. Two of the options entailed use of chemical treatment methods to either higher or lower technical levels. A third option was to irrigate through naturally growing vegetation.

In general terms soluble iron and manganese can usually be partly treated by oxidation of the raw water and pH elevation to above pH 9. Precipitation, followed by removal of insoluble oxides by filtration or sedimentation completes the process. Control of manganese is a fairly common practice in the treatment of groundwater for potable supplies, however the process is often difficult and the chemistry complex.

Laboratory studies to investigate methods of manganese removal from the drainage waters were carried out. It was shown that to precipitate 95% or more of the manganese from solution it was essential to raise the pH to 9.5. With the facilities available it was not possible to establish the necessary oxygenation required but it appeared that 50% saturation would be adequate. Land irrigation treatment does not necessitate pH elevation to high values; however Boken (1958) concluded that of all the physical and chemical factors involved in manganese reduction and oxidation, pH was the most important with low pH favouring solubility.

In a full scale treatment plant the three stage chemical treatment process of oxidation, pH elevation and precipitation is usually achieved by the following means. Oxidation is achieved by either chlorine gas or sodium hypochlorite and pH elevation by sodium carbonate or sodium hydroxide. Precipitation is often achieved through various configurations of rapid gravity or pressure filters.

The refurbishment of an existing 'Aquapac 25' treatment plant was considered. Investigations by Kennicott Ltd enabled them to conclude that the use of hypochlorite and sodium carbonate was insufficient to effect manganese reduction to below the guideline value of 0.14 mg/l. Kennicott recommended conversions of the Aquapac which they would guarantee would enable compliance with the manganese guidelines.

On the basis of the evidence presented the proposals for the 'Aquapac' conversion were considered as likely to be effective. However there were disadvantages. The capital cost was high, unlikely to be less than £60,000 for a plant unable to treat more than 0.6 Ml/day. The operation is complex especially with regard to media regeneration. There was also the problem of disposal of regeneration and backwash liquor. The additional risk to the aquatic environment from chemicals required for the treatment process, which by necessity would be close to the watercourse was considered to be unacceptably high.

Low technology chemical treatment methods were also considered. Plans for a pilot study were drawn up utilizing cascade aeration to enhance natural oxidation processes and limestone cobbles to elevate pH sufficiently to precipitate insoluble manganese. Laboratory studies had already indicated that a pH in excess of 9.5 and settlement for over 32 hours would be required to reduce the manganese in a drainage water sample from 3.3 to 0.2 mg/l.

A pilot on-site land treatment plant was established in October 1989. This entailed pumping dam drainage up into a steep wooded slope on the east flank downstream of the embankment (Figure 1). Dispersal of the drainage waters would be through a 10 m section of perforated flat-hose into the surrounding vegetation. It was envisaged that a periodic shifting of the discharge location would be required.

Total iron concentration was 0.56 mg/l (78% soluble) before irrigation treatment and decreased by 30% to 0.39 mg/l (32% soluble) after treatment. Total manganese concentration was 2.05 mg/l (89% soluble) before irrigation treatment and decreased by 70% to 0.62 mg/l (95% soluble). The results for post-irrigation drainage water quality were sufficiently encouraging to consider development toward a full irrigation system.

During operation of the pilot scheme a strip of sloped woodland only 50 m wide was used. A more permanent system was commissioned at the end of June 1990, utilising a larger area of woodland. This system entails pumping the drainage waters by underground pipeline to a head tank above the woodland area. Distribution is then carried on into the woodland by 150 m of pipeline and then through 11 equally spaced regulating valves (Figure 1).

It was envisaged that, by reducing the irrigation rate, improved percolation would result in an increased treatment efficiency. It was also believed that by spreading the load the system would be capable of treating volumes greater than 1.2 Ml/day.

Several physical and chemical mechanisms were identified as being potentially involved with the irrigation treatment. Improving the oxidizing conditions should increase the insoluble compared to soluble fraction. This would apply particularly to iron where the decay of soluble Fe^{2+} to insoluble Fe^{3+} typically has a half-life of 4 min upon exposure to the atmosphere (Stumm and Lee, 1961). By improving percolation through the soil rather than saturation and run-off, removal of particulate iron could be increased. Iu et al (1981) reported that waterlogging was significantly implicated in the chemical reduction (solubilization) of soil manganese. It was thus considered that improving percolation would probably reverse the trend, toward oxidation and precipitation instead of reduction and solubilization.

The amount of metal that could be removed by biological processes was uncertain. It is known that considerable amounts can be passively removed from the atmosphere by species of moss (Goodman and Roberts, 1971). It was expected that an unknown proportion of iron and manganese applied to the irrigation area would be actively absorbed by higher plants during the growing season as reported by Cheng (1972). However the efficiency of the irrigation system for both iron and manganese removal did not vary significantly during the year. As the woodland was deciduous it was concluded that any adsorption or absorption mechanisms were likely to be passive.

Results for composite embankment drainage and treated drainage before and after the adoption of the permanent irrigation system are shown in Table III. Total iron, manganese and suspended solids are illustrated in Figure 5. This shows that the treatment system was especially effective after commissioning the permanent distribution system. Occasions when the suspended solids, manganese and iron were observed to increase intermittently have been attributed to local erosion problems.

The woodland irrigation scheme has thus proved a highly effective, low technology method of treating the embankment drainage to an acceptable standard.

INVESTIGATION OF MEMBRANE DRAINAGE

SOURCES OF INFORMATION

A drainage layer was provided under the asphalt membrane on the upstream face of the dam with pipes leading into the inspection gallery. During February 1991 four groups of membrane drains were identified where the flow of each individual drain exceeded 2,000l/day. A representative drain from three of these groups was selected for routine chemical analysis.

VOLUME OF FLOW

Membrane drains were first observed flowing in November 1989 shortly after impounding commenced. The flow was then observed to increase coincidently with increased reservoir capacity during the winter of 89/90 and 90/91 to a maximum of 119,000 and 270,000 l/day respectively. A decline from peak flow during each of the following spring/summer periods was observed even though reservoir head remained stable (Figure 4). The rainfall profile was one of dry summer conditions followed by periods of moderate and heavy rainfall during autumn and winter. Consequently it was considered that the origin of the high winter flows from all the membrane drains was unlikely to be directly from the reservoir.

BACTERIOPHAGE INVESTIGATION

In May 1990 a bacteriophage tracer survey similar to that previously used at Colliford Lake was carried out (Martin, 1988). An area of the reservoir between the draw-off tower and the embankment was dosed with a suspension of the harmless bacteriophage *Enterobacter cloacae* and

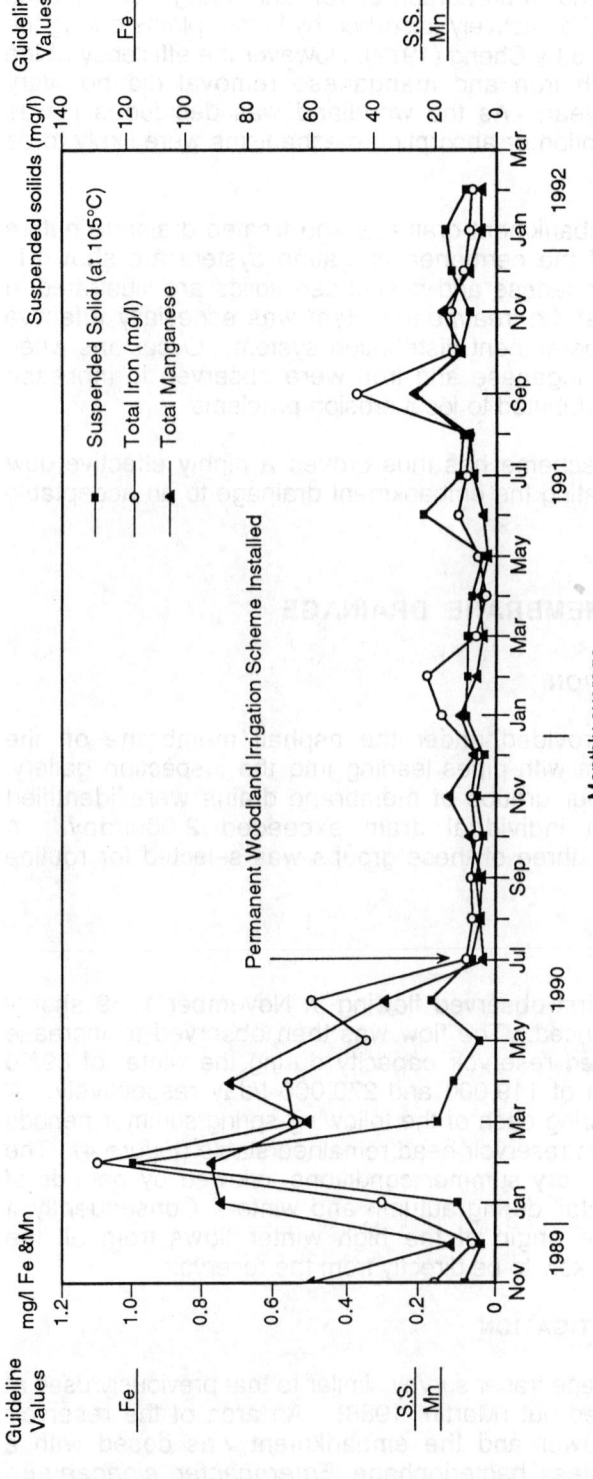

Figure 5: Quality of treated embankment drainage versus time.

during the following two months samples were taken from the reservoir and from various waters around the embankment including membrane drainage.

The bacteriophage concentration within the reservoir was initially over 1,000 plaque forming units per ml (PFU/ml) declining to 19 PFU/ml after 6 days. No bacteriophage was detected within the membrane drainage waters until day 6 when it was detected at only very low concentrations. The results suggest that seepage through the membrane is slight and that it is diluted by water from at least one other source.

POLYAROMATIC HYDROCARBONS

Analysis of membrane drainage for polyaromatic hydrocarbons (PAH) was undertaken to identify possible residues of bitumen that may leach from either the asphalt membrane or the crest road. No such residues were detected.

WATER CHEMISTRY

Membrane drainage sampled during the high flow period of winters 1989/90 and 90/91 was observed to be chemically similar to reservoir water or surface water in general but not to groundwater (see Tables I and IV). Only pH shows a significant variation from surface water quality and this may be attributed to dissolved carbon dioxide.

During summer when flow is lower a different water quality is observed. Higher concentrations of sulphate, silicate, potassium and magnesium indicate a residual seepage of drainage from the embankment. Lower concentrations of nitrate indicate that these waters have not recently been surface waters. The high calcium and bicarbonate concentration as well as the neutral pH are considered as evidence of reactions with calcite contained in the crushed dolerite of the membrane drainage blanket material.

The precise origin of the membrane drainage is still uncertain but would appear to involve components of reservoir water, embankment drainage and surface water.

CONCLUSIONS

The Roadford Dam project has illustrated how geochemical studies can be used to predict the nature and extent of reactions resulting from the use of natural materials in civil engineering works. Monitoring of the chemistry of drainage waters has been of value in the development of a satisfactory low-technology treatment method for the embankment drainage and an investigation into possible leakage through the upstream membrane.

Table IV: Chemistry of membrane drainage

	16W		26E		13W	
	Winter 21/0/91	Summer 14/10/91	Winter 21/2/91	Summer 14/10/91	Winter 21/2/91	Summer 02/9/91
Flow	Max	Min	Max	Min	Max	Min
pH	5.7	7.3	5.5	7	6.7	6.5
SO_4^{2-}	15.8	25	11.8	12.	15.2	22
HCO_3^-	45	249	24	43	23	145
NO_3^-	1	0.4	1.2	0.6	1.2	0.5
Ca^{2+}	21.5	90	12.7	19.2	13.3	57.4
Mg^{2+}	3.9	8	3.4	3.5	3.6	4.5
K^+	2.8	15.5	2.6	3.6	2.5	8.3
Fe (soluble)	0.33	0.01	0.33	0.07	0.34	0.02
Fe (total)	0.37	0.01	0.43	0.17	0.36	0.02
Mn (soluble)	0.12	0.01	0.04	0.04	0.07	0.03
Mn (total)	0.12	0.01	0.6	0.05	0.08	0.03
Al (soluble)	0.06	0.01	0.11	0.03	0.1	0.02
Al (total)	0.15	0.01	0.21	0.17	0.2	0.02
SiO_2	4.5	15.4	3.9	3.6	3.9	10.3
O_2 % sat	67	88	66	91	80	69
Dissolved CO_2	69	24	89	8.3	8.9	89
Conductivity	201	520	158	177	159	368

19/02/91   Max Total   270 000 l/day
23/10/91   Min Total   3 887 l/day
All parameters as mg/l except pH, O_2 % saturation and conductivity (μS/cm)

ACKNOWLEDGEMENTS

The authors would like to acknowledge the permission of South West Water Services Limited to publish this paper.

REFERENCES

BOKEN, E. (1958). Investigations on the determination of the available manganese content of soils. *Plant and Soil*, **9**, 269-285.

CHENG, B.T. (1972). Dynamics of soil manganese. *Ahi-Simp Int. Agrochem*, **9**, 180-191.

CLARK, C.J. & CRAWSHAW, D.H. (1979). A study into the treatability of ocherous minewater discharges. *Journal of the Institution of Water Pollution Control*, **78**, 446-462.

EVANS, J.D. & WILSON, A.C. (1992). The instrumentation, monitoring and performance of Roadford dam. *British Dam Society Conference, Stirling, June 1992*.

GILKES, P.W., MILLMORE, J.P. & BELL, J.E. (1991). The Roadford Scheme: Planning, reservoir construction and the environment. *Journal of the Institution of Water and Environmental Management*, **5(6)**, 659-670.

GOODMAN, G.T. & ROBERTS T.M. (1971). Plant and soil as indicators of metals in the air. *Nature*, **231**, 287-292.

IU, K.L., PULFORD I.D. & DUNCAN, H.J. (1981). Influence of waterlogging and lime on organic matter additions on the distribution of trace metals in an acid soil. *Plant and soil*, **59 (1)**, 317-325.

JOHNSTON, T.A. & EVANS, J.D., (1985). Colliford dam sand waste embankment and asphaltic concrete membrane. *Proceedings of the Institution of Civil Engineers, Part 1*, **78**, 689-709.

KENNARD, M.F., KNILL, J. & VAUGHAN, P.R. (1967). The geotechnical properties and behaviour of Carboniferous shale at Balderhead dam. *Quarterly Journal of Engineering Geology*, **1**, 3-24.

KRAUSKOPF, K.B. (1967). *Introduction to Geochemistry*. McGraw-Hill, New York.

MACDONALD, A. & REID, J.M. (1991). Embankment dam behaviour: the contribution of geochemistry. In: *The Embankment Dam*, Thomas Telford, London, 185-192.

MARTIN, C. (1988). The application of bacteriophage tracer techniques in South West Water. *Journal of the Institution of Water and Environmental Management 1988*, **2(6)**, 638-642.

SCULLION, J. & EDWARDS, R.W. (1980). The effects of pollutants from the coal industry on Fish. *Environment Pollution*, **Series A21**, 141.

SPEARS, D.A., TAYLOR, R.K. & TILL, R. (1971). A mineralogical investigation of a spoil heap at Yorkshire Main Colliery. *Quarterly Journal of Engineering Geology*, **3**, 239-252.

STUMM, W. & LEE, G.F. (1961). Oxygenation of ferrous iron. *Industrial Engineering Chemistry*, **53**, 143.

TAYLOR, R.K. (1984). *Composition and engineering properties of British colliery discards*. British Coal Mining Dept., London.

WILSON, A. & EVANS, J.D. (1991). The use of low grade rockfill at Roadford dam. In: *The Embankment Dam*, Thomas Telford, London, 21-28.

2-6 THE ASSESSMENT OF VOLUMETRIC LOSS FOR CLAY/SAND AND CLAY/CLAY MIXES

S.H. AL-JASSAR & A.B. HAWKINS
Engineering Geology Research Group, University of Bristol, Wills Memorial Building, Queens Road, Bristol BS8 1RJ.

ABSTRACT

The paper reports volumetric loss tests carried out on sand/clay mixes and clay/clay mixes. The feasibility study indicated that testing of samples at their liquid limit does not provide a realistic indication of the shrink/swell potential of natural soils. Work to date is showing that the best assessment can be made when the volumetric loss is established on samples with moisture contents of half their liquid limit.

INTRODUCTION

Although Cooling & Ward (1948) had written on the seasonal swelling and shrinking of soils, little detailed work was available until the mid 1970s when significant distress to structures resulted in large insurance claims for many buildings constructed on either shrinkable/swellable soils or in the proximity of existing or recently removed vegetation. The extent of the problem is indicated in Building Research Establishment Report 240 (1991) which published a graph showing the increase in claims in the post-1970 period (Figure 1). A number of important papers have been presented since 1980, including those in the Geotechnique Symposium in Print (1983) and by Chenney (1988) and Chandler (1992). The National House Builders Council have issued their own guidelines (Guidance Note 3).

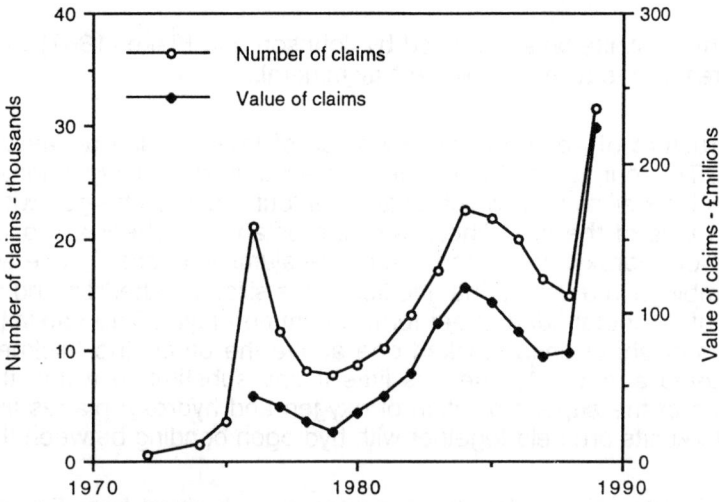

Figure 1: Annual value of insurance claims for subsidence and heave damage to housing (after BRE digest - 240, 1991).

Swelling occurs when the constituent parts of the soil absorb or adsorb sufficient water that the increase in pressure within the soil is greater than that of any overburden load. Such minerals as quartz, feldspar or calcite will absorb only minor volumes of water hence in soils where these are dominant, any expansion is likely to be minimal. In soils in which clay minerals are the dominant material, however, the electrically charged constituent parts play a major role in the change in volume. Indeed, it is only when clay minerals are present that significant volumetric changes take place, although the rapidity with which soils wet and dry will be influenced by their capillary characteristics.

In standard procedures, linear and volumetric shrinkage are used to provide an assessment of the potential volumetric change. However, it is considered that the mathematically derived values for volumetric shrinkage recommended by ASTM and British Standard may have little practical significance and for this reason the experiments reported here were carried out to establish the volumetric loss. This is defined simply as the difference between the initial wet volume and the dry volume.

The paper describes experiments to assess the potential for shrink/swell carried out on both clay/clay mixes and clay/sand mixes. It briefly reviews the applicability of the British Standard and ASTM testing procedures in providing data for use in constructional design.

MATERIALS USED IN THE EXPERIMENTS

Kaolinite, illite and montmorillonite (calcium and sodium) were used for the experiments, mixed with a commercially obtained silty fine to medium sand.

KAOLINITE (K)

The term kaolinite was first used by Johnson and Blake (1867) to describe the purest of the alumina silicate clay minerals.

The structure of each crystal is a stack of layers, each an association of sheets. The four tips of the tetrahedra are occupied by oxygen ions and their centres by a silicon ion which shares its four positive charges with the four oxygen ions of the tips. The tips of the octahedral sheet are occupied by oxygen or hydroxyl and in the centre are aluminium ions. These two sheets are combined in a unit so that the tips of the silica tetrahedron and one of the layers of the octahedral sheet form a common layer. The kaolinite mineral consists of sheet units stacked one above the other, the thickness of the units being about 7 Å. There is little if any substitution within the lattice. Because of the superimposition of oxygen and hydroxyl planes in adjacent units, the units are held together with hydrogen bonding between the layers.

The samples used in the experiments were obtained from English China Clay Loveland Ltd.

ILLITE (I)

This term was first proposed by Grim, Bray and Bradley (1937) for the least pure of the common alumina silica clay minerals, the main impurities being quartz and calcite.

Illite is similar to micas in its general structure, being composed of an octohedral sheet between two tetrahedral sheets. The tips of all the tetrahedrons in each silica sheet point towards the centre of the unit and are contained with the octahedral sheet in a single layer, with a suitable replacement of hydroxyls by oxygen. The unit is the same as that for montmorillonite except that less than one Si^{3+} out of four was replaced by Al^{3+}. This results in a proportionate diminution of alkali ions between layers. In the octahedral sheet some of the Al^{3+} is replaced by Mg^+, Fe^{2+} and Fe^{3+}.

The mineral frequently occurs in extremely small particles mixed with other clay mineral constituents. The structure is relatively fixed in position so that polar ions cannot enter and cause expansion. In addition the interlayer balancing cations are not exchangeable except where they occur at the edges of the layers. The substantially complete removal of K^+ from illites removes the bonding force holding the silicate layers together so that expansion might be possible causing degraded illite. This process, known as depotassification, changes the illite clay mineral into an interstratified illite:montmorillonite mineral. (Grim, 1962).

Natural/pure illite is difficult to obtain; the best British source being the Silurian mudstones at Woodbury on the Welsh Borderlands. The X-ray trace of the glycolated sample implied it had some minor mixed layer component; the details being given by Morgan (1978) and Srodon (1986) who report the illite content to be 70-80%.

MONTMORILLONITE (MC AND MS)

The term "montmorillonite" was first proposed by Damour and Salvetat (1847).

Hoffman, Endell and Wilm (1933) suggested the structure of this mineral is made up of a central alumina octahedral sheet surrounded by two silica tetrahedral sheets. The tips of the tetrahedrons point towards the centre of the unit so that each silica sheet and one of the hydroxyl layers of the tetrahedral sheet forms a common layer. In the stacking of silica-alumina-silica units, oxygen layers of each unit are adjacent to oxygens of the neighbouring units which results in a very weak bond and excellent cleavage between the layers. As a consequence of this structure water can enter between the unit layers causing the lattice to expand. This expansion is perpendicular to the layers and causes the stacking periodicity to have variable values, in many cases close to 14 Å. The thickness of the water layers depends in part on the nature of the exchangeable cations at a given water vapour pressure. The lattice is always unbalanced by substitution of cations with the prominent feature of the structure being a partial substitution of Al^{3+} and Si^{3+} in the tetrahedral sheet. Most montmorillonite has some Mg^{2+} substituting Al^{3+}. This clay mineral is relatively easily dispersed into extremely small particle sizes. This is particularly true when Na^+ is the exchangeable cation (Grim, 1962).

The calcium montmorillonite used in the study was from the Fuller's Earth Works at Redhill; the wet chemistry implies approximately 8% calcium oxide is present. The sodium montmorillonite was obtained as a commercial bentonite.

SAND (S)

The sand was obtained commercially and found to comprise 75% fine sand and 15% coarse silt, the remainder being medium and fine silt. It was considered that such sand would be typical of the quartz present in most natural soils.

WET CHEMISTRY

The wet chemistry of the four clays and the sand used in the experiments was obtained by atomic absorption. The results are given in Table I.

Table I: Wet chemistry of the sand and four clay minerals; the water being obtained by heating the materials to 900°C.

	CaO%	MgO%	K_2O%	Na_2O%	Al_2O_3%	SiO_2%	H_2O%	FeO%
S	1.0	0.2	0.8	1.1	5.32	92.85	0.09	0.02
K	0.12	0.5	0.10	0.00	37.65	49.01	11.9	0.31
I	2.27	6.62	6.36	0.05	23.2	53.3	7.9	1.08
Mc	7.6	4.2	0.52	0.27	20.7	59.0	7.7	2.80
Ms	1.2	3.05	0.62	4.72	22.1	62.4	6.3	1.74

PUBLISHED METHODS OF ASSESSING VOLUMETRIC SHRINKAGE

In 1952 the Transport and Road Research Laboratory in the book Soil Mechanics for Road Engineers, included a method for establishing volumetric shrinkage. This involved determining the change in volume between a standard wet sample and a completely dried material by determining the volume of the dry material in a mercury bath. In order to reduce the problem of air pockets remaining within the filled mould, it was recommended that for convenience the material should be slightly wet of the liquid limit.

In 1990, ASTM and BS 1377 suggested similar test procedures and calculations for volumetric shrinkage and shrinkage limit, including the use of moulds and a mercury bath for measuring the dry soil peds. They advocated remoulding the soil to its liquid limit and gave calculations by which the shrinkage limit and volumetric shrinkage could be obtained. It is considered that this is an unnecessarily complicated procedure, hence prone to error. As a consequence, the experiments reported here determine volumetric loss, which is defined as the percentage difference between the initial wet volume and the dry volume of the samples.

In order to assess natural ground conditions, the tests on the sand/clay and clay/clay mixes have been undertaken at three moisture contents:

a) the liquid limit
b) half the liquid limit
c) one third of the liquid limit.

RESULTS

The results of the index analyses on the pure clays are given in Table II and those for the mixes in Table III.

Table II: The index limits for the four clay minerals used in the tests.

	w_l	w_p	I_p
Kaolinite (K)	57	28	29
Illite (I)	72	32	40
Calcium Montmorillonite (Mc)	102	45	57
Sodium Montmorillonite (Ms)	440	44	396

SAND/CLAY MINERAL MIXES

The mixes were prepared with the proportion of the sand to clay decreasing in 20% steps. The samples were moistened to the liquid limit condition following the procedure recommended by BS 1377: 1990. The index limits for these mixes, given in Table III, are also shown graphically in Figure 2a; the percentage indicated on the horizontal axis representing the proportion of clay. The plasticity indices for the same samples are presented in Figure 2b. As expected, for samples with more than 80% sand, no index limits could be obtained. Figure 2c shows the volumetric loss for the samples with a moisture content at the liquid limit; Figure 2d with a moisture content of a half the liquid limit and Figure 2e, at one third of the liquid limit.

The results of the linear shrinkage tests, undertaken as recommended in BS1377: 1990, are given in Tables III and IV. Figures 2f to 2h give the linear shrinkages for the sand/clay mixes tested at the liquid limit, a half and one third of the liquid limit.

Table III (cont'd on next page): Index properties for the various sand/clay and clay/clay mixes, their linear shrinkage (W_L), volumetric shrinkage (V_S) and volumetric loss (V_L) tested at the liquid limit.

		w_l%	w_p%	I_p%	W_L%	V_S%	V_L%
K	**S**						
100	0	57	28	29	9	27	34
80	20	47	27	20	8	24	30
70	30	42	24	18	8	20	29
60	40	37	19	18	6	18	28
50	50	34	18	15	6	18	24
40	60	28	17	11	6	15	23
37	70	25	17	8	5	17	17
20	80	24	16	8	2	17	14
I	**S**						
100	0	72	32	40	18	7	19
90	10	64	29	35	17	7	56
70	30	53	26	27	15	7	49
50	50	40	22	18	11	13	37
30	70	33	22	11	8	16	26
20	80	31	21	10	3	21	14
Mc	**S**						
100	0	101	49	53	25	13	64
80	20	84	37	47	22	15	57
70	30	74	35	40	19	19	51
60	40	68	31	37	16	21	46
50	50	59	22	36	12	24	39
40	60	53	21	32	10	27	32
30	70	44	21	24	6	28	27
20	80	39	18	21	4	29	18
Ms	**S**						
100	0	439	42	397	46	2	91
90	10	388	41	377	45	5	90
70	30	294	40	253	37	7	87
50	50	176	36	140	30	15	78
40	60	154	31	123	26	16	75
20	80	85	26	59	16	27	47
I	**K**						
100	0	72	32	40	18	7	59
90	10	69	32	37	17	8	55
70	30	63	31	32	15	13	51
50	50	60	31	29	13	14	47
30	70	59	30	29	10	20	41
10	90	57	30	27	9	24	35
0	100	57	28	29	9	27	34
Mc	**K**						
0	100	101	49	53	25	13	64
90	10	95	47	48	25	19	61
70	30	80	40	40	20	22	53
50	50	71	32	39	14	24	47
30	70	63	31	32	9	26	41
10	90	57	29	28	8	26	34
0	100	57	28	29	9	27	

Table III (cont'd): Index properties for the various sand/clay and clay/clay mixes, their linear shrinkage (W_L), volumetric shrinkage (V_S) and volumetric loss (V_L) tested at the liquid limit.

		w_l%	w_p%	I_p%	W_L%	V_S%	V_L%
Ms	**K**						
100	0	439	42	397	46	2	91
80	20	339	41	298	42	7	88
70	30	304	37	277	39	8	87
50	50	210	32	178	31	12	83
30	70	136	31	105	23	17	72
10	90	82	31	50	13	22	51
0	100	57	28	29	9	27	34
Mc	**I**						
100	0	101	49	53	25	13	64
90	10	101	49	52	24	13	62
70	30	91	46	44	22	13	60
50	50	80	39	44	21	12	60
30	70	76	36	40	20	10	59
10	90	73	33	40	18	9	60
0	100	72	32	40	18	7	59
Ms	**I**						
100	0	439	42	397	46	2	91
80	20	358	39	319	42	4	90
60	40	264	36	221	39	6	85
50	50	220	36	184	36	7	83
30	70	142	35	107	29	7	76
10	90	84	32	50	20	7	62
0	100	72	32	40	18	7	59

Table IV: Summary of the volumetric loss and linear shrinkage for the sand/clay and clay/clay mixtures, also shown in Plate 1.

		Clay Mix (%)			Volumetric Loss (%) at			Linear Shrinkage (%) at		
S	K	I	Mc	Ms	w_l	50% w_l	33% w_l	w_l	50% w_l	33% w_l
	100				34	15	1	9	4	1
		100			59	42	26	18	11	8
			100		64	36	26	25	14	8
				100	91	81	74	46	39	32
50	50				24	7	0	6	1	0
50		50			37	17	9	11	6	1
50			50		39	12	4	12	4	1
50				50	78	60	49	30	17	15
	50	50			47	22	10	13	6	2
	50		50		47	12	0	14	3	0
	50			50	83	64	54	31	24	18
		50	50		60	33	15	21	12	5
		50		50	83	70	62	36	27	25

CLAY/CLAY MIXTURES

The results of the liquid limit tests on the various clay mineral mixes given in Tables III and IV are shown graphically in Figure 3a; the values given on the horizontal axis indicating the proportion of the clay mineral first named in the legend. The linear shrinkage for the clay mineral mixes is given in Figures 3f to 3h.

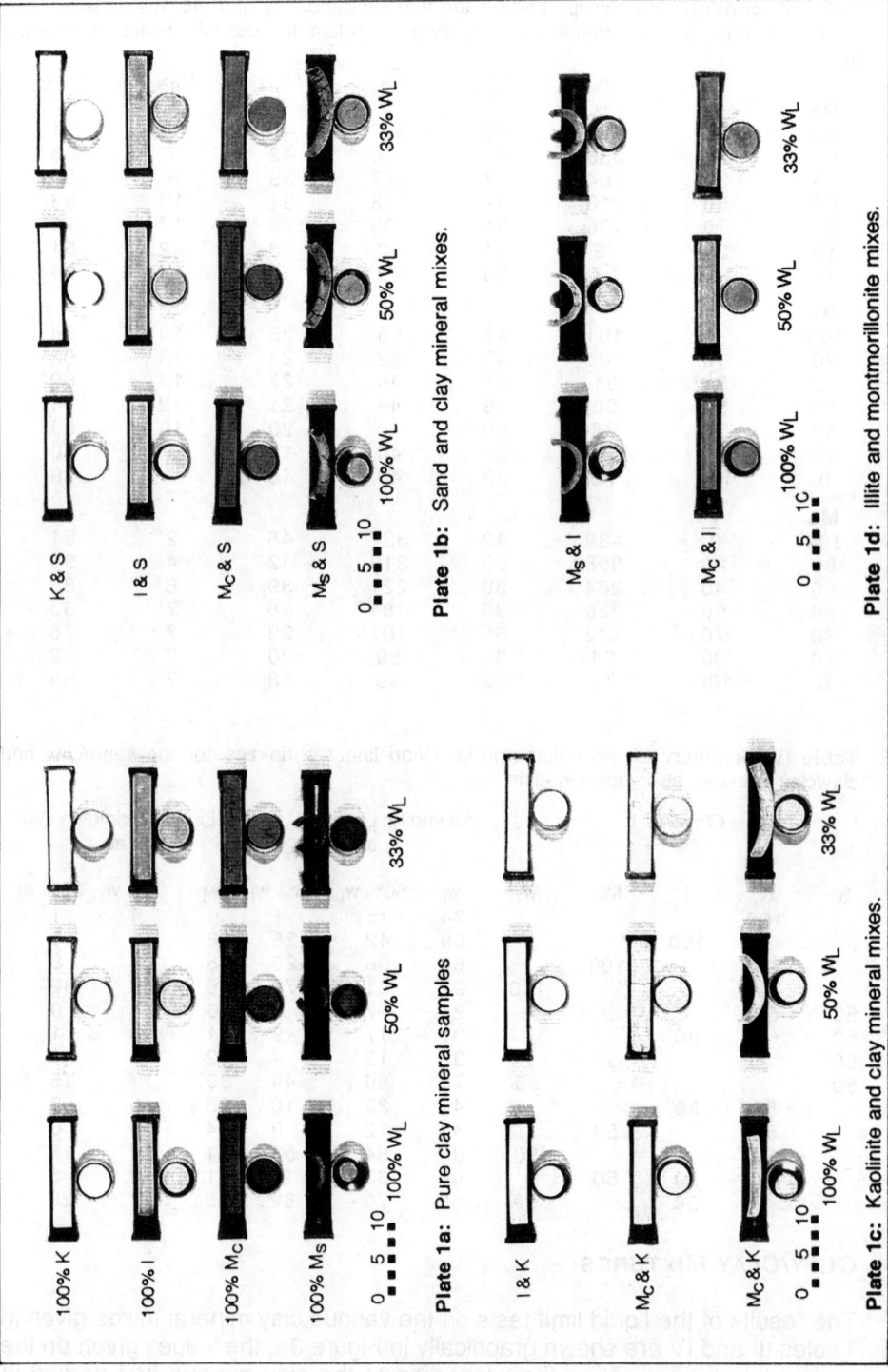

Plate 1: Changes in volumetric loss and linear shrinkage for clay/clay and sand/clay mixes.

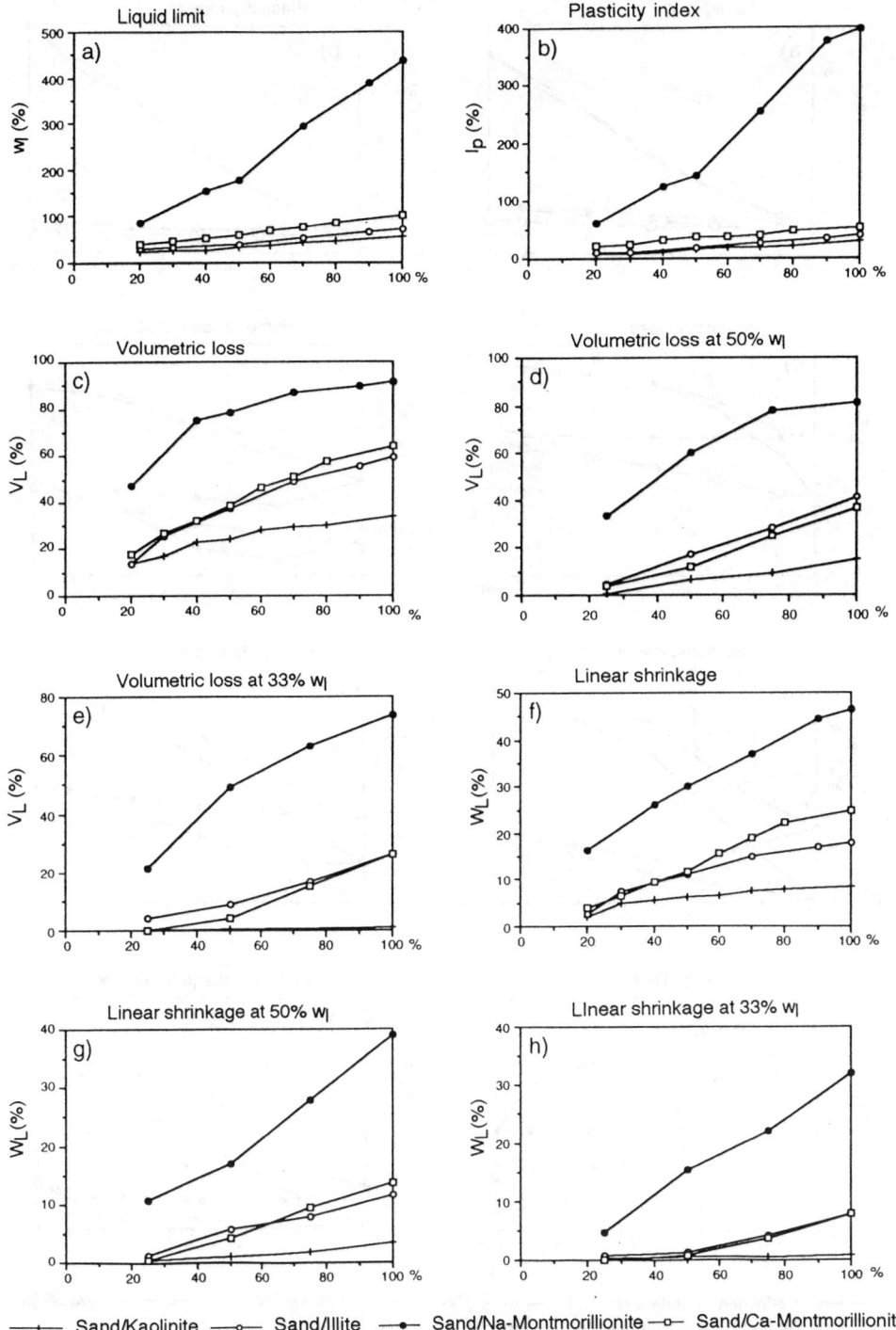

Figure 2: Liquid limit (w_l), plasticity index, volumetric loss (V_L) at w_l, 50% and 33% of w_l; linear shrinkage (W_L) at w_l, 50% and 33% of w_l for clay/sand mixes.

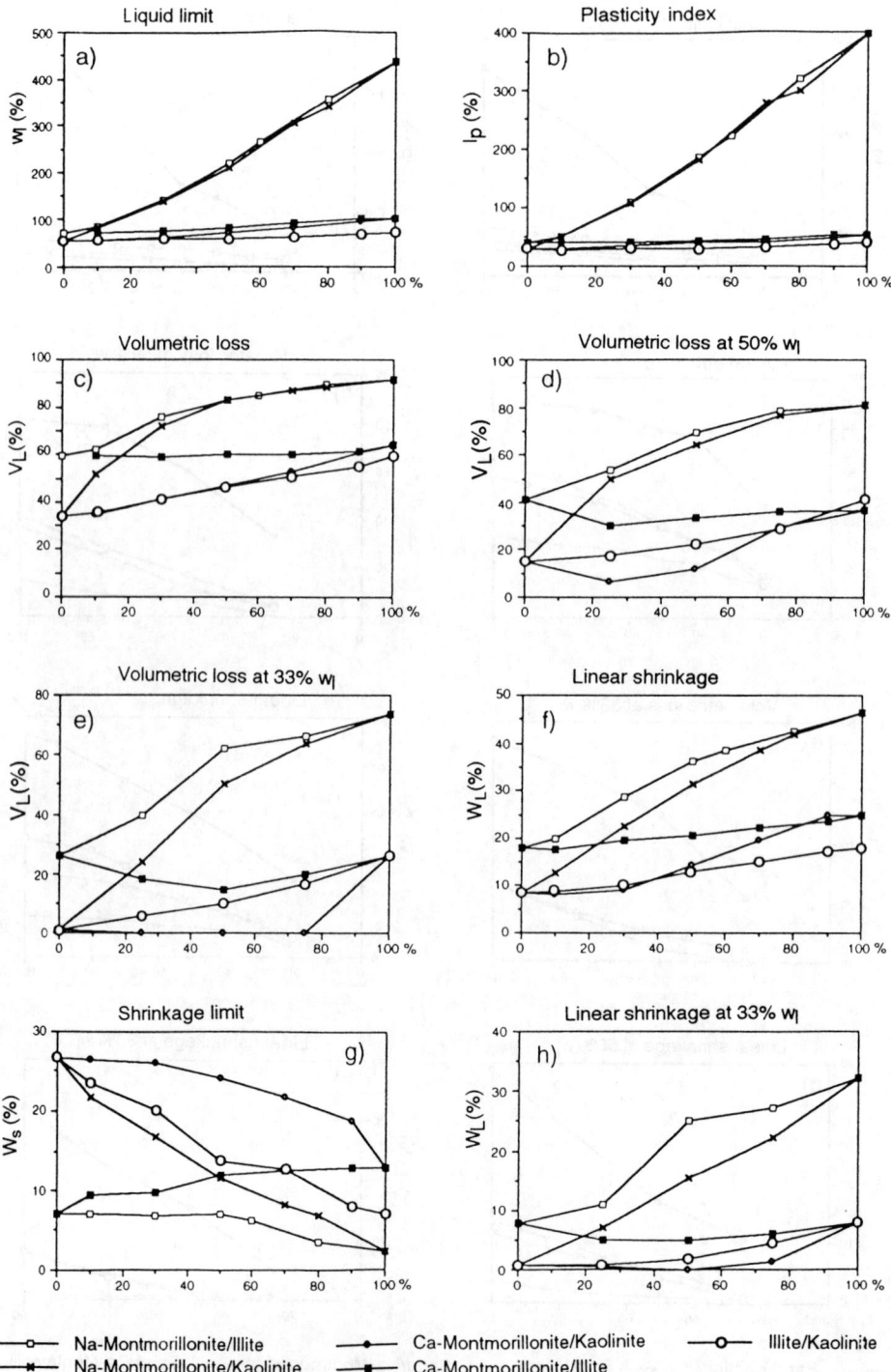

Figure 3: Liquid limit (w_l), plasticity index, volumetric loss (V_L) at w_l, 50% and 33% of w_l; linear shrinkage (W_L) at w_l, 50% and 33% of w_l for clay/clay mixes.

When the shrinkage limit (defined as the moisture content at which there is no further volume shrinkage with decreasing moisture content) is plotted for the clay/sand and the four clay mineral mixes in Figures 4 and 5, it can be seen that the moisture content at the shrinkage limit decreases progressively. The kaolinite/sand sample with more than 40% kaolinite (Figure 4) does not follow the same trend as the other materials; a phenomenon not yet explained.

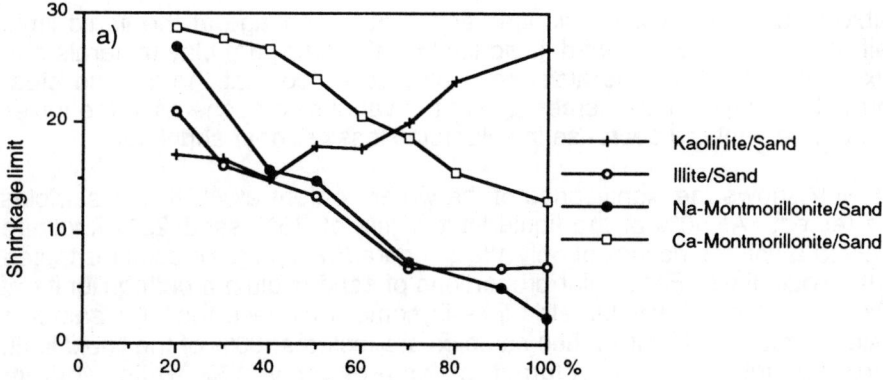

Figure 4: Shrinkage limit for clay/sand mixes.

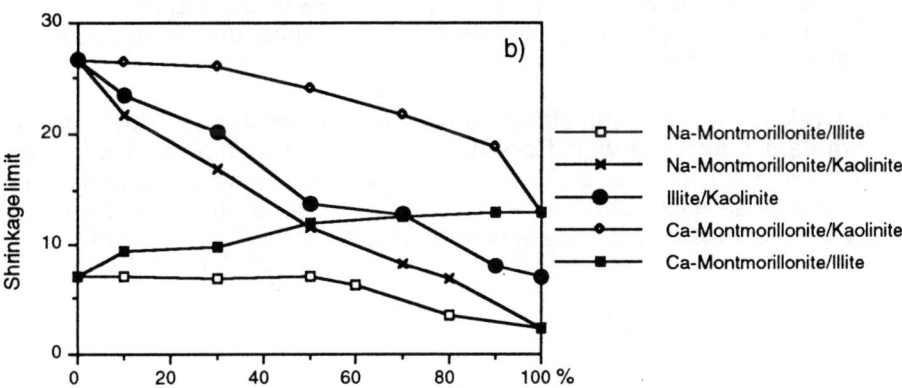

Figure 5: Shrinkage limit for clay/clay mixes.

DISCUSSION

Using the five components: sand, kaolinite, illite and sodium/calcium montmorillonite, a large number of mixes are possible and can be prepared to simulate different soil conditions. To date, however, the work has only been undertaken on mixes of various proportions of two of the five components.

Table III indicates that at the liquid limit a 80:20 sand:kaolinite mix shows a volumetric loss of only 14% while for the same ratios the sand:sodium montmorillonite mix at its liquid limit has a volumetric loss of 47%, emphasising the significance of sodium montmorillonite even when only a minor proportion is present. Indeed, for pure sodium montmorillonite, the volumetric loss is 91% implying major volume changes could be expected for soils dominated by this mineral. In contrast, pure kaolinite showed a volumetric change of 34%.

Table IV gives the volumetric loss and linear shrinkage at the liquid limit, half liquid limit and one third liquid limit for the four pure clay minerals and mixes of 50% clay minerals/sand. It will be noted that there is no clear correlation between the percentage of the various components in the mixes and the percentage decrease in volumetric loss or linear shrinkage.

Table V shows the significance of the water content at which the samples are tested. At 50% of the liquid limit a mix of 75% sand 25% kaolinite showed a volumetric loss of only 1% compared with 14% for samples tested at the liquid limit. For similar proportions of sand:sodium montmorillonite at 50% liquid limit, the volumetric loss is some 14% less than for samples tested at the liquid limit. When samples are tested at 33% of the liquid limit, pure kaolinite has a volumetric loss of only 1% while sodium montmorillonite has 74% and illite and calcium montmorillonite 26%.

The results obtained for the calcium montmorillonite are slightly anomalous. The samples have been shown to contain some 8% CaO hence it is likely that the calcium content has acted either as a bonding cement or a coating to some of the clay mineral aggregates. This may explain why the volumetric loss is less than might be expected, although the liquid limit is similar to that anticipated.

The work is not yet sufficiently advanced to determine which of the parameters measured would be the most reliable in indicating the likely volumetric loss in field situations. At this stage, it would appear that volumetric loss established on a mix tested at 50% of the liquid limit may allow the most realistic assessment to be made.

CONCLUSIONS

The nature of the mineral structure is the most important factor influencing the volumetric loss of a soil. In the case of the montmorillonite used in these tests, however, it is likely that some minor cementation or coating of the clay mineral aggregates has occurred. The illite clay mineral used in this research produced a higher than anticipated volumetric loss, possibly due to the presence of some mixed layer clay mineral.

Undertaking tests at the liquid limit clearly provides a volumetric loss which is not representative of normal ground conditions. It is recommended that soils with a moisture content of a half their liquid limit are used, rather than the wetter samples advocated by British Standard and ASTM.

REFERENCES

ASTM (1992). *Soil and Rock; Dimension stone; geosynthetics.* Volume 04.08, USA.

BRINDLEY, G.W. & ROBINSON, K. (1946). The structure of kaolinite. *Mineralogy Magazine,* **27**, 242-253.

BRINDLEY, G.W. (1951). The Kaolin Minerals, "X-ray Identification and crystal structures of clay 'minerals'. *Mineralogical society of Great Britain Monograph, Chapter 2,* 32-75.

BRE DIGEST 240, (1991). *Low rise buildings on shrinkable clay soils: Part 1.*

BRITISH STANDARDS INSTITUTE, (1990). *Methods of test for soils for civil engineering purposes. BS1377 Parts 1 and 2.*

CHANDLER, R.J., CRILLY, M.S. & MONTGOMERY-SMITH (1992). A low cost method of assessing clay desiccation for low rise building. *Civil Engineering,* 82-89.

CHENEY, J.E. (1988). 25 years' heave of a building constructed on clay, after tree removal. *Ground Engineering,* July, 13-27.

COOLING, L.F. & WARD, W.H. (1948). Some examples of foundation movements due to causes other than structural loads. *Proceedings 2nd International Conference on Soil Mechanics, Rotterdam,* **2**.

GRIM, R.E., BRAY, R.H. & BRADLEY, W.F. (1937). The mica in argillaceous sediments. *American Mineralogy,* **22**, 813-829.

GRIM, R.E. (1962). *Applied Clay Mineralogy.* McGraw Hill Book Co, New York.

HOFMANN, V., ENDELL, K. & WILM, D. (1933). Kristallstuktur und Quellung von Mont morillonite. *Zeitschr. Krist,* **86**, 340-348.

JOHNSON, S.W. & BLAKE, J.M. (1867). On Kaolinite and Pholerite. *American Journal of Science,* **43**, 351-361.

MORGAN, D.J. (1978). The clay bentonites at Woodbury quarry, Worcestershire. *6th International Clay Conference, Field Excursions,* 13-16.

PAULING, L. (1930). The structure of micas and related minerals. *Proceedings National Academcy of Science,* **16**, 123-129.

SRODON, J., MORGAN, D., ELSINGER, E.V., EBERL, D.D. & KARLINGER, M.R. (1986). Chemistry of Illite/Smectiite and end member Illite. *Clays and Clay Minerals,* **34**, 368-378.

TRANSPORT & ROAD RESEARCH LABORATORY, (1952). *Soil Mechanics for Road Engineers,* HMSO, London.

2-7 EFFECT OF SOME CHEMICAL ADDITIVES ON THE STRENGTH DEVELOPMENT OF SOIL-CEMENT

R. ANGELOVA
Geotechnical Laboratory, Bulgarian Academy of Sciences, Sofia, Bulgaria

ABSTRACT

The use of cement or lime for soil stabilisation is widespread in Bulgaria; in foundation works on collapsing loess soils using a soil-cement cushion and in water irrigation by the construction of impermeable screens. One way of controlling the technological parameters and regulating the properties of the soil-cement is the addition of various chemical reagents but there are almost no data in the literature concerning the effect of chemical activators on the long-term strength of soil-cement.

This paper describes the influence of the most common inorganic chemical additives (NaOH, CaO and $CaCl_2$) on the strength kinetics and the fabric formation in loess-cement. The investigation was carried out over a period of two years and included strength tests, X-ray diffraction and scanning electron microscopy as well as the determination of carbonate content, free CaO, pH and the composition of soluble salts.

The nature of the fabric formed and the strength parameters depend on the composition of the loess, the percentage of cement and the amount of activator. It has been shown that the effect of the additives varies over time and that some of the additives may reduce the long term strength of the soil-cement.

INTRODUCTION

In the early 1950s, pioneering research work began in the field of soil stabilisation by mixing cement and lime. Investigations included the effect of some chemical additives (Nicholls, 1952; Handy et al, 1959; Lambe et al, 1959; Moh, 1962; Laguros and Davidson, 1963; Bezruk, 1958; Lybimova and Iagodovskaya, 1961; Evstatiev, 1965 etc).

The application of different chemical additives in soil-cement is one of the ways of optimising and regulating its technological parameters. Additives may improve the strength and deformation properties of soil-cement, change its structure or allow a reduction in the cement content hence a more economical mix.

Despite the considerable number of publications on the effect of additives on the strength of soil-cement, their long-term influence has not yet been thoroughly explored. The problem is of major importance as soil-cement is a material which will be in place for decades in such structures as foundation cushions, piles, screens, building blocks and bricks etc. Since 1966 in Bulgaria alone more than 500,000 m^3 of soil-cement cushions have been emplaced in collapsible loess beneath buildings and structures and sixteen water irrigation levellers with a total area exceeding 160,000 m^2 have been lined by soil-cement screens.

This paper discusses current work on three of the most commonly used chemical additives ($CaCl_2$, CaO and $NaOH$) in relation to their influence on strength kinetics and fabric formation in loess-cement. The choice of additives was made considering their availability, low cost and well-known effect at the standard one-month period of curing.

INVESTIGATION METHODS

The study uses three varieties of loess from Northern Bulgaria - sandy, silty and clayey loess; for details of their composition and general properties see Angelova and Evstatiev, 1990. The basic binding agent is Portland cement PC 35 with a constant mineral content of C_3S - 50 to 60%; C_2S - 17 to 18%; C_3A - 8 to 9%; C_4AF - 12% and specific surface area 3500 to 3800 cm^2/g.

The additives of calcium oxide and sodium hydroxide are chemically reactive and meet the requirements for analytical reagent grade (AR). The calcium chloride is technologically pure and is obtained from the by-product liquids of soda-ash production. It corresponds to the regulations in Bulgarian Standards (BDS 3885-74), the content of $CaCl_2$ being not less than 72%.

The samples were prepared with cement proportions of 4, 7 and 10% of the dry weight of the loess soil and the amount of additives varied from 0 to 2%.

Cylindrical samples of 50 mm height and diameter were used in establishing the uniaxial compressive strength (R_c). They were moulded at optimum water content (W_{opt}) and compacted to maximum dry density (ρ_{dmax}). The dry soil and cement were mixed together. Calcium oxide was added directly but calcium chloride and sodium hydroxide were first dissolved in water corresponding to the optimum water content. $CaCl_2$ was added to the soil-cement mixture with sandy loess; NaOH to the silty loess mix and CaO to the clayey loess samples. This choice reflected the different granulometric and mineral composition of the loess soils and was based on former investigations (Evstatoeve, 1965; Angelova, 1987).

Strength tests were performed on six parallel samples, kept for varying periods of time (from 30 to 720 days) in air-moisture medium at a temperature of 20°C and water-soaked for 24 hrs. The crushed samples were dried in a vacuum oven at 40°C and the following tests performed: X-ray diffraction, scanning electron microscopy (SEM), determination of pH and the composition of soluble salts, free CaO and carbonate content.

TEST RESULTS AND DISCUSSION

CALCIUM CHLORIDE ADDITIVE

The preliminary tests after a year of curing showed that calcium chloride only increases the strength of soil-cement mixtures with sandy loess and more than 4% cement (Angelova, 1987). The mixtures with silty and clayey loess exhibited the negative effect of the admixture, which is more pronounced at greater percentages of $CaCl_2$ and lower cement content. Strength decrease in these cases varied from 5-35% for a cement content of 7% to 20-50% for a cement content of 4%. Further research is required into the influence of $CaCl_2$ admixtures on the strength development with samples of sandy loess and 10% Portland cement.

The effect of admixture on strength depends not only on its quantity but also on the curing time - t (Figure 1). The most substantial increase in strength (25 to 30%) was registered after three months curing of samples with 1 to 2% $CaCl_2$. After the sixth month all samples with the additive showed a strength decrease of between 9 and 11% compared with those without $CaCl_2$. Thereafter a strength increased was noted and after a year of curing the $CaCl_2$ samples gave between 8 and 15% higher strength results than the control ones. This difference persisted to the end of the two year investigation period.

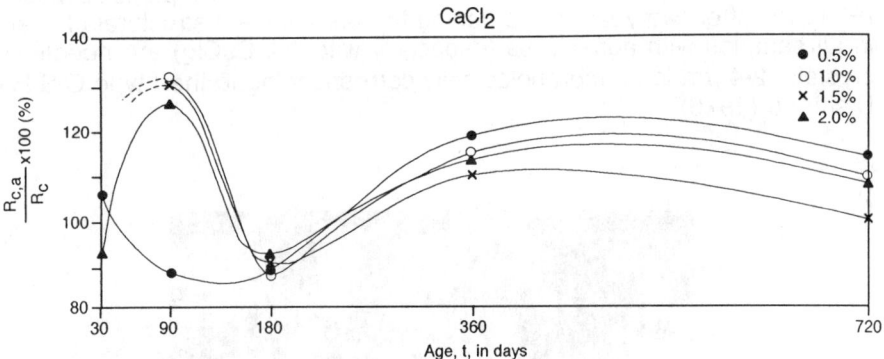

Figure 1: Effect of curing time on strength.

The calcium chloride admixture induces changes in the soil-cement microstructure. According to Diamond (1976), the network-like II type calcium silicate hydrates C-S-H appear after three months in the mixtures with $CaCl_2$ while in pure soil-cement two years of hardening is required (Plate 1). These data confirm the accepted concept of the accelerating effect of $CaCl_2$ on cement hydration, especially on the basic clinker mineral alite C_3S (Singh and Ojha, 1981). Hexagonal crystals with dominant aluminium content (2-4 μm to 10-12 μm in diameter and 0.5 to 1.5-2.0 μm thick) can be observed in mixtures with $CaCl_2$ between the third and sixth months (Plate 1). It is likely that these are calcium aluminate hydrates (C-A-H) of different

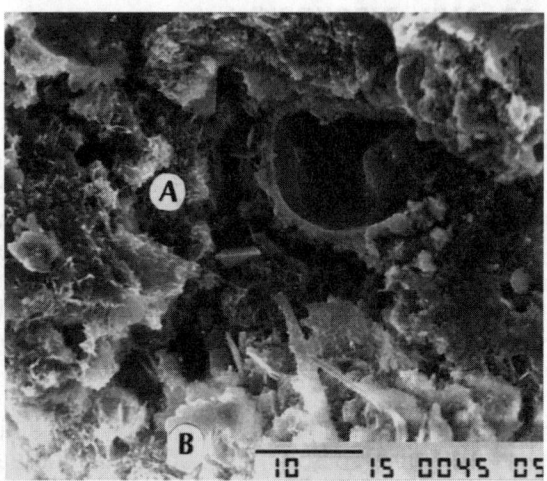

Plate 1: Mixture with $CaCl_2$ between 3 and 6 months.

composition and structure and calcium chloraluminate hydrate ($C_3AH_{10}.CaCl_2$). After two years of curing these crystals are observed less frequently, only occurring in single regions. The $CaCl_2$ also causes the formation of crystals 1-3 μm in diameter, which are of undefined composition. They usually coat the surface of the larger silt particles of loess (Plate 2). After two years of hardening the predominant structural elements in all samples with admixtures (especially with 2% $CaCl_2$) are needle-like crystals, 2-4 μm long, morphologically corresponding to the I type C-S-H of Diamond (1976).

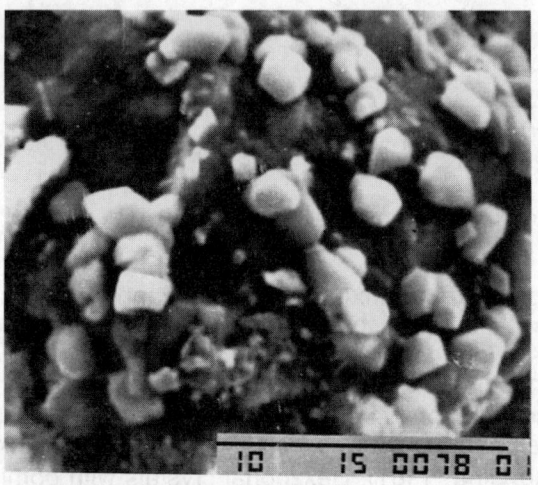

Plate 2: $CaCl_2$ crystals coat loess particles.

The addition of calcium chloride in soil-cement increases the degree of crystallinity of the basic binding mass and differences in the composition and the structure of the calcium silicate hydrates. After three months of cure mixtures with 2% $CaCl_2$ exhibited distinct outline peaks even of C-S-H, which are known to form under raised temperature and pressure conditions. Even earlier the X-ray diffraction charts show reflections of different calcium aluminate hydrates: CAH_{10} (d = 1.443; 0.3715; 0.269 and 0.237 nm), C_4AH_{19} (d = 0.278 and 0.253 nm) etc. Calcium chloride, joining the aluminate phase of soil-cement, forms complex compounds as $C_3AH_{10}.CaCl_2$ (most intensive lines at d = 0.774; 0.391; 0.2596 and 0.237 nm) and $3Ca(OH)_2.CaCl_2.12H_2O$ (d = 0.9237; 0.419 and 0.2774 nm).

Calcium chloride exerts a considerable influence on the free CaO content in the soil-cement. In pure soil-cement, free CaO is established up to the sixth month (0.2% in the first month and 0.02% in the sixth). In mixtures with 0.5% $CaCl_2$, free CaO is present up to the third month (0.03-0.06%) but with 2% $CaCl_2$ free CaO is absent even during the first month.

Carbonate content is only slightly increased (0.5-1.5%) by the addition of $CaCl_2$ to soil-cement. In all samples (with and without the admixture) carbonate content diminishes by about 3-4% over long term curing of one to two years.

Calcium chloride considerably decreases the soil-cement's alkalinity (pH), the effect becoming more pronounced in the course of time. The pH reduction for loess-cement is from 11.9 after one month of curing to 9.4 after two years; for the mixtures with 0.5% $CaCl_2$ it decreases from 11.8 to 8.6 and for the mixtures with 2% $CaCl_2$ from 11.6 to 7.8 (Table I). The reduced values of pH in the system are not conducive to C-S-H formation (especially of tobermorite as it is a stable compound only at pH > 9 (Kuznetsova et al, 1989).

Calcium chloride leads to alteration in the composition and content of the water extracted soluble salts (Table I). The mixtures without additive have a total salts content of 0.33-0.51% compared with 1.35-1.55% for samples with 2% $CaCl_2$. Naturally, the amount of Ca^{2+} and Cl^- ions increases with higher percentages of the admixture. After two years of hardening the mixtures with $CaCl_2$ manifest a decrease in the concentration of Ca^{2+} ions. The increased admixture and curing time lead to decreased quantities of the CO_3^{2-} and HCO_3^- ions. The sulphate ions concentration is marginally reduced over the investigation period.

The analyses showed that the calcium chloride activated and accelerated the hydration of the basic clinker minerals in loess-cement mixtures. This is confirmed by the formation of a voluminous mass of network-like C-S-H at the early stages of curing, before the third month; by the higher crystallinity of the basic binding mass of calcium silicate hydrates; by the number of hexagonal crystals with a predominantly aluminium content; by the rapid decrease of pH and by the extinction of free CaO in the system.

Table I: Changes in water extracted soluble salts content with time.

Type of mixture	Age months	pH	Total salts (%)	Composition of the soluble salts					
				Ca^{2+}	Na^+	CO_3^{2-}	HCO_3^-	SO_4^{2-}	Cl^-
Sandy loess + 10% Portland cement + $CaCl_2$ in the following percentages:									
0%	1	11.9	0.33	3.9	-	2.7	2.1	1.6	0.5
	24	9.4	0.51	4.4	-	0	0.4	2.7	2.0
0.5%	1	11.8	0.53	7.0	-	2.2	1.8	1.2	5.2
	24	8.6	0.82	7.5	-	0	0.4	2.2	6.5
2.0%	1	11.6	1.55	18.0	-	1.4	1.4	1.6	16.3
	24	7.8	1.35	15.0	-	0	0.4	1.5	14.9
Silty loess + 10% + Portland cement + NaOH in the following percentages:									
0%	1	11.4	0.29	3.2	0.1	1.8	1.9	2.3	-
	24	9.6	0.26	2.2	0.2	0.8	0.9	1.9	-
0.2%	1	11.8	0.31	0.8	2.9	3.2	2.5	2.8	-
	24	9.7	0.28	0.7	3.2	1.0	1.4	1.9	-
2.0%	3	12.1	1.25	0	1.8	24.9	17.7	4.3	-
	24	11.0	0.99	0	16.5	15.0	15.0	3.1	-
Clayey loess + 10% Portland cement + CaO in the following percentages:									
0%	1	11.7	0.35	3.9	-	1.9	1.8	1.8	-
	24	9.7	0.31	3.0	-	0.6	0.9	2.9	-
0.5%	1	11.9	0.35	4.4	-	2.5	2.5	2.8	-
	24	10.0	0.36	3.5	-	1.1	1.2	2.8	-
1.5%	1	11.8	0.25	4.6	-	3.0	2.2	2.3	-
	24	10.1	0.32	3.3	-	1.3	1.3	-	-

SODIUM HYDROXIDE ADDITIVE

A number of publications confirm the positive effect of sodium hydroxide (NaOH) on the strength of soil-cement after short periods. The preliminary investigations reported here prove that after a year of curing NaOH in quantities of less than 0.5 to 0.75% increases the strength of the mixtures with silty and clayey loess by about 5 to 30%. The effect is markedly negative for the sandy loess mixtures.

The influence of NaOH on the strength development and fabric formation of soil-cement over time was investigated on samples of silty loess with 10%

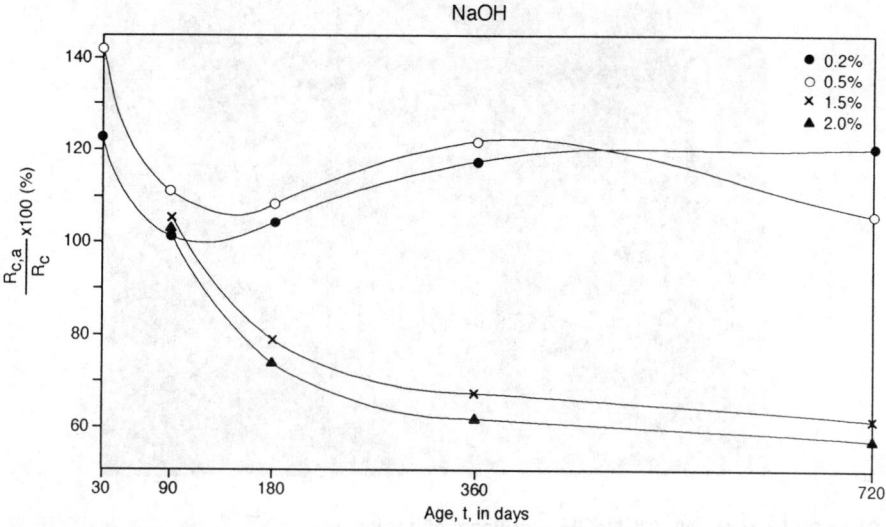

Figure 2: Strength changes with NaOH.

Portland cement. The results for strength changes with time show that the positive effect of NaOH is displayed only when the proportion of admixture is less than 1% but this effect was not constant over the investigation period (Figure 2).

The mixtures of 0.2 to 0.5% NaOH showed the greatest strength increase (20 to 40%) at up to one month curing. Thereafter this positive effect decreased between the third and six months and after a slight increase the strength stabilised for the remainder of the two year period of hardening. More than 1% of NaOH produced an overall net decrease in strength over the period of investigation, reaching 40 to 45% for the mixtures containing 1.5 to 2.0% NaOH after two years of hardening (Figure 2).

The differences in microstructure with lower (0.2 %) and higher (2%) quantities of the additive are substantial. The microstructure of the 0.2% NaOH samples is similar to those without additive, being characterised by needle- and network-like C-S-H in the early stages and gel-like ones during later periods. Some of the silt grains in soil-cement appeared to be free from the coating of the clay particles. Considerable alterations in the microstructure of the samples containing 2% NaOH were also noted, especially after two years of hardening. These included the partial destruction of some crystals and microaggregates, resulting in an overall less compacted fabric (Plate 3).

Some strong reflections can be observed on the diffraction charts of loess-cement with 0.5% NaOH: calcium aluminate hydrates CAH_{10} and C_4AH_{13}; calcium silicate hydrates $C_5S_6H_3$, $C_3S_2H_3$, CS_2H_2 etc; lime $Ca(OH)_2$ and some hydrogarnets. The characteristic reflections of $Na_2CO_3.10H_2O$, Na_2SO_4, $C_3AH_{12}.CaCO_3$, some calcium silicate and aluminate hydrates were noted in the mixtures with 2% NaOH additive (Angelova, 1987).

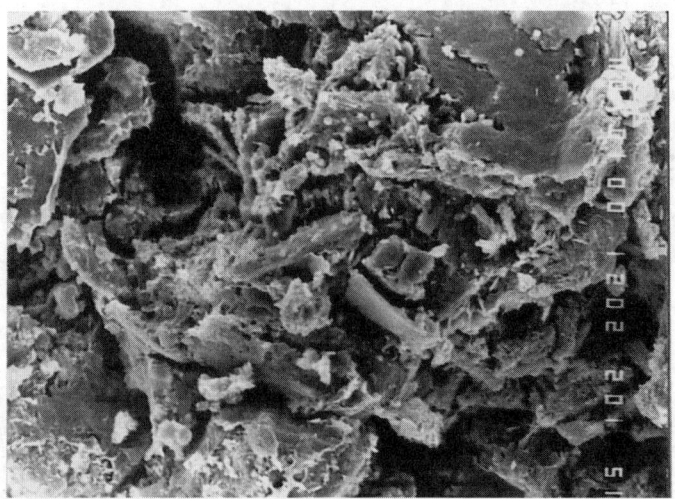

Plate 3: Sample with 2% NaOH after years of hardening.

During the whole period of investigation no free CaO was registered in soil-cements with either no NaOH or with 0.2 and 0.5% NaOH. In contrast, the mixtures with 1.5 and 2.0% NaOH showed free CaO in the course of a year; its content being almost constant from the first to the twelfth month at 0.12 to 0.15% for the samples with 1.5% NaOH and 0.16 to 0.18% for those with 2% NaOH. After two years of hardening free CaO is absent in all of the samples.

A slight increase of about 2% in carbonate content has been registered in loess-cement with and without additive after a period of one or two years.

The alkalinity increases with greater quantities of additive while the rate of pH diminution decreases over time (Table I). The total salts content increases in proportion to the amount of additive; in the case of 2% NaOH it is about 1% higher than for pure soil-cement. The concentration of Na^+ ions also increases while that of Ca^{2+} decreases. In the water extract of mixtures with 0.2 and 0.5% NaOH, Ca^{2+} ions are registered even after two years of curing while in the mixtures with 2% NaOH these ions disappear before the third month. The admixture also raises the quantity of CO_3^{2-} and HCO_3^- ions. A small increase of sulphate ions has been observed with greater quantities of sodium hydroxide (Table I).

The data demonstrate the influence of sodium hydroxide in cement hardening by the raised alkalinity as well as the chemical reaction with loess and cement minerals. Small quantities of NaOH (0.2 to 0.5%) are more favourable for cement hydration. They lead to the formation of an additional gel-like binding mass, rich in sodium, which enhances contact between particles and thus increases the strength. With greater quantities of NaOH (1 to 2%) however there is a reaction with the finely dispersed quartz, carbonates and clay minerals. This process is accompanied by destruction of microaggregates and particles, resulting in a less compacted fabric. Furthermore, in such quantities NaOH may react with the dolomite $CaMg(CO_3)_2$ within the loess, producing water soluble Na_2CO_3. Clearly such alterations in the composition and fabric of loess-cement with larger quantities of NaOH will result in a considerable decrease in strength values.

CALCIUM OXIDE ADDITIVE

Experience in Bulgaria shows that it is difficult to mix clayey loess with cement and the strength of this soil-cement is usually lower than that obtained under laboratory conditions. The introduction of small amounts of quicklime CaO before the mixing with cement improves the soil workability, the more regular distribution of cement and the homogeneity of the mixture.

The effect of CaO on strength development has been investigated for a clayey loess and 10% Portland cement. The addition of up to 0.5% CaO has almost no effect on the soil-cement strength (Figure 3) but with 1.5 to 2.0% of lime the strength shows a greater increase (20 to 50%) during the early stages of curing, up to the third month. A slight decrease of this positive effect is noted between the third and the twelfth months, followed by a further but more moderate increase during the second year. At the end of the investigation period the mixtures containing 1.5 to 2.0% of lime displayed 10 to 20% higher strength than the pure soil-cement (Figure 3).

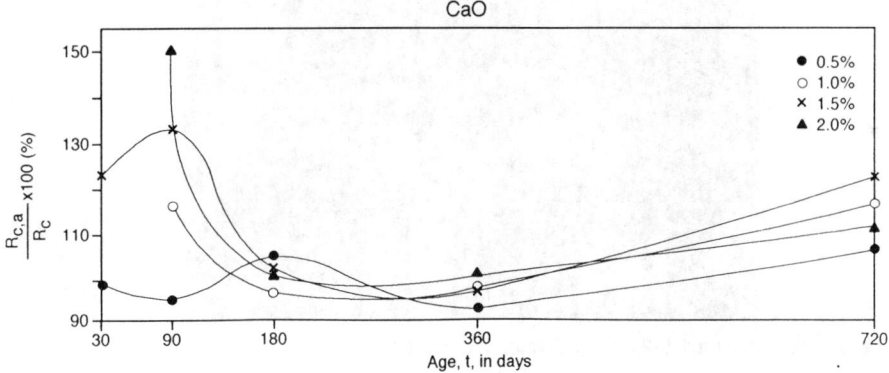

Figure 3: Strength changes with time.

Small amounts of CaO (0.2 to 0.5%) do not substantially change the loess-cement microstructure. The basic binding mass consists of needle-like C-S-H up to 1 μm long and gel-like C-S-H. Flat irregularly shaped particles rich in calcium are also present. The structure seems more aggregated than that of pure soil-cement. The alterations are greater in the case of 1.5 to 2.0% CaO content. As a consequence of the physico-chemical interaction with lime, the clayey component of the loess is aggregated and the microstructure has a granular appearance (Plate 4). Needle-like I type C-S-H, 3 to 4 μm long are present in single regions. At about the sixth month, the needle-like C-S-H (5 to 7 μm long) significantly increase their volume and are densely interwoven (Plate 5).

Plate 4: Mixture with 1.5-2.0% CaO.

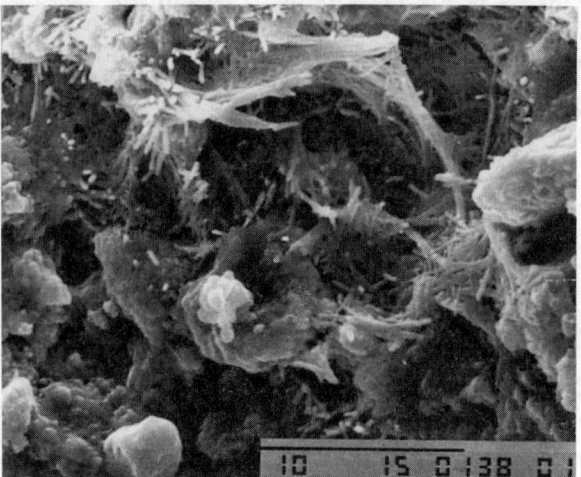

Plate 5: Mixture with 1.5-2.0% CaO after six months.

The intensive growth of these new formations in the comparatively compact structure most probably causes internal stresses in the soil-cement, thus reducing its strength. This fact has been established only for the mixtures with a higher cement content of 7 to 10%. After a period of two years of hardening the densely interwoven needle-like crystals are no longer observed. The structure is compact and granular; accretions of fine (up to 1 μm long) needle-like crystals can be seen only in isolated positions.

The small quantities of CaO (up to 0.5%) result in the appearance of some new reflections, which undergo various phase transformations with time. More reflections of C-S-H can be observed in the mixtures with greater CaO content (2%) after a period of three months. The strongest peaks of $Ca(OH)_2$ (d = 0.264 and 0.481 nm) are noticeable until the end of the first year. New reflections with d = 0.756; 0.279 and 0.272 nm appear at that time probably related to calcium carboaluminate hydrate $C_3AlH_{12}.CaCO_3$.

Free CaO is absent in soil-cement with and without additive throughout the period of investigation. Carbonate content increases negligibly with higher quantities of additive from 25 to 26% for pure soil-cement and 26.5 to 28.0% for mixtures with 1.5 to 2.0% CaO. Calcium oxide raises the alkalinity pH insignificantly (Table I), having little influence on the total salts content. The concentration of Ca^{2+} and SO_4^{2-} ions does not depend on the amount of CaO in soil-cement. The higher quantities of additive slightly increase the quantity of CO_3^{2-} and HCO_3^- ions in the water extract (Table I).

The results obtained show that calcium oxide reacts with the clay minerals of loess. In addition CaO changes the conditions of cement hydration and of new phases formation. It does not substantially alter the microstructure, composition and strength of loess-cement in quantities of up to 0.5%, but the addition of greater amounts of CaO (1.5 to 2.0%) leads to noticeable differences. Initially, the clay fraction is strongly aggregated and the microstructure acquires a granular appearance. The larger quantity of admixture contributes to the formation of high-alkaline C-S-H at the early stages up to the third month. This affects the strength, which is 30 to 50% higher than that of pure soil-cement. At a later stage, however, between the sixth and twelfth months, the intensive growth of the needle-like C-S-H in the dense fabric of loess-cement induces an increase in internal stresses and a fall in strength values. This effect is noted only in mixtures with a greater cement content, 7 to 10%. Smaller quantities of cement (4 to 7%) probably do not allow the intensive growth of needle-like C-S-H which would result in volumetric increase. Towards the end of the twelfth month, phase transformations of C-S-H occur which change their morphology and lead to a greater rate of decrease of pH. At the end of the two year period the strength increase is 10 to 20% for mixtures with 1.5 to 2.0% CaO.

CONCLUSIONS

1. The influence of chemical admixtures $CaCl_2$, NaOH and CaO in loess-cement depends on:

    a) the granulometric and mineral composition of the mix;

    b) the quantity of the basic binding agent and of the admixture

    c) the duration and

    d) the conditions of curing etc.

    Under similar initial conditions the effect of the admixtures on strength development vary with time. This process is accompanied by changes in the chemical-mineral composition and microstructure of the soil-cement.

2. Calcium chloride in quantities of 1 to 2% exerts a positive effect only on sandy loess soils stabilised with more than 8% of Portland cement. The greatest strength increase for these mixtures (30%) is observed up to the third month. The more intensive strength growth is due mainly to the activation and acceleration of hydration of basic cement clinker minerals.

3. Sodium hydroxide in amounts of 0.2 to 0.5% has a positive effect on the strength of the mixtures of silty and clayey loess and 4 to 10% Portland cement. The addition of more than 1% of NaOH sharply decreases the strength of loess-cement and this effect is more pronounced with time. The considerable strength reduction is associated with partial destruction of separate grains and micro-aggregates and an increase of water soluble salts which results in a less compacted fabric. Clearly therefore, the amount of sodium hydroxide used in a mixture must be carefully considered.

4. Between 1 and 2% calcium oxide improves the workability and has a long-term positive effect on the strength of clayey loess with low cement quantities (up to 10%). As a result of the physico-chemical interaction with calcium oxide, the clayey component of loess is aggregated and this aids homogenisation of soil-cement. In greater proportions (1.5 to 2.0%) the CaO contributes to the formation of high-alkaline calcium silicate hydrates at the early stages of curing.

5. When considering the application of various chemical activators for soil stabilisation with cement it is essential to establish their long term effect on strength. The results of the present investigation prove that some admixtures which increase soil-strength after the standard one month of hardening, lead to a significant decrease in strength over longer periods of time.

REFERENCES

ANGELOVA, R. (1987). *General relationships of kinetics of structure formation of cement stabilized loess from North Bulgaria.* Ph.D. Thesis, University of Moscow, 230 pp (in Russian).

ANGELOVA, R. & EVSTATIEV, D. (1990). Strength gain stages of soil-cement. *Proceedings 6th International Congress of the International Association of Engineering Geology*, Balkema, Rotterdam, **4**, 3147-3154.

BEZRUK, V. (1958). Methods of complex soil stabilization by cement and admixtures of lime and electrolytes. *Autotransizdat*, Moscow, 17 pp (in Russian).

DIAMOND, S. (1976). Cement paste microstructure - an overview at several levels. *Proc. Conf. on hydraulic cement pastes: their structure and properties*, University of Sheffield, 2-30.

EVSTATIEV, D. (1965). Investigations on Portland cement stabilized loess from Bulgaria. *Papers on Geology of Bulgaria - Engineering Geology and Hydrogeology*, **IV**, 131-153 (in Bulgarian).

HANDY, R., JORDAN, J., MANFRE, L. & DAVIDSON, D. (1959). Chemical treatments for surface hardening of soil-cement and soil-lime-fly ash. *Highway Research Board Bulletin*, **241**, 49-66.

KUZNETSOVA, T., KUDRIASHOV, I. & TIMASHEV, V. (1989). Physico chemistry of binding materials. 384 pp, *Vishaya shkola*, Moscow (in Russian).

LAGUROS, J. & DAVIDSON, D. (1963). Effect of chemicals on soil-cement stabilization. *Highway Res. Rec.*, **36**, 172-203.

LAMBE, T., MICHAELS, A. & MOH, Z.C. (1959). Improvement of soil-cement with alkali metal compounds. *Highway Research Board Bulletin*, **241**, 67-108.

LYBIMOVA, T. & IAGODOVSKAYA, T. (1961). On the structure formation processes in cement stabilized soils. *Coll. Mag.*, **5**, 596-604 (in Russian).

MOH, Z.C. (1962). Soil stabilization with cement and sodium additives. *Journal of the Soil Mechanics and Foundation Division*, **88(SM6)**, 81-105.

NICHOLLS, R. (1952). *Preliminary investigation for the chemical stabilization of loess in Southwestern Iowa*. M.Sc. Thesis, Iowa State University Library.

SINGH, N. & OJHA, P. (1981). Effect of $CaCl_2$ on the hydration of tricalcium silicate. *Journal of Materials Science*, **16**, 2675-2681.

2-8 METHANE GENERATION FROM VOID FORMERS USED IN FOUNDATIONS ON CLAY SITES

R.W. JOHNSON
NHBC, 9-10 Clevedon Triangle Centre, Avon BS21 6HX

ABSTRACT

During 1990/91, there were reports that void formers used in the foundation construction of buildings on clay sites were a source of methane gas generation. A small explosion occurred in the basement of a building in London. This incident initiated a review by BRE and NHBC, together with the manufacturers of the void former on the use of these materials.

INTRODUCTION

A significant proportion of Southern England is underlain by clays which possess a swelling/shrinkage potential with changes in moisture content. The primary cause of moisture fluctuation is seasonal and hence in addition to rainfall and evaporation it is affected by the presence of vegetation. The main influence of seasonal shrinkage and swelling is restricted to depths of about 1m, but with vegetation significant volume changes may occur to depths of 3.5 m or more.

Designing and constructing foundations of buildings in these areas must take account of the potential movements. Shrinkage due to the reduction in moisture content is simply accommodated by deepening the foundation to a depth where the soils are stable. The extra depth can also accommodate swelling as the soils rehydrate. However, swelling is not solely vertical but also horizontal. Therefore lateral pressures must either be taken into account in the design of the foundations or materials introduced into the foundation construction to create voids or compress and absorb the swelling.

Similarly if a portion of the foundation is constructed at a depth where vertical swelling may occur, e.g. beneath the ground beam in a pile and beam foundation solution, then again providing a void or compressible material is necessary if damage to the building is to be avoided.

COMPRESSIBLE MATERIALS AND VOID FORMERS

The most commonly used compressible material is low density expanded polystyrene. It possesses sufficient strength to offer support to wet concrete but will compress to approximately 50% of its thickness in response to a swelling clay. There are however limitations of use. For example, it is unsuitable under most ground floor slabs because the uplift pressure generated in compressing polystyrene up to 50% thickness reduction is usually greater than the dead-load from the slab.

A popular alternative is a permanent void former called "Clayboard". Clayboard is constructed from two hardboard faces separated with paper honeycomb. It is manufactured in thicknesses of 50, 75, 100 and 150mm.

The Clayboard is designed to support the weight of wet concrete. However, when the honeycomb is exposed to ground water, there is a dramatic reduction in strength, enabling any clay swelling to occur without stressing the concrete foundation or floor slab. In soils where there is a low ground water level, there is a facility for "wetting" the Clayboard honeycomb after the concrete has cured.

The product has been used extensively for many years in the foundations for housing, industrial and commercial developments. However, Clayboard has an organic origin and is therefore biodegradable. It has long been recognised that under anaerobic conditions (absence of oyxgen) methane can be generated from the decomposition of biodegradable substances. Methane when mixed with air in volumes between 5 and 15% is combustible. There are recorded instances of explosions occurring causing loss of life and injury in addition to building damage as a result of the ignition of methane.

METHANE GENERATION FROM CLAYBOARD

During the summer of 1990 it was reported in the technical press that methane gas had been detected in the basement of two commercial buildings in London. In one of the basements, an ignition source caused a minor explosion. The subsequent investigation identified Clayboard, which had been used below ground level beneath the basement floor, as the methane source.

ACTION BY NHBC/BRE

Following the incidents in London, BRE were asked to advise DOE Building Regulations Division upon appropriate action.

Similarly NHBC who, up until that time accepted Clayboard without restriction, were anxious to assess the risk and advise the housebuilding industry accordingly.

A tripartite meeting was arranged between the manufacturers of Clayboard, BRE and NHBC during the Autumn of 1990. The purpose of the meeting was to examine the following:-

1. Review the state-of-the-art knowledge on methane generation from Clayboard.

2. Assess when the results of the current research project and recommendations will be published.

3. Anticipating the research results would not be available for some time, discussing possible interim precautions for the private housebuilding industry and to enable BRE to suitably advise DOE Building Regulations.

It was reported at the meeting that the state-of-the-art knowledge of methane generation from Clayboard when related to construction is in its infancy and clearly no definitive recommendations will be available in the immediate future.

Therefore as an interim policy, it was decided to consider the circumstances when anaerobic conditions were most likely to occur with a view to trying to eliminate significant gas generation occurring. It was agreed that if Clayboard is used in large quantities in saturated conditions (at or below ground water level) without adequate ventilation, then there was a risk that methane could accumulate to volumes in air where combustion was possible.

As a result of these discussions, it was agreed Clayboard under floor slabs, if allowed to become saturated, could generate methane and find access through service entry points into the building. It was considered Clayboard in this location was inappropriate, especially in housing.

Many low-rise houses are constructed with pre-cast beam and block suspended ground floors. Below is a void, passively ventilated by airbricks in the external walls. It is well known that small gas accumulations that collect in the void will eventually disperse. Therefore it was decided that if Clayboard was to be used in foundations, there must be a pre-cast concrete ground floor with a passively ventilated void. A passively ventilated void has a limited capacity to accommodate the accumulation of gas. Therefore, a further restriction was placed upon the use of Clayboard with foundations. Clayboard was limited to depths of 2m on the side of trenchfill (Figure 1). This restricts the total volume of Clayboard and limits the quantity of Clayboard likely to become submerged and saturated due to ground water.

Similarly Clayboard was permitted to the side and underside of ground beams where piles and piers are used (Figure 2). Again, usually the ground beam remains above ground water level and the volume of material used is limited.

The NHBC published these conclusions in their Newsletter to the private housebuilding industry in January 1991:

> "CLAYBOARD
>
> This product, manufactured by Dufaylite Developments Ltd., is designed to absorb clay heave thereby preventing forces generated by swelling clay imposing pressure on the substructure of a building.

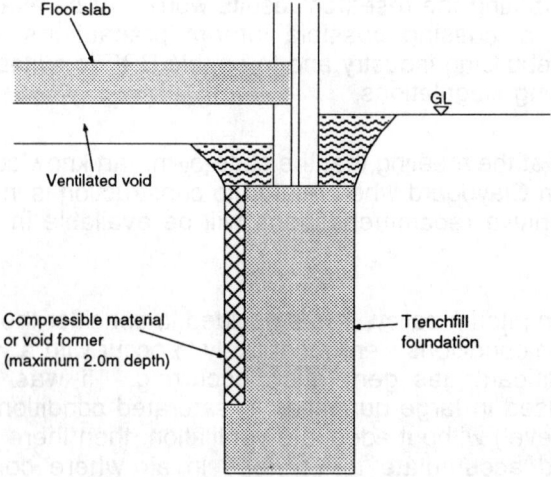

Figure 1: Restrictions for the placement of Clayboard on the side of trenchfill foundations.

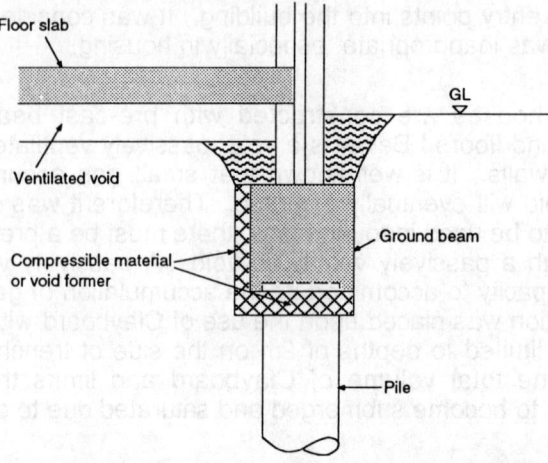

Figure 2: Restrictions for the placement of Clayboard around ground beams where piles and piers are used.

The material is formed of hardboard facing with a paper honeycomb infill.

The constituents of the product are biodegradable. The decomposition of the materials may produce methane gas. The risk from methane is reduced if the following precautions are taken when using Clayboard:

1. Clayboard is only used in foundations if the ground floor is suspended precast concrete or beam and block

construction with a ventilated void in accordance with NHBC Handbook requirements Fo32 and NHBC Standards 5.2 - D10.

2. Clayboard is not used on the inside face of trench fill or strip foundations which are greater than 2m deep.

3. Clayboard is only placed in dry excavations.

Clayboard should NOT be used under suspended in-situ reinforced concrete slabs.

These precautions follow discussions with the manufacturer and have their agreement.

NHBC field staff will be looking for these precautions immediately but with common sense and good judgement."

BRE have published their views in BRE News. In addition, they invited building owners (from the commercial sector) to allow them with the aid of instrumentation to check if any methane was present and if so the volume in basements and other poorly ventilated spaces where Clayboard had been used.

With regard to existing buildings, including housing, it was agreed that the risk to health and safety was very low and it was agreed that any attempt to generate anxiety amongst building owners, occupants and homeowners was unfounded.

FUTURE ACTION

The manufacturers of Clayboard have suffered a reduced market share as a result of the two incidents in London, the BRE published advice and the NHBC interim policy.

Their first action has been to introduce a Mark II Clayboard with polypropylene faces. The biodegradable core remains but the potential volume of gas-generating material is reduced. They consider this is an interim step towards introducing a totally non-biodegradable void former in the future.

2-9 GASEOUS POLLUTANTS AND THEIR IMPLICATIONS FOR SITE SAFETY

J.S. BUTTERWORTH
4 Ridgeside, Bledlow Ridge, High Wycombe, Bucks
D.P. CREEDY
J.S. EDWARDS

ABSTRACT

The paper reviews the occurrence of natural and pollutant gases which may be of significance to the safety of construction site workers and others. The properties of gases, including some of their toxic and asphyxiant properties are discussed together with a summary of methods of measurement and the conventions in use for expressing their concentrations. Reference is made to current legislation and guidelines related to safe working and structural design in respect of potential gas hazards.

INTRODUCTION

The implications of the presence of gases in construction relate both to safety and to the integrity of any development. The consequences of gas occurrences may be experienced at any time from the initial site investigation through to the post-development period.

In terms of safety, the effects of gases may range from the sub-clinical to the lethal. The effects of gases on structures can be as disparate as the superficial crazing of concrete caused by carbonation of the cementitious matrix or the destruction of a structure by gas explosion.

In many branches of engineering the composition of the ambient atmosphere may differ considerably from that of normal air, due either to the incursion of gas from an external source or as a result of the removal of oxygen. The result may be an atmosphere that is asphyxiating, explosive or toxic. In this paper the important properties of the gases most likely to be encountered are described; these include their flammability characteristics, solubility in water, mixing behaviour and physiological properties.

The sources of gas are manifold and, in the case of construction activity, many will be present as a direct result of site activity; welding fumes and vapours associated with adhesive solvents are obvious examples. These sources are predictable and controllable; other sources occur less expectedly and may be a result of previous use, or abuse, of the site.

HAZARDOUS GASES FOUND ON CONSTRUCTION SITES

A wide variety of hazardous gases and vapours can be encountered at any stage of construction activity, from site investigation to occupancy of the completed building. Some of these substances may be flammable; others may be toxic or asphyxiating.

The principal flammable gases are methane, carbon monoxide and, less commonly, hydrogen. Methane occurs in a wide range of geological and biochemical environments (Creedy, 1989) and its presence should be considered on any site. Carbon monoxide, a common oxidation product of carbonaceous material, is likely to manifest its presence due to its toxicity before becoming an explosive problem. Hydrogen sulphide is another highly toxic gas that could be encountered.

Any non-toxic gas that displaces oxygen can be considered an asphyxiant. Carbon dioxide is the most important asphyxiant and high concentrations are not uncommon in soils even in the absence of landfill gas.

Oxygen deficient atmospheres are invariably encountered in badly ventilated underground openings, often as a result of oxygen being consumed in biological and chemical oxidation processes.

Ancient shallow coal mine workings will contain varying proportions of methane, oxygen deficient air (predominantly nitrogen) and products of oxidation (chiefly carbon dioxide). Lesser concentrations of ethane and higher hydrocarbons may be present together with carbon monoxide and hydrogen. Of the hydrocarbons, only methane occurs in sufficiently high concentrations to constitute an explosion hazard.

Oxygen has an affinity for both coal and ironstone. Partial oxidation of coal leads to the production of carbon dioxide and carbon monoxide. The action of groundwater on siderite (iron carbonate) nodules within ironstone rakes may also lead to the production of carbon dioxide.

BASIC PROPERTIES OF GASES

FLAMMABILITY OF COMBUSTIBLE GASES

In a flammable gas/air mixture a flame becomes self-propagating when sufficient of the flammable gas is present to cause combustion to spread through the mixture from a point of ignition.

There are two limits of flammability, a higher and a lower, for each pair of flammable gas and supporter of combustion. The lower limit corresponds to the minimum amount of flammable gas that will liberate sufficient heat for the combustion process to be self-sustaining. The addition of more flammable gas to a lower-limit mixture causes an increase in the liberation of heat with a corresponding increase in explosive violence when the mixture is ignited.

The mixture will be at its most explosive at the stoichiometric point, i.e. when there is just sufficient oxygen to completely burn the gas. For a methane/air mixture the stoichiometric composition is 9.48%. If still more flammable gas is added to the mixture the amount present becomes so large that the oxygen concentration is reduced and the amount of heat liberated is only just sufficient to sustain combustion. This is the upper limit of flammability; at greater flammable gas concentrations the gas mixture will not support combustion.

The limits of flammability, in air, of the gases of most concern are given in Table I.

Table I: Limits of flammability of pure gases (NTP).

Gas	LEL (%)	UEL (%)
Methane	5.0	15.0
Ethane	3.0	12.5
Propane	2.2	9.5
Butane	1.9	8.5
Hydrogen	4.0	74.0
Carbon monoxide	12.5	74.0
Hydrogen sulphide	4.4	45.0

The limits of flammability of gas mixtures are affected by the composition of the mixture, strength of the ignition source, temperature, pressure and nature of the surroundings.

In some circumstances the flammable gas under consideration may not be a single gas but a combination of flammable gases and in addition there may be inert gases present, eg fumes from fires and explosions, landfill gas, spillages of petroleum, and natural gas.

The upper and lower limits of flammability will be affected by the presence of other combustible gases. The presence of higher hydrocarbons such as ethane, propane and butane in a natural gas, for example, can produce a gas mixture having limits of flammability considerably different from the oft-quoted values of 5%-15% for pure methane.

The ratio of inert gas to the flammable components is also an important factor in determining the limits of flammability of gas mixtures. The combustion characteristics of landfill gas, for example, are modified considerably by the presence of gases other than methane - particularly carbon dioxide and hydrogen. The flammability limits of three typical landfill gas compositions are illustrated in Figure 1, together with those of natural gas. This highlights the wide variation in flammability limits that can be encountered and reinforces the necessity for careful assessment of any other gases present.

The flammability limits of a mixture of methane in air containing various concentrations of oxygen can be determined by means of the well-known COWARD diagram (Coward and Jones, (1931), Figure 2.

Ignition of gas mixtures usually occurs through contact with an open flame, hot surface, or electric spark. Hot surfaces ignite methane at a temperature much higher than that of a flame, while in the case of electric sparks or arcs a minimum spark energy is required.

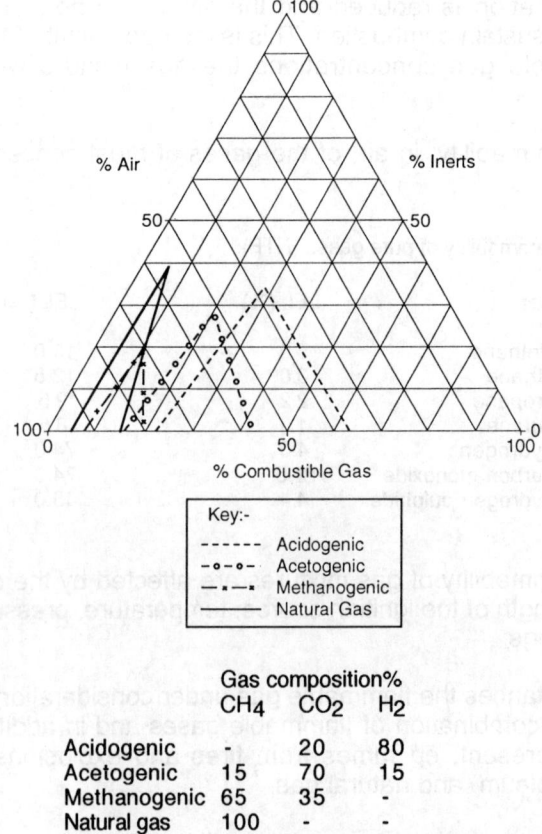

	Gas composition%		
	CH₄	CO₂	H₂
Acidogenic	-	20	80
Acetogenic	15	70	15
Methanogenic	65	35	-
Natural gas	100	-	-

Figure 1: Variation of flammability limits with changes in landfill gas composition.

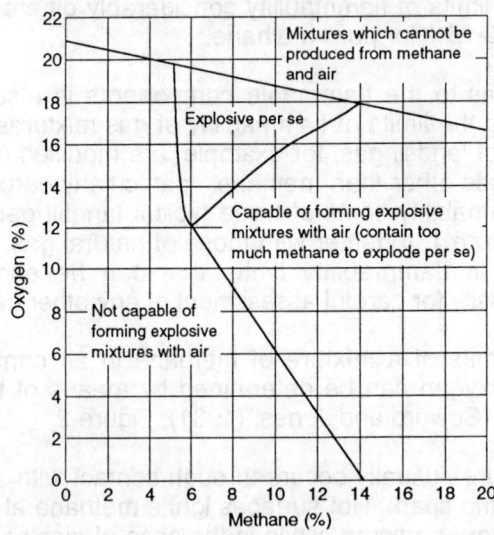

Figure 2: Limits of flammability of mixtures of methane, air and nitrogen.

The limits of flammability of methane/air mixtures are dependent upon the temperature at constant pressure; as temperature is increased the lower limit decreases and the upper limit increases. Thus a non-flammable mixture may become flammable if its temperature is raised sufficiently. The effect of pressure on the limits of flammability of gas mixtures is insignificant within the normal range of atmospheric pressure variations.

ADSORPTION OF GASES

Methane and higher hydrocarbons may be found as a free gas in pores, fractures and cavities in rocks of all types and also as bubbles in water in modern sediments. The methane contents of non-carbonaceous Jurassic and Carboniferous rocks range from 0.1 to 150 l/tonne.

Methane is found highly concentrated in coals and other carbonaceous materials as a result of adsorption. The quantity of gas adsorbed depends upon the composition, rank and moisture content of the coal substance and also on the temperature and partial pressure of the gas. The relationship between the gas pressure and adsorbed gas content at the lower pressures encountered on surface sites is approximately linear:

$$q = kP(C/100)$$

where q is the quantity (m^3/t) of a gas of concentration C% adsorbed at a final total gas pressure of P kPa. A typical value for k is 0.012 m^3/t per kPa.

SOLUBILITY OF GASES

All gases dissolve in water to some extent and therefore most naturally occurring waters will contain dissolved gases, the most abundant of which are N_2, O_2, CO_2, CH_4, H_2S and N_2O. The first three are derived from the atmosphere whilst the remainder, when encountered on surface sites, are usually the result of biogenic processes involving the decomposition of organic matter in moist anoxic conditions.

At a given temperature the solubilities of gases which do not react with the solvent to any appreciable extent are directly proportional to the partial pressures of the gases above the solution. This relationship is known as Henry's Law which may be expressed mathematically in the form,

$$P = HX$$

where,
- P = partial pressure of the solute in the gas phase (mm Hg).
- X = mole fraction of the solute in the liquid phase.
- H = Henry's constant (mm Hg/mole fraction).

Although Henry's Law is exact only in the infinitely diluted state, it is a good approximation in dilute solutions. If the solubility of a gas is known for one pressure the Henry's Law constant may be calculated and used to calculate the solubility at any other gas pressure.

The solubilities of some of the more common unreactive gases in pure water are given in Table II.

Table II: Solubility of gases at 1atm and 25°C

Gas	Solubility (mg/l)
Nitrogen	17.90
Oxygen	40.40
Hydrogen	1.58
Methane	21.47

Methane is frequently found in solution in groundwater. In some circumstances the presence of methane in a dissolved state in groundwater can give rise to a serious explosion hazard and indeed has resulted in a number of serious incidents, eg Abbeystead, Carsington, Furnas etc.

The concentration of dissolved methane that would be required to create a given concentration of methane in the headspace above a body of water may be calculated thus:

$$\% \; CH_4 \text{ in air at equilibrium} = 3.417 \times 10^2 \; Q/P$$

where Q = dissolved methane concentration (mg/l)
P = absolute pressure (kPa).

For a concentration of 5% methane in air by volume, (i.e. equal to the lower explosive limit) and assuming that water containing dissolved methane flows through an unventilated airtight chamber at atmospheric pressure, it can be shown that a 5% mixture would build up if the water contained as little as 1.5 mg/l dissolved methane.

From the above analysis therefore it is likely that safe conditions would not be found in an unventilated chamber. Anywhere that methane-bearing water comes into contact with air in a poorly ventilated space should be considered a potential hazard point and a combination of methane removal and ventilation should be adopted in such circumstances.

GAS MIXING

A knowledge of the mixing behaviour of gases is essential to an understanding of the problems associated with the entry and subsequent possible accumulation of gases in buildings or other confined spaces.

Mixing of an intrusive gas with the ambient air takes place under the actions of diffusion, turbulent jet mixing, buoyancy or turbulent interaction with ventilating air. Of these various processes, gas mixing by molecular diffusion is extremely slow in comparison and can often be ignored. The effect of all mixing processes is to dilute any release of gas progressively as the gas travels away from the point of leakage. Gases once mixed will not separate due to density differences.

The factors which influence the build-up of gas concentrations in an enclosure are: gas density, source characteristics and degree of ventilation. The specific gravity of the gas affects the tendency of the gas to form a layer within an enclosure, either at the roof in the case of a lighter-than-air gas such as methane or at the floor in the case of heavier-than-air gases such as petrol vapour, propane, butane, and some landfill gases. The formation of a layer inhibits mixing and can affect the time taken for an explosive mixture to develop following an ingress of gas.

ASPHYXIATING AND TOXIC BEHAVIOUR OF GASES

Some of the gases that may be encountered in construction are either asphyxiating or toxic. In addition, the presence in air of substantial amounts of an inert gas may give rise to an oxygen-deficient atmosphere.

OXYGEN DEFICIENCY

Oxygen is necessary for the support of life and for the processes of combustion.

The physiological effects of an oxygen deficient atmosphere vary between individuals, level of activity and with length of exposure. A person at rest is not significantly affected until the oxygen concentration falls to 14% and under these conditions it is possible to survive without loss of consciousness. However activity and the inability of oxygen to diffuse into the blood at a sufficiently high rate will result in anoxia. On entering an extremely oxygen deficient atmosphere, a person will collapse in a very short time (40 seconds) without warning or prior distress and if revived will have no recollection that anything happened.

CARBON DIOXIDE

Carbon dioxide may occur in concentrations higher than that of normal air as a result of oxidation and combustion of organic materials and from respiration. It may also be given off from certain types of strata. It is one of the constituent gases given off by fires and explosions, blasting and diesel engines and can also be produced by the action of acid waters on carbonate rocks. A major constituent of landfill and sewage gas, carbon dioxide is noncombustible, has a specific gravity of 1.53 and is very soluble in water, its solubility increasing with pressure. The high density results in CO_2 rich gas mixtures accumulating in low areas.

The main physiological effect of carbon dioxide is a stimulation of the respiratory and central nervous systems. At low to medium concentrations (<10%) the symptoms are mainly increased breathing with progressive headache and exhaustion, but very high concentrations (>15%) may result in loss of consciousness and death. Its high solubility results in rapid diffusion and physiological effects are almost instantaneous.

If elevated carbon dioxide concentrations are accompanied by a reduction in oxygen concentration the effects will be more severe. Generally there is no permanent disabling effect at concentrations up to 10% and exposed individuals recover quickly following the administration of oxygen, although this may be accompanied by nausea and vomiting.

CARBON MONOXIDE

Carbon monoxide is produced by the incomplete combustion of carbonaceous material. It commonly occurs after explosions of flammable gas or coal dust and is present during fires. Carbon monoxide is a common constituent of blasting fumes and may also be generated by poorly maintained diesel engines and overheated air compressors. It is extremely toxic and for this reason it is one of the most feared gases in underground excavations.

Carbon monoxide has a much greater (about 250 times) affinity for haemoglobin than oxygen and forms carboxyhaemoglobin, thus preventing oxygen from being transported in the body leading to a condition of anoxaemia. Although carboxyhaemoglobin is more stable than oxyhaemoglobin, dissociation readily occurs in an atmosphere of oxygen or ordinary air and most of the gas disappears from the bloodstream in a few hours. Red blood cells do not appear to be destroyed by saturation with the gas and readily resume their function: acute carbon monoxide poisoning is thus a reversible process, although many organs, tissue and cells may suffer irreversible damage as a result of oxygen deprivation.

It is important to appreciate that poisoning may not always produce an orderly progression of symptoms, particularly at high concentrations. One cannot rely on symptoms of headache, dizziness or nausea to warn of approaching disaster.

OTHER GASES AND VAPOURS

Chlorinated solvents may be transformed by 'heat', for example by inhalation through a lighted cigarette, to even more toxic species such as phosgene. The range of possible solvents which may be encountered on such sites as waste tips and derelict chemical plant is virtually unlimited, hence the first priority must be the identification of the species present.

In the case of toxic or radioactive gases such as radon the physiological aspects of the gas in question should be considered. The hazard associated with a particular gas may also involve such factors as length of exposure and other contributing aspects such as age, health etc.

UNITS OF MEASUREMENT

CONCENTRATION

Gases will expand to fill the space available, which makes the expression of concentration a confusing issue, exacerbated by the profusion of units in common use. An ideal unit would take account of the effects of temperature and pressure of the gas or mixture of gases. For this purpose the most useful unit is "partial pressure", defined as the pressure each gas would exert if it alone occupied the volume of the mixture at the same temperature. In practice the most useful units are volume percent (sometimes replaced by molar percent) and mass per unit volume (mg/m^3).

In using the latter two units the assumption is made that the gas is at standard temperature and pressure and behaves as an ideal gas with a molar volume of 22.4 l. In practice none of these assumptions is likely to be true but deviations are of little or no practical significance under normal circumstances. Conditions under which the above considerations would be significant would include high pressure working conditions, eg compressed air, or at high altitude.

Volume percent effectively expresses the proportion of molecules of the gas in question in the gas mixture. For low concentrations the unit more commonly used is parts per million (ppm), where:

$$\text{parts per million} = \text{percent} \times 10000$$

Unless otherwise specified it may be assumed, in the case of gases, that parts per million means parts per million by volume.

Units based on mass are related to units based on volume by the molecular weight of the gas thus:

$$\text{Concentration (gm/l)} = \frac{\text{volumetric proportion} \times \text{molecular weight}}{22.4}$$

$$\text{Concentration (mg/l)} = \frac{\text{ppm (volume)} \times \text{molecular weight}}{22.4}$$

A further correction factor of 293/273 is applied to the main concentration where measurements are referred to the more usual temperature of 20°C.

EXPLOSIVE LIMITS

The limits of flammability of gases that combine with oxygen in the air are usually expressed in terms of the lower and upper limits of concentration that will just sustain combustion under the conditions specified. It should be made clear, however, that limits of flammability refer to clearly defined conditions of test and that in any given set of circumstances there may be deviations from expected behaviour.

DETECTION AND MEASUREMENT

Monitoring of hazardous atmospheres, whether in buildings, boreholes or excavations, can be achieved using a variety of methods including hand-held instruments (eg methanometers), reaction tubes, pressurised samples for later analysis by gas chromatography or infra-red spectrography, fixed remote sampling using tube bundle systems and remote transducers with telecommunication facility.

Equipment for use in hazardous atmospheres must be safe under all conditions and the associated power supplies and control units are inevitably heavy, bulky and relatively expensive. In some circumstances it may be possible to adopt an emergency shut-down approach in which 'intrinsically safe' or 'flame proof' detection systems activate circuits when a dangerous condition is detected, triggering shut down of the electrical systems.

Regular maintenance and calibration of all monitoring devices is essential if meaningful results are to be obtained. It is important that personnel using such equipment are made fully aware of its limitations The primary function of a detector is to give warning of conditions that differ significantly from normal. It is axiomatic that data should be treated with a degree of caution and that informed judgement can only be made on the basis of samples taken with the correct equipment by experienced personnel.

Detector instruments may be classified into several broad categories:

(i)   personal monitoring for safety - hand held instruments responding to a specified gas for a given range of hazard/s.

(ii) remote monitoring using fixed monitors for a defined hazard.

(iii) preliminary investigations for unknown processes and hazards.

(iv) full investigation - monitoring of known phenomena using hand-held or remote monitors.

Instruments designed for one task are not necessarily suited to others. Unfortunately, all too often they are used in entirely inappropriate circumstances without consideration of the application and the limitations of the instrument. Instruments used in situations where flammable gases are expected must have been certified as suitable by an appropriate authority such as the British Approvals Service for Electrical Equipment in Flammable Atmospheres (BASEEFA). Flame ionisation detectors (FIDs) should be used with particular care as very few makes are certified for use in hazardous atmospheres. If it is planned to use such instruments in these circumstances it is important that a 'cautious entry' approach be adopted, which entails measurement of the concentration of flammable gases by means of a certified instrument prior to entering or advancing further into a potentially hazardous area.

It is important to recognise the limitations of gas detectors when making investigations. For instance, pellistor (catalytic oxidation) devices used in low-concentration methanometers, (0-5%) effectively measure heat of combustion. The instrument response thus depends on the availability of oxygen. The presence of other interfering gases may also affect the response of such instruments, for example carbon dioxide in landfill gas. Again detector tubes are not always entirely specific and may respond to other gases.

Wherever there is any uncertainty as to the reliability of instrument readings or where the composition of the gas in question is open to doubt, tube samples of the gas should be taken for more reliable laboratory investigation. Whenever possible it is recommended that duplicate samples should be obtained.

EXPOSURE LIMITS

GENERAL

Occupational exposure limits for gases, vapours and airborne particulates have been established by the Health and Safety Executive and are published annually by the Health and Safety Executive (EH40). Limits are given for both long-term and short-term exposure, 8 hour and 10 minute time-weighted averages (TWA) respectively. Values for gases and vapours are quoted as both ppm (by volume) and mg/m^3.

Table 1 in EH40 lists the Maximum Exposure Limits (MELs) for substances falling within Schedule 1 of the COSHH regulations 1988. These limits should be regarded as mandatory and not to be exceeded. The presence of a substance in Table 1 implies a requirement for a programme of monitoring to be established by the employer in accordance with regulation 10.

Table 2 in EH40 lists approved Occupational Exposure Standards (OESs). These are effectively targets rather than mandatory values. Nevertheless, the prudent employer will ensure compliance with these standards if for no other reason that they may be used as criteria in assessing compliance with the Health & Safety at Work Act.

A further set of values, Indicative Limit Values (ILVs) has been published in EEC Directive 91/322/EEC. The substances listed in the Directive all fall within Table 2 of EH40. The values given (in mg/m^3 only) correspond to the values in Table 2 for the 8-hour TWA. Again these values are not mandatory but are advisory levels to be taken into account in setting national levels.

MIXED EXPOSURE

EH40 is equivocal about assessing exposure to mixed gases and vapours, except where the exposure limit is itself set for a mixture. Where mixtures exhibit synergistic effects then the recommended procedure is to sum the individual partial fractional exposure. A total fractional exposure of one or less is then acceptable. For non-synergistic mixtures the recommendation is merely that the appropriate level of exposure should not be exceeded for any of the individual components.

In the real situation, except in the case of the simplest mixtures of the simplest gases, the possibility of synergistic reactions can never be discounted. On sites contaminated with solvents, for example, then there is every likelihood that each would produce similar physiological effects. It would seem sensible in these circumstances to assume the worst and to assess exposure by the sum of the fractional exposures method. There is a good precedent for using this approach where gassing landfills are being investigated or developed.

NON-OCCUPATIONAL EXPOSURE

Those most likely to experience non-occupational exposure are residents on adjacent sites, both during and after development. In developing badly-contaminated sites it is probable that adjacent residents will be at greater risk during rather than after development. There is no simple method for evaluating non-occupational exposure. Indeed, EH40 states that, with the exception of limits for vinyl chloride and cotton dust, "limits cannot readily be extrapolated to give indications associated to non-occupational exposure".

In residential situations an important factor which must be taken into account is the perception of risk. The obvious presence of a pollutant gas, detected by smell, will arouse concern even if the concentration is within an acceptable range. Tables are published of so-called odour thresholds and in many, perhaps most cases, these are below the 8-hour TWA. Unfortunately, odour thresholds are not absolute values; the sensitivity of individuals may vary by a factor of ten or more. Furthermore, some of the most toxic species have little or no odour. Where odour thresholds are very much lower than the 8-hour TWA the probability is that the species has a foul smell and no detectable level will be acceptable.

The answer to the dilemma is perhaps to ensure that the occupational exposure at source falls within acceptable limits and thereafter to rely on dispersion to further protect peripheral residents unless odour nuisance dictates additional control measures,

ACTION LEVELS FOR HAZARDOUS GASES

There is no absolute safe action level for a hazardous gas. This can only be determined in the context of the gas properties, expected flow rates, the nature of the site, ventilation conditions and monitoring and control provisions.

For working coal mines, reporting and action levels are clearly defined within mining legislation (Mines and Quarries Act 1954 and Regulations). However, what constitutes a safe gas concentration in a monitoring borehole on a development site cannot be unequivocally defined. A risk of explosion only arises, for instance, if methane is allowed to accumulate in an enclosed space forming a mixture of at least 5% by volume in air. This in turn will depend on the balance between the inflow rate, the gas composition and the degree of ventilation.

Practical action levels must be set lower than maximum exposure limits for toxic and asphyxiant gases and at concentrations below lower explosive limits for combustible gas mixtures to permit controls to be applied sufficiently early to be effective.

The occupational maximum exposure standards published by the Health and Safety Executive for carbon dioxide are 0.5% and 1.5% for the 8 hour and 10 minute time weighted averages respectively. The minimum acceptable oxygen level in a workplace is 18% but different rules apply in coal mines.

The problems associated with the lateral migration of hazardous gases from landfills have been recognised for some time and evidence of no significant methane migration risk from landfills is generally required prior to approval of a development scheme.

The effectiveness of any landfill gas control scheme is determined by the concentrations of gas measured in monitoring boreholes at the perimeter of the site in which the concentration of flammable gas should not exceed 1% by volume. An action-level of 0.5% for carbon dioxide was proposed by Waste Management Paper 27 but this value has been revised upwards to 1.5% in the second edition of this paper.

Various ground treatment techniques and structural methods have been developed to enable land affected by landfill gas migration to be developed safely. Although the magnitude of the problem and hence the efficacy of a solution, is largely dependent on gas flow rate, most precautions are designed on a gas concentration criteria.

PROTECTION OF INDIVIDUAL BUILDINGS

Draft revisions to the building regulations made reference to the use of a suspended concrete floor with a ventilated void for dwellings where methane concentrations lie between 0.5% and 1%, and the use of a concrete floor slab and membrane on blinded fill when methane is present but in concentrations less than 0.5% The published revision (DOE) retains only the 1% trigger threshold for methane (1.5% and 5.0% for carbon dioxide) but makes reference to guidance available from BRE publications (BRE).

If any concentration of flammable gas is found in a building which could be attributable to a gas leak the regulations require that British Gas should be informed immediately (Gas Safety, Installation and Use, Regulations 1985). Relevant statutory bodies must be informed of occurrences of flammable gas in buildings or confined spaces in excess of 5% of the LEL or carbon dioxide greater than 0.5% by volume. If 1% flammable gas by volume, or 20% LEL, is detected within a building, then that building should be evacuated.

SOURCES AND METHODS OF FORMATION

GAS SOURCES

The more common sources of hazardous gases in construction include:

Strata - gases may be contained within the rock mass either in an absorbed state, filling pores and fissures, or adsorbed onto the internal surfaces of rocks such as coal or shale. Alternatively, volatile liquids may be absorbed within permeable rocks and soils to form a vapour source.

Groundwater - many gases can be found dissolved in groundwater and can be transported and liberated as infiltrating groundwater enters an excavation. Similarly, leachate from landfill sites can also provide a means of transporting dissolved gases, especially methane.

Landfill - the biodegradation of waste materials liberates significant quantities of a wide range of gases including methane, carbon dioxide, hydrogen and nitrogen.

Blasting - the detonation of explosives produces several gases, most of which are toxic. These include carbon monoxide, oxides of nitrogen, sulphur dioxide and carbon dioxide.

Fires and explosions - incomplete combustion produces gases similar to the above but also includes hydrogen and toxic gases associated with the combustion of PVC belting and electric cables, e.g. hydrogen cyanide, phosgene and chlorine.

Internal combustion engines - substantial amounts of dangerous gases are produced by the engines of construction equipment, especially if poorly maintained. Gases include carbon monoxide, oxides of nitrogen, sulphur dioxide, carbon dioxide and aldehydes.

Chemical reactions - the action of acid waters on carbonates, sulphides and cyanides in the strata can liberate hydrogen sulphide, carbon dioxide and cyanide gases.

Bacterial activity - bacteria are widespread in nature and can be encountered in underground situations as well as in landfill operations. Methanogenic bacteria will liberate methane and carbon dioxide whilst sulphate reducing bacteria may produce hydrogen sulphide. In recent times the existence of methane-oxidizing bacteria has been recognised. Aerobic bacteria and other fauna and flora produce carbon dioxide as a result of respiration.

Oxygen-deficient atmospheres - these can be created in several ways:

1. High temperature oxidation in fires, explosions and internal combustion engines.
2. Low temperature oxidation of wood and other carbonaceous materials, and iron pyrite.
3. Respiration.
4. Absorption in acid mine water.
5. Dilution with other gases.
6. Biological oxidation.

Oxygen deficient atmospheres have been associated with compressed air tunnelling. Oxygen may be removed from air travelling through ground containing organic or oxidizable inorganic material. In some instances air has migrated considerable distances (up to half a mile) from a compressed air excavation to appear in an adjacent underground structure. Oxygen deficiency may also result when pressure is reduced; air which has been forced into the surrounding soil and rock, and which may be reduced in oxygen content or contaminated with extraneous gases, will expand and enter the excavation.

GAS MIGRATION AND ACCUMULATION PROCESSES

Hazardous gases can reach a site by many different routes. They may be brought onto site directly by introducing fill containing adsorbed gas, chemical waste or by landfilling with putrescible waste which may release gases during anaerobic decomposition. Alternatively, gases could originate outside the site boundary and then migrate onto the site along high permeability pathways. Gas flow rates will be limited either by the production rate of gas or by the resistance of the pathways through which it is transmitted.

A number of gas-related processes may be active on a particular site and although the primary source of gas may be obvious it is important that no new hazards are introduced as a consequence of any ill-considered activity on or adjacent to the site.

By analogy with hard rock geology there are five basic requirements for the accumulation and retention of gases in near surface strata, namely the presence of a source, a migration mechanism, a reservoir, a seal and a trap configuration. The gaseous source may be geological, industrial waste or domestic waste. Sources to be investigated may include - soils, groundwater, rock fabric, old mineral workings, sewers, contaminated ground, refuse.

GAS MIGRATION PATHWAYS

Migration between source and reservoir may occur as a result of pressure gradients driving fluids through bedding planes, joint networks, intergranular porosity, fracture planes or in fact through any prospective reservoir material that does not trap the fluid. Circulating groundwaters can provide an important mechanism for transporting gases from source to trap in solution. In some instances the body of liquid itself may be considered a gas reservoir.

Examples of the range of possible migration pathways that might be encountered include:

(i) unconsolidated strata - compacted waste, permeable soils and sediments.
(ii) settlement cracks - backfill, spoil, perimeter of landfills, mining-induced subsidence.
(iii) consolidated strata - faults, fractures, joints, bedding planes, blast-induced fractures, subsidence fractures, permeable sandstones.

(iv) underground cavities - limestone caves, old mineral workings, shafts, exploration or old monitoring boreholes.
(v) services - ground adjoining sewers, water pipes, drains, cables, land drains.

GAS RESERVOIRS

Examples of suitable reservoir rocks include porous sandstones, well-jointed limestones and fractured igneous rocks. The suitability of the latter may be enhanced by weathering. Sandstones, consisting of coarse, well-rounded, lightly cemented quartz grains provide the optimum porosity in sedimentary rocks. Within unconsolidated sediments, sandy lenses in muds or clay provide an ideal reservoir, trap and seal configuration.

An essential component of any trap is a seal or cap rock lying on top of the reservoir thus preventing vertical migration and loss of gases directly to the surface. Clay caps on restored landfills fit this criterion. However, restrictions on vertical gas movement can lead to pressure gradients developing which can promote lateral migration.

Soft rock equivalents of fault traps can be encountered in miniature on construction sites where a permeable sediment or infill abuts against an impermeable horizon, the discontinuity being an artefact of historical activity, natural soft sediment slip feature, or a mining subsidence-induced break.

Abandoned coal mines can be considered a special case of free gas reservoir, the source being the continuing but decaying emission of methane from old waste areas. High methane pressures and purities have occasionally been experienced on intersecting old mine workings, even at shallow depths, presumably as a result of hydraulic pressurisation caused by partial flooding in the absence of provision for venting.

In the case of vapours the reservoir may be the liquid vapour source absorbed into permeable materials.

GASES IN SOIL LAYERS

Above the water table, intergranular voids are occupied by moisture and "ground air" (von Pettenkofer,1871) the composition of which differs from that of the atmosphere due, in particular, to biological activity being deficient in oxygen and enriched in carbon dioxide. Gases that have escaped from groundwater may also accumulate.

Where methane is released near the surface, natural methane removal may also be taking place by microbial oxidisation in soils. Such effects have been demonstrated around leaks in natural gas pipes and in soils above and close to landfill sites (Adamse, Hoeks, DeBont and Van Kessel, 1972; Adams and Ellis, 1960; Mancinelli, Shulls and McKay, 1981). Jones and Nedwell (1990) measured methane concentration profiles above a restored landfill site and found that surface emissions only occurred where methane concentrations reached the soil surface. When sampling gases in soils the absence of atmospheric levels of methane (about 1.7 ppm) should not be totally unexpected.

LEGAL ASPECTS AND SAFETY PROCEDURES

GENERAL

Gaseous pollutants have implications for safety for the site investigator, developer, end-user and neighbour alike and therefore merit consideration. It is not the intention of this paper to delve into the legal complexities pertaining to the occurrence, treatment and control of gaseous pollutants on construction sites, however, but merely to outline the framework within which constructors are required to operate.

HEALTH AND SAFETY OF THE CONSTRUCTOR

Statutory protection of the health, safety and welfare of the site investigators and constructors is encapsulated by the provisions of the Health & Safety at Work Act 1974. The various Regulations drawn under the umbrella of the Act include requirements to ensure that reasonable practicable means are taken to prevent workers inhaling dust and fumes, provision of adequate ventilation and procedures for entering confined spaces in which hazardous gases may be present.

The Control of Substances Hazardous to Health (COSHH) is a statutory instrument of the Health & Safety at Work Act which sets out "maximum exposure limits" for certain substances which must not be exceeded together with "occupational exposure standards" which, insofar as is reasonably practicable, must be considered as upper limits. A more detailed discussion of the action levels is given in section 5. The Health & Safety Executive (HSE) publish guidance notes to assist in the interpretation of the various Regulations.

Arguably, one of the most important hazards facing constructors is posed by the accumulation of asphyxiating gases in confined spaces. Guidance on this matter can also be obtained from the HSE.

Precautions relating to the use and storage of LPG on construction sites, another potentially serious hazard, must be taken seriously because of the relatively large volumes of highly inflammable material involved. Consideration should also be given to unusual hazards, for instance where structures are capable of behaving as efficient receiving aerials in proximity to an intense source of electromagnetic radiation and where flammable gases may be present.

The principles of site safety inherent within regulations and recommended procedures require:

    (i)   responsible site management and supervision
    (ii)   adequate initial and refresher training

(iii) effective emergency procedures
(iv) provision of appropriate detection, monitoring and protection equipment
(v) common sense

"HEALTH AND SAFETY OF THE ENVIRONMENT"

Environmental Impact Assessment is an important part of planning procedure within which consideration may need to be given to the effect or interaction of a development with a natural or anthropogenically related gaseous hazard on or near the site. Such assessments are not directed at the health and safety of the constructors but at the welfare of the neighbourhood and also, if an industrial development, at the end-users' workforce.

The provisions of the Environmental Protection Act are relevant to the end-use of the site. If derelict land is being developed then the project may be seen as an enhancement of the environment. Should the renewed site then support a "controlled process", hopefully the provisions of the EPA will prevent the continuation of the "pollution cycle". The philosophy which underpins regulation of the Environment is BATNEEC (Best Available Technology Not Entailing Excessive Cost) which presumably relies on man's incessant drive towards finding better technological solutions to the environmental problems created by new technological developments.

HEALTH AND SAFETY OF THE END-USER

Draft and in-force Building Regulations are aimed at ensuring gaseous hazards discovered during the investigation of a site do not lead to chronic problems within the structures which could create unacceptable risks for the occupants. The concern that the construction industry feels for the hazardous gas problem as it impacts on the end-user is reflected in CIRIA's (Construction Industry Research and Information Association) research programme which is currently addressing the problem of "Methane and Associated Hazards to Construction".

CONCLUSIONS

The circumstances in which gases may be found are many and varied. Any hazardous gas flow, however small, is potentially dangerous if conscious efforts are not made to dilute and disperse it. A wide range of techniques exist which can be used to assess the magnitude of likely gas problems and thus assist in selection of remedial and control measures.

REFERENCES

ADAMS, R.S. & ELLIS, R. (1960). Some physical and chemical changes in soils brought about by natural gas. *Soil Society Proceedings*, 41-44.
ADAMSE, A.D., HOEKS, J., DEBONT, J.A.M. & VAN KESSEL, J.F. (1972). Microbial activities in soils near natural gas leaks. *Archiv fur Mikrobiologie*, **83**, 31-35.

BUILDING RESEARCH ESTABLISHMENT, (1990). BRE Radon: *Guidance on protective measures for new dwellings.*
BUILDING RESEARCH ESTABLISHMENT, (1991). BRE *Construction of new buildings on gas contaminated land.*
COWARD, M.F. & JONES, G.W. (1931). Limits of flammability of gases and vapours. *US Bureau of Mines Bulletin* , **279**.
CREEDY, D.P. (1989). Geological sources of methane in relation to surface and underground hazards. *Methane - Facing the Problems Symposium, Nottingham.*
DEPARTMENT OF THE ENVIRONMENT, (1992). The Building Regulations: "Site preparation and resistance to moisture, Approved Document C".
DEPARTMENT OF THE ENVIRONMENT, (1991). Landfill Gases. *Waste Management Paper*, **27**, HMSO.
HEALTH AND SAFETY EXECUTIVE. Occupational Exposure Limits. Guidance Note EH40, HMSO.
JONES, H.A. & NEDWELL, D.B. (1990). Soil atmosphere concentration profiles and methane emission rates in the restoration covers above landfill sites: equipment and preliminary results. *Waste Management and Research,* **8**, 21-31.
MANCINELLI, R.L., SHULLS, W.A. & MCKAY, C.P. (1981). Methane oxidising bacteria used as an index of soil methane concentration. *Applied and Environmental Microbiology,* **42**, 70-73.
VON PETTENKOFER, M. (1871). Uber den Kohlensaurenegehalt der Grundluft im Gerollboden von Munchen in Verschiedene. *Zeiten. Z. Biol.,* **7**, 395-417.

2-10 HYDROGEN SULPHIDE GENERATION BENEATH A DEEP BASEMENT IN GAULT CLAY

D.F.T. NASH
Department of Civil Engineering, University of Bristol, Bristol BS8.
M.L. LINGS
Department of Civil Engineering, University of Bristol, Bristol BS8.
S.A. JEFFERIS
Golder Associates, 54 Moorbridge Road, Maidenhead, Berks., SL6 8BN.

ABSTRACT

During the final stages of construction of an underground car park founded in Gault Clay, substantial quantities of groundwater were unexpectedly found to be entering the excavation at one location beneath the suspended basement slab. To allow for long term heave of the base of the excavation, the slab had been cast on a collapsible degradable void former and the water was percolating through this. Although the water could readily be collected and pumped away, it was found to be severely contaminated with hydrogen sulphide - an unpleasant and highly poisonous gas.

The source of the water was traced to an ungrouted unrecorded borehole. Chemical analysis showed that the groundwater came from the underlying Lower Greensand aquifer, and on entering the excavation it had a significant concentration of sulphate but was uncontaminated by sulphide. Investigations indicated the presence of sulphate reducing bacteria in the void former, and it is believed that these were reducing the sulphate in the groundwater to produce the hydrogen sulphide. The borehole has now been grouted up and the water entry stemmed.

The paper describes the investigations undertaken to solve this unexpected and unusual problem.

INTRODUCTION

A three level underground car park has recently been constructed in the centre of Cambridge, with a five-storey hotel complex above ground. The car park measures approximately 65m x 45m, and a section through it is shown in Figure 1. The structure is founded on large diameter bored piles, and the 10m deep excavation, which is retained by a 17 m deep perimeter diaphragm wall 0.6m thick, was constructed using top-down methods. The car park floors which prop the diaphragm wall consist of in situ reinforced concrete waffle slabs supported on steel columns connected to the tops of the bored piles.

The lowest floor slab is solid, and was constructed with a 150mm void beneath it to allow for clay heave in the long term. This space was constructed using a proprietary collapsible void former, which comprised

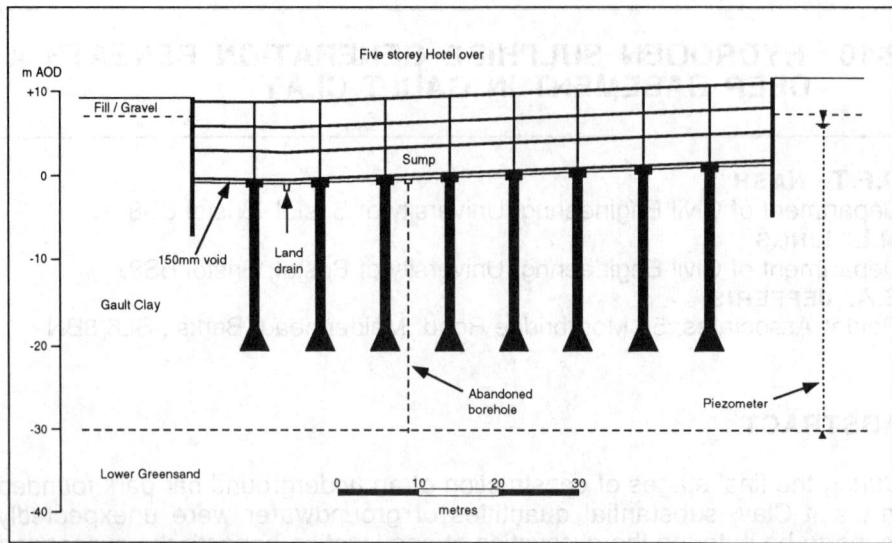

Figure 1: Indicative section.

two layers of hardboard separated by a cellular arrangement of cardboard, and was sufficiently strong to support the weight of wet concrete during construction. The void former was designed so that after subsequent wetting, its structure could collapse to a small residual volume, and it is usual to flood such a void former at the end of construction to accelerate its collapse. In this case flooding of the void former was not carried out until these investigations were completed.

The ground conditions at the site comprise some 3m of made ground and gravel above 38m of Gault Clay which overlies the Lower Greensand. The Gault Clay was laid down under marine conditions during the Cretaceous, and was overlain by up to 400 m of Chalk which has since been eroded; it is thus heavily overconsolidated. It is a stiff to hard silty clay of high plasticity, and is closely fissured and jointed. Within the depth of the excavation, some small pockets of softer structureless clay were noted, perhaps reworked by frost action.

Ground level is at approximately +10m OD. Initially the pore pressures were hydrostatic below a level of about +7m OD, a condition controlled both by water present in the made ground just over the surface of the clay, and by the piezometric level in the Greensand aquifer at depth.

THE PROBLEM

To facilitate drainage, all floors of the car park, and the underlying excavated surface of the Gault Clay, slope across the site as indicated in Figure 1. At the lowest level of the car park there are petrol interceptor pits which are designed to receive surface water run-off from all the car park levels, and collect any ground water entering the space beneath the basement slab. Shortly after construction was completed it was found that an unexpectedly large volume of water was entering the interceptor pits. The water was found to have high sulphide concentrations, which led to the production of substantial quantities of hydrogen sulphide, an unpleasant and highly poisonous gas. On account of the smell, the contractor was required to take the water off site by tanker, incurring considerable inconvenience and expense.

INVESTIGATIONS CARRIED OUT

SOURCE OF WATER

Water entering the interceptor pits can come from two main sources. One is from the surface water drainage system, which collects surface water run-off from all the car park levels. During the later stages of construction, this system collected water arising from various different site activities, and on some occasions this included rain water. The other source is ground water which enters the void beneath the slab and collects in a land drain at the lowest point, and flows via silt pits into the interceptor pit. While small quantities of water were visible weeping through the joints of the diaphragm wall it was not thought likely that this contributed significantly to the water entering the interceptor pits.

In the period soon after the end of car park construction, it had not been anticipated that any ground water would be found entering the excavation. Piezometers within the site still showed that pore water pressures were generally depressed following the major unloading of the ground that had taken place. A very small general seepage of water into the excavation, predicted as the long term steady state condition, was not expected for many years, on account of the low permeability of the Gault Clay.

During an early stage of the investigations, discussions were held with the contractor's site staff in order to obtain a first hand history of the problem. They reported that during construction there had been some water ingress from a "spring" at one location during construction of the lowest slab, and that a sump had been dug to enable the water to be collected and periodically removed by pumping. The sump had been covered over when the slab was cast; its location is indicated on Figure 1.

The investigations then focussed on the continuing water ingress into this sump. Vent holes, which had already been drilled through the slab, were systematically logged, and the presence of any water or smell was recorded. Manhole covers were lifted and a comprehensive set of measurements taken of drainage positions, invert levels and standing water levels. Additional holes were drilled through the slab to locate the sump precisely. It was found that the void former was only wet at positions downslope from the sump. These observations suggested that the majority of ground water entering the excavation was coming from the sump. Water overflowing from the sump was then flowing downhill through the void former beneath the slab, and eventually reached the interceptor pits via the land drainage system.

In order to confirm that the majority of the water was indeed coming from the sump, tests were made during which water was syphoned directly from the sump beneath the slab into the interceptor pits via a hosepipe, thereby preventing it from flowing beneath the slab. Such tests had a threefold purpose: firstly to assess the quantity of ground water entering the sump; secondly to enable a check to be made on whether any other significant sources of ground water existed; and thirdly to assess the water quality.

During the investigations the only access to the sump was through small diameter holes drilled through the basement slab so that it was difficult to assess how the water was coming in. The water could have been entering through a man-made feature, such as an old borehole or the result of an earlier in-situ test, although there were no records of any known holes in the area of ingress. Alternatively it might have been due to some unusual geological feature within the Gault Clay which had led to a zone of higher permeability. A combination of these two possibilities could also have explained the feature.

RATE OF INFLOW

At the time when water from the pits was being taken off site by tanker, the water was first allowed to build up in the pits and in the voids beneath the slab before being pumped right down during a tanker visit. By collating all the information on tanker trips and quantities taken off site, it was concluded that the amount of water removed during this period amounted to an average of around 2,000 litres per day. This water was of course a combination of surface water, arising chiefly from site operations, and ground water.

The syphon tests, in which water was syphoned from the sump and transferred directly into the interceptor pits, enabled a better estimate to be made of the quantity of ground water entering the sump. After there had been continuous flow through the syphon for 19 hours, the quantity was measured at 0.43 litres/min. This was assumed to represent a steady state condition, and implied an average ingress of water into the sump of just over 600 litres/day. During the period of these tests there was no observable flow of water from the land drain into the interceptor pits, so it was concluded that there were no other substantial sources of ground water ingress apart from that found at the sump.

WATER QUALITY

Early in the investigations it was suspected that the void former was playing some part in the generation of the sulphides. The syphon tests showed that the ground water from the sump, when prevented from coming into contact with the void former and transferred instead directly to the pits, was of good quality. It had no unpleasant taste, and gave off no smell even when strongly agitated.

CHEMICAL TESTING OF WATER SAMPLES

Chemical tests were carried out on various samples of water collected during an early site visit. These included samples of ground water taken from a deep piezometer installed outside the site into the top of the underlying Lower Greensand aquifer at a depth of about 40 m, water syphoned from the sump via the hosepipe, water taken from a vent hole in the slab located downslope from the sump, and water taken from the silt pits. In addition tests were carried out on a specimen of Gault Clay taken from beneath the slab. The results of these chemical analyses are given in Table I. Before considering the results at each location in detail it is appropriate to make some general observations.

From the results shown in Table I it can be seen that the sulphate concentrations are in the range 30 to 300 mg/litre of sulphate as SO_3. The water ranks as Class 1 (upper limit is equivalent to 330 mg/litre of SO_3) of BRE Digest 363 (1991), and thus posed minimal threat to structural concrete. The sulphide concentrations are in the range < 0.1 to 22 mg/litre. The highest value could give an equilibrium concentration of hydrogen sulphide in air of order 7000 ppm (by volume) in a confined or unventilated space which is well above the lethal concentration.

The ammoniacal nitrogen contents are in the range < 0.1 to 2.2 mg/litre. For drinking water ammoniacal nitrogen concentrations above 0.02 mg/litre would be questioned, while for sewage the actual concentration might be in the range 10 to 50 mg/litre. The higher site figures suggest some slight contamination from organic matter.

The iron and manganese contents are in the range 0.7 to 6.4 and 0.1 to 1.7 mg/litre respectively, except for the sample from the vent hole which shows much higher values of 148 and 17 mg/litre. The presence of iron and manganese in the waters suggests reducing conditions and a strongly reducing environment.

Sodium, potassium, calcium and magnesium analyses were carried out only on two of the samples and the results are unexceptional for a groundwater. The potassium value is slightly higher than for many waters but not significant.

It should be noted that all the samples contained some suspended solids. The analyses were carried out on the supernatant fluid after a period of settlement. However, there may have been some contribution to the measured values from colloidal matter, for example iron, manganese and ammoniacal nitrogen, which may have influenced the results.

CHEMICAL TEST RESULTS AT EACH LOCATION

With just a single sample from each location it was not appropriate to try to explain the fine detail of the results, but some general observations can be made.

Table I: Results of chemical analyses

Water Samples   Location

(mg/l)	Piezometer	Sump/Hosepipe	Vent Hole	Silt Pit 1	Silt Pit 2
Sulphate	300	250	150	30	40
Sulphide	1.4	< 0.1	22.0	12.7	0.2
Ammon. Nitr	0.1	0.4	2.2	1.6	0.1
Iron	6.4	2.3	148	0.7	1.6
Manganese	1.7	0.2	17	0.3	0.5
Sodium	114	103	-	-	-
Potassium	22	21	-	-	-
Calcium	132	119	-	-	-
Magnesium	25	27	-	-	-

Gault Clay Soil Sample

Total Sulphate	0.42 %
Sulphate in a 2:1 water extract	450 mg/l
Sulphide	1.4 mg/l
Moisture content	23.5 (%)

a) Sump/Hosepipe and Piezometer samples

The results for the sample from the piezometer are unexceptional for groundwater under slightly anaerobic conditions such as may occur in a deep aquifer. The ammoniacal nitrogen content of 0.1 mg/litre would have to be explained if the water were to be used for drinking. The sample of water taken from the sump via the hosepipe was obtained at the start of the first syphon test, after the syphon had been running for about half an hour. This was probably not long enough for the sump water, itself in contact with the void former, to be completely flushed out. The sample was thus a mixture of the water entering the sump with the stagnant water present previously.

The sodium, potassium, calcium and magnesium results for the two samples were very similar suggesting that they came from a common source. The sulphate results for the two waters were also similar, although there was apparently a small reduction of sulphate between the piezometer and the sump. The sulphide results for both samples were low, with that of the piezometer sample slightly higher than for the sump sample. This suggests that hydrogen sulphide was not being generated from the sulphate in the water during its passage through the ground, and that some sulphide had been lost.

Loss of hydrogen sulphide probably occurs as the water arrives at the soil surface and is exposed to air although it is possible that sulphide could be precipitated by reaction with the soil. Loss of sulphide on exposure to air is normal since the concentration of hydrogen sulphide in water when in equilibrium with free air is less than 0.01 mg/litre. Thus in time effectively all hydrogen sulphide is lost from a hydrogen sulphide solution left in free air.

The lower contents of iron and manganese in the sump water suggest that there had been some oxidation of the water, and that as a result iron and manganese had been precipitated. This oxidation was probably the result of exposure to air. The sump water appears to have had a slightly higher ammoniacal nitrogen content than the piezometer water. This was probably due to incomplete flushing out of the sump before the sample was taken from the hosepipe.

b) Vent Hole

This vent hole was located approximately 6m downslope from the sump. The sulphate content of this water was 100 mg/litre lower than that of the sump/hosepipe water. 100 mg/litre of sulphate would produce 40 mg/litre of sulphide on reduction. The sulphide content of the water was 22 mg/litre and thus some sulphur had been lost from the water, presumably as hydrogen sulphide.

The contents of iron and manganese in the water from the vent hole were very high suggesting strongly anaerobic conditions resulting in the enhanced solubility of iron and manganese. The source of the iron and manganese could not be confirmed from the available data. It could have come from re-solution of material previously precipitated from the flowing groundwater or there might have been some selective bacterial concentration. Also there might have been some contribution from the clay at the ground surface or from the void former.

The high ammoniacal nitrogen content appeared to come from the void former - there was a very substantial amount of cardboard debris in the hole.

c) Silt Pits 1 and 2

Water from the land drains is discharged into silt pits before flowing to the interceptor pits. Water taken from silt pit 1 showed a high sulphide content and a low sulphate content. This water had lost 220 g/litre of sulphate relative to the sump/hosepipe water, which is equivalent to 88 mg/litre of sulphide of which only 12.7 mg/litre remained in the water. Thus once again it seems that some hydrogen sulphide had been evolved. It appeared that the water in silt pit 2 had been stagnant for some time, allowing reactions to proceed towards completion and for the hydrogen sulphide content to move towards equilibrium with air (that is effectively towards zero).

The relatively high ammoniacal nitrogen content in pit 1 was probably from the suspension/solution of organic material from the void former, whereas the low ammoniacal nitrogen in pit 2 suggested relatively complete breakdown of the organic matter in the stagnant water.

The moderate contents of iron and manganese in silt pits 1 and 2 are reasonable for water moving towards aerobic conditions.

GAULT CLAY SAMPLE

The total sulphate content of the soil was 0.42%. If all this sulphate were present as soluble sulphate then the concentration of sulphate in the 2:1 water:soil extract would have been 2100 mg/litre. The actual concentration was 450 mg/litre and thus it would seem that some of the sulphate was present as gypsum which is of limited solubility. The sulphate content of the piezometer water was 300 mg/litre and that of the sump/hosepipe sample 250 mg/litre. Assuming that the sump water and the piezometer water both originated in the Greensand, then the sump water had lost rather than gained sulphate on flow to the surface. This suggests that the groundwater was not in equilibrium with the soil sulphate, since equilibrium would have substantially enhanced the concentration. Thus it appeared that the groundwater flow was along a preferred path which had been leached of sulphate.

CONCLUSIONS FROM CHEMICAL TESTS

The chemical tests showed close agreement between the quantities of sodium, potassium, magnesium and calcium found in the piezometer and sump/hosepipe samples, which strongly suggested that the water appearing in the sump originated from the underlying Lower Greensand aquifer. This water had a high sulphate and low sulphide concentration, but by the time the water had reached the land drainage system, the sulphate content had reduced and the sulphide content increased.

GAS GENERATION

There are various areas beneath the basement slab such as beneath pile caps and interceptor pits, where water can collect and stagnate, there being nowhere for it to drain away to. Any ground water that enters the excavation collects there and remains permanently in contact with the void former. This is thought to have resulted in the generation of gases by a number of possible mechanisms.

The chemical tests provided clear evidence that sulphate in the groundwater entering the site was being reduced to sulphide, and hence hydrogen sulphide gas. Based on a flow rate from the sump of 600 litres per day of water with sulphate content of 300 mg/litre, then the maximum quantity of hydrogen sulphide that could be reduced from the water is 77 grams/day. The most likely mechanism for this reduction process was the presence of sulphate reducing bacteria growing and feeding on the wet void former. In certain cases samples of wet cardboard recovered from the void former beneath the slab showed black spotting, which helped to confirm this view. Other possible mechanisms for the production of hydrogen sulphide gas were considered, for example from the iron sulphide known to be present within the Gault Clay, but it was concluded that the bacterial mechanism was the dominant effect.

It was noticeable that the smell was at its worst when water was pumped after being allowed to stand for a long time. Under the pumping regime originally adopted, sulphate-rich water was remaining in contact with the void former for long periods of time, causing large quantities of sulphide to be produced. Without adequate ventilation, this had allowed high concentrations of sulphide to build up in the water. When this water was eventually pumped out, hydrogen sulphide gas was released into free air, assisted by the agitation process.

With such a large quantity of wet cardboard beneath the slab, it seemed inevitable that methane would also be produced. Although no direct evidence was found for the presence of methane on site, it was assumed that it could be present in the future. The quantities of methane produced could be significantly greater than the quantities of hydrogen sulphide. Ultimately all methane production will cease when all available carbon etc. has been removed from the void former, and the quantity of methane that may be evolved is related to the quantity of organic matter beneath the floor slab. The period over which methane generation will continue cannot reliably be predicted but could be very many years.

REMEDIAL MEASURES

Once the source of the water and gas had been established, various measures were considered for reducing or eliminating the problem. These included :

a) attempting to stop the inflow of water at the sump, for example by drilling and grouting,

b) attempting to kill the bacteria which were causing the problem,

c) regularly pumping the water away with the hope that concentrations of sulphide arriving at the interceptor pits might be reduced to acceptably low levels, and

d) diverting the water directly from the sump into the pits so that the water would then no longer come into contact with the void former, and would not have a chance to become contaminated.

It was initially proposed that the water be diverted, and at that stage the concrete slab was broken out so that further investigations of the sump could be carried out. These eventually revealed the presence of what appeared to be an old borehole approximately 125mm in diameter from which the water was emerging. It was then decided to attempt to grout this up to stem the flow, despite the presence of a 6 m artesian head of water. This work has now been carried out, and it is understood that the flow of water has ceased and that sulphide rich water is no longer flowing into the interceptor pits.

In addition permanent vents have been installed through the slab to allow any gases generated to escape. The car park ventilation system is operated whenever the car park is in use, and can safely remove any gas.

CONCLUSIONS

The investigations reported here revealed an unusual example of the interplay of microbiology and groundwater chemistry on a construction project. The main features may be summarised as follows :

Groundwater was found to be entering the void beneath the basement slab. The water was coming up from the sub-artesian Lower Greensand aquifer below the Gault Clay via an abandoned ungrouted borehole and was found to have a high sulphate content. The quantity of water entering at this point was measured at over 600 litres per day. By the time the water reached the interceptor pits via the land drainage system the water was found to have high sulphide concentrations, which led to the production of significant quantities of hydrogen sulphide, an unpleasant and highly poisonous gas.

The chemical tests provided clear evidence that sulphate in the groundwater entering the site was being reduced to sulphide, and hence hydrogen sulphide gas. The most likely mechanism for this reduction process was the presence of sulphate reducing bacteria growing and feeding on the wet void former, and there were clear signs of black spotting on samples of the wet cardboard which support this hypothesis. It was considered that hydrogen sulphide generation would continue as long as there was a source of sulphur and organic carbon.

It was originally proposed that the sulphate-rich water be diverted in order that it would not flow through the void former, so as to reduce the hydrogen sulphide generation substantially. However it proved possible to eliminate the ingress of water by grouting the borehole.

It is anticipated that there could be some residual production of hydrogen sulphide from the clay or void former, and that bacterial activity could also result in the generation of methane. A permanent ventilation system has been installed to eliminate any hazard arising from gas generation.

REFERENCES

BUILDING RESEARCH ESTABLISHMENT (1991). *BRE Digest No 363 : Sulphate and acid resistance of concrete in the ground.*

SESSION 3 - TRANSPORT

3-1	Influence of groundwater chemistry and motion on highway construction materials *W.J. French*	199
3-2	Physico-chemical aspects of soluble salt damage to thin bituminous road surfacing *B. Obika, R.J. Freer-Hewish & D. Newill*	211
3-3	Generation of acid groundwater beneath City Road, London *N.S. Robins, D.G. Kinniburgh & M.J. Bird*	225
3-4	Biological stability of geosynthetics in critical earth structures *D.G. Bright*	233
3-5	The physical and chemical characteristics of geotextiles and their effects on in situ degradation *P.R. Rankilor*	253
3-6	The physical environments of geotextiles in civil engineering structures *P.R. Rankilor*	277
3-7	Evaluation of soil chemistry and its role in highway embankment erosion - a case study *R.K. Srivastava, A.V. Jalota, R.P. Tiwari, T. Nath & A.K. Sahu*	283
3-8	The influence of calcite on the index properties of Triassic mudrocks *J. Collins & A.B. Hawkins*	293
3-9	Contribution of sulphides and carbonates to the weathering of sedimentary rocks *T. Oyama, M. Chigira & T. Sidahara*	309
3-10	Strength characteristics of root systems in mudstone *J-W. Chen & D-H. Lee*	321
3-11	Engineering properties of mudstone-lime-slag mixtures *D-H. Lee & Y-M. Tien*	329

3-1 INFLUENCE OF GROUNDWATER CHEMISTRY AND MOTION ON HIGHWAY CONSTRUCTION MATERIALS

W.J. FRENCH
Geomaterials Research Services, 1 Falcon Park, Crompton Close, Basildon, Essex.

ABSTRACT

Damage to road substrates, bases and surface materials, whether soil, fill, concrete or bitumin-based, can be rapidly and seriously affected by chemical reactions with ground water. The principal influences of water reflect the presence and abundance of water, its composition in terms of hydrogen ions, oxygen and dissolved ionic species, its rate of flow and the porosity and other properties of the constructional materials and substrate.

Concrete is damaged by pure water through the solution of calcium hydroxide and the conversion of the hydrates into calcium hydroxide and calcium sulpho-aluminate. With the introduction of CO_2 or other acidifying compounds, the rate of solution is greatly enhanced and surface concrete and aggregate may be dissolved to a depth of some tens of millimetres. Sewage water can be particularly aggressive. Compounds in the water may lead to substantial chemical reactions including the well known sulphate attack, formation of magnesium silicate hydrates and formation of calcium chloroaluminate. Cement and lime stabilized soils may be profoundly affected in that the stabilization produces calcium silicate aluminate hydrates which, if sulphate ions are available, are subject to rapid conversion into calcium silicate aluminate sulphate hydrates in the presence of abundant water, with enormous expansion.

Salt crystallization through evaporation of the carrier leads to expansion of rock and concrete with the development of cracks and disintegration of rock and cement paste. This process is particularly damaging in arid regions, but can occur from time to time in less aggressive environments. In arid regions damage due to salt crystallization causes substantial deterioration of the substrate, road base, and road surfacing materials. The effect is particularly damaging where rising ground water occurs, since the process and deposits migrate with the rise in the ground water.

INTRODUCTION

Factors influencing the magnitude of potential chemical damage to highway constructions include the following:

(i) The abundance and changes in abundance of water in contact with the construction in all its aspects.

(ii) The composition of the water in terms of hydrogen ions, oxygen, and other dissolved species.

(iii) The rate of flow of water and changes in flow rate.

(iv) The porosity size distribution of the constructional materials and the potential for chromatographic separation (French, 1976 and 1978).

Ground water and infiltrating surface water reach equilibrium with many rocks and soils fairly rapidly in engineering terms with respect to a number of important constituents so that the magnitude of the damage to be expected will vary substantially with the potential for continual water transference in the vicinity of construction. Motion of water, even the relatively slow transference of moisture through the capillary fringe by evaporation or transpiration will substantially affect the rate of deterioration and chemical change in the constructional zone. Both growth of crystalline phases and solution of crystalline phases are accelerated by increased flow rates. Berner (1968) has given equations appropriate to crystal growth for calcite from flowing ground water. Slow flow rates of, say, up to 3 m per year have little effect on the rate of growth, but as the flow rate is increased by each order of magnitude, the rate of growth (or solution of crystalline phases) is doubled. James and Lupton (1978) have reviewed the kinetics of the solution of crystals and illustrated the contrasts that occur between the development of the solution front in fissure flow and in pore water flow and have shown how rates of solution can be derived. The magnitude of damage must therefore be seen in terms of the life expectancy of the structure in combination with potential rates of moisture transference.

The hydrogen ion concentration of the solution is obviously significant for the deterioration of many soils, rocks, and constructional materials although the indication of potential aggressiveness is usually rather crude. For concrete, various standards have been erected (e.g. see Bartholomew, 1979 and Harrison, 1977, 1987). The following provides an illustration of standard conditions specified for particular pH value ranges:

Aggressiveness	German Standard	BRE Digest 174	French Standard
Negligible to slight	>6.5	6.0-9.0	5.5-6.5
Moderate	5.5-6.5	3.5-6.0	4.5-5.5
High	4.5-5.5	3.5-6.0	4.0-4.5
Very high	<4.5	<3.5	<4.0

Such values may also be appropriate for soils containing carbonates, but for constructional materials based on bitumen, the table is possibly the inverse of that for concrete and the most aggressive conditions require pH values in excess of 10.

The availability of oxygen is important in effecting decomposition of soil and rock minerals, especially sulphides. Growth of particular solid phases is also encouraged by availability of oxygen and the now traditional Eh-pH diagrams such as those given by Garrels and Christ (1965) provide a basis for the interpretation of soil mineral stabilities, in particular aqueous environments and allow prediction of the damaging species likely to be present in various Eh-pH regimes.

Other ions in solution accentuate or reduce aggressiveness in particular circumstances and may cause physical damage by crystallization or solution, or by specific chemical reactions. The effects of sea water as ground water and more highly saline ground water contrast with those of terrestrial water of low salinity.

Prevention of chemical damage requires elimination of water or at least prevention of flow and evaporative or transpirational loss and prevention of direct contact between the constructional materials and the water.

Principal aspects of damage to constructional materials for highways can be considered as follows:

(i) The influence of water on soils and rock substrate and fill.

(ii) The influence of water on buried surfaces such as drainage and other pipes including flow within the pipes.

(iii) The effects of water and dissolved ionic species on concrete and related constructional materials.

(iv) The effect of water and dissolved ionic species on bituminous binders.

Interaction of these various aspects is also important in that for example changes may occur in concrete as a result of the solution or reaction of phases in fill material. Since reviews covering the full breadth of this subject and dealing with the differing water types and the range of constructional materials would be considerably long, an arbitrary selection has been made.

DAMAGE TO CONCRETE

The principal effects of ground water attack on portland cement concrete and allied materials are as follows:

(i) Leaching of cement paste compounds.

(ii) Reactions of the cement paste with components in the water.

(iii) Alteration of the cement hydrates by addition of water.

(iv) Expansive chemical reactions such as those seen in sulphate attack and alkali-aggregate reaction.

(v) Physical damage through the formation of salt crystals within the paste and sometimes aggregate particles.

LEACHING

Damage to major concrete structures by leaching and the principles involved have been reviewed in detail by Mason (1989), see Smith (1984). Leaching occurs simply through the passage of neutral water through concrete, along cracks and joints. Calcium hydroxide, which makes some 25% of cement paste and often occurs in zones on aggregate surfaces and may be precipitated in voids and cracks, is soluble in water to about 1.2 g per litre. It is therefore readily leached from concrete where pure water can percolate through capillary pores or along cracks. The models of James and Lupton (1978) may be relevant to this dissolution but the deterioration is exacerbated by the breakdown of the chemical equilibrium between the calcium hydroxide and the calcium silicate hydrates. The result is that further breakdown of the calcium silicate hydrates occurs, leading to the formation of silicates of low Ca/Si ratio. These form plate-like crystals of little cementitious value. The most obvious evidence of this process occurs in the formation of deposits of calcium hydroxide or calcium carbonate on the surface of concrete, in surrounding fill or soils, and in cracks and voids in the concrete. The surface of the concrete also becomes etched and aggregate exposed. First the surface laitance of the concrete, then the paste with fine aggregate and finally coarse aggregate may be loosened on the surface of the concrete. The rate of deterioration depends mainly on flow rate, which may accelerate as damage progresses and on the composition and density of the concrete. In rare instances complete disintegration of the concrete can result, simply from the movement of what is in effect clean water and this is the most common cause of minor to moderate damage of concrete used beneath pavements, in pavements, and in bridge decks.

Along with the removal of portlandite there is often formation of abundant ettringite and a pseudo-sulphate attack results. In this reaction the sulphate is entirely derived from the cement, but the effect is such that very substantial quantities of ettringite can be produced which can represent some 10% by volume of the cement paste. This effect can also occur where portland cement is used to stabilise clay soils.

LEACHING BY DILUTE ACIDS

If the water is acidified then the leaching process can obviously become extremely aggressive. This process is of such importance that a remarkable body of research has been developed defining standard quantities for the use of concrete in particular aqueous environments and in assessing the potential degree of aggressivity. Ronben (1975) measured the rate of attack on concrete pipes at various values of pH showing that the depth of corrosion increased by an order of magnitude with a reduction of pH of some units. Aggressive CO_2, making for reduction in pH, i.e. that in excess of the CO_2 required to stabilize calcium bicarbonate, has been defined in diverse ways - so many in fact, that it may be doubted whether any can be relied upon entirely. Such definitions have been developed for several decades and some of the earlier data remain of value. Lea (1970), for example, gives the following guide:

Aggressiveness	Aggressive CO_2 parts/100 000	Temporary hardness $CaCO_3$ parts/100 000
Very low	<15	>35
Slight	15 - 40	>35
Slight	<15	3.5 - 35
Marked	40 - 90	>35
Marked	15 - 40	3.5 - 35
Marked	<15	<3.5
Very high	>90	>35
Very high	>40	3.5 - 35
Very high	>15	<3.5

Lea also provides data leading to a corrosion evaluation chart which has been modified and developed by various authorities over the years. Other useful charts have been produced subsequently such as that by Baronio & Berra (1986). Locker & Sprung (1975) investigated the variation in depth of attack on concrete with pH and concrete compositional variations. Their conclusions were as might be expected, limestone aggregate leads to eight times the rate of attack compared with quartz-based aggregate: a W/C ratio of 0.7 is three times as bad as a W/C ratio of 0.5; portland cements are three times worse than portland blast furnace slag cements, and high cement content is substantially worse than low to moderate cement content. Several indices of aggression of water containing CO_2 have also been derived over the years with again earlier indices retaining considerable value. The well known Langelier index and the effectively identical driving force index are widely used, and relate closely to the Ryznor index defined in 1944. An aggressiveness index defined as $pH + log10[Ca^{2+}][Alk]$ has been defined where the brackets signify concentrations in mg per litre. Waters with this index of less than 10 are considered very aggressive. The potential for solution or precipitation has been reviewed in detail by Rossum and Merrill (1983), who use the driving force index, pH, and alkalinity of the solution in assessing the aggressiveness.

Acidity produced from decomposition of vegetation and oxidation of sulphides is as effective as that induced by solution of CO_2 and both are effective in causing a substantial solution of surface layers of concrete. Very serious damage also commonly occurs from the oxidation of hydrogen sulphide in sewers and concrete in the vicinity of leaking sewers may be seriously damaged by solution. The presence of CO_2 in water may lead to the precipitation of vaterite and calcite in soils that have been stabilized with lime or portland cement. If this occurs, it could lead to considerable shrinkage and failure to develop full strength in the stabilization process. This has been considered by Bagonza et al (1987) and by Sampson et al (1987). The primary process investigated by these authors relates to the prevention of carbonation of stabilization products by atmospheric CO_2

rather than dissolved CO_2. The carbonation of stabilization products through dissolved CO_2 is much less likely to occur and it seems more likely that the principal source of damage from this source will be the expansive formation of thaumasite. Further work is clearly required in this area.

CHEMICAL ATTACK ON CONCRETE

Sulphates

Possibly the most common source of damage to concrete from ground water is sulphate attack and this has been well known and described in detail over decades (BRE Digests 174 and 250 and for example Harrison and Teychenne (1981)). The effect of the presence of sulphate in ground water is however strongly dependent on the quality of the concrete. At any given sulphate level, damage is maximized if the concrete is porous and generally of low density. Concrete below ground that has some 15% of void and is highly porous has been found to be thoroughly deteriorated throughout in a matter of a few years where dense concrete in the same environment has altered for only a few millimetres in the same time.

Particularly damaging is the local release of sulphates from fill or soil substrate to concrete roads and other pavements. Certain forms of pyrite (but not all), pyrrhotite, and marcasite, may be oxidized with the liberation of sulphate ions. Reaction of these ions with the cement paste hydrates leads to the formation of gypsum and subsequently ettringite and possibly thaumasite. Heave is generated that leads to serious cracking. Another source of this form of deterioration derives from the presence of brick rubble in the substrate, which provides sodium sulphate in solution in the ground water.

In arid areas the ground water commonly has both high sulphate and chloride. There is often alteration of surface layers of concrete which become very friable. If the water is derived from sea water, then Mg ions are likely to be important in addition to the sulphate and chloride. Magnesium ions penetrate into the concrete to form a layer of magnesium silicate hydrates a few millimetres from the surface. This material has little cementitious value and leads to a breakdown of surface layers. Sulphate damage is most prevalent in the under surfaces of pavements and along joints and cracks. Damage can be mitigated in some cases by the use of a sulphate resisting portland cement or a slag cement, but consideration needs to be given to the possible damage from chlorides which may be exacerbated by the use of a cement with low alumina.

Chloride

Penetration of chloride ions into concrete often leads to the development of a zone of altered paste some 10 to 30 mm from the surface. This zone is characteristically white in reflected light and is dark and lacks the normal portland cement structure when seen in thin section. Penetration of the chloride leads to some weakening of the paste and there may be spalling of surface layers. The layer appears to contain chloroaluminates which may form as well shaped crystals in cracks and on some aggregate surfaces. Needles of calcium chloroaluminate have been found in damaged concrete from below ground in arid regions where the occurrence of the chloroaluminate is very similar to that of ettringite and where the material can be confused with ettringite unless chemical analysis is carried out. This damage is of course additional to the profound damage caused by corrosion of steel in the concrete.

ALTERATION OF STABILIZED SOILS

Stabilization of clay soils using lime or cement is also substantially affected by the presence of sulphates. The stabilization is typically dominated by reaction between calcium hydroxide and smectite with the production of complex calcium silicate aluminate hydrates. The presence of sulphates is probably innocuous unless sulphate ions are present in solution when the wet (normally saturated) environment leads to the formation of calcium silicate aluminate sulphate hydrates of needle-like habit. The density of these phases is very low (c. 1700 kg/m^3) and their needle-like form leads to rapid and sometimes extreme expansion with the development of substantial amounts of space in the soil. If the water contains carbonate ions or the material is later carbonated then crystalline forms similar to thaumasite may be produced. Examples of this process have been described over some years from various regions. The process may go through two stages with sulphate ions being produced initially through the oxidation of iron sulphides in clay soils.

CRYSTALLIZATION DAMAGE

It is common to find crystals of gypsum, and particularly sodium chloride, grown in the surface layers of concrete. This growth has been generated in concrete placed in experimental sabkha systems at Queen Mary & Westfield College. The crystallization occurs towards the top of the capillary fringe and in the almost dry zone immediately above the fringe. The crystallization causes spalling of surface layers and a general disruption of the outer few centimetres of the concrete. Particularly serious damage has been found where saline waters migrate through concrete to precipitate out in the surface layers of the concrete. In some circumstances up to thicknesses of more than 100 mm of the concrete have been virtually totally destroyed by this crystallization process. The process of transference of salts into and through concrete has been reviewed by French (1978). The process also takes place in aggregate particles and porous coarse aggregate may act to transmit the crystallization process through concrete. This process is described in more detail in dealing with salt precipitation in aggregates in bituminous roads.

BITUMINOUS STRUCTURES

CHEMICAL DETERIORATION

Deterioration of bituminous materials by ground water is mainly brought about through the deterioration of substrate and aggregate materials rather than direct alteration of the organic binder. However in arid regions the rise of salts into the bituminous layers can cause deterioration of the binder itself. The effect noticed is that the binder becomes dry and powdery and loses its adhesive qualities. Bitumen pieces caught up in concrete typically remain stable, but where the paste has high alkalies and water is abundant the bituminous material tends to become soluble and dispersed. Bituminous rocks occurring in aggregate lead to staining of cement paste and the surface of concrete. These processes occur because bituminous materials are partially soluble in alkaline solutions and in rare cases where ground water may be alkaline, perhaps through contamination, the chemical effects may be expected to lead to severe damage to bituminous binders.

SALT PRECIPITATION

A significant amount of literature has been published concerning salt damage to bituminous roads from the presence of salt in solution in the ground water. This is mostly related to construction in arid areas, though a few examples occur in temperate regions where very high salinity is derived from contamination of the ground water or from surface addition of salts for de-icing purposes. Papers providing an introduction to this type of deterioration include Fookes (1976), Fookes & French (1977), Obika et al (1989), French et al (1982), Cooke (1981), Netterberg (1984, 1979, 1974, 1970) and Sperling & Cooke (1985). The damage that occurs relates essentially to salt crystallization processes and to the presence of saline ground water. An index of the aggressiveness of such water has been given by Doornkamp et al (1980).

Moisture containing salt can migrate through soils and rocks of almost any type (French 1978). Experiments have shown for example that sandstones and limestone blocks encased in impermeable plastics allow moisture to rise through them by capillary transfer for a metre or more in a matter of a few months. When evaporation takes place, salt crystallization produces a crack transverse to the direction of transport of the moisture. Continued precipitation of salts causes the rock to become swollen and barrel shaped and to develop cracks parallel to vertical surfaces. These cracks contain abundant salt crystals. Where the crystals are sodium chloride these usually have a fibrous form with the fibres grown with their length normal to the crack surfaces. The salt crystallization is controlled by the environmental humidity and crystals can be both resorbed and reprecipitated over daily cycles. In the field such salt crystals have been observed to form within road base and road surface materials variously, depending on local conditions. They may disappear overnight only to reappear next day.

While the rate of sodium chloride crystallization and solution can be extremely rapid, the solution and precipitation of gypsum is slower and typically needs months for substantial crystals to grow. In addition, the gypsum tends to form at a higher humidity or moisture content than the sodium chloride so that it often forms below the location at which sodium chloride would be precipitated. Crusts of desert rose have been observed to form on piles of sand where the growth produced represented about 300 mm of a mixture of gypsum crystals some 100 mm in length embedded in sand where the gypsum made some 80% of the mixture. This layer grew over a period of 18 months to two years. Alternate layers of gypsum and algal material have been found developed on the sands of reclaimed ground from sea water where the period required for the development of these materials was two to three years.

Changes in ground water level whether progressive or alternating can lead to solution and redeposition of salt crystals with the site of crystallization moving from level to level in the road structure. Rising ground water has led to the severe deterioration of bituminous roads. Initially the water may be sufficiently deep for abundant crystallization to occur in the substrate or immediately below the road surface. With time, the site of precipitation rises into the surface layers and subsequent rises may cause saturation of the surface and removal of the salts. The result is virtually complete disintegration of the road surface. Further rise in the ground water level may sometimes have been accompanied by the development of intermittent surface flow leading to destruction of the road surface by erosional processes.

The most common forms of damage occur where the road surface materials are close to the top of the capillary fringe. Here intermittent salt precipitation occurs within the road base and surface layers. Precipitation is greatly affected by the quality of the bituminous surface layer and is much reduced where this layer is essentially impermeable. However, the presence of only a few pieces of porous rock in the surface layer or relatively small areas of lower quality can lead to progressive disruption of the better material. Some of the features observed through the precipitation of salt in bituminous materials and in fill are disintegration of porous rocks and the accumulation of lenticular masses of salt crystals, which cause blistering or localized heave of the road surface.

Experiments have shown that the transmission of ground water can take place through slightly porous rocks, even where the rocks are in point contact only. The rate of transmission of the moisture is scarcely affected by the space between the rock particles and rate of transmission is governed mainly by the porosity of the rock particles. It can therefore be difficult to design structures that will prevent the movement of the saline ground water by capillarity and various experiments have been carried out in order to evaluate the qualities of geotextiles in preventing the movement of salts (French et al, 1982; Obika et al, 1989). Some of these experiments have now been running for almost a decade. They have shown suitably designed geotextiles to be effective in preventing salt movement and incidentally have provided a good deal of information on the rate of damage of concrete and aggregate in rock materials where these are embedded in different parts of the ground water regime. The ability of salt crystallization, without chemical attack, to cause disruption of rock particles, rock masses, and even well made concrete, is plain in these experiments. A wide range of soluble phases are reported to have caused damage. The soluble salts are most commonly sodium chloride and calcium sulphate, although magnesium and the various sodium sulphates are also important. Nitrates and borates have occasionally been reported to have caused distress.

The influence of the environment and location of the road is particularly important. Where the constructional materials lie above the top of the capillary fringe, damage is minor with only slight transference through the vapour phase or along grain surfaces. Where the capillary fringe occasionally reaches the constructional material, abundant salts are precipitated, aggregates may be decomposed, the bitumen develops small closely spaced cracks and becomes detached from the aggregate. There may be salt crystallization on the road surface and exposed aggregate often has salt crust upon it. Small depressions in the road surface or potholes may have salt crystallization developed within them from time to time. Where the road materials are essentially within the capillary fringe, but where the saturation may be intermittent, there is usually rapid salt decomposition of the road materials with surface potholes being abundant, aggregate being decomposed and the bituminous binder being converted into sand. Where the road materials are saturated in moisture the extent of damage to fill or bituminous materials tends to be less, though of course there is a greater risk of physical damage from water movement.

POLYMORPHIC TRANSITION

The presence of carbonate in sub-base or substrates where ground water is changing may lead to volume changes in the soil. Changes from gypsum to anhydrite or calcite to aragonite and their reversals need to be considered. Aragonite, for example, may be produced and remain metastable when the ground water is saline or sea water. Here it may co-exist with high-Mg calcite (Berner 1971). Changing the environment to fresh water may cause both phases to convert to low magnesium calcite. This represents an increase in volume for the solid bases but because of the possibility of higher porosity in the materials containing aragonite, the conversion is more likely to cause settlement. This process may occur over a period of a few years.

REFERENCES

BAGONZA, S.E., PECTE, J.M., FREER-HEWISH, R. & NEWILL, D. (1987). Carbonation of stabilized soil-cement and soil-lime mixtures *in Proceedings of Seminar H, PTRC Transport and Planning Seminar Annual Meeting, University of Bath, London PTRC, Education and Research Services*, 29-48.

BARTHOLOMEW, R.F. (1979). *The protection of concrete piles in aggressive ground conditions: An international appreciation.* London.

BARONIO, B. & BERRA, M. 1986. Concrete deterioration with the formation of thaumasite - analysis of the causes. *Il Cemento*, **B83 (3)**, 169 - 184.

BERNER, R.A. (1968). Rate of concretion growth. *Geochimica et cosmochimica acta*, **32**, 477-483.

BERNER, R.A. (1971). *Principles of chemical sedimentology.* McGraw-Hill.

COOKE, R.U. (1981). Salt weathering in deserts. *Proceedings of the Geologists' Association*, **92**, 1-16.

DOORNKAMP, J.C., BRUNSDEN, D. & JONES D.KC. (1980). *Geology, Geomorphology, and Pedology of Bahrain.* Geo abstracts, Norwich UK

FOOKES, P. G. (1976). Road geotechnics in hot deserts. *Journal of the Institution of Highway Engineers*, **23 (10)**.

FOOKES, P.G. & FRENCH, W.J. (1977). Soluble salt damage to surfaced roads in the Middle East. *Journal of the Institution of Highway Engineers*, **24 (12)**.

FOOKES, P.G., FRENCH, W.J. & RICE, M.M. (1985). The influence of ground and ground water geochemistry on construction in the Middle East. *Quarterly Journal of Engineering Geology*, **18 (2)**, 101-128.

FRENCH, W.J. (1976). The role of solvent migration in alkali-silica reactivity. In: *The effects of alkalies on the properties of concrete*. Cement and Concrete Association, 177-193.

FRENCH, W.J. (1978). The migration and precipitation of water soluble ions in concrete. In: *Effects of alkalies in cement and concrete*, Perdue USA, 47-67.

FRENCH, W.J., POOLE, A.B., RAVENSCROFT, P. & KHIABANI, M. (1982). Results of preliminary experiments on the influence of fabrics on the migration of ground water-soluble minerals in the capillary fringe. *Quarterly Journal of Engineering Geology*, **15**, 187-199.

GARRELS, R.M. & CHRIST, C.L. (1965). *Solutions, Minerals, and Equilibria*, Harper, New York.

HARRISON, W.H. (1987). Durability of concrete in acidic soils and waters. *Concrete*, **21 (2)**, 18-24.

HARRISON, W.H. (1977). Chemical resistance of concrete. *BRE publication CPZ3/77*, pp 3.

HARRISON, W.H. & TEYCHENNE, (1981). Sulphate resistance of varied concrete. *Second interim report on long term investigation at Northwick Park. BRE publication HMSO*, pp 171.

JAMES, A.N. & LUPTON, A.R.R. (1978). Gypsum and anhydrite in foundations of hydraulic structures. *Geotechnique*, **28**, 249-272.

LEA, F.M. (1970). *The chemistry of cement and concrete*. Arnold

LOCHER, F.W. & SPRUNG, S. (1975). The resistance of concrete to aggressive carbonic acid. *Beton Technische Berichte, Forschungsinstitut der Zementindustrie*, 91-104.

MASON, E. (1989). Exposure of dam concrete to special aggressive waters. *Bulletin of the International Commission of Large Dams*, **71**, 1-177.

NETTERBERG, F. (1970). Occurrence and testing for deleterious salts in road construction materials with particular reference to calcretes. *Symposium on soil and earth structures in arid climates, Adelaide*. 87-92.

NETTERBERG, F., BLIGHT, G.E., THERON, P.F. & MARAIS, G.P. (1974). Salt damage to roads with bases of crusher run, Witwatersrand quartzite. *Proceedings of the Second African Conference on Asphalt Pavements for Southern Africa, Durban*, 34-53.

NETTERBERG, F. (1979). Salt damage to roads - an interim guide to its diagnosis, prevention and repair. *IMIESA, Official Organ. of the Institute of Municipal Engineers of Southern Africa*, **4 (9)**.

NETTERBERG, F. (1984). Salt damage to some pavements in Zimbabwe. *Proceedings of the Eighth Regional Conference for African Soil Mechanics and Foundation Engineering, Harari*, **1**, 311-319.

NETTERBERG, F., PAIGE-GREEN, P., MEHRING, K. & VON-SOLMS, C. L. (1987). Prevention of surface carbonation of lime and cement stabilized pavement layers by more appropriate curing techniques. *Proceedings of the Annual Transportation Convention (ATC 1987), Pretoria, Paper No. 4A/X,* pp 45.

OBIKA, B., FREER-HEWISH, R.J., & FOOKES, P.G. (1985). Soluble salt damage to thin bituminous roads and runway surfaces. *Quarterly Journal of Engineering Geology,* **22**, 59-73.

RONBEN, C. (1979). Testing the chemical resistance of concrete to non-corrosive acid attack. *Quality Control of Concrete Structures - RILEM Symposium, Swedish Cement and Concrete Research Institute, Stockholm,* **1**, 267-274.

ROSSUM, J.R. & MERRILL, D.T. (1983). An evaluation of the calcium carbonate saturation indexes. *Research and Technology Journal AWWA,* 95-100.

SAMPSON, L.R., NETTERBERG, F. & PROLANA, S.F. (1987). A full-scale road experiment to evaluate the efficiency of bituminous membrane for the prevention of in-service carbonation of lime and cement-stabilized pavement layers. *Proceedings of the Annual Transportation Convention (ATC 1987), Pretoria, Paper No. 4A/1X.,* pp 23.

SMITH, M.A. (1984). Identification and assessment of ground conditions aggressive to concrete. *Concrete in the Ground - The Concrete Society,* 13-31.

SPERLING J. & COOKE, R.U. (1985). Laboratory simulation of rock weathering by salt crystallization and hydrating in hot arid environments. *Earth Surface Processes and Landforms,* **10**, 542-555.

3-2 PHYSICO-CHEMICAL ASPECTS OF SOLUBLE SALT DAMAGE TO THIN BITUMINOUS ROAD SURFACING

B. OBIKA
Sir Owen Williams and Partners Ltd., Consulting Engineers.
R.J. FREER-HEWISH
Civil Engineering, University of Birmingham.
D. NEWILL
Overseas Unit, Transport and Road Research Laboratory.

ABSTRACT

Soluble salt contamination of highway construction materials occurs in climatic zones where evaporation exceeds precipitation. This results in an upward migration of moisture to the surface, where salts present in solution are precipitated. These climatic zones cover large areas of Australia, Africa, The Middle East and North and South America.

Existing guidelines and recommendations for road design and construction in saline areas are based mainly on experience of local materials and conditions without a full understanding of the damage mechanism. This has resulted in conflicting advice on how to prevent salt damage.

A laboratory simulation approach to understand the salt damage process has been reported. This paper is concerned with an appreciation of the geochemical processes, which are important for current laboratory and field testing programmes.

INTRODUCTION

Damage to bituminous surfaced roads and runways in arid, semi arid and warm coastal environments has, in certain cases, been attributed to the action of soluble salts. The existing literature on the subject does not explain the fundamental mechanisms, however, and this has led to varying advice on how to prevent salt damage. There is a need to appraise the existing recommendations for the design and construction of roads and runways in saline environments, based on a better understanding of the salt damage process.

Soluble salt damage to bituminous surfacings is caused by upward migration of salts under an evaporation gradient. At or near the surface the salts in solution become supersaturated and crystallise. This can cause a physical degradation of the bituminous surfacing, often in the form of blistering, doming or general powdering/disintegration (Plate 1). Bituminous prime coats and surface dressings are particularly susceptible to salt damage.

(a)

(b)

Plate 1: a) Doming of cap seal coat, Sua Pan Airstrip, Botswana,
b) Blistering of a bituminous cutback prime.

Salt damage will only occur in climatic zones where evaporation exceeds precipitation. Under such conditions, moisture is drawn to the surface where soluble salts present in solution are precipitated. These climatic zones cover large areas of Australia, Africa, The Middle East and North and South America. A detailed review of the extent of occurrence of salt damage and existing recommendations in these areas is given in Obika, Freer-Hewish and Fookes (1989).

The work reported in this paper is part of an ongoing study at the University of Birmingham initially sponsored by the British Overseas Development Administration (ODA) following the occurrence of salt damage to the bituminous surfacing of a runway in the West Indies. The work has included extensive laboratory climatic cabinet simulation and field studies in the West Indies and Chile. Experimental road sections have also been constructed in Botswana under a two year research programme sponsored by the Overseas Unit, Transport and Road Research Laboratory. The experimental road sections are currently being monitored and will be reported in a later paper.

This paper describes the pertinent aspects of salt solubility, movement, crystallisation and crystal pressures which can lead to a better understanding of the salt damage mechanism.

EXISTING RECOMMENDATIONS

The few published papers on salt damage deal with local environments and materials and this has resulted in a variety of recommendations for damage prevention which are often conflicting. Early published reports in Australia (Cole and Lewis 1960) and in South Africa (Weinert and Clauss 1967) suggested upper limits of sodium chloride (NaCl) content and sulphate content (as SO_3) of 0.2% and 0.05% respectively. Subsequent work (Netterberg et al 1974; Fookes and French 1977, Januszke and Booth 1984, Obika et al 1989) have shown that these limits are probably not applicable universally. In addition to imposing maximum salt limits the recommendations for salt damage prevention also include: rolling the bituminous surface, preferred use of bituminous emulsion binders rather than bituminous cutbacks for prime coats and minimising time duration between base construction and surfacing. The relative effectiveness of these measures is discussed in Obika et al (1989).

Horta (1985) provided the first detailed examination of the damage mechanisms and suggested possible approaches to developing preventative measures. He described the occurrence of salt damage in various parts of North Africa and noted the presence of NaCl filamentous crystals in many damaged surfacings. This work also drew attention to some critical crystal growth factors and the need for a rational approach to understanding the damage mechanism.

SALT DAMAGE MECHANISMS AND CRYSTAL PRESSURES

Clearly, the development of methods of prevention and repair of salt damaged pavements should be based on consideration of the fundamental damage mechanism. Evaporation, salt migration, solubility, crystallisation and crystal pressures are discussed below.

EVAPORATION AND MIGRATION OF DELETERIOUS SALTS

Salt migration occurs by capillary movement of saline solution due primarily to evaporation. The effect of evaporation is to provide a suction gradient which is greater than the opposing gravitational gradient encouraging the movement of moisture towards the pavement surface.

There are two stages in the drying process (Ward 1975). The first is characterised by a constant rate of evaporation governed largely by climatic and pavement surface conditions and the second by a declining rate of evaporation governed mainly by the ability of the pavement layers to transmit moisture to the evaporating surface. Thus at first the suction gradient increases as the pavement becomes drier, but at the same time the moisture films through which water movement occurs become thinner and fewer in number. The rate of moisture movement therefore, decreases as the pavement dries. This reduced moisture movement and reduced evaporation is further encouraged by the decreasing moisture gradient. In addition, if salt crystals are precipitated at the surface they will retain moisture at the pavement surface and reduce the moisture gradient further.

Thus the most rapid increase in salt content at the pavement surface may occur soon after pavement construction, followed by a second stage of declining rate. The time required to reach this second stage will depend on the initial salt content of the material, the moisture condition of the pavement and the climatic conditions.

A typical relationship between salt content at the unsealed pavement surface and time is shown in Figure 1. This has been developed from laboratory climatic cabinet simulations of a West Indian climate using crushed West Indian oolitic limestone. As noted above, the relationship may vary but it is important to recognise that rapid upward salt migration can occur within 48 hours after construction or ground excavation. Design parameters based on pre-construction or initial salt content determinations may not always be appropriate.

SOLUBILITY AND CRYSTALLISATION OF DELETERIOUS SALTS

Only those salts which are soluble in water can migrate to the surface of the pavement. The solubility of a given salt is related to its crystallisation thresholds. The solubility of sodium chloride increases only slightly with increase in temperature, whereas sodium sulphate, magnesium sulphate and sodium carbonate show a rapid initial increase with temperature. At typical ambient temperatures of $30°C$, common in hot climates, the solubility of the salts is near its maximum.

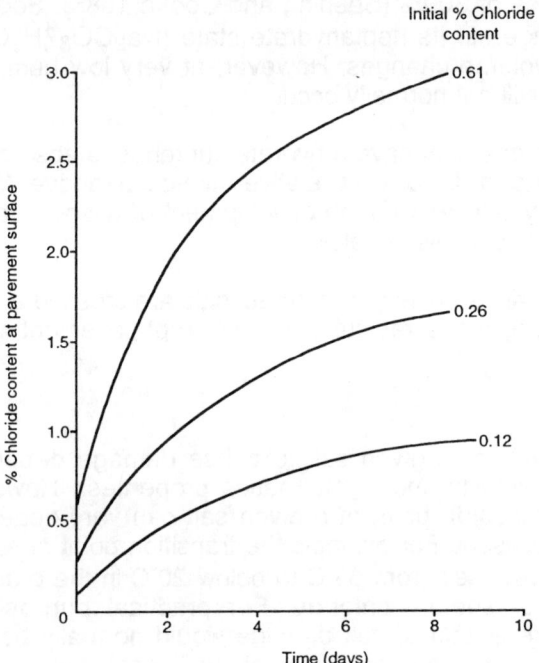

Figure 1: Relationship between initial chloride content and chloride content at pavement surface with time.

Salt crystallisation evolves through three stages:
- attainment of supersaturation;
- formation of crystal nuclei;
- growth of crystals.

When a solution contains more dissolved salts than the equilibrium concentration it is said to be supersaturated. Supersaturation is the most important variable influencing the magnitude of disruptive pressures for a given crystal. It can be induced in a number of ways including cooling, evaporation of solvent and reduction of the solubility of one salt by another.

SALT CRYSTAL TRANSITION POINTS

Certain salts, such as sodium sulphate and sodium carbonate, have hydrates whilst others such as sodium chloride do not. A hydrate is a salt which contains water of crystallisation. Hydration can occur when the temperature is at or below the transition point of a given salt. At the transition point there is a reversible change from an anhydrous salt to a hydrate or from a lower to a higher hydrate. For a given temperature of hydration, at or below the transition point, there is also a critical ambient relative humidity which must be exceeded if hydration is to take place.

In the presence of excess moisture (r.h. >60%) sodium sulphate (Na_2SO_4) forms a decahydrate ($Na_2SO_4 10H_2O$) at 32°C with an associated volume change in excess of 300% (Sperling and Cooke 1985). Sodium carbonate (Na_2CO_3) changes to its heptahydrate state ($Na_2CO_3 7H_2O$) at 31°C with similar crystal volume changes. However, at very low humidities, such as 30%, hydration will not normally occur.

Sodium chloride does not have a hydrate but tends to absorb moisture from the atmosphere and dissolve at relative humidities above 76%. Below this humidity it recrystallises with the development of associated crystallisation pressures. This is discussed below.

In many hot dry lands these crystal thresholds are crossed at least once in a single day creating pressures sufficient to disrupt pavement surfacings.

Salt Mixtures

The relative ability of a given salt to cause damage depends to a large extent on its solubility and crystallisation properties. However, both the solubility and transition point of a given salt can vary depending on what other salts are present. For example the transition point of sodium sulphate decahydrate is reduced from 32°C to below 20°C in the presence of a low concentration of sodium chloride. For practical purposes the design approach for prevention of salt damage would normally be based on the dominant salt but an awareness of the above is essential.

Sodium Chloride Threshold

The presence of sodium chloride crystals, known as halite, has been reported in several cases of salt damage. It is also the most widely occurring natural salt and requires particular attention.

Increasing dissolved salt concentration decreases the chemical activity of H_2O in a solution and as a result lowers the equilibrium water vapour pressure. This arises because when salts are added to water, some fraction of the water molecules are structurally bound to the salt ions, reducing the chemical activity of H_2O. The extent of lowering of the equilibrium water vapour pressure depends on the particular type of dissolved salt present. For NaCl solutions, the vapour pressure lowering is approximately 23% at normal ambient temperature (Kinsman 1926). Thus in a concentrated NaCl solution such as a brine, the chemical activity of H_2O is 0.76 (the chemical activity of H_2O in pure water is unity) and the equilibrium vapour is achieved at a relative humidity of 76% as opposed to 100% vapour pressure. Thus for the NaCl solution to be evaporated, relative humidities less than 76% must prevail in the overlying atmosphere.

It follows, therefore that NaCl crystals, can only be precipitated and accumulated in those areas where the mean relative humidity of the atmosphere is less than 76%. Mean relative humidities of many low latitude coastal and arid regions, commonly fluctuate between 70 and 80% (Kinsman 1926). At all relative humidities above 76%, there will be a net flux of water vapour from the atmosphere into the salt solution. At all relative humidities below 76% evaporation of the solution and precipitation of halite will occur.

Halite may crystallise and redissolve at least once in a single day creating high heaving pressures at the pavement surface.

In the sabkhas of Abu Dhabi, Trucial Coasts and Persian Gulf, unsealed roads which are dry at mid-day, frequently become wet at dawn as a result of moisture attracted by halites present on the unsealed road surface.

It is concluded that in climatic zones where the relative humidity remains high (above 76%) all year round, salt damage due to halites is unlikely as crystals will not precipitate from capillary solution in substantial quantities. Where the daily fluctuations of relative humidity cross the 76% threshold, damage due to halites should be anticipated.

Filamentous Crystals

The occurrence of filamentous crystals (whiskers) in cases of salt damaged roads and runways has been documented (Horta 1985, Cole and Lewis 1960). Detailed studies of damaged field and laboratory surfacings using a scanning electron microscope (Obika and Freer-Hewish 1988) have shown that whiskers of sodium chloride salts were responsible for the damage to bituminous surfacings.

Whiskers, often referred to as filamentous or fibrous are high super-saturation, elongated crystals with higher than normal strength and crystallisation pressures. There are several known properties and features of this type of crystal which are fundamental in understanding the mechanism and prevention of salt damage. For example it has been shown that sodium chloride(NaCl) whiskers will grow preferentially in substrates of finer porosity. Thus base courses with high fines content are particularly susceptible to salt heave. Horta (1985) noted that the Old Adrar Airport in Algeria with a base course high in clay content, showed early heave caused by NaCl whisker growths. It is known that the growth of whiskers can be inhibited by the presence of low supersaturation crystals of the same chemical composition or by impurities (Nabarro and Jackson 1958); for example sodium chloride whiskers will not normally grow from solutions when sodium chloride cubic crystals are also present.

SALT CRYSTAL PRESSURES

Opinions differ on the relative importance of stresses produced during salt crystallisation, growth, hydration and thermal expansion on the breakdown of materials. Evans (1970) has compiled a review of salt crystallisation relevant to salt weathering of rocks and lists the various arguments concerning the origin and nature of salt crystal pressures. Theoretical and observed pressures indicate crystal growth and hydration may be important (Correns 1949, Wellman and Wilson 1965, Winkler 1975).

Crystal Growth Pressures

The pressure (P) exerted by crystal growth can be expressed as (Correns 1949):

$$P = \frac{RT}{V_s} \log c/c_s \quad [1]$$

where R is the gas constant of the ideal gas law, T is the temperature in degrees Kelvin, V_s is the volume of the solid salt, c is the actual concentration of solute during crystallisation and c_s is the concentration of the solute at saturation.

The major factor influencing the magnitude of crystal growth pressure is the level of supersaturation c/c_s. Pressures in excess of 500 atmospheres can develop at low supersaturation (Obika and Freer-Hewish 1988).

Hydration Pressures

The hydration pressure (P_h) is given by (Winkler 1975):

$$P_h = \frac{(nRT)}{V_h - V_a} \times 2.3 \log \frac{P_w}{P_{w1}} \quad [2]$$

where n is the number of moles of water gained during hydration, V_h is the volume of hydrates in cubic centimetres per gram-mole, V_a is the volume of the original salt before hydration in cubic centimetres per gram-mole, P_w is the vapour pressure of water in millimetres of mercury at a given temperature and P_{w1} is the vapour pressure of hydrated salt in similar units to P_w.

The rate of hydration is also important in determining the overall destructive effect of a particular type of salt. The hydration of sodium sulphate heptahydrate to the decahydrate may repeat several times in a single day (Wellman and Wilson 1965). The hydration takes about 20 minutes. The ability of magnesium sulphate to cause more damage than sodium sulphate in laboratory testing is attributed to the faster hydration rate of magnesium sulphate.

Thermal Expansion Pressures

Cooke and Smalley (1968) suggested that thermal expansion of salts within rock pores may create pressures sufficient to cause rock weathering. These pressures arise because of the lower thermal coefficient of expansion of rock minerals when compared with that of certain salt crystals. The importance of pressures due to the above mechanism has not been investigated in any detail and further work is required.

Very few measurements of crystal growth pressures have been recorded. The earliest known observations are those of Becker and Day (Evans 1970). The authors noted that alum and $CaSO_3$ crystals growing between two plates of glass can lift one kilogram (kg) through several tenths of a millimetre. Taber (1916) observed that growth of loaded crystals was slower than unloaded crystals. However, the growth of the former could be increased by raising the level of supersaturation. Mosebach (1951) measured pressures in excess of 47 atmospheres by growing copper sulphate ($CuSO_4 5H_2O$) crystals. The author also demonstrated that different crystal faces have different abilities to grow against pressure. For example, the (010) face of potassium chromate ($K_2Cr_2O_7$) was found to grow against a pressure twice the limit for the (001) face. Correns (1949) also found that certain faces of the alum crystal did not show any growth whereas other faces grew against pressures up to 42 atmospheres.

MECHANICS OF CRYSTALLISATION AT BITUMEN - BASE INTERFACES

Correns (1949) compared observed and calculated pressures using equation 1 and found significant variations at high supersaturation which he explained in terms of the phase boundary tensions. Thus if the phase boundary tension between crystal (a) and solution (b) is σ_{ab}, that between the substance of the plates (c) (under and above the crystal) and the solution (b) is σ_{bc} and that between crystal (a) and substance (c) is σ_{ac} then the following relation must hold for the crystal to grow:

$$\sigma_{ac} > \sigma_{bc} + \sigma_{ab} \qquad [3]$$

Correns found that although the crystal grew against pressures when sandwiched between two glasses, the same crystal (with identical face orientation) did not grow when sandwiched between two mica plates. In the latter case the author assumed $\sigma_{ac} = \sigma_{bc} + \sigma_{ab}$. The surface tension theory may be extended to the case of salt crystallisation at the bitumen-base interface of a road pavement, see Figure 2. If the bitumen and the soil form the upper and lower plates respectively, the phase boundary tensions can be expressed as follows:

Between: bitumen (b) and crystal (c) σ_{bc}
          bitumen and solution (s) σ_{bs}
          crystal and solution σ_{cs}
          crystal and soil (p) σ_{cp}
          solution and soil σ_{sp}

In order that the crystal might grow the following relation must prevail:

$$\sigma_{bc} + \sigma_{cp} > \sigma_{bs} + \sigma_{cs} + \sigma_{sp} \qquad [4]$$

The above relationship demonstrates the importance of the chemical properties of the pavement materials. For example, any additive or inherent property of the bitumen or the soil which is likely to increase the phase boundary tension between the materials and the salt crystal will tend to

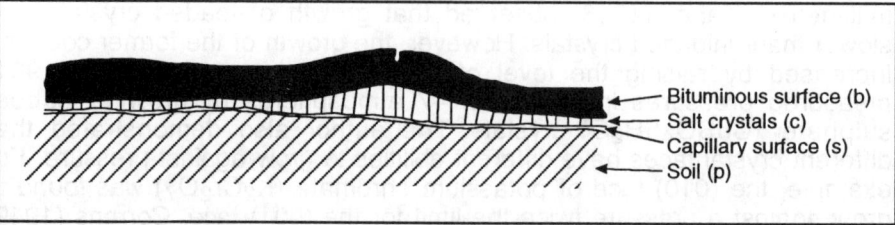

Figure 2: Salt crystallisation at bitumen-base interface.

increase the potential for the salt crystal to grow. Additives may also be used to reduce the boundary tensions involved, thereby reducing the potential for crystal growth.

INFLUENCE OF BASE POROSITY ON CRYSTALLISATION AT BITUMEN-BASE INTERFACES

For materials of equal strength, those with large pores separated from each other by microporous regions will be the most susceptible to salt weathering (Wellman and Wilson 1965).

The work required to be done during crystal growth on one face of a crystal is equal to $(P_1 - P_s)dV$; where P_1 is the pressure in the solution, P_s is the solid and dV is the increase in volume. This must equal the work done in extending the surface, σdA, where σ is the boundary surface tension between the crystal face and its saturated solution and dA is the increment of area.

Thus
$$P_1 - P_s = \sigma \frac{dA}{dV}$$

Consider salt crystals growing at the bitumen-base interface where the micropores formed by the interface are larger than those in the base. The crystals will grow first at the interface as a result of evaporation. The growth continues until the interface pores are filled. As $P_1 - P_s = \sigma dA/dV$, for the crystals to extend by growth into the smaller pores of the base would involve a high increase in area (dA) relative to volume (dV). This would require an unduly large amount of work to be done. The crystal will therefore continue to grow at the bitumen-base interface, until the pressure is sufficient to heave or rupture the bituminous surfacing. The above mechanism indicates that heaving pressures will be greater in a base with high fines content and this is consistent with field observations.

The mechanism whereby salt crystals grow at the bitumen-base interface rather than the top (exposed side) of the bitumen requires some consideration. A possible explanation is that the bituminous layer may act as a semi-permeable membrane. Upon reaching the bitumen-base interface, moisture only is allowed to pass through the microporous bitumen generally in the vapour phase.

An alternative explanation, however, may be that nucleation is favoured at the interface due to the existence of crystals accumulated on the base prior to bitumen application. These crystals continue to grow until the interface pores are filled. As a large chemical potential would be required for the crystals to extend into the small micropores of the bitumen, the crystals continue to grow at the bitumen-base interface causing an upward heave of the bituminous layer.

Crystallisation under applied load

By considering the relation between pressure solution and the force of crystallisation, Weyl (1959) developed an expression for the pressures involved during crystallisation. Thus, if the level of supersaturation and the stress coefficient of solubility are known, the pressure against which a given crystal will grow can be determined.

The author reasoned that the phenomenon of pressure solution and the force of crystallisation is the result of removal or deposition of mineral matter in the region of contact between two mineral grains. The force of crystallisation is therefore the antithesis of pressure solution. In developing the theory, it is assumed that a film of solution exists, a few atoms thick, between two crystal grains through which diffusion occurs. Evidence for the existence of this film is given by Correns (1949), Weyl (1959) and Evans (1970).

The rate of diffusion through this film depends on the grain size, the effective normal stress (total normal stress minus hydrostatic pressure) between the grains, the diffusion constant in the solution film, the film thickness and the stress coefficient of solubility; the latter being related to the tendency of crystals to dissolve under stress.

Weyl (1959) demonstrated that for a given set of parameters, as the average effective stress is increased, the force of crystallisation occurs by growth with a hollow centre. Further increases in effective stress decreases the rate of crystal growth as the radius of the hollow centre decreases until it disappears. As the effective stress is further increased, the rate of crystal growth continues to decrease until the stress across the film is constant and there is, therefore, no diffusion gradient. A further increase in effective stress changes the direction of solute diffusion and pressure solution occurs.

The force against which crystallisation can occur increases with the degree of supersaturation and with decreasing stress coefficient of solubility. Weyl gives a practical example; calcite has a fractional change of solubility with hydrostatic pressure of approximately 10^{-3} per atmosphere, therefore at 1% supersaturation, calcite can be expected to crystallise against a force of 10 atmospheres. If supersaturation is higher it will crystallise against a proportionately higher force.

THE EFFECT OF CLAY ON CRYSTALLATION PRESSURES

If a clay film is present within the solution film bordering the two crystal grains discussed above, the effect will be to increase the force of crystallisation. The clay film consists of a series of platelets with associated water films. These water films provide significantly increased rates of solute diffusion which result in increased rates of crystallisation.

THE HALITE DAMAGE PROCESS

The foregoing may be related to the halite damage process. The damage process is initiated by the accumulation of cubic halite crystals on the surface of the base before the bituminous surfacing is applied. Under these conditions evaporation is moderate but continuous. The capillary solution reaching the surface is never allowed to reach the high levels of supersaturation as cubic halite crystals nucleate and crystallise out instantaneously on reaching the surface. The cubic crystals have comparatively low surface free energy, exert lower pressures and form at lower levels of supersaturation. Cubic crystals are the stable form of halites.

Although when the bituminous surfacing is applied the evaporation rate is immediately reduced, a moisture gradient still exists through the pavement and further migration of saline moisture to the surface occurs. The moisture is unable to escape rapidly and the salt crust on top of the base may be partially redissolved. The resulting salt solution is now at a higher level of supersaturation. The high supersaturation combined with the confined volume between the base and surfacing provide conditions favourable to filamentous crystal growth. The filamentous crystals (or whiskers) have high surface free energy and exert higher crystal pressures. Evaporation still occurs but at a much slower rate through the bituminous surfacing, aided by drying shrinkage cracks. The filamentous crystals nucleate and grow perpendicular to the surface of the base between the bitumen and the base. As it is energetically less feasible (Wellman and Wilson 1965) for the crystal to extend into the micropores of the base or bitumen, it continues to grow at the interface, lifting the bituminous surfacing by exerting pressures along the crystal longitudinal axis.

Further growth of filamentous crystals takes place slowly as more salt migrates to the surface. Where the surfacing cracks, increased growth occurs as evaporation accelerates, drawing more salt to the surface. Early trafficking is often beneficial as this development of domes and blisters created by the salt crystals is suppressed by tyre pressures. This prevents increased evaporation and hence crystallisation.

CONCLUSIONS

Certain climatic conditions are essential for salt damage of bituminous surfacings to occur. Damage is most likely where there is a large diurnal change in temperature and relative humidity and where evaporation is generally high. In such environments repeated crossing of the crystal formation thresholds may lead to the development of destructive pressures at the pavement surface. The pressures developed by salt crystallisation can be high and depend to a large extent on the level of supersaturation of the saline solution.

Where contaminated pavement material contains more than one type of salt the crystallisation characteristics are difficult to determine. Existing recommended salt limits require careful interpretation in such cases and a small scale laboratory or field simulation (Obika and Freer-Hewish 1988) may be useful in predicting whether damage may occur.

An improved understanding of the heaving mechanism caused by salt crystallisation between bituminous surfacing layers and base layers is assisting in the development of guidelines for preventative measures to avoid salt damage for wide ranging conditions.

ACKNOWLEDGEMENTS

The study described in this paper was sponsored by the British Overseas Development Administration (ODA). The paper is also published by permission of the Director of Transport and Road Research Laboratory, UK. The authors are grateful for information provided by Wallace Evans and Partners and the facilities provided by Sir Owen Williams and Partners Geotechnical Limited.

REFERENCES

COLE, D.C.H. & LEWIS, J.G. (1960). Progress report on the effect of soluble salts on stability of compacted soils. *Proceedings of the 3rd Australia - New Zealand Conference Soil Mechanics and Foundation Engineering, Sydney, Australia*, 29-31.

COOKE, R.U. & SMALLEY, I.J. (1968). Salt weathering in deserts. *Nature*, **220**, 1226-7.

CORRENS, C.W. (1949). Growth and dissolution of crystals under linear pressure. *Discussion of the Faraday Society*, **5**, 267-271.

EVANS, L.S. (1970). Salt crystallisation and rock weathering: a review. *Revue Geomorphologie, Dyn*, **19**, 153-177.

HORTA, J.C. & DE, O.S. (1985). Salt heaving in the Sahara. *Geotechnique*, **35(3)**, 329-337.

JANUSZKE, R.M. & BOOTH, E.H.S. (1984). Soluble salt damage to sprayed seals on the Stuart Highway. *ARRB Proceedings*, **Part 3, 12**. 18-30.

KINSMAN, D.J.J. (1926). Evaporites: relative humidity control of primary mineral facies. *Journal of Sedimentary Petrology*, **46(2)**, 273-279.

MOSEBACH, R. (1951). Neue Ergebnisse zur frage des Wachstums von Kristallen unter einseitigem Druck. *Fortschritte de Mineralogie*, **29 & 30**, 25-33.

NABARRO, F.R.N. & JACKSON, P.J. (1958). Growth of crystal whiskers, a review in growth and perfection of crystals. *In:* R.H.Doremus, B.W. Roberts, D. Tunbull (eds.) *Growth and perfection of crystals*, Wiley, New York; 14-91.

NETTERBERG, F., BLIGHT, G.E., THERON, P.F. & MARAIS, G.P. (1974). Salt damage to roads with bases of crusher-run Witwatersrand quartzite. *Proceedings of the 2nd Conference on Asphalt Pavements for Southern Africa, Durban, August*, 34-53.

OBIKA, B. & FREER-HEWISH, R.J. (1988). Study of salt damage of bituminous surfaces for highway and airfield pavements. *Final Contract Report to Overseas Development Administration, U.K.*

OBIKA, B., FREER-HEWISH, R.J. & FOOKES, P.G. (1989). Soluble salt damage to thin bituminous road and runway surfaces. *Quarterly Journal of Engineering Geology*, **22**, 59-73.

SPERLING, C.H.B. & COOKE, R.U. (1985). Laboratory simulation of rock weathering by salt crystallisation and hydration processes in hot arid environments. *Earth Surface Processes and Landforms*, **10**, 541-555.

TABER, J. (1916). The origin of veins of the asbestiform minerals. *National Academy of Science, Proceedings*, **2**, 659-664.

WARD, R.C. (1975). *Principles of Hydrology*. McGraw Hill, London.

WEINERT, H.H. & CLAUSS, M.A. (1967). Soluble salts in road foundations. *Proceedings of the 4th Regional Conference for Africa on Soil Mechanics and Foundation Engineering, Cape Town*; 213-218.

WELLMAN, H.W. & WILSON, A.T. (1965). Salt weathering: a neglected agent in coastal and arid environments. *Nature*, **205**, 1097-1098.

WEYL, P.K. (1959). Pressure solution and the force of crystallisation: a phenomenological theory. *Journal of Geophysical Research*, **64(11)**, 1098-2001.

WINKLER, E.M. (1975). *Stone properties, durability in man's environment*. Springer-Verlag, New York.

3-3 GENERATION OF ACID GROUNDWATER BENEATH CITY ROAD, LONDON

N.S. ROBINS, D.G. KINNIBURGH & M.J. BIRD
Hydrogeology Group, British Geological Survey, Wallingford, Oxfordshire, OX10 8BB.

ABSTRACT

Corrosion of railway tunnel linings in the London Underground beneath City Road is associated with acid groundwater. The groundwater is contained in a 3 to 4 m deep, 60 m wide sand channel situated between the London Clay and the underlying Woolwich and Reading Beds. The water typically has a pH of less than 3, a sulphate concentration greater than 100,000 mg/l and metal (cobalt, nickel, zinc and aluminium) concentrations greater than 100 mg/l. The most likely source of the acid is the oxidation of pyrite. The production of acid will probably continue as long as water drains from the London Clay above, oxygen is supplied via the railway tunnels and pyrite is available within the sand. Core analysis indicates that there is sufficient pyrite remaining in the strata to maintain this reaction for many years.

INTRODUCTION

Routine inspection of the London Underground Northern Line tunnels in September 1962 revealed water ingress south of Old Street station. Reports of water in the tunnel had also been made in 1945 and in 1958. The new ingress occurred only eight months after major track renewals and an investigation concluded that seepage of highly acidic water had caused serious deterioration of the tunnel linings. The origin of the corrosive water was attributed to both natural and anthropogenic sources (Follenfant, 1975).

Remedial work was carried out in December 1962. Caustic soda (sodium hydroxide) was injected behind the linings to neutralise the acid. The cause of the tunnel deterioration and the origin of the acid water was not determined. Subsequent investigations suggested that the tunnels pass through a sand channel trending NNE-SSW in which acid water collects (Rainey and Rosenbaum, 1989), but during tunnel widening in 1922/24 the sand was found to be dry and loose.

Hydrogeological investigations have been undertaken to attempt to delineate the extent of the sand and to identify the source of the acid water (Bird et al, 1989). The work comprised a detailed sampling and analytical programme on material recovered from purpose-drilled boreholes. Further tunnel repairs have now been successfully completed.

GEOLOGY

The generalised geological sequence in the study area is Chalk, Thanet Beds, Woolwich and Reading Beds and London Clay with river terrace gravels at the surface.

Sands occur at the top of the Woolwich and Reading Beds in the vicinity of City Road (Lake, 1987). Borehole records show that the sands fine upwards in the manner of a channel deposit. The channel, which trends NNE - SSW, is of marine or coastal origin despite the presence of woody debris and some leaves. The presence of frequent burrows in the finer grained sediments indicate a brackish or marine environment similar to present day coastal tidal channel and bank complexes along the North Sea coast of Germany. There, tidal channels fill with migrating sand and pass laterally into laminated intertidal and shallow subtidal clays.

In terms of the regional history of the London basin the channel is situated on the northern fringe of a complex of tidal channels, lagoons and sand banks. These resulted from a rise in sea level, the precursor to the London Clay transgression.

EXPLORATORY DRILLING AND SAMPLING

To estimate the channel width (Figure 1) six boreholes were drilled, four in the suspected NNE - SSW axis of the channel and two at right angles. Continuous U100 core sampling was carried out through the zone of interest. The U100 cores provided undisturbed samples with good recovery from the lower part of the London Clay, as well as through the sand channel and from about 3 m into the underlying Woolwich and Reading Beds.

The boreholes were completed with acid resistant piezometer tubing (32 mm internal diameter). A 1 m long mesh filter (200 μm) over a slotted pipe (10% open area) was set at the base of the sand channel over bentonite backfill. Rounded 10 mm gravel was used as a formation stabilizer with coarse sand and a bentonite seal placed above.

Each U100 sample was sealed with wax to prevent evaporation of pore water and to minimise possible contamination and oxidation. The samples were later extruded from the U100 tube using a hydraulic extruder and the central core of the sample retained. This was divided for analysis of grain-size, the gravimetric moisture content and for pore-water extraction. Pore water samples were obtained by spinning 300 g of core at 14000 rpm for 30 minutes in a high speed centrifuge. Yields varied considerably depending on the degree of saturation of the sand. Typically about 8 ml was collected and filtered. The sample was passed through a 0.45 μm filter and split into two, one part being acidified to a final concentration of 1% HNO_3.

Major and minor cations were determined by aspirating acidified samples into an ARL 34000C inductively coupled plasma optical emission spectrometer (ICP- OES). The more saline waters required dilution. Corrections were made for the spectral interferences caused by unusually

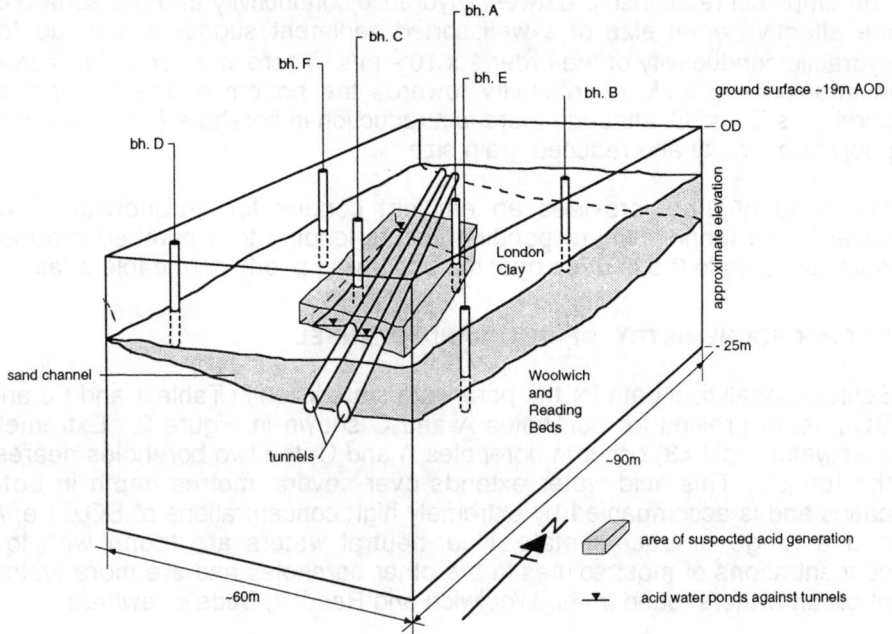

Figure 1: Schematic model of railway tunnels and the sand channel.

high concentrations of iron and aluminium. Automated colorimetry was used to determine chloride and nitrate. The pH of the waters was determined in the laboratory.

RESULTS

PHYSICAL PROPERTIES OF THE SAND CHANNEL

Descriptive logging of recovered core suggests that the sand channel rests on the undulating surface of the Woolwich and Reading Beds. The thickness of the channel ranges from 4.3 m at borehole A to 3.1 m at borehole B, but there was no evidence of the channel in boreholes D or F which lie beyond the western edge of the channel (Figure 1). Although there are insufficient data to define the area covered by the channel with precision, it is most probably an isolated pocket with dimensions some 170 m north-south and 60 m east-west. The effective area of the channel within which corrosion and consequent leakage have been observed in the tunnels is approximately 40 m along the tunnels by 30 m wide between boreholes A and C.

The channel contains fine to medium grained sands, ranging from 0.23 mm to 0.47 mm median grain size. There are abundant pyrite and black pyritic horizons. The lower channel sands in boreholes B and C are a uniform medium-grained sand which becomes a clayey-sand in borehole A and borehole E.

The empirical relationship between hydraulic conductivity and the square of the effective grain size of a well sorted sediment suggests a value for hydraulic conductivity of the order 1×10^{-3} m/s. There is a general increase in inferred hydraulic conductivity towards the bottom of the channel in boreholes B and C, although there is a reduction in borehole A as a result of poorer uniformity and reduced grain size.

The sand channel provides an efficient conduit for groundwater flow towards the tunnels where ponding occurs leading to a perched groundwater body up to 0.5 m deep over the underlying poorly permeable strata.

HYDROGEOCHEMISTRY OF THE SAND CHANNEL

Selected analytical data for the pore waters are listed in Table 1 and Fe and SO_4 depth profiles for boreholes A and C shown in Figure 2. Extremely acid waters (pH <3) occur in boreholes A and C, the two boreholes nearest the tunnels. This acid water extends over several metres depth in both cases and is accompanied by extremely high concentrations of SO_4, Fe, Al and a range of trace metals. Near-neutral waters are found with low concentrations of most solutes in the other boreholes and are more typical of groundwaters found in the Woolwich and Reading Beds elsewhere.

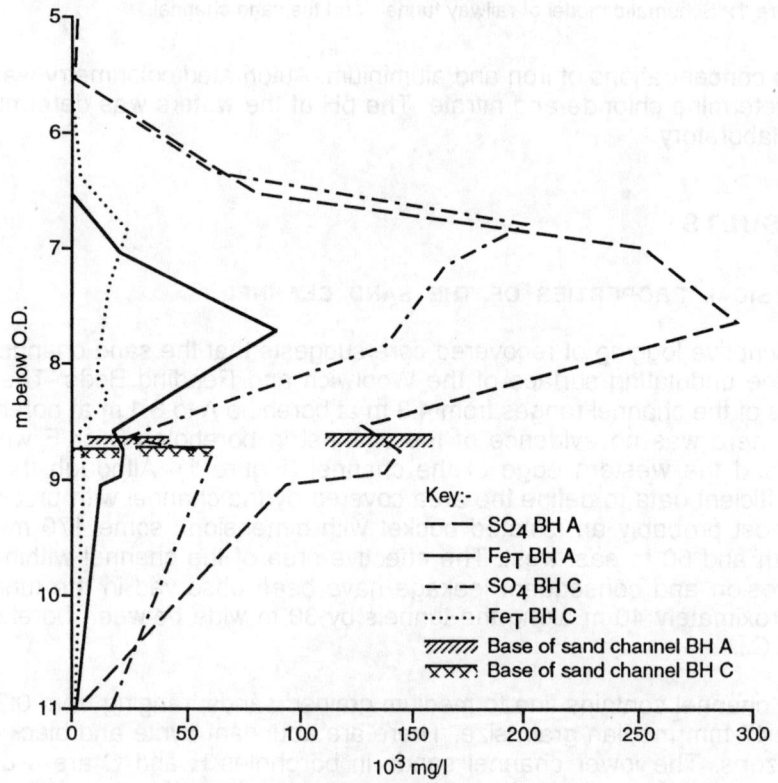

Figure 2: SO_4 and Fe_T profiles in boreholes A and C.

Table I: Concentration of selected inorganic determinands in the sand channel pore waters (mg/l)

Borehole	Depth (m below ground)	pH	Na	K	Ca	Mg	SO$_4$	Cl	FeT	AlT	Co	Ni	Zn
A	24.3	8.0	105	51	500	173	2080	65	0.08	0.05	0.2	0.25	0.61
	26.5	0.7	138	<35	348	2930	296000	<30	91700	20600	371	504	989
	27.5	1.8	97	<35	549	3850	145000	<30	23500	15200	199	257	484
B	25.9	8.4	107	23	92	41	539	72	0.04	<0.15	<0.03	<0.1	0.03
C	25.5	2.2	220	<40	670	675	25300	30	1630	2020	84	107	1090
	26.5	–	130	<40	425	3100	20400	30	25000	22400	2390	2380	9230
	27.7	1.8	130	<40	540	2850	10900	30	10400	13100	53	87	120
E	25.3	8.2	61	11	23	9	38	27	0.03	<0.1	<0.02	<0.05	<0.02

The extremely acid waters apparently derive from the oxidation of pyrite. This is illustrated by the high SO_4 and Fe concentrations in the pore water and the presence of *Thiobacillus ferrooxidans*. The high concentrations of the other solutes, particularly Al, Co, Ni and Zn, are probably of a secondary nature and have resulted from the dissolution of minerals or other materials (including pulverized fuel ash used as a tunnel lining) in the prevailing acid environment. The high sulphate concentrations are extremely aggressive towards cement (Robins and Milowdowski, 1986).

It is interesting to note that the most acidic and sulphate-rich waters in boreholes A and C do not always contain the highest concentrations of Zn and other metals, as the highest concentrations tend to be found towards the edge of the acid zone. This poor correlation suggests that the trace metals are not derived from the pyrite *per se* but have come from other minerals. The metals within the acid zone may have been removed by leaching and the highest concentrations are now found at the periphery where acid is first coming into contact with the fresh matrix. The high solute concentrations reflect slow groundwater movement. Indeed Edmunds and Cook (1987) reported the tunnel seepage to have a tritium activity of only 4.4 TU, suggesting a slow moving body of water several decades old.

Looking at anthropogenic sources, the only acid spill that is consistent with the data is one of sulphuric acid. This could not have entered the aquifer without being neutralized on passing through the drift and the London Clay. The amount of acidity is considerable and would have required a major leakage, perhaps via an old well or borehole. However it is of note that the acidity in the two boreholes adjacent to the tunnels is absent from the four boreholes at a greater distance from the tunnels, strongly suggesting that the tunnels play a direct role in the production of acidity. It seems unlikely, therefore, that the acidity has resulted from a spill. The possibility of an oxidizing agent (other than air) being spilled also seems unlikely as there are no other reaction products present: sodium hypo-chlorite, for example, would have given high concentrations of chloride and sodium.

Pore water seeping from the bottom of the London Clay shows no deterioration in quality compared with that higher up the sequence.

Evidence of humic acids was found in the pore water in borehole A adjacent to the tunnels. This probably derives from timber packing used during earlier civil engineering works on the tunnels. These acids are weak and not significantly aggressive.

In February 1989 outgassing was observed from borehole B. Analysis revealed a gas depleted in oxygen and somewhat enriched in carbon dioxide and methane relative to air. Such compositions are not unusual in London basin boreholes (MacLean, 1966). Nevertheless the depletion of oxygen is consistent with oxidation processes taking place within the aquifer.

SCALE OF ACID GENERATION

The area of saturated sand is some 40 m long, 30 m wide and 0.5 m deep amounting to a volume of 600 m^3. Of this the railway tunnels occupy some 140 m^3 leaving a saturated volume of aproximately 460 m^3. If the porosity of the sand is about 0.40 then the volume of water in the saturated zone is about 180 m^3.

In order to estimate the weight of pyrite in the acid generating zone of the sand channel, selected sand samples were refluxed for two hours with concentrated nitric acid (1 g wet weight plus 1 ml HNO$_3$), diluted to 100 ml with distilled water and the dissolved Fe and SO$_4$ measured. The Fe and S released were converted to percentage pyrite giving an average pyrite concentration of 0.2% by volume. As the volume of pyritic sand is estimated to be 40 m by 30 m and an average 3.7 m deep, it follows that the total volume of pyrite is about 9 m^3, equivalent to 45 000 kg in weight.

Assuming there is 200 m^3 of acidic water in the channel, or lost previously to seepage into the tunnel, with an average sulphate concentration of 100 000 mg/l, this contains

$$\frac{200 \times 1000 \times 100\,000}{100 \times 100 \times 96} = 208 \text{ kmol SO}_4 = 104 \text{ kmol pyrite}$$

as each mole of SO$_4$ is derived from the oxidation of 0.5 mole of pyrite.
At pH > 4:

$$FeS_2(s) + \frac{15O_2}{4} + \frac{7}{2}H_2O = Fe(OH)_3(s) + 4H^+ + 2SO_4^{2-}$$

At lower pH, some of the Fe(OH)$_3$(s) will dissolve producing Fe^{3+} which can oxidize more pyrite (releasing yet more acid). The reaction is autocatalytic.

The weight of pyrite oxidized is given by 104 x 120 kg = 12500 kg, i.e. just over a quarter of the total available pyrite in the sand channel. Rainey and Rosenbaum (1989) estimated the rate of leakage from the London Clay above to be 2 x 10^{-4} m/a, i.e. 0.24 m^3/a for the area of interest. With this low rate of drainage it is likely that highly acid water will persist for many years to come.

DISCUSSION

The evidence presented by the exploratory drilling supports a number of assertions made by Rainey and Rosenbaum (1989). Until 1820 groundwater levels in Central London were about 5 m AOD and the sand horizon at the top of the Woolwich and Reading Beds was naturally fully saturated. Groundwater abstraction induced a steady decline in water levels so that the sand became dewatered some time between 1850 and 1875 (Marsh and Davis, 1983).

The tunnels allowed oxygen to enter the sands. This resulted in the oxidation of pyrite which was present in the sand and changed the naturally reducing environment into an unstable aerated environment. Natural slow percolation combined with water released by consolidation provided seepage water low in dissolved oxygen which moved from the London Clay into the sand.

The piston effect of passing trains and natural changes in barometric pressure pumped air into and out of the sands adjacent to the tunnels. The three ingredients: pyrite, air and water were thus available for the generation of highly acid waters. The acid became particularly concentrated as a result of the near stagnant conditions and the lack of readily weatherable minerals. There is sufficient pyrite remaining to continue the reaction at a similar rate for a considerable time.

ACKNOWLEDGEMENTS

The authors thank Mr I L J Chudleigh (London Underground Limited) for his enthusiastic support of this project. Colleagues within BGS who contributed to the study include J M Cook, W G Darling, R P McKittrick and M A Perkins. The paper is published by permission of London Underground Limited, Civil Engineering Department, and the Director, British Geological Survey (NERC).

REFERENCES

BIRD, M.J., COOK, J.M., DARLING, W.G., KINNIBURGH, D.G., MCKITTRICK, R.P., PERKINS, M.A. & ROBINS, N.S. (1989). An investigation into the occurrence of acidic groundwater beneath City Road, London. *British Geological Survey Technical Report WD/89/34C.*

EDMUNDS, W.M. & COOK, J.M. (1987). Old Street Station: groundwater investigations. *British Geological Survey, Technical Report 87/3.*

FOLLENFANT, H.G. (1975). *Reconstructing London's Underground*, 2nd edition, London Transport, London.

LAKE, R.D. (1987). Old Street Station: regional geological appraisal. *British Geological Survey, Technical Report.*

MACLEAN, R.D. (1966). Foul air in wells and boreholes in the London area. *Proceedings of the Society for Water Treatment and Examination*, **15(4)**, 271-283.

MARSH, T.J. & DAVIES, P.A. (1983). The decline and partial recovery of groundwater levels below London. *Proceedings of the Institution of Civil Engineers* **74**, 273-276.

RAINEY, T.P. & ROSENBAUM, M.S. (1989). The adverse influence of geology and groundwater on the behaviour of London Underground railway tunnels near Old Street Station. *Proceedings of the Geological Association*, **100(1)**, 123-132.

ROBINS, N.S. & MILODOWSKI, A.E. (1986). Borehole cements and the downhole environment - a review. *Quarterly Journal of Engineering Geology*, **19**, 175-181.

3-4 BIOLOGICAL STABILITY OF GEOSYNTHETICS IN CRITICAL EARTH STRUCTURES

D.G. BRIGHT
The Tensar Corporation, Morrow, Georgia, U.S.A.

ABSTRACT

Critical earth structures are commonly designed for long lifetimes of about 100 years and are expected to remain stable over this duration. As these structures frequently incorporate geosynthetics, it is necessary to know the mode, mechanism and profile of potential biological deterioration with time to facilitate an appropriate initial design to ensure long term stability.

The primary polymers (PE, PP, PET, PVC) in geosysnthetics are organic compounds of sufficient molecular weight to be considered non-digestible and consequently of no nutritional value to micro-organisms. However, recent information shows that polymers possessing a specific chemical group are susceptible to microbial attack. Some of these polymers experience problems in certain soil chemistries common in the earth structures, thus restricting their use.

Geosynthetics composed of multiple polymers with additives exhibit a more complex behaviour. The microbial growth of algae, fungi and bacteria on buried geosynthetics is primarily due to the attack of these micro-organisms on the low molecular weight additives of plasticizers, lubricants, emulsifiers and fillers. These additives can be digested by micro-organisms through enzymatic reactions; the products of which can in turn initiate and propagate reactions deteriorative to the primary polymers in the geosynthetic. This can result in a deterioration of the whole geosynthetic product, thus affecting its long term durability.

Biocides are sometimes incorporated into geosynthetics to retard microbial attack. However the best biocide cannot function effectively if it is immobile; this suggests that a biocide must migrate to be effective. Migration can eventually deplete the biocide's concentration and render the additives susceptible to microbial attack. Thus the effectiveness of biocides is short term.

The designer and practitioner will discover that geosynthetics composed of a single polymer present the fewest problems affecting durability. Geosynthetics of multiple polymers requiring additives present a greater number of problems affecting durability that are compounded by the synergisms between these different polymers and their respective additives.

INTRODUCTION

Civil engineering structures involving large earth works are usually critical soil structures with long life expectancies. Over the past ten years, these structures have been designed to incorporate polymeric products in a grid configuration to facilitate soil stabilization and reinforcement. Plate 1 shows a steep embankment with embedded polymeric grids together with a wrap-around face treatment. Figure 1 is a typical cross-section of such a structure showing the arrangement of these grids within the soil embankment. The introduction of polymeric grids has initiated the construction of reinforced soil structures with slopes steepened to angles of 27 to 70 and with life expectations of 75 to 120 years. These structures are critical in nature; that is, failure of the polymeric grid tensile elements would result in catastrophic failure of the soil structure and potential loss of life.

The dependency placed on these polymeric grids in the soil environment of a critical earth structure has led to new concerns regarding their stability and durability over the long-term. The grids, called geogrids, are fabricated either of a homogeneous polymer or copolymer or a heterogeneous composite of several polymers. Geogrids fabricated of a single polymer or copolymer are most commonly of a high molecular weight, high density polyethylene (HMW HDPE) or polypropylene (PP). The most common composite format for a geogrid is the lamination of a plasticized polyvinyl chloride (PVC) coating on a fibre substrate of polyethylene terephthalate (PET).

Plate 1: A steep embankment soil structure with polymeric grid reinforcement.

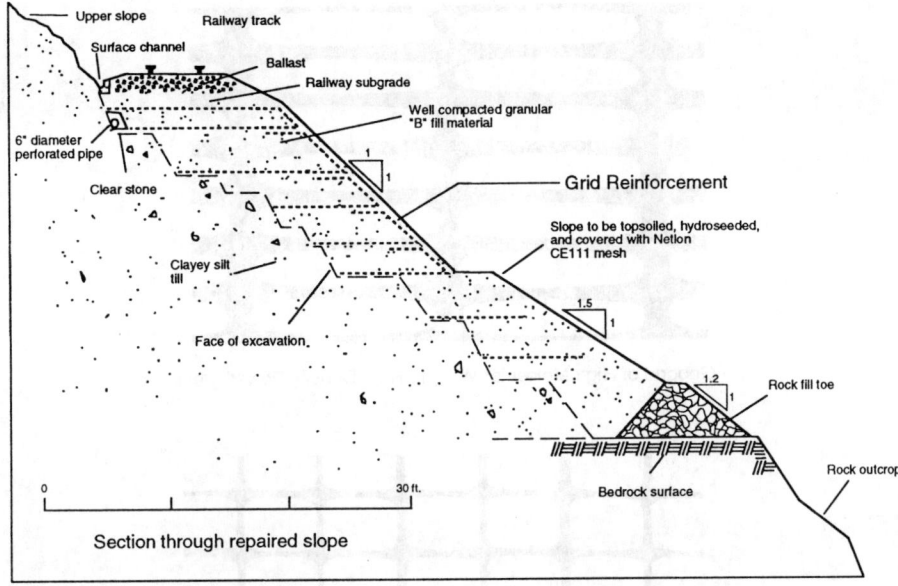

Figure 1: A cross-sectional view of a soil structure with polymeric grid reinforcement.

Geogrids of a single polymer or copolymer are typically 97+% by weight HMW HDPE or PP with about 2+% carbon black and <1% processing stabilizers. One manufacturer fabricates these geogrids by first extruding a solid sheet which is then perforated with a precise arrangement of apertures. The perforated sheet is mechanically drawn to orient its molecular structure, significantly improving creep resistance. Sheets drawn in just one direction are principally of HMW HDPE whereas sheets drawn in two directions are principally of PP. Examples of these geogrids are shown in Figure 2.

For a composite geogrid of a laminate configuration, the coating is applied to the substrate principally to protect it against deteriorative forces in the geotechnical environment. The performance requirements of a coating include adherence to and waterproofing of the substrate (Seaman and Venkataraman, 1976), prevention of wicking of liquids into internal voids (Seaman and Venkataraman, 1976) and protection from installation damage and field service including the deteriorative effects of weathering and chemical and biological attack. The ideal performance criteria are that both the coating and the substrate are resistant to all deteriorative modes (Seaman and Venkataraman, 1976) over the composite's anticipated service lifetime and that there are no deteriorative synergisms between components.

The interaction between polymers and the natural environment has been of concern ever since materials (e.g. polymers) were fashioned into products (e.g. geosynthetics such as geogrids) and exposed to the environment.

Geogrid of High Molecular Weight, High Density Polyethylene

Geogrid of Polypropylene

Figure 2: Examples of geogrids.

The exposure of geosynthetics to soil constituents can cause their biodeterioration and biodegradation (Bessems, 1979). Biodeterioration is an undesirable change or deterioration in the properties or value of a material or product caused by organisms (Bessems, 1979; Eggins and Oxley, 1980; Seal and Eggins, 1981; Seal, 1990); whereas biodegradation is a breaking down of the molecular structure of the material caused by organisms resulting in a loss of its structural integrity (Seal, 1990; Huang et al,1990; Hueck, 1974). The principal concern today is just how stable and durable are geosynthetics in a critical earth structure over a life time of 75 to 120 years.

SOIL CONSTITUENTS

CHEMISTRY

Within the United States there are over two hundred different soil types (Connolly, 1972). These soils contain both inorganic chemicals such as mineral acids and alkalis, salts, minerals, clays, sand, gases and water and also organic chemicals from plants, animals and macro- and micro-organisms (Rankilor, 1981).

ORGANISMS

Macro-organisms include insects and rodents. Micro-organisms of importance in biodeterioration processes are bacteria, fungi, actinomycetes, algae (Seal and Eggins, 1981; Koerner, 1990; Shapira and Magier, 1991) and yeast (Osman, 1972). Their concentration in the soil varies by class: bacteria (1000 pounds/acre or about 1 billion/gram of soil, or about a trillion/cubic inch of soil), fungi (10-100 m of mold filament/gram of soil), actinomycetes (0.1-36.0 million/gram of soil) (Connolly, 1972; Rankilor, 1981; Koerner, 1990). These micro-organisms are found all over the world, in environmental conditions (Seal and Eggins, 1981) ranging from the tropics to the polar regions (Butler and Eggins, 1966).

Biodeterioration/biodegradation processes are effected by chemical composition, surface characteristics and physical state of the geosynthetic as well as the environmental factors. Temperature, humidity, pH, osmotic pressure and redox potential are very important in initiating biodeterioration/biodegradation and encouraging subsequent development (Seal and Eggins, 1981, Seal, 1990; Shapiro, 1991). Ambient and higher temperatures, high humidities, and the absence of UV light favour the growth of fungi while bacteria frequently need the presence of moisture for reproduction (Seaman and Venkataram, 1976; Heap and Morrell, 1968; Tirpak, 1970). Fungi are more adaptive to growing under adverse conditions (Cadmus, 1977). The most favourable temperature conditions are 30°C, but they can survive below 0°C and up to 65°C. Relative humidities of 95-100% are the most conducive to growth, but they still thrive at 70% (Tirpak, 1970). Both bacteria and fungi require a source of carbon for growth and obtain it from enzymatic degradation of organic materials (Seaman and Venkataraman, 1976; Heap and Morrell, 1968; Tirpak, 1970) such as polymers.

MECHANISM OF ACTIVITY BY MICRO-ORGANISMS

As micro-organisms grow, they excrete enzymes into their surrounding medium. These enzymes cause degradation of the host material by breaking down its large molecular units into much smaller units to serve as nutrients for the parent microbial (Tirpak, 1970; Cadmus, 1977; MacLachlan, 1966). The net effect is a reduction in molecular weight, a deterioration of physical properties and eventual disintegration of the host material (Heap and Morrell, 1968). One micro-organism may have an initial advantage by virtue of the immediately available nutrients, but it may be replaced or joined by other micro-organisms leading to a succession of other micro-organisms (Seal and Eggins, 1981) so that a particular micro-organism initiating degradation is not necessarily that which perpetuates and/or concludes the process. Given the inherent propensity for rapid mutation, micro-organisms can produce enzymes specifically suited to the host substrate (Huang, et al, 1990). The level of enzymatic activity varies according to the micro-organism, culture medium and the presence of inhibitors (biocides) (Lazar, 1975). Microbial enzymes can promote several other highly degradative reactions: oxidation-reduction, decarbonylation, and hydrolysis (Seal, 1990; Huang et al., 1990; Tirpak, 1970; Cadmus, 1977; Fields and Rodriguez, 1975; Frobisher, 1968).

BIOLOGICAL TEST PROTOCOL

Although there are no uniform or formalised test methods for determining biodegradability of polymers (Bessems, 1979; Koerner, 1990; Aminabhavi, 1990), two elements are generally important (Bessems, 1979; Aminabhavi, 1990):

a) incubation in environments conducive to microbial attack of the polymers and

b) the measurement of the degree of degradation through change in both chemistry and properties.

Historically, the testing is divided into two categories: laboratory testing and field studies. Laboratory testing permits greater experimental control, but at the expense of being able to actually duplicate exposure conditions while field studies involve burial in the soil for several years. However, soil burial tests have a problem with reproducibility, due to the difficulty in controlling environmental conditions and microbial population (Bessems, 1979; Aminabhavi, 1990). The solution to this dilemma would be to use the right micro-organisms with the right conditions (Aminabhavi, 1990), but the conditions under which micro-organisms grow in a controlled laboratory are radically different from those prevailing in the field test (Bessems, 1979).

In the laboratory, biodegradation testing on polymer is usually conducted in solid agar media. The agar media contains all the nutrients necessary for microbial growth except the carbon source, which is provided by the polymer. Tests usually run for at least three weeks, with growth rates being classified by fraction of surface covered. Changes in physical and chemical properties are also monitored (Huang, et al. 1990). However, this time frame may not necessarily be sufficient to recognize or detect changes in properties as a result of microbial activity.

It is a common fallacy to suppose that the presence of growing micro-organisms on a material is in itself sufficient proof of the biodegradability of the material. Trace amounts of impurities can stimulate growth. Rigorous proof of biodegradability requires clear change in an essential property of the polymer in which the suspected causative micro-organism is the sole effective agent (Hueck, 1974).

SOIL CHEMISTRY SUSCEPTIBILITY OF POLYMERS

Inorganic chemicals include mineral acids and alkalis, salts, minerals, clays, sand, gases and water. The corrosion of polymers can result from the interaction with such inorganic chemicals which may be present in substantial quantities in certain aggressive soils such as acid soils (e.g. acid sulphate soils), organic soils, saline alkaline soils and calcareous soils (Rankilor, 1981). The susceptibility of polymers used in geosynthetics to these soil chemistries and others is illustrated in Table I prepared by Elias (1989) for the Federal Highway Administration.

Table I: General guide to anticipated resistance of polymers to specific soil environments (after Elias, 1989).

Soil environment	PETP	PA	PE	PP	PVC
Acid Sulphate Soils - characterised by low pH and considerable amounts of Cl^- and SO_4^{2-} ions (e.g. pyritic soils in the Appalachian region	?	X	NE	?	?
Organic Soils - characterised by high organic contents and susceptibility to microbiological attack (e.g. dredged fills)	NE	?	NE	NE	X
Salt Affected Soils - in areas of seawater saturation or in dry alkaline areas as the Southwestern, US	?	NE	NE	NE	NE
Ferruginous - contains Fe_2SO_3	NE	?	NE	X	X
Calcareous - in dolomitic areas	X	?	NE	NE	?
Modified Soils - soils subject to deicing salts, cement stabilised or lime stabilised	X	?	NE	NE	?

KEY: NE = No Effect; ? = Questionable Use; X = Not Recommended; PETP = Polyester; PA = Polyamide; PE = Polyethylene; PP = Polypropylene; PVC = Poly(vinyl chloride)

BIOLOGICAL SUSCEPTIBILITY OF POLYMERS AND ADDITIVES

Polymers contain carbon and micro-organisms require a source of carbon to survive and reproduce (Seal and Eggins, 1981; Koerner, 1990; Pankhurst, Davies and Blake, 1971). Thus, the biological susceptibility of polymers used in stabilization and reinforcement of critical earth structures over very long periods is a major concern.

Microbes produce enzymes that attack many polymers (Huang, et al., 1990), but the resulting degree of degradation is controlled by the following conditions:

1. Molecular weight level (Huang, et al., 1990; Koerner, 1990),
2. Availability and accessibility of molecular chain end groups (Huang, et al., 1990; Koerner, 1990),
3. Hydrophobicity (Huang, et al., 1990) and
4. Low surface area to volume ratio (Huang, et al., 1990).

Enzymology can explain this biodegradation. First, enzymes require accessibility to the sites where the polymer is susceptible to attack (Aminabhavl, 1990). Low surface area to volume ratio and hydrophobicity can control accessibility. Second, there must be chemical groups susceptible to enzymatic attack (Aminabhavl, 1990) for the generation of nutrients such as carbon. Polymer chemistry and molecular weight level are controlling factors here. Micro-organisms tend to attack the end groups of large molecular chains, but the number of end groups is inversely proportional to the molecular chain length or weight (Aminabhavi, 1990). Thus, lengthening the molecular chains by physical and/or chemical means will decrease the susceptibility of the molecule to microbiological attack (Seal and Eggins, 1981).

The susceptibility to microbial attack is also dependent on the chemical structure of the polymer's monomeric repeat unit. Polymers with mixed linkages, such as -COO-, -CO- and -CN- (e.g., polyesters, polyamides, polyeurethanes), show a much greater susceptibility to biodegradation due to hydrolytic attack than polymers with simple aliphatic linkages such as -C-C- (e.g., polyethylene, polypropylene), (Huang et al 1990; Aminabhavi 1990).

The biological susceptibilities of specific polymers and compounds, as used in geosynthetic reinforcements, are addressed individually below: polyethylene (PE), polypropylene (PP), unplasticized polyvinyl chloride, polyethylene terephthalate (PET), plasticizers, and plasticized polyvinyl chloride.

Polyethylene

Polyethylenes have molecular chains of simple aliphatic linkages, -C-C. PE molecular weights (MW) range from 2000 to 6,000,000. High molecular weight, high density polyethylene (HMW HDPE) used in high strength geogrids has a typical MW range of 250,000 to 1,500,000 (Juran, (ed.), 1989). HMW HDPE inherently has a very low number of methyl (CH^{3-}) end groups available for microbial attack, and these are not readily accessible (Aminabhavi, 1990). As PE is hydrophobic, there is no easy access to internal molecular structure, (Hueck, 1974). Thus the principal points of attack are at the ends of carbon chains (the methyl end groups), and any activity is confined to the lower molecular weight oligomers (MacLachlan, 1966; Aminabhavi, 1990) of about three carbon lengths (Seal and Eggins, 1990). Any gradual breakdown of these oligomers from their chain ends inward will be cut short by any side chains (Hueck, 1974). Commercial grades of PE have molecular chains consisting of more than 1000 monomeric units so that no biodegradation can be expected (Kuster, 1979).

As early as 1944, Brown (1946) had assessed that PE did not serve as a source of carbon for the growth of fungi and thus was resistant to fungal growth. In later years, others have substantiated these findings as well as bacteriological resistance (Heap and Morrell, 1968; MacLachlan, 1966; Pankhurst et al, 1971). The Bell System (Connolly, 1971; Connolly, 1972; De Coste, 1972; Miner, 1972) buried molded samples of PE for eight years in Georgia soils (acidic with a 5.2 pH and high rain fall) and New Mexico soils (alkaline with an 8.2 pH and low rainfall). These MW grades of PE, commonly used in wire and cable structures, exhibited excellent resistance to physical degradation (no significant evidence of biological attack) and very minimal property deterioration (retention of electrical and mechanical properties) (Connolly, 1972; Connolly, 1971; De Coste, 1972). Burial depth and location did not affect the biological resistance of the PE (Miner, 1972). HDPE has been used as the principal resin in pipe for natural gas distribution throughout the USA since the early 1950s and the same generic resin and pipe are still being used today. As recently as 1972, a study for the Environmental Protection Agency (EPA) by Potts, Clendinning and Ackart (1972) on the biodegradability of various packing plastics assessed the microbial resistance of PE packaging films as very high.

The MW grades of HDPE, therefore, exhibit sufficient resistance to biological degradation and property deterioration from attack by micro-organisms (Connolly 1972; Rankilor 1981; Heap and Morrell 1968; Cadmus 1977; MacLachlan 1966; Lazar 1975; Aminabhavi 1990; Pankhurst 1971; Brown 1946; Connolly 1971; De Coste 1972; Miner 1972; Potts et al 1972) to warrant their use for long-term applications in critical earth structures.

Polypropylene

Polypropylenes also have molecular chains of simple aliphatic linkages, -C-C-, but with a methyl group (CH^{3}-) attached to one of the carbons. PP molecular weight ranges from 200,000 to 600,000 (Juran (ed.), 1989). The same reasons and logic for biological resistivity pertain to PP as for PE. With PP being hydrophobic, there is no easy access to its internal molecular structure and the methyl groups within each monomeric repeat unit. Thus, the principal points of attack are at the ends of the carbon chains (the methyl end groups), and any activity is confined to the lower molecular weight oligomers. Any gradual breakdown of these oligomers from their chain ends inward will be cut short by the side chains of the PP. Thus, PP exhibits resistance to biological degradation and property deterioration from attack from micro-organisms (Rankilor, 1981; Cadmus, 1977; MacLachlan, 1966; Aminabhavi, 1990; Pankhurst, et al., 1971; Potts et al, 1977) sufficient for long-term applications in critical earth structures.

Unplasticized polyvinyl chloride

As early as 1944, Brown (1946) had assessed that the rigid, unplasticized grades of polyvinyl chloride (PVC) did not serve as a source of carbon for the growth of fungi and was resistant to fungal growth. In later years, others have substantiated the same findings along with bacteriological resistance (MacLachlan, 1966; Lazar, 1975; Pankhurst, et al., 1971). The Bell System (Connolly, 1972; Connolly, 1971; De Coste, 1972; Miner, 1972) buried molded samples of PVC for eight years in Georgia soils (acidic with a 5.2 pH and high rain fall) and New Mexico soils (alkaline with an 8.2 pH and low rainfall). The rigid, unplasticized grades of PVC exhibited excellent resistance to physical degradation (no evidence of biological attack) and no property deterioration (retention of electrical properties and a slight increase in mechanical properties) (Connolly, 1972; Connolly, 1971; De Coste, 1972; Miner, 1972). Burial depth and location did not affect the biological resistance of unplasticized PVC (Miner, 1972). As recently as 1972, a study for the Environmental Protection Agency (EPA) by Potts et al (1972) on the biodegradability of various packing plastics put the microbial resistance of PVC packaging films as being very high.

Thus, rigid, unplasticized grades of PVC exhibit resistance to biological degradation and property deterioration from attack from micro-organisms (Seal, 1990; Connolly, 1972; Rankilor, 1981; MacLachlan, 1966; Lazar, 1975; Aminabhavi, 1990; Pankhurst, 1971; Brown, 1946; Connolly, 1971; De Coste, 1972; Miner, 1972; Potts et al, 1972).

Polyethylene terephthalate

As mentioned earlier, the susceptibility to microbial attack is dependent on the chemical structure of the polymer. Polymers with mixed linkages within their monomeric repeat units, such as the ester group (-COO-), exhibit a greater susceptibility to biodegradation due to hydrolytic attack on the monomeric repeat units along the chain backbone (Huang, et al., 1990; Aminabhavi, 1990; Allsopp and Seal, 1986). Microbials mutate rapidly to produce enzymes specifically suited to the host substrate (Lazar, 1975). For example, esterase is an enzyme which is excreted by fungi (Seal and Eggins, 1981; Aminabhavi, 1990) and accelerates the hydrolysis of polyesters (Fields and Rodriguez, 1975; Allsopp and Seal, 1986; Klausmeier, 1961; Mish, 1986). The esterase enzyme is excreted by fungi (Seal, 1981; Aminabhavi, 1990) commonly found in the soil (Fields, 1975; Klausmeier, 1961). Both fungi and bacteria have exhibited this ability to use polyesters through this particular enzyme (Fields and Rodriguez, 1975; Allsopp and Seal, 1986). This attack is referred to as enzymatic hydrolysis (Seal, 1990; Frobisher, 1968). The esterase enzyme does not exhibit specificity towards a particular ester (Klausmeier, 1961); thus enzymatic hydrolysis can occur at the ester group (-COO-) (Aminabhavi, 1990) within the monomeric repeat unit along the backbone (Huang, et al., 1990) of PET. Functionally, the esterase enzyme seeks out an ester group and connects to it. Using water, this enzyme converts the ester into an alcohol and acid that the parent microbial can then digest (Gibbons, 1989). The use of antihydrolysis additives to combat conventional hydrolytic degradation is not effective in preventing enzyme catalyzed hydrolysis (Allsopp, 1986). Thus, the presence of ester groups makes an ester based polymer susceptible to biodeterioration in the short-term (Seal, 1990).

Aminabhavi (1990) lists PET fabrics as exhibiting only fair microbial resistance (however he states that aromatic polyesters, such as PET, are resistant to such attack. No explanation is given for this contradiction). Although Potts et al (1972) state that PET is resistant, so much contrary information (Seal and Eggins, 1981; Huang, et al., 1990; Rankilor, 1981; Fields and Rodriguez, 1975; Gibbons, 1989) has been published since the Potts et al paper that it must now be considered suspect. Field studies by various researchers investigating the biological susceptibility of PET have produced mixed results.

PET geotextile samples were exposed to accelerated soil burial testing for up to seven years. The soil was a moist, organically rich soil at a pH of 6.7 and maintained at 29 ± 1°C and 85-90% relative humidity. The PET showed negligible loss of strength, but there were large standard deviations in the data (Colin, et al. 1986). Interestingly, Hoechst-Celanese (1988) ran soil burial tests on high tenacity PET in a woven fabric under the same conditions for just five years and reported a 15% loss in (burst) strength. Sotton, Leclercq, Paute and Fayoux (1982) buried PET fabrics in dam, road and drainage sites in Europe for 12 years; exhumed samples exhibited <30% loss in mechanical properties. It is not clear as to what proportion of these property losses is attributable to installation and/or exhumation damage versus classical or enzymatic hydrolysis, but Sotton et al (1982) do state that losses of >30% in mechanical properties can occur due to light exposure and punctures.

A review of the basic chemistry involved clearly indicates that biological susceptibility of PET is a real phenomenon. Biological degradation has been documented by some, although not all, researchers. Clearly, the use of PET in critical earth structures is questionable for long-term applications, and its use should, therefore, be limited to short-term applications.

Plasticizers

Plasticizers are added to polymers to increase flexibility. They function by separating the polymer's molecular chains sufficiently to weaken the intermolecular forces between them. The susceptibility of these plasticizers to microbial attack (Koerner, 1990; Mills and Eggins, 1953; Stahl and Pessen, 1953; Williams and Dale, 1983) has a profound influence on the susceptibility of the whole polymeric composite (Mills and Eggins, 1953; Stahl and Pessen, 1953; Williams and Dale, 1983). It has been known for some time that plasticizers may provide a prime source of nutrient for microbial growth on polymeric composites (Bessems, 1979; Klausmeier, 1962). Loss of plasticizer by any means - chemical extraction (leaching), physical removal, biological consumption or modification - results in changes in physical and performance properties of the polymeric composite (Lazar, 1975; Klausmeier, 1961; Mills and Eggins, 1974).

The most common plasticizer used with vinyl polymers, particularly for PVC, is dioctyl phthalate (DOP) (Klausmeier, 1961; Klausmeier, 1972; Menzel, Rohlfing and Rohlfing, 1990). DOP is frequently, and erroneously, referred to as diisooctyl phthalate (DIOP). They are not the same (Sax and Lewis, 1987). More specifically, DOP is di(2-ethylhexyl)phthalate and is sometimes abbreviated as DEHP (Kaplan, 1977).

DOP is an aromatic (phthalate) ester and consequently is subject to enzymatic hydrolysis (Seal and Eggins, 1981; Seal, 1990; Connolly, 1972; Tirpak, 1970; Williams, Kanzig and Klausmeier, 1969). Although DOP is considered resistant to mold growth, bacteria have been found to use DOP as a sole source of carbon and energy. Its degradation can occur by esterase enzymes converting it back to phthalic acid (Allsopp and Seal, 1986; Williams and Dale, 1983); further deterioration can be catalyzed by adaptive enzymes (Williams and Dale, 1983). Thus, phthalates are not as inert as was previously thought (Seal, 1990). Mathur and Rouatt (1975) report that DOP supports growth of bacteria known to occur in soil and industrial waste disposal systems. Some plasticizers are resistant when present as the sole carbon source, but the presence of other organic nutrients (such as stabilizers, pigments, fillers, hydrolytic stabilizers) in the same composite may stimulate the utilization of these plasticizers by microbials leading to the eventual deterioration of the whole composite (Seal, 1990; Allsopp and Seal, 1986). However, the additive carbon black is not susceptible to microbial attack (Tsuchii, Hayashi, Hironiwa, Matsunaka and Takeda, 1990) and may be used as a stabilizer against UV degradation.

The physical aspects of the susceptibility of a plasticizer are important. It must be available at the surface of the polymer and not completely immobilized within the polymeric structure which would prevent its migration. The presence of other additives will encourage surface growth which may then extend to other components in the composite (Seal, 1990). Plasticizer diffusion studies have demonstrated a correlation between migration of biodegradable plasticizers and deterioration of plasticized vinyl formulations (Osman, Klausmeier and Jamison, 1971). Generally, many components in a polymeric composite will migrate continuously to the surface and then be leached into the environment. This provides a constant fresh supply of nutrients to the colonizing microbes (Seal and Eggins, 1981) on the surface.

Laboratory and soil burial tests have produced mixed and contradictory results. Preparations of DOP in an industrial environment may support the growth of various fungi after prolonged incubation despite the fact that preparations of DOP in a laboratory environment do not (Kaplan, 1977). Thus, industrial preparations of plasticized vinyl polymers using DOP are likely to be susceptible to microbial attack.

The basic chemistry of plasticizers commonly used with vinyl polymers indicates that a problem with microbial attack exists in the short-term. As the field results are inconclusive/variable, caution should be exercised in the use of any vinyl polymers containing plasticizers in long-term applications.

Plasticized polyvinyl chloride

Rigid PVC is plasticized to improve toughness and flexibility for use as a coating on woven fabric substrates for tarpaulins, automobile and pool liners and geogrids. As plasticized PVC is often used in a wet or damp environment as a waterproof barrier, its resistance to microbiological attack is critical (Upsher and Roseblade, 1984). The most common plasticizer used with PVC is dioctyl phthalate (DOP) (Klausmeier, 1961; Klausmeier,

1972; Menzel, et al., 1990). This plasticizer and other plasticizers and additives - such as antioxidants, pigments, fillers and hydrolytic stabilizers - constitute 50 to 60% of the total weight of a formulation (Seal and Eggins, 1981; Klausmeier, 1972; Williams, et al., 1969; Upsher and Roseblade, 1984). These components render plasticized PVC susceptible to microbial attack (Seaman and Venkataraman, 1976; Bessems, 1979; Seal and Eggins, 1981; Seal, 1990; Rankilor, 1981; Cadmus, 1977; Williams, et al., 1969; Osman, et al., 1971; Upsher and Roseblade, 1984; Kaplan, Greenberger and Wendt, 1970; Griffin and Urbie, 1984).

Under conditions of high humidity (80-100%), typical of ground atmospheres (Jailloux and Verdu, 1990), additives used to make a PVC plastisol can absorb water and carry it into the plastisol. The most commonly used plasticizer, DOP, is among the most highly adsorptive plasticizer of the twelve commonly used and tested by Marshall (1990). This adsorbed moisture facilitates the migration through leaching of plasticizers like DOP and DIOP to the surface of the PVC plastisol (De Coste, 1972; Griffin and Uribe, 1984) where micro-organisms can cultivate in the presence of moisture and ground atmospheric oxygen (Bessems, 1979; Seal and Eggins, 1981; Seal, 1990; Tirpak, 1970) using the PVC plastisol as the host substrate (Bessems, 1979). A high surface-to-volume ratio ensures a rapid and continuous loss and degradation of the plasticizer, thus hastening deterioration of performance properties, such as tensile strength, flexibility, toughness and elasticity (Tirpak, 1970; Upsher and Roseblade, 1984).

Formulations of PVC plasticized with phthalate based plasticizers, including DOP or DIOP, were buried for four years in soils in Georgia (5.2 pH acidic, humid climate) and New Mexico (8.2 pH alkaline, dry climate). The PVC showed no evidence of disintegration from microbial attack. However, the plasticizers were the principal target of attack by micro-organisms. Migration, extraction and deterioration resulted in a 4 to 8% weight loss in plasticizer over the four years. This caused the original formulation to become brittle with higher tensile and modulus and net lower elongation (De Coste, 1972). These formulations were found to be susceptible to attack by macro-organisms (insects) as well (Connolly, 1972; De Coste, 1972). Any loss of plasticizer sufficient to change mechanical properties and cause brittleness is considered a deterioration of the PVC plastisol formulation.

Samples of PVC plasticized with DIOP were inoculated with 13 yeast cultures isolated in uncultivated soils in Minnesota. The media were mineral agar plates and liquid culture for 15 days at 30°C. Although sample weight loss was only 2% over the 15 days, the results show that yeasts found in the soil environment can degrade PVC plasticized with a common plasticizer (Osman, et al., 1972). Hueck (1974) reports a plasticized PVC buried in soil for 16 weeks; the plastisol had a 78% increase in residual strength reflecting a decrease in plasticizer as was apparent from considerable microbial growth. Seal and Pantke (1988) buried specimens of plasticized PVC in English soil for up to 18 months. The plastisol registered at 15+% and 20% weight loss after 6 and 18 months, respectively, which was attributed to microbial attack (Seal and Pantke, 1988). Thus, plasticized formulations of PVC are susceptible to microbial attack and degradation, even in the short-term of just weeks.

Biological susceptibility of plasticized PVC is well documented in the literature, therefore the use of this plastisol to provide protection to underlying materials in a soil environment, should be limited to short-term applications.

BIOLOGICAL STABILITY OF THE COMPOSITE OF TWO GEOSYNTHETIC POLYMERS

The most common format for a composite geogrid is the lamination of a plasticized polyvinyl chloride coating on a fibre substrate of polyethylene terephthalate. Analysis of the individual components (plasticizers, plasticized PVC and PET) to biological deterioration has shown that the most common plasticizers (DOP and DIOP) and plasticized PVC are susceptible to microbial attack and deterioration, even in the short time span of just weeks. Due to its inherent chemistry, PET is also susceptible to biological deterioration.

The components in composite geogrids have exhibited an individual susceptibility to biological deterioration. The productions of degradation of one component can directly, or indirectly, initiate and propagate the degradation of other components by modes other than biological. Degradation of the aromatic ester plasticizers by the esterase enzyme via hydrolytic reactions produces the phthalic acid (Allsopp and Seal, 1986; Williams and Dale, 1983). Deterioration of the plasticized PVC coating exposes the underlying PET fibre substrate. Due to the void created by the deteriorating plasticizer and presence of an acid, water is now more accessible to the PET fibres. This facilitates maximum adsorption of water by the PET fibre, which has a plasticizing effect and reduces tensile and modulus strength (Hoechst-Celanese Corporation, 1988; Davis, 1988). The adsorption of moisture could be sufficient for classical hydrolysis to be initiated (Schneider and Groth, 1987). The presence of an acid can catalyze (accelerate) this degradative reaction. With time, the hydrolytic reaction can become autocatalytic (Risseeuw and Schmidt, 1990), further accelerating degradation of the PET fibre substrate. Although this mechanism of classical hydrolysis is different from that of enzymatic hydrolysis, the two mechanisms could easily function in concert, reducing the stability and durability of a PVC plastisol coated PET fibre geogrid to the short-term.

Thus, the biological degradation of individual components within a composite structure creates the potential for synergistic effects between components to accelerate the degradation of others and deterioration of the whole composite. Field studies to date have not been of sufficient scope or duration to elucidate this phenomenon.

FUNCTIONALITY, DURABILITY AND LIFE EXPECTANCY OF BIOCIDES

From the above, it can be seen that plasticized formulations of PVC are susceptible to microbial attack and degradation. A biocide may therefore be added to a PVC formulation for one of two reasons: to impede the growth of or to destroy completely the micro-organism (Bessems, 1979; Shapiro and Magier, 1991). Biocides impede or destroy because they interfere with the metabolism of the micro-organisms, destroying their vitality and preventing amitosis. They differ widely in the extent of their effectiveness, however, as different biocides block different enzymatic reactions (Bessems, 1979; Jakubowski, Gyuris and Simpson, 1983). In addition, a plasticized composite containing a fungicide adequate to protect against fungal growth remains susceptible to bacterial attack; the converse is also true of a bactericide (Stahl and Pessen, 1953).

Biological stability of a plastisol system is strongly influenced by its plasticizer and the effectiveness of a biocide must be related to the plasticizer. The best biocide cannot function effectively if it is immobilized within the polymer structure while the biologically susceptible plasticizer is migrating or leaching to the surface and supporting microbial growth. Nor can an immobile biocide repel a microbial invasion which feeds on the coating's surface (Cadmus, 1977; Kaplan, et al., 1970). Ideally, a biocide should provide protection for all components of a composite even when it is applied to only one component (Kaplan, et al., 1970). Thus a composite with static components will not be likely to survive the environment (Cadmus, 1977); the mobilised biocides are the most effective.

Assessment of biocide effectiveness may be undertaken in several ways, individually or collectively; by visual observation, oxygen consumption or uptake, sample weight loss and change in sample flexibility or elasticity (Bessems, 1979). Loss of biocide service may be attributable to a number of factors, including, for example, aqueous leaching, physical migration, hydrolysis, thermal oxidation, auto-oxidation and microbial decomposition as well as general incompatibility with the existing microbials (Gabriele and Iannucci, 1984).

After a certain period of incubation, each sample will be surrounded by a static zone, the extent of which acts as an indicator of the actual biocide effectiveness. The width of this static zone indicates how the biocide is released by the PVC formulation. A wide zone indicates rapid leaching; a small or narrow static zone indicates slow leaching (Bessems, 1979). This implies that the biocide must be mobile and in concert with the migrating plasticizer, to be effective in protecting that particular plasticizer from a particular micro-organism. Thus a given loading of biocide has only a specified lifetime of effectiveness before it is depleted. The question must therefore be raised as to how much of each of how many biocides must be formulated into a plasticized PVC for sufficient protection from micro-organisms for 75 to 120 years.

CONCLUSIONS

1. Fungi and bacteria deteriorative to some polymers exist in all parts of the world.

2. Fungi and bacteria excrete enzymes that can initiate and propagate several reactions, such as oxidation-reduction, decarbonylation, and hydrolysis, that are highly degradative to some polymers, particularly polyesters.

3. Due to simple chemistry and minimal additives, high density grades of polyethylene, polypropylene, and rigid unplasticized polyvinyl chloride are resistant to soil chemistries and biological deterioration over the long-term.

4. The plasticizers commonly used in vinyl composites are susceptible to biological degradation in the short-term.

5. Due to the inherent presence of ester groups and the generation of the esterase enzyme by common fungi and bacteria, the biological susceptibility of polyethylene terephthalate is a real phenomenon, and its application in earth structures should be confined to the short-term.

6. Plasticized vinyl composites are susceptible to biological degradation in the short-term.

7. The synergistic effects between biodegrading components of a composite structure can accelerate the deterioration of the whole composite.

8. Biocides exhibit specificity, have a limited time span of effectiveness due to loading quantity and are leachable.

9. Based on an extensive literature survey, polymers having the simplest of chemistries, such as HMW HDPE and PP, and requiring minimal additives are resistant to microbial attack.

REFERENCES

ALLSOPP, D. & SEAL, K.J. (1986). *Introduction to Biodeterioration*, Edward Arnold, London; 39-49.

AMINABHAVI, T.M. & BALUNDGI, R.H. (1990). A Review on Biodegradable Plastics. *Polymer-Plastics Technology Engineering*, **29(3)**, 234-62.

BESSEMS, E. (1979). Some Microbiological Problems of Fabrics Coated with Plasticized PVC. *Journal Coated Fabrics*, **9**, 26-37.

BROWN, A.E. (1946). The Problem of Fungal Growth. *Modern Plastics*, **23(8)**, 189-195.

BUTLER, N.J. & EGGINS, H.O.W. (1966). Microbiological Deterioration and the Tropical Environment. In: *Microbiological Deterioration in the Tropics*, Society Chemical Industry, London, 3-12.

CADMUS, E.L. (1977). The Biological Stability of Polymers. *Journal Coated Fabrics,* **7**, 33-42.

COLIN, G., MITTON, M.T., CARLSSON, D.J. & WILES, D.M. (1986). The Effect of Soil Burial Exposure on Some Geotechnical Fabrics. *Geotextile and Geomembrane,* **4(1)**, 1-8.

CONNOLLY, R.A. (1971). Soil Burial of Materials and Structures. *Biodeterioration of Materials, Volume 2,* Walters, A.H., Hueck-van der Plas, E.H. (eds.), Wiley & Sons, New York, 168-78.

CONNOLLY, R.A. (1972). Soil Burial Tests: Soil Burial of Materials and Structures *The Bell System Technical Journal,* **51(1)**, 1-21.

DAVIS, G.W. (1988). Aging and Durability of Polyester Geotextiles. In: *Durability and Aging of Geosynthetics,* Geosynthetic Research Institute, Philadelphia.

DE COSTE, J.B. (1972). Effect of Soil Burial Exposure on the Properties of Plastics for Wire and Cable. *The Bell System Journal,* **51(1)**, 63-86.

EGGINS, H.O.W. & OXLEY, T.A. (1980) Biodeterioration and Biodegradation. *International Biodeterioration Bulletin,* **16(2)**, 53-56.

ELIAS, V. (1989). Durability/Corrosion of Soil Reinforced Structures FHWA/RD-89/186.

FIELDS, R.D. & RODRIGUEZ, F. (1975). Microbial Degradation of Aliphatic Polyesters. *Proceedings of the Third International Biodegradation Symposium,* Sharpley, J.M., Kaplan, A.R. (eds.) Applied Science Publishers, Kingston, Rhode Island, 775-84.

FROBISHER, M. (1968). *Fundamentals of Microbiology,* W.B. Saunders Co, Philadelphia, 66-68.

GABRIELE, P.D. & IANNUCCI, R.M. (1984). Protection of Mildewcides and Fungicides from Ultraviolet Light Induced Photo-oxidation. *Journal Coatings Technology,* **56(712)**, 33-48.

GIBBONS, A. (1989). "Making Plastics that Biodegrade",*Technology Review,* 69-73.

GRIFFIN, G.J.L. & URBIE, M. (1984). Biodegradation of Plasticized Polyvinyl Chloride. In: *Biodeterioration 6,* Barry, S., Houghton, D.R. (eds.) The Biodeterioration Society, Washington; 64857.

HEAP, W.M. & MORRELL, S.H. (1968). Microbiological Deterioration of Rubbers and Plastics. *Journal Applied Chemistry,* **18**, 189-94.

HOECHST-CELANESE CORPORATION (1988). High-Performance Geosynthetics from Trevira™ High Tenacity Polyester Yarns. *Technical Fibers Group*:1-8.

HUANG, J-C., SHETTY, A.S. & WANG, M-S. (1990). Biodegradable Plastics: A Review. *Advances in Polymer Technology,* **10(1)**, 23-30.

HUECK, H.J. (1974). Criteria for the Assessment of the Biodegradability of Polymers. *International Biodeterioration Bulletin,* **10(3)**, 87-90.

JAILLOUX, J-M. & VERDU, J. (1990). Kinetics Models for the Life Predictions in PET Hygrothermal Ageing: A Critical Survey. *4th International Conference on Geotextiles, Geomembranes and Related Products,* G. den Hoedt (ed.), The Hague, Netherlands, 727.

JAKUBOWSKI, J.A., GYRUIS, J. & SIMPSON, S.L. (1983). Microbiology of Modern Coatings Systems. *Journal Coatings Technology,* **55(705)**, 49-53.

JURAN, R. (ed.) (1989). *Modern Plastics Encyclopaedia,* McGraw-Hill, New York; 50-85.

KAPLAN, A.M. (1977). Microbial Degradation of Materials in Laboratory and Natural Environments. *Developments in Industrial Microbiology*, **18**, 203-210.

KAPLAN, A.M., GREENBERGER, M. & WENDT, T.M. (1970). Evaluation of Biocides for Treatment of Polyvinyl Chloride Film. *Polymer Engineering Science*, **10(4)**, 241-6.

KLAUSMEIER, R.E. (1972). Results of the Second Interlaboratory Experiment on Biodeterioration of Plastics. *Int. Biodeterioration Bulletin*, **8(1)**, 3-7.

KLAUSMEIER, R.E. & JONES, W.A. (1961). Microbial Degradation of Plasticizers. *Developments in Industrial Microbiology*, **2**, 47-53.

KOERNER, R.M. (1990). *Designing with Geosynthetics*. Prentice Hall, Englewood Cliffs, N.J.; 106,303,393-5.

KUSTER, E. (1979). Biological Degradation of Synthetic Polymers. *Journal Applied Polymer Science: Applied Polymer Symposium*, **35**, 395-404.

LAZAR, V. (1975). The Study of Microbiological Corrosion of Plastics in Romania. *International Biodeterioration Bulletin*, **11(1)**, 16-23.

MACLACHLAN, J., HEAP, W.M. & PACITTI, J. (1966). Attack of Bacterial and Fungi on Rubbers and Plastics in the Tropics. *Microbiological Deterioration in the Tropics, Monograph 23, Society Chemical Industry, London*, 185-99.

MARSHALL, R.A. (1990). Moisture Absorption by PVC Plastisol Components. *SPE Annual Technical Conference, Society of Plastics Engineers, Dallas, Texas*, 592-4.

MATHUR, S.P. & ROUATT, J.W. (1975). Utilisation of the Pollutant Di-2-Ethylhexyl Phthalate by a Bacterium. *Journal Environmental Quality*, **4(2)**, 273-5.

MENZEL, B., ROHLFING, M. & ROHLFING, W.H. (1990). Plasticizers. *Kunststoffe German Plastics*, **80(7)**, 30-33,810-815.

MILLS, J. & EGGINS, H.O.W. (1974). The Biodeterioration of Certain Plasticisers by Thermophilic Fungi. *International Biodeterioration Bulletin*, **10(2)**, 39-44.

MINER, R.J. (1972). Soil Burial Tests: Effect of Soil Burial Exposure on the Properties of Molded Plastics. *The Bell System Journal*, **51(1)**, 23-42.

MISH, F.C. (1986). *Webster's Ninth New Collegiate Dictionary*, Merriam-Webster, Springfield, Mass.

OSMAN, J.L., KLAUSMEIER, R.E. & JAMISON, E.I. (1972). The Ability of Selected Yeast Cultures to Degrade Plasticized Polyvinyl Systems. *Developments in Industrial Microbiology*, **11**, 447-51.

OSMAN, J.L., KLAUSMEIER, R.E. & JAMISON, E.I. (1971). Rate-Limiting Factors in Biodeterioration of Plastics. In: *Biodeterioration of Materials*, Walter, A.H., Hueck-van der Plas, E.H. (eds.) Wiley & Sons, New York, 66-75.

PANKHURST, E.S., DAVIES, M.J. & BLAKE, H.M. (1971). The Ability of Polymers or Materials Containing Polymers to Provide a Source of Carbon for Selected Micro-organisms. *Proceedings of the Second International Biodeterioration Symposium*, Walters, A.H., Hueck-Van Der Plas, H. (eds.), Halsted Press, Lunteren, The Netherlands, 76-90.

POTTS, J.E., CLENDINNING, R.A. & ACKART, W.B. (1972). *An Investigation of the Biodegradability of Packaging Plastics*, EPA-R2-72-046, Contract No. CPE-70-124.

RANKILOR, P.R. (1981). *Membranes in Ground Engineering*, John Wiley, New York, 65-83.

RISSEEUW, P. & SCHMIDT, H.M. (1990). Hydrolysis of HT Polyester Yarns in Water at Moderate Temperatures. *4th International Conference on Geotextiles, Geomembranes and Related Products*, G. den Hoedt, editor, Balkema, Rotterdam, The Netherlands, 691-7.

SAX, N.I. & LEWIS, R.J. (1987). *Hawley's Condensed Chemical Dictionary*, 11th edition, van Nostrand Reinhold, New York.

SCHNEIDER, H. & GROTH, M. (1987). An Analysis of the Durability Problems of Geotextiles. *Geosynthetics '87, IFAI, New Orleans*, 434-41.

SEAL, K.J. & EGGINS, H.O.W. (1981) The Biodeterioration of Materials. *Essays in Applied Microbiology*, Norris, J.R., Richmond M.H. (eds.), John Wiley & Sons, London, 39-52.

SEAL, K.J. & PANTKE, M. (1988). Microbiological Testing of Plastics: Ongoing Activities of IBRG Plastics Project Group to Improve Standard Test Procedures. *International Biodeterioration*, **24**, 313-9.

SEAMAN, R.N. & VENKATARAMAN, B. (1976). Utilization of Vinyl Coated Synthetic Fabrics in Industrial Applications. *Journal Coated Fabrics*, **5(4)**, 225-4.

SHAPIRO, O. & MAGIER, L. (1991). Biocide Application Plays Many Roles in Adhesives. *Adhesives Age*, **34(2)**, 22-4.

SOTTON, M., LECLERCQ, B., PAUTE, J.L. & FAYOUX, D. (1982). Some Answers' - Components on Durability Problem of Geotextiles. *Second International Conference on Geotextiles, Las Vegas, Nevada*, 553-8.

STAHL, W.H. & PESSEN, H. (1953). The Microbiological Degradation of Plasticizers. *Applied Microbiology*, **1**, 30-5.

TIRPAK, G. (1970). Microbial Degradation of Plasticized PVC. *SPE Journal*, **26(7)**, 26-30.

TSUCHII, A., HAYASHI, K., HIRONIWA, T., MATSUNAKA, H. & TAKEDA, K. (1990). The Effect of Compounding Ingredients on Microbial Degradation of Vulcanized Natural Rubber. *Journal Applied Polymer Science*, **41(5&6)**, 1181-1187.

UPSHER, F.J. & ROSEBLADE, R.J. (1984). Assessment by Tropical Exposure of Some Fungicides in Plasticized PVC. *International Biodeterioration*, **20(4)**, 243-52.

WILLIAMS, G.R. & DALE, R. (1983). Shorter Communications: The Biodeterioration of the Plasticiser Dioctyl Phthalate. *International Biodeterioration Bulletin*, **19(1)**, 37-8.

WILLIAMS, P.L., KANZIG, J.L. & KLAUSMEIER, R.E. (1969). Evaluation and Production of Esterases on Plasticizer Substrates by Fungal Species. *Developments in Industrial Microbiology*, **10**, 177-82.

3-5 THE PHYSICAL AND CHEMICAL CHARACTERISTICS OF GEOTEXTILES AND THEIR EFFECTS ON IN SITU DEGRADATION

P.R. RANKILOR
Manstock Geotechnical Consultancy Services Ltd.,1 North Parade, Manchester, U.K.

ABSTRACT

This paper defines some of the terminology associated with geotextiles, geogrids and related materials. It then addresses the questions most frequently posed by engineers when considering the potential use of geotextiles in earthworks designs, discussing types of degradation and time scales, the susceptibility of particular polymers, the influence of the environment and the implications for different engineering structures.

The various polymers are described with particular attention being paid to the properties which are of prime importance in different applications. Agencies of degradation are discussed in the light of a long term international weathering program comparing fourteen different geotextiles and geogrids. In conclusion, specific attention is drawn to a number of points with particular implications for design and construction and guidance offered on permissible light exposure periods during site construction works.

INTRODUCTION

The 'degradation' of geotextiles in civil engineering is of concern to all engineers. However, most engineers will not be sure exactly what this means. Which type of degradation might be a problem? Are any particular polymers particularly susceptible? Are any particular environments particularly arduous? What kind of time scales are relevant? What are the implications on different engineering structures?

This paper describes the various structure types which comprise geotextiles before addressing these very relevant and frequently posed questions.

TERMINOLOGY

Firstly it is necessary to appreciate that the term 'geotextiles' is sometimes used to include geogrids while the word 'geosynthetics' is used to incorporate both geotextiles and geogrids - leading to much debate on the fate of natural fibre geotextiles which arguably cannot be classified as 'synthetic' materials. In order to cover the subject adequately, however, all types of geo-products are included in this paper using the following terminology.

Geotextile

A planar sheet of permeable textile made from either synthetic or natural materials, with the appearance commonly recognised as a textile.

Geonet

A planar sheet of material made from either synthetic or natural materials, with the appearance of a netting. The holes between the fibres may be rectangular, diamond or other, but the product is not a rectangular grid with fixed intersections. Such a product may sometimes be referred to as a 'mesh' structure.

Geogrid

A planar sheet of material made from synthetic polymers, with the appearance of a rectilinear grid. The holes between the fibres are rectangular and the product has fixed intersections. It is not necessarily stiff but usually has a high ultimate strength and modulus for soil reinforcement purposes.

In the context of geotextiles, the term 'fibre' can vary considerably in its meaning. Some products are made with fibres considerably finer than a human hair whilst geogrids, for example, may have structural 'fibre' elements up to 10 mm across.

In the case of geotextiles, not only is the specific surface area of the material extremely large, but the true dimension of fibre cross sections in relation to ultra violet light penetration is small. It is therefore likely that given any two geotextiles made from an identical polymer, the relative dimensions of the individual component fibres will be important in governing the way in which the overall fabric weathers and decomposes. It has been interesting to observe from experimental work that the actual methods of construction of the individual fibres and fabrics also appear to govern the way in which they respond to stress and thus open themselves up to additional degradation.

The term 'polymer' is generally used by engineers to mean the generic chemical group from which most geotextiles are manufactured. It rarely means a pure polymer as manufacturers almost universally add other chemicals to act as stabilisers and fillers. This is important as papers have been published stating certain properties of pure polymers which are misleading, since they are quite different from those found in geotextile 'polymers'. Furthermore, there is considerable ignorance in the engineering world concerning the number of different polymers produced commercially under global names. "Nylon" is a classic example. Does one mean Nylon 6 or Nylon 66 for example? Both come under the general heading of 'Polyamide' but their properties are very different. Global terms tend to be used in engineering discussions when referring to industrial raw materials made predominantly from a particular group of polymers. Much of the ignorance is specifically due to the natural secrecy maintained by commercial manufacturers about their formulae, additives and manufacturing processes, in order to protect their product speciality.

A fuller study of the terminology of geotextiles can be obtained from a number of textbooks (Rankilor, 1981; Koerner, 1986) and papers (Rankilor, 1988) on the subject.

CONSTITUENT MATERIALS

Allowing for the factors mentioned above, the main 'polymers' from which geotextiles and geogrids are made are:-

> Polypropylene
> Polyester
> Polyethylene
> Polyvinylidene
> Polyamide (Nylon)

Various manufacturers frequently add 'carbon black' and other additives to the polymer melt before manufacture in order to enhance the UV stability of the geotextile. Anti-oxidant chemicals and stabilisers are often added, but these do not usually change the colour. Geotextiles are therefore generally black or 'white' (translucent), with very few having other colours.

In order to study the process of deterioration, observations have been made of the effects of both UV exposure and buried degradation on fourteen commercially available types of geotextile and geogrid. The early stages of this work were reported by Rankilor (1986) and some specific examples discussed by him in a subsequent paper (1990).

The fourteen selected samples were representative of virtually all of the several hundred different types of geotextiles and geogrids available on the world market. They are each commonly used world-wide.

Rather than use commercial names, each geotextile has been identified by an appropriate code lettering system:

> NP = Needle Punched,
> HB = Heat Bonded,
> W = Woven,
> NS = Needle Scrim,
> GG = Geogrid

Table 1 summarises the properties of the geotextiles and geogrids chosen and provides a brief description sufficient for the purposes of the present study. A total of 1024 individual tensile tests were reported, ie:

NP1	83	NP2	99				
HB1	85	HB2	76				
W1	66	W2	97	W3	88	W4	73
NS1	94	NS2	63	NS3	90		
GG1	22	GG2	45	GG3	43		

Table 1: Geotextiles Selected for Weathering Research

Code	Weight g/m^2	Thickness	Colour	Construction	Polymer	Fibre Size
HB1	140	Thin	White	Heat-bonded	Polyethylene (33%)/ Polypropylene (6-7%)	Very Fine
HB2	136	Thin	Grey	Heat-bonded	Polypropylene	Very Fine
NP1	150	Medium Felt	Grey	Needle Punched	Polyester	Very Fine
NP2	450	Thick Felt	Grey	Needle Punched	Polypropylene	Very Fine
NS1	600	Thick	White	Needle Punched. Resin bonded on to a scrim	Polyamide	Fine
NS2	1000	Thick Felt	Black	Needle Punched on to a scrim	Polypropylene	Fine
NS3	00	Thick Felt	White on Black Scrim	Needle Punched on to a scrim	Polyester	Fine
W1	130	Thin	Black	Woven Tape	Polypropylene	Coarse
W2	393	Thin	Green	Woven tape calendered after weaving	Polyvinylidene (85%) / Stabilisers (15%)	Medium
W3	243	Thin	Black	Woven	Polypropylene (85%) / Stabilisers (15%)	Medium
W4	50	Medium	White	Woven Multi-filaments	High Tenacity Polyester	Fine
GG1		Thick	Black	Punched sheet extended rectangular Geogrid	Polypropylene	Macro-Fibre
GG2		Thick	Black	Punched sheet extended asymmetrical Geogrid	Polyethylene	Macro-Fibre
GG3	660	Thick	Black	Hexagonal hot extruded self welded mesh Geogrid	Polyethylene	Macro-Fibre

AGENCIES OF DEGRADATION

SHORT-TERM EXPOSURE TO LIGHT DURING INSTALLATION.

One of the main purposes for studying the effects of UV exposure on geotextiles is to assess the possible damage caused by such exposure either during the designed period of installation or through accidental exposure owing to non-compliance with contractual specifications.

For many years engineers have asked the question "how long can I leave the geotextile exposed without damaging it?" Unfortunately, they tend to overlook, or are unaware of, the fact that the question is meaningless unless the environment is specified. Current work shows conclusively that the length of time will vary, depending primarily upon the climatic region in which the geotextile is weathered.

To study this aspect further it was decided to establish a limited number of weathering stations in widely differing climatic and environmental conditions. It took approximately two years to obtain suitable sites in each of these areas so that a full testing programme could be established. The greatest difficulty was not in actually locating suitable areas or making the appropriate contacts but in finding sites which were both exposed and at the same time would be undisturbed and vandal-proof for the necessary research period of up to ten years. Plate 1 shows a typical site array of UV exposure samples in Sweden.

The location of the weathering sites is given in Figure 1. The first to be set up was situated in peat-lands on the exposed Pennine Hills near Buxton in the U.K. The second was established in Indonesia, in the sea in a private

Plate 1: Photograph of the Swedish weathering site.

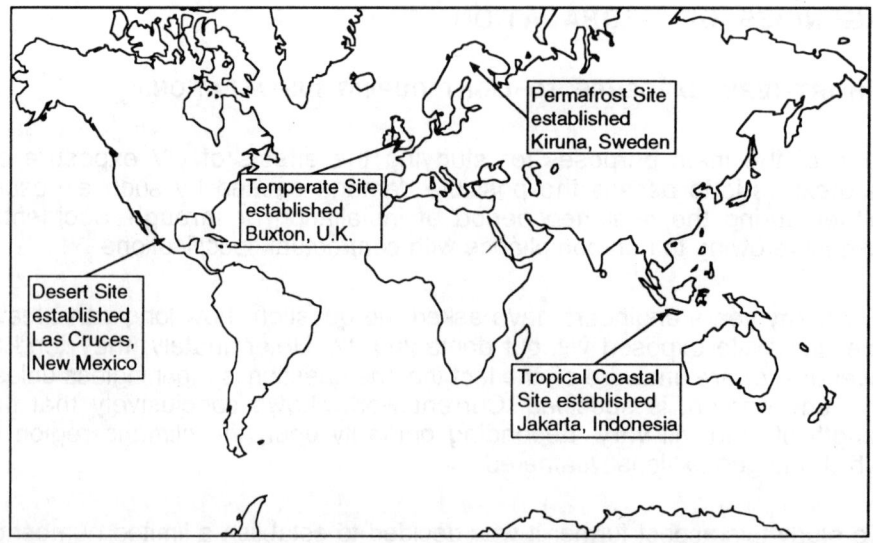

Figure 1: Geographical distribution of environmental weathering sites.

harbour just north east of Jakarta city. The third was north of the Arctic Circle near a small town called Kiruna in Sweden where test specimens were set up in a permafrost peat-land environment. Finally, the last installation was at an altitude of approximately 2500 m in the sand desert of New Mexico, U.S.A. where ultra violet radiation would be extremely strong and where the diurnal temperature range is high. Suffice it to say that in this part of the world it is common to have to replace house roofs at regular intervals owing to the repeated thermal expansion and contraction caused by the hot days and freezing nights.

These different environments were chosen to provide a widely varying set of conditions which were nevertheless representative of areas where substantial civil engineering earthworks involving the use of geotextiles are currently taking place.

The selection therefore represented:

1. Temperate climate conditions - Europe, northern United States, Japan, etc.
2. Tropical countries - the broad span of countries around the equatorial region including Central Africa, South East Asia, North Australia and much of South America.
3. Hot desert regimes - the development work in the Middle East, the hot desert countries of the southern United States, Mexico and the heavy industrial development along the North African coast.
4. Arctic tundra type climate - of civil engineering construction work taking place in high latitude areas such as Canada, North Europe, U.S.S.R., the Falkland Islands and southern South America.

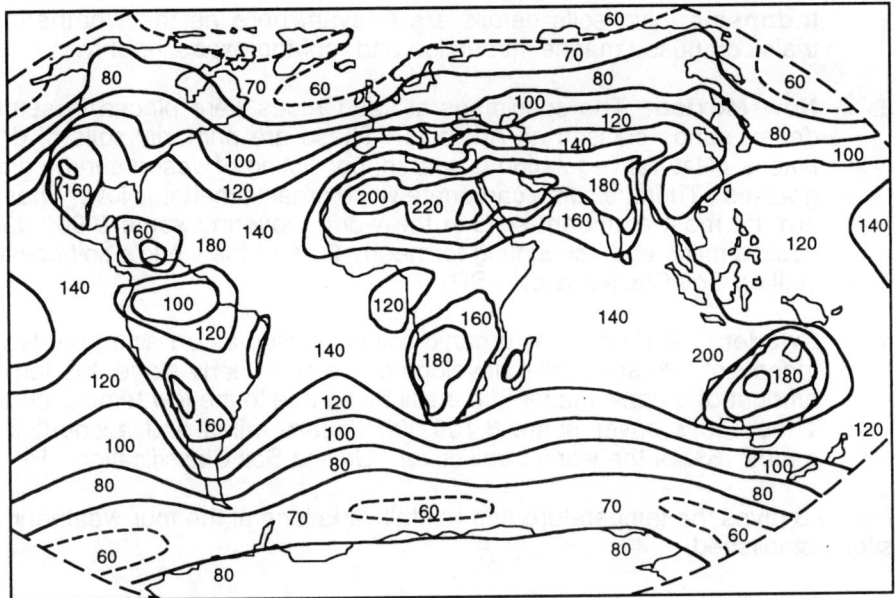

Figure 2: Generalised isolines of global insolation in units of kcal/cm^2/year.

Figure 2 (from Schneider, 1987) shows the insolation levels for most of the globe. It can be seen that the sites chosen have two fundamentally different levels of light intensity as follows:-

1.	U.K. -	70 kcal/cm^2/year
2.	Sweden - Kiruna -	70 kcal/cm^2/year
3.	Indonesia - Java -	140 kcal/cm^2/year
4.	New Mexico desert - Las Cruces -	160 kcal/cm^2/year

As the energy levels of Indonesia and New Mexico are approximately double those of the U.K. and Sweden, approximate comparisons may be made of the order of 'equal' or 'double' exposures, whilst considering the impact of the substantially different temperatures between the U.K. and Sweden and the substantially different humidities of New Mexico and Java.

The soil types at the four chosen sites may be described briefly as follows:

1. **U.K:** The soil at the Buxton site is a peaty soil classified under the New Soil Order System as a 'moist' Histosol. Histosols are 'organic' soils which normally remain saturated with water for at least 30 consecutive days a year; however the Buxton site, although subject to frequent precipitation, is too elevated to become flooded. The soil overlies Carboniferous Limestone calcium carbonate bedrock at a depth of less than 500 mm, grading down from a brown organic soil into a humic sub-soil horizon which grades indeterminately into rockhead. Unified Soil Classification - Pt-Topsoil.

2. **Indonesia:** No soils details are relevant here as the weathering trials comprised marine immersion and roof top exposure only.

3. **New Mexico:** The specimens at Las Cruces were placed in semi-desert sands classed as Aridisols. These are primarily soils of dry places. Natural vegetation is sparse, consisting of desert shrubs and grasses. There is some carbonate while organic matter is low. These are the most abundant soils in the world, covering some 9,900,000 square miles and accounting for nearly 20% of the world's soil cover. Unified Soil Classification - SP.

4. **Sweden:** At Kiruna, the organic soil is classed as an acid peat-land Inceptisol. Its soil profile development is at an early stage, the large amount of organic matter in the soil being due to the low temperature. These soils cover some 8,100,000 square miles and account for nearly 16% of the world's soil cover. Unified Soil Classification - Pt.

Figure 3 gives the temperature and rainfall variations at the four weathering sites established.

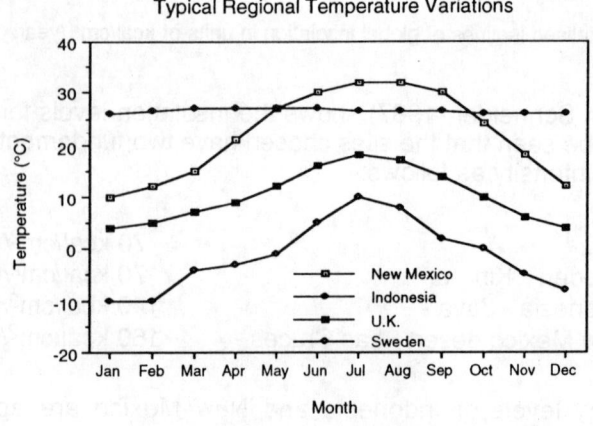

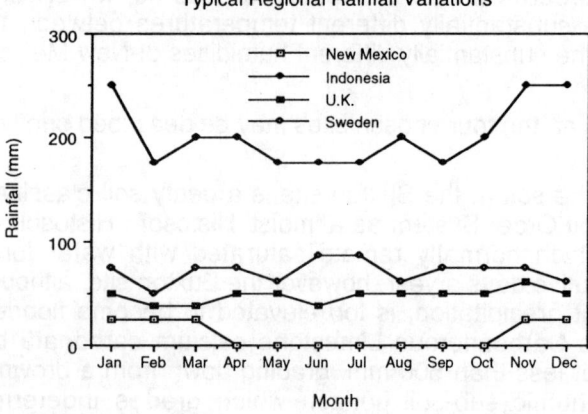

Figure 3: Temperature and rainfall variations at the four weathering sites.

When selecting the various research sites it was not the detail of each local situation that was important but rather that the areas were representative of certain climatic regimes in order to allow a pragmatic, empirical assessment of variations in weathering, which could in future guide more accurate and more finely tuned research programmes.

By using the particular four sites chosen it was considered it would be possible for an assessment to be made of the difference, if any, between the effects of the tropical environment as against the desert or an extremely cold climate against the temperate U.K. It was found that over a period of three years, for example, there was no discernible difference in the ultimate tensile strength of many geotextiles exposed to sunlight in the U.K. or the Scandinavian permafrost region. On the other hand, ultra violet exposure in Jakarta, Indonesia, had a greater effect on the same geotextiles than the drier desert environment of New Mexico.

Much more significant than the insolation difference, however, was the size of fibre making up the geotextile. It was found that fine fibre geotextiles in the temperate and cold exposure sites were not significantly affected, but in the tropical and desert sites dramatic fibre damage was noted within days rather than months. Plate 2 shows a large polypropylene fibre exposed for one year to tropical sunlight and Plate 3 shows a fine poly-propylene fibre exposed to desert sunlight for one year. It can be seen that although the large polypropylene fibre has been degraded with lateral serrations and some longitudinal splitting and proto-fibrillation, the overall physical integrity of the fibre remains. Conversely, the fine polyester fibre exposed for one year to the desert light, has totally degraded into a series of non-coherent disks.

As a result of the research work, a simple subdivision of allowable exposure times for different climatic regions has been drawn up (Table II). Permitted exposure could be varied in accordance with this chart, without any noticeable increase in deterioration of the fabric, irrespective of the general type. Such a generalised approach may be useful when preparing contractual documents when the geotextile to be offered is unknown.

Table II: Recommended time limits for geotextile UV exposure.

	Temperate	Arctic	Desert	Tropical
Summer	8 weeks	4 weeks	2 weeks	1 week
Winter	12 weeks	6 weeks	2 weeks	1 week

Comparative inter-environmental exposure tests are rare and whilst the above table invites improvement, nonetheless it is subjectively based upon the author's experience and more than one thousand tensile tests from samples exposed to UV over a period of four years.

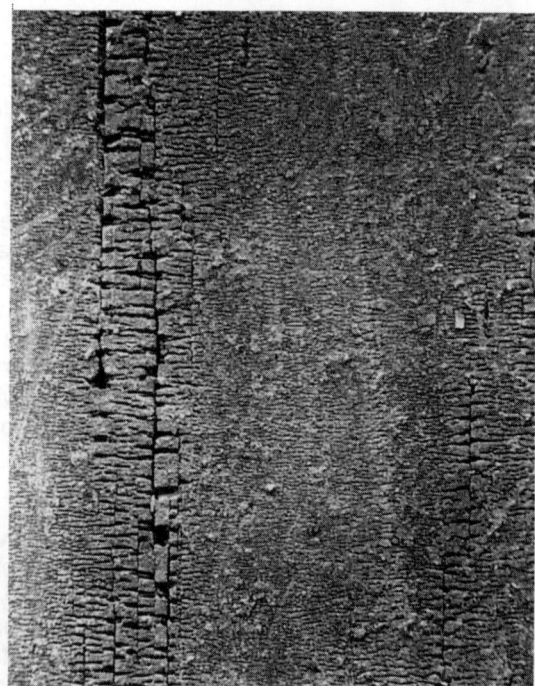

Plate 2: Scanning electron micrograph of weathered Polypropylene Tape geotextile from the Jakarta weathering site.

Plate 3: Scanning electron micrograph of weathered Polypropylene Fine Round Fibre geotextile from the Las Cruces, New Mexico weathering site.

LONG TERM EXPOSURE TO LIGHT AFTER INSTALLATION

There are a number of products - almost exclusively geogrids - which are designed for long term exposure to the natural light. In particular they are required for snow fences, sand fences, highway separation visibility barriers, gabions for the support of steep slopes or for erosion control, rock fall control meshes, ice and snow fall control meshes and soil erosion prevention textiles.

Although the majority of these products are intended to last as long as possible, erosion control textiles, such as 'Geojute', are designed to degrade rapidly as part of their function. The others are predominantly made from large diameter mesh, extruded plastic or plastic sheeting, thus giving them the longest survival period before deterioration sets in.

In terms of quantitative geotextile usage, the proportion of geotextiles subjected to long-term exposure is very small - possibly less than one percent of all applications.

SOIL EXPOSURE

There are a wide variety of different soil types in the world and various maps have been published (Rankilor 1981). However, as far as geotextiles and engineers are concerned, the main subdivisions of interest are more simple:

a) Is the soil/water which will be in contact with the textile highly acidic?

b) Is the soil/water which will be in contact with the textile highly alkaline?

c) Does the soil/water which will be in contact with the textile contain any unusual content which will attack the geotextile, such as industrial chemical pollutants or vehicle oils/fuels?

In the context of geotextile degradation, most soils are neither highly acidic nor highly alkaline. Clay soils, limestones and chalks naturally tend to be alkaline while peat soils in particular are acid, but rarely to an extreme level, with the possible exception of those where the presence of iron pyrites can give rise to strong acid levels, when natural groundwater dries up and the concentration is increased. Textiles are rarely if ever in contact with such waters however and the author is not aware of any recorded cases of damage as a result of such conditions.

It is recognised that all the geotextiles are highly resistant to mid-range pH levels, with polyethylene and polypropylene being the most stable. Polyester is for all intents and purposes equally stable, but is more susceptible to attack at high alkalinity levels. In particular the by-products given off by setting cement will attack polyester vigorously which precludes the use of that polymer in direct contact with concrete and cement during the setting process.

Clearly, the kind of soil in an area may be of no direct importance to the consideration of 'soil exposure' for a civil engineering geotextile. Only a proportion of applications actually use the geotextile in contact with the local soil. In a significant number of uses, the geotextile is only in contact with artificially-imported fill materials and is laid in an environment which does not experience the flow of natural ground waters. In a reinforced soil wall, for example, the reinforcing geotextile or geogrid is placed in a matrix of imported granular materials which will usually bear no relationship to the local soils. Furthermore, the retaining wall is constructed of free draining materials and so the environment of the geotextile is one of fresh surface water percolation (if any) and a 'non-soluble' granular material such as crushed sandstone. It is likely that the environment will therefore be as near neutral as possible, although the wall itself might be built in an area of naturally extreme soil conditions.

In the case of a geotextile used in the blacktop asphalt layer of a road, there will be no contact with the local soils but it will experience an extreme form of pollutant attack from diesel, petrol, carbon-dioxide generated acid and other vehicle related chemicals.

EXPOSURE TO WATER

From the point of view of damage related to soil exposure, the field situation is often better than the theoretical one. Conversely, with pure water the attack on most polymers is expected to be slight, but in some geotextile installations, the water in contact with the geotextile turns out to be not so innocuous.

In terms of mechanical properties, exposure to water will immediately weaken most geotextile polymers to some degree. Polyamide is by far the most susceptible and, possibly for this reason, is rarely used for geotextile manufacture. Simple weakening does not in itself comprise long term degradation if it is a single-stage phenomenon and does not operate continuously. In theory this would appear to be the case. Unfortunately, in practice, a further element is present: the wetting of the polymer followed by drying. Wet-dry cycling does appear to be detrimental to the life of the geotextile as the polymers are subjected to alternate swelling and shrinking.

Table III: Mechanical Properties

	Tenacity (cN/tex)	Breaking Extension (%)	Initial Modulus (cN/tex)	Work of Rupture (cN/tex)
Nylon	75	16	450	6.0
Polyester	55	8	1400	0.7
Polypropylene	65	17	700	7.0

Wet-dry cycling is found in many practical situations, for example, in ground drains subject to intermittant water flow. Probably the most extreme example is that of a geotextile placed in a marine tidal zone. Diurnal tidal movement and wave action ensures that frequent wetting and drying takes place.

In addition to the direct effects of water the more important effects of the materials contained in the water must be considered. In the sea there are salts; dissolved oxygen (plentiful in the wave zone); organic life and pollutants from shipping (oil spills and similar). Further, the temperature of the water is not insignificant in supporting the processes which can attack geotextiles and it would appear likely that warm tropical sea water will have a more rapid degrading effect than cold stream waters.

SOME SPECIFIC PROPERTIES OF GEOTEXTILE POLYMERS

The property of tenacity is not often recognised in engineering considerations. Although of significance where high loads are to be experienced, it is not as important as initial modulus which specifies the designable stress/strain response of the polymer. For a different reason the work of rupture is important in choosing a geotextile for a particular purpose. In particular, a high work of rupture is advantageous in such applications as haul road construction where the geotextile is to be severely stretched and almost ruptured. In simple terms, a polypropylene fabric will require considerably more energy to break than a polyester one of the same ultimate strength. The figures in Table III are based on information presented in Ford (1986).

Clearly, polyester has the most useful initial modulus for reinforced soil applications, providing a very rapid absorption of stress with minimal strain.

Abrasion resistance is a quite different factor, to be considered for different purposes. In particular, polyamide is by far the most abrasion resistant polymer used for geotextile construction. However, its low modulus and poor water tolerance mean that it is rarely used. The weight loss in a standard test is the relative indication of a polymer fabric's resistance to abrasion. As can be seen in Table IV, nylon is superior by a great margin.

Table IV: Abrasion Resistance Test Results.

	Weight Loss (g)
Nylon (Polyamide)	0.02
Polyester	0.6
Polypropylene	0.9

The effect of immersion in water at 20°C is quite marked on the common polymers. In particular, there is a very large fall in modulus for polyester when soaked. When used in dry reinforced bank and wall environments, the modulus may well be high, but the design use of polyester in wet environments such as embankment and roadway bases merits careful consideration as to the values chosen for initial modulus. On the other hand, the ultimate strength of the polymer is little affected and on the whole, in terms of being unaffected by water, polypropylene is without doubt the best (Table V).

Table V: Effects of water immersion.

	Retained Tenacity (%)		Breaking Extension (%)		Initial Modulus (cN/tex)	
	Water 20°C	Dry	Water 20°C	Dry	Water 20°C	Dry
Polyamide	82	80	22	20	230	440
Polyester	100		27	25	900	1420
Polypropylene	100		17	15	720	710

Thermal behaviour is generally of limited importance although it may be significant in extreme desert or high latitude environments and in-asphalt applications may also call for high thermal endurance. The melting point of polyester is 247°C with a glass transition point of 70-80°C. Thus for all normal geotextile end uses, the polymer behaves consistently without passing through a transition. Indeed, polyester actually increases in tenacity if heated to temperatures of 100-200°C. Alternatively, polypropylene, with a melting point of 162°C has a glass transition point of only 0°C and so can behave somewhat variably around that temperature. It is also a highly brittle polymer and may suffer excessive damage from sharp impact during installation.

Creep is also an important consideration for design purposes. Of the common polymers, polyester is by far the most creep resistant, with a creep coefficient of 0.15 against 0.28 for polyamide and 1.3 for polypropylene slit film yarn. However, with polypropylene, it is believed that adjustment of the extrusion conditions may help to minimise creep tendencies in the polymer.

In terms of chemical stability, polypropylene is highly stable and virtually inert to acid or alkali immersion at normal levels. Experiments have shown 100% retention of tenacity in polypropylene after ten hours of immersion in 10% solutions of acid and caustic soda at 99°C.

As indicated above, there is a tremendous variation in the intrinsic chemical and physical properties of the common polymers and in their responses to different physical and chemical environments.

SOME SPECIFIC EFFECTS ON GEOTEXTILES AND THEIR ENGINEERING IMPLICATIONS

During the course of the study it was noted that the geotextiles deteriorated noticeably with age when kept at room temperature in a UV free environment. The engineer should therefore consider the probable age of the product, particularly in overseas situations where shipping and storage for years in extreme hot or cold conditions are likely to cause a significant reduction in expected properties.

As the shapes of stress/strain curves have been found to be very different for different geotextiles (see Figure 4), the current pre-occupation with the ultimate tensile strength of geotextiles represents a major potential hazard

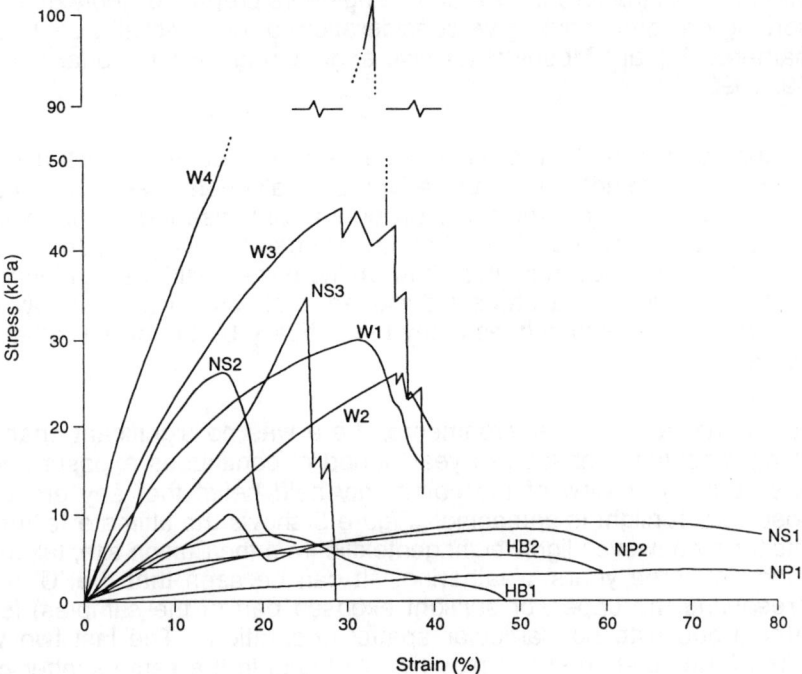

Figure 4: Typical stress/strain curves obtained from more than one thousand tensile tests on fourteen representative geotextiles and geogrids.

to present design philosophy. Not only may textiles have markedly different stress/strain curves from those published by the manufacturers, but in many cases weathering further alters the shape of the stress paths followed by the materials in being strained to ultimate failure.

The range of shapes of stress/strain curves is much greater than most engineers envisage and it is likely that textiles have higher practical ultimate strain levels than those advertised by the manufacturer. Figure 4 combines the results of the many tests enumerated above and may be used as an approximate 'likely' guide in the absence of graphical data from a manufacturer.

It is imperative that design engineers should view the full stress/strain graph for design purposes rather than make assumptions about the shape of the stress/strain response curve yet it is surprising how rarely this is done. Having obtained the stress/strain curve for a particular product, the potential influence of weathering in the proposed environment should be considered - not just in respect of its ultimate tensile stress, but also as regards potential alterations to the shape of the stress/strain curve in the first 5% of strain, ie the range used for stress absorption purposes in soil reinforcement designs.

The variety of stress/strain curve shapes clearly indicates that care must be taken to assess and design the working range of strain for geotextiles in relation to the structure in question. Engineers preparing geotextile stress absorbing designs should give consideration to the possibility of using the 'Weathered Secant Modulus' for civil engineering design works (Rankilor 1989, 1990).

It is important to note that the changes experienced during the first few years of weathering are not necessarily adverse for all geotextiles in terms of the design equations into which the properties are inserted. For example, although some graphs show that UV-exposed polymers decrease in ultimate strength, they may also show an increase in stiffness and modulus. If reinforced soil applications are designed to stress polymers at low strain levels, this increase in stiffness may theoretically be beneficial rather than adverse.

In temperate and cool environments, there was no significant change of working modulus over a three year period of continuous exposure, which was surprising in view of the commonly held belief that any amount of exposure to sunlight is damaging. Figure 5 shows the ultimate failure test points for a nonwoven lightweight geotextile from the Buxton site, taken over a period of three years' weathering. It can be seen that the 'U' results (representing the upper, or sunlight exposed part of the samples) form a scatter group with no particular spatial orientation. The last two years results of the soil-buried samples (L) are found in the same scatter cloud, indicating that there is no effect from either sunlight or soil burial over that period of time.

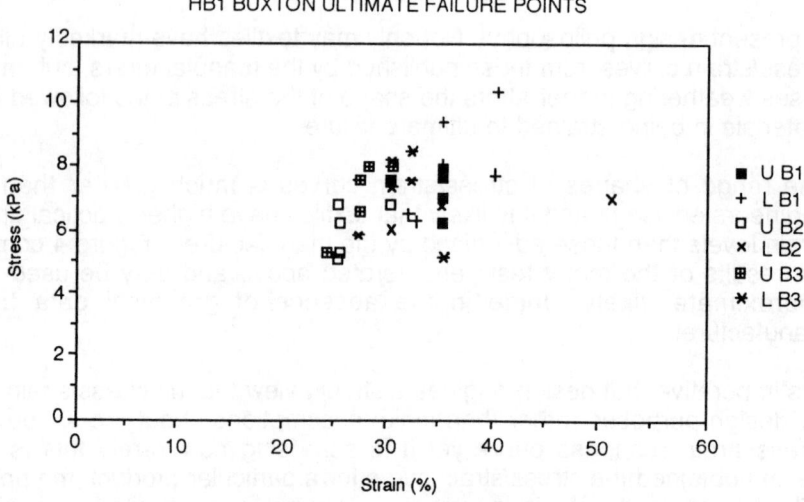

Figure 5: Tensile strip test result showing lack of deterioration of a thin nonwoven geotextile. Three years temperate exposure (U - upper light exposed: L - lower soil; buried).

In the case of this nonwoven fabric, variation of strength is not attributable to the fibres themselves, which are as uniform as those of a woven product, but to the distribution of the fibres within the textiles. Substantial variations of strength occur in accordance with the variation of 'thick' and 'thin' patches

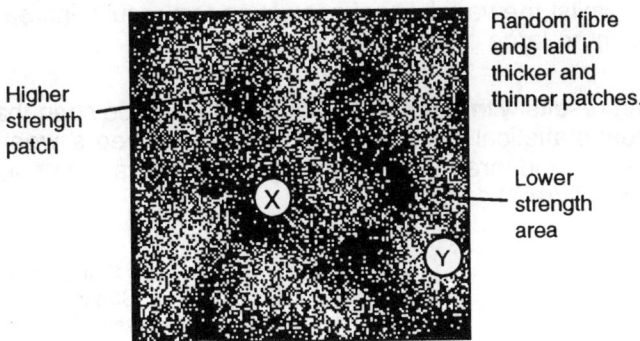

Figure 6: Diagrammatic illustration of a nonwoven fabric with patches of variable thickness and strength.

in the textile (Figure 6). Wide testing (200 mm test grip) tends to smooth out localised variations in the textile properties. In particular, as weathering effects are more significant on weaker parts of the textile, wide testing will mask the increasing scatter of test results and hide their potential impact on design considerations. For this pragmatic index study, therefore, narrow strip tensile testing (25 mm test grip) was adopted. The research has shown that the nonwovens suffer from an increase in variability of property with time, as the agents of deterioration attack the weaker parts with greater success than the stronger.

Cluster groups of data become cluster strings as the weaker points deteriorate more rapidly. This extending of the scatter group would necessitate engineers undertaking a large number (say 20 to 25) of index tests per sampling session in order to establish a coherent group for marine exposed samples of nonwoven fabrics over any sensible period of time.

It is of vital importance that the variability of geotextiles is taken into account by design engineers when they are considering the application of design equations. This is particularly emphasised because it is generally purported that narrow width tests, cone tests and grab tests can be used as index tests for checking on site-delivered quality. When original variability is masked by wide width testing statistics, or even accidentally overlooked by designers, the interpretation of isolated index test results becomes unreliable.

The statistical scatter effects obtained using small samples, give an indication of the likely variation in results and therefore the number of small scale index tests needed to assess the true overall property of various geotextile product types by construction and polymer. This may be used as a first guide in the specification and interpretation of index test results for site acceptance.

In summary, there are a wide range of responses to environmental exposure, depending upon the different environment, the polymer concerned, its additives and its fibre dimensions. On the whole, the woven fabrics, with their more uniform initial constructions, maintained a more uniform response, whilst the variability of nonwoven test results increased with time, even though no large overall trend was apparent.

Comparing results with graph HB1 Control F, it can be seen that in fact there was no real statistical difference between three years weathering at the temperate site and three years standing in the laboratory under UV free, dry conditions (Figure 7).

Textile HB 1		Mean Stress (kN/m^2)	Strain (%)	Standard Deviation Stress (kN/m^2)	Strain (%)
3 yr old	Control	7.1	36.0	1.3	4.1
1 yr	Buxton U	7.4	34.0	0.7	2.2
2 yr	Buxton U	6.3	26.0	0.8	2.2
3 yr	Buxton U	7.3	27.2	1.1	2.2
1 yr	Buxton L	8.5	37.0	1.6	2.7
2 yr	Buxton L	7.0	31.0	0.7	2.2
3 yr	Buxton L	7.3	31.2	1.4	3.0

Figure 7: Statistics for the ultimate peak level 'Failure Points' of HB1 geotextile in Buxton, derived from graphs HB1 B1F, HB1 B2F & HB1 B3F.

It is clear that continuous deterioration must have taken place, highlighting the fact that mass variation in nonwoven quality can mask the results of field index testing, even over a period of three years. The plotted results suggest that index testing with about ten samples for each buried or exposed set would be advantageous in terms of statistical description reliability.

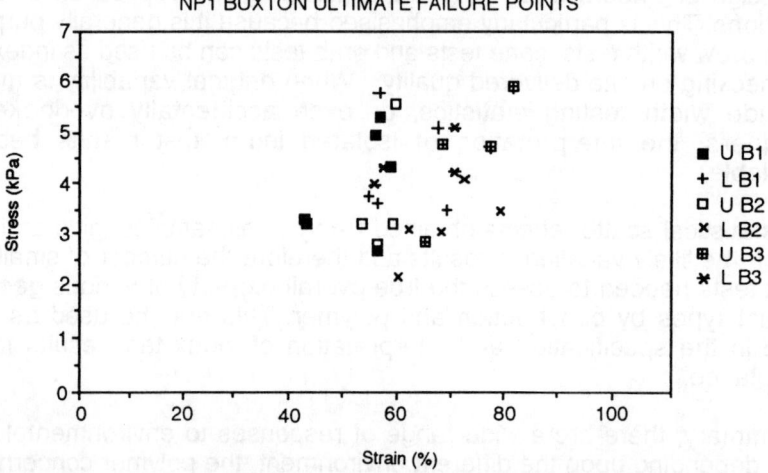

Figure 8: The NP1 ultimate failure points in Buxton, derived from graphs NP1 B1F, NP1 B2F & NP1 B3F.

The buried specimens (LB1, 2 and 3) showed no sign of deterioration throughout the three years of testing, in terms of the scatter of ultimate failure points (Figure 8).

The UV exposed samples can be seen to move to the right (becoming more extensible) with each successive year. The relatively large scatter indicates the order of specimen-to-specimen variability with each sample. The respective overlap of the sample scatter indicates the reliability and meaningfulness of the test method.

The increasing extensibility of the UV exposed parts of the NP1 samples is indicated by the increase of mean percentage strain from 50.8 to 56.6 to 72.7% (Figure 9). Further exposure may show this trend to continue. However, it is entirely possible, if not likely, that this 'trend' may be reversed in future years.

Textile		Mean		Standard Deviation	
NP 1		Stress (kN/m^2)	Strain (%)	Stress (kN/m^2)	Strain (%)
3 yr old	Control	5.2	66.8	1.0	12.3
1 yr	Buxton U	4.2	50.8	0.9	7.6
2 yr	Buxton U	3.5	56.6	1.7	2.5
3 yr	Buxton U	4.5	72.7	2.5	7.5
1 yr	Buxton L	4.3	60.4	1.0	7.0
2 yr	Buxton L	3.2	68.2	1.4	7.7
3 yr	Buxton L	4.4	63.0	2.1	8.1

Figure 9: Statistics for the ultimate failure points in Buxton, derived from graphs NP1 A1, NP1 A2.

In terms of ultimate tensile strength, no trends can be observed in terms of the 'means', but both the upper and lower samples show remarkably similar distinct increases in the scatter of points with age. The 'Upper' Standard Deviation of ultimate tensile strength increases from 0.9 to 1.7 to 2.5; the 'Lower' equivalent is 1.0 to 1.4 to 2.1. It is apparent that the weaker parts of the variable product are being more rapidly degraded leading to increased variability of performance.

In contrast with the preceding nonwoven products, a remarkable uniformity of result can be observed with the woven fabric. The scatter plot shown in Figure 10 is typical of the kind produced by woven fabrics and the closely grouped points can be seen to spread along the line of the deformation stress/strain curve.

The third year results show that the last W1 specimen had a higher ultimate strength than the one tested after one year (Figure 11). Further years of testing will provide information on trends, but bearing in mind the relative

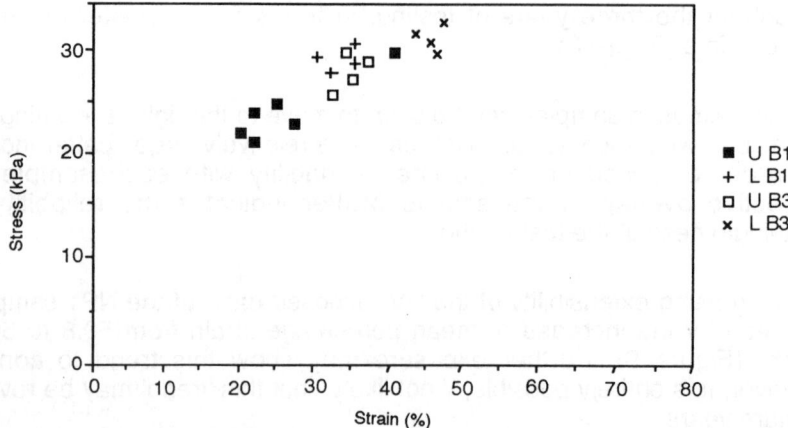

Figure 10: The W1 ultimate failure points in Buxton, derived from graphs W1 B1F, W1 B2F & W1 B3F.

uniformity of a woven product, the apparent increase in ultimate strength with time might well prove real. As discussed elsewhere, polymers can increase in mechanical strength before decreasing as degradation sets in. Certainly the response obtained in this study does not support the commonly anticipated rapid deterioration of physical properties with time.

Textile		Mean		Standard Deviation	
W 1		Stress (kN/m^2)	Strain (%)	Stress (kN/m^2)	Strain (%)
3 yr old	Control	28.2	32.6	2.2	4.4
1 yr	Buxton U	23.0	23.2	1.5	2.7
2 yr	Buxton U				
3 yr	Buxton U	28.5	25.6	1.7	3.0
1 yr	Buxton L	29.5	33.2	1.1	2.1
2 yr	Buxton L				
3 yr	Buxton L	31.2	44.2	1.3	2.7

Figure 11: Statistics for the ultimate peak level 'Failure Points' of W1 geotextile in Buxton, derived from graphs W1B1 & W1B2.

In contrast to the results from the nonwoven products, Figure 11 shows small values of Standard Deviation for strain for the geotextile in Buxton although the stress level variations are more similar. This confirms the ineffectiveness of simply describing a fabric by its ultimate tensile strength. The shape of the curve, as represented by variations in ultimate strain levels, is clearly important.

It can be observed that in both cases, the upper exposed specimens have a lower average and overall strength and extensibility than the soil exposed samples. On the whole, however, there is no substantial shift between them with time. For example, the difference between the first year lower and upper specimen mean is 6.5 kN/m^2, whereas the difference between the second year lower and upper specimen mean is only 2.7 kN/m^2. This difference, together with an examination of the test plot positions, suggests that the 'strengthening' from UV exposure is still continuing at three years whilst the strengthening from soil exposure was slowing down in the third year.

SUMMARY OF WHAT THE ENGINEER NEEDS TO KNOW

The engineer needs to recognise that all polymers are deteriorating with time. However, under some circumstances, their rate of deterioration is no greater than under dry laboratory storage. In other situations exposure to the environment can deteriorate the textile at such a rate that it will have virtually no strength after one year. Some guidance has been given as to the different environments for each condition.

Essentially, exposure to sun or soil in temperate or cold conditions has little effect on geotextiles beyond their storage deterioration, over a period of at least several years. However the study indicated that the inherent manufactured variability of nonwoven fabrics is exacerbated by weathering as weaker parts of the fabric deterior-ate faster than stronger ones. Exposure to tropical and desert sunshine rapidly attacked the geotextiles, with a clearly more detrimental effect on geotextiles with fine fibres - again usually the nonwoven type. As exposure to sunlight can be restricted to short periods by contractual specification, this should not present a working limitation on these kinds of fabrics. However, there was also a corresponding reaction of fine fibre textiles to tropical marine immersion. This gives some cause for concern and needs further urgent investigation. The present results may be considered to be only indicative due to the restricted nature of the samples; a full marine immersion research program is clearly required.

When specifying geotextiles for filtration work, consideration should be given to the inherent variability of nonwoven geotextiles, which should be balanced against their apparent overall functional superiority for filtration purposes. This is a subject of great debate, but the variability of non-wovens is not in doubt and must be borne in mind when the nominal weight of textile is being specified.

With textiles to be used for soil reinforcing purposes, ie to carry mid- and long-term loads, the engineer should study the full stress-strain curve and assume that this curve will lessen slightly with time around the 5% strain range. As working strains are generally in this order, the theoretical design modulus of the textile should be adjusted accordingly. For a life span of 100 years, it may be advisable to assume that the modulus ultimately available may be little more than a tenth of that originally tested. This will require large initial safety factors but it is considered that the current long term laboratory test data of twelve continuous loaded years is still too short and too 'clean' to provide a realistic indication of the likely 100 year behaviour of these polymers.

CONCLUSIONS

This research work has highlighted a number of points with implications for the civil engineering design and construction industry, in relation to geotextiles. These include:-

1. Products deteriorate with age, even if kept at room temperature in a UV free environment.

2. It has been found that stress/strain curves frequently exhibit considerably more strain and may be quite different in shape from that indicated by the manufacturers.

3. In some cases weathering does not affect the ultimate stress levels of geotextiles but does depress the modulus, particularly around the important 5% strain levels.

4. It is recommended that larger numbers of index tests are undertaken during on-site testing than may be current practice in order to assess the variability of products arriving at site.

6. Exposure to sunlight in cool climates has virtually no detrimental effects over periods of many months. The current concern regarding site exposure may therefore be unwarranted in temperate countries. A guidance table is included in this paper.

7. The two main polymers currently in use for soil reinforcement (polyester and polyethylene) have shown adverse weathering reactions in the desert soil test environment. This raises important questions with regard to their suitability in hot environments such as Saudi Arabia and to the possible effects of storage in such countries.

8. There is a need for small scale weathering control sites on major installations in order to monitor the specification performance of geotextiles with time.

REFERENCES

FORD, J.E. (1986). *Durability and Environmental Resistance of Geotextile Materials.* An in-house document, Shirley Institute, Didsbury, Manchester.

KOERNER, R.M. (1986). *Designing with Geosynthetics.* Prentice-Hall.

RANKILOR, P.R. (1981). *Membranes in Ground Engineering.* John Wiley & Sons Ltd, Chichester, England.

RANKILOR, P.R. (1986). Problems relating to light degradation and site testing of geotextiles - interim results of an international weathering programme. *Durability of Geotextiles - RILEM Conference,* Saint-Remy-les-Chevreuse, near Paris, France.

RANKILOR, P.R. (1988). The range and function of geotextiles. *International Textile Engineering Technology Conference*, The Textile Institute, Manchester.

RANKILOR, P.R. (1989). *The weathering of fourteen different geotextiles in temperate, tropical, desert and permafrost conditions.* Unpublished Ph.D. thesis, University of Salford, UK.

RANKILOR, P.R. (1990). The comparative weathering of fourteen different geotextiles and geogrids in desert, tropical, permafrost and temperature weathering stations. *Proceedings 4th International Conference of the Geotextile Society,* The Hague, Holland.

SCHNEIDER, H. (1987). An analysis of the durability problems of geotextiles. *IFIA Conference Geosynthetics '87,* New Orleans.

3-6 THE PHYSICAL ENVIRONMENTS OF GEOTEXTILES IN CIVIL ENGINEERING STRUCTURES

P.R. RANKILOR
Manstock Geotechnical Consultancy Services Ltd, 1 North Parade, Manchester, U.K.

ABSTRACT

The paper describes the range of environmental conditions likely to be experienced by geotextiles and geogrids in civil engineering earthworks structures. Various polymers are discussed and the properties of products relevant for different applications. The implications of the highly varied environments on different polymers and particularly on the long term life expectancy of geotextiles and geogrids are explored. Potential functional time periods are suggested for different types of engineering works.

For convenience, the information is also represented in tabular form and an initial reference list is provided for those who may wish to extend their study and further explore this field.

INTRODUCTION

In civil engineering, geotextiles are used in an extremely wide variety of environments; they are also subjected to greatly differing conditions of installation and end usage (Rankilor, 1981).

Environmental parameters include:-
    Wetness of the environment.
    pH level of the environment.
    Temperature regime within the structure.
    Level of organic attack during the structure's life.
    Special chemical conditions pertaining at the site.

Conditions of installation and use include:-
    Stress during installation.
    Damage during installation
    Stress levels experienced in use.
    Length of time of stress experienced in use.

At present there is no system for deducing a categorisation of geotextiles based on a sub-classification of these major divisions. Nevertheless, it is possible to consider some of the major conditions above in terms of their implications for the use of different types of geotextiles for various civil engineering structures.

Perhaps the simplest case is "wetness". As can be seen in Figure 1, geotextiles are expected to perform in a range of environments from totally

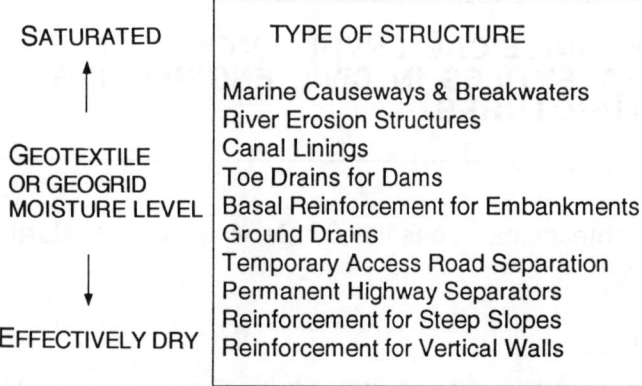

Figure 1: Range of moisture conditions for working geotextiles.

saturated to totally dry. Other things being equal, geotextiles will generally suffer less degradation if they are dry than if they are wet.

There is also an obvious demand on the geotextile which varies directly in accordance with the physical stress which is likely to be placed on it. For example, in a vertical retaining wall the geotextile is likely to have both a high and a long-term stress placed upon it. In a site access road, the stress on the geotextile is high, but the time requirement is short. In a ground drain, the stress on the geotextile is very low but it needs to continue functioning for a long-term period.

Clearly therefore, whether the civil engineering structure is above ground, in the ground, or under water will have a distinct effect on the potential life of the geotextile. It is possible to make the simple statement that the overall "demand" on the geotextile will be a function of the stress and the amount of moisture to which the geotextile will be subjected. In simple mathematical terms, this can be represented as:-

$$\text{Demand} = f.\sigma.W.$$

The intensity of demand on a geotextile can be seen, inter alia, to be a function (f) of the stress (σ) and wetness (W) of the environment. A highly stressed, very wet environment will undoubtedly present the greatest demand on a geotextile while a low stress, dry environment will present the least demand.

An examination of the range of working temperatures for geotextiles shows that they are currently being incorporated into structures in every climatic règime. Figure 2 classifies end uses by virtue of the amount of stress experienced in the geotextile during use in relation to wetness of environment. It shows that at least four major types of structure are being built in arid desert environments where the air temperature can exceed 40°C.

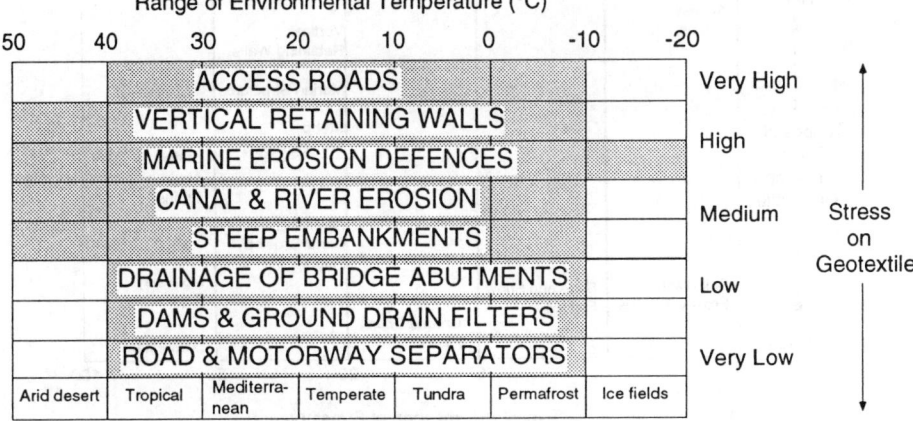

Figure 2: Diagram illustrating that geotextiles are used in virtually every climatic environment, irrespective of stress implications.

It is clear that there is no simple way to establish a "grade" of geotextile by simply multiplying the two different parameters σ and W. Even if multiplication factors could be scientifically allocated, this would not take account of the many additional factors which may be present; for example, exposure to light/temperature and such other environmental influences as soil acidity or possible organic attack. As can be seen from Figure 2, wetness of application and environmental temperature are not related - for example, a considerable number of canal and marine works are constructed in the Saudi Arabian coastal areas as well as the Russian steppes or on the coasts of Alaska and Newfoundland in areas where the sea freezes!

Consideration of the conditions of installation is also important. For example, stress can damage a geotextile either by causing immediate excessive strain, or by generating long–term creep failure. Before designing a structure, therefore, the imposed stresses must be assessed. As illustrated in Figure 3, not all civil engineering structures impose the same stress levels for the same periods of time.

The total "stress demand" to be placed on a geotextile varies from a minimum at the bottom left hand corner of the table to a maximum at the top right hand corner. In general, most polymers are suitable for the left hand side applications on the graph and for the low stress applications but the high stress, long-term applications demand the use of polymers which can stabilise under stress and not creep to failure.

In considering the types of environment in which geotextiles can be used, there would appear to be two major classification requirements. The first is the degree of stress imposed by the civil engineering structure and its immediate environment; the second is the climatic region in which the particular structure is to be constructed.

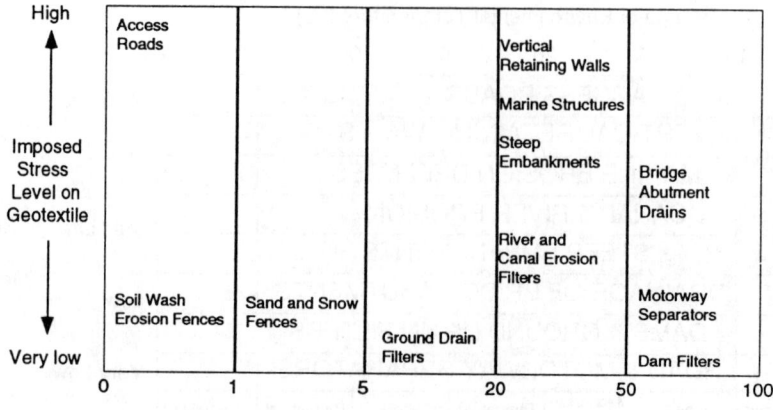

Figure 3: Expected life spans for a number of different structures.

In the case of stress, research has been conducted over the last fifteen years by various parties such as Netlon Ltd (1987). There has therefore been a realistic start on the assessment of the response of different geotextiles to high stress levels and in particular to long-term induced stresses which could lead to creep failure.

The long-term functional effects of using geotextiles in wet environments have been well documented by authors such as Heerten (1986) but only very limited work has been done on the degradative effects of water exposure itself. The early emphasis was necessarily on the technical effectiveness of geotextiles as filters but it is clear that the long-term exposure of various polymers to water will have an adverse effect on their physical structure. Polyamide in particular suffers a substantial strength loss when submerged in water and, possibly for this reason, is rarely used in geotextile construction.

Although some work has been undertaken by the author (Rankilor, 1986, 1989 and 1990) the long term degradation of geotextiles is an extremely complex subject which clearly warrants further research, particularly in the context of civil engineering applications. There is little doubt that water temperature, dissolved oxygen, dissolved chemicals and organic activity levels will all have a bearing on the potential life of different polymers within the simply-defined framework of a 'wet' environment. Furthermore, it is likely that the marine geotextile environment will be additionally demanding as desiccation and hydration of the polymer with alternating tidal cycles induces more rapid deterioration than either desiccation or hydration alone as long-term environments.

The reaction of geotextiles to their environment is particularly difficult to analyse as different polymers react differently to the various levels of certain categories of agent; eg the pH level of the immediate contact environment. As some polymers behave considerably better under high pH conditions whilst others show improved performance under low pH conditions, it is impossible to suggest a linear division whereby one extreme of environmental agent is considered more aggressive than the other for all polymers.

This observation leads to the logical conclusion that, ideally, any environmental classification system relating to geotextiles should be defined in terms of individual polymer types. The polymers most commonly used in civil engineering fabrics include;

>polypropylene
>polyester
>polyethylene
>polyamide
>polyvinylidene

It might be realistic to take each polymer in term and assess the order of degradative impact of the installation environment on that particular polymer. Beyond this, recent research work (Rankilor, 1989) suggests environmental attack will be substantially greater on polymers that are spun as fine filaments and incorporated into textiles in that form, than on polymers that are extruded as coarser elements and thus present a far smaller specific surface area for attack.

In summary, it is suggested that any system for analysing and classifying the degradation of geotextiles will need to consider the following:-

1. Sub-division by individual polymer types.
2. Sub-division by specific surface area available for attack.
3. Sub-division based upon utilisation end-use.
    - a) Wetness
    - b) Temperature
    - c) Stress
    - i) Installation
    - ii) Length of service
4. Sub-division by chemistry of environment.

These sub-divisions are by no means exhaustive but simply propose a means of subjective assessment. They indicate clearly the complexity of the likely permutations and combinations which may exist in civil engineering environments and which makes the overall assessment of suitability of geotextiles for different purposes so difficult.

It is clear that further study is required into both the many different factors which influence the deterioration of geotextiles and the long term effect of the environment before the utilisation of geotextiles can confidently be adopted in major civil engineering structures. In addition to soil exposure, there is a real need for research into the effect of marine immersion on geotextiles, particularly in warm tropical seas.

ACKNOWLEDGEMENTS

The author wishes to thank UCO Technical Fabrics NV, Lokeren, Belgium, for their support and permission to use the diagrams herein, which were originally developed by the author for a Design Manual for Geosynthetic Materials yet to be published at the time of writing this paper.

REFERENCES

HEERTEN, G. (1986). Functional design of filters using geotextiles. *Third International Conference on Geotextiles,* Vienna, Austria.

RANKILOR, P.R. (1981). *Membranes in Ground Engineering.* John Wiley & Sons Ltd, Chichester, England.

RANKILOR, P.R. (1986). Problems relating to light degradation and site testing of geotextiles - interim results of an international weathering programme. *Durability of Geotextiles - RILEM Conference,* Saint-Remy-les-Chevreuse, near Paris, France.

RANKILOR, P.R. (1989). *The weathering of fourteen different geotextiles in temperate, tropical, desert and permafrost conditions.* Unpublished Ph.D. thesis, University of Salford, UK.

RANKILOR, P.R. (1990). The comparative weathering of fourteen different geotextiles and geogrids in desert, tropical, permafrost and temperature weathering stations. *Proceedings 4th International Conference of the Geotextile Society,* The Hague, Holland.

WRIGLEY, N. (1987). Durability and long-term performance of Tensar polymer grids for soil reinforcement. Materials Science and Technology, *Journal of the Institute of Metals,* **3**, 3-5.

3-7 EVALUATION OF SOIL CHEMISTRY AND ITS ROLE IN HIGHWAY EMBANKMENT EROSION - A CASE STUDY

R.K. SRIVASTAVA
Civil Engineering Department , MNR Engineering College Allahabad, India
A.V. JALOTA
Civil Engineering Department , MNR Engineering College Allahabad, India
R.P. TIWARI
Civil Engineering Department , K N I T, Sultanpur, U P, India
T. NATH
Public Works Department, U P, India
A.K. SAHU
Airport Authority of India, New Delhi, India

ABSTRACT

The paper presents chemical and geotechnical analyses (including X-ray diffraction and SEM) of soil samples collected from highway embankments which are prone to erosion. The problem of soil erosion has become a recurring feature in India, resulting in both accidents and economic loss. The conventional measures of embankment protection, eg stone pitching and tree and shrub plantation, have not proved successful. The study has indicated that the soil chemistry itself is largely responsible for the susceptibility of the fill to erosion; a factor which had not been taken into consideration during the design of the embankments. The samples tested show high percentages of exchangeable calcium cations.

INTRODUCTION

In developing countries, especially India, the soil used in embankment construction is usually that which is available in the vicinity of the proposed structure. During the design stage the engineering properties of the soil are considered and any necessary measures for soil improvement suggested. However, the conventional geotechnical tests do not always reveal all the characteristics of the soils and this may give rise to problems during the life time of the structure. One such type of problematic fill is known as dispersive soil.

These erosion-prone soils are not identifiable in the routine geotechnical analyses. However, when used for the construction of earth embankments, earth dams, irrigation canal banks etc, they may cause extensive damage to the structure as a result of surface erosion, piping, slaking and gully erosion etc. This causes not only economic loss due to recurring maintenance costs but also accidents and even failure of the structures. As earth dams and embankments are costly structures and require very large amounts of soil, identification tests to determine the susceptibility of the material to erosion are important. For this reason, chemical analyses are required as well as conventional geotechnical analysis as it has been noted that the amount and type of exchangeable cations is largely responsible for the erodible behaviour of soils.

LITERATURE REVIEW

The literature available on studies carried out on erodible soils is meagre and most of these comprise largely case studies of observed erosion. Pioneering work in this area has been carried out by Sherard et al (1972, 1976a, 1976b) and Flanigan and Holmgren (1977) have described a field procedure for identification of soil erodibility. Decker and Dunnigan (1977) have concluded that the erodible soils generally have alkaline pore water with pH values higher than 8.5. Some important studies on various aspects of erodible soil behaviour have been reported by Chandra and Chen (1985), Chandra and Garcia (1984), Cole, Ratanasen, Maiklad, Liggins and Chirapunta (1977), Coumoulos (1977), Forsythe (1977), Melvill and Mackellar (1980). An excellent state of the art report on erodible soils has been prepared for India by the Central Soils and Materials Research Station, New Delhi, India (1986). The limitations of standard soil testing procedures for evaluating their suitability as core materials in regard to dispersivity and internal erosion have been discussed by Narula and Sharda (1990) for medium height embankment structures being constructed in the outer Himalayas for watershed management. Recently Leonards et al (1991) have discussed the results of the tests carried out to assess the filtration characteristics of the chimney drain and on the erodibility of the upstream clay blanket at Conner Run Dam located in West Virginia, USA. The studies reported clearly indicate that some soils are prone to erosion while others with similar geotechnical properties are not. Susceptibility to erosion has been observed to be dependent on the mineralogy and chemistry of the soil, the pore water and the eroding water.

CASE STUDY

The present study discusses the problems of soil erosion encountered on three stretches of road, the Varanasi-Bhadohi Road, the Lumbini-Dudhi Road and the Rewara Phatak-Radheypur Road which are major arterial roads in the state highway system and major district roads. The recurring problem of soil erosion has resulted in traffic difficulties due to the reduced embankment widths and increase in accidents. The magnitude of the problems is such that at critical locations the width has been reduced to 5 m after only one or two rainy seasons. The conventional methods of control, eg repair, compaction, stone pitching etc, have not proved successful and the problem persists.

The geotechnical tests carried out indicate that the soil used in embankment construction is predominantly silty soil, classified as 'ML' type. A study of the borrow areas from where the soil had been taken suggested that the land was not suitable for agricultural purposes because of its high alkaline nature (ie it was barren land). Visual observation of the eroded embankment (Plates 1 to 4) and study of the borrow area implied the soil

SOIL CHEMISTRY AND ITS ROLE IN HIGHWAY EMBANKMENT EROSION

Plate 2: Patterns of soil erosion.

Plate 4: Failure of stone pitching.

Plate 1: Patterns of soil erosion.

Plate 3: Embankment erosion.

used in the stretches of highway embankment might be susceptible to erosion. As a consequence soil samples were collected from several locations and geotechnical and chemical analyses carried out.

GEOTECHNICAL PROPERTIES OF THE SOILS

Table I presents the properties of soil samples from twenty locations. The samples were disturbed soil which had actually been eroded. Particle size analysis, consistency limits, specific gravity, optimum moisture content and maximum dry density have been determined from the disturbed soil samples. Strength parameters, cohesion and angle of internal friction were also determined on the remoulded samples.

It can be seen that the soil used for construction of the highway embankment is predominantly silty, with a very low percentage of clay and a small amount of sand. As anticipated these soils have a very low plasticity index and in some cases they are non-plastic. A typical particle size variation curve of these ML soils is shown in Figure 1. The results of the X-ray diffraction analysis indicated the minerals to be predominantly kaolinite and quartz. Scanning electron microscopy for typical samples is shown in Plates 5 and 6.

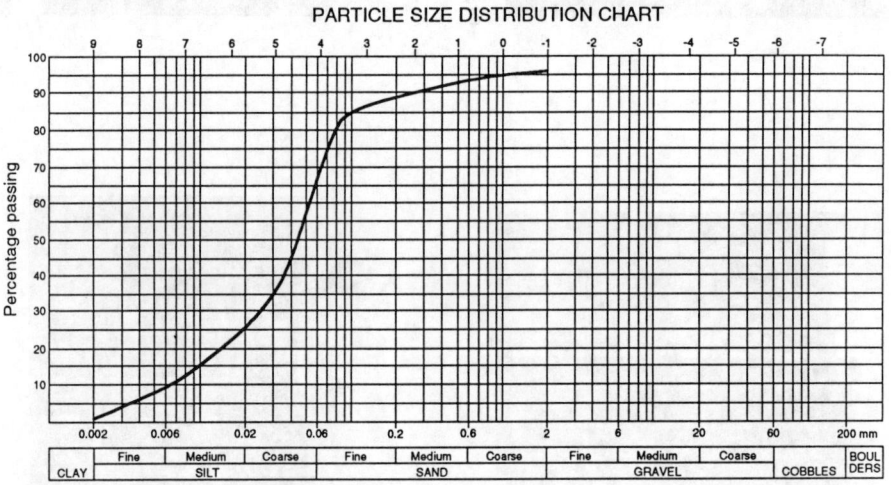

Figure 1: Particle size distribution curve of a typical soil sample.

CHEMICAL ANALYSIS OF SOILS

Chemical analyses of the soil samples have been carried out to determine their pH value, conductivity and total and exchangeable cation content.

The acidic or alkaline nature of the soil (pH) was determined using a pH meter (Beckmeen, USA). Ten grams of soil sample were added to 25 ml of deionized water and left standing for six hours before the solution was placed in the pH meter. The pH values of various soil samples are included

Table I: Properties of soils at various locations.

Sample No.	Particle size distribution				Consistency limits			Soil Type	S.G.	OMC %	MDD gm/cc	Strength parameters	
	Clay %	Silt %	Sand %	Gravel %	w_l %	w_p %	I_p %					c_u kg/cm^2	ϕ_u°
1.	8	85	7	0	23	20	3	ML	2.65	16.9	1.68	0.16	28
2.	7	85	8	0	23	20	3	ML	2.66	16.8	1.68	0.16	27
3.	1	89	10	0	Non plastic			ML	2.66	16.0	1.69	0.00	39
4.	2	77	12	0	23	20	3	ML	2.65	15.0	1.70	0.14	30
5.	0	88	12	0	Non plastic			ML	2.67	15.5	1.75	0.00	38
6.	3	85	12	0	22	20	2	ML	2.65	14.2	1.81	0.15	30
7.	2	86	12	0	22	20	2	ML	2.67	14.0	1.80	0.14	30
8.	4	86	10	0	23	20	3	ML	2.66	15.5	1.75	0.15	30
9.	0	89	11	0	22	20	2	ML	2.65	16.0	1.74	0.10	33
10.	0	90	10	0	22	20	2	ML	2.65	16.0	1.73	0.09	33
11.	5	88	7	0	23	20	3	ML	2.66	15.8	1.70	0.14	31
12.	4	89	7	0	23	20	3	ML	2.67	15.5	1.71	0.14	31
13.	4	88	8	0	23	20	3	ML	2.67	15.3	1.69	0.14	31
14.	2	89	9	0	22	20	2	ML	2.65	14.5	1.79	0.13	31
15.	1	88	11	0	22	20	2	ML	2.65	15.8	1.79	0.11	32
16.	2	85	13	0	22	20	2	ML	2.66	14.7	1.72	0.12	32
17.	1	79	12	8	Non plastic			ML	2.67	15.9	1.69	0	38
18.	0	73	17	10	Non plastic			ML	2.67	15.4	1.75	0	38
19.	3	85	12	0	22	20	2	ML	2.65	14.5	1.79	0.13	31
20.	2	76	13	9	Non plastic			ML	2.66	14.0	1.73	0	38

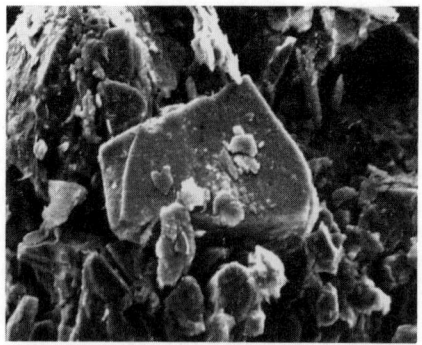 

Plates 5 and 6: SEM analysis of typical samples.

in Table II. The conductivity of the soil samples was ascertained by use of a conductivity meter after the solution had been prepared as for the pH determination. The results are also given in Table II. The total cation and exchangeable cation (Na, Km, Ca and Mg) content was determined for all the soil samples using the Integrated Current Plasma Analyser (ICP - 8410 Plasma Scan, Labtum, France). To determine total cation content of the soil 10 grams of sample is added to 30 ml of concentrated HCl and the solution boiled until full evaporation takes place. The residue is then dissolved in 100 ml of 1:1 HCl and the solution thus obtained is filtered. After filtration the solution is put into the ICP analyser and the total cation content of the soil determined. The results are presented in Table II.

The amount of exchangeable cations was ascertained by adding 10 grams of soil sample to 100 ml of 1 M Ammonium Acetate. The solution was stirred and left for 24 hrs, after which it was filtered and put into the ICP analyser to ascertain the exchangeable cations content. The results are included in Table II.

It can be seen from Table II that in general the pH of all the soil samples is high, ranging between 8.25 and 8.75 with an average pH value of 8.5. This indicates the alkaline nature of the soil which has rendered it unsuitable for agricultural purposes. The conductivity value (umho) ranges from 135 to 386 with an average value of approximately 200. This value again is in the higher range and indicative of the potentially erodible nature of the soil.

The total dissolved salts present in the soil samples ranged from 300 to 558 ppm with an average of 389. Comparing the percentage of exchangeable cations with the total cations present for all four cations it can be seen from Table III that the ratio is highest for calcium ions, followed closely by sodium ions but comparatively small for potassium and magnesium ions. This indicates that in the present case the erodible nature is dominantly controlled by the number of exchangeable calcium ions present, although the sodium ions also play a significant role.

Table II: Chemical analysis of soil samples obtained from various locations.

Sample No.	pH	Conductivity (umho)	Na Total	Na Exchangeable	K Total	K Exchangeable	Ca Total	Ca Exchangeable	Mg Total	Mg Exchangeable
1	8.75	136	14	9	115	16	219	176	41	7
2	8.25	148	12	6	131	21	182	128	38	6
3	8.75	165	16	10	103	22	175	144	47	6
4	8.75	135	14	9	114	15	220	176	41	8
5	8.75	125	11	6	94	16	150	136	45	7
6	8.55	206	19	13	86	20	261	125	45	7
7	8.45	204	28	20	76	16	179	121	46	8
8	8.30	215	19	7	81	19	188	136	45	7
9	8.25	320	72	49	146	28	291	224	49	8
10	8.25	148	13	6	132	21	184	129	38	6
11	8.70	164	16	11	108	22	173	144	47	6
12	8.75	125	11	6	94	16	153	136	45	7
13	8.50	205	27	20	77	16	180	121	46	8
14	8.45	228	21	15	92	13	276	163	48	7
15	8.55	187	24	16	121	9	215	150	46	7
16	8.45	230	22	16	90	14	271	164	46	7
17	8.55	205	19	12	85	21	262	130	44	7
18	8.50	185	24	16	122	9	220	153	46	7
19	8.25	380	66	44	140	26	293	220	48	7
20	8.30	215	20	8	80	18	190	137	45	7

Table III: Cation percentage in total dissolved salts.

Sample No.	Total Dissolved Salts (ppm)	Na %	K %	Ca %	Mg %
1.	389	3.6	29.5	56.3	10.5
2.	363	3.3	36.1	50.1	10.5
3.	341	4.6	30.2	51.3	13.8
4.	389	3.6	29.3	56.5	10.8
5.	300	3.7	31.3	50.0	15.0
6.	411	4.6	20.9	63.5	10.9
7.	329	8.5	23.1	54.4	13.9
8.	333	5.7	24.3	56.4	13.5
9.	558	12.9	26.2	52.1	8.8
10.	367	3.5	35.9	50.1	10.4
11.	344	4.6	31.4	50.3	13.6
12.	303	3.6	31.0	50.4	14.8
13.	330	8.2	23.3	54.5	13.9
14.	437	4.8	21.1	63.1	10.9
15.	406	5.9	29.8	52.9	11.3
16.	429	5.1	20.9	63.2	10.7
17.	410	4.6	20.7	63.9	10.7
18.	412	5.8	29.6	53.4	11.2
19.	547	12.1	25.6	53.5	8.8
20.	335	5.9	23.9	56.7	13.4

CONCLUSIONS

A brief case study of some embankments prone to erosion has been presented. Field observations and laboratory tests have indicated that the soil erosion is dominantly related to the chemistry of the soil, although other geotechnical factors may be contributary. The soil used for embankment construction is generally silty and although when placed it may be satisfactory from a geotechnical point of view, its resistance to erosion may be aggravated because of the presence of a significant amount of exchangeable calcium cations. The conventional means of stabilising such soils is by stone pitching and tree plantation. These methods have not proved successful and chemical additives may be the only practical and economic means to control soil erosion in the present case.

The main observations from this study are:

1. X-ray diffraction and SEM analysis indicate that the soil contains minerals which do not have significant attractive forces.

2. The pH values of the soil samples range from 8.25 to 8.75. Such alkaline soils are known to be easily erodible.

3. The conductivity values range from 136 to 390 (umho) indicating the presence of high concentrations of total dissolved salts.

4. In all the soils tested the dominant cation is calcium, ranging between 150 to 293 ppm; of which the exchangeable cation ranges from

120 to 226 ppm. It is suggested that the erodible nature of the soils in the present case is related to the high percentage of exchangeable Ca ions which are effectively electrically inert.

5. The percentage of exchangeable sodium ions is also high, but as the total concentration is less this is not considered to be the dominant influence.

REFERENCES

CHANDRA, S. & CHEN, G.L. (1985). Improvement of Dispersive Soil by Using Different Additives. *Indian Geotechnical Journal*, **14(3)**, 202-216.

CHANDRA, S. & GARCIA, E.B. (1984). Chemical and Mineralogical Study of Collapsible and Dispersive Soil. *Indian Geotechnical Conference, Calcutta, India.*

COLE, B.A, RATANASEN, C., MAIKLAD, P., LIGGINS, T.B. & CHIRAPUNTU, S. (1977). Dispersive Clay in Irrigation Dams in Thailand. *ASTM STP-623*, 25-41.

COUMOULOS, D.G. (1977). Experience with Studies of Clay Erodibility in Greece. *ASTM STP-623*, 42,53.

CENTRAL SOILS AND MATERIAL RESEARCH STATION, (1986). State of the Art Report on Dispersive Characteristics of Fine Grained Soils. *Report No. RDI-86/ASE*, New Delhi, India.

DECKER, R.S. & DUNNIGAN, L.P. (1977). Development and Use of the Soil Conservation Service Dispersive Test. Dispersive Clays, Related Piping and Erosion in Geotechnical Projects, *ASTM STP-623*, 94-109.

FORSYTHE, P. (1977). Experience in the Identification and Treatment of Dispersive Soils in Mississippi Dams. *ASTM STP-623*, 135-155.

FLANAGAN, C.P. & OLMGREN, G.G.S. (1977). Field methods for determination of soluble salts and percent sodium from extract for identifying dispersive clay soils. *ASTM STP-623*, 121-134.

LEONARDS, G.A., HUANG, A.B. & RANOSA, J. (1991). Piping and Erosion Tests at Conner Run Dam. *Journal of Geotechnical Engineering, ASCE*, **117(1)**.

LOGANI, K.L. & HECTOR, M. (1979). Techniques Developed During Foundation Treatment of the Ullum Dam Constructed of Dispersive Soils. *Proceedings of the 13th International Congress on Large Dams*, **1**, 729-748.

MCDANIEL, T.N. & DECKER, R.S. (1979). Dispersive Soil Problem at Los Esteros Dam. *Journal of the Geotechnical Engineering Division, ASCE*, **105(GT9)**, 1017-1029.

MELVILL, A.L. & MACKELLAR, D.C.R. (1980). The Identification and Use of Dispersive Soils at Elandsjagt Dam, South Africa. *Proceedings of the VIIth Regional Conference for Africa on SM&FE, Accva*.

NARULA, P.L. & SHARDA, Y.P. (1990). Dispersive Soils - A constraint for embankment dams in outer Himalaya. *6th International IAEG Congress, Rotterdam.*

SHERARD, J.L., DECKER, R.S. & RAKER, N.L. (1972). Piping in Earth Dams of Dispersive Clays. *Proceedings Conference Performance of Earth and Earth-Supported Structures, ASCE*, **1**, 589-626.

SHERARD, J.L., DUNNIGAN, L.P. & DECKER, R.S. (1976a). Identification and Nature of Dispersive Soil. *Journal of the Geotechnical Engineering Division ASCE*, **102(GT4)**, 287-302.

SHERARD, J.L., DUNNIGAN, L.P., DECKER, R.S. & STEELE, E.F. (1976b). Pinhole test for identifying dispersive soils. *Journal of the Geotechnical Engineering Division, ASCE*, **102(GT1)**, 69-87.

3-8 THE INFLUENCE OF CALCITE ON THE INDEX PROPERTIES OF TRIASSIC MUDROCKS

J. COLLINS & A.B. HAWKINS
Engineering Geology Research Group, University of Bristol, Wills Memorial Building, Queens Road, Bristol BS8 IRJ.

ABSTRACT

The paper reports analyses of samples taken from three boreholes in the east Devon Triassic mudrocks. The data on the engineering properties highlight the variation that occurs within what appears to be consistent red mudrocks. The test results indicate an almost linear correlation between the calcite content and the index and residual shear strength properties.

INTRODUCTION

In England, Triassic mudrocks occupy approximately 15% of the outcropping solid geology. Audley Charles (1970) described their lithology and Meigh (1976) discussed many of the geotechnical properties of the Triassic mudrocks in several different parts of the UK. In addition to the outcrop as seen on the geological maps (Figure 1), the Trias extends under a number of the British estuaries and into the North Sea. It is therefore a very important strata as regards foundations and one of the lithologies often referred to as a "weak rock". As such it may contain some bedding features but in most situations when exposed at the surface the Triassic appears as a stiff fissured clay (weak mudrock), with the grain size and degree of cementation controlling the individual peds/discontinuity bounded blocks.

Most of the British Triassic deposits are continental in character, laid down either as rock fans adjacent to the mountains, as sandy pediments or in environments characteristically known as playas or sabkhas. This paper is concerned only with the latter lithologies which until 1980 were referred to as the Keuper Marl but following the recommendations of the Stratigraphic Committee of the Geological Society, are now frequently termed Mercia Mudstone. In some ways this is useful as the mudrocks were frequently not calcareous, particularly at shallow depths, as a consequence of weathering rather than original deposition.

Following the work of Hawkins, Lawrence and Privett (1986, 1988) and Hawkins and McDonald (1992), it is known that the index properties of the Fuller's Earth are inversely related to the calcite content. These authors have shown that where the calcite content is low, high liquid limits and low ultimate residual shear strengths values are obtained. It is also possible to observe an almost linear relationship between calcite, the index limits and residual shear strength.

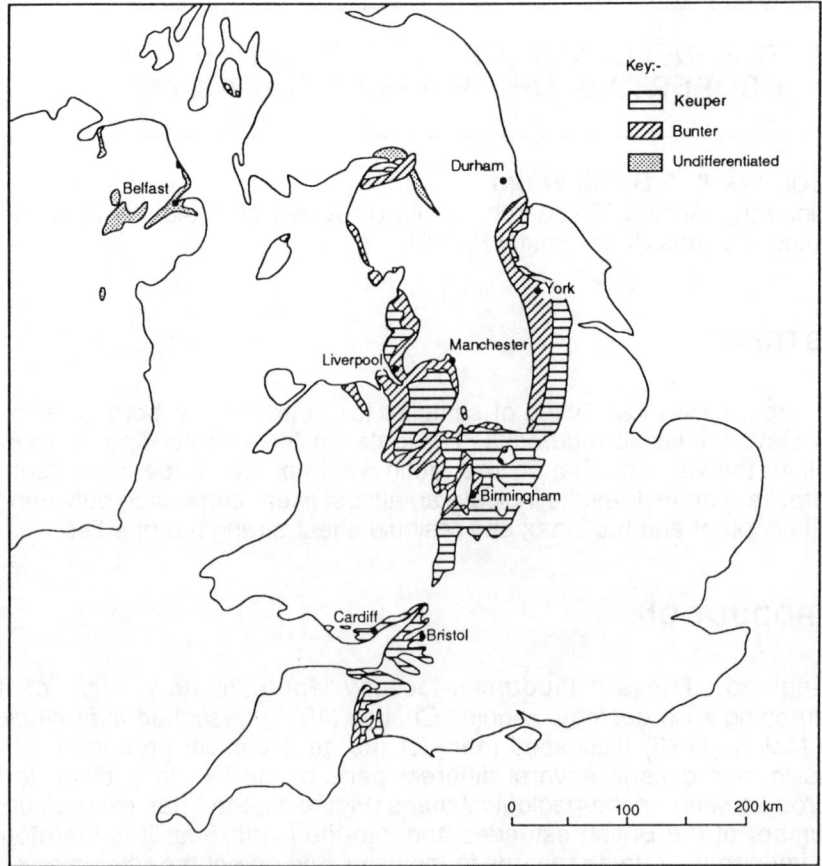

Figure 1: Outcrops of Triassic mudrocks in Britain, after Meigh (1976).

This paper describes the relationship between the calcite content and the index and residual shear strength properties for three boreholes near Axminster in South West England. The data presented are from samples taken from the borehole core at one metre intervals.

BACKGROUND GEOLOGY

The geology of the east Devon area was described in the geological Memoir by Woodward and Ussher (1911). This included an important description of the borehole drilled by the Geological Survey in 1901 at Lyme Regis; the log produced by Jukes Brown (1902) is given in Figure 2. This was used for the field description of the Triassic of the area until the borehole was re-logged and described in modern nomenclature by Warrington and Scrivener (1980). Their more detailed log is also included in Figure 2 to facilitate comparison between the two stratigraphic terminologies. From the more recent interpretation it is believed that the Mercia Mudstone along the south Devon coast is in the order of 450 m thick. Being a continental deposit, it is frequently not well bedded and

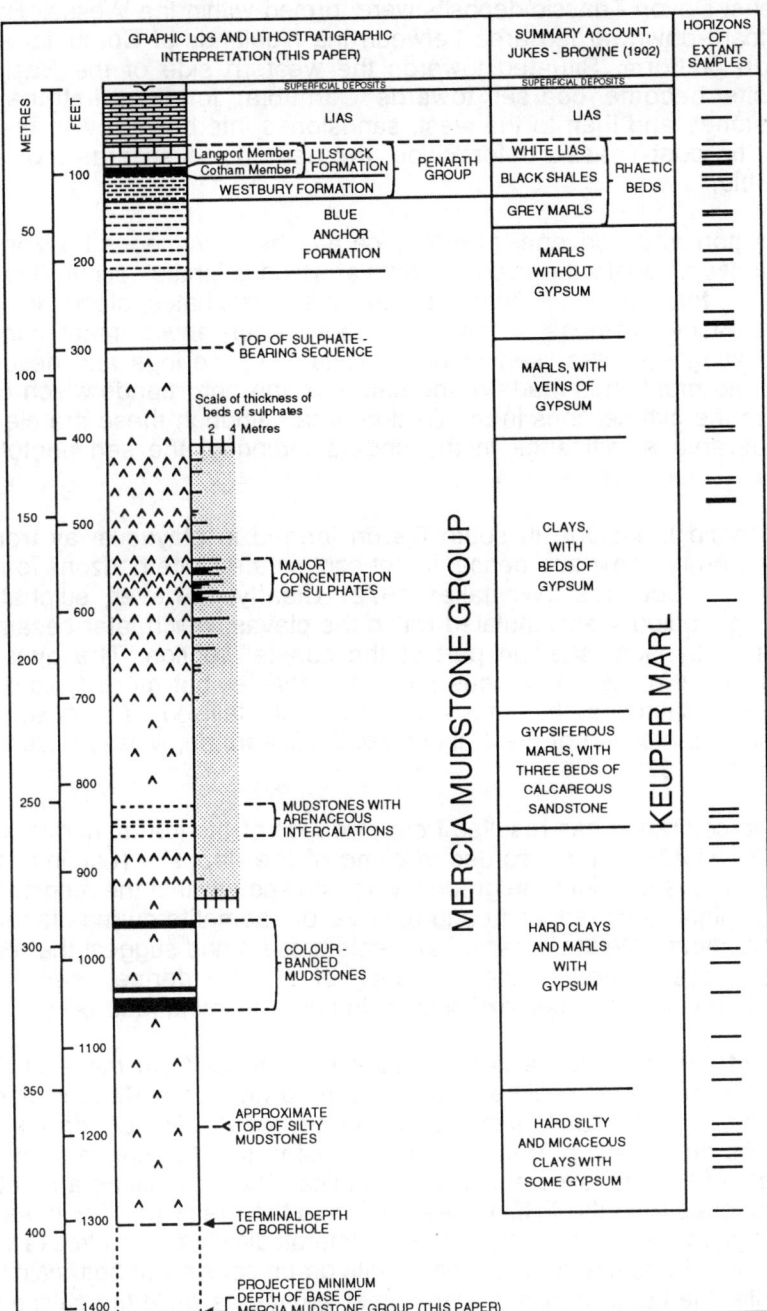

Figure 2: Comparison of the Lyme Regis borehole log by Jukes - Browne (1902) and Warrington & Scrivener (1980).

almost without fossils. As a consequence, in the faulted coastal area it is not possible to be confident of the thickness of these deposits.

The east Devon Triassic deposits were formed within the Wessex Basin, a significant downwarped area between the highlands of Cornubia and the London platform. Situated towards the western side of the Basin, the deposits become coarser towards Cornubia, forming first the Otter Sandstones and then to the west, sandstones interbedded with breccias. With little post-Triassic deformation, the beds dip eastwards, away from Cornubia.

Warrington and Scrivener (1980) separate the arenaceous horizons and also show pictorially the main horizons at which sulphates occur. It is noted in the log that the upper 30 m have few if any sulphates, clearly indicating an increasing wetness of the contemporaneous environment such that evaporation was less likely to be significant. In the logs and descriptive paper, no mention is made of the nature of the hard bands which can be seen in the cliff sections in the Seaton area, although these are clearly of considerable significance in the understanding of the sedimentological regime.

The Mercia Mudstone in south Devon formed in playas away from the contemporary coast and hence do not contain the halite horizons found for instance in Somerset (Whittaker, 1972). Clearly, however, sulphate-rich waters periodically accumulated within the playas, which later became the bands of gypsum seen in part of the coastal section. The gypsum is occasionally in the amorphous form of alabaster but more typically it is present as satin spa, the fibrous form which crystallised as a consequence of hydraulic injection of mineral-rich fluids (Shearman, Mossop, Dunsmore and Martin, 1972).

The red coloration has resulted from oxidation of the iron minerals with up to 9.1% Fe_2O_3 being recorded in some of the deposits (Laming, 1982). Some authors consider the green layers and spots within the mudstones to have originated by reduction and removal of haematite during diagenesis. Other authors note the organic fragments present and suggest the reduced horizons are related to the existence of a more dense, semi-desert vegetation which inhibited oxidation of the accumulating mudrocks.

Jeans (1978) collected a large number of samples from the east Devon coast and prepared sedimentological logs, dividing the Keuper Marl into three facies: mudstone, sandstone and carbonate. He undertook X-ray analyses on the $<2\mu m$ fraction and suggested that the common minerals throughout the Marls were chlorite and mica. Jeans identified a number of clay minerals from the XRD traces and noted the presence of both sepiolite and palygorskite. As seen in Figure 3, mineralogically the mudrocks vary. It will be noted that the dominant carbonate group consists of both calcite and dolomite, the latter occurring in the older sediments while the calcite is the main carbonate in the younger rocks.

GEOLOGICAL INTERPRETATION

In the present study, samples were taken from boreholes drilled in the Axe valley area, north of Seaton. X-ray analyses were undertaken on both the

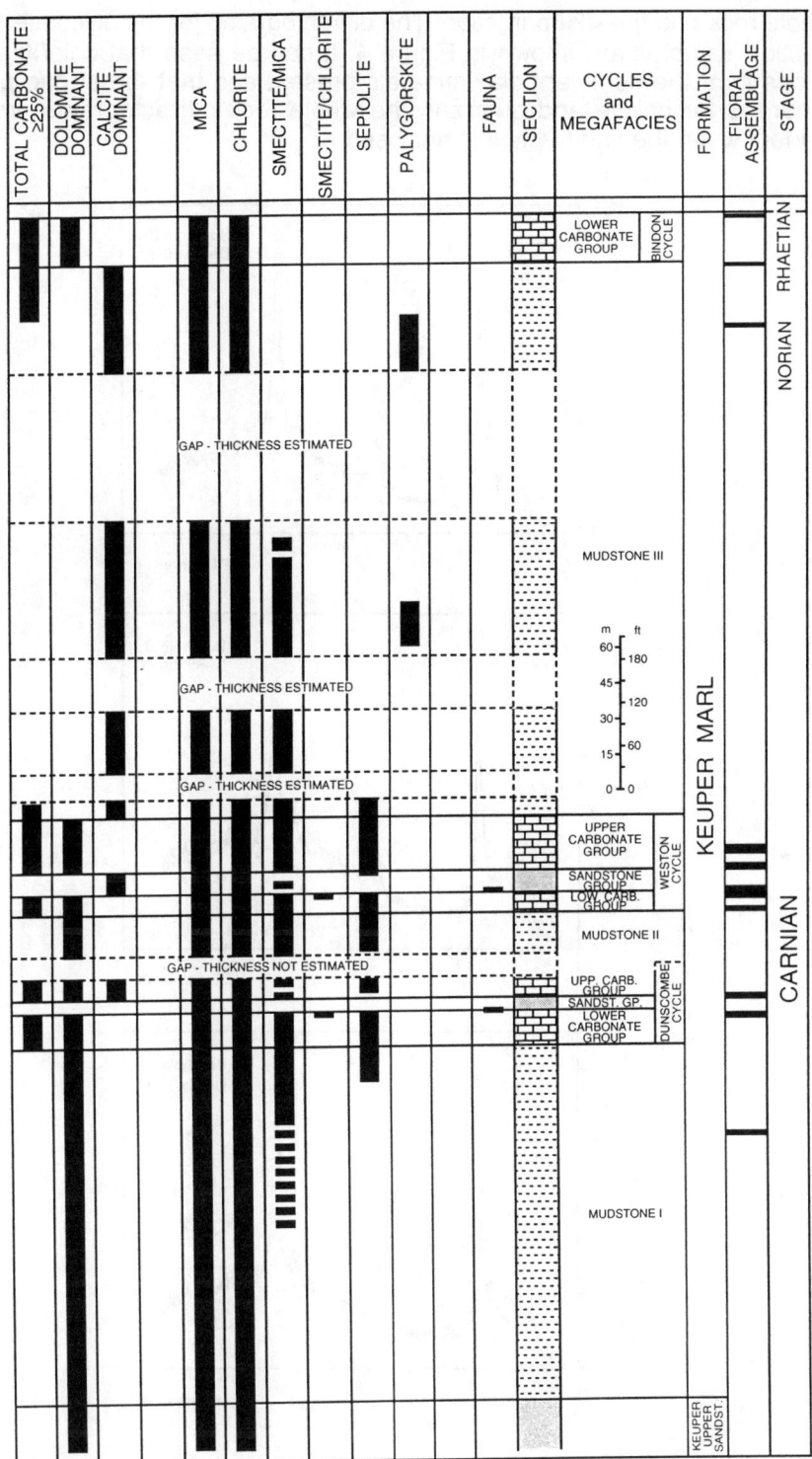

Figure 3: Summary of the Keuper Marl, south Devon coast.

whole rock and the <2μm fraction. The diffractograms for the glycolated clay fraction samples are shown in Figure 4. It can be seen that chlorites and micas form the dominant clay minerals present and that the strong peaks recorded for calcite and quartz in the whole rock diffractograms are not evident when the fine fraction is analysed.

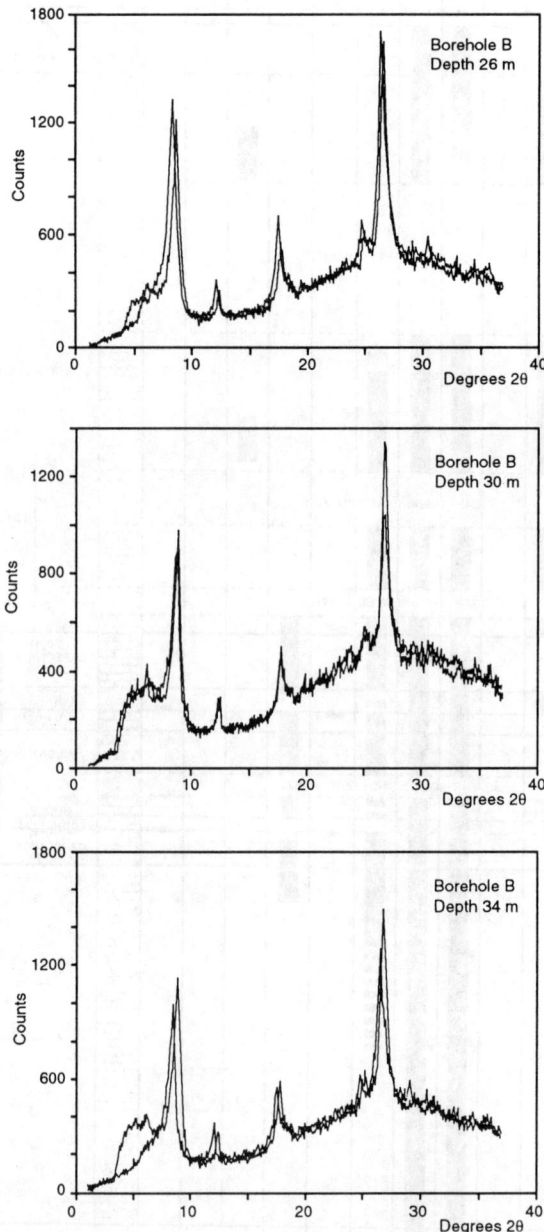

Figure 4: X-ray diffraction traces of the <2μm fraction including glycolated runs from borehole B at 26, 30 and 34 m.

INDEX PROPERTIES

The index properties were measured as recommended in BS 1377: 1990. The liquid limit was undertaken using the cone method as this is considered more reliable and reproducible than the Casagrande cup. As with other workers, some difficulty was experienced in obtaining the plastic limit and hence these results should be taken as ±2%.

The results obtained for the index properties are presented in Tables I, II and III and shown graphically in Figure 5.

The results from BH B show a very narrow range in liquid limits, only 8% over 21 samples, while for BH C the range is 9% for 32 samples. It is of note that in BHs B and C the material is effectively non-plastic and would be likely to be susceptible to frost heave, a process which may well have aided the weathering of the formation.

The calcite content for each of the boreholes varied between 11-40% in BH A; 11-33% in BH B and 12-39% in BH C.

Table I: Index properties and calcite content from BH A.

Depth (m)	Plastic Limit %	Liquid Limit %	Plasticity Index %	Calcite Content %
2	20	46	26	17
3	20	47	27	24
4	21	40	19	22
6	23	48	26	11
7	26	45	19	19
9	25	44	19	17
10	19	36	17	23
11	19	32	12	17
12	26	44	18	19
13	23	41	18	22
14	20	47	27	20
15	20	36	16	23
16	20	40	21	19
17	17	29	12	16
18	15	37	22	21
19	14	35	21	15
20	18	33	15	22
21	19	37	18	21
22	17	36	19	19
24	19	37	18	31
25	19	33	14	39
26	19	37	18	24
27	20	37	17	25
28	16	29	13	30
29	16	34	19	35
30	18	44	25	19
31	20	37	17	16
32	16	34	18	26
33	14	35	21	22
34	16	35	19	21
35	14	27	14	20

Table II: Index properties, calcite content and ultimate residual shear strength for BH B

Depth (m)	Plastic Limit %	Liquid Limit %	Plasticity Index %	Calcite Content %	ϕ'_r (ult)
26	19	34	15	26	22
27	17	29	13	33	25
28	18	33	15	22	23
29	20	30	10	22	27
30	21	32	11	15	24
31	21	35	14	16	19
32	18	27	9	25	29
33	21	30	9	24	27
34	19	28	10	26	25
35	20	30	10	21	21
36	21	30	9	32	23
37	19	28	10	25	29
38	22	33	11	11	14
39	19	29	10	24	27
40	20	29	9	24	24
41	22	32	11	29	24
42	19	39	10	20	25
43	18	29	11	21	27
44	21	31	11	14	23
45	21	31	11	11	24
46	22	34	13	11	23

Table III: Index properties, calcite content and ultimate residual shear strength for BH C.

Depth (m)	Plastic Limit %	Liquid Limit %	Plasticity Index %	Calcite Content %	ϕ'_r (ult)
8	25	34	9	30	26
9	22	33	11	39	25
10	24	35	10	28	23
11	21	31	10	30	25
12	24	35	11	21	21
13	21	32	11	32	25
14	19	29	10	34	28
15	23	33	11	23	21
16	21	32	10	31	24
18	19	28	10	37	26
19	20	30	10	36	27
20	24	39	15	20	17
21	20	31	10	31	26
22	21	31	10	30	25
23	24	36	12	15	17
24	23	35	11	20	22
25	23	32	10	19	23
26	22	33	11	19	27
27	24	36	12	12	19
28	22	32	10	20	24
29	23	34	11	20	22
30	21	34	13	24	22
31	22	35	13	16	20
32	21	33	12	22	24
33	23	36	13	17	22
34	22	36	14	17	24
35	22	35	13	16	20
36	22	34	12	20	23
37	21	34	12	19	22
38	24	37	13	16	18
39	24	37	13	13	19
40	21	34	12	21	23

INFLUENCE OF CALCITE ON THE INDEX PROPERTIES OF TRIASSIC MUDROCKS

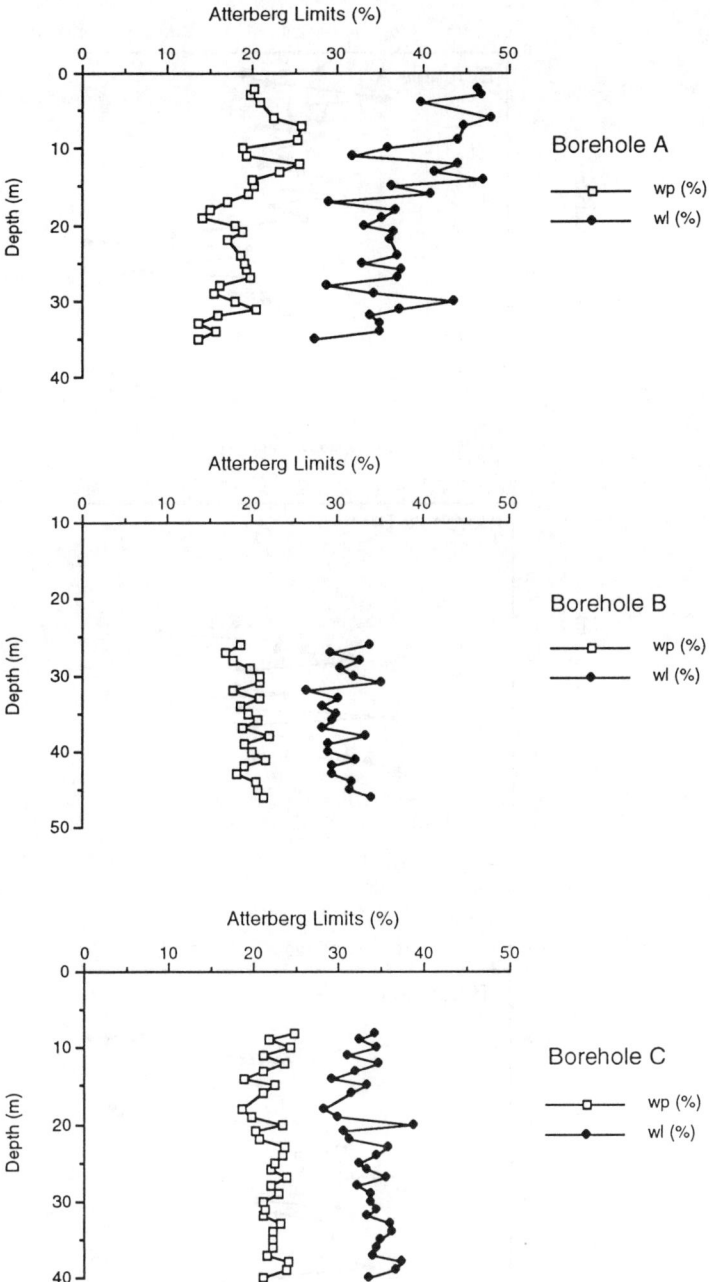

Figure 5: Atterberg limit profiles for boreholes A, B and C.

As can be seen from Figure 6, the range of calcite contents is very similar for each of the boreholes. In the case of BH A, the proportion of calcite is generally 20% ± 5% to a depth of 24 m where it increases suddenly to

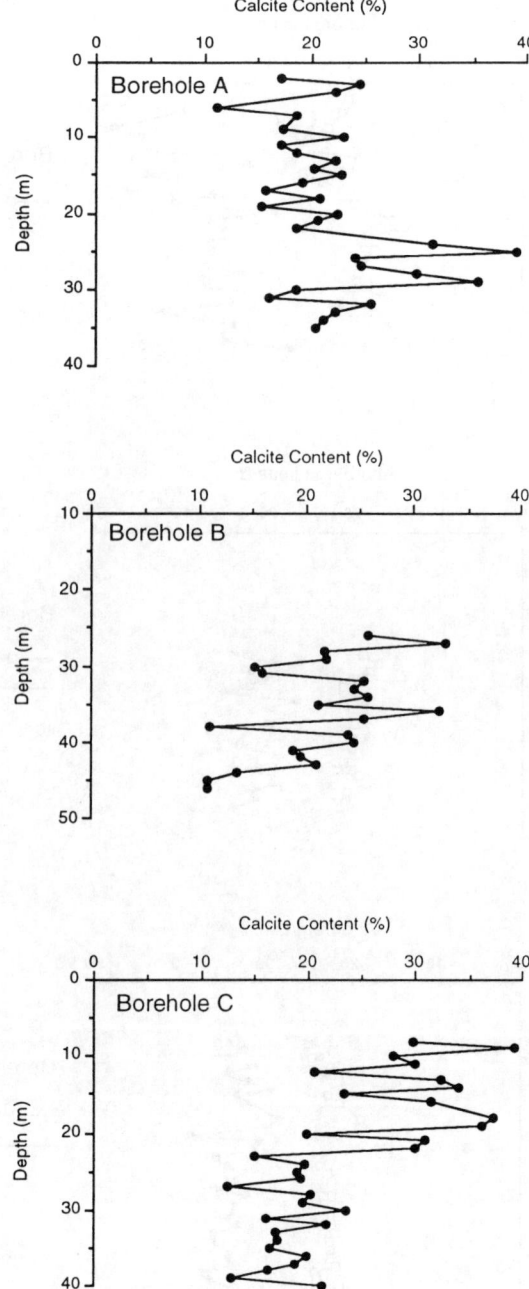

Figure 6: Calcite profiles for boreholes A, B and C.

39%. The more calcareous horizon extends to 29 m while below 30 m the calcite content again drops to approximately 20% (Table I). In BH B the calcite profile has distinct peaks and troughs, indicating a variable carbonate content, probably related to depositional characteristics. Borehole C indicates a variable calcite content in the upper 8 to 23 m where distinct peaks and troughs are present but below 24 m the variability is much less pronounced.

Each of the samples from BH B were centrifuged to separate the calcite. Microscopic examination indicated that subangular and angular fragments were dominant although some rhombic crystals were seen. Table IV gives the silt fraction separated in the samples with the highest, intermediate and lowest proportions of calcite.

Table IV: Silt fraction in the samples from BH B with the maximum, minimum and intermediate proportions of calcite.

	Maximum Calcite	Intermediate Calcite	Minimum Calcite
Coarse silt	14	11	5
Medium silt	12	8	4
Fine silt	7	3	2

It is clear from Table IV that the calcite is dominantly medium to coarse silt size and hence the inert silt grains and any sand grains present will have a significant influence on the index properties.

The relationship between the calcite content and the liquid and plastic limits is given in Figure 7. Although there is a considerable scatter of points for BH A, in BH B an almost linear relationship is seen which is even more pronounced for BH C. The data in Table I indicate that in BH A there is a change in the calcite content and the index properties above and below 20 m. As can be seen in Figure 8, there is a more linear relationship for the lower samples while no apparent trend can be discerned in the material above approximately 20 m.

With weak to moderately weak material (1.25 to 12.5 MPa), obtaining an appropriate peak shear strength is always very difficult and the results have little relevance for a slope stability problem. For this research, therefore, it was considered more appropriate to use remoulded material. It has already been appreciated by Hawkins and McDonald (1992) that calcite dissolved along a shear surface will have a dramatic effect on the shear strength, hence it was decided to obtain the residual shear strength values using the Bromhead ring shear apparatus. The values obtained on samples from BHs B and C using a notional speed of 0.024° per minute are given in Tables II and III.

As the residual shear strength envelope is invariably curved, following the work of Hawkins and McDonald (1992) it was decided to plot the $\phi'_{r\ ult}$ results ignoring any residual cohesion value. It should be noted that the ultimate residual shear strength ($\phi'_{r\ ult}$) is taken as the straight line part of the failure envelope which usually occurs when samples are tested above

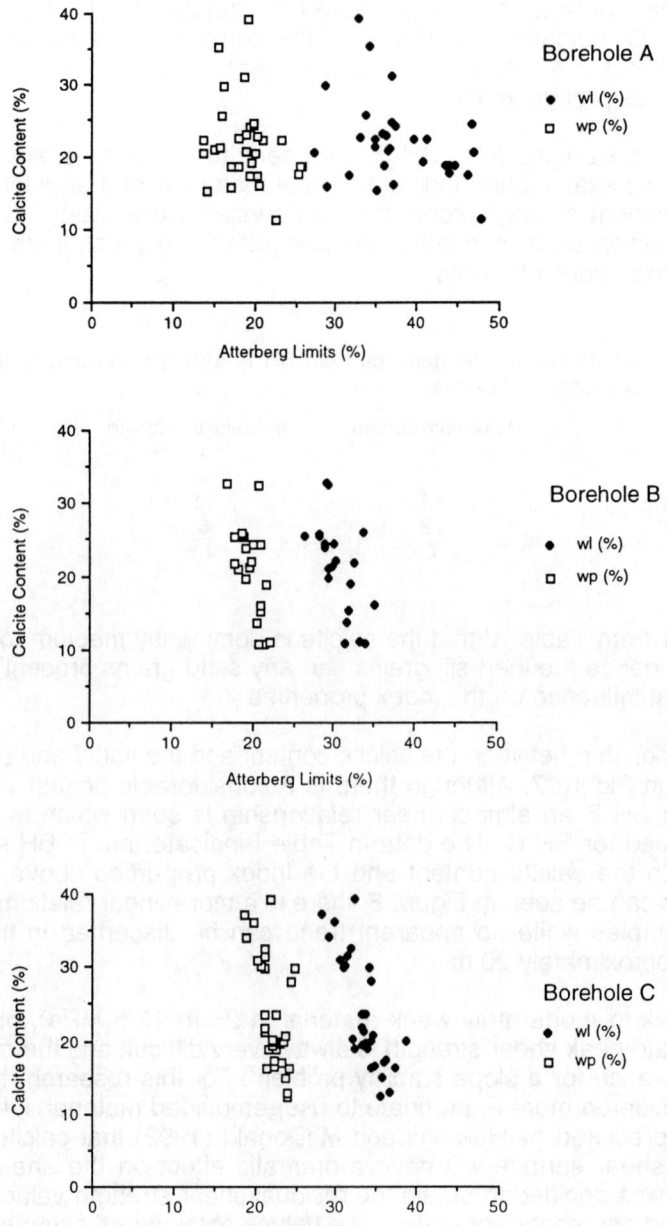

Figure 7: Relationship between the liquid and plastic limits and calcite contents for boreholes A, B and C.

an effective normal stress (σ'_n) of 100 to 200 kPa. It is appreciated that the use of $\phi'_{r\,ult}$ should be accompanied by a $c'_{r\,ult}$. Such a combination of uses, however, becomes very complex and consequently only the $\phi'_{r\,ult}$ values are given.

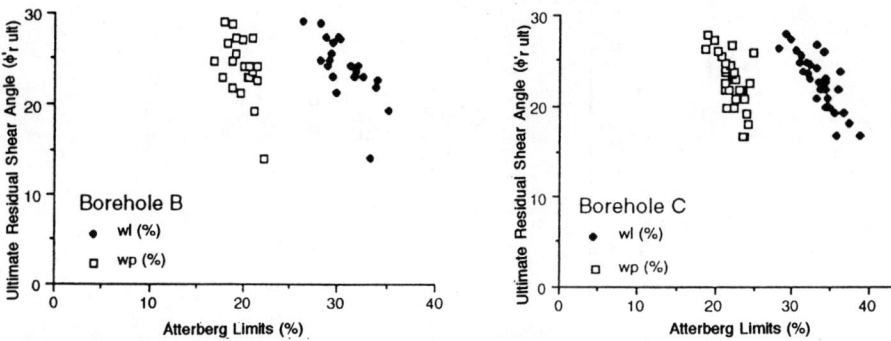

Figure 8: Relationship between the liquid and plastic limits and the calcite content for the upper and lower parts of borehole A.

When the ultimate residual shear strength is plotted against the index values (Figure 9) it can be seen that an acceptable straight line is produced for the BH B samples while in the case of BH C a good linear trend is present. When the $\phi'_{r\,ult}$ is plotted against the calcite content (Figure 10) a positive trend is indicated. In the lower of the three diagrams the $\phi'_{r\,ult}$ values from both the BH B and BH C samples are plotted against the calcite content. It is noted that the ultimate residual shear strength for the samples from BH C are on average 5° higher than those for BH B for the equivalent calcite contents. The reason for this is not yet known but is probably related to the variable percentage of fine quartz in the samples.

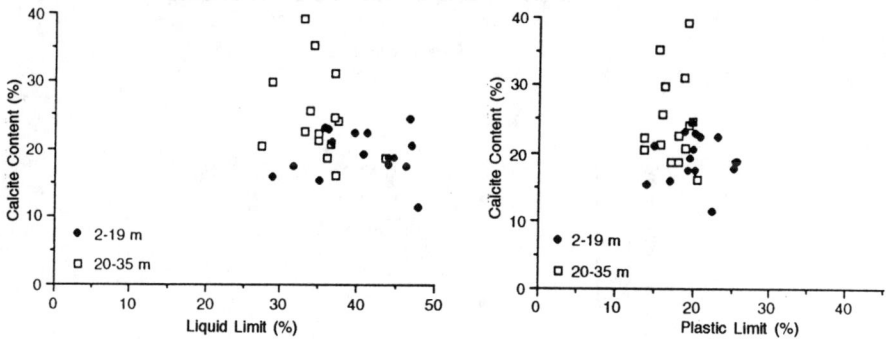

Figure 9: Relationship between the liquid and plastic limits and the ultimate residual shear strength for boreholes B and C.

DISCUSSION

The Mercia Mudstone is an atypical British mudrock as it was formed in a semi-desert environment where the oxides contained within the clay mineral complex suffered syndepositional oxidation. Although the XRD traces of the whole rock sample indicate the dominant minerals to be quartz and calcite, the presence of illite and some kaolinite as well as other minerals is important. Microscopic examination shows the silt-sized

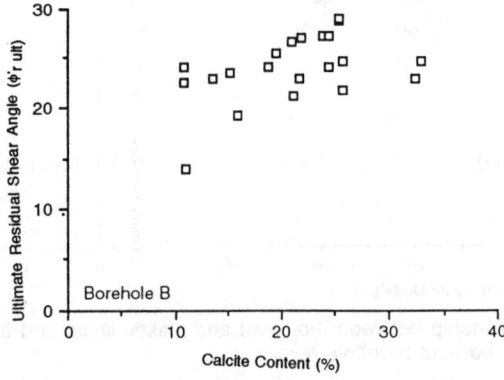
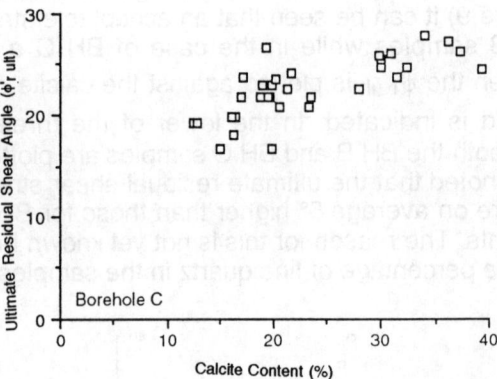
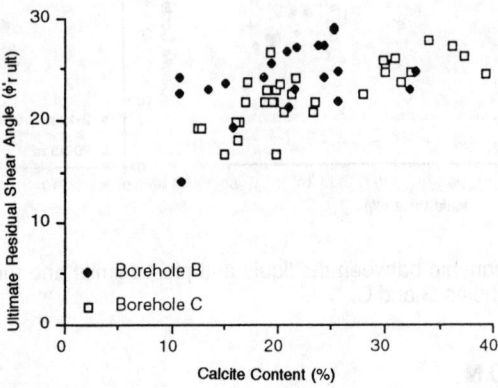

Figure 10: Relationship between the calcite content and the ultimate residual shear strength for boreholes B and C.

particles to be both rounded and subangular to angular in nature. When the grains from the same batch were quartered and leached of their calcite, it was noted that the quartz grains were more rounded than the calcite fragments, indicating that much of the quartz was incorporated as abraded grains while the calcite generally formed within the accumulating sediment.

In addition to the angularity and inert nature of the calcite grains, the aggregate structure of the Triassic deposits (Chandler and Davis, 1973) also has a major influence on the index properties, being one reason why the liquid limit is lower than might otherwise be expected. It was noted, for instance, that in a number of trial pits adjacent to BH A, there were significant changes in the ground conditions with depth. These are recorded in Table V which shows a pronounced increase in the clay fraction in the near-surface zones with generally more plastic material closer to the ground surface.

Table IV: The change in geotechnical properties with depth.

Trial Pit Number	Depth (m)	Clay Fraction %	w_l %	w_p %	MDD Mg/m^3	OMC w %
T1	0.85	15	34	21	1.75	17
	1.95	6	30	16	1.86	12
T2	1.55	31	34	21	1.66	21
	3.95	8	37	25	1.72	19
T3	1.50	21	54	32	1.53	25
	4.15	6	38	25	1.67	20
T4	1.05	26	40	25	1.70	19
	2.95	14	34	22	1.75	17

In a semi-desert environment it is likely that the nature of the deposit will vary depending on its location within the accreting playa. It is noted, for instance, that higher liquid and plastic limits are obtained for the samples from BH A compared with those from BHs B and C, approximately 5 km to the east. However it is likely that BH A may be located at a lower stratigraphic horizon than the other two boreholes, which are known to be near the top of the Mercia Mudstone. The differences emphasise the variations which may occur in this stratigraphic unit, which despite Meigh's Rankine Lecture (1976), is still too often considered a consistent geotechnical material.

CONCLUSIONS

The Triassic mudrocks found in east Devon formed as playa lake accumulates. Although in some cases the calcite content is relatively homogeneous, at other levels it is clear that past post-sedimentation enrichment has taken place. This is probably related to capillary rise with selective cementation producing an early form of a calcrete duricrust.

Boreholes B and C, near the top of the Mercia Mudstone Group, have very small ranges of liquid and plastic limits compared with BH A from lower in the sequence. When plotted against the calcite content an acceptable straight line is obtained for BHs B and C but in the case of BH A, there is no linear relationship.

When the calcite content is plotted against the ultimate residual shear strength a positive trend is evident. From a straight line through the points it can be seen that the residual shear strength increases at 1° for each 3% increase in calcite. Consequently, the proportion of calcite content present will have a significant effect on the shear strength, which is controlled largely by the angularity of the coarse fraction (see Hawkins and McDonald, 1992). With dissolution of the calcite however, the angularity would decrease and with a smaller proportion of inert grains, the shear strength is likely to change dramatically.

REFERENCES

AUDLEY-CHARLES, M.G. (1970). Stratigraphic correlation of Triassic rocks of the British Isles. *Quarterly Journal of the Geological Society*, **126**, 19-47.

CHANDER, R.J. & DAVIS, A.G. (1973). Further work on the engineering properties of Keuper Marl. *CIRIA Report 47*.

HAWKINS, A.B., LAWRENCE, M.S. & PRIVETT, K.D. (1986). Clay mineralogy and plasticity of the Fuller's Earth Formation, Bath, UK. *Clay Minerals* **21**, 293-310.

HAWKINS, A.B., LAWRENCE, M.S. & PRIVETT, K.D. (1988). Implications of weathering on the engineering properties of the Fuller's Earth Formation. *Geotechnique*, **38**, 517-532.

HAWKINS, A.B. & MCDONALD, C. (1992). Decalcification and residual shear strength reduction in Fuller's Earth clay. *Geotechnique*, **42**.

HAWKINS, A.B. & MCDONALD, C. (1992). The influence of particle shape on residual shear strength. *Geotechnique*, **42**.

JEANS, C.V. (1978). The origin of the Triassic clay assemblages of Europe with special reference to the Keuper Marl and Rhaetic of parts of England. *Philosophical Transactions of the Royal Society*, **A 289**, 549-639.

JUKES-BROWN, A.J. (1902). On a deep boring at Lyme Regis. *Quarterly Journal of the Geological Society*, **58**, 279-289.

LAMING, D.J.C. (1982). The New Red Sandstone. In: *The Geology of Devon*, Durrance E.M. & Laming D.J.C. (eds), University of Exeter, 148-178.

MEIGH, A.C. (1976). The Triassic rocks, with particular reference to predicted and observed performance of some major foundations. *Geotechnique* **26**, 391-452.

SHEARMAN, D.J., MOSSOP, G., DUNSMORE, H. & MARTIN, M. (1972). Origin of gypsum veins by hydraulic fracture. *Proceedings of the Institution of Mining and Metallurgy*, **81**, B149-B155.

WARRINGTON, G. & SCRIVENER, R.C. (1980). The Lyme Regis (1901) Borehole succession and its relationship to the Triassic sequence of the east Devon coast. *Proceedings of the Ussher Society*, **5**, 24-32.

WHITTAKER, A. (1972). The Somerset Saltfield. *Nature*, **238**, 265-266.

WOODWARD, H.B. & USSHER, W.A.E. (1911). The geology of the country near Sidmouth and Lyme Regis. *Memoir of the Geological Survey of Great Britain, London*.

3-9 CONTRIBUTION OF SULPHIDES AND CARBONATES TO THE WEATHERING OF SEDIMENTARY ROCKS

T. OYAMA, M. CHIGIRA & T. SIDAHARA
Central Research Institute of Electric Power Industry, 1646 Abiko, Abiko, Chiba, 270-11, Japan

ABSTRACT

Weathering profiles in mudstones have two weathering fronts, an oxidation front in approximately the middle part and the dissolution front at the base. The brown oxidized zone exists between the ground surface and the oxidation front while the dissolved zone is between the oxidation front and the dissolution front.

At the oxidation front pyrite is oxidized and chlorite is transformed into smectite while gypsum formed in the dissolved zone is moved by leaching or capillary rise. At the dissolution front calcite is taken into solution, partly due to sulphuric acid generated by the oxidation of pyrite. Depending on the hydrologic conditions, particularly ground water level, gypsum may be formed.

The physical properties of rock which deteriorate in the dissolved zone as a consequence of the dissolution of calcite may be improved again in the oxidized zone by the cementation of grains with iron oxide and/or hydroxide.

Secondary weathering from a cut slope surface over several years may also have a deleterious effect depending on the mineral composition, fabric and hydrologic properties of the rocks.

INTRODUCTION

The weathering of soft sedimentary rocks, such as the Neogene and Quaternary rocks in Japan, is an important geological and engineering geological process as it has a more significant influence on the physical properties than fracture frequency. Chemical weathering in particular has recently been found to have an important role in the weathering of the soft sedimentary rocks as, in geological time, it develops characteristic weathering profiles as deep as tens of metres (Spears and Taylor 1972; Russell and Park, 1979; Dixon, Hossner, Senkayi and Egashira1982; Hawkins, Lawrence and Privett, 1988; Chigira, 1990; Chigira and Sone, 1991). In addition, artificial slopes of soft rocks may be weathered rapidly by chemical processes (Pye and Miller, 1990).

This paper considers the long term weathering of soft sandy mudstone and discusses the secondary weathering from a cut slope surface during a four year period.

GEOLOGICAL SETTING AND METHODS

The site investigated was a cut slope at the edge of a hill in south Chiba Prefecture, central Japan (Figure 1). The Pleistocene Kakinokidai Formation (Mitsunashi et al, 1976) underlies the site, which consists mainly of a massive sandy mudstone with a marine origin. At the slope, the intercalated thin sandstone and conglomerate beds (Figure 2) are north-south trending with a dip of 10° to the west. The colour of the rocks was brown within several metres of the shoulder of the cut surface and grey below it. Top soil was approximately 0.2 m thick.

Five inclined drill holes with a length of about 3 m were made from the slope surface and continuous cores obtained (Figures 2 and 3). Drill hole No 1 was made in the lower part of the slope (grey) while Nos 2, 3, 4 and 5 were made in its upper (brown) part. The cored rocks consisted of homogeneous sandy mudstone. After drilling, no groundwater flowed from the drill holes.

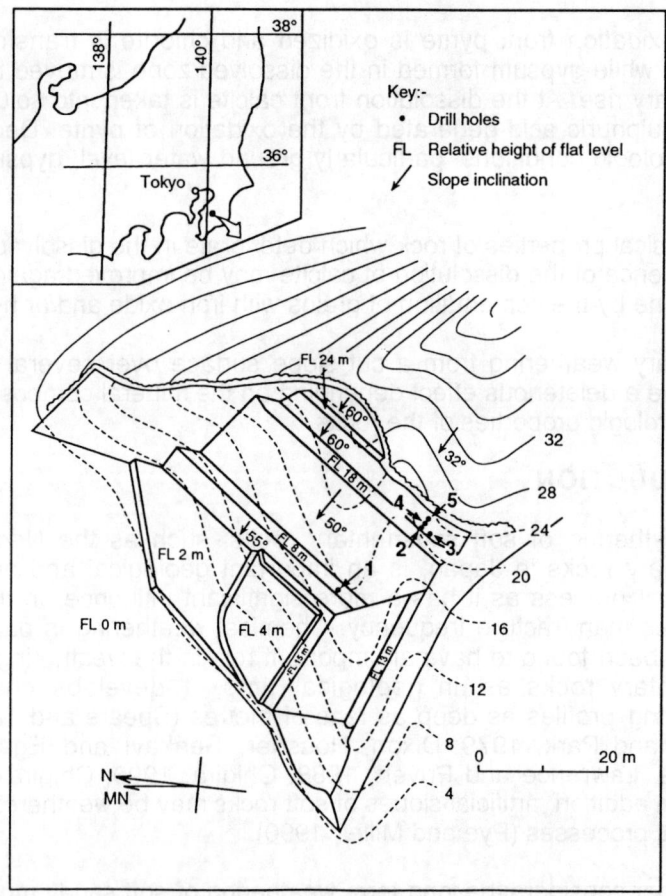

Figure 1: Map of investigated area.

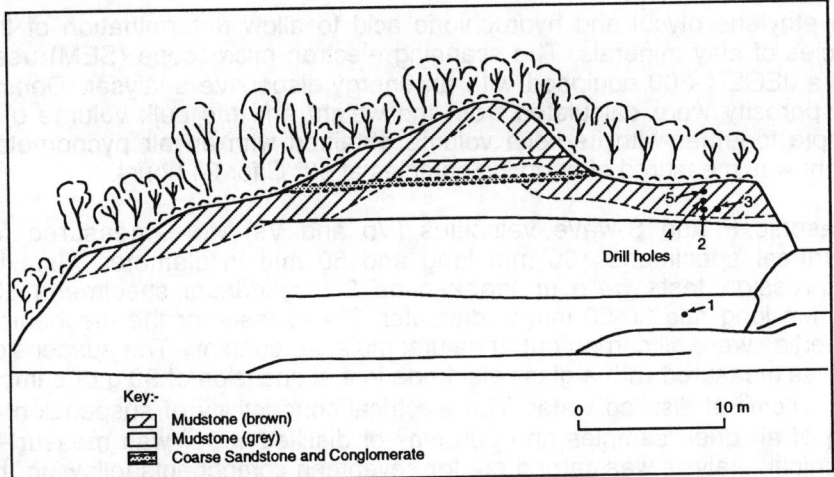

Figure 2: Section of slope.

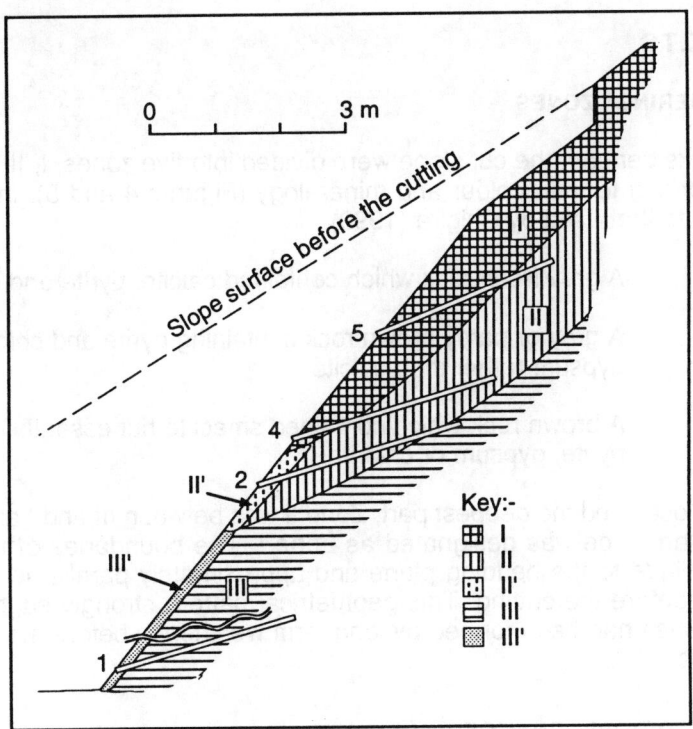

Figure 3: Locations from which cores were obtained.

More than twenty samples were selected from the drilled cores and immediately sealed with paraffin for subsequent mineralogical, chemical and physical analyses. Mineral analyses were done with a JEOL JDX-8200T X-ray diffractometer (XRD). The analyses were performed for the bulk powder samples and for the <2μm fractions. The samples were treated with ethylene glycol and hydrochloric acid to allow determination of the species of clay minerals. The scanning electron microscope (SEM) used was a JEOL T-300 equipped with an energy-dispersive analyser. Density and porosity were calculated from the weight and the bulk volume of a sample together with its solid volume obtained with an air pycnometer; weight was measured after drying samples at 60° C for 24 hours.

Ultrasonic P and S-wave velocities (Vp and Vs) were measured for cylindrical specimens 100 mm long and 50 mm in diameter. Uniaxial compression tests were undertaken on 2:1 cylindrical specimens 60-100 mm long and 30-50 mm in diameter. These tests for the mechanical properties were all carried out at natural moisture content. The suspension pH was measured with a glass electrode in a suspension of 30 g of sample and 50 cm^3 of distilled water. The electrical conductivity of suspension of 30 g of air dried samples and 200 cm^3 of distilled water was measured. Chemical analysis was carried out for seventeen components following the procedures described by Chigira (1990).

RESULTS

WEATHERING ZONES

The rocks beneath the cut slope were divided into five zones, I, II, III, II' and III' according to their colour and mineralogy (Figures 4 and 5), in a similar manner to that used by Chigira (1990).

Zone III   A grey fresh rock which contained calcite, pyrite and chlorite.

Zone II   A grey or partly brown rock containing pyrite and chlorite and gypsum rather than calcite.

Zone I   A brown rock which contained smectite but essentially no pyrite, gypsum or chlorite.

Zone III occupied the deepest part, Zone II was between III and I and the top part of the slope was designated as Zone I. The boundaries of the zones were oblique to the bedding plane and approximately parallel to the slope surface before the cutting. This geometrical pattern strongly suggests that these zones had been formed by long term weathering before the cutting of the slope.

MINERALOGICAL CHANGE

The minerals detected by XRD and microscopic observation were quartz, feldspar, illite, chlorite, smectite, calcite, gypsum and pyrite; the mineralogical changes presumed to be made by weathering occurred in the chlorite, smectite, calcite, gypsum and pyrite.

In Zone III, which is presumed to be the original fresh rocks, calcite was present as microfossils and as fine cement materials (Plate 1a). Calcite

WEATHERING OF SEDIMENTARY ROCKS BY SULPHIDES AND CARBONATES 313

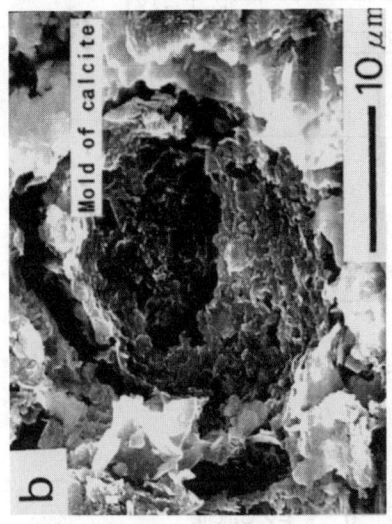

Plate 1a: Zone III, containing calcite as microfossils and fine cement.

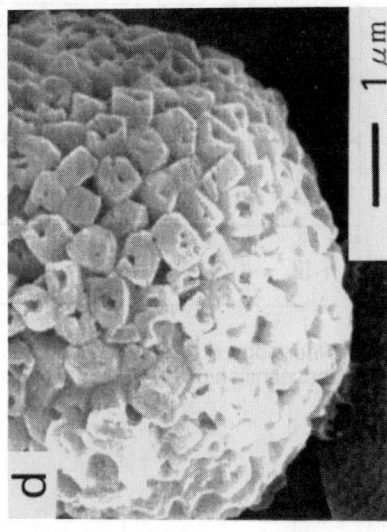

Plate 1b: Zone II, containing mulch of microfossils.

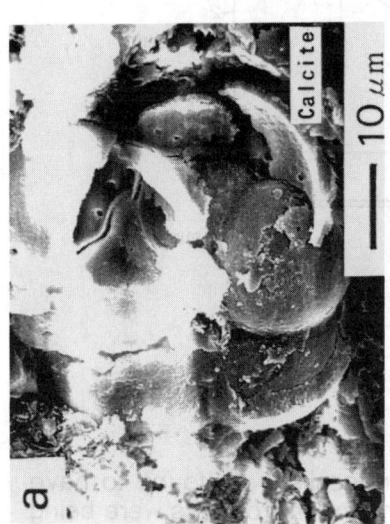

Plate 1c: Zone II, containing gypsum laths.

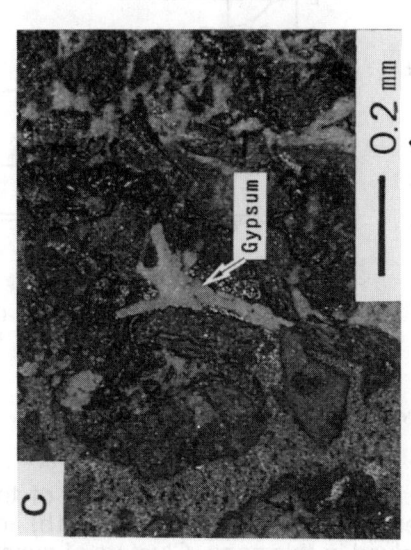

Plate 1d: Zone I, containing iron hydroxide with the pseudomorph of framboidal pyrite.

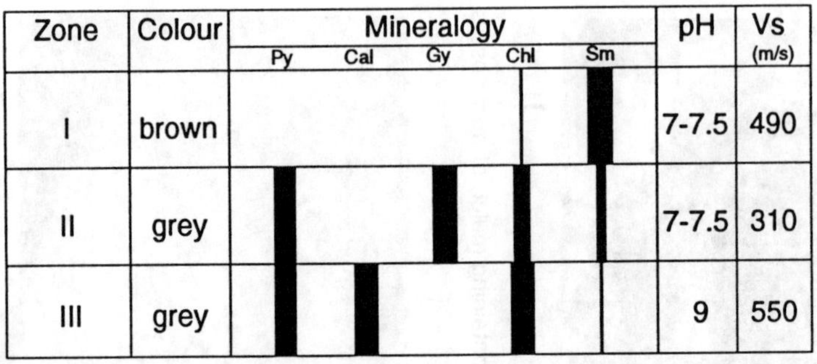

Zone	Colour	Mineralogy					pH	Vs (m/s)
		Py	Cal	Gy	Chl	Sm		
I	brown				▮	▮	7-7.5	490
II	grey	▮		▮	▮		7-7.5	310
III	grey	▮	▮		▮		9	550

Figure 4: Mineralogy of recovered cores; Py, pyrite: Gy, gypsum: Chl, chlorite: Sm, smectite: pH, suspension pH: Vs, velocity of ultrasonic S-wave.

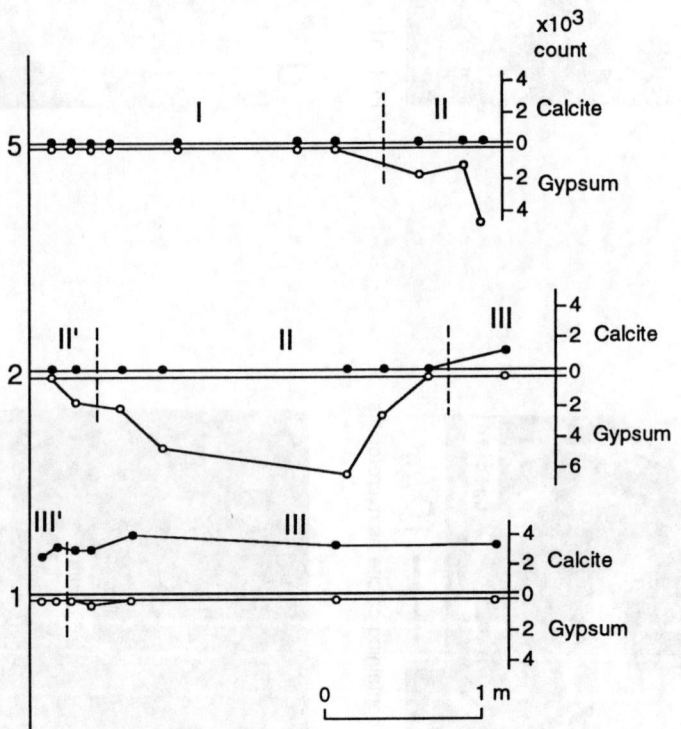

Figure 5: XRD analysis of recovered cores.

was lost in Zone II and moulds of the microfossils were found to have suffered some dissolution (Plate 1b), indicating the microfossils were being dissolved in this zone. Clusters of gypsum laths were also identified in Zone II using the staining method (Tucker, 1988; Plate 1c). Sulphide was contained as framboidal pyrite in Zones III and II. In Zone I, however, pyrite was essentially absent and iron oxide or hydroxide occurred with the pseudomorph of framboidal pyrite (Plate 1d). In Zone I, the grains were cemented by the iron oxide and/or hydroxide.

CHEMICAL CHANGE

The samples from this site probably varied significantly in density before weathering, therefore volume-constant assumptions cannot be used to examine the change in chemical composition. However, there were several chemical components which seemed to be influenced by the weathering: C, S and Fe. Figure 6 is a diagram of carbonate versus non-carbonate carbon which may be contained as organic carbon, showing that carbonates were lost in Zones II and I. This would be consistent with the disappearance of calcite in Zones II and I noted above.

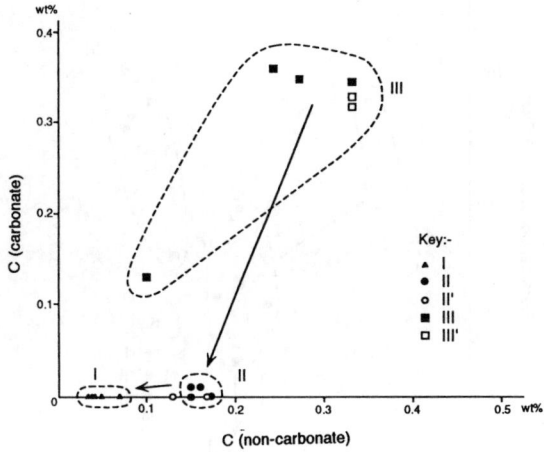

Figure 6: Carbonate carbon versus non-carbonate carbon.

Figure 7 shows the relationships between the amounts of sulphate sulphur and non-sulphate sulphur, assumed to be contained as pyrite. Sulphate increased from Zone III to Zone II and was depleted in Zone I.

The values of $Fe^{2+}/(Fe^{3+} + Fe^{2+})$ were approximately 0.4m 0.4-0.2 and 0.2-0.1 in Zones III, II and I respectively.

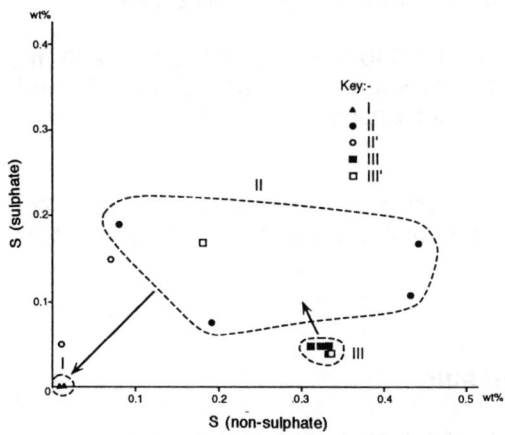

Figure 7: Sulphate sulphur versus non-sulphate sulphur.

SUSPENSION PH AND CONDUCTIVITY

Figure 8 is a diagram of suspension pH versus electrical conductivity (EC) which characterizes the interstitial waters in the rocks and soluble components. The pH values were approximately 9 in Zone III, 7 to 7.5 in Zones I and II and 4 to 5 in Zone II. The conductivities decreased from Zone III to II and were substantially less in Zone I.

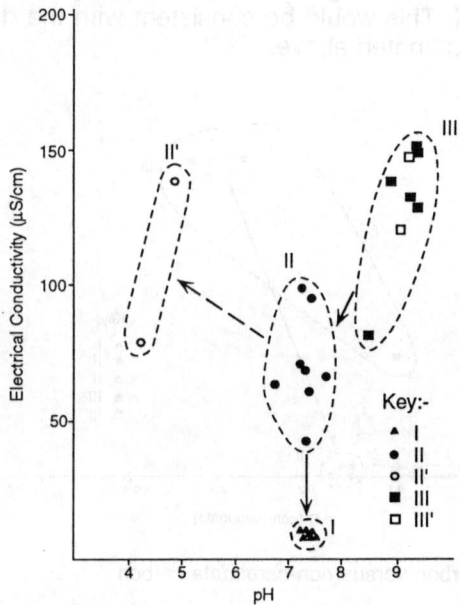

Figure 8: Suspension pH versus electrical conductivity (EC).

PHYSICAL PROPERTIES

Densities of the rocks were 1.5-1.7 Mg/m^3 in Zone I and 1.4-1.5 Mg/m^3 in Zones II and III. Porosities were 40-50% in all zones.

Average Vp values were 1,040 m/s, 730 m/s and 1,220 m/s in Zones I, II and III respectively and average Vs values 490 m/s, 310 m/s and 550 m/s (Figure 9). Vp and Vs decreased from Zone III to Zone II and increased from Zone II to Zone I.

Average uniaxial compressive strengths were 20 kPa in Zone I, 10 kPa in Zone II and the upper part of Zone III and as high as 40 kPa in Zone II.

DISCUSSION

LONG TERM WEATHERING

The weathering profile of the mudstone studied in this paper was shown to have resulted from both long term and secondary short term weathering.

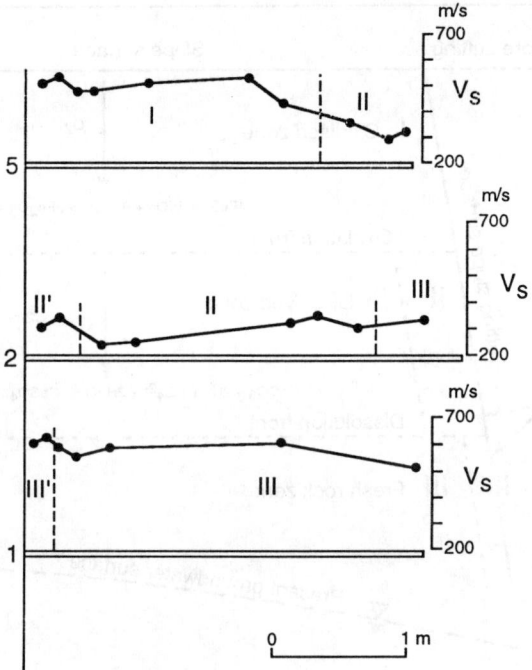

Figure 9: V_S values in Zones I, II and III.

That due to long term weathering consisted of three Zones; I, II and III corresponding respectively to the oxidized zone, the dissolved zone and the fresh rock zones of Chigira (1990). The boundary between Zones I and II is the oxidation front and that between Zones II and III the dissolution front. The distribution patterns of the mineralogical, chemical and physical properties in the weathering profile elucidate the processes operating at this study site as discussed below (see Figure 10).

Dissolution of calcite and the formation of gypsum were clearly observed at the dissolution front in this study site. At the oxidation front, loss of gypsum, decrease of the suspension conductivities and rock hardening were noted in addition to the transformation of chlorite to smectite and the depletion of pyrite, as has been recorded in other locations (Chigira, 1990).

The dissolution of calcite and formation of gypsum is explained as follows. At the base of the dissolved zone (dissolution front) calcite is dissolved by the sulphuric acid generated by the oxidation of pyrite at the base of the oxidized zone (oxidation front) and gypsum is formed by

$$4FeS_2 + 15O_2 + 8H_2O \rightarrow 2Fe_2O_3 + 16H^+ + 8SO_4^{2-}$$
(oxidation front)

$$CaCO_3 + 2H^+ + SO_4^{2-} + 2H_2O \rightarrow CaSO_4 \cdot 2H_2O + H^+ + HCO_3^-$$
(dissolution front)

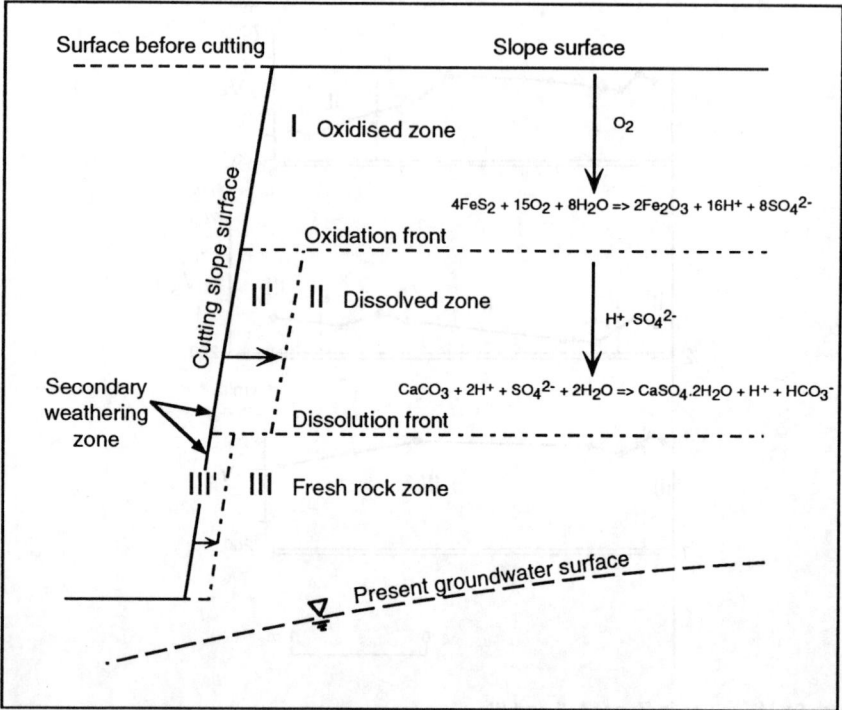

Figure 10: Processes operating at the study site.

Gypsum in Zone II was not a pseudomorph of calcareous microfossils hence this gypsum was not formed by calcite replacement but precipitated from interstitial water in the rock. Whether gypsum is formed in the dissolved zone will depend on the ground water level. If the dissolved zone is saturated, sulphate irons may remain in the interstitial water in the rock and may not appear as a sulphate mineral; this is assumed to be the case studied by Chigira (1990). At this site, however, the ground water level was lower than the No 1 drill hole in the fresh rock zone, at least after the slope was cut. The ground water level in general is likely to be at about the oxidation front (Chigira, 1990; Chigira and Sone, 1991). It is not known whether the formation of gypsum at this site occurred before or after the slope was cut.

Gypsum in the dissolved zone may be leached at the oxidation zone (perhaps near the oxidation front). The reduced electrical conductivity of the suspension in the oxidized zone (Zone I) indicates lesser amounts of soluble components than are present in the dissolved and fresh rock zones.

At the oxidation front chlorite was transformed into smectite, as was found by Chigira (1990). The iron (Fe^{2+}) in the chlorite may be oxidized to Fe^{3+} at the oxidation front.

Physical properties, such as uniaxial compressive strength and ultrasonic velocity of rocks, deteriorated at the dissolved zone by the dissolution of calcite and improved again in the oxidized zone by the cementation of grains with iron oxide and/or hydroxide.

SECONDARY WEATHERING ALONG THE CUT SLOPE

Secondary weathering from the slope surface after the cutting resulted in the formation of subzones II' and III'. Zone III' had the same mineralogical, chemical and physical properties as Zone II but its colour changed to pale brown to a depth of 0.25 m from the slope surface. In Zone II' to a depth of 0.45 m the gypsum content was less than that in Zone II and the suspension was acidic. However the ultrasonic velocities did not decrease from Zone II to II' or from Zone III to III'.

The weathering effects apparent in Zones II and III at this site during the four year study period were therefore a slight change in colour and mineralogy.

CONCLUSIONS

Long term weathering of soft sandy mudstone was studied as well as short term and secondary weathering from a cut slope surface over a period of four years. The following results were noted:

1. The geological weathering profile had oxidized and dissolved the fresh rock zones from the ground surface.

2. Reactions occurred at the oxidation front, the base of the oxidized zone where the oxidation of pyrite, transformation of chlorite to smectite and dissolution of gypsum took place in the dissolved zone as well as the cementation of grains by iron oxide and/or hydroxide.

3. The reaction at the dissolution front at the base of the dissolved zone was the dissolution of calcite by sulphuric acid which was generated by the oxidation of pyrite at the oxidation front and the formation of gypsum. The gypsum, however, may have been formed after the ground water level was lowered by the cutting of the slope.

4. The sandy mudstone was weakened at the dissolved zone and strengthened at the oxidized zone by the dissolution of calcite and the cementation with iron precipitates respectively.

5. Due to secondary weathering the fresh rock changed colour to pale brown to a depth of 0.25 m from the cut slope surface without distinguishable change in the mineralogy of physical properties.

6. The dissolved zone became slightly acid and its constituent gypsum decreased to a depth of 0.45 m from the cut surface.

REFERENCES

CHIGIRA, M. (1990). A mechanism of chemical weathering of mudstone in a mountainous area. *Engineering Geology*, **29**, 119-138.

CHIGIRA, M. & SONE, K. (1991). Chemical weathering mechanisms and their effects on engineering properties of soft sandstone and conglomerate cemented by zeolite in a mountainous area. *Engineering Geology*, **30**, 195-219.

DIXON, J.B., HOSSNER, L.R., SENKAY, A.L. & EGASHIRA, K. (1982). Mineralogical properties of lignite overburden as they related to mine spoil reclamation. In: *Acid sulphate weathering*, Kittric, J A., Fanning, D.S. & Hossner, I.R. (eds), Soil Science Society of America Special Publication 10, 169-191.

HAWKINS, A.B., LAWRENCE, M.S. & PRIVETT, K.D. (1988). Implications of weathering on the engineering properties of the Fuller's Earth formation. *Geotechnique*, **38**, 517-532.

MITSUNASHI (1976). *Geological map of Tokyo Bay and adjacent areas, 1:100,000*. Geological Survey of Japan, 91pp.

MILLER, J. (1988). Microscopical techniques: 1. Slices, slides, stains and peels. In: *Techniques in Sedimentology*, Tucker, M. (ed). Blackwell Scientific Publications, 86-107.

PYE, K. & MILLER, J.A. (1990). Chemical and biochemical weathering of pyritic mudrocks in a shale embankment. *Quarterly Journal of Engineering Geology*, **23**, 365-381.

RUSSELL, D.J. & PARKER, A. (1979). Geotechnical, mineralogical and chemical interrelationships in weathering profiles of an overconsolidated clay. *Quarterly Journal of Engineering Geology*, **12**, 107-116.

SPEARS, D.A. & TAYLOR, R.K. (1972). The influence of weathering on the composition and engineering properties of in situ Coal Measures rocks. *International Journal of Rock Mechanics and Mining Science*, **9**, 729-756.

3-10 STRENGTH CHARACTERISTICS OF ROOT SYSTEMS IN MUDSTONE

J-W. CHEN & D-H. LEE
Department of Civil Engineering, National Cheng Kung University, Tainan, Taiwan

ABSTRACT

The mudstones which form the foothills of south western Taiwan characteristically swell/slake after water absorption and shrink/fracture when their water content decreases. The weathering of the mudstone is accelerated by the wet-dry cycles, resulting in disasters such as severe slope erosion. In addition during the rainy season landslides occur frequently in this area. In order to prevent such disasters, different kinds of root systems have been adopted for stabilisation of slopes in the area.

The paper evaluates the effectiveness of root systems in improving the resistance of the mudstone to failure.

MUDSTONE IN SOUTH WESTERN TAIWAN

The Pliocene rocks in south western Taiwan are composed of a thick mudstone series which has been variously named as the Kutingkeng Formation, the Nahua Mudstone or the Yuching Shale. These mudstones were deposited in a comparatively deep marine environment and may well be related to turbidity currents (Ho, 1975). According to Wu (1967), the area covered by this series is approximately 1014 km^2. The total thickness of the mudstone series is about 5000 m. Generally the bedding is not obvious due to the uniform lithology, although some sandy interlayers do occur.

The Pliocene mudstones possess shrinkage/swelling potential and low to medium compressibility. The pore water has high exchangeable sodium potential and a pH between 8 and 9. The multivalent cations in the pore water, such as Ca^{2+} and Mg^{2+} are precipitated as hydroxides and carbonates (Yen and Tsai, 1985). As a result, clay minerals cannot provide an adequate binding force, exacerbating the sheet erosion and weathering degradation of the mudstone slopes. As a consequence, the land surface is typical of badland morphology (Plate 1).

SOIL ROOT SYSTEMS

Root systems have been used to stabilise landslides and to control erosion on slopes for many years. Greenway (1987) summarised the mechanisms by which vegetation affects slope stability while Gray and Ohashi (1983) and Wu, Mcomber, Erb and Beal (1988) established theoretical models to

Plate 1: Mudstone in south western Taiwan.

describe stress-strain behaviour of soil-root systems under shear. However, only a limited number of studies (Erdo and Tsurata, 1969; Waldron, 1977; Wu, Beal and Lan, 1988) have been conducted on the influence of root systems on the shear strength of the soil.

In this study, four species of vegetation were planted on mudstone slopes in order to determine the effect of these root systems on the shear strength and swelling potential of the mudstone. The roots of these species, Bahia grass, Rhodes grass, Bermuda grass and Leucaena glauca are shown in Figure 1.

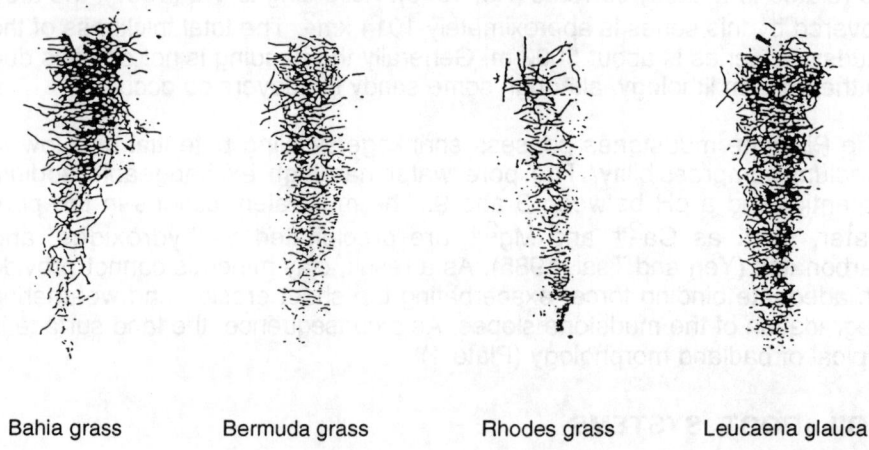

Bahia grass   Bermuda grass   Rhodes grass   Leucaena glauca

Figure 1: Root systems used in this study.

SAMPLING

A special sampler was designed in order to reduce disturbance (Plate 2). This consisted of a 100 mm long cutting shoe of 50 mm internal diameter, connected by a thread at one end to a 240 mm long sleeve of 54 mm internal diameter. Inside this are 50 mm internal diameter rings, each 25 mm in length, to take the cutting shoe and a 50 mm internal diameter inner ring, 25 mm in height. After the sampler was hammered into the soil perpendicular to the ground surface to the desired depth it was pulled out and sealed at both ends. In the laboratory the cutting shoe and sleeve are separated and the inner rings removed. Thus each 25 mm ring provides a single specimen.

Plate 2: Mudstone-root system sampler.

DIRECT SHEAR TEST

In order to study the variation of shear strength of the mudstone-root system with depth, each sample was divided into three parts; a top, middle and bottom representing the mudstone-root systems from 0 to 80 , 80 to 160 and 160 to 240 mm depth respectively. Each part contained three inner rings from which specimens were prepared for direct shear testing. The three tests from each part allowed a determination of one set of c' and ø' values for the top, middle and bottom layers, representing the shear strength of the mudstone with the particular species of root system at each depth zone. Figure 2a gives the failure envelope for the natural mudstone and Figures 2b, c, d and e those for the mudstone with each of the different root systems The c' and ø' values of the mudstone and mudstone-root systems are given in Table I.

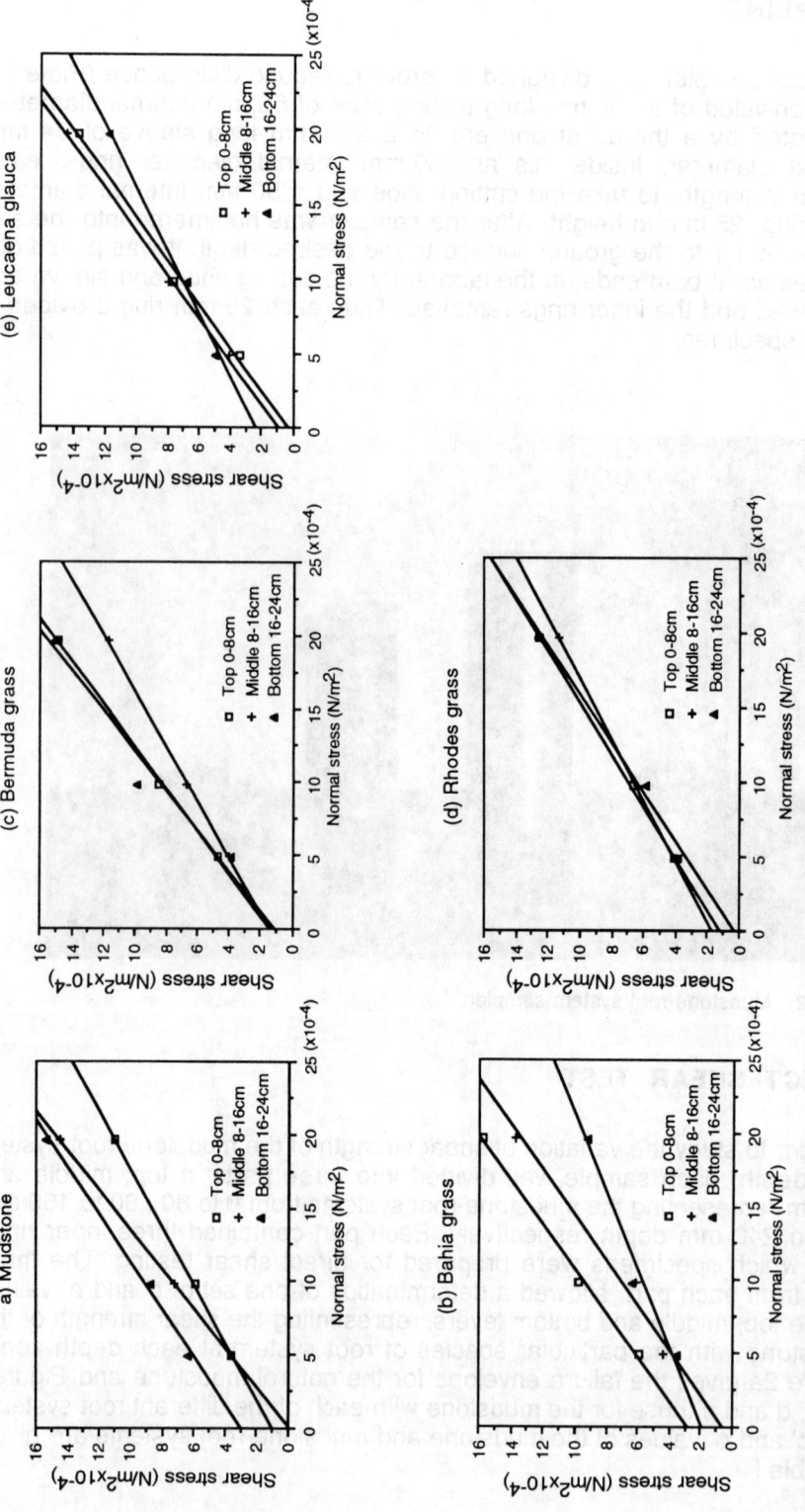

Figure 2: Failure envelopes of mudstone and mudstone-root systems.

Table I: Shear strength parameters of mudstone with root systems.

Shear Strength Parameters	Depth (m)	Mudstone	Bahia Grass	Rhodes Grass	Bermuda Grass	Leucaenu Glauca
c' value (kN/m^2)	Top	13	31	16	34	5
	Middle	0	9	22	16	11
	Bottom	32	22	4	12	26
ø' value (°)	Top	27	34	29	26	35
	Middle	37	34	26	29	35
	Bottom	32	22	32	37	26

Clearly, a root system may not be uniformly distributed with depth, hence the root content in each specimen was also measured. After the direct shear tests had been completed, the same specimens were dried in an oven for 24 hours and the roots sieved out and weighed. The variation of root content with depth is given in Table II. It should be noted that root content is presented as the mass of root material contained in an inner ring of the sampler.

As can be seen from Table I, with the exception of the Leucaenu Glauca in the top layers the mudstone-root systems increased in c' value while all increased in ø' value except the mudstone with the Bermuda grass. In the middle layers all mudstone-root systems increased in c' value but decreased in ø' value. In the bottom layer both c' and ø' values of all the mudstone-root systems decreased, with the exception of the ø' value of Bermuda Grass. From Table II it can be seen that the root content for each species is concentrated in the top layer. Generally the root mass was small below 105 mm and hence not surprisingly, beneath this depth no improvement in the shear strength of the mudstone was noted.

Table II: Amount of root systems with depth variation (in grams)

Depth (mm)	Bahia Grass	Rhodes Grass	Bermuda Grass	Leucaenu Glauca
5-30	0.63	0.12	0.11	0.30
30-55	0.52	0.07	0.07	0.23
80-105	0.16	0.02	0.03	0.20
105-130	0.13	0.01	0.02	0.15
130-155	0.05	<0.01	0.01	0.08
160-185	0.03	<0.01	<0.01	0.04
185-210	0.05	<0.01	<0.01	0.02
210-235	0.03	<0.01	<0.01	0.01

FREE SWELL TEST

In his work on the swelling characteristics of mudstones in south western Taiwan, Chow (1988) indicated that the swelling strain of remoulded mudstone reached 15% after 24 hours saturation. In the present study mudstone and mudstone-root system specimens of 50 mm diameter and 25 mm height were placed in a fixed ring consolidometer for free swell testing. The test was similar to the one-dimensional consolidation test but without any load on the specimen. Figure 3 shows the swelling strain of the natural mudstone, the mudstone-Bahia grass and mudstone-Rhodes grass root systems respectively. The final swelling strain of the mudstone-root systems is reduced by about one third compared to that of the weathered mudstone.

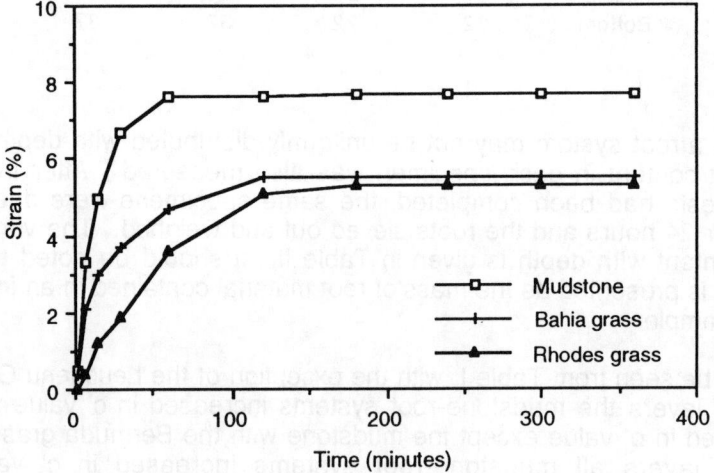

Figure 3: Swelling strain characteristics of mudstone and root systems.

CONCLUSIONS

From the results of the tests carried out the following conclusions are drawn:

1. Weathered mudstone with a root system has a larger c' value than that with no root system.

2. Root systems only improve the c' value component of the shear strength of weathered mudstone to depths of less than 80 mm.

3. Root systems reduce swelling potential of the weathered mudstone but do not appear to affect the ø' values of the weathered mudstone.

4. Bahia grass is the most effective of the four tested root systems.

5. Root systems reduce the swelling potential of weathered mudstones.

REFERENCES

CHOW, D.C. (1988). *Study on swelling property and shear strength of mudstone*. M.Sc. Thesis, National Cheng Kung University (in Chinese).

ERDO, T. & TSURATA, T. (1969). The effect of tree roots upon the shearing strength of soil. *Annual Report No. 18 of the Hokkaido Branch, Tokyo, Forest Experiment Station, Tokyo, Japan*, 168-179.

GRAY, D.H. & OHASHI, H. (1983). Mechanics of Fibre Reinforcement in Sand. *Journal of Ground Engineering, ASCE*, **109(3)**, 335-353.

GREENWAY, D.F. (1987). Chapter 6 - Vegetation and Slope Stability. In: *Slope Stability*, Anderson, Richard (eds.), John Wiley & Sons Inc.

HO, C.S. (1975). *An Introduction to the Geology of Taiwan; Explanatory Text of the Geology Map of Taiwan.* The Ministry of Economic Affairs, R.O.C.

WALDRON, L.J. (1977). Shear Resistance of Root-Permeated Homogeneous and Stratified Soils. *Soil Science Society of America Proceedings*, **41**, 843-849.

WU, C.M. (1967). Study on the erosion problem in mudstone area in southwestern Taiwan. *Hydraulics Journal*, **3** (in Chinese).

WU, T.H., BEAL, E. & LAN, C.C. (1988). In situ shear test of soil-root systems. *Journal of Ground Engineering., ASCE*, **114**, 1376-1394.

WU, T.H., MCOMBER, R.M., ERB, R.T. & BEAL, P.E. (198). Study of Soil-Root Interaction. *Journal of Ground Engineering, ASCE*, **114**(12), 1351-1375.

YEN, F.S. & TSAI, Y.H. (1985). On the Physio-Chemical Properties of the Mudstones in the Foothills of Southwestern Taiwan. *Final Report on Research of Disaster Prevention*, **No. 74-09**, N.S.F. (in Chinese).

3-11 ENGINEERING PROPERTIES OF MUDSTONE-LIME-SLAG MIXTURES

D-H. Lee & Y-M. Tien
Department of Civil Engineering, National Cheng Kung University
Tainan, Taiwan

ABSTRACT

This paper reports on a study undertaken to evaluate the relative merits of various mudstone-lime-slag mixtures for use as a subgrade material during the construction of a freeway in Taiwan. The results suggest that the addition of 3% lime to mudstone leads to a significant enhancement of the engineering properties at a reasonable cost. Uncrushed slag did not have a significant effect during the maximum curing period of ninety days, but this may be due to its large particle size and the limited duration of the study.

INTRODUCTION

A new freeway is being constructed in western Taiwan, in part to encourage the development of new towns in this area. For this reason it was necessary to position the road some distance from the existing highway. As a consequence, a 100km section of the new freeway passes through a mudstone area in the south of the island. This mudstone is known to be highly shrinkable/swellable and both erosion and slope failure have taken place.

It was appreciated that the construction of this section would involve excavation of the mudstone, which in its natural state would not form a good material for subgrade or backfill. A study was therefore undertaken to ascertain whether the addition of lime and/or blast furnace slag could improve the engineering properties of the excavated mudstone, such that the stabilised material could be utilised during the construction contract.

GEOLOGY AND TESTING MATERIALS

MUDSTONE

The mudstone used for this study was taken from the highlands in south western Taiwan. The Pliocene mudstone formation has a thickness of about 4,000 m (Ho, 1986) and covers an area of some 1,000 km^2. The mudstone consists of illite, chlorite, kaolinite, gypsum and quartz (Wang, 1970); Table I shows its physical properties. For the testing, the mudstone should be crushed and ground and the portion passing the 425 μm sieve selected. In the dry state the bearing capacity and shear strength of the mudstone are high but it loses much of its strength and is prone to swelling and slaking after experiencing water absorption.

Table I: Physical properties of untreated mudstone in the southern part of Taiwan.

Grain size		
	Sand (> 0.06 mm)	3 - 9%
	Silt (0.07-0.002 mm)	62 - 65 %
	Clay (<0.002 mm)	27 - 34 %
Liquid Limit (w_l)		38%
Plastic Limit (w_p)		24 %
Plasticity Index (I_p)		14 %
Specific Gravity (Sp.G)		2.74 - 2.75
Void Ratio (e)		0.51

BLAST FURNACE SLAG

Blast furnace slag is a by-product of the steel manufacturing process. Two different types of blast furnace slag may be obtained depending on whether the cooling method uses air or water. When the temperature is reduced gradually by air cooling, the slag has a larger particle size and better crystallisation, hence it forms a good aggregate for concrete despite its poor cementation. The water cooling process is very rapid, however, hence the slag contains much glassy material of granular shape. In the presence of an activator, such as lime, alkali or alkali sulphate, granulated slag reacts with water to form a cementitious material. It is not considered a pozzolan, however, as it requires only very small amounts of the activator to induce hydration.

In this study water-cooled granulated blast furnace slag was used, supplied by the China Steel Company in Kaohsiung. As can be seen from the particle size distribution curve presented in Figure 1, most of the particles have a diameter of between 0.1 and 5 mm. The main chemical components of the slag are given in Table II.

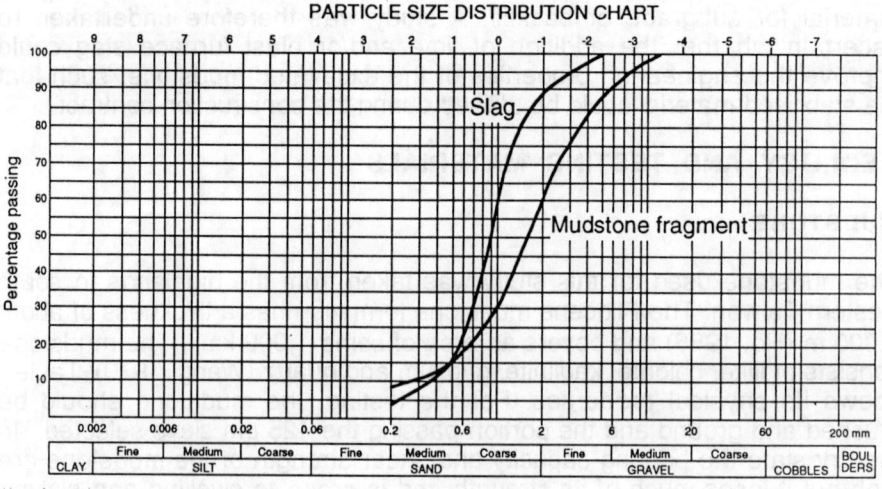

Figure 1: Particle size distribution curves of mudstone fragment and slag.

Table II: Chemical composition of China Steel Company water quenched blast furnace slag.

Constituent	Weight (%)
SiO_2	34.6-36.7
MgO	7-8.2
CaO	38-42
Al_2O_3	13.5-15.5
S	0.8-1.0
K_2O	0.4-0.7
FeO	0.3-0.7
MnO	0.4-0.7
TiO_2	0.4-0.7

TEST PROCEDURE

The slag was mixed directly with the mudstone material and lime without being ground. Ten combinations of mudstone-lime-slag were tested (Table III). A modified Proctor compaction test (ASTM) was carried out on a sample from each mixture and the optimum moisture content (OMC) and maximum dry density (MDD) obtained. The samples were securely wrapped in vinyl bags and subjected to a curing process at room temperature (c. f27°C) for 0, 28 and 90 days. Swelling tests, CBRs, direct shear and consolidation tests were undertaken on each of the samples, following the ASTM procedures set out in ASTM 1990.

DISCUSSION OF RESULTS

It will be appreciated that to obtain the OMC and MDD on samples of each of the ten mixtures cured for 0, 28 and 90 day durations using the modified Proctor compaction test would be extremely time-consuming. It was therefore decided to follow this procedure for only three of the mudstone-lime-slag mixtures with combination ratios by weight ($w_m:w_s:w_l$) of 1:0:0, 0.95:0.05:0 and 0.84:0.10:0.06 respectively. Using Lagaro and Moh's (1970) triangular co-ordinates method and the Proctor compaction results from these three mixtures, it was possible to calculate the γ_{dmax} and OMC of the other mixes.

To check the reliability of these calculations, Proctor compaction tests were undertaken on two other mixes with $w_m:w_s:w_l$ ratios of 0.92:0.00:0.03 and 0.91:0.03:0.06 respectively. The full results are given in Table III. As can be seen, the calculated and estimated figures are very similar and it was therefore considered reasonable to use triangulated results for the purposes of this study.

Figure 2 shows the relationship between the dry density and water content for five of the mixtures. Although with a lower water content the γ_d decreases as the proportion of lime increases, where the water content is higher the influence of the lime decreases and the compaction curve for all five samples follows an almost identical, steep dip.

Table III: The O.M.C. and γ of mudstone-lime-slag mixtures.

	\multicolumn{10}{c}{Mixture}
	A	B	C	D	E	F	G	H	I	J
mudstone	100	97	95	90	94	92	87	91	89	84
slag (%)	0	3	5	10	3	5	10	3	5	10
lime (%)	0	0	0	0	3	3	3	6	6	6
γ_d max (g/cm³) (experimental)	1.95		1.95			1.85		1.83		1.86
γ_d max (g/cm³) (calculated)	1.95	1.95	1.95	1.95	1.91	1.91	1.91	1.86	1.86	1.86
O.M.C.(%) (experimental)	13.2		13.7			13.8		14.39		14.70
O.M.C.(%) (calculated)	13.2	13.12	13.7	13.17	14.0	13.95	13.82	14.88	14.83	14.70
Remark	calc. base		calc. base							calc. base

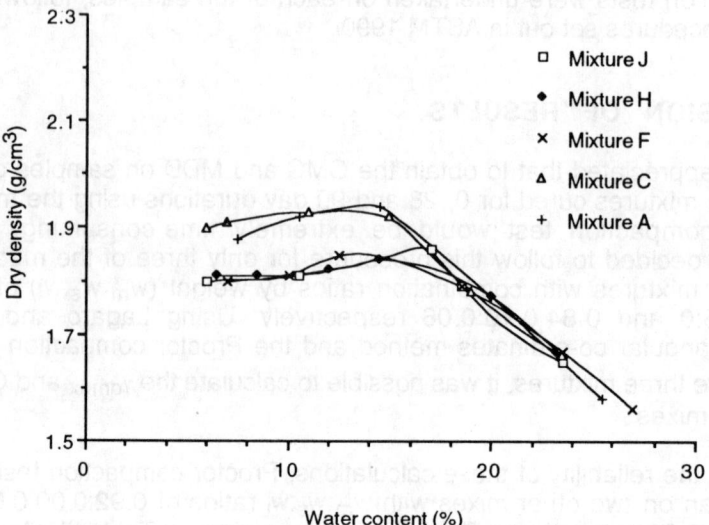

Figure 2: Compaction curves for five mudstone-lime-slag mixtures.

With a lower water content the OMC increases and the γ_d decreases with higher proportions of lime. This may be due to the small size of the lime particles and the increased thickness of the adsorbed water.

The swelling strains, CBR values and compression indices of the ten mudstone-lime-slag mixtures taken at the end of each of the duration periods are given in Table IV. It can be seen that, irrespective of curing time or slag content, the engineering properties of the samples without lime remain relatively consistent. This suggests that slag by itself does not form a suitable stabilizing additive. On the other hand, the engineering properties of the mixtures containing 3 or 6% lime are significantly enhanced and the longer the period of curing the greater the improvement. It is of note that the proportions of slag used with the lime had little effect on the engineering properties of the mixture, despite the fact that it is known that slag can use lime as a catalyst to initiate cementation. It is likely that in this study the particle size of the slag used was too large to produce a good hydro-cement effect.

Table IV: Engineering properties of mudstone-lime-slag mixtures.

Property	Curing time	A	B	C	D	E	F	G	H	I	J
CBR (%)	0 day	1.2	2.2	1.7	2.0	31.2	37.3	40.8	54.9	35.1	44.0
	28 days	1.2	1.9	1.2	2.3	61.8	54.4	70.0	55.9	73.6	70.2
	90 days	2.0	2.0	2.7	2.3	96.1	74.5	89.3	95.2	99.1	87.5
Swelling Strain (%)	0 day	3.8	4.1	4.7	3.9	0.7	0.4	0.4	0.4	0.4	0.3
	28 days	4.3	4.0	3.7	3.6	0.2	0.14	0.17	0.14	0.11	0.13
	90 days	3.8	3.7	3.3	2.5	0.07	0.08	0.04	0.04	0.03	0.07
C_c	28 days	0.16	0.14	0.13	0.18	0.12	0.12	0.100	0.08	0.089	0.089
	90 days	0.12	0.11	0.10	0.12	0.05	0.05	0.07	0.06	0.06	0.05
Shear Strength c (MPa)	28 days	0.18	0.19	0.17	0.16	0.16	0.17	0.19	0.17	0.16	0.17
	90 days	0.16	0.15	0.18	0.18	0.17	0.20	0.21	0.22	0.25	0.21
ϕ (deg)	28 days	22.8	21.1	23.0	23.1	35.1	34.8	33.5	37.5	35.3	36.1
	90 days	22.7	23.1	22.3	23.5	39.1	38.7	37.6	38.5	37.5	38.9

After 90 days curing the mixtures with 3% lime gave a compression index C_c of 0.06, shear strength parameters c = 190 kPa, ϕ' = 38.5°, a CBR of 80% and a swelling strain of 0.07%. This represents a significant improvement over the properties of the natural mudstones and with such a high CBR means excavated material treated in this way could be used as subgrade for the construction of the new freeway (Ingles and Metcalf, 1972). It is of note that after 90 days curing there is relatively little additional improvement when 6% lime is used. It is considered that a mix containing 3% lime cured for 90 days provides a satisfactory and economic solution to enhancing the properties of mudstones for use as road subgrade.

CONCLUSIONS

1. The mudstone-slag mixes did not form good subgrade material as regards volumetric stability, bearing capacity and shear strength.

2. Mudstone-lime-slag mixes containing 3% lime showed a low swelling strain and low compression index after 90 days curing, with CBRs of 80%, cohesion c of 190 kPa and ϕ' of 38.5°. Such material would therefore form adequate subgrade material for road construction.

3. In this study, slag was not seen to enhance the engineering properties of the mixes significantly. However this may be due to the large particle size of the slag which was not crushed/ground before incorporation into the mixture. This would inhibit the onset of hydrocementation and hence adversely influence possible stabilising effects.

REFERENCES

HO, C.S. (1986). *An introduction to geology of Taiwan - explanatory text of the geology map of Taiwan, 2nd edition*, The Ministry of Economic Affairs, Taiwan; 71-78.

INGLES, O.G. & METCALF, J.B. (1972). Soil stabilization - principles and practice. *Chapter 9, Stabilization in Road Construction*, 225-246.

LAGARO, R.C. & MOH, Z.C. (1970). Stabilization of deltaic clay with lime rice hull ash admixtures. *Proceedings of the Second Southeast Asia Conference on Soil Engineering, Singapore*, 215-223.

LEE, D.H. & TEN, Y.M. (1987). Engineering properties of mudstone. *Sino-Geotechnical*, **19**, 26-34.

RAMACHANDRAN, V.S., FELDMAN, R.F. & BEAUDOIN, J.J. (1981). *Concrete Science*, Heyden and Son, 286.

WANG, Y. (1970). Clay mineralogy of Gutingkeng mudstone, Southern Taiwan. *Science Reports of National Taiwan University, ACTA Geological Taiwanica*, **14**, 9-19.

SESSION 4 - MARINE & ESTUARINE

4-1	An overview of the factors responsible for the degradation of materials exposed to the marine/estuarine environment *A.K. Tiller*	337
4-2	Biofouling and microbial corrosion in estuarine waters *R.G.J. Edyvean & H.A. Videla*	355
4-3	Susceptibility of steel piles to corrosion in the marginal lands of the Niger Delta *S.C. Teme*	369
4-4	The influence of microbial assemblages on sediment erodibility: an engineering perspective *D.M. Paterson*	383
4-5	Organic residues as a factor in the plasticity of clay soil from the SERC soft clay research site, Bothkennar *M.A. Paul*	401
4-6	Chemical stabilisation of estuarine alluvium *A. B. Hawkins, J.C. Lonsdale & R.W. Narbett*	413
4-7	Laboratory study of soil-fluid interaction behaviour *R.K. Srivastava, A.V. Jalota & B. Singh*	431

4-1 AN OVERVIEW OF THE FACTORS RESPONSIBLE FOR THE DEGRADATION OF MATERIALS EXPOSED TO THE MARINE/ESTUARINE ENVIRONMENT

A.K. TILLER
National Corrosion Service, National Physical Laboratory, Teddington, Middlesex

ABSTRACT

The paper presents a review of the factors responsible for the degradation of materials used in the marine/estuarine environment. The important aspects which should be considered by design and structural engineers if problems are to be avoided and the life expectancy of the structure is to be achieved are discussed. The data presented are either widely available or can be readily obtained from a variety of sources.

The importance of the chemistry of the marine/estuarine environment, its variability in degradation processes and mechanisms are discussed. Attention is also drawn to the influence of the hydrodynamics and ecology of estuarine waters on materials of construction. Simple procedures for assessing the corrosiveness of the environment are also highlighted, including the range of preventative measures which are available.

INTRODUCTION

The aim of this paper is to review the data available on a variety of topics which have a bearing on the degradation of engineering structures and related equipment exposed to the marine/estuarine environment. Materials exposed to a entirely seawater environment tend to be subjected mainly to an oxygen saturated system with a well defined chemistry, such as salinity, chlorinity and oxygen concentration. This is not the case in the estuarine situation which usually has a variable water chemistry with marked fluctuations in chloride, sulphate, nitrate, carbonate and dissolved gases and in the nature and type of biofouling which develops. Estuarine muds and sediments are seldom simple systems and frequently contain heavy metal ions such as iron, copper and manganese, etc. These elements are often present as a result of effluent dumped from vessels operating in the region or from industrial activities where discharges occur upstream. Such variables in environmental conditions present a challenge to the materials engineer and must be addressed irrespective of the nature and type of structure being developed. This is not necessarily a difficult or daunting task provided the appropriate philosophy is adopted, ie "prevention is better than cure". By understanding these variables and the role they play in the degradation process, improved performance and reliability will be achieved.

FACTORS AFFECTING DEGRADATION IN ESTUARINE ENVIRONMENTS

The factors affecting the degradation of materials in estuarine conditions may be classified as chemical, physical and biological. Alone or in combination, these factors determine the behaviour of materials and the form of degradation which occurs.

In comparison with land-derived waters, seawater contains a relatively high concentration of salts which remains substantially constant. It is often considered to be equivalent to a 3.5 wt% solution of sodium chloride, but there are also significant levels of calcium, magnesium, potassium and sulphate together with traces of many other constituents. For convenience, the concentration of salts in seawater is expressed as salinity; defined as total weight in grams of solid matter dissolved in 1 kg of water as shown in Table I.

Table I: Composition of seawater.

	gms/kg of water of salinity 35 gms/kg
Chloride	19.4
Sodium	10.8
Sulphate	2.7
Magnesium	1.3
Calcium	0.41
Potassium	0.39
Bicarbonate	0.14
Bromide	0.07
Shartium	0.008
Boron	0.004
Fluoride	0.001

Even more remarkable than this near-constancy of total salt content is the fact that the relative proportions of the major constituents of seawater are virtually consistent whatever its source. Exceptions to both of these rules occur however, particularly at estuarine sites where the seawater is diluted with varying quantities of land-derived water of different compositions. Natural seawater is also well-buffered to pH-values between 8.1 and 8.3, although in stagnant areas and in areas of estuarine pollution the pH may fall to 7.

In addition to the dissolved salt content, the dissolved gases in estuarine waters play an important part in determining its corrosivity. The most important of these is oxygen. In the same way as for land-derived waters, the oxygen controls the corrosion rate of metals and unpolluted seawater can contain up to 12 ppm dissolved oxygen depending on temperature. On-shore and at estuarine sites where fixed structures exist, this value may be lower or reduced by pollution and under these conditions the oxygen level can fall to zero. The implications of this and the role of pollution are discussed later but it is important to note that the presence of hydrogen sulphide in these environments will have a marked effect on the rate of degradation of the structure. The other important gaseous constituent is carbon dioxide which contributes to the chemistry of carbonate scale formation.

A useful, albeit broad, generalisation about estuarine conditions is that there is a relatively high concentration of aggressive anions such as chloride, sulphate and nitrate (Table II). However, there are other factors in addition to the chemical content which facilitate corrosion and determine its type and severity; these are discussed below.

Table II: Example of variation in composition of estuarine water with time, Alexandra Docks, Grimsby (after Tiller, 1988).

Time hrs	Dissolved O_2 (% Satn)	Chloride ppm	Nitrate ppm	Ammonia NH_4 ppm
0900	85	3200	3.05	0.84
1200	40	2800	1.47	1.95
1600	36	10800	2.37	3.18
1800	76	12600	3.60	0.35

OTHER FACTORS ASSOCIATED WITH CORROSION

FLOW VELOCITY AND TIDAL CONDITIONS

The influence of flow velocity on the corrosion of metals has been widely studied and is well documented in the literature. Much of this is relevant to the process chemical industry or to situations where cooling water systems are employed, ie marine condenser units. Although hydrodynamic parameters are well defined for these systems, they may not necessarily be directly related to the type of flow characteristics which occur in estuarine conditions as a result of tidal effects. Nevertheless, the overall rate of a corrosion reaction will be determined by either the rate of transport of oxygen to the metal surface (ie diffusion control), transport of corrosion products from the metal, or the rate of reaction at the surface. The effect of flow parameters is primarily on reactions under diffusion control where the reduction of the diffusion boundary layer thickness can directly affect the rate of access or egress of reactants or products. Thus in the corrosion of ferrous metals the rate of corrosion will be controlled by the rate of supply of oxygen from the water to the metal surface. In the practical situation there is always some movement of the water and even in quiescent conditions some natural convective movement will occur.

At low rates of forced convection, such as those caused by passing propeller driven vessels, the laminar flow regime will change to a more turbulent system. In a laminar flow regime the rate of corrosion is increased somewhat but this may be beneficial in overcoming any tendency to local attack such as pitting. Under these conditions the rate of attack (k) can usually be expressed as a function of velocity (V) as follows:

$$k = K_o V^n$$

where K_o and n are constants, n varying typically between 0.5 for flat plate geometries and 0.33 for flow through irregular geometrics, eg in harbours, berth installations, etc.

As the flow increases, turbulent conditions are generated; the transition will be governed by the nature and magnitude of the fouling and surface roughness as well as the system geometry. Variation in the corrosion rate with turbulent regimes usually depends on the type of environment - for example clean and unpolluted sea and estuarine waters or fresh rivers. The corrosion rate usually increases more rapidly with increase in flow due to increased oxygenation of the system. In polluted systems, particularly those containing high total suspended solids either as fine mud, sand particles or domestic sewage, the aggressiveness of the system increases due not only to the increase in flow but also because of impingement effects leading to erosion corrosion. This is particularly important for the part of a structure just above the mud or sea bed, including the buried section. Movement of estuarine mud or sea sediments will increase the erosion corrosion effect.

HYDROGRAPHIC FACTORS

The effect of tidal levels on fresh water flow and the extent of oxygenation or pollution varies seasonally. In the Tees, Tyne and Wye estuaries there is a greater tidal range during spring tides compared to neaps and this affects the composition of the water. In addition to the tidal movement of water in the estuary, there is usually a continuous seaward displacement caused by the entry of land water thus stratification of sea- and fresh-water can occur, encouraging variable oxygen gradients and salinity and hence the rate of corrosion.

Studies carried out in the Thames and Afon Dyfi estuaries showed clearly that the factors discussed above play an important role in the rate of corrosion. The results indicated that in a relatively unpolluted estuary such as the Afon Dyfi, the rate of corrosion was about 2-3 times greater than that experienced in the Thames (Table III). This increase could be attributed to such factors as retention time of fresh water, flow velocity stratification and hydrographic aspects of the two rivers and estuaries.

POLLUTION AND ACIDIFICATION

Reference has already been made to the effect of domestic sewage and industrial effluent discharges on the overall corrosivity of the estuary. Where chemical discharges occur the role of heavy metal ions such as manganese, copper and iron become important. The influence of ferrous ion levels on corrosion is in aiding the precipitation of metallic sulphides which are in themselves corrosive. Copper ions in solution plate out on ferrous material which could encourage galvanic corrosion and acidification

Table III: Ratios of corrosion for mild steel at sites of comparable salinity.

Compared Sites	Ratios of Corrosion rate							
	6 summer months		6 winter months		1 year		2 years	
	Pre-dicted	Exper-imental	Pre-dicted	Exper-imental	Pre-dicted	Exper-imental	Pre-dicted	Exper-imental
Dovey Junction with Putney	12	8.5	6	4	6	6	6	3.5
Pant Eidal with Tilbury	7	2	10	3.5	10	3.5	10	1
Aberdovey Jetty with Southend	1	0.8	1	-	1	-	1	-

* Predicted values obtained from the equation:

$$\frac{\text{Rate A}}{\text{Rate B}} = \frac{(\%\ \text{saturation with } O_2)_A \times (\text{rate of flow})_A}{(\%\ \text{saturation with } O_2)_B \times (\text{rate of flow})_B}$$

of the region will exacerbate the problem. Sludge discharges released during certain tidal conditions may not be adequately diluted and act in a manner similar to that of fresh water and non-mixing situations.

Pollution may also be caused by indiscrete dumping from passing vessels, particularly of oil and oil-type materials. Although deposited oil films on metals may initially provide some protection, this will deteriorate with time. In nearly all cases the degradation is brought about by a combination of oxidation and the activity of aerobic bacteria capable of utilising hydrocarbon sources. This usually results in the precipitation of aggressive metabolites and subsequent corrosion of the underlying metal.

More recently, an additional factor has been recognised: the effect of acid rain. Much has been written about the susceptibility of UK surface and ground waters to acid deposition, particularly with regard to ecological changes. From the research carried out it is now clear that the susceptibility of soils and natural waters to acid results in marked changes in the soil chemistry and in some cases large water catchment areas are now more acidic than hitherto.

To date the ecological effects have not been quantified with respect to the degradation of materials subject to estuarine conditions. Obviously where the buffering capacity of fresh water is insufficient to counter these changes, acidic environments will predominate and over a period of time this would be expected to increase the overall potential corrosiveness of the environment. An increase in the rate of corrosion of mooring chains and shackles used in a number of estuaries around the UK is a recent example of this problem.

It is interesting to note that out of thirteen locations examined eight were within regions where contamination due to acid rain was classified as medium to high, whilst in other there were rivers with sources where the risk of acid rain contamination was high (Rackley & Tiller, 1988).

FORMS OF DEGRADATION IN ESTUARINE CONDITIONS

For metallic structures in estuarine conditions several forms of corrosion can occur, viz, general, pitting, crevice, galvanic, erosion-corrosion and microbially-induced. In non-metallic materials such as concrete and reinforced concrete structures degradation is still mainly a chemical effect with diffusion of corrosive species into the body of the material. However, as with metallic systems, concrete is also vulnerable to biofouling and microbial corrosion as well as the erosion effect.

UNIFORM CORROSION

Corrosion is an electrochemical process in which a metal reacts with its environment to form an oxide or other compound. The cell developed in this process has three essential constituents: an anode, a cathode and a suitable conducting solution. Simply, the anode is the site at which the metal is corroded; the electrolyte solution is the corrosive medium; and the cathode forms the other electrode in the cell and is not consumed in the corrosion process. At the anode the corroding metal passes into the electrolyte as positively charged ions, releasing electrons which participate in the cathodic reaction. Hence, the corrosion current between the anode and the cathode consists of electrons flowing within the metal and ions flowing within the electrolyte.

The whole surface of a component may become the anode, and the surface of another component in contact with it the cathode. Usually, however, corrosion cells will be much smaller and more numerous, occurring at different points on the surface of the same piece of metal. For example, anodes or cathodes may arise from differences in the constituent phases of the metal itself, from variations in natural or protective coatings on the metal, or from variations in the electrolyte (caused for instance, by differential aeration). The electrolyte may be in bulk form (as when a component is immersed in or is handling a liquid during service) or it may be present only as a thin condensed or absorbed film on the metal surface.

The overall rate of the process under these circumstances may depend on a number of factors, but the important criterion is the balance between the rates of the anodic and cathodic reactions, required to maintain electroneutrality of the metal, and the rate of corrosion is controlled by the slower of the two. Frequently, the availability of oxygen governs the rate at which the cathodic process can absorb electrons and thus controls the corrosion rate.

Although many metals display active corrosion behaviour under immersed conditions others, in particular many engineering alloys, form a stable film of solid corrosion product, which is frequently an oxide. If this oxide has good cohesive strength and adhesion to the metal and low ion conductivity, the metal dissolution rate is limited to the rate at which the metal ions can pass through the film, which is often a factor of 10^3 to 10^6 lower than the rate of dissolution of film-free metals. This is the phenomenon of passivity, exhibited by stainless steels, nickel alloys, titanium and many similar materials, which enables many engineering materials to be used in estuarine environments.

Although uniform corrosion is responsible for the greatest destruction of metal, it is not technically a major cause for concern. The rate of corrosion is measurable, usually by relatively simple tests, and the lifetime of a component can be predicted with confidence. It is the more complex forms of corrosion which present problems by encouraging localised or accelerated rates of attack, so that failure may occur with only little total metal loss, frequently as a result of the formation of microscopic pits or cracks which may not be readily detected.

GALVINIC (BIMETALLIC) CORROSION

When two dissimilar metals are immersed in a conducting solution they usually develop different potentials and if the metals are in contact this potential difference provides the driving force for corrosion. The less resistant of the two metals will corrode, while the more resistant is often completely protected. An electrochemical series based on the thermodynamic behaviour of metals in solutions of their own ions is

Table IV: Galvanic series of metals in seawater.

BASE	Magnesium		
	Zinc		
	Aluminium	(commercial)	
	Cadmium		
	Duralumin	(Al with 4.5% Cu)	
	Mild steel		
	Cast iron		
	Stainless steel		ACTIVE
	(Type 430; 18% Cr)		
	Stainless steel		ACTIVE
	(Type 304; 18% Cr 10% Ni)		
	Lead-tin solders		
	Lead		
	Tin		
	Nickel		
	Brasses		
	Copper		
	Bronze		
	Monel		
	Silver solders	(70% Ag 30% Cu)	
	Nickel		PASSIVE
	Stainless steel	(Type 430)	PASSIVE
	Stainless steel	(Type 304)	PASSIVE
	Silver		
	Titanium		
	Graphite	(Carbon) (non-metal)	
	Gold		
NOBLE	Platinum		

frequently used as the basis for ranking metals under these conditions but this is misleading since it is not usual to find pairs of metals exposed under these conditions, and in any event the engineering materials most often considered are alloys rather than pure metals. A more practical ranking is obtained from a galvanic series based on corrosion measurements in specific corroding environments. Examples are given in Table IV for a seawater environment.

Although tables of galvanic properties will show which alloy of a couple will become the anode and which the cathode, no indication is given of the rate of corrosion to be expected from the anodic material. For example, many of the active/passive materials such as stainless steel behave well in estuarine water and would be expected to be cathodic in many couples, but are in fact not very good substrates for the oxygen-reduction reaction and readily become polarised. Under these conditions acceleration of attack of the less noble member of a couple is often much less than would be expected. The prediction of actual rates of galvanic attack is therefore more complex than simple prediction of the polarity of the couple, and measured values are of far more practical significance. Published tables of galvanic behaviour (BSI PD 6484: 1979), refer to measurement at ambient temperature, and are a useful source of data.

ELECTROLYTIC- STRAY CURRENT CORROSION

The previous section discussed galvanic or bimetallic corrosion which is concerned mainly with the combination of dissimilar metals. However, in addition to this form of corrosion there is also the problem of electrolytic corrosion caused by stray current effects which is caused by a current from an external source. This source may be from a vessel's battery, shore supply, electronic equipment or other electrical installations such as welding equipment.

Under the influence of stray currents the rate of corrosion can be extremely high and current values can vary from a few microamps to several amps. Prevention of this type of corrosion is a matter of good electrical installation and all underwater metallic fittings should be electrically isolated. This in practice requires measures such as fitting an insulating coupling on the system. If such insulation is not implemented then the structure could be subjected to a high rate of corrosion and degradation.

This problem can also arise when cathodic protection systems are installed having inadequate earthing and bonding.

CREVICE CORROSION

The existence of crevices, either as a result of design (eg under laps, bolt and rivet heads, gaskets etc) or as a result of operation (eg by deposition of sand, dirt, corrosion deposits etc) can lead to occluded regions in which oxygen cannot be replenished. Thus, whereas both anodic and cathodic reactions would initially occur uniformly over the surface, after only a short interval poor availability of oxygen in the crevice can lead to cessation of the cathodic reaction in the occluded region, with the anodic reaction continuing, and balanced by cathodic reactions outside the region of attack.

With active metals such as iron, anodic reaction also continues outside the crevice and so although there may be some acceleration of corrosion within the crevice, the effect will be to increase the attack by a modest factor only. With stainless steels and similar materials however, the effect may be very different. Consumption of oxygen within the crevice may then mean that the passive film cannot be maintained at that location whereas outside the crevice, the film will remain intact so that enormous ratios of cathodic to anodic area are available. As dissolution of metal proceeds, an excess positive charge can build up due to the presence of metal ions in the crevice solution, which is balanced by diffusion of anions from the bulk of the solution. These are primarily the small and very mobile (and highly aggressive) chloride ions and under these circumstances the crevice solution becomes more aggressive than the bulk. In addition, as the concentration of metal ions builds up, a point is reached at which the solubility limit is exceeded and metal hydroxide is precipitated, leaving an excess of hydrogen ions in the crevice which results in a reduction in pH, and the environment becomes yet more aggressive.

The reaction rate in the crevice is thus further increased and indeed the whole process becomes autocatalytic. The crevice may contain up to 10 times the chloride content of the bulk and the pH inside the crevice usually falls to values of pH 1 or lower.

PITTING CORROSION

Pitting corrosion is another form of extremely localised attack which can seriously affect the overall life of a metallic structure. In many cases the incubation period for pit initiation is very long but once it has started it is usually driven by an autocatalytic reaction which leads to an increase in the rate of corrosion. However in some cases pitting attack may be stifled due to build-up of corrosion product or lack of oxygen, as for example when thick encrustations develop on the metal surface. Tubercles formed as a result of biofouling are less likely to be benign. Usually the pitting attack beneath the fouling continues due to bacterial activity by microrganisms such as the sulphate-reducing bacteria.

EROSION-CORROSION

Erosion-corrosion is the acceleration of the rate of deterioration of a metal resulting from relative movement between the corrosive fluid and the metal surface. The phenomenon is characterised by the development of a surface morphology of grooves, gulleys, waves, rounded holes, etc, which usually exhibit a directional pattern. These features are produced as a result of the flow velocity of the liquid and the mechanical removal either of corrosion products or of solid metal or cement. Most metals and alloys including concrete are susceptible to erosion-corrosion damage, including many which depend upon the formation of a surface oxide film for resistance to corrosion. Erosion-corrosion occurs when such protective films are damaged by the flow and are then attacked at a rapid rate.

The entrainment of solids such as sand in flowing fluid markedly increases the erosion effects observed, although when the removal of materials is primarily controlled by the solids the phenomenon is strictly termed "impact erosion". With most metals the impact of the particle removes the protective film mechanically, whereupon subsequent corrosion repairs the film which can then be removed again by impact, in a cyclic manner. Impact erosion is particularly pertinent to the degradation of concrete installations. All materials of construction are vulnerable to this effect which can also be caused by powered craft manoeuvring in harbour environs which subsequently disturb the sand and sediment, releasing particulate matter.

MICROBIAL-CORROSION

Microbially induced corrosion is defined as the deterioration of the material, usually a metal, by processes which occur either directly or indirectly as a result of bacterial metabolic activity. The importance of the phenomenon has increased considerably, not least because of the now widely recognised association with biofouling phenomena. Familiar examples are the growth of marine algae, barnacles, mussels, molluscs and seaweed or marine structures. These growths either produce aggressive metabolites or create micro-habitats suitable for the proliferation of other bacterial species, eg anaerobic conditions favouring the activity of the well known sulphate-reducing bacteria. In addition, the presence of growths or deposits on the metal surface encourages the formation of differential aeration and concentration cells between the deposit and the surrounding environments, which may stimulate existing corrosion processes. Although the sulphate reducing bacteria are the more important of those involved in corrosion, micro-organisms from a wide range of genera and species are associated with the problem. These can be classified into three categories:

   (i) fungi,
   (ii) algae and
   (iii) aerobic and anaerobic bacteria.

It is not intended here to discuss the microbiological characteristics of these organisms since these will be discussed and reviewed by Edyvean (1992).

In order to appreciate the immediate relevance of the more important aspects of microbial corrosion it must be emphasised that microbial involvement in the corrosion mechanism does not involve any new form of corrosion process.

Corrosion mechanisms involving bacteria are also associated with the following processes.

(a) production of differential aeration and concentration cells due to the absorption of nutrients (oxygen) by microbial growths adhering to the metal surface.

(b)   production of a corrosive metabolite, ie inorganic and/or organic acids, or the end products of fermentation.

(c)   interference in the cathodic process under oxygen free conditions by the obligate anaerobic sulphate reducing bacteria and their metabolite sulphide.

Although the above is useful it is more often appropriate to discuss the problem in terms of either aerobic or anaerobic environmental conditions.

Aerobic Corrosion

The two forms of microbial corrosion experienced under this heading involve the colonization of the material by microbial growths and deposits or the formation of tubercles. The former is due to the presence of organisms such as marine algae, seaweed, barnacles etc. Tubercle formation occurs on marine structures encouraging the formation of the differential corrosion cells described earlier. Hence all types of microbes or growths which colonize a surface must be considered potentially dangerous.

For metals the mechanism of the corrosion process is the uptake of oxygen by the fouling organisms which creates a depletion of oxygen below the mass. The poorly aerated region becomes the anode of the corrosion cell whilst the cathodic region is the more aerated area away from the deposit.

Where corrosion can be attributed to the formation of inorganic and/or organic acids, it is nearly always caused by the activity of the *Thiobacillus thio-oxidans* or the presence of fungal growths. In addition to these, certain sea weeds and other similar types of organisms can produce acids at the growth-metal interface. However, as far as we are aware few if any case histories of this type of corrosion have been reported for marine structures.

Anaerobic Corrosion

In a chemically neutral and oxygen-deficient aqueous environment neither of the two cathodic reactions discussed earlier can take place at a substantial rate because the cathodes of the corrosion cell are polarised. However, in practice, corrosion under these conditions can be very severe and several orders of magnitude greater than that experienced under normally aerated conditions. The primary factor responsible for this situation is the activity of the sulphate reducing bacteria and their metabolic product, sulphide. The original theory postulated that these organisms utilised the cathodic hydrogen formed during the corrosion process thus depolarising the system and stimulating the anodic reaction.

Unfortunately this mechanism does not explain the very high rates of corrosion observed in practice. These now seem to be due to a number of environmental factors and in particular to the ferrous ion content of the solution, the metabolic product sulphide and the nature of the sulphide film formed on the metal surface. Environments low in ferrous ion encourage the formation of protective sulphide films. These become less protective when the ferrous ion content is high and the system retains the anaerobic conditions which occur in sediments and muds. This situation brings about a major chemical change resulting in the formation of non-protective iron sulphide scales. The combined effect of these sulphides, plus entrapped bacterial cells, increases the corrosiveness of the environment, encouraging rapid depolarisation of the cathode by removing the absorbed hydrogen from the surface of the metal. This situation is exacerbated by the fact that iron sulphide scales are themselves corrosive to ferrous materials.

The rate of corrosion due to sulphate reducing bacteria activity is predominantly controlled by the action of these sulphides. Sulphide corrosion also occurs beneath the tubercles and mounds of corrosion caused by marine growths and other microbial species colonizing the surface of the structure. It is also prevalent where metals are buried in estuarine mud or sea sediments.

Typical corrosion rates are usually not excessive because of limitation of growth of the organisms. A typical pitting rate is 1-2 mm per year, although rates of up to 6 mm per year can occur. The intrinsic corrosiveness towards steel of the various iron sulphides which can be formed is given in the Table below (Tiller, 1990).

H_2S alone		12.8 mm/yr
$FeS_{(1-x)}$	(Mackinawite)	5.3 mm/yr
Fe_2S_4	(Greigite)	119.6 mm/yr

In practice, however, many environments are dynamic and stagnant conditions seldom prevail with most being subjected to changes which result in aeration of the system. Fluctuating anaerobic/aerobic cycles can produce elemental sulphur arising from the oxidation of the sulphides. When this occurs the rate of corrosion can increase very rapidly up to values of between 270 mm/yr and 800 mm/yr.

TYPES OF ESTUARINE EXPOSURE

The behaviour of a material and/or structure depends on the exposure conditions which can be divided into distinct types; atmospheric, splash zone, tidal zone, immersed zone and mud - of which the last four are the most important.

SPLASH ZONE

The splash zone is the region above the mean high tide level which is wetted by wave action and spray. As this is also exposed to the atmosphere the film of water on the metal surface is fully aerated and corrosion in this region can be severe.

On a steel structure extending from the atmosphere down to the mud line, the highest corrosion rate occurs in the region immediately above high tide level. Figure 1 is a schematic of the corrosion profile on a steel pile. The maximum rate of corrosion in the splash zone has been measured as 5-10

OVERVIEW OF MATERIAL DEGRADATION IN THE MARINE/ESTUARINE ENVIRONMENT

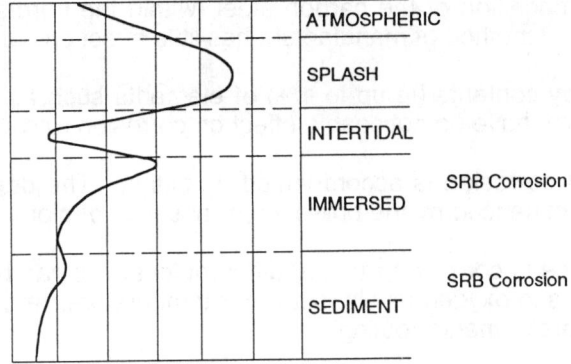

Figure 1: Zones of corrosion attack.

times that for the fully immersed parts. Values of 1.4 mm/yr have been reported. These high rates are due to continuous contact with highly aerated seawater and to the erosive effect of sea spray, tidal and wave action.

The extent of the splash zone varies with location. In areas of relative calm for most of the year, the zone extends only a short distance (1 metre approx) above mean high tide level. In areas where there is vigorous wave action for long periods, the splash zone is much more extensive and can extend for tens of metres above the high tide level.

TIDAL ZONE

This is the zone between the low tide and high tide levels and its height will vary in different locations according to the tidal range. Corrosion rates are usually lower in this region compared to the splash zone. On a metal structure the corrosion rates are lower than on the fully immersed parts. Reinforced concrete structures in this zone normally remain saturated with water so that the rate of oxygen diffusion through the concrete to the reinforcement is reduced. As corrosion of the reinforcement proceeds the build-up of corrosion products will cause cracking and spalling of the cover concrete and loss of sections of the reinforcement may affect the structural integrity.

FULLY IMMERSED ZONE

In the fully immersed zone corrosion of steel is dependent on diffusion of oxygen to the metal surface so that corrosion rates depend on the oxygen contact at the depth of immersion, current flow, extent of marine fouling and other factors influencing the availability of oxygen at the surface of the structure. Typical rates of corrosion are given below together with other useful data.

a) Corrosion rates under fully immersed conditions lie in the range 0.05 - 0.015 mm/yr.

b) The composition of the carbon steel (within the normal specification limits) and method of manufacture has little effect on corrosion rate.

c) Low alloy contents (ie up to 1%) of elements such as copper, nickel, chromium, have no noticeable effect on corrosion rate.

d) General corrosion is accompanied by pitting. The depth of pitting is greatly influenced by the presence or absence of biofouling.

e) Difference in corrosion rates at different locations can be explained by variations in oxygen levels, seawater temperature, rate of current flow and layers of macro-fouling.

MUD ZONE

This is the zone beneath the sea bed. This is often made up of sedimentary deposits of sand, clay and mud, although rocks and stones are also found. Bottom sediments, both shallow and deep, usually contain bacteria and are usually anaerobic. Sulphate reducing bacteria are often present and can produce sulphides which affect the corrosion behaviour of steel as discussed.

The oxygen levels and hence corrosion rates are normally low but the difference in oxygen level on an immersed metal surface which extends into the sea bed can set up differential cells, causing attack at the mud line.

ASSESSMENT AND PREDICTION OF THE CORROSIVENESS OF THE ENVIRONMENT

Where structures are to be erected or installed the site must be assessed using both site and laboratory investigations. Many of these are relatively simple procedures utilising well established analytical and biological techniques.

For estuarine waters a site survey and monitoring of the chemical parameters will provide valuable information on the chemical characteristics and flow conditions which prevail. Further information on the hydrographic characteristics of the estuary are readily available from the coastal authorities, in particular harbour masters. The river authority for the region is also a valuable reference source and will provide information on pollution aspects. Where piled structures are being considered and ground anchors may be involved borehole sampling and testing will provide useful data. The main factors responsible for the corrosiveness of soils are anaerobic bacteria, soil chemistry, redox potential, resistivity, differential aeration and the formation of concentration cells. These are not independent features but often have to be considered as a whole in order for the assessment to be made.

Table V is an example of a simple scheme which has proved successful for disturbed and undisturbed soils (Tiller, 1990).

Table V: Assessment of corrosiveness of soils (after Tiller, 1990).

Parameter	Aggressive	Non Aggressive
Soil resistivity (Ωcm)	<2000	>2000
Redox potential at pH 7.0 [V.nhe*]	<0.40V	>0.40V
	<0.43V if clay	>0.43V if clay
Water content (Wt %)	>20%	<20%

*nhe = normal hydrogen electrode.

For sediments a more complex system has been used which relies mainly on analysis of the sediment as shown in Table VI (King, 1981).

In addition to these assessments it is necessary to implement a microbiological assay which will indicate the extent of biofouling which can occur including the potential risk for microbial corrosion.

Table VI: A system to assess the corrosiveness of a sediment (after King, 1981).

	Factor	Corrosiveness Rating
1.	Sediment;	
	Mud	4
	Sandy-mud	2
	Sand or rock	0
2.	Organic content;	
	If Mud:	
	high	3
	medium	2
	low	0
	If Sand:	
	high	1
	medium/low	0
3.	Water depth; (includes disturbance)	
	shallow (60 m)	2
	deep (60 m)	0
4.	Availability of Nitrogen (N) and Phosphorous (P);	
	high organic content + high N&P	2
	low organic content + high N&P	1
	low N&P	0
5.	Temperature;	
	above 10°C	2
	below 10°C	0

A total rating of above 10 indicates a sediment potentially more aggressive than the sea environment.

PREVENTION AND CONTROL

The prevention and control of corrosion and/or degradation of materials in estuarine environments should be considered at the design stage. In general the most popular protection methods are the use of surface coatings alone or in combination with cathodic protection. Modern protective coating systems will provide a very high standard of corrosion protection provided the surface preparation of the material is carried out in accordance with the appropriate specification. Similarly coating selection should be based on a combination of practical experience or BS 5493 1977.

For partially immersed systems cladding with metal such as copper nickel alloys or plastic sheet have been reasonably successful on offshore structures and should be acceptable for some estuarine conditions. In some instances metallic coatings are available such as flame sprayed aluminium and hot dipped or sprayed zinc.

Hot dipped galvanizing produces a thick coating which seals crevice regions, edges seams and welds and therefore reduces some of the problems associated with corrosive attack. For the marine environment most materials are coated in accordance with BS 729 or its equivalent. Their performance is a function of coating thickness ie a coating of 305 g/m^2 (0.0017 in) exposed to the fully immersed, tidal and splash zone, should have a life expectancy of 2.6 years and in mud the life expectancy could be between 3 and 20 years.

Most metallic systems including ground anchors and reinforced concrete structures can be protected using cathodic protection. Detailed accounts of this technique are outlined in the conference paper by R Hoffman(1981). Provided the system is designed and operated as outlined in the specification BS 7361 part 1, problems should not be encountered. The method utilises either a sacrificial anode scheme which corrodes preferentially to protect the structure or an impressed current system utilising an external D.C. electrical source. It is therefore possible by adapting the schemes outlines above together with those discussed above to design and protect structures from extensive degradation.

REFERENCES

BOOTH, G.H., COOPER, A.W. &.TILLER, A.K. (1967). Corrosion of mild steel in the tidal waters of the Afon Dyfi and a comparison with the River Thames. *Brit. Corr. Journal,* 2.

EDYVEAN, R.G.J. & VIDELA, H.A. (1992). Biofouling and microbial corrosion in estuarine waters. *Proceedings of the International Conference on the Implications of Ground Chemistry and Microbiology for Construction, Bristol.*

HOFFMAN, R. (1992). Cathodic Protection, Practice and Theory. *Proceedings of the International Conference on the Implications of Ground Chemistry and Microbiology for Construction, Bristol.*

KING, R.A. (1981). Prediction of corrosiveness of sediments. In: *Marine corrosion on offshore structures*, Society of Chemical Industry, London, p.23.

RACKLEY, F.A. & TILLER, A.K. (1988). *Guideline document on aspects of corrosion of mooring chains*. Brit. Ports Federation London.

TILLER, A.K. (1988). Examination and assessment of the corrosivity of the water and soil in the Alexandra Dock Grimsby. *NCS Report No.0065* Humberside Country Council Technical Dept. Grimsby.

TILLER, A.K. (1990). Biocorrosion in Civil Engineering. In: *Microbiology in Civil Engineering FEMS Symposium No.59*, Howsam, P (ed) E. and F.N.Spon, Chapter 3.

4-2 BIOFOULING AND MICROBIAL CORROSION IN ESTUARINE WATERS

R.G.J. EDYVEAN
Chemical Engineering Department, University of Leeds, LS2 9JT. UK.

H.A. VIDELA
Bioelectrochemistry Section, INIFTA. University of La Plata, La Plata, 1990 Argentina.

ABSTRACT

Biofouling and microbiological corrosion is mediated by micro-organisms embedded in a gelatinous organic matric (the biofilm) attached to the metal surface. Microbial adhesion processes lead to an important modification of the metal/solution interface inducing changes in the type and concentrations of ions, pH values, oxygen levels, flow velocity, etc. These alter the classical concept of the electrical interface as used in inorganic corrosion studies. Metal dissolution in an aggressive environment like estuarine or seawater is thus conditioned by two different processes occurring at the interface: corrosion products forming from the metal into the solution, and biofouling settlement from the bulk solution towards the metal surface. This paper emphasizes the importance of biofilms and their effects on corrosion on different metal surfaces such as stainless steel, carbon steel and copper-nickel alloys in the marine environment and uses examples to demonstrate the points.

INTRODUCTION

It is widely acknowledged in the literature (Characklis 1981; Videla, 1989) that metals immersed in a biologically active liquid media like seawater, undergo a sequence of biological and inorganic changes that result in an important modification of the metal/solution interface. The first stage of the sequence of biological changes is the development of a thin film (approximately 20 to 80 Å thick) on the metal surface. This film is formed by inorganic ions and high molecular weight organic compounds that alter the electrostatic charge and wettability of the metal surface (Dexter, 1976). In a second stage, bacterial colonization, facilitated by the conditioning film, leads to the development of a biofilm, consisting mainly of water, bacterial cells and their extracellular polymeric substances (EPS). In open waters there are a considerable number of sessile and sedentary marine organisms, both microscopic and macroscopic, which will follow and complement this bacterial settlement and readily colonize any surface. This "fouling" can thus range from as little as a bacterial biofilm a few microns thick to one including larger organisms and accumulated debris some tens of centimetres on the metal surface.

To consider just a "simple" bacterial biofilm:

The settlement and growth of the bacteria and their extracellular products forms a layer between the metal and the environment. This layer is biologically active, dynamic and heterogeneous. In relation to the electrochemical reactions of the corrosion process even this bacterial biofilm is thick, thus the metal surface is no longer exposed to the bulk, or even a uniform, environment.

The high water content of the extracellular polymeric substances (Geesey, 1982) suggests that reactions in this interfacial system will be governed by the different transport processes taking place through the EPS. This radically changes the conventional electrochemical concept of electrical interface used in inorganic corrosion studies as important concentration gradients, changes in the type of ions, pH values, oxygen levels, and redox conditions, will modify the behaviour of the metal surface and its corrosion products.

The sequence of inorganic changes (corrosion, and passivation by the formation of stable metal oxide layers) occurs simultaneously with biofilm formation but in the opposite direction. Whereas corrosion or passivation is a process directed from the metal surface to the solution, biofouling is produced from the bulk towards the metal/solution interface (Videla, 1989) producing a complex "biologically conditioned" metal/solution interface which is not necessarily uniform in time or space (Characklis and Marshall, 1990).

Electrochemical concepts and parameters used to assess inorganic corrosion (in the absence of biofilms) have thus to be adapted to the characteristics of the biofouled metal surface. This means that a bioelectrochemical approach is required to understand the complex interactions at the metal/solution interface.

BIOFOULING AND MICROBIAL CORROSION ON DIFFERENT METAL SURFACES IN SEA WATER

The different kinds of interactions can be illustrated with reference to various metal surfaces: carbon steel, stainless steel and copper-nickel alloys. Carbon steel exhibits a complex behaviour in sea and estuarine water as a consequence of its active dissolution in the presence of aggressive anions like chlorides. A boundary layer of corrosion products, EPS and bacterial cells is always present at the metal/solution interface. Conversely, stainless steel, and other corrosion resistant alloys, is characterized by a good passive behaviour giving it more resistance in seawater (depending on the type of stainless steel), and by the lack of corrosion products on the metal surface. This feature facilitates closely adherent bacterial colonization that can lead to modification of the corrosion resistance of the metal rendering it more susceptible to attack by chlorides etc. Copper-nickel alloys exhibit a complex intermediate behaviour due to the interaction of the toxic effects of copper ions leached into solution and the formation and removal of passive layers as biofilm detachment occurs.

An assessment of corrosion during the early stages of biofouling settlement was recently made for several metals and alloys in polluted harbour/estuarine seawater by using scanning electron microscopy and electrochemical corrosion measurements (Videal, de Mele & Brankevich, 1988). It was found that the nature of the metal surface plays a relevant role in facilitating or inhibiting microbial adsorption. Thus, a corrosion-resistant metal or alloy presents an ideal substratum for microbial colonization, similar to inert non-metallic surfaces such as glass or different kinds of plastics. Bacterial colonization and biofilm formation increased according to the following sequence: copper < 70/30 copper-nickel alloy < brass < aluminium < stainless steel < titanium. On the last two metal surfaces the biofilm formation and its structure were more easily observed by SEM. It was found that on a metal-solution interface like aluminium, where copious layers of different types of corrosion products are formed, microbial colonization occurs on an unstable and continuously growing layer of inorganic products. On this kind of metal surface, detachment of parts of the corrosion product and their associated biofilm readily takes place.

Corrosion on iron and iron based alloys occurs with the formation of rust. The reactions are:-

$$Fe \longrightarrow Fe^{2+} + 2e \text{ (ANODE)}$$

and

$$O_2 + 2H_2O + 4e \longrightarrow 4OH^- \text{ (CATHODE)}$$

The metal ions react in the electrolyte environment to produce a complex mixture of hydrated iron oxides and hydroxides with a general composition:

$$Fe(OH)n(H_2O)$$

In the case of stainless steel, the alloying of iron with other elements (mainly chromium, nickel, or molybdenum) gives surfaces which show spontaneous passivity in many aggressive environments such as seawater. However, stainless steels are sensitive to pitting (Videla, de Mele & Brankevich, 1988) and other types of localized corrosion in chloride-containing media and bacterial metabolic activity can produce conditions favourable to such attack. For example, it was recently reported (Dexter et al, 1986) that the initial step in crevice corrosion (the depletion of oxygen in the crevice solution), becomes the dominant mechanism in the presence of bacteria and may accelerate the initiation of the crevice attack. The rate of oxygen utilization by bacteria mainly depends on the density of microbial cells in the biofilm and on the species of bacteria present but oxygen depletion in crevice corrosion can be as fast as electrochemical mechanisms.

Bacterial colonization can occur within a period of 24 to 72 hours when stainless steel samples are exposed to flowing natural seawater depending on conditions (Videla et al, 1988). The lack of corrosion products on the stainless steel surfaces allows biofouling settlement over an almost uniform passive layer of oxide. Copious microfouling deposits of bacteria, EPS, and particulate material are generally formed, leading to a patchy distribution of the biofilm. Because of its technological importance, several interpretations of the effects of biofouling on the corrosion behaviour of high resistance stainless steel exposed to seawater have recently been made using open circuit potential vs. time measurements (Mollica, Trevis, Traversso, Ventura, Scotto, Alabisio, Marcenaro Montini, de Carolis and Dellepiane, 1984; Johnsen and Bardal, 1986; Scotto, Di Cintio and Marcenaro, 1985). A continuous increase of the open circuit potential to more noble values with time was found by different authors for several kinds of stainless steel in seawater. However, the presence of bacterial biofouling causes an increased scattering in this potential when compared with results obtained in sterile artificial seawater, sodium chloride solutions or membrane filtered seawater. Although the interpretation of these effects is still the subject of controversy, it is thought that this "noise" is an indication of the stimulation of pitting corrosion of stainless steel by the biofouling. This can occur in two ways (Dexter & Gao, 1987):

i) by the formation of differential aeration cells, due to the patchy distribution of the biofilms. (Differences in oxygen concentration set up corrosion cells in which the areas with the lowest oxygen availability corrode).

ii) by increasing the rate of the cathodic oxygen reduction reaction, due to the alteration of oxygen gradients within the biofilm.

A third effect due to adsorption processes altering electrode capacitance has also been mentioned (Mansfield, Tsai, Shih, Little, Ray & Wagner, 1990).

The corrosion behaviour of carbon steel in seawater is affected by the abundant deposits of corrosion products of varied chemical composition. External layers of corrosion product containing lepidocrocite (gamma-Fe_2O_3, H_2O) and wustite have been reported (Edyvean, 1984). After extended periods of exposure, film layers of goethite (alpha-$Fe_2O_3.H_2O$) and magnetite (Fe_3O_4) are generally found. In marine environments, a fouling layer consisting of bacteria and microalgae embedded in EPS consolidates corrosion products on carbon steel (Lynch & Edyvean, 1988). This kind of cohesive effect of EPS in a biofilm depends on several environmental and biological factors and will finally determine the extent of corrosion products/biofilms interactions.

An important effect of micro-organisms on carbon steel corrosion behaviour is due to the microbial dissolution of ferric oxides and hydroxides by the reducing capacity of iron (III) species present in certain types of bacteria. It was recently reported that a marine vibrio was able to expose the metal surface (by the metabolic dissolution of ferric deposits) to the direct action of aggressive species (i.e. sulphides, chlorides) present in the medium (Gaylarde & Videla, 1987).

Biofilm effects on the corrosion behaviour of carbon steel induce:

i) differential aeration between metal areas covered and uncovered by microbial colonies and,

ii) a "depassivation" or activation of the metal surface produced by the microbial reduction of insoluble ferric deposits into soluble ferrous compounds.

Micro-organisms inside pits and crevices can create redox conditions that favour metal attack.

Copper-nickel alloys, in spite of their well-documented antifouling properties, show biofilms formed by bacteria entrapped between corrosion products and EPS after a few weeks exposure to seawater. The structure of the passive layer can be altered by EPS leading to a "sandwich" of biological and non-biological material. This is disrupted when biofilm detachment occurs owing to the shear stress effects of seawater. As a consequence, a patchy distribution of the biological and inorganic deposits is produced, which accounts for an increase in the corrosion rate through differential aeration. Biogenic sulphides produced by sulphate-reducing bacteria (SRB) can alter the cuprous oxide passive layer on copper-nickel alloys increasing corrosion (Videal, de Saravia & de Mele, 1989).

THE ROLE OF HYDROGEN SULPHIDE IN CORROSION

THE PRODUCTION OF HYDROGEN SULPHIDE

Biogenic hydrogen sulphide is produced by sulphate-reducing bacteria (SRB) and this group are probably the greatest single cause of microbial corrosion problems.

SRB are strict anaerobes, that is they can only grow in the absence of oxygen. However, they exist in a dormant form in most aqueous environments and readily become active when the conditions become favourable. Such conditions can be found under deposits, under fouling by other organisms, in estuarine muds etc. A biofilm need only be 15 to 20 microns in thickness for anaerobic conditions, and hence SRB activity, to exist at the metal surface. As their name implies SRB derive energy from the reduction of sulphates to sulphide. The end result is the production of hydrogen sulphide. This produces a characteristic "bad egg" smell when disturbed and often a black discoloration at metal surfaces and in muds due to the formation of sulphide compounds such as iron sulphide.

The overall reaction can be described (albeit very simply) in the following equation:-

$$4H_2 + SO_4^{2-} + 2H^+ \longrightarrow H_2S + 4H_2O$$

Sulphate-reducing bacteria are always found in close association with other micro-organisms which play key roles in the creation of the necessary physical conditions and the provision of the relatively simple nutrient compounds required by SRB (Hamilton & Sanders, 1984).

This close association creates difficulties in measuring the amount of hydrogen sulphide that can be produced biologically. Levels of between 15 and 100 ppm have been measured on offshore oil rigs (Wilkinson, 1983); up to 800 ppm in decomposing seaweed (Efird and Lee, 1979; Thomas, Edyvean and Brook, 1988); and over 2000 ppm in pure cultures (Miller, 1949). Despite this very wide spread there is a consensus emerging that between 50 and 200 ppm may be representative of SRB active conditions at a metal surface.

CORROSION BY SRB

While there is evidence of other mechanisms of corrosion enhancement by SRB (depolarisation of cathodic sites, creation of differential cells and the production of a corrosive phosphorus compound) perhaps most of the damage caused is due to the production of sulphide ions, the formation of elemental sulphur and galvanic corrosion due to the formation of iron sulphide films (Tiller, 1988).

Hydrogen sulphide is generated in an anaerobic environment and thus the main corrosion reaction that it affects is also anaerobic. The surface electrochemical reactions are thus (for a ferrous metal):-

$$Fe \longrightarrow Fe^{2+} + 2e \text{ (ANODE)}$$

and

$$2H^+ + 2e \longrightarrow H_2 \text{ (CATHODE)}$$

This cathodic reaction can be split into two components:-

$$H^+ + e \longrightarrow H$$

and

$$H + H \longrightarrow H_2$$

the latter being the rate limiting step in the overall reaction.

As the cathodic reaction in anaerobic conditions is far slower than its counterpart under aerobic conditions (oxygen reduction), the rate of corrosion in anaerobic environments is usually very slow. In fact a "polarisation" effect has been described in which hydrogen remains adsorbed on the metal surface and acts as a barrier to further corrosion.

When hydrogen sulphide is present the ferrous ions are presented with a reactant and the basic corrosion reaction becomes:

$$Fe^{2+} + H_2S \longrightarrow FeS + 2H^+$$

While this reaction is a very simplified statement of the process, it does serve to show the two important components of hydrogen sulphide corrosion of ferrous metals, i.e. iron sulphide and hydrogen (the role of hydrogen will be discussed in more detail below).

The reaction is virtually autocatalytic in that iron sulphide is cathodic to steel and will thus "drive" an alternative corrosion reaction between the metal (anode) and the iron sulphide (cathode) rather than the reaction between anodic and cathodic sites on the metal surface.

However, this does not explain how any initial "polarisation" (adsorbed hydrogen on the surface) is overcome nor the likely "barrier" nature of a uniform iron sulphide film. In fact protective iron sulphide films are often found in hydrogen sulphide environments. These are always described as "thin and adherent" (King, Smith and Miller, 1973; King, Dittmer and Miller, 1976) and it seems that, for corrosion to occur, either these films have to be disrupted (by bacterial action or by other reactants such as oxygen) or there has to be sufficient free ferrous ions present near the metal surface for a very loose iron sulphide film to form (thus creating cathodic and anodic sites). The new corrosion reaction may be too vigorous for the iron sulphide cathode itself to become polarised or, as suggested by King and Miller (1971) the molecular hydrogen is removed from the iron sulphide by SRB. Hamilton, Moosavi and Pirrie (1988) give as key elements in the corrosion process the relative proportions of iron sulphides such as mackinawite ($FeS_{(1-x)}$) and pyrite (FeS_2) in the corrosion product, discontinuities within these sulphide films, physical disturbance and access to oxygen.

That the rate of corrosion of ferrous metals in the presence of hydrogen sulphide is highly dependant on the sulphide films formed is shown by the following figures (King, Smith and Miller, 1973);

1) in the presence of a thin adherent mackinawite film the rate is 5.3 mm/yr,

2) in the presence of hydrogen sulphide alone the rate is 12.8 mm/yr,

3) in the presence of greigite (Fe_2S_4) it is 119.6 mm/yr and

4) in the presence of elemental sulphur it is 1110 mm/yr.

The importance of anaerobic-aerobic transitions is emphasised by Hardy and Bown (1984) who also cite several more examples of high corrosion rates for steel in "sour" systems exposed to air. In their own work Hardy and Bown found corrosion rates of 1.45 mg/dm/day for SRB cultures but 129 mg/dm/day when the cultures were exposed to air. The attack was found to be confined to areas beneath tubercles consisting of loosely adherent corrosion product and not in areas of thick tightly adherent films. The corrosion was thus highly localised pitting.

HYDROGEN SULPHIDE-INFLUENCED FAILURE BY CRACKING

Perhaps even more important than the surface corrosion effects of hydrogen sulphide are its effects on the formation and propagation of cracks.

The mechanism by which hydrogen sulphide enhances the failure of metals by cracking is to poison that part of the cathodic reaction which allows hydrogen to leave the metal surface. This leaves hydrogen atoms at the metal surface and thus allows them more time to diffuse into the atomic structure of the metal. Hydrogen entry into a metal causes damage to the interatomic bonds and is known as hydrogen embrittlement. Hydrogen embrittlement can reduce the fracture toughness (resistance to failure) of steel by more than half and produces failure in several forms, all due to the development of cracks. These forms depend on whether there is an internally or externally applied stress, whether that stress is static or cyclic and whether a corrosive environment is present. Under corrosion-fatigue conditions (a corrosive environment and a cyclic load) dramatic increases in crack growth rates have been found in hydrogen sulphide saturated seawater when compared to results from air or seawater alone (Thomas, Edyvean & Brook, 1988).

EXAMPLES OF MICROBIAL CORROSION

Two typical examples can be used to illustrate biofouling and MIC effects on metal biodeterioration.

There is an important distinction between the different environments and the type of biological activity in each. The first environment is that of the external surfaces, exposed to open water and marine fouling ranging from bacteria to macroscopic seaweeds. The second, mainly fouled by bacteria, is the interior of plant and pipework.

EFFECTS ON STATIC STRUCTURES

The external surfaces of "offshore" structures, which can be considered to include jetties and bridges as well as structures such as offshore oil platforms, are exposed to a range of environments from the highly aerated "splash zone" influenced by tides, currents, wind and weather to the highly anaerobic sediments at the base of the structure, influenced by the chemical and organic nature of the sediments themselves.

ENHANCING CORROSION

The partial barrier that marine fouling places between the seawater and the metal surface of the submerged structure allows a considerable range of macro- and micro-biological environments to be formed. Marine macrofouling grows in intimate association with bacteria and the interactions between the two are of major importance in seawater environments. It is likely that the close interactions found between different species of bacteria extend at least to bacterial-microalgal interactions. Also, the continual cycle of growth and decay of the macrofouling is a major source of nutrients for bacteria. One result of the provision of organic material and suitable anaerobic environments at the metal surface is the activity of sulphate-reducing bacteria (SRB) and the consequent detrimental production of hydrogen sulphide (H_2S).

Fouling will influence the corrosion of metals in several ways:

1. By a physical presence creating differential aeration, pH or other concentration cells on both macroscopic and microscopic scales;

2. By the production of corrosion promoting metabolites such as acids (both organic and inorganic) and sulphur compounds (especially H_2S);

3. By stimulation of cathodic (reduction) reactions; and

4. By stimulation of anodic (oxidation) reactions.

Several reviews have been published on corrosion enhancement by marine fouling in general (Edyvean, 1984), algae in particular (Terry and Edyvean, 1986), and by the sulphate-reducing and sulphur-oxidizing bacteria (Cragnolino and Tuovinen, 1984) and aerobic biofilms (Scotto, Di Cintio and Marcenaro, 1985).

The general, biologically enhanced, corrosion mechanisms occurring on the external surfaces of offshore structures are the same as in any other biocorrosion situation. In areas of trapped water and crevices under either macrofouling or bacterial biofilms considerable changes in pH, dissolved oxygen and ionic balance can occur. These changes can enhance corrosion by the formation of differential pH, aeration and chemical concentration cells. The daily pH change under healthy algae, due to photosynthetic action tends to be over 2 pH units from a baseline of 8 to over 10, and when the buffering capacity is broken the pH can fall to around 6 during the dark (Terry and Edyvean, 1981). pH levels between 4.5 and 7.5 have been measured beneath animal fouling (Woolmington and Davenport, 1983) and, if bacterial acid decomposition takes place, pH levels as low as 1.8 can be created (Woolmington and Davenport, 1983). Thus, the total difference in pH between two adjacent sites can be up to 9 pH units, and this can form highly corrosive concentration cells.

Oxygen concentration cells can lead to crevice and pitting corrosion. These forms of corrosion are very important in seawater environments where extreme differences in oxygen concentration can be created between portions of bare surface in contact with oxygenated water and those areas of metal covered by fouling or even thin films of bacteria.

Corrosion by metabolic products of bacteria and the direct or indirect stimulation of anodic and cathodic reactions are similar to other biocorrosion situations except that there is likely to be a wider range of environments created under and around fouling. Enhancement of the cathodic oxygen reaction by general aerobic bacterial films has been well documented, especially to the pitting corrosion of stainless steels (Dexter, Lucas and Gao, 1986; Scotto, Di Cintio and Marcenaro, 1985). Enhancement of the corrosion reaction through the production of H_2S by sulphate-reducing bacteria has also been widely reported, while anodic enhancement has been demonstrated for nitrate-reducing bacteria and blue-green algae. H_2S production is still an unknown factor on external surfaces offshore. High SRB activity has been detected (Hamilton and Sanders, 1984) but levels of H_2S near the metal surface are still reliant on laboratory simulation.

HYDRODYNAMIC LOADING AND CORROSION FATIGUE

Fixed offshore structures are susceptible to the effects of cyclic stress loading due to wave action. The first 30 m of the structure below the water line carries some 70% of the total loading due to wave action (Heaf, 1981) and the presence of fouling, especially large algae which can dominate this zone (eg the kelp seaweeds in the North Sea) will considerably enhance this loading by increasing both the diameter and roughness of tubular members of the structure. Both of these factors increase the drag coefficient and hence the drag forces on the structure. It has been calculated that a layer of fouling 150 mm thick will decrease the fatigue-life (the predicted life to failure) by 54% (Heaf, 1981). Recent studies summarized by Vaux (1988) indicate that the force coefficient can be increased by the presence of fouling by up to a factor of 3.

The combination of loading and corrosion - corrosion-fatigue - is an important factor in the integrity of offshore structures. Thomas, Edyvean and Brook (1988) tested structural steels for their corrosion-fatigue behaviour using seawater in which various levels of H_2S were bacterially generated. Their results show that crack growth rates are higher than in unsoured seawater, even at low levels of H_2S (increasing crack growth rate by a factor of 10 at 73 ppm H_2S and 20 at 400 ppm H_2S).

SEA BED PROBLEMS

At the sea bed, oxygen availability tends to be lower, fouling growth is less and cyclic loading not as marked. However, the water, mud/sand junction provides possibilities for bacterially assisted differential cells and corrosion problems may be enhanced by organically rich muds. In addition the presence of iron ions tends to increase the microbial activity around the legs of the structure with subsequent corrosion problems.

COASTAL POWER PLANT COOLING WATER SYSTEMS

Biofouling and microbial corrosion are major problems in power plant cooling water systems. Nearly all problems in this area arise from bacterial growth fouling the surfaces of pipes and plant, causing blockage, decreasing heat exchange efficiency, enhancing corrosion and "souring" the water with hydrogen sulphide. Microbial corrosion is widespread and occurs on the majority of the constructional materials used, even in highly corrosion resistant metals or alloys like titanium or high molybdenum austenitic stainless steel (Licina, 1989). Corrosion damage may result from direct or indirect action by micro-organisms. Direct action from the organisms growing on the metal surface will produce effects such as differential aeration. Indirect action can result from the growth of organisms remote from the corroding metal surfaces by the production of acids and "sour" conditions. Corrosion rates of up to 1 mm per year have been measured together with significant pitting.

Most cases of biofouling and MIC in the power industry present similar characteristics: the formation of discrete deposits of bacterial cells, EPS and corrosion products on the internal wall of heat exchanger tubes, water conduits or storage tanks (Pope, 1986). Bacterial activity under the deposits produces deep pits filled with sulphide corrosion product. There are several groups of micro-organisms which will thrive in any of the areas described above. These include:- sulphate-reducing bacteria (SRB), iron bacteria, slime forming bacteria, sulphur oxidizing bacteria, nitrogen utilizing bacteria, as well as some fungi and micro-invertebrates and micro-algae.

It is important to realise that the attached, sessile bacterial population is more numerous, and more important in terms of corrosion enhancement, than the planktonic one sampled from the water and misleading results can be obtained if this factor is overlooked. For example, sulphate-reducing bacteria can be found in large numbers in biofilms on pipe walls when they are in very low numbers in the water.

Biofouling and corrosion studies have recently been reported for a power plant cooling water system located in Mar del Plata harbour, Argentina (Brankevich, de Mele and Videla, 1990). The water intake of the power plant is highly polluted due to a lack of water renewal and waste discharges by the fishing industry into the harbour. These features cause low levels of dissolved oxygen and high sulphide content. Thus, biofouling in the cooling systems of this plant consists of species and groups resistant to pollution. However, a total of 46 taxa, belonging to 6 phyla, were identified in taxonomic studies (Bastida, 1971).

The corrosion of samples of different types of stainless steel and copper-nickel alloy exposed for several weeks to flowing seawater in the water intake canal of the power plant was studied. The conclusions drawn from this case history are:

a) stalked ciliates, observed as predominant biofouling species, can facilitate passive layer detachment through adhesion effects developed at the fixation points of their peduncles, and assisted by water velocity effects.

b) biofilms affect the passive behaviour of copper-nickel alloys by reducing the corrosion layers' adhesion, facilitating the removal of passive layers and leading to differential aeration effects.

c) biofilm formation facilitates the initiation of corrosion on stainless steel by altering the oxygen concentration gradients at the metal/solution interface.

d) open circuit electrical potential oscillations (characteristic at the onset of localized corrosion) are observed with stainless steel samples. Blistering of the metal surface and shallow micropits were observed in all cases.

e) sulphide anions present in polluted seawater lead to the formation of an oxide layer of poor protective characteristics, which facilitates the initiation of corrosion attack.

f) biofilm and corrosion product interactions enhanced biodeterioration effects markedly, on the different metal surfaces tested.

CONCLUSIONS

Seawater is a highly specialised environment, both for corrosion and for biological activity. The interactions between a corroding metal surface and a developing biofilm in such a medium are complex and cannot be described solely by one methodology. An interdisciplinary and integrated approach is required.

ACKNOWLEDGEMENTS

The authors wish to thank Engs. M.M.S.Freitas, R.A.Silva, M.F.L.de Mele and G.J.Brankevich for some of the experimental results mentioned in this paper. Both authors acknowledge and thank The Royal Society (UK) and CONICET (Argentina), for a joint research agreement, which enabled this paper to be written.

REFERENCES

BASTIDA, R. (1971). Las Incrustaciones Biologicas en el Puerto de Mar del Plata, periodo 1966/67. Rev. Muc.Cs. Nat.3.-Dexter, S.C. 1976. Influence of Substrate Wettability on the Formation of Bacterial Slime Films on Solid Surfaces Immersed in Natural Seawater, *in Proc. 4th Int. Congr. on Marine Corrosion and Fouling,* Juan-Les-Pins, Antibes, France, 137-144.

BRANKEVICH, G.J., DE MELE, M.F.L. & VIDELA, H.A. (1990). Biofouling and Corrosion in Coastal Power Plant Cooling Water Systems. *Marine Technology Society Journal*, **24**, 18-28.

CHARACKLIS, W.G. (1981). Fouling Biofilm Development: A Process Analysis. *Biotech. Bioeng.*, **23**,1923-1960.

CHARACKLIS, W.G. & MARSHALL, K.C. (1990). Biofilms: A Basis for an Interdisciplinary Approach. In: *Biofilms*, W.G.Characklis, K.C.Marshall eds.), John Wiley & Sons, New York, 3-15.

CRAGNOLINO G. & TUOVINEN, O.H. (1984). The role of sulphate-reducing and sulphur oxidising bacteria in the localised corrosion of iron-based alloys - a review. *International Biodeterioration*, **20**, 9-27.

DEXTER, S.C. (1976). Influence of Substrate Wettability on the Formation of Bacterial Slime Films on Solid Surfaces Immersed in Natural Seawater. *in Proc. 4th Int. Cong. on Marine Corrosion and Fouling,* Juan-Les-Pins, Antibes, France, 137-144.

DEXTER, S.C. & GAO, G.Y. (1987). Effect of Seawater Biofilms on Corrosion Potential and Oxygen Reduction of Stainless Steel. *Corrosion 87*, paper **No.377**, NACE, Houston, TX.

DEXTER, S.C., LUCAS, K.E. & GAO, Y. (1986). The Role of Marine Bacteria in Crevice Corrosion Initiation. In: *Biologically Induced Corrosion*, S.C. Dexter, ed.), NACE, Houston TX, 144-153.

EDYVEAN, R.G.J. (1984). Interactions Between Microfouling and the Calcareous Deposit Formed on Cathodically Protected Steel in Seawater. *Proc. 6th International Congress on Marine Corrosion and Fouling*, Athens, 469-483.

EDYVEAN, R.G.J. (1987). Biodeterioration problems of North Sea oil and gas production. *International Biodeterioration*, **23**, 199-231.

EFIRD, K.D. & LEE, T.S. (1979). Putrid seawater as a corrosive medium. *Corrosion*, **35**, 79-83.

GAYLARDE, C.C. & VIDELA, H.A. (1987). Localised Corrosion Induced by a Marine Vibrio. *International Biodeterioration*, **23**, 91-104.

GEESEY, G.C. (1982). Microbial exopolymers: Ecological and Economic Considerations. *Am.Soc. Microbiol. News*, **48**, 9-14.

HAMILTON, W.A. & SANDERS, P.F. (1984). Sulphate-reducing bacteria and anaerobic corrosion. In: *Corrosion and marine growth on offshore structures*, Lewis, R.J. & Mercer, A.D. (eds). Ellis Horwood, Chichester ISBN 0-85312-564-3 23-30.

HAMILTON, W.A., MOOSAVI, A.N. & PIRRIE, R.N. (1988). Mechanisms of anaerobic microbial corrosion in the marine environment. In: *Microbial Corrosion*, Sequeira C.A.C. & Tiller A.K. (eds), Elsevier Science Publishers, London & New York ISBN 1-85166-299-5, 13-19.

HARDY, J.A. & BOWN, J.L. (1984). Sulphate-reducing bacteria: their contribution to the corrosion process. *Corrosion,* **40**, 650-654.

HEAF N.J. (1981). The effect of marine growth on the performance of fixed offshore platforms in the North Sea. In: *Marine fouling of offshore platforms. Vol. 1.* London, Society for Underwater Technology.

JOHNSEN, R. & BARDAL, E. (1986). The Effect of a Microbiological Slime Layer on Stainless Steel in Natural Seawater, *Corrosion 86*, paper **No.227**, NACE, Houston, TX.

KING, R.A. & MILLER, J.D.A. (1971). Corrosion by the sulphate reducing bacteria. *Nature*, **233**, 491.

KING, R.A., DITTMER, C.K. & MILLER, J.D.A. (1976). Effect of ferrous iron concentration on the corrosion of iron in semicontinuous cultures of sulphate-reducing bacteria. *Br. Corr. J.*, **11**, 105-107.

KING, R.A., SMITH, S.S. & MILLER, J.D.A. (1973). Corrosion of mild steel by iron suphides. *Br. Corr. J.*, **8**, 138-147.

LICINA, G.J. (1989). An Overview of Microbiologically Influenced Corrosion in Nuclear Power Plant Systems. *Materials Performance*, **28**, 55-60.

LYNCH, J.L. & EDYVEAN, R.G.E. (1988). Biofouling in Oilfield Water Systems: A review. *Biofouling*, **1**, 147-162.

MANSFIELD, F., TSAI, R., SHIH, H., LITTLE, B., RAY, R. & WAGNER, P. (1990). Results of Exposure of Stainless Steels and Titanium to Natural Seawater. *Corrosion 90*, paper **No.109**, NACE, Houston TX.

MILLER, L.P. (1949). Rapid formation of high concentrations of hydrogen sulphide by sulphate-reducing bacteria. *Contributions from the Boyce Thompson Institute*, **15**, 437-65.

MOLLICA, A., TREVIS, A., TRAVERSSO, E., VENTURA, G., SCOTTO, V., ALABISIO, G., MARCENARO, G., MONTINI, U., DE CAROLIS, G. & DELLEPIANE, R. (1984). Interaction Between Biofouling and Oxygen Reduction Rate on Stainless Steel in Seawater. *Proc. 6th International Congress on Marine Corrosion and Fouling*, Athens, 269-281.

POPE, D.H. (1986). *A Study of Microbiologically Influenced Corrosion in Nuclear Power Plants and a Practical Guide for Countermeasures.* EPRI NP-4582, Project 1166-6 Report, 90 pp.

SCOTTO, V., DI CINTIO, R. & MARCENARO, G. (1985). The Influence of Marine Aerobic Microbial Film on Stainless Steel Corrosion Behaviour. *Corros. Sci.*, **25**, 185-194.

SZKLARSKA-SMIALOWSKA, Z. (1986). Effects of Environmental Factors on Pitting. In: *Pitting Corrosion of Metals*, NACE, Houston, TX, 201-253.

TERRY L.A. & EDYVEAN, R.G.J. (1981). Microalgae and corrosion. *Botanica Marina*, **24**, 177-183.

TERRY L.A. & EDYVEAN, R.G.J. (1986). Recent investigations into the effects of algae on corrosion. In: *Algal Fouling*, L.V.Evans and K.D. Hoagland (eds), Elsevier, Amsterdam, 211-230.

THOMAS, C.J., EDYVEAN, R.G.J. & BROOK, R. (1988). Biologically enhanced corrosion-fatigue. *Biofouling*, **1**, 65-77.

TILLER, A.K. (1988). The impact of microbially induced corrosion on engineering alloys. In: *Microbial Corrosion.*, Sequeira C.A.C. and Tiller, A.K. (eds), Elsevier Science Publishers, London & New York ISBN 1-85166-299-5. 3-12.

VAUX, R. (1988). Kelps add to wave loading. *Offshore Research Focus*, **62**, 12. London, Department of Energy.

VIDELA, H.A. (1989). Biological Corrosion and Biofilm Effects on Metal Biodeterioration. In: *Biodeterioration Research 2*, C.E.O'Rear, G.C. Llewellyn (eds.), Plenum Press, New York, 39-50.

VIDELA, H.A. (1989). Metal Dissolution/Redox in Biofilms. In: *Structure and Functions of Biofilms*, W.G. Characklis, P.A. Wilderer (eds), John Wiley & Sons, New York, 301-320.

VIDELA, H.A., DE MELE, M.F.L. & BRANKEVICH, G.J. (1988). Assessment of Corrosion and Microfouling of Several Metals in Polluted Seawater. *Corrosion*, **44**, 423-426.

VIDELA, H.A., GOMEZ DE SARAVIA, S.G. & DE MELE, M.F.L. (1989). Relationship Between Biofilms and Inorganic Passive Layers in the Corrosion of Copper-Nickel Alloys in Chloride Environments. *Corrosion 89*, paper **No. 185**, NACE, Houston TX.

WILKINSON, T.G. (1983). Offshore monitoring. In: *Microbial Corrosion*, The Metals Society London, Book 303, 117-22.

WOOLMINGTON, A.D. & DAVENPORT, J. (1983). pH and pO_2 levels beneath marine macrofouling organisms. *Journal of Experimental Marine Biology and Ecology*, **66**, 113-124.

4-3 SUSCEPTIBILITY OF STEEL PILES TO CORROSION IN THE MARGINAL LANDS OF THE NIGER DELTA

S.C. TEME
Geotechnical Division, Institute of Geosciences and Space Technology, Rivers State University, P.M.B. 5080, Port Harcourt, Nigeria

ABSTRACT

The area of the Niger Delta can conveniently be divided into three principal zones; freshwater, mangrove (or transition) and coastal. Within the mangrove or transition zones the land-masses are surrounded by saline mangrove vegetations. The boundaries between these landmasses and the surrounding mangroves have been defined as "marginal lands". Although, in terms of physiography and vegetation, the characteristics of each zone are different, a feature common to all three is the predominance of iron (Fe) in the groundwater.

This paper discusses a case of corroded cylindrical steel piles used for the foundation of a 20-year old mineral processing plant situated within the marginal lands of Port Harcourt city, Nigeria. It presents the results of the subsurface investigations carried out and discusses the substructure design for the new mineral plant to be constructed adjacent to the old one, drawing attention to the remedial measures taken to prevent corrosion.

INTRODUCTION

The Niger Delta lies at the southernmost portion of Nigeria and is bounded by latitudes 4°15' and 5°45' north of the equator and longitudes 5°25' and 7°35' east of the Greenwich Meridian (Figure 1).

For the purposes of environmental studies, the Niger Delta has been subdivided into three basic hydrometeorological zones, namely freshwater, mangrove (or transition) and coastal zones.

Within the mangrove or "transition" zone of the Niger Delta are landmasses, most of which support human habitation. These landmasses are usually surrounded by saline mangrove vegetation; the boundaries between the landmasses and the surrounding mangrove vegetations have been defined by the writer as "marginal lands". In recent years, owing to the population explosion that has characterized most of the developing nations of the world, there has been need to utilize these "marginal lands" for housing and industrial developments but degradation of metals used in the building sub-structures has proved a major problem with diurnal and seasonal fluctuations of groundwater levels and the abundance of various ions and minerals within the subsurface water regime encouraging not only corrosion but also encrustation of metals used in substructural constructions.

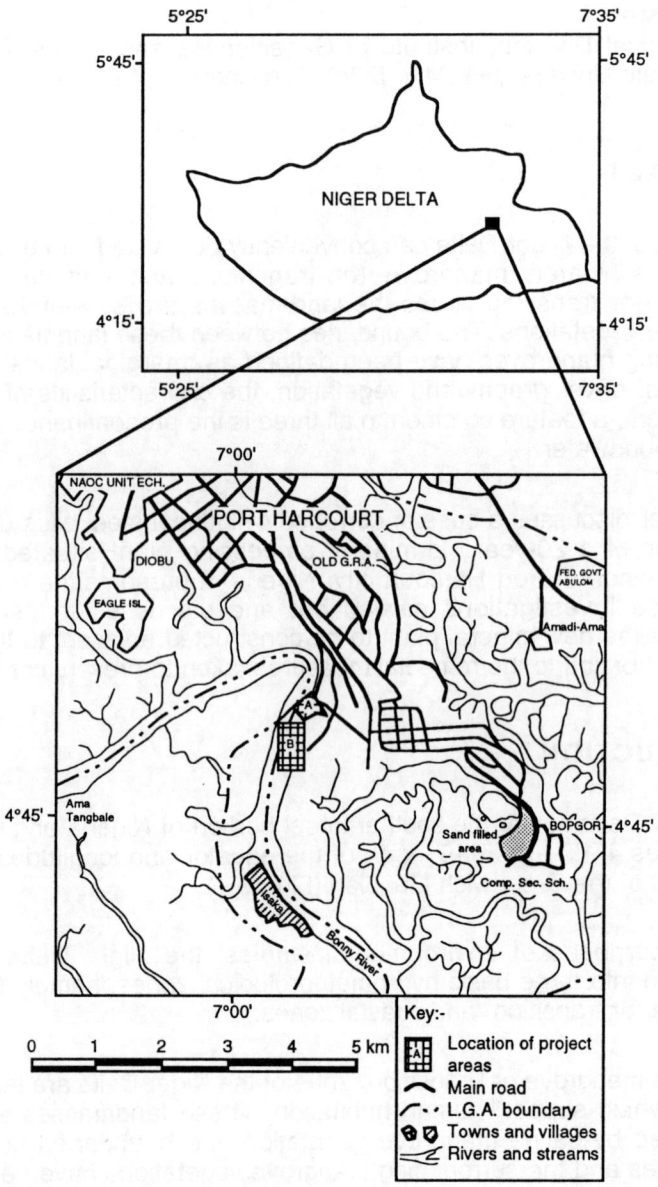

Figure 1: Location of case study area within the Niger Delta, Nigeria.

DESCRIPTION OF CASE STUDY SITES

The sites discussed in this paper are located approximately between latitude 4°45' north and longitude 7°00' east of the Greenwich Meridian, within the marginal lands of Port Harcourt city in the Niger Delta (Figure 1). They occupy an area of approximately 9,000 m² along the left bank of the Bonny River.

The generalised local geology is the coastal plains sands of Pleistocene-Oligocene age, part of the uppermost sections of the Benin Formation. Materials in this Formation include friable salts, whitish to light brown sands, dark greyish shales and clays. The water table is generally at a depth of about 0.15m below ground level. The general topography of the sites is flat-lying and almost at sea level and mangrove vegetation of a secondary nature is predominant.

The subsurface lithology of the site areas consists of a topmost layer of silty sands and sandy gravels (sandfill) underlain by sandy clays, medium plastic clays, silty-clayey sands and whitish-grey sands and gravels at depth. The water table was generally 0.15 m below ground level. The plasticity indices (I_p) of the subsoils ranged from non-plastic to 76.5%, residual friction angles ($ø'_r$) from 8.0° to 34.5°, residual cohesion values (c_r) from 0.70 to 12.8 kPa; coefficients of compressibility (Mv) from 1.42 to 2.44 m²/MN and coefficients of consolidation (C_v) from 4.30 to 9.8 m²/year.

The pH of the subsurface water ranged from 7.1 to 7.3 at temperatures of 26.8°C to 29.0°C, while the redox potential (Eh) was 150-200 mV. The chloride (Cl^-) concentration varied from 2,840 to 16,773 mg/l; sulphate (SO_4^{2-}) concentrations from 0.83 to 4.33 mg/l and Fe^{2+} concentration varied from 0.69 to 1.36 mg/l.

THE TWENTY-YEAR OLD MINERAL PROCESSING PLANT

This facility, denoted "A" in Figure 1, was built on reclaimed land and supported on cylindrical steel piles (Figure 2). Incorporated within the reinforced concrete foundation beam at 5 m intervals are 200 mm x 100 mm x 6.0 m steel H-beams which serve as supporting columns for the walls of the plant. Due to erosion of fill materials adjacent to the building, some of the cylindrical steel piles have been either completely or partially corroded leaving the original concrete fills as the only supporting media for the buildings. In addition to the corrosive activities on the steel piles there is the problem of sulphate attack on the foundation concrete.

NEW MINERAL PROCESSING PLANT

This facility, denoted "B" in Figure 1, occupies a land mass of approximately 4119 m² adjacent to the old plant and was intended to both update the works and increase the production capacity.

As a result of the observed corrosion of steel piles and the sulphate attack on foundation concrete members in the old plant, a detailed subsurface investigation programme was carried out for the new mineral processing plant. This comprised soundings using a 2.5 tonne Dutch Cone Penetrometer and borings using both shell-and-auger rigs and hand augers. In total, one shell-and-auger boring to a depth of 19.8 m and five hand augers borings to depths of 3.0-4.0 m were made in addition to three Dutch Cone Penetrometer soundings to depths ranging from 20.0 m to 20.8 m. The locations of the borings and CPT are given in Figure 3. Figure 4 shows a typical profile of cone penetration resistance versus depth at the site. Both disturbed and undisturbed samples were obtained from the hand auger and shell-and-auger holes for laboratory testing.

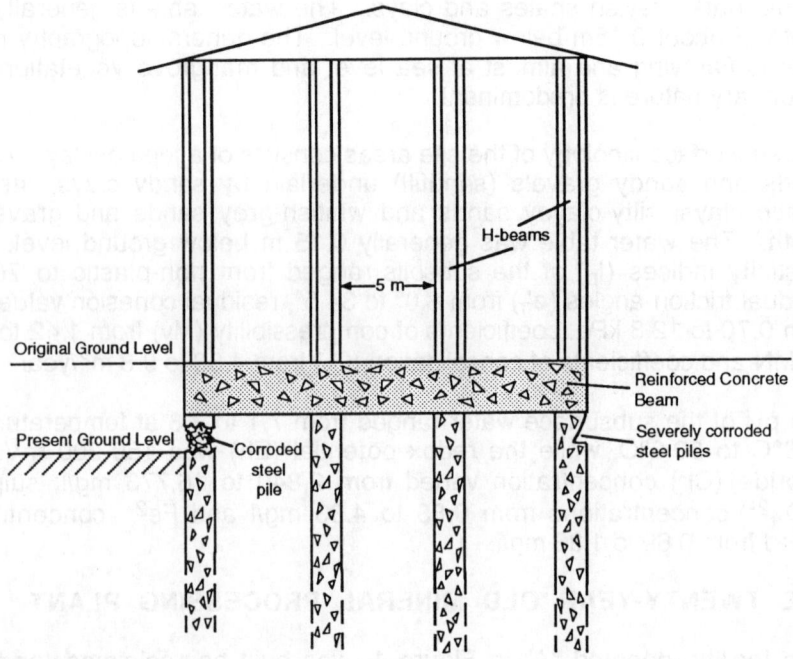

Figure 2: Schematic representation of the 20 year old mineral processing plant showing corrosion of supporting steel structures.

SUBSURFACE PROFILES AND DESCRIPTIONS

The geology at the new mineral processing plant site is shown in Figure 5 and consists of:

i) Greyish sands and gravels (hydraulic sandfill layer)
ii) Sandy-clay layer
iii) Soft, dark-grey organic (marine) clay layer.
iv) Greyish brown silty clayey sand layer
v) Whitish-grey sands and gravelly sand layer.

CORROSION OF STEEL PILES IN THE NIGER DELTA 373

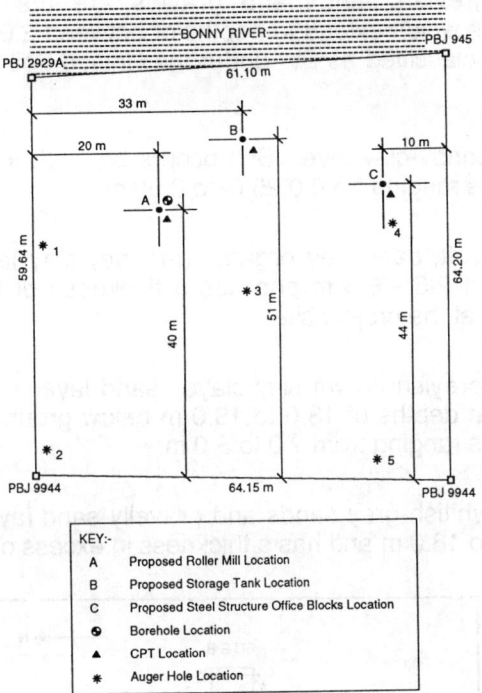

Figure 3: Schematic layout of the boring and CPT locations at the new mineral processing plant site.

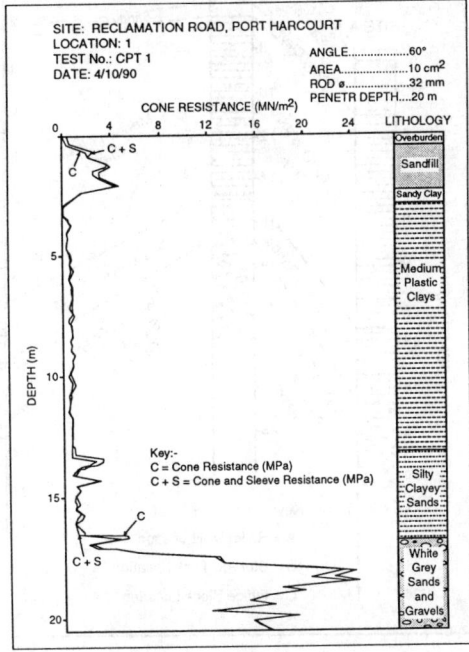

Figure 4: Typical cone resistance versus depth profile at the new mineral processing plant site, Port Harcourt.

i) The greyish sands and gravels are the uppermost sandfill materials which extend to depths of between 2.0 and 6.0 m in places and are classified as SW under the Unified Soil Classification (USC) system.

ii) The sandy-clay layer (SC) occurs beneath the sandfill and has a thickness ranging from 0.25 m to 0.50 m.

iii) The soft, dark-grey organic (marine) clay layer (OH) occurs at depths of 2.5 - 6.5 m and has a thickness of between 8.0 m and 12.50 m at the project site.

iv) The greyish brown silty clayey sand layer (CL-ML; SM and SC) occurs at depths of 13.0 to 18.0 m below ground surface and has a thickness ranging from 2.0 to 3.0 m.

v) The whitish-grey sands and gravelly sand layer occurs at depths of 17.0 to 18.0 m and has a thickness in excess of 10.0 m.

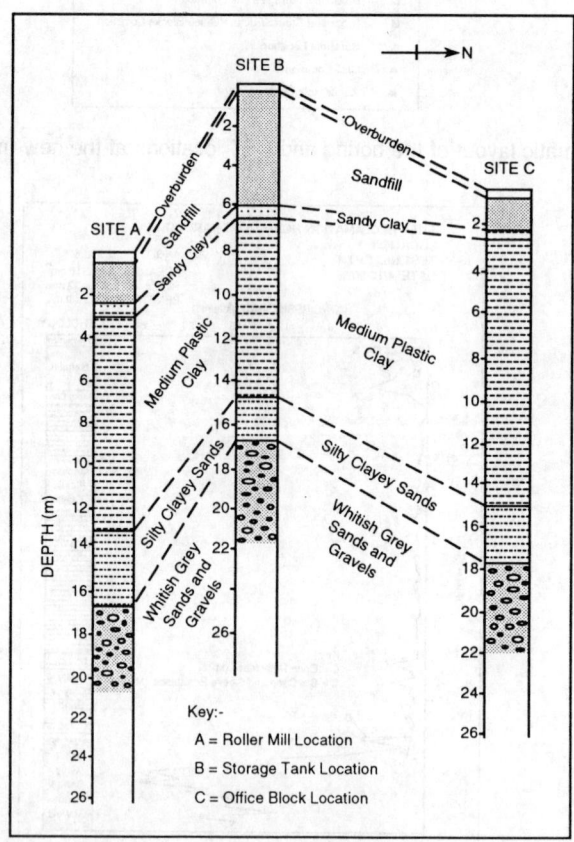

Figure 5: Fence diagram of the subsurface at the new mineral processing plant site.

LABORATORY INVESTIGATIONS AND RESULTS

Laboratory investigations were undertaken on soil samples obtained from the boring and ground water samples. The tests carried out on soil samples included grain size distribution analysis; consistency limits (Atterberg limits); direct shear tests; unconsolidated-undrained (u-u) triaxial tests and oedometer consolidation tests. Analyses of the ground water determined pH; chloride (Cl^-); sulphate (SO_4^{2-}) and iron (Fe^{2+} and Fe^{3+}) contents.

SOIL GRADATION PATTERNS

The gradation patterns based on the Unified Soil Classification (USC) system range from well-graded for the upper hydraulic sandfill and the lower whitish-grey sands and gravels through gap-graded for silty-clayey sands. Typical gradation curves for subsurface materials from the project site are shown in Figure 6.

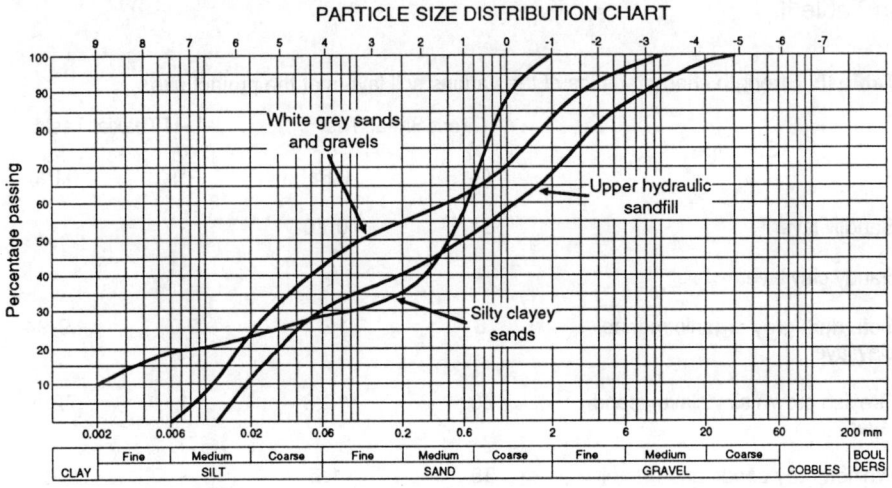

Figure 6: Typical particle size distribution curves for the subsurface materials from project site.

SOIL CONSISTENCY

The consistency limits of the various subsoil layers at the new mineral processing plant site are given in Table I.

As can be seen, the upper sandfill and sandy clay layers are non-plastic to slightly plastic respectively, while the organic marine clays and silty-clayey sands beneath are moderately to highly plastic. The whitish grey sands and gravelly sands at depth are non-plastic.

Table I: Consistency limits of the various subsoil layers at the project site.

Soil type	w_l (%)	I_p (%)	w_n (%)
Sandfill layer	NP	NP	51-75
Sandy clay layer	NP-46	NP-48	51-70
Soft, dark-grey organic marine clay layer	106-145	45-76	69-72
Greyish brown silty clayey sand layer	37-46	19-22	48-84
Whitish-grey sands and gravelly sand layer	NP	NP	53-54

SOIL STRENGTH CHARACTERISTICS

The strength properties of the subsurface materials at the project site were assessed by means of both direct shear tests on remoulded materials and undrained-unconsolidated triaxial compression tests. The results are given in Table II.

Table II: Strength characteristics of the various soil layers at the project site.

Soil type	Direct Shear Test		U-U Triaxial Tests	
	ϕ'_r (°)	c'_r (kPa)	ϕ_u (°)	c_u (kPa)
Sandfill layer	34	0.70	-	-
Sandy clay layer	-	-	-	-
Soft, dark-grey organic marine clay layer	8	12.4	2-7	69-94
Greyish brown silty clayey sand layer	23	12.8	1	72.5
Whitish-grey sands and gravelly sand layer	35	1.5	-	-

SHEAR STRENGTHS

The average value of the residual friction angles for the upper sandfill layer was 34° while the corresponding residual cohesion was 0.70 kPa. For the organic (marine) clays, the average values of the residual and undrained friction angles were 8° and 5° respectively while the residual cohesion intercept and undrained cohesion were 12.4 and 83.5 kPa respectively. The silty clayey sands gave average residual and undrained friction angles of 22.6° and 1.0° respectively with residual and undrained cohesion values of 12.8 and 72.5 kPa. These values, although low, are typical of the organic clays and silty-clayey-sands within the Niger Delta (Teme, 1988, 1990a, 1990b) and could easily support similar allowable bearing pressures

(Tomlinson, 1980). The variation of shear strength with depth at the project site is shown in Figure 7.

SOIL BEARING CAPACITY

In addition to the field Dutch Cone Penetration Tests (CPT) and the Standard Penetration Tests (SPT) undertaken to determine or evaluate the bearing capacities of the various subsoil units, laboratory one-dimensional oedometer consolidation tests were carried out on selected undisturbed U_{100} tube samples.

COEFFICIENT OF VOLUME COMPRESSIBILITY (M_V)

The coefficients of volume compressibility (m_V) for the silty-clayey sands and organic (marine) clays at the project site varied from 1.42 to 2.44 m^2/MN as would be expected for typical Niger Delta marine organic clays and silty-clayey-sands.

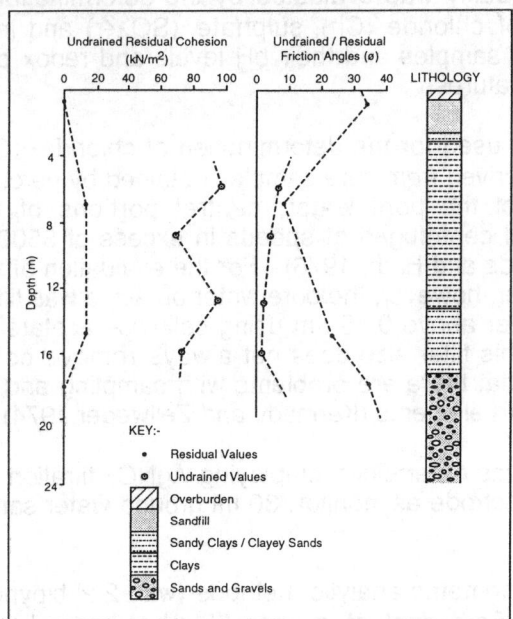

Figure 7: Variation of shear strength properties with depth at the project site.

COEFFICIENT OF CONSOLIDATION (C_V)

The coefficients of consolidation (c_V) for the organic (marine) clays and silty-clayey-sands varied from 4.30 m^2/year to 9.80 m^2/year, which are again normal for such soils within the Niger Delta. The bearing capacities of the various subsoil units at the project site based on Dutch Cone Penetration Test results are given in Table III.

Table III: Consolidation and bearing characteristics of subsoil units at the project site.

	Depth Range (m)	Coefficient of Compression m_V (m²/MN)	Coefficient of Consolidation c_V (m²/yr)	Soil bearing pressures from CPT (MPa)
Hydraulic sandfill layer	0.0-5.8	N/A	N/A	3.0-6.0
Sandy clay layer	2.10-6.4	2.44	6.80	0.75-1.80
Organic clays (marine clays)	2.5-15.0	1.42-2.44	4.30-9.80	0.0-0.8
Silty clayey sands	13.2-17.80	-	-	1.5-3.5
Sand and gravelly sands	16.8 & below	-	-	above 8.0

GROUND WATER QUALITY

Groundwater quality was evaluated by the determination of the concentration levels of chloride (Cl^-), sulphate (SO_4^{2-}) and iron (Fe^{2+}) in the borehole water samples and their pH levels and redox potentials (Eh) at ambient temperatures.

Water samples used for the determination of chloride (Cl^-) and sulphate (SO_4^{2-}) were derived from core samples obtained by percussion drilling. In order to extract the pore water, central portions of the cores were fragmented and centrifuged at speeds in excess of 3500 revolutions per minute (Edmunds and Bath, 1976). For the evaluation of the Fe content in the ground water, however, the pore water obtained was filtered to separate particulate matter above 0.45 μm using cellulose acetate filters. It should be noted that this filter size does not always remove colloidal materials, however, and that there are problems with sampling and filtration for iron and other related elements (Kennedy and Zellweger, 1974).

Chloride (Cl^-) was determined employing $AgNO_3$ titration with the Ag/AgS ion selective electrode as monitor; 30 ml ground water samples were used for this analysis.

The spectrophotometric analytic methods (with 2,2'-bipyridyl) was used to determine the Fe^{2+} content in circa 55 ml of ground water sample, in accordance with the technique recommended by Rainwater and Thatcher (1960).

For the sulphate (SO_4^{2-}) determination, the analytical method used was titration with $Ba(ClO_4)_2$ using thorin as the indicator (Fritz and Yamamura, 1955).

Both pH and the redox potential (Eh) were measured at the site after the percussion drilling, in order to avoid carbon dioxide loss and aeration. A battery operated pH meter was used to measure the pH level of the ground water in the newly drilled borehole to minimize the problems of obtaining meaningful, accurate pH values (Barnes, 1964; Whitefield, 1971).

The redox potential (Eh) was measured by attaching a simple flow cell to the discharge line from an existing borehole at the adjacent premises of the old mineral processing plant. The water samples from this discharge line were obtained via a submersible pump and both the pH and Eh were evaluated in order to obtain a better understanding of mineral stability patterns and corrosion potentials of the ground water at the project site. Similar approaches have been reported by Garrels and Christ (1965) and Barnes and Clarke (1969).

A summary of the results of the chemical analyses carried out on the ground water samples is given in Table IV. Values of pH and Eh obtained from field testing have been superimposed on Eh-pH diagrams based on thermodynamic data from Garrels and Christ (1965) in order to make correlations, as has been done by workers such as Hem (1961); Langmuir (1969) and Edmunds (1973). The field values were found to lie in the Fe^{2+} region (Figure 8), indicative of reducing conditions which could lead to serious corrosion of metals. This would explain why the steel cylindrical piles used as foundation supports for the adjacent 20-year old mineral processing plant were severely corroded.

Table IV: Summary of results of chemical analyses on groundwater samples from project site.

Ion	Range of Concentration (mg/l)	Parameter	Range of Values
Cl^-	2,840.0-16,773.0	pH	7.1-7.3
SO_4^{2-}	0.83-4.33	Eh	150-200 mV
Fe^{2+}	0.69-1.36		

Note: Range of temperatures at which pH and Eh were measured was 26.9°C to 29.0°C.

REMEDIAL MEASURES

In order to reduce the effect of corrosion on the support system of the new mineral processing plant building, precast reinforced concrete piles were specified rather than the cylindrical steel piles used for the old plant. Each reinforced concrete pile had a length of 20 m with a top cross-section of 300 mm^2 and an angle of pile taper (ω) equal to 1.5°, as shown in Figure 9. Average depths of pile penetration were between 19.0 and 20.0 m; the water table being very close to the ground surface (0.15 m). In addition, the base of the raft-on-pile foundation system used for the new plant building was waterproofed to avoid the ingress of ground water into the system.

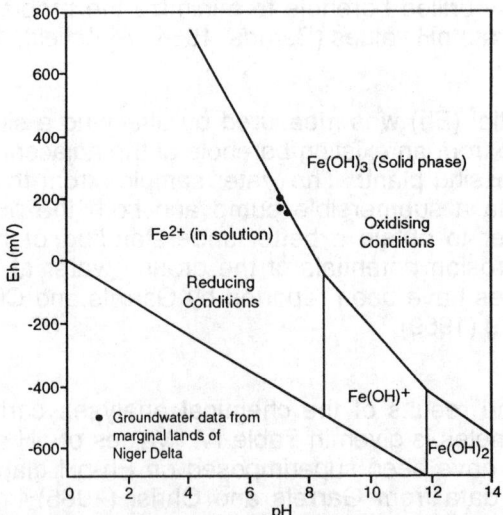

Figure 8: Eh-pH stability field diagram showing superimposed plots of data from groundwaters in the marginal lands of the Niger Delta.

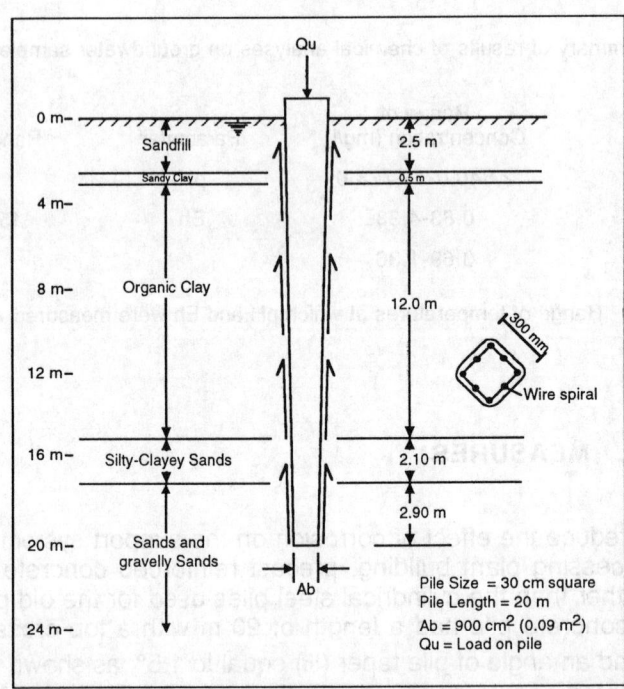

Figure 9: Schematic cross-section of tapered reinforced concrete pile used at the project site.

SUMMARY AND CONCLUSIONS

The long-term performance of steel piles within the marginal lands of the Niger Delta has been observed at the site of a 20-year old mineral processing plant building at Port Harcourt. Detailed geotechnical and geochemical investigations were undertaken to assess both the soil characteristics and the ground water quality prior to the construction of a new mineral processing plant adjacent to the existing buildings. On the basis of results obtained from these studies, it can be concluded that reducing conditions exist which could explain the serious corrosion of metals evident in the 20-year old mineral processing plant.

In order to minimize such corrosion within the marginal lands of the Niger Delta, the use of pre-cast reinforced concrete piles is recommended. These have been used in the foundation of a new mineral processing plant building. As a further preventative measure, the foundations of the buildings have also been waterproofed.

REFERENCES

BARNES, I. (1964). Field measurement of alkalinity and pH. *Water Supply Paper*, **No. 1535-H**, U.S. Geological Survey.

BARNES, I. & CLARKE, F.E. (1969). Chemical properties of groundwater and their corrosion and encrustation effects on wells. *U.S. Geological Survey*, **Paper 498-D**.

EDMUNDS, W.M. (1973). Trace element variations across an oxidation-reduction barrier in a limestone aquifer. In: *Proceedings of the Symposium on Hydrogeochemistry and Biogeochemistry*, Toyota, Ingerson, E. (ed), 500-526.

EDMUNDS, W.M. & BATH, A.N. (1976). Centrifuge extraction and chemical analysis of interstitial waters. *Environmental Science and Technology*, **10**, 467-472.

FRITZ, J.S. & YAMAMURA, S.S. (1955). Rapid micro-titration of sulphate. *Analytical Chemisty*, **27**, 1461-1464.

GARRELLS, R.M. & CHRIST, C.L. (1965). *Solutions, minerals and equilibria*. Harper and Row, New York.

HEM, J.D. (1961). Stability field diagrams in the aid of iron chemistry studies. *Journal of the American Water Works Association*, **53**, 211-228.

KENNEDY, V.C. & ZELLWEGER, G.W. (1974). Filter pore size effects on the analysis of Al, Fe, Mn and Ti in water. *Water Resources Research*, **10**, 785-790.

LANGMUIR, D. (1965). Geochemistry of iron in a coastal plain groundwater of the Camden, New Jersey area. *U.S. Geological Survey*, **Paper 650-C**, 224-235.

RAINWATER, F.H. & THATCHER, L.L. (1960). Methods for the collection and analysis of water samples. *Water Supply Irrigation*. **Paper 1454**.

TEME, S.C. (1988). Report on subsurface investigations for ferry terminal, Abam-Ama, Okrika. *Technical Report, No. 1 for the Federal Ministry of Transport, Inland Waterways Department*, Lagos, 68 pp.

TEME, C.C. (1990a). *Report on subsurface investigations for the Nigerian National Petroleum Corporation (NNPC) Pollution Control Centre, Marine Base, Port Harcourt. Technical Report*, 101 pp.

TEME, S.C. (1990b). *Report on subsurface investigations for Nishan Industries Limited, Mineral Processing Plant, Port Harcourt. Technical Report*, 95 pp.

TOMLINSON, M.J. (1980). *Foundation design and construction*. Pitman Advanced Publishing Programme, London, 4th edition, 793 pp.

WHITFIELD, M. (1971). *Iron selective electrodes for the analysis of natural waters*. Australian Marine Sciences Association. Sydney, 130 pp

4-4 THE INFLUENCE OF MICROBIAL ASSEMBLAGES ON SEDIMENT ERODIBILITY: AN ENGINEERING PERSPECTIVE

D.M. PATERSON
Department of Botany, University of Bristol, Bristol, BS8 1UG.

ABSTRACT

The influence of microbial populations and assemblages on the erodibility of sediments is reviewed and some of the techniques used to measure sediment stability briefly described. Evidence is presented for the biogenic stabilisation of sediments by a variety of microbial systems under controlled conditions within the laboratory and in field studies. The mechanisms (network formation, cohesive effects, and boundary layer influences) of biogenic stabilisation are described and steps towards the prediction of sediment stability through the use of biological parameters such as biomass and biochemical markers discussed. The progress in current research towards the use of *in situ* technology for interdisciplinary studies of sediment behaviour is supported with examples from natural ecosystems. The lack of seasonal data and the variability of biological systems is recognised as a limitation to our current knowledge of the biological mediation of sediment behaviour. Further data must be gathered in order to test the use of the parameterisation of biogenic stabilisation within models of sediment behaviour.

INTRODUCTION

The erosion, transport and deposition of marine and estuarine sediments is of environmental and engineering significance in coastal waters. Anthropogenic alteration to natural flow environments can lead to massive redistribution of sediments with consequential environmental and economic effects. The stability of the substratum is also of direct significance in coastal engineering. In order to predict the influence of anthropogenic alterations in coastal waters, it is necessary to understand the complex of parameters and processes which control the behaviour of sediments (Figure 1). Without the breakdown and understanding of these factors it will be impossible to produce useful predictive models of sediment dynamics. The process of erosion, the loss of particles or aggregates from the surface of the sediment, is central to this issue. This short review concentrates on the influence of micro-organisms on the process of erosion and outlines some of the evidence for microbial mediation of sediment behaviour. The measurement of biological effects is discussed and the potential of biological indices for sediment stability considered.

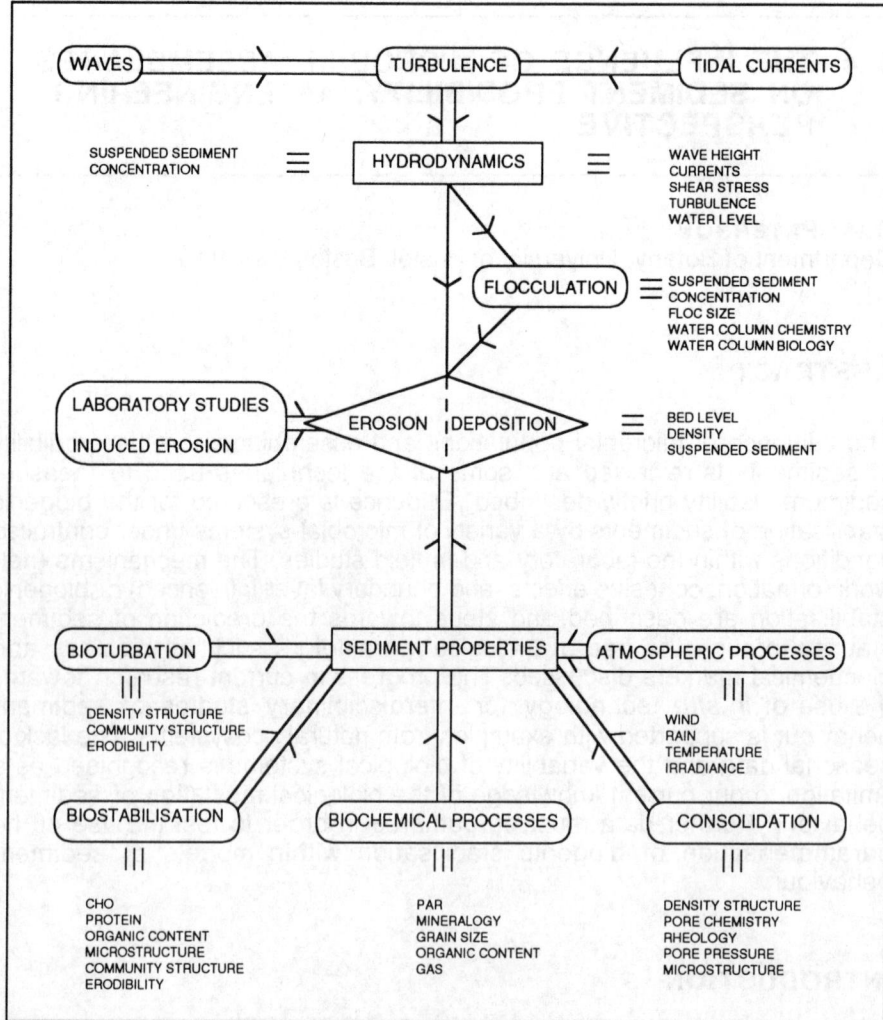

Figure 1: Schematic representation of the parameters and processes which interact to control the erosion and deposition of intertidal sediments. Biological effects influence flocculation, deposition, and consolidation in addition to the processes of biostabilisation and bioturbation. (Diagram prepared for the LISP-UK proposal in association with Dr A. Delo (HR Wallingford) and members of the LISP steering group).

HISTORIC PERSPECTIVE

The most common approach to the prediction of sediment behaviour was based on a purely physical paradigm. The response of selected particles to physical forcing was examined under laboratory conditions and the results used to determine the relationship between grain size and critical erosion stress, giving rise to predictive relationships such as the Shields curve (Figure 2) or Yallin parameter (Miller, McCave and Komar, 1977). While it was found that non-cohesive sediments behave in a broadly predictable manner, the smaller cohesive particles do not. These small particles have a

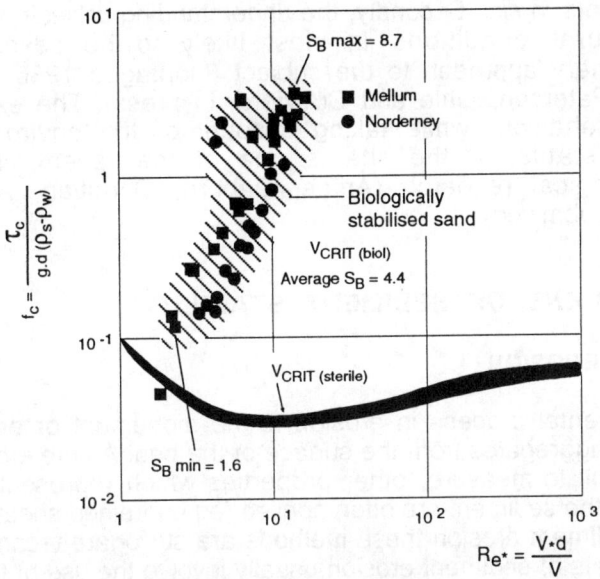

Figure 2: The results of critical threshold erosion measurements of biogenically stabilised sands plotted in a Shields' diagram. (Redrawn from Manzenrieder 1983).

large surface area to volume ratio and their behaviour is affected by the electrostatic properties of the particle surface (eg. Van der Waals forces). The behaviour of cohesive sediment beds is therefore not simply related to particle size (Amos, van Wagganer and Daborn, 1988) but is also influenced by the physico-chemical micro-environment surrounding the particles. Time-dependent changes also take place as the particles become more closely aligned in the process of sediment compaction. Investigation of the response of cohesive sediments to physical forcing must therefore recognise such complex factors as bed history and physico-chemical conditions.

In addition to physical variables, it has now been shown beyond doubt (Dade, David, Nichols, Nowell, Thistle, Trexler and White, 1990; Meadows and Tait 1989; Meadows, Tait and Hussain, 1990; Montague 1986; Paterson 1989; Paulic and Montague and Mehta, 1986) that under natural conditions the behaviour of intertidal and subtidal sediments is greatly influenced by biotic factors. De Boer (1981) demonstrated *in situ* that the destruction of sediment biota led to the subsequent erosion of the experimental area. Such biogenic influences extend to both cohesive and non-cohesive sediments although the phenomenon is more easily demonstrated with the more predictable non-cohesive sediments, as shown by Figure 2.

These results have two important consequences. Firstly, measurements made in the laboratory and intended to represent the erosion of sediments *in situ* must be interpreted with caution as extrapolation from the laboratory to the field may be misleading. It is difficult to maintain or duplicate natural conditions within test sediments and it is therefore preferable to make measurements *in situ*. Secondly, the understanding of sediment behaviour under natural conditions is most likely to be advanced by an interdisciplinary approach to the subject (Montague 1986, Underwood, McArthur, Paterson, Little and Crawford, in press). The examination of sediment behaviour while taking account of the environmental and biological status of the site is one of the recent challenges in sedimentological research (Amos, Daborn, Christian, Atkinson and Robertson, submitted).

MEASUREMENT OF SEDIMENT STABILITY

SEDIMENT EROSION

The fundamental process in erosion is the movement or entrainment of particles or aggregates from the surface of the bed. As the erodibility of the bed is difficult to measure, other properties which represent the general "stability" of the sediment are often applied (eg undrained shear strength). In terms of sediment erosion these methods are surrogate techniques. Direct measurements of sediment erosion usually involve the use of flow channels (flumes). The behaviour of cohesive sediments in response to erosive stress is not yet fully understood.

Two types of erosion have been identified by Mehta and Patheniades (1982) and supported by Amos (1989). These take the form of "aggregate by aggregate" erosion (Type I erosion) and "mass erosion", a more extensive bed failure (Type II erosion). In most flume studies Type I erosion is observed although Type II erosion has been found to occur within a Carousel flume under high flow conditions (azimuthal current > 0.45 m/sec; Amos 1989). Results of flume studies tend to be expressed either as the bed shear stress (τ) at which erosion began (τ_c) or a shear velocity (U^*) at the same point (U^*_c). The former has units of N/m^2 while U^* is a velocity measurement (m/s); these terms are related by the expression:

$$U^* = (\tau/q)^{1/2} \quad [1]$$
where τ = Bed shear stress (N/m^2)
q = Fluid density (kg/m^3)

SURROGATE METHODS

Surrogate methods used as a measure of sediment "stability" and sometimes to infer erodibility include fall cone apparatus (Hansbo 1957) which measures surface undrained shear strength averaged over the upper centimetre of sediment; shear vane techniques (Kravitz 1970) which measure the resistance of the sediment bed to shearing forces at a variety of depths (Lambe and Whitman 1970) and, more recently, the "INSIST" system (*in situ* simple shear test), a friction plate device developed by Christian, Gillespie and Amos (1990). INSIST measures the *in situ* drained shear strength in the upper millimetre of sediment by measuring horizontal failure stress under different vertical compressive loadings. While these devices offer the advantages of speed and ease of use and produce valuable data, care must be taken in relating these results to the erodibility of sediments. The reliability of the correlation between such measures and the actual force required to entrain or move particles *in situ* is uncertain and results may be highly variable (Kravitz 1970). However, under controlled conditions, (Meadows and Tait 1989, Meadows et al. 1990) surrogate methods clearly demonstrate alterations in sediment properties brought about by biological action.

DIRECT METHODS

The most common method of determining sediment erodibility is by the use of laboratory flow channels or "flumes". These devices vary greatly in size, design and in the technology applied to their operation (see Nowell and Jumars 1987). At the most basic level, a flume may consist of a simple cylindrical chamber in which water flow is driven by a paddle or propeller mechanism. Examples include one of the earliest laboratory studies of microbial stabilisation which was carried out by Holland, Zingmark and Dean, (1974). More sophisticated devices include straight channels powered by large pumps and race track or carousel flumes (Nowell, Jumars and Eckman, 1981; Nowell, Jumar, Self and Southard, 1989; Amos, 1989).

By whatever method the erosive stress is applied to the sediment, the determination of the point at which erosion begins - the point of incipient scour - is still problematic. As the initial events of erosion may cause very little sediment to be thrown into suspension, detection by remote devices is difficult. Many workers still rely on direct observation which has none of the problems associated with high technology systems although the results are more subjective (cf. Dade et al. 1990). An alternative method is to use a mathematical approach to extrapolate the critical erosion stress from the subsequent erosion of sediment from the bed after the point of incipient scour has been exceeded (Mehta and Partheniades 1982). In brief, stress is applied to the bed until erosion ceases. It is assumed erosion continues until critical bed shear stress of the newly exposed bed exceeds the flow-induced stress. The level of flow-induced stress is then increased and the process repeated in a series of steps. The resultant erosion of sediment from the bed at each step is calculated. From this information and knowledge of the bed shear stress at each successive step, a regression relationship between depth and bed shear stress can be produced. The intercept of this regression with the y-axis represents the bed shear stress at zero depth eg the critical bed shear stress for erosion (Amos 1989).

Many flume systems are now deployed in the field and several are reviewed by Black (1989). The logistic problems of field deployment are often great but the use of *in situ* systems is a genuine advance towards the understanding of the behaviour of natural sediment systems (see Young and Southard 1978, Amos 1989 and Amos et al. submitted). Practical considerations for the use of flumes to simulate benthic environments have been reviewed by Nowell and Jumars (1987).

MECHANISMS OF MICROBIAL STABILISATION

An increase in the resistance of a sediment to erosion caused by the direct or indirect influence of biota can be termed biogenic stabilisation. Organisms responsible for biogenic stabilisation range from mangrove trees (Scoffin 1970), to seagrasses (Fonseca 1989) and microbes (Coles 1979, Dade et al. 1990), including filamentous algae, diatoms, cyano-bacteria and bacteria (Plates 1-3). Although larger organism may have a more noticeable effect on sediment erosion and transport, microbes occur on virtually all submerged surfaces and their influence is potentially far more pervasive than macroscopic species (Plate 2). The influence of microbes on sediment stability has in part been recognised because of advances in scanning electron microscopy which have helped to visualise the natural structure at the boundary interface (Paterson et al 1986, Plate 1).

There are several mechanisms by which microbes can stabilise the substratum that they inhabit (Table I). Some specific examples and the influence of these mechanisms on sediment erosion are given below. These mechanisms should not be regarded as mutually exclusive for microbial assemblages or even for individual species (Plates 1D & E). The complexity of benthic microbial assemblages, particularly in mat forming systems comprised of a complex assemblage of organisms (Oppenheim & Paterson 1990), ensures that many mechanisms of stabilisation will occur in parallel. This an important factor that hinders the development of a single biological/biochemical parameter of biogenic stabilisation for use in simulation modelling. The major mechanisms of biogenic stabilisation are network formation (Plate 1), cohesive effects (Plates 1-3) and boundary layer influences (Plates 1 and 3). These influences have been measured in the field and reproduced in the laboratory (Plate 3) where control systems can be more easily established (Holland et al. 1974, Paterson 1990).

Table I: Mechanisms of biogenic stabilisation and some examples of representative organisms.

Mechanism	Organisms
Physical network	Cyanobacteria (eg *Oscillatoria, Microcoleus, Lyngbya*), Fungi, Green algae (*Vaucheria*), Filamentous bacteria
Boundary interface roughness	Diatoms (eg *Amphora, Surirella*), Green algae, Cyanobacteria (*Lyngbya*), Bacteria
Cohesive effects: Matrix	Diatoms (Epipelic eg. *Gyrosigma, Navicula*), Bacteria, Polychaete worms (*Nereis diversicolor*), Snails (*Hydrobia ulvae*)
Particle cohesion	All producers of extracellular polymeric material (eg algae, cyanobacteria, bacteria, protozoa, etc)

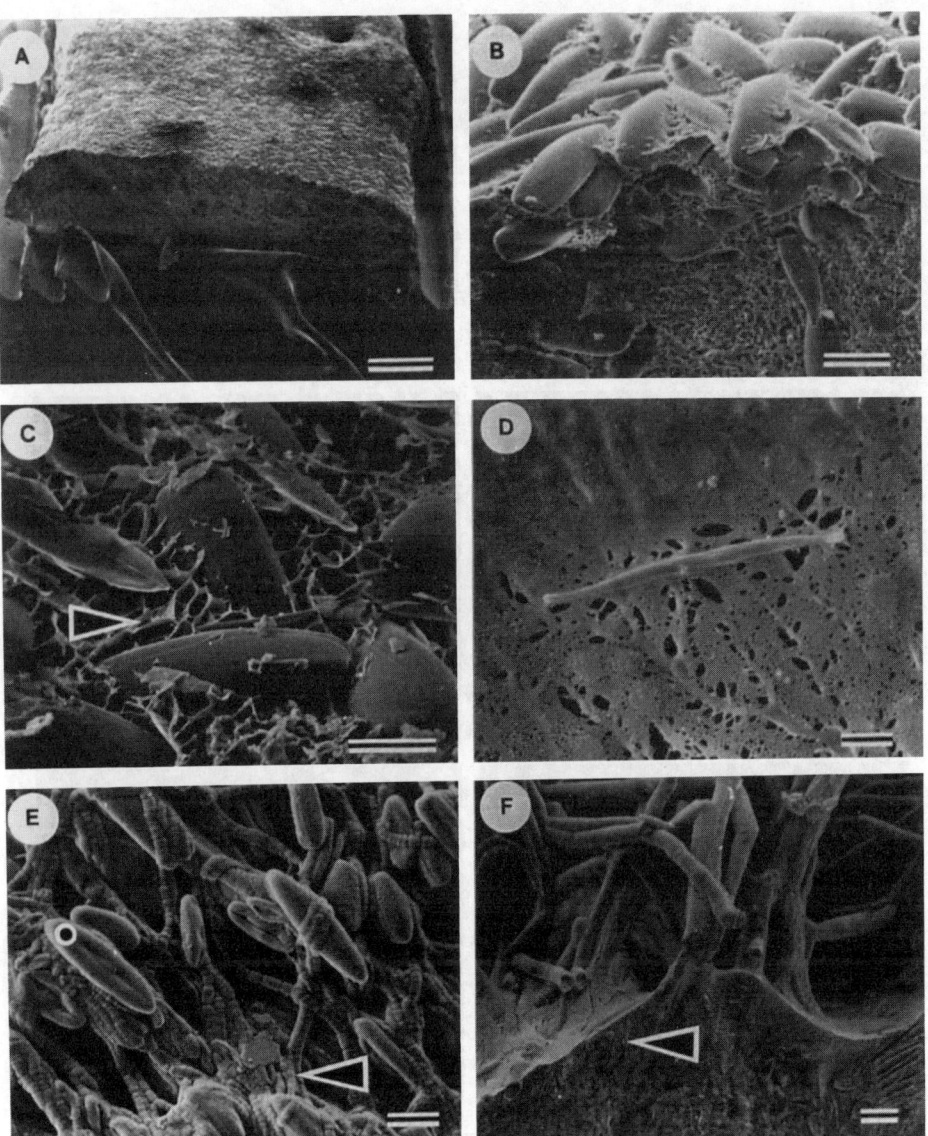

Plate 1: Low-temperature scanning electron micrographs of cohesive sediments: A) A sediment sample at low magnification within the SEM. Bar marker = 1 mm: B) Detail of the freeze-fractured sample edge showing the accumulation of diatoms at the sediment surface. The diatoms form a cohesive matrix at the sediment surface. Bar marker = 50 μm: C) A similar sample after complete freeze-drying showing damage to cells (above arrow) and shrinkage of the mucilage into defined strands (arrow). Bar marker = 50 μm: D) Mucilage on the surface of cohesive sediment under a single diatom (Cylindrotheca signata). The influence of the diatom on sediment erodibility is through the cohesive matrix and the smoothing of the sediment surface. Bar marker = 10 μm: E) Mixed matrix of cyanobacteria (arrow) and diatoms (spot) on the surface of intertidal cohesive sediment demonstrating both network formation and cohesive effects. Bar marker = 10 μm: F) The boundary layer colonised by the intertidal green alga Vaucheria sp. Filaments of the Vaucheria can be seen penetrating into the sediment. Fractured filaments (arrow) are visible below the sediment surface forming a binding matrix. Bar marker = 1 mm.

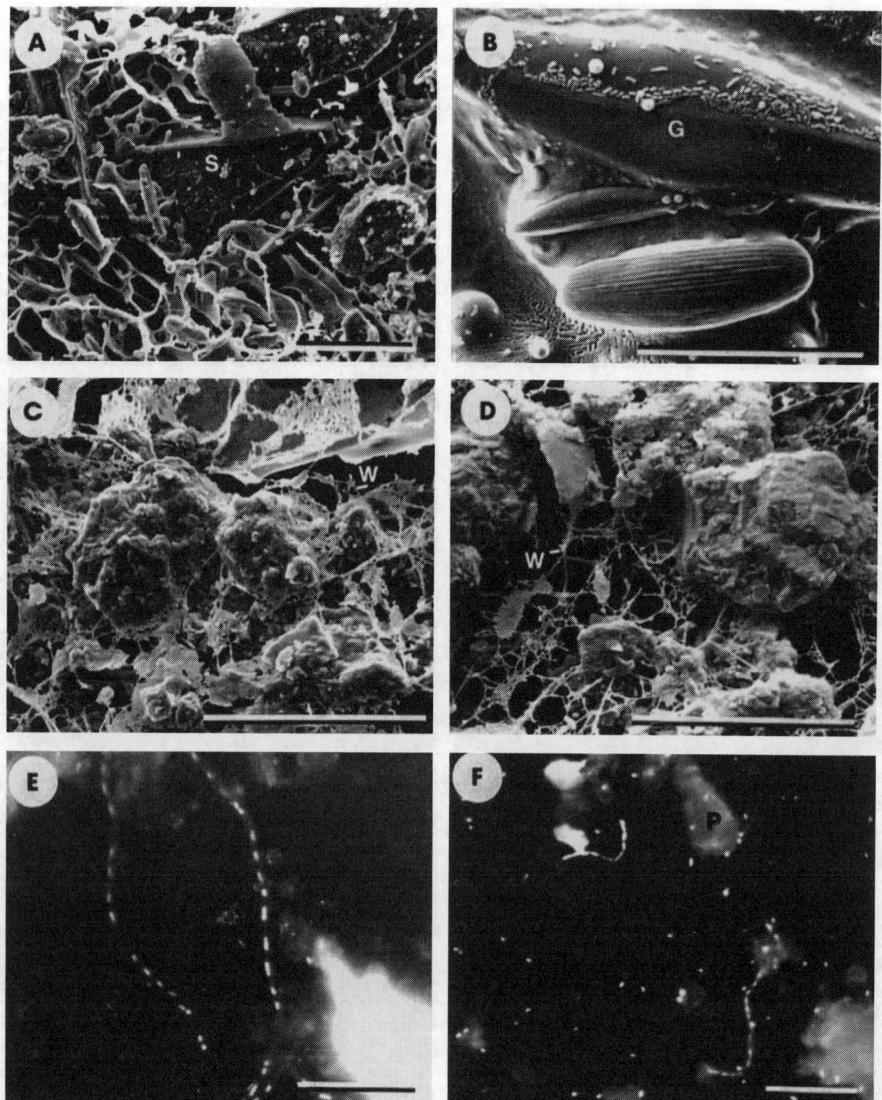

Plate 2: Bacteria within sediment systems. A-D) Low-temperature scanning electron micrographs of intertidal sediment from the Severn estuary: A) The surface sediment inhabited by a population of the diatom Surirella gemma (S). A network of mucilage extends over the sediment around the diatoms and bacteria can be seen on the surface of the valve. Bar marker = 50 µm: B) Bacteria on the surface of an epipelic diatom (G = Gyrosigma sp.). The sediment surface is obscured by an organic matrix. Bar marker = 50 µm: C-D) Organic matrix formed in sediments in which diatom growth has been prevented by the addition of an inhibitor of photosynthesis (DCMU): C) Web-like strands of organic matrix (W) can be seen among the particles. Bar marker = 40 µm: D) These strands are thought to be of bacterial origin and can form extensive fine networks. Bar marker = 40 µm: E-F) Epifluorescence micrographs of sediments stained with acridine orange: E) Filamentous bacteria clearly visible despite the disturbance of sediment during preparation for microscopy. Bar marker = 10 µm: F) Single bacteria in suspension and in association with sediment particles. Bar marker = 10 µm.

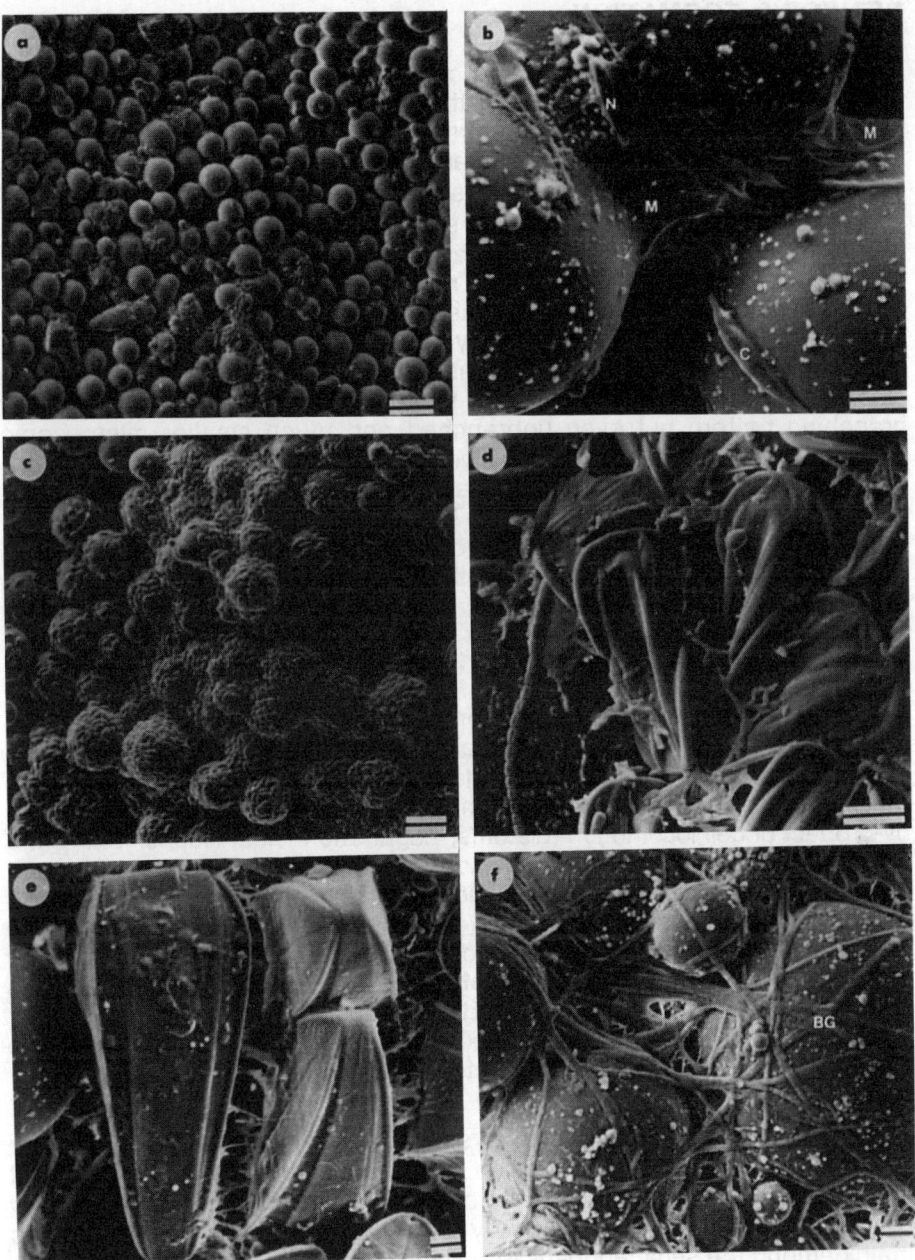

Plate 3: Low-temperature scanning electron micrographs of laboratory glass bead cultures of diatoms and cyanobacteria: A) Glass bead substratum after the addition of diatom inoculum. Bar marker = 100 μm: B) The development of cohesion between the glass beads. Bar marker = 10 μm: C) Biogenic stabilisation by the attachment of epipsammic diatoms. Bar marker = 50 μm: D-E) Biogenic stabilisation by epipelic diatoms forming inter-cell networks. Bar marker = 10 μm: F) Cyanobacterial filaments traversing and binding the glass beads. Bar marker = 10 μm.

NETWORK FORMATION

One of the most direct mechanisms of stabilisation is by the formation of a physical network above or among the surface layers of sediments. The filamentous green alga Vaucheria forms an effective network in the surface sediments of intertidal mudflats in many estuaries (Plate 1F). The filaments can increase the critical bed shear stress from 2-4 N/m^2 for uncolonised sediment to 6-13 N/m^2 for colonised sediment (George, unpublished data). In addition to the fibrous network, filaments projecting above the sediment surface act to reduce the flow in the region of the boundary layer and may enhance the deposition of fine sediment. Vaucheria beds are clearly visible and individual filaments are relatively large on a microbial scale (Plate 1F) but many smaller organisms, such as red algae of the genus *Audouinella* also form effective networks. Both of these forms show some complexity of branching pattern but even unbranched and much smaller microscopic filaments are capable of imparting significant erosion resistance to sediment systems. Several species of cyanobacteria (microscopic blue-green algae) are effective stabilisers (Plate 3F). Trichomes (cyanobacterial filaments) may act individually or be associated in mucilage sheaths (eg *Microcoleus* sp) adding to the strength of individual binding elements.

COHESIVE EFFECTS

Cohesive effects arise from the production of extracellular polymeric substances (EPS) such as the mucilages secreted from microbial cells (Plates 1-3). These microbial mucilages may fulfil several functions such as: attachment of cells to the substratum (bacteria, cyanobacteria, diatoms); a by-product of cell locomotion (diatoms, Edgar and Pickett-Heaps, 1984); the formation of a protective capsule around the cell (some bacteria and cyanobacteria); protection from desiccation; and perhaps even protection against pollutants such as heavy metals which are collated by these organics. Cohesive influences on sediment erodibility can be subdivided by the specific mechanism of their action but are broadly treated here as belonging to matrix effects and particle cohesion (Table I). The relative influence and longevity of these effects are related to the properties of the EPS produced, whether highly refractory or easily dispersed.

The influence of EPS in marine systems has been reviewed by Decho (1990). In terms of sediment erosion, the influence of EPS on the sediment arises from sediment particles being effectively stuck together, either directly by microbial cells or more indirectly through the formation of an organic matrix among the sediment particles (Plates 1-3).

The influence of EPS on sediment erosion depends on the concentration and type of EPS produced. The nature of the EPS spectrum from a sediment is not easily predicted. For example, EPS production from a population of bacteria is influenced not only by the type of bacterium but also by the physiological condition of the cell and the nutrient status of the surrounding medium (Dade et al. 1990). Despite this, general relationships have been produced between the concentration of EPS in the sediment and the critical erosion stress for particular sediments (Grant and Gust 1987, Dade et al. 1990). Most work has been concentrated on sandy, non-cohesive sediments with less information as yet available on cohesive sediments (Paterson et al. 1989 and 1990, Little et al. 1992).

BOUNDARY LAYER INFLUENCES

Boundary layer influences arise from changes in the characteristics of the boundary interface brought about by biological activity. When fluid flows over a surface, the stress at the surface is related to several factors of which one is the "roughness" of the boundary itself. A hydraulically smooth boundary will experience less surface stress than a rough surface under similar flow conditions. Therefore the smoothing of a surface by the development of a microbial film over the natural roughness of the sediment tends to reduce the range of stress experienced at the boundary interface (Plates 1D, 3C & F). In addition to this, EPS in solution within the boundary layer may also act to reduce friction stresses within the boundary layer through viscous effects.

RESPONSE OF THE SEDIMENT TO STRESS

Whereas abiotic sandy sediments respond in a broadly predictable fashion to applied erosive stress, cohesive sediments are more variable. As described by Paterson 1989, the surface of cohesive sediment may deform in an elastic manner in response to stress. When a threshold stress was exceeded the sediment "burst" throwing particles into suspension. This study supported earlier conclusions (Moore and Masch 1962) that the critical stress for erosion exceeds the stress required to cause subsequent erosion of the upper millimetre of sediment. This suggests that the interface was "armoured" with respect to immediately underlying layers, which may reflect greater microbial activity within the interface region. Parchure (1984), in laboratory experiments with cohesive beds in which microbial growth was encouraged, interpreted the influence of the microbes as a surface region in which shear strength increased with depth above a bed of uniform shear strength. The region of influence in these fairly uncompacted sediments was within the upper 5 mm of the sediment.

The behaviour of fine non-cohesive sediments may become more like their cohesive counterparts under the influence of biological action. Particles no longer act individually but become bound within an organic matrix at the surface of the sediment. Erosion may then occur by the shedding of aggregates from the bed (Grant and Gust, 1987). Scanning electron microscopy of natural or artificial particles shows the development of network or cohesive bonds between the grains (Vos 1988, Paterson 1990).

Where the surface of the sediment is completely overgrown by a microbial mat (Plate 3 C-F.) the response of the surface to stress depends on the characteristics of the mat (Neumann et al. 1970). The surface of well developed mat systems can withstand forces in excess of 20 N/m^2 as demonstrated by seasonal intertidal mat systems found on the island of Texel, Netherlands (Paterson, unpublished data).

PREDICTION OF SEDIMENT STABILITY

The contribution of interdisciplinary study of the erodibility of sediments is to improve our understanding and ability to simulate or predict the behaviour of natural sediments. The first task, to establish that the biology of the system had a significant influence on the erosion threshold of sediments, has been conclusively demonstrated in a number of laboratory and field studies of which a few have been discussed. A common expression of the influence of the biogenic effect is by a biological stabilisation coefficient. This usually takes the form:

$$S_b = \frac{F_{crit}\,(biol)}{F_{crit}\,(ster)} \qquad [2]$$

where S_b = Biological stabilisation coefficient
F_{crit} (biol) = Erosion threshold force of natural sediment
F_{crit} (ster) = Erosion threshold force of sterile (control) sediment

Values of Sb found by Manzenrieder (1983) ranged from 1.0 to 8.7. and demonstrated that the erosional behaviour of the non-cohesive sediment measured *in situ* was related not to sediment particle size but determined by the degree of biogenic stabilisation (Figure 2). Many other studies have demonstrated the influence of biogenic stabilisation (Heinzelmann and Wallisch 1991) on the behaviour of a number of different substrata (Table II) but such data is of limited practical use to modellers since each study is highly localised. This information must be further developed in order to gain more generic measures of the influence of biogenic stabilisation.

Table II: The effect of microbial communities on the critical threshold for the onset of erosion represented as the % increase in stability over control values.

Study	Date	Material	Organisms	% Stabilisation[1]
Non-cohesive Sediments				
Neumann	1970	Sand	Algae	500
Holland	1974	Sand	Diatoms	(100)
Rhoads	1978	Beads	Microbial	25-60
Grant (W)	1982	Sand	N.A.	100
Manzenrieder	1983	Sand	Bact/algae	300-770
Grant (J)	1987	Sand	Bacteria	25
Vos	1988	Sand	Diatoms	100
Dade	1990	Sand	Bacteria	100
Paterson	1991	Beads	Diatoms	400
Silt and Cohesive Sediments				
Holland	1974	Silt	Diatoms	(>350)
Rhoads	1978	Silt	Microbial	56
Parchure	1984	Kaolinite	Microbial	300
Paulic	1986	Natural	Autotrophs	200
Paterson	1989	Mud	Diatoms/Bact	200

[1]Figures derived from the relative increase in sediment stability over control measurements;
( ) = best estimate from available data

BIOLOGICAL PARAMETERS

If biogenic stabilisation can be correlated with a measurable biological parameter then two questions can be addressed for a particular situation. Firstly, is biogenic stabilisation likely to enhance the stability of this sediment and, secondly, what is the predicted threshold stress for the measured value of the predictive parameter? Some progress has been made toward this goal. Grant and Gust (1987) demonstrated a correlation between the quantity of photopigment and the critical erosion velocity for sediments colonised by purple sulphur bacteria. In a sense, this is an indirect measure as the binding action of the bacteria occurs through the cohesive effect of bacterial EPS. In more recent work, Dade et al. (1990) found that direct measurement of a component of bacterial EPS (uronic acid) was proportionally related to the erosion resistance of quartz grains. Further, the addition of purified EPS induced stabilising effects when re-applied to acid-washed sand. This work demonstrated that a biological parameter gave more accurate predictions for sediment behaviour than calculations based on particle properties.

Despite recent progress there are a number of reasons why care must be taken in the parameterisation of biological stabilisation using selected indicators. These stem from the complexity of the stabilisation process and the variability of biological systems. As outlined above, biogenic stabilisation occurs by a number of different mechanisms and the parameter used must reflect the dominant stabilising effect. This requires some knowledge of the biology of the system and more work is required to demonstrate suitable parameters for different microbial systems. The most promising candidates are photopigment content (a measure of autotrophic biomass) and some measure of EPS (cf uronic acid, colloidal carbohydrate, total carbohydrate).

BIOLOGICAL VARIABILITY

Biological effects are influenced by many factors including season (Frostick and McCave 1979), nutrient availability and ecological interactions such as grazing and competition. The nutrient status of the medium in experimental studies has been shown to influence the production of EPS, and hence sediment stability (Dade et al. 1990). The determination of the overall effects of biogenic stabilisation is not straightforward therefore, yet biological factors clearly mediate the behaviour of sediment systems. From further study it should be possible to develop seasonal functions to increase the accuracy of sediment transport models. The experiment of de Boer (1981) showed that disruption of biogenic stabilisation can cause large scale changes in the environment and work by Canadian scientists has shown a surprising ecological "cascade" effect (Daborn 1989, Daborn et al. submitted) linking the influence of wading birds, grazing amphipods, microbial autotrophs and sediment cohesion/stability. The latter study demonstrated that ecological effects have a direct effect on the cohesion of sediment particles and hence their erodibility.

FUTURE DEVELOPMENTS

The practical application of our increased knowledge is likely to be in the form of more accurate parameterisation for predictive models of sediment behaviour. As stated by Montague (1986):

> "The physical and chemical properties of sediments that contain a significant biotic community are radically different from those of abiotic sediments".

We have not yet reached the stage where enough data have been gathered to make widespread predictions about the influence of biogenic stabilisation on sediment transport. The development and use of *in situ* techniques is a major advance in this area which is already generating new data (Amos, in press). Laboratory work is confirming the potential of biological parameters (Dade et al). Seasonal surveys are required to measure the variability of biological factors (Little et al. 1992) for the development of new or adjusted models.

More exotic possibilities, such as the deliberate enhancement of microbial activity in order to prevent sediment erosion may seem far fetched but in fact may be realised as a natural consequence of the development and proper management of ecosystems. For example, since EPS produced by microbial metabolism enhances sediment stability, any alteration in conditions which influences microbial assemblages may have a feed-back effect on stability. Where photo-autotrophs can develop, a significant increase in sediment stability may be the result. The current development of *in situ* techniques and interdisciplinary studies should lead to significant advances and an answer to some of these question within the next few years.

ACKNOWLEDGEMENTS

The author would like to thank A.E Delo and the steering group of LISP-UK (Littoral Investigation of Sediment Properties) for advice on compiling Figure 1 which is taken from the LISP-UK proposal and G J C Underwood for Plate 2. Little kindly agreed to comment on the manuscript. The financial support of the Royal Society is also gratefully acknowledged.

REFERENCES

AMOS, C.L., VAN WAGGONER, N.A. & DABORN, G.D. (1988). The influence of subaerial exposure on the bulk properties of fine-grained intertidal sediment from Minas Basin, Bay of Fundy. *Est. Coast & Shelf Sci.*, **27**, 1-13.

AMOS, C.L. (1989). Sea carousel: A benthic annular flume. In: *Littoral Investigation of sediment properties*, G.R. Daborn (ed), Acadia Centre for Estuarine Research, Nova Scotia, Publication 17, 145-182.

AMOS, C.L., DABORN, G.R., CHRISTIAN, H.A., ATKINSON, A. & ROBERTSON, A. (1992). *In situ* erosion of fine-grained sediments from the bay of Fundy. *Marine Geology*. (submitted).

BLACK, K. (1989). *The in situ measurement of sediment erodibility: a review*. Unpublished Report submitted to ETSU, Department of Energy, Harwell, U.K.

CHRISTIAN, H.A., GILLESPIE, D. & AMOS, C.L. (1990). Results of a new device to measure the *in situ* shear strength of cohesive sediments, Minas Basin, Bay of Fundy. *Abstract Volume, 13th Int. Sedimentological Congress, Nottingham, U.K.*

COLES, S.M. (1979). Benthic microalgal populations on intertidal sediments and their role as precursors to salt marsh development. In: *Ecological processes in coastal environments: The first European Symposium of the British Ecological Society*, Jeffries, R.C. & Davies, A.J. (eds), Blackwell Scientific. Oxford, 25-42.

DABORN, G.R. (1989). Animal-sediment interactions. In: *Littoral Investigation of sediment properties*, G.R. Daborn (ed), Acadia Centre for Estuarine Research, Nova Scotia, Publication 17, 205-226.

DABORN, G.R., AMOS, C.L., BRYLINSKI, H.C., DRAPEAU, G., FAAS, R.W., GRANT, J., LONG, B., PATERSON, D.M., PERILLO, G.M.E. & PICCOLO, M.C. (1992). An ecological "cascade" effect: Migratory birds affect stability of intertidal sediments. *Limnol. Oceanog.* (submitted).

DADE, B.W., DAVIS, J.D., NICHOLS, P.D., NOWELL, A.R.M., THISTLE, D., TREXLER, M.B. & WHITE, D.C. (1990). Effects of bacterial exopolymer adhesion on the entrainment of sand. *Geomicrobiology Journal*, **8**, 1-16.

DE BOER, P.L. (1981). Mechanical effects of micro-organisms on intertidal bedform migration. *Sedimentology*, **28**, 129-132.

DECHO, A.W. (1990). Microbial exopolymer secretions in ocean environments: Their role(s) in food webs and marine processes. *Oceanogr. Marine Biology Annual. Review*, **28**, 73-153.

EDGAR, L.A. & PICKETT-HEAPS, J.D. (1984). Diatom locomotion. *Progress in Phycological Research*, **3**, 47-88.

FONSECA, M.S. (1989). Sediment stabilization by Halophita decipiens in comparison to other seagrasses. *Estuarine Coastal & Shelf Science*, **29**, 501-507.

FROSTICK, L.E. & MCCAVE, I.N. (1979). Seasonal shifts of sediment within an estuary mediated by algal growth. *Estuarine Coastal & Mar. Sci.*, **9**, 569-576.

GRANT, J. & GUST, G. (1987). Prediction of coastal sediment stability from photopigment content of mats of purple sulphur bacteria. *Nature*, **330**, 244-246.

HANSBO, S. (1957). A new approach to the determination of the shear strength of clay by the fall-cone test. *Proceedings Royal Swedish. Geotechical Institution*, **14**, 1-49.

HEINZELMANN, C. & WALLISCH, S. (1991). Benthic settlement and bed erosion. A review. *Journal of Hydraulic Research*, **29**, 355-371.

HOLLAND, A.F., ZINGMARK, R.G. & DEAN, J.M. (1974). Quantitative evidence concerning the stabilization of sediments by marine benthic diatoms. *Marine Biology*, **27**, 191-196.

KRAVITZ, J.M. (1970). Repeatability of three instruments used to examine the undrained shear strength of extremely weak saturated cohesive sediments. *Journal Sedimentary Petrology*, **40**, 1026-1037.

LAMBE, T.W. & WHITMAN. R.V. (1970). *Soil mechanics*. Standard edition. Wiley & Sons, New York.

LITTLE, C., PATERSON, D.M., CRAWFORD, R.M., UNDERWOOD, G.J.C. & MCARTHUR, J. (1992). *Algal stabilisation of Estuarine sediment.* Report to the Energy Technology Support Unit (DoE).

MANZENRIEDER, H. (1983). *Retardation of initial erosion under biological effects in sandy tidal flats.* Leichtweiss, Inst. Tech. University Braunschweig, 469-479.

MEHTA, A.J. & PARTHENIADES, E. (1982). Resuspension of deposited cohesive sediment beds. *Eighteenth Conference Coastal Engineering,* 1569-1588.

MEADOWS, P. & TAIT, J. (1989). Modification of sediment permeability and shear strength by two burrowing invertebrates. *Marine Biology,* **101**, 75-82.

MEADOWS, P., TAIT, J. & HUSSAIN, S.A. (1990). Effects of estuarine infauna on sediment stability and particle sedimentation. *Hydrobiologia,* **190**, 263-266.

MILLER, M.C., MCCAVE, I.N. & KOMAR, P.D. (1977). Threshold of sediment motion under unidirectional currents. *Sedimentology,* **24**, 507-527.

MOORE, W.L. & MASCH, F.D. (1962). Experiments on the scour resistance of cohesive sediments. *Jour. of Geophysical Res.,* **67**, 1437-1446.

MONTAGUE, C.L. (1986). Influence of biota on erodibility of sediments. In: *Lecture notes on coastal and estuarine studies.,* **14**, Mehta, A.J. (ed.), 251-269.

NEUMANN, A.C., GEBELEIN, C.D. & SCOFFIN, T.P. (1970). The composition, structure and erodibility of subtidal mats, Abaco, Bahamas. *Journal of Sedimentary Petrolology,* **40**, 274-297.

NOWELL, A.R., JUMARS, P.A. ECKMAN, J.E. (1981). Effects of biological activity on the entrainment of marine sediments. *Marine Geology,* **42**, 133-153.

NOWELL, A.R.M., JUMARS, P.A., SELF, R.F.L. & SOUTHARD, J.B. (1989). The effects of sediment transport and deposition on infauna: Results obtained in a specially designed flume. In: *Lecture Notes on Coastal and Estuarine Studies,* 31. Ecology of Marine Deposit Feeders, Lopez, G., Taghon, G. & Levington, J. (eds), 245-266.

NOWELL, A.R.M. & JUMARS, P.A. (1987). Flumes: Theoretical and experimental considerations for simulation of benthic environments. *Oceanography Marine Biology Annual Review,* **25**, 91-112.

OPPENHEIM, D.R. & PATERSON, D.M. (1990). The fine structure of an algal mat from a freshwater maritime Antarctic lake. *Canadian Journal of Botany,* **68**, 174-183.

PARCHURE, T.M. (1984). *Erosional behaviour of deposited cohesive sediments.* Ph.D. Thesis, University of Florida.

PATERSON, D.M. (1989). Short-term changes in the erodibility of intertidal cohesive sediments related to the migratory behaviour of epipelic diatoms. *Limnol. Oceanogr.,* **34(1)**, 223-234.

PATERSON, D.M. (1990). The influence of epipelic diatoms on the erodibility of an artificial sediment. *Proceedings of the 10th International Symposium on Living and Fossil Diatoms, 1988, Joensuu* , 345-355.

PATERSON, D.M., CRAWFORD, R.M. & LITTLE, C. (1986). The structure of benthic diatom assemblages: A preliminary account of the use and evaluation of low-temperature scanning electron microscopy. *J. Exp. Mar. Biol. Ecol.*, **96**, 279-289.

PATERSON, D.M., CRAWFORD, R.M. & LITTLE, C. (1990). Sub-aerial exposure and changes in the stability of intertidal estuarine sediments. *Estuarine Coastal & Shelf Science*, **34(1)**, 223-234.

PAULIC, M., MONTAGUE, C.L. & MEHTA, A.J. (1986). Influence of light on sediment erodibility. *Third International Symposium on River Sedimentation, Mississippi*, 1758-1764.

SCOFFIN, T.P. (1970). The trapping and binding of subtidal carbonate sediments by marine vegetation in Bimini Lagoon, Bahamas. *Journal of Sedimentary Petrology*, **40**, 249-273.

UNDERWOOD, J.C., MCARTHUR, J., PATERSON, D.M., LITTLE, C. & CRAWFORD, R.M. (1992) Biogenic stabilisation and altered tidal range. In: *The changing coastline. The 20th Estuarine and Coastal Science Association International Symposium.* (In press).

YOUNG, R.N. & SOUTHARD, J.B. (1978). Erosion of fine-grained marine sediments: Sea-floor and laboratory experiments. *Geological Society of America Bulletin*, **89**, 663-672.

VOS, P.C. DE BOER, P.L. & MISDROP, R. (1988). Sediment stabilization by benthic diatoms in intertidal sandy shoals. In: *Tide-influenced Sedimentary Environments and Facies*, De Boer, P.L. (ed), Reidel, 511-526.

4-5 ORGANIC RESIDUES AS A FACTOR IN THE PLASTICITY OF CLAY SOIL FROM THE SERC SOFT CLAY RESEARCH SITE, BOTHKENNAR

M.A.PAUL
Department of Civil and Offshore Engineering, Heriot-Watt University
Edinburgh, Scotland EH14 4AS

ABSTRACT

The UK Soft Clay Research Site at Bothkennar is founded on Postglacial silty clays of estuarine to shallow marine origin. X-ray diffraction analyses have shown that the clay fraction is dominated by illite, intermixed with quartz and felspar flour. Smectites appear to be absent. Although it might be expected that such a mineral suite would lead to a soil of low plasticity, it is in fact moderately plastic and the clay fraction is of medium to high activity (0.7 to 1.4).

The soil contains on average about 5% organic material in the form of both macroscopic (largely terrestrial) remains and soft-tissue residues. The distribution of this organic material is believed to be facies related. Engineering tests have shown that when the organic material is removed from the soil, its behaviour becomes consistent with that expected from the clay mineralogy. For this reason, it is suggested that the enhanced soil plasticity results from the presence of organic residues which modify soil behaviour by interaction with the clay mineral platelets. The processes by which this is effected are discussed using evidence from artificial mixtures of soil and organic additives.

INTRODUCTION

During recent investigations into the engineering geology of the Carse Clay at the SERC Soft Clay Research Site at Bothkennar, it has become apparent that this soil is more plastic than would be expected from its clay mineralogy alone. It is suggested this is attributable to the presence of organic residues, for example from the partial decomposition of soft bodied organisms or from a variety of *in vivo* secretions. Although geotechnical engineers have long recognised the importance of organic materials in the behaviour of engineering soils, the majority of studies have been concerned with the effects of macroscopic remains such as plant tissues (e.g.Skempton & Petley, 1970; Al-Khafaji,1981; Hobbs, 1986). The purpose of this paper is therefore to report the initial findings of an investigation into the effects of non-fibrous organic material on the plasticity of clay soil, using the Bothkennar Carse Clay as a case study.

GEOLOGICAL BACKGROUND

The SERC Soft Clay Research Site at Bothkennar lies near Grangemouth on the west bank of the Forth Estuary. The engineering geological setting has been described by Hawkins, Larnach, Lloyd and Nash (1989) and by Paul, Peacock and Wood (1992). The site lies within the outcrop of the Carse clays, which here occupies the upper part of the Quaternary sequence. This sediment is of Postglacial age and formed at a time of higher relative sea level in an extension of the estuary which inundated the Forth valley as far west as Stirling. It is considered to correlate with the Claret Beds of Browne, Graham and Gregory (1984) and thus to date from around 5,000 years BP.

Figure 1 shows the generalised geological profile at the site. It is possible to subdivide the sediments into three broad facies: bedded, mottled and laminated. In the first, original bedding surfaces and other primary sedimentary structures remain visible. In the second, these structures have been destroyed to a greater or lesser extent and the visible fabric is dominated by a mottling on a scale of a few millimetres to about one centimetre. In the third facies, silt laminae of a few millimetres thickness are interbedded with either bedded or mottled sediments at a spacing of a few centimetres.

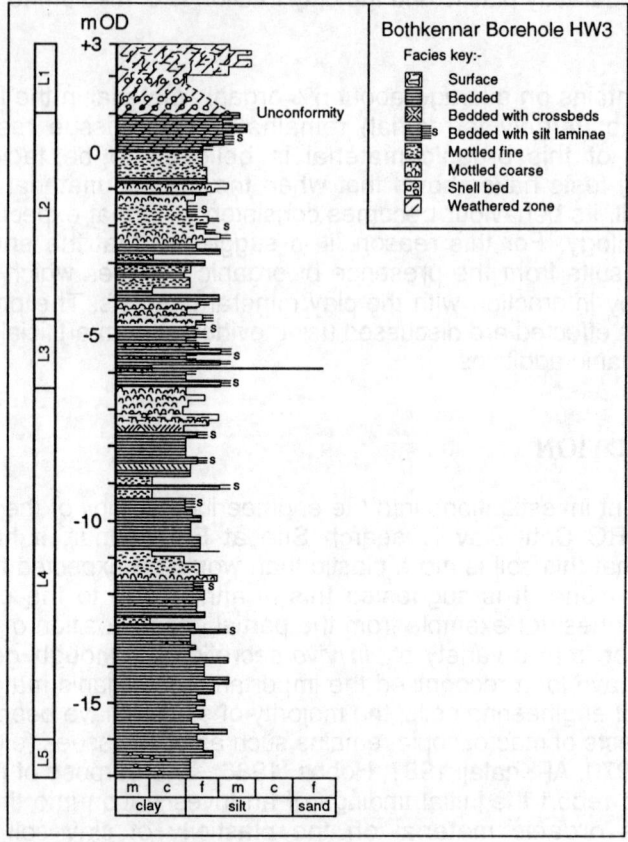

Figure 1: General geological profile at the Bothkennar site.

The sediments are interpreted as the product of quiet water deposition by tidal processes in water depths of around 10 to 20 m. The bedded facies is the 'normal' sedimentary product of this environment and the laminated facies is the result of periodically more energetic conditions. The mottled facies arises from bioturbation of the sediments of either of the other two facies.

SOIL COMPOSITION AND PLASTICITY

PARTICLE SIZE AND MINERAL COMPOSITION

The particle size distribution has been determined by both pipette and Sedi-Graph. The clay sized fraction usually comprises 35-50% of the material: the proportion of < 1 μm particles is greater in the mottled than in the bedded facies, but otherwise the two facies are similar in grading. Sediment from the bedded facies has an average particle size of 4-5 μm and a clay sized content of 35-45%. Sediment from the mottled facies is slightly finer, having an average size of around 3 μm and a clay sized content of 40-50%. In the laminated facies, individual laminae are composed of well sorted medium to coarse silt, the interbedded sediments being typical of their parent facies.

The mineralogy of the clay fraction has been determined by X-ray powder diffraction using variously air dried, furnace dried and glycolated samples. Figure 2 shows a typical diffractogram. The results suggest that the mineral assemblage throughout the sediment is of almost constant composition, although some variation occurs in the proportions of the minerals present due to changes in the silt:clay ratio. In the clay sized fraction the dominant minerals are quartz, illite and chlorite, with subsidiary feldspar, probable biotite and possible minor smectite. Kaolinite was recognised in minor amounts. The assemblage as a whole is typical of a glacially derived, rock flour dominated soil formed by erosion of bedrock in the southern Highlands.

ORGANIC MATERIAL

The majority of natural sediments contain organic material in some form. Their chemistry has been well studied and a wide variety of compounds identified (cf Degens and Mopper, 1975). The exact composition depends on the geological setting in which the sediment has accumulated and on the post-depositional history to which it has been subjected. In particular, a distinction can be made between those sediments whose organic component is mainly or wholly of terrestrial origin and those in which it is of marine origin. In the former, the organic material is derived largely from plant tissues whereas in the latter there will be some input from soft bodied planktonic and benthonic organisms. The contribution from the latter increases with increasing distance offshore.

Figure 3 profiles the organic content at Bothkennar as determined both by loss on ignition and by oxidation with hydrogen peroxide. As would be expected, there is a systematic difference between the results of the two methods. Loss on ignition normally overestimates the organic content, due

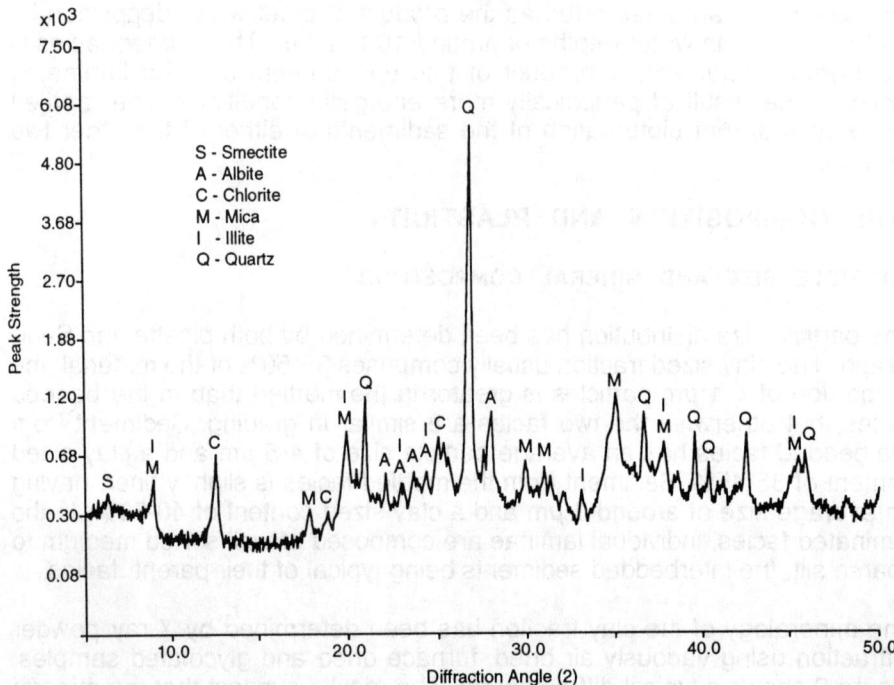

Figure 2: Typical diffractogram.

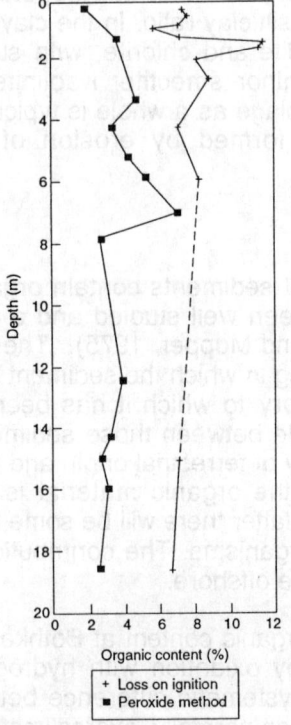

Figure 3: Profile of organic content.

to clay and mica dehydration and carbonate decomposition. The latter accounts for the spike at around 1.5m depth. Treatment by hydrogen peroxide normally underestimates organic content, due to partial oxidation and resistance of some tissues to this process. For this study, however, the peroxide method has been preferred in order to avoid possible damage to the clay minerals which might invalidate later geotechnical tests. As illustrated by Figure 3, the organic content (peroxide method) increases steadily with depth until a peak value of around 6% is reached at 7m depth: there is a sharp fall, and below this depth the value then remains at around 2-3%. Comparison with the lithofacies profile (Figure 1) suggests that the change at around 7m corresponds to the laminated/mottled facies boundary. Below this boundary, in the bedded/mottled region, the organic productivity may have been lower than in the mottled region above. This is an attractive concept as it relates the present organic content to the extent of bioturbation. Some caution is required however as the sample interval is fairly wide. A detailed study within part of the mottled facies (Hawkins, Lloyd and Nash, 1991) indicates some (albeit poor) correlation between the organic content (ignition loss) and percentage cover of mottles.

The nature of the organic material at Bothkennar is as yet unclear and is the subject of ongoing research. Only a small amount of detrital vegetable matter has been recovered and little organic material is visible on SEM micrographs. On the basis of other investigations on marine organic material (e.g. Degens and Mopper, 1975), it is considered likely that much of the organic material present is in the form of residues from the organisms responsible for the mottling. These residues are attached directly to the clay minerals. Their exact nature is unknown but it is reasonable to speculate that, as in other nearshore marine sediments, they will consist mainly of cellulose, polysaccharides, proteins and lipids, together with various breakdown and condensation products. The studies of Degens and Mopper (1975) for example have shown that 40-80% of organic material in marine sediments may be carbohydrate and protein.

SOIL PLASTICITY

Atterberg limits have been determined on soil at the natural water content (ie on soil which has not been dried) and on samples which have been treated with hydrogen peroxide to measure and remove organic matter. As can be seen in Figure 4, it is clear that the plasticity of the soil is lower after it has suffered a peroxide treatment. The results also suggest there is a broadly quantitative relationship between the plasticity of the untreated soil and its percentage organic content. This is illustrated further in Figure 5 which compares the activity of the samples (Skempton, 1953) and shows that the peroxide treatment also reduces the activity. It is believed that this effect is due to loss of the organic material and not to particle aggregation as the peroxide treatment was combined with ultrasonic disaggregation to ensure dispersion.

Figure 5 also suggests that before treatment the activity of the clay sized fraction is anomalous in terms of its composition. It would be expected that an illite/chlorite/rock flour mixture (cf. Figure 2) would possess a relatively

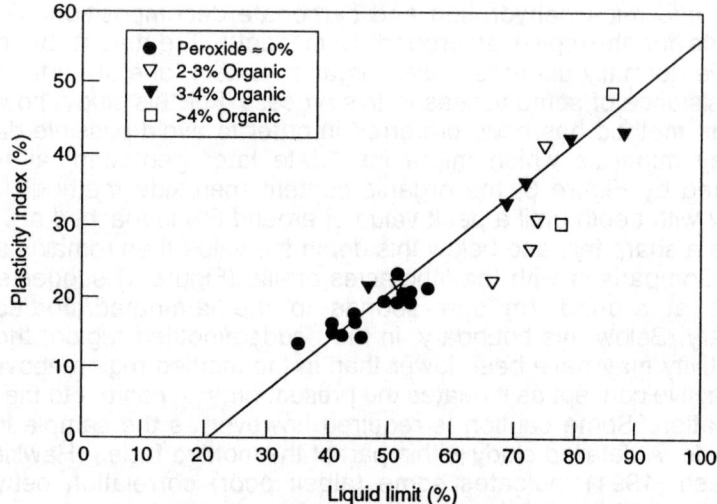

Figure 4: Plasticity chart for natural and treated samples.

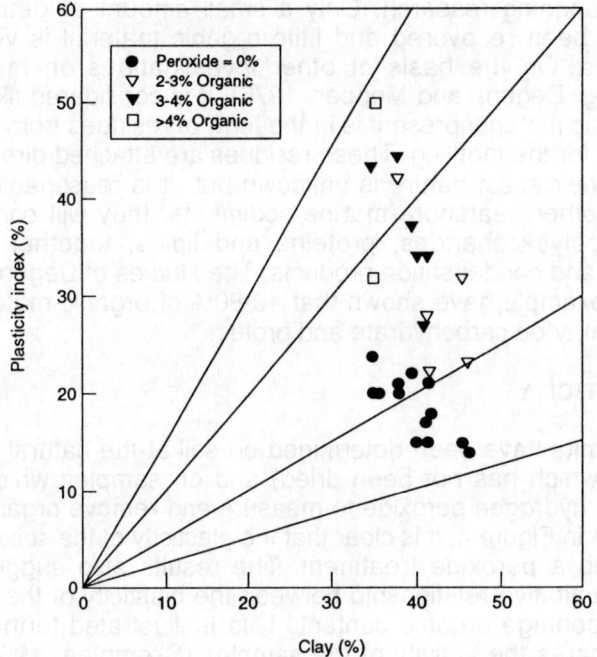

Figure 5: Activity chart for natural and treated samples.

low activity, perhaps around 0.5 (Skempton, 1953). However, as Figure 5 shows values ranging up to about 1.50, there appears to be a broad relationship with the percent organic content. Such high values of activity are normally due to smectite (swelling) minerals but from the XRD analysis these do not appear to be quantitatively significant at Bothkennar. In contrast, however, the activity of the peroxide treated soil is about 0.4, ie much closer to that expected from the mineralogy. Figure 6 shows a depth profile of activity before and after treatment. Comparison with Figure 1 shows that the (untreated) higher values are usually associated with sediment from the mottled facies, as would be anticipated from the organic percentage. After treatment the activity is almost uniform with depth, reflecting the uniform clay mineralogy.

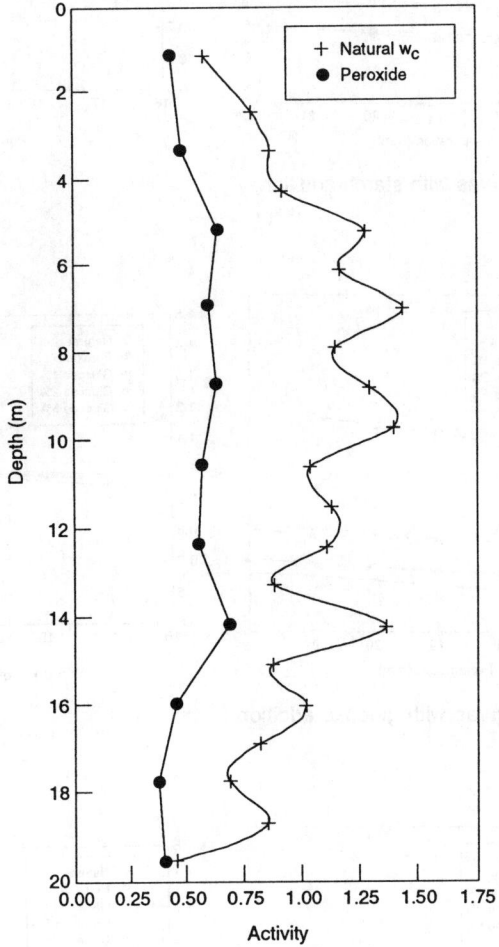

Figure 6: Depth profile of activity before and after treatment.

EFFECTS OF SIMPLE ORGANIC ADDITIVES

A series of tests has been conducted to determine the effects of simple organic compounds on soil plasticity. Glucose, soluble starch and glycerine were chosen in order to simulate possible degradation products of polysaccharides and lipids although other materials could equally well have been used. In each case the compounds were added in aqueous solution to

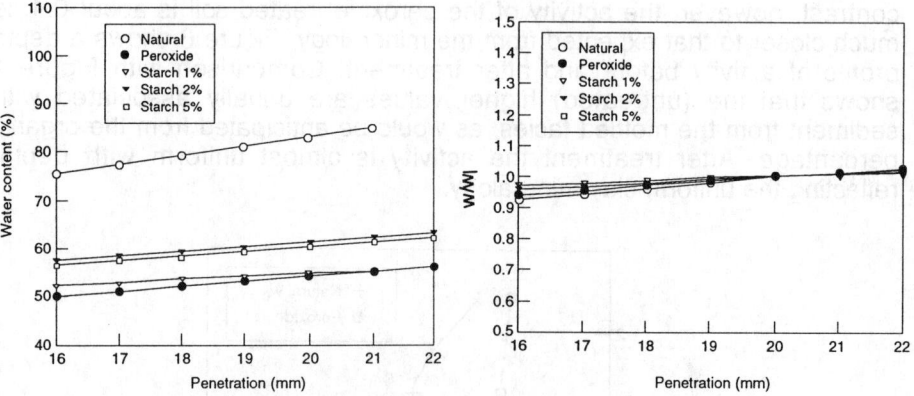

Figure 7: Flow curves with starch addition.

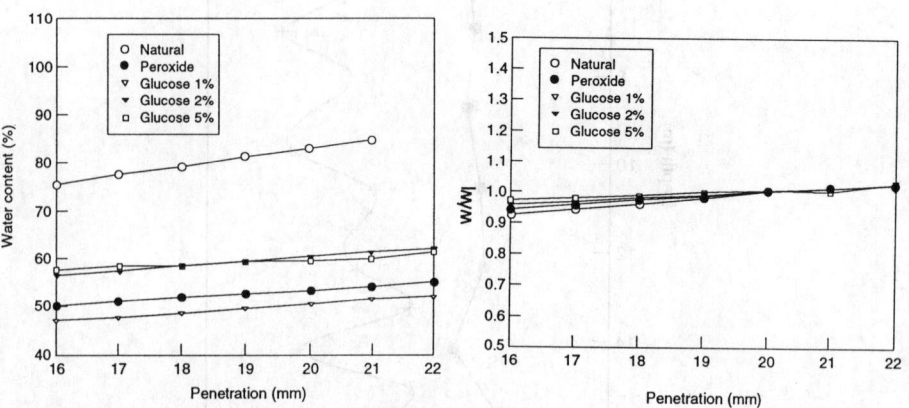

Figure 8: Flow curves with glucose addition.

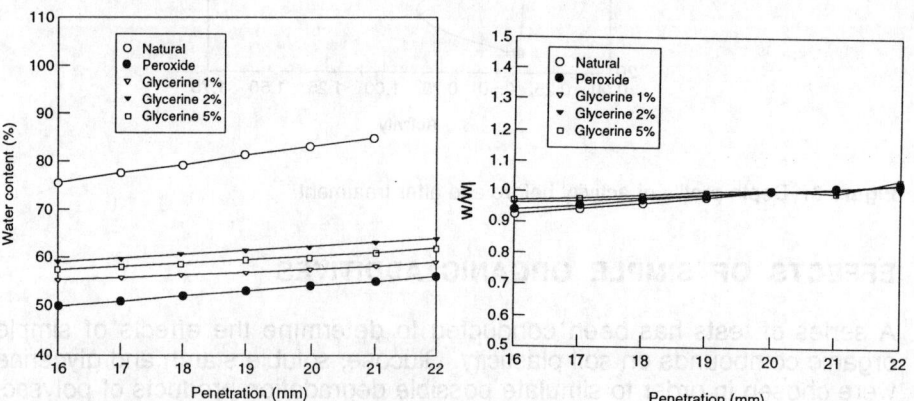

Figure 9: Flow curves with glycerine addition.

peroxide treated soil in strengths of 1%, 2% and 5% by weight. The liquid and plastic limits were then determined by conventional methods. Figures 7 to 9 show the results of cone penetrometer tests on these mixtures. These flow curves illustrate that in each case the water content at a given penetration is increased by the presence of the organic additive, and that the greatest increase is achieved at low concentrations. In absolute terms there is very little difference between the different compounds: all increase the liquid limit by about 10% water content. The figures also show the normalised flow curves. As expected, they are all almost on an unique curve; which suggests a common relationship between water content and yield strength similar to that seen in natural clay soils.

Figures 10 and 11 show the plasticity results. In all cases the organic additive increases the conventional plasticity of the soil although, as in the penetrometer tests, the continued increase in their effectiveness becomes less as higher concentrations are reached. Figure 10 shows that on the conventional plasticity chart the additives cause the soil to fall close to and below the A line, which corresponds to the behaviour of 'natural' organic soils. It is of note, however, that these samples fall some distance from the untreated Bothkennar clay. Figure 11 illustrates the same effect in terms of the clay fraction activity, which is raised from around 0.5 for the peroxide treated soil to around 0.75 at 2% additive. This is below the mean value of 0.9 for the untreated soil and well below the maximum values of around 1.25-1.50. The implication is that although the additives used here bring about some increase in clay activity, there are naturally occuring compounds, as yet unidentified, which are still more effective.

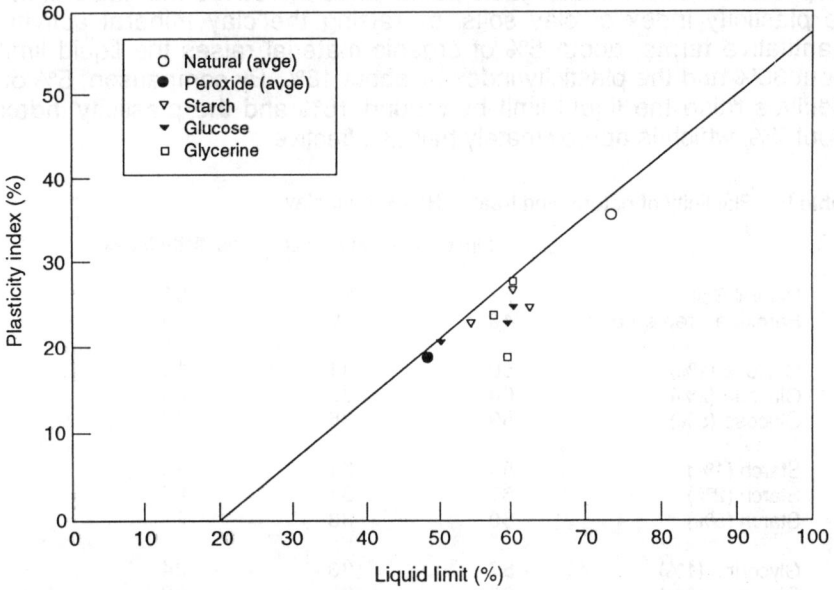

Figure 10: Plasticity chart - effect of additives.

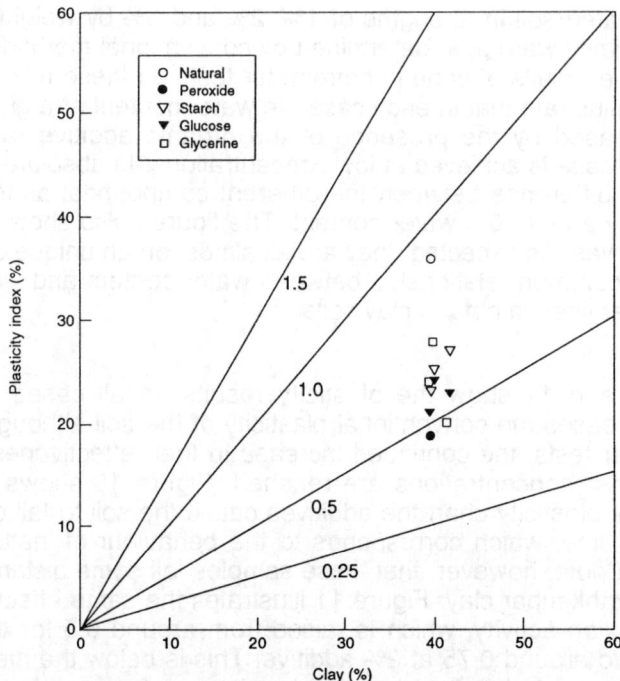

Figure 11: Activity chart - effect of additives.

DISCUSSION

The results are summarised in Table 1. Taken as a whole, they show that the presence of various organic compounds increases the liquid limit and the plasticity index of clay soils, by raising the clay mineral activity. In quantitative terms, about 5% of organic material raises the liquid limit by about 30% and the plasticity index by about 16%. By comparison, 5% of the additives raise the liquid limit by around 15% and the plasticity index by about 7%, which is approximately half as effective.

Table I: Plasticity of natural and treated Bothkennar clay.

	Liquid Limit	Plastic Limit	Plasticity Index
Natural Soil	73	37	36
Peroxide Treated Soil	48	29	19
Glucose (1%)	50	31	19
Glucose (2%)	60	35	25
Glucose (5%)	59	36	23
Starch (1%)	54	31	23
Starch (2%)	62	37	25
Starch (5%)	60	33	27
Glycerine (1%)	57	33	24
Glycerine (2%)	60	32	28
Glycerine (5%)	59	40	19

These results raise two fundamental questions: which compounds in the natural soil are able to effect this increase in plasticity, and by what mechanisms do they achieve it? Preliminary implications may be drawn from these results, although elucidation of the details of the chemical mechanisms involved are beyond the scope of this study; for possibilities see for example Mortland (1970), Theng (1974), Burchill, Greenland and Hayes (1981)).

The chemistry of the additives suggests that compounds which contain hydroxyl groups may be of importance, as they are capable of bonding to the clay surfaces or their adsorbed cations by a number of mechanisms. Other candidates are compounds with amino groups (eg proteins, peptides and amino acids), which may also co-ordinate with cations on the clay surfaces. In both cases, further water molecules can then bond to these molecules, increasing the weight of water that a given weight of the clay can hold. Taking plausible values for the molecular weights of these candidates suggests that around eight water molecules must be bonded in this way to each organic molecule. It is also possible that polymeric molecules such as polysaccharides may bind clay platelets together and so enable them to retain a higher weight of water at a given rheological behaviour. In all these cases the exact details of the mechanisms involved and of their effects on the rheology of the clay are not yet clear and are the subject of continuing research.

CONCLUSIONS

This work has shown that the plasticity of the Carse clay at Bothkennar is enhanced by the presence of non-fibrous organic material, which possibly originates from soft-bodied organisms. Candidate compounds include polysaccharides, proteins and lipids, together with their various degradation products. This material increases the liquid limit and plasticity index by around 30% and 15% respectively. The addition of related organic compounds (glucose, starch, glycerine) increases the liquid limit and plasticity index of pre-treated 'inorganic' clay by about half these amounts. It is suggested that the basic mechanism involves the bonding of polar molecules to the clay surfaces, which in turn allows the adsorption of further water.

ACKNOWLEDGEMENTS

This research arose from SERC grant GR/E/74748 into the engineering geology of the Bothkennar Soft Clay Research Site. Particular thanks are due to Mr H.Barras for generous field and laboratory assistance. Helpful discussions with H.Barras, B.F.Wood and D M Paterson are gratefully acknowledged.

REFERENCES

AL-KHAFAJI, A.W.N. & ANDERSLAND, O.B. (1981). Compressibility and strength of decomposing fibre clay soil. *Geotechnique*, **31**, 497-508.

BROWNE, M.A.E., GRAHAM, D.K. & GREGORY, D.M. (1984). Quaternary deposits in the Grangemouth area, Scotland. *British Geological Survey Report*, **16/3**.

BURCHILL, S., GREENLAND D.J. & HAYES, M.H.B. (1981). Adsorption of organic molecules. In: *The Chemistry of Soil Processes*, D.J. Greenland, M.H.B. Hayes (eds.), John Wiley and Sons, London.

DEGENS, E.T. & MOPPER, K. (1975). Early diagenesis of organic matter in marine soils. *Soil Science*, **119**, 65-72.

HAWKINS, A.B., LARNACH, W.J., LLOYD, I.M. & NASH, D.F.T. (1989). Selecting the location, and the initial investigation of the SERC soft clay test bed site. *Quarterly Journal of Engineering Geology*, **22**, 281-316.

HAWKINS, A.B., LLOYD, I.M. & NASH, D.F.T. (1991). The significance of mottling in the description of Quaternary estuarine soils. In: *Quaternary Engineering Geology*, A. Forster, M.G. Culshaw, J.C. Cripps, J.A. Little, C.F. Moon (eds.), Geological Society Engineering Geology Special Publication 7; 637-42.

HOBBS, N.B. (1986). Mire morphology and the properties and behaviour of some British peats. *Quarterly Journal of Engineering Geology*, **19**, 7-80.

MORTLAND, M.M. (1970). Clay-organic complexes and interactions. *Advances Agron.*, **22**, 75-117.

PAUL, M.A., PEACOCK J.D. & WOOD, B.F. (1992). The engineering geology of the Carse Clay at the National Soft Clay Research Site, Bothkennar. *Geotechnique*, (submitted).

SKEMPTON, A.W. (1953). The colloidal 'activity' of clays. *Proceedings. 3rd International Conference of Soil Mechanics and Foundation Engineering*, **1**, 57-61.

SKEMPTON, A.W. & PETLEY, D.J. (1970). Ignition loss and other properties of peats and clays from Avonmouth, King's Lynn and Cranberry Moss. *Geotechnique*, **20**, 343-356.

THENG, B.K.G. (1974). *The Chemistry of Clay-Organic Reactions*, John Wiley and Sons, New York, 343 pp.

4-6 CHEMICAL STABILISATION OF ESTUARINE ALLUVIUM

A.B. HAWKINS, J.C. LONSDALE & R.W. NARBETT
Engineering Geology Research Group, University of Bristol, Wills Memorial Building, Queens Road, Bristol BS8 1RJ.

ABSTRACT

The firm to stiff weathered crust on which the construction of buildings and highways across estuarine levels takes place, varies in thickness depending on the tidal regime of the area. Construction would be significantly aided if the thickness and strength of this crust overlying the soft, normally consolidated deposit could be thickened. The paper reports experiments using chemical additives to enhance the strength of the estuarine soils and concludes that addition of a mixture of $CaO.CaSO_4$ provides the most effective solution.

INTRODUCTION

Approximately three percent of Britain's land mass is covered by estuarine alluvial sediments. While in the past these were used only for agriculture, in the present century the increasing demand for industrial and housing development has resulted in a need to use these geotechnically marginal areas. Unlike most engineering situations, in coastal alluvial sites a firm to stiff weathered crust overlies sediments which are normally consolidated and frequently retain a very soft to soft consistency. The thickness of the desiccated crust is influenced mainly by the depth to which weathering has taken place within the soft sediments. This physical process results in the strengthening of the deposits by point contact cementation by mobilised iron together with calcium carbonate.

Although the rate of sea level rise has decreased progressively in the last five to six thousand years (Hawkins, 1984), the main control on the depth to which weathering has affected the sediments is the relationship between present and recent past ground surface levels and associated ground water regimes. The difference between the ground surface elevation and the level to which water is retained within the alluvium is controlled by the tidal cycle. In the case of the Severn Estuary, the significant tidal levels for Avonmouth are given in Figure 1. With the twice daily tides always inundating levels above the mean high water neap tide (MHWN), below this level the sediments will remain fully saturated. Evidence to date suggests that the saturation level is approximately half way between MHWN tides and the level of the mean high water spring (MHWS) tides. As seen in Table I, such a level at Avonmouth would be approximately 5 m AOD. Since much of the sediment accretion has reached approximately 7.5 m AOD on the estuary side of the sea wall, the strength profile has been modified by chemical reactions which have taken place above the saturation level. Here, in the unsaturated zone, oxidation and calcium dissolution has occurred allowing the movement and re-precipitation of irons and calcium, increasing point contact cementation (Hawkins, 1984). Clearly, in areas such as the Severn Estuary where an aerated zone up to 2.5 m thick may be present, continued weathering results in the formation of a thick crust. In other estuaries where the tidal ranges are much smaller (Table I) the depth of the weathered, strengthened crust would be anticipated to be correspondingly less.

The presence of many housing and light industrial complexes on alluvial lowlands confirms that construction can take place on the weathered, stiffened crust although for heavier structures pile foundations are required. This paper examines the possible methods of thickening/stiffening the alluvial crust which could reduce the cost of engineering development on these coastal, alluvial lowlands. Such a scheme would have significant financial advantages, although care would have to be taken to ensure the long term integrity of the induced strength.

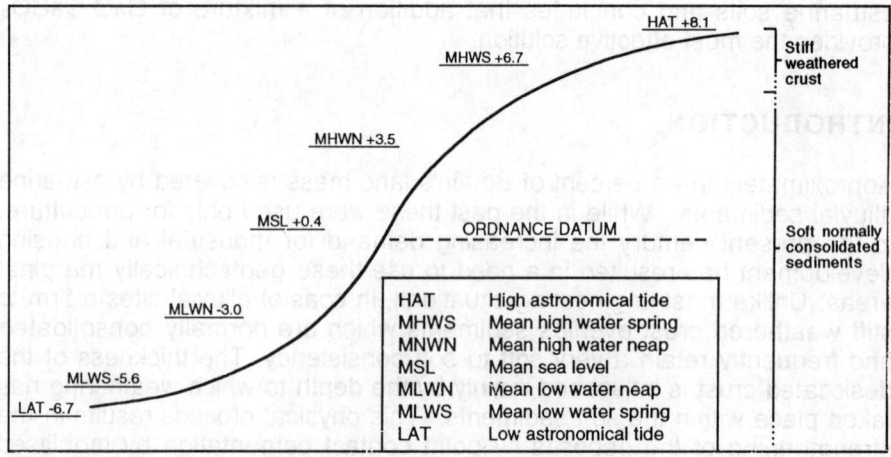

Figure 1: Levels of various tidal stages at Avonmouth taken from the Admiralty Tide Tables and the nature of the corresponding alluvium.

Table I: Various tidal heights relative to Ordnance Datum for a number of ports around the coasts of Britain and the English Channel.

	Tidal Range (m)	MHWN	MHWS	HAT	half MHWN to half MHWS	half MHWS to HAT	Difference
Avonmouth	14.8	4.5	6.7	8.1	5.6	7.4	1.8
Southampton	4.9	0.96	1.76	2.16	1.36	1.96	0.6
Shearness	6.5	1.9	2.8	3.5	2.35	3.15	0.8
Immingham	8.1	1.9	3.4	4.1	2.65	3.75	1.1
Rosyth	6.4	1.75	2.85	3.45	2.30	3.15	0.75
Liverpool	10.5	2.47	4.37	5.37	3.42	4.87	1.45
Wilhelmshaven	5.1	1.4	1.9	2.3	1.65	2.1	0.55
Calais	7.2	1.83	3.03	3.43	2.43	3.23	0.8

CHEMICAL STABILISATION

One method of improving the geotechnical properties of the near-surface alluvial soils is the process of chemical stabilisation. According to Mitchell (1981) the main objectives of mixing chemical additives to soils are the improvement or control of:-
    a)   volume stability,
    b)   strength and stress-strain properties,
    c)   permeability and
    d)   durability.

LIME

The most commonly applied chemical additive is calcium oxide (lime) which has been used to strengthen soils in many parts of the world since at least Roman times. Extensively adopted in the USA and Scandinavia, lime stabilisation has not been widely developed in the UK although it has recently been accepted for highway design.

With the addition of lime the calcium ion replaces other metallic cations on the surface of the clay mineral. Grim (1968) cites the order in which cations are commonly replaced as $Na^+ < K^+ < Ca^{2+} < Mg^{2+}$. As a result of both clay mineral structure and chemistry therefore, the nature of the cation makes little difference in kaolinitic soils but in the case of smectite-rich soils the exchange of cations is more significant.

Bell and Coulthard (1990) considered that all clay minerals are attacked by lime but that those containing a higher proportion of silica are likely to experience a more pronounced reaction. As a consequence, the two layer clay minerals are less susceptible than the three layer clay minerals in which lamellae expose silica faces on both sides. However, a silica face cannot be considered "available" for reaction if it is bound to a similar face by ions which are not readily exchangeable. Consequently illite and chlorite are less reactive with lime than is montmorillonite.

The addition of lime results in the flocculation of clay minerals due to either cation exchange or the reduction in the thickness of the hydrous double layer which surrounds clay minerals. Hilt and Davidson (1960) suggest that both mechanisms alter the density of the electrical charge around the clay particles, strengthening the attractive forces which in turn increases the bonding of the clay minerals. These authors state that when >1% lime is added a more permanent improvement in the soil strength is produced. In addition it is considered that calcium silicate and aluminate which are formed as a consequence of the longterm breakdown of clay minerals through pozzolanic reactions, assist flocculation by bonding the soil particles.

It is not surprising that the particle size of the original soil is significant in the success of lime stabilisation, not only because of the importance of the clay mineral content but also as it affects the rate and extent of diffusion. Van Impe (1989) states that at least 35% of the soil mass must be of clay or silt grade in order for the addition of lime to produce a significant strengthening. Kézdi (1979) produced a figure showing the soils he considered to be suitable for improvement by lime stabilisation (Figure 2), with sandy silty clays being the most likely to benefit from such treatment.

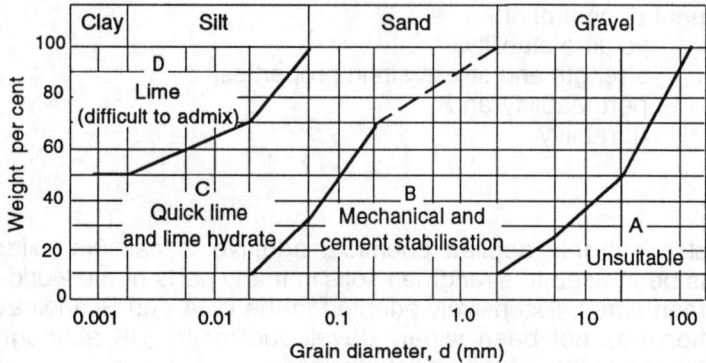

Figure 2: Particle size distribution classes for the suitability of lime stabilisation, after Kézdi (1979).

Organic matter has a high exchange capacity and is believed by Gillott (1987) to compete with the clay minerals for the available calcium ions in the soil. Dumbleton (1962) suggests that the precipitate produced from the reaction of lime and the organic matter surrounds the lime particles thus preventing further reaction within the soil. Other authors, such as Aly Sabry and Parcher (1979), believe that when the organic content is < 1.5% its presence does not significantly hinder the reaction between the soil and the lime; such a low organic percentage is typical of estuarine alluvium except where peat layers are present.

There are three main methods by which lime can be applied to a soil for the purpose of subsurface stabilisation.

i) Lime columns were developed in Sweden using a specially shaped auger (Broms and Boman, 1979). When the auger has reached the maximum required depth its rotation is reversed and quicklime is forced into the soil by compressed air as the auger is withdrawn. In this way a column of mixed lime/soil is created.

ii) Lime piles are another commonly used technique for lime stabilisation. These are created by drilling or using driven tubes to create voids into which quicklime can be placed and compacted (Tsytovich, Abelev and Takhirov, 1971). The benefits of this method include:

    a) The process of inserting the tubes initially compacts the adjacent soil.

    b) As pore water is attracted into the pile, the lime hydrates causing a volume increase. Bell (1988) and Rogers and Bruce

(1991) have indicated that the effective diameter of the pile may increase by 30-70%, resulting in an enhancement of the strength of the soil as a consequence of the induced lateral stresses.

c) The heat generated by the hydration of the lime causes some of the surrounding pore water to evaporate, reducing the saturation of the soils.

iii) Lime slurry or lime-fly ash slurry injection processes are used to treat expansive clays and other soft soils in the USA. Hydrated or anhydrous lime is mixed with water up to a maximum concentration of 40% solids (Rogers and Bruce, 1991) and then injected into the ground under pressure to depths of up to 40 m (Joshi, Natt and Wright, 1981). The injected slurry preferentially penetrates natural planes of weakness within the soil, such as bedding, laminations and fissures. If the injection points are close enough an interlocking framework of lime seams is created. Such seams stabilise the adjacent soil by contact with and diffusion of lime and hence act as moisture barriers, reducing the potential for shrink/swell. Excess slurry at the surface can be mixed into the upper soil and compacted, providing a weatherproof capping in addition to a "stabilised raft" which can redistribute swell pressures in the event of differential movement (Bell, 1988).

EFFECTS OF LIME ADDITION TO REMOULDED ESTUARINE ALLUVIUM

Samples of intertidal sediments from Northwick Warth between Avonmouth and Aust (ST 560 887) were considered to be representative of the soft, laminated deposits beneath the stiffened weathered crust, ie the level where ground treatment could be most beneficial. The samples were remoulded in an electrical mixer for one minute and various percentages of lime (by weight) were added to the remoulded samples before mixing again for one minute. The lime used was anhydrous calcium oxide manufactured by BDH Chemical Limited. Anhydrous lime was chosen for the experiments as the drying effect in estuarine material due to the exothermic hydration of quicklime would be beneficial in the stabilisation process. The specimens were then remixed to ensure an even consistency of the chemical admixture and to simulate the processes involved in the emplacement of a lime column.

Initially it was considered that aluminium foil and cling film could be used to maintain the moisture content of the samples but within five hours the aluminium foil placed over lime stabilised material was both discoloured and disintegrating. Clearly a reaction was taking place between the foil and the treated samples. A possible explanation of this affinity of lime stabilised samples for aluminium foil may be found in the work of Queiroz de Carvalho (1981) who stated that amorphous silica and aluminium are the most important soil properties affecting soil-lime reactions. As a consequence, the soil lime mixtures were placed into plastic tubes 100 mm in diameter and 70 mm high which were sealed with a screw top and tape to provide an air-tight environment. The samples were stored at a room temperature of about 22°C.

The moisture contents of all samples were tested at intervals as set out in Table II. As expected, those with the higher percentages of anhydrous lime showed the most significant reduction in moisture content.

Table II: Moisture contents of different alluvial-lime mixtures and their variation with time.

Time	Moisture content (%)					
Initial Sample	66.4	65.2	58.8	64.9	63.1	62.4
Mixtures	No CaO	+1% CaO	+2% CaO	+4% CaO	+6% CaO	+8% CaO
1 hr	66.4	62.3	57.7	57.3	53.3	49.6
1 day	66.3	63.2	57.9	57.2	53.4	49.6
2 day	66.6	62.3	57.9	-	-	51.2
4 day	66.4	62.7	-	57.0	52.9	52.0
6 day	-	-	59.0	56.7	53.0	-
8 day	67.1	63.1	57.5	55.5	52.7	53.5
16 day	64.9	62.8	58.1	56.4	54.4	52.5

As with the moisture content, the effect of the addition of lime on the Atterberg limits was also seen to be almost instaneous.

The liquid limit of alluvial samples showed an initial increase and then appeared to decrease progressively with time while the plastic limit increased for all samples, supporting the work of Hilt and Davidson (1960). Bell and Tyrer (1987) noted similar increases in liquid limit with kaolinite and fine grained quartz which Croft (1964) suggested was probably due to the action of hydroxyl ions modifying the affinity of the surfaces of clay particles for water. With expansive clay minerals, however, there is a decrease in the liquid limit as calcium is absorbed into the clay mineral structure. It is of note that estuarine samples, dominantly illitic with subsidiary kaolinite, chlorite and smectite, show a similar behaviour to stabilised London Clay. There is a pronounced reduction in the plasticity index due to diagenetic flocculation and as seen in Table III the addition of 1-4% lime can cause significant changes in plasticity.

Table III: The variation of plasticity index with time of alluvial-lime mixtures.

Time	Plasticity Index (%)				
	1% CaO	2% CaO	4% CaO	6% CaO	8% CaO
Initial Sample	30	28	22	24	24
1 day	30	27	19	21	18
2 day	31	-	19	-	-
4 day	27	28	19	19	19
8 day	28	27	18	21	19
16 day	25	26	20	20	16
32 day	25	26	16	16	15
64 day	24	25	20	15	15
128 day	24	25	17	16	15

As no sophisticated cone penetrometer system was available, the change in strength of the admixtures was assessed using the fall cone test which can be correlated to shear strength by the formula produced by Hansbo (1957). This method allows rapid and multiple test measurements on small scale samples. The penetrations of the 30°/80 gm cone with time and varying CaO concentrations are given in Table IV.

Table IV: Cone penetrations of alluvial-lime mixtures expressed as a percemtage of the penetration (20.5 mm) obtained in untreated remoulded samples.

Time	Relative cone penetration in treated material (%)					
	0% CaO	1% CaO	2% CaO	4% CaO	6% CaO	8% CaO
1 hour	93.8	59.8	49.8	47.9	42.9	47.9
1 day	90.0	52.1	41.5	39.4	36.6	38.8
2 day	89.1	46.1	40.8	-	-	36.8
4 day	90.2	43.8	-	33.3	27.3	33.4
6 day	-	-	34.7	29.5	27.3	-
8 day	90.2	40.4	31.3	28.7	25.7	28.5
16 day	89.4	37.1	26.5	25.0	25.1	28.5

It can be seen that there is a significant and almost instantaneous decrease in the cone penetration following the addition of lime. While the depth of penetration into the untreated sample decreased by only 4% from that noted for the initial material over a period of sixteen days, the samples with 1% CaO decreased by 63% and those with 6 and 8% CaO recorded larger decreases in penetration of 75 and 71% respectively. Compared with the initial, untreated sample, it is clear that the maximum effect is achieved with the addition of a relatively small proportion (4-6%) of CaO.

Calculating the shear strength from fall cone results using the constants provided by Wood (1985), the strengths obtained with the 30°/80 gm cone are given in Table V. From this table it can be seen that even with 1% CaO there is a pronounced increase in strength with time while the maximum increase is still related to the addition of 6% CaO. As a consequence specimens were prepared with 6 and 8% lime and tested using the 33 mm hand shear vane. The results are plotted in Table VI and it can be seen that the shear strengths are lower than the calculated values from the fall cone tests. Some anomalous results were recorded which have not yet been explained but it is likely that the shear vane penetrated too far into the soil and there was an inter-reaction between the shearing soil and the base of the container.

To establish the extent to which lime diffusion took place, undisturbed samples were placed into similar sized containers and a 10 mm diameter cork borer used to remove a central column from the soil. Lime was placed into this 10 mm cylinder and tamped gently to produce an effective lime pile.

After two weeks the sample was split open and the conductivity of the soil away from the lime pile was measured together with the pH. As seen from

Figure 3, the two pH readings close to the lime column were just above 9 while those away from the column were slightly above 8, similar to that for untreated estuarine alluvium. The conductivity measurements indicate a significant increase between about 4 and 16 mm from the lime column.

Table V: Calculated shear strengths of alluvial-lime mixes obtained from fallcone tests using the constants reported by Wood (1985).

Time	Calculated Shear Strength (kPa)				
	1% CaO	2% CaO	4% CaO	6% CaO	8% CaO
1 day	-	4.3	7.3	17.1	21.5
2 day	9.2	-	9.2	-	-
4 day	11.8	10.6	12.1	20.7	30.6
8 day	10.8	11.2	13.6	25.2	21.9
16 day	20.3	13.9	12.6	29.6	49.5
32 day	17.5	16.5	20.9	23.5	29.0
64 day	19.8	13.3	15.6	35.3	23.2
128 day	22.5	17.9	26.7	32.5	25.0

Table VI: Shear strengths of alluvial-lime mixes obtained by hand vane testing.

Time	Shear strength (kPa)	
	6% CaO	8% CaO
1 day	-	14
4 day	13	15
8 day	14	12
16 day	15	21
32 day	13	14
64 day	17	14
128 day	15	18

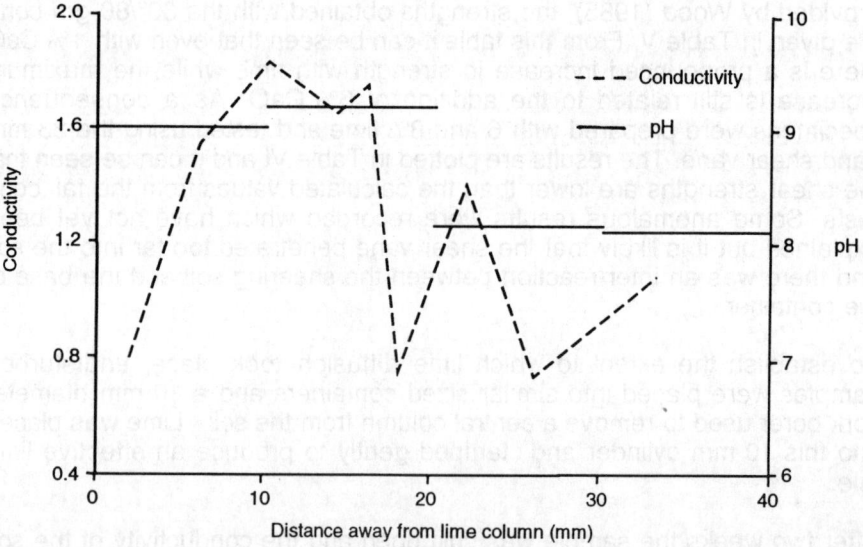

Figure 3: Variation of conductivity and pH away from a small scale lime pile.

This suggests that the influence of the lime reaction extended to between 15 and 20 mm from the edge of the lime source. Clearly this method of determining the width of the diffusion zone is not precise enough and consequently the electron microprobe was used to differentiate changes in soil conditions.

The graph of calcium levels determined by the electron microprobe is plotted in Figure 4 and indicates a change at a sampling interval of 20, approximately 10 mm from the original cylinder. This method appears to provide a better system of monitoring the diffusion of chemical additives such as lime.

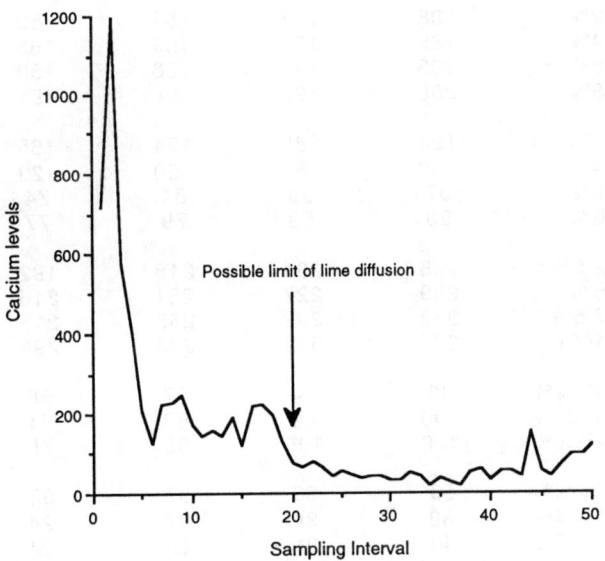

Figure 4: Profile of calcium levels away from a lime pile as measured by an electron microprobe in a 25 mm long alluvial sample.

OTHER CHEMICAL ADMIXTURES

The number of techniques used in the chemical stabilisation of soils is extensive. The most successful compounds reported in the literature were chosen as potential alternatives to lime for the stabilisation of estuarine alluvium. The additives and the change in fall cone penetration are given in Table VII. The main departure from the established practice in the literature was the use of anhydrous calcium sulphate; previous work having used either hydrated or semi-hydrated gypsum. However it was decided that the ability of anhydrous compounds to take in water would be beneficial to the stabilisation process. To ensure that the control specimen best represented the change compared with the chemically mixed soils, the inert material polytetrafluorethylene (PTFE) was added to create a secondary control. When combinations of KCl and $CaSO_4$ additives were used in conjunction with lime, the percentage lime in the mixture was kept

Table VII: Fallcone results of remoulded estuarine alluvium mixed with various chemical additives.

		\multicolumn{5}{c}{Cone penetration (1/10th mm)}				
		2 days	4 days	8 days	16 days	32 days
CONTROLS						
Remoulded Alluvium		194	175	167	155	149
PTFE	4%	207	223	196	178	162
CHEMICAL ADDITIVE						
CaO	2%	59	63	53	53	49
	4%	59	54	55	54	49
	6%	52	50	45	49	44
	8%	36	28	36	38	33
KCl	2%	188	161	150	150	141
	4%	188	197	169	165	164
	6%	193	146	166	150	141
	8%	206	199	20	191	156
$CaSO_4$	2%	194	188	184	165	149
	4%	152	151	150	129	123
	6%	97	99	84	74	66
	8%	93	93	79	77	68
OH-Al	2.5%	248	234	218	192	179
	5%	289	227	251	216	214
	7.5%	333	292	255	215	213
	10%	375	332	338	296	203
CaO:KCl	4%:2%	90	84	77	68	54
	4%:4%	100	89	82	71	51
	4%:6%	109	105	89	75	54
$CaO:CaSO_4$	4%:2%	53	36	37	32	23
	4%:4%	38	30	27	24	16
	4%:6%	46	31	26	22	13
CaO:OH-Al	2%:5%	115	103	91	84	84
	4%:10%	151	136	125	108	101
	8%:20%	110	146	157	138	139

constant in order to assess the influence of the other component on the soil stabilisation.

EXPERIMENTS

In these experiments the samples were prepared in a similar manner to those described above except the mixing time in each case was three minutes. Samples were placed in containers 160 mm in diameter and 90 mm in height and then sealed using cling film and tape. By using the larger samples it was possible to undertake a number of fall cone or hand shear vane tests in the same sample. Every attempt was made to open and re-seal the samples as quickly as possible but as seen from the results in Table VII, there was some change in the controls with time which may be partly attributed to the desiccation of samples.

a) CaO

As seen in Table VII there is a decrease in penetration with both increased proportion of CaO and with length of curing. The cone penetration of the 2% lime mixture after two days (5.9 mm) is clearly a fast positive response compared with that of the remoulded alluvium (19.4 mm). With 8% CaO the mixture cured for two days has a penetration of only 3.6 mm dropping to 3.3 mm after 32 days. The 3.3 mm penetration is only 20% of that attained for the remoulded control, clearly showing a significant increase in strength.

b) KCl

Moum, Sopp and Løken (1968) used potassium chloride in an investigation of the stabilisation of quick clays. In a similar manner Eggestad and Sem (1976) reported the stabilisation of soil beneath embankments in the Oslo area.

The addition of potassium chloride to the alluvial samples resulted in only a small decrease in cone penetrations compared with the controls. Clearly the lack of strength enhancement indicates that this chemical additive is not a viable alternative to lime for the stabilisation of estuarine alluvium.

c) $CaSO_4$

Although there is little difference between the 2% sample and the control results after two days, for the same curing time the 8% sample has only a half the penetration compared to the control and a similar relationship is recorded after 32 days. The decrease in cone penetrations for $CaSO_4$ mixes is not as significant as that obtained with CaO admixtures. However as the decrease appears to be progressive, over the long term similar strength improvements may be achieved.

d) OH-Al

Foster and Gazzard (1975) investigated the influence of polymeric hydroxy aluminium cations on the consolidation, shear strength and index limits of clays. Soils stabilised with hydroxy aluminium were shown by Bryhn et al (1983) to have equivalent increases in strength to those stabilised by quick lime.

Hydroxy aluminium produced the worst results of the tests with no short-term beneficial stabilisation. The cost of applying this chemical additive to large scale stabilisation problems on estuarine alluvium would also be prohibitive.

e) CaO:KCl

Improvements in strength were not as significant as those obtained by lime alone. The detrimental effect of potassium chloride on the ability of lime to stabilise estuarine alluvium is confirmed with greater penetrations achieved by mixtures containing a higher ratio of potassium chloride.

f) $CaO:CaSO_4$

Soils stabilised by gypsum and lime mixtures have been shown by a number of authors to achieve higher strengths than those stabilised by lime alone (Nieminen, 1978; Holm, Trank and Ekstrom, 1983; Kujala, 1983; Kujala and Nieminen, 1983). A consequence of the gypsum/lime/soil reaction is the formation of ettringite. This develops strong expansion pressures which can significantly improve the soil strength.

This appeared to be the most effective admixture. After two days with a 2:1 ratio of $CaO:CaSO_4$ the penetration was less than a third of the cone penetration obtained from the remoulded sample which decreased with time. The 1:1 mixture recorded only 20% of the penetration measured for the controls after two days, decreasing to 11% after 32 days. After two days the 2:3 ratio showed a penetration of only 4.6 mm compared with 19.4 in the remoulded control sample while after 32 days the penetration was only 1.3 mm compared with 14.9 mm for the control. The improved stabilisation of alluvial samples can be observed by directly comparing either the 2:1 mixture with the results of the 6% CaO or the 1:1 mix with 8% CaO.

g) CaO:OH-Al

No record of previous work using this mix has been found in the literature but it was included in these experiments for the sake of completeness. Samples containing both lime and hydroxy aluminium showed some improvement compared with the controls although these were smaller with increasing proportions of hydroxy aluminium. In practice this combination has no advantage as a potential stiffener of soft alluvial soils. As seen from Table VIII there is an improvement but the enhancement is much less than that noted for $CaO:CaSO_4$.

Hand shear vane strengths were measured directly on samples of $CaO:CaSO_4$ for 8, 16 and 32 days for the same mixture ratios. These are presented in Table VIII and show a marked increase in strength both in time and against the remoulded control sample. The strength improvement for the 2:3 mixture is approximately double that attained by the 2:1 mix. Again direct comparisons can be made with admixtures containing only CaO, which illustrate that the strengths of the $CaO:CaSO_4$ mixture are in excess of three times those achieved by the addition of lime alone. In addition, as the $CaO:CaSO_4$ showed an appreciable increase after 32 days it was re-read after 64 days and values of 50, 90 and 100+ kPa obtained (Figure 5).

When the $CaO:CaSO_4$ mixtures were tested for changes in index properties (Table IX), the results showed more significant increases in both the liquid and plastic limits than those obtained with CaO admixtures although the lower plasticity indices of the stabilised mixes were similar.

Table VIII: Hand vane results of remoulded estuarine alluvium mixed with various chemical additives (values >30 kPa measured with 19 mm vane).

		Shear Strength (kPa)			
		2 days	8 days	16 days	32 days
CONTROLS					
Remoulded Alluvium		0.7	-	1.0	1.0
PTFE	4%	0.8	-	1.0	0.9
CHEMICAL ADDITIVE					
CaO	2%	8.0	-	8.9	7.9
	4%	7.4	-	8.5	8.3
	6%	10.1	-	10.2	10.6
	8%	14.7	-	13.5	14.1
KCl	2%	1.6	-	1.7	1.6
	4%	1.0	-	1.2	1.3
	6%	1.4	-	1.2	1.5
	8%	1.2	-	1.0	1.2
CaSO4	2%	-	1.0	1.3	1.3
	4%	-	1.7	1.8	1.7
	6%	-	3.6	3.9	3.7
	8%	-	4.6	4.7	4.6
OH-Al	2.5%	0.6	-	0.8	0.7
	5%	0.3	-	0.5	0.6
	7.5%	0.2	-	0.4	0.5
	10%	0.2	-	0.3	0.3
CaO:KCl	4%:2%	-	5.6	7.8	11.5
	4%:4%	-	5.0	7.8	11.4
	4%:6%	-	4.0	6.3	9.9
CaO:CaSO4	4%:2%	-	19.8	28.1	36
	4%:4%	-	30	32	53
	4%:6%	-	33	42	70
CaO:OH-Al	2%:5%	3.6	-	4.3	4.2
	4%:10%	1.7	-	2.4	2.3
	8%:20%	1.6	-	1.5	1.6

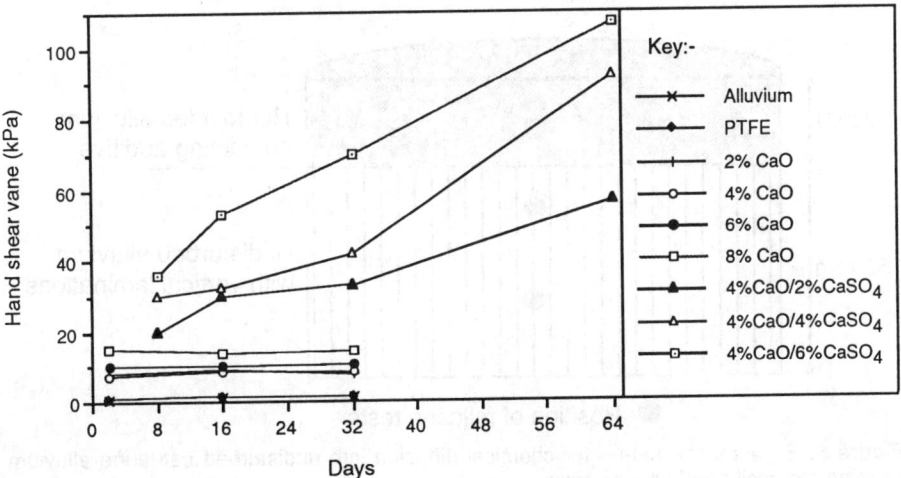

Figure 5: Hand shear vane readings for various mixes up to 64 days.

Table IX: Atterberg limits of lime and CaO:CaSO$_4$ mixtures measured after 32 days.

		Liquid Limit (%)	Plastic Limit (%)	Plasticity Index (%)
Remoulded alluvium		48	24	23
CaO	2%	54	33	21
	4%	54	33	21
	6%	53	33	20
	8%	52	35	17
CaO:CaSO$_4$	4%:2%	66	39	27
	4%:4%	61	42	19
	4%:6%	63	43	20

DIFFUSION TESTS

In order to assess the extent of the diffusion zone from various chemical sources a series of tests were performed in which the top 20 mm of collected samples was removed, mixed with the additives and then replaced. As it was required to simulate natural conditions the container was pressed into the soil horizontally such that when the upper part of the contained sample was removed and replaced with treated soil it would simulate in a horizontal mode a column/pile situation. The experimental set-up is shown in Figure 6. In order to obtain quantitative results fall cone tests were performed after 20 days, half-way into the upper mixed zone and at depths in the undisturbed material equivalent to 10 and 30 mm. It can be seen in Figure 7 that all the mixed materials gave a reduced strength compared with that 10 mm into the undisturbed sample below the base of the additive zone. This lower penetration implies that in addition to the cone entering undisturbed material, there was some diffusion from the chemical admixtures, especially those containing CaO. At a greater depth most samples showed an increased penetration (lower shear strength) indicating the natural state of the alluvium away from the influence of the admixtures.

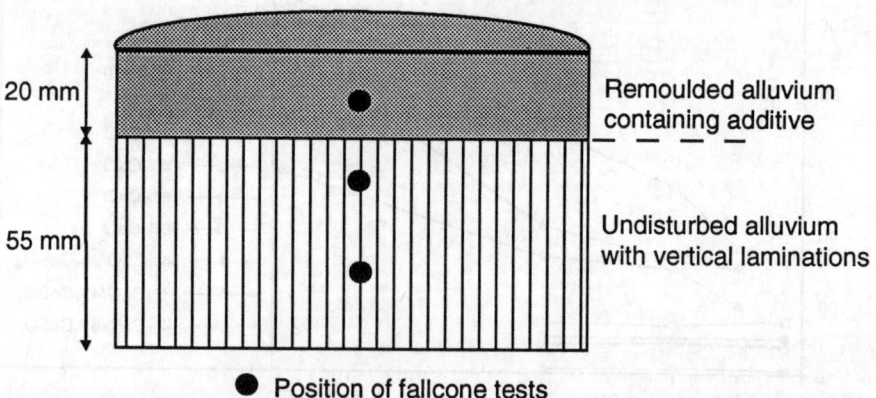

Figure 6: Experimental set-up for chemical diffusion into undisturbed estuarine alluvium showing the position of fallcone tests.

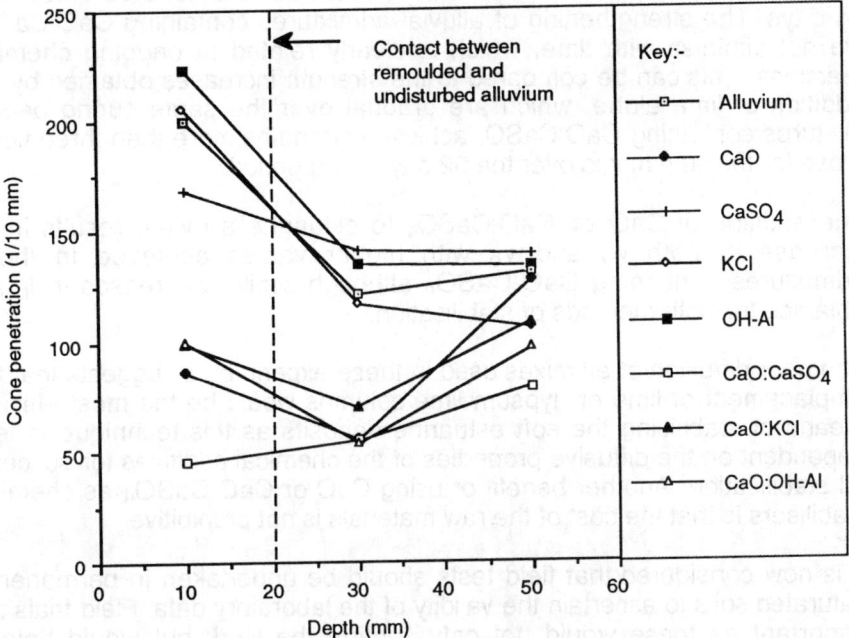

Figure 7: Cone penetrations from an experiment measuring diffusion of chemical additives into undisturbed estuarine alluvium.

The effects of the different chemicals on the undisturbed alluvium are indicated by comparison with the control samples. Only those samples containing CaO showed a significant difference. The higher strengths of the material at 30 mm below the chemical sources may not only be related to the diffusion of the chemical additive but also to the exothermic nature of the CaO hydration causing drying out of the samples. At 30 mm below the contact with the additive mixture the strength had remained constant or fallen slightly indicating the limit of stabilisation effects.

CONCLUSIONS

Experiments were performed to determine whether the strength of saturated alluvial soils could be sufficiently increased using some form of chemical additive to provide a possible alternative treatment for sub-surface problems.

The addition of CaO resulted in a significant increase in strength compared to the control sample; samples containing 2 to 8% CaO gave cone penetration values of approximately 70% less than those of the control sample

The best chemical admixture for the stabilisation of estuarine alluvium appeared to be a $CaO:CaSO_4$ combination with which after eight days the undrained shear strength measured using the hand vane was thirty times that of the control specimen, increasing to almost one hundred times after 64 days. The strengthening of alluvial admixtures containing $CaO:CaSO_4$ did not diminish with time, which is clearly related to ongoing chemical reactions. This can be compared to the strength increases obtained by the addition of lime alone, which are gradual over the same curing period. Mixtures containing $CaO:CaSO_4$ achieved strengths more than three times those for the lime mixes over the 32 day curing period.

The addition of CaO or $CaO:CaSO_4$ to estuarine alluvium results in an increase in both w_l and w_p with higher values achieved in those admixtures containing $CaO:CaSO_4$ although similar decreases in I_p are obtained for both methods of stabilisation.

The poor diffusion of all mixes used in these experiments suggests that the emplacement of lime or gypsum lime columns would be the most efficient means of stabilising the soft estuarine deposits as this technique is less dependent on the diffusive properties of the chemical additives for successful stabilisation. Another benefit of using CaO or $CaO:CaSO_4$ as chemical stabilisers is that the cost of the raw materials is not prohibitive.

It is now considered that field tests should be undertaken in permanently saturated soils to ascertain the validity of the laboratory data. Field trials are important as these would not only confirm the work but would help in determining the ideal spacing of the chemically stabilised soil columns in order to achieve the optimum effects. In addition it would be essential to establish the diameter of columns in order to minimise the cost of such stabilisation.

REFERENCES

ALY SABRY, M.M. & PARCHER, J.V. (1979). Engineering properties of soil-lime mixes. *Proceedings American Society of Civil Engineers*, **105**, TE, 59-70.

BELL, F.G. (1988). Stabilisation and treatment of clay soils with lime, Part 2, some applications. *Ground Engineering*, **21**, 22-30.

BELL, F.G. & TYRER, M.J. (1987). Lime stabilisation and clay mineralogy. *Proceedings of the International Conference on Foundations and Tunnels, London*, **2**, 1-7.

BELL, F.G. & COULTHARD, J.M. (1990). Stabilisation of clay soils with lime. *Municipal Engineer*, 7, 125-140.

BROMS, B.B. & BOMAN, P. (1979). Stabilisation of soil with lime columns. *Ground Engineering*, **12**, 23-32.

BRYHN, O.R., LØKEN T. & AAS, G. (1983). Stabilisation of sensitive clays with hydroxy-aluminium compared with unslaked lime. *Proceedings 8th European Conference on Soil Mechanics and Foundation Engineering, Helsinki*, **2**, 885-896.

CROFT, J.B. (1964). The processes involved in the lime stabilisation of clay soils. *Proceedings Australian Road Research Board*, **2**, 1169-1203.

DUMBLETON, M.J. (1962). Investigations to assess the potentialities of lime for soil stabilisation in the United Kingdom. *Road Research Technical Paper 64*, 76pp.

EGGESTAD, Å. & SEM, H. (1976). Stability of excavations improved by salt diffusion from deep wells. *Proceedings 6th European Conference on Soil Mechanics and Foundation Engineering, Wien*, **1**, 211-216.

FOSTER, R.H. & GAZZARD, I.J. (1975). The influence of polymeric hydroxy-aluminium cations on the consolidation, shear strength and consistency limits of clay soils. *Geotechnique*, **25**, 513-525.

GILLOTT, J.E. (1987). *Clays in Engineering Geology*, Elsevier, Oxford.

GRIM, R.E. (1968). *Clay mineralogy*. McGraw-Hill, New York. 2nd edition

HANSBO, S. (1957). A new approach to the determination of the shear strength of clay by the fallcone test. *Proceedings Royal Swedish Geotechnical Institute*, **14**, 47.

HAWKINS, A.B. (1984). Depositional characteristics of estuarine alluvium: some engineering implications. *Quarterly Journal of Engineering Geology*, **17**, 219-34.

HILT, G.H. & DAVIDSON, D.T. (1960). Lime fixation on clayey soils. *Highway Research Board, Washington, Bulletin 262*, 20-32.

HOLM, G., TRANK, R. & EKSTROM, A. (1983). Improving lime column strength with gypsum. *Proceedings 8th European Conference on Soil Mechanics and Foundation Engineering, Helsinki*, **2**, 903-7.

JOSHI, R.C., NATT, G.S. & WRIGHT, P.J. (1981). Soil improvement by lime-fly ash slurry injection. *Proceedings 10th International Conference on Soil Mechanics and Foundation Engineering, Stockholm*, **3**, 707-712.

KÉZDI, A. (1979). Stabilised earth roads. *Developments in geotechnical engineering*, **19**, Elseveir.

KUJALA, K. (1983). The use of gypsum in deep stabilisation. *Proceedings 8th European Conference on Soil Mechanics and Foundation Engineering, Helsinki*, **2**, 925-928.

KUJALA, K. & NIEMINEN, P. (1983). On the reactions of clays stabilised with gypsum lime. *Proceedings 8th European Conference on Soil Mechanics and Foundation Engineering, Helsinki*, **2**, 929-932.

MITCHELL, J.K. (1981). Soil improvement - state of the art report. *Proceedings 10th International Conference on Soil Mechanics and Foundation Engineering, Stockholm*, **4**, 509-565.

MOUM, M.J., SOPP, O.I. & LØKEN, T. (1968). Stabilisation of undisturbed quick clay by salt wells. *Vag-och vatten-bygarren*, **14**, 23-29.

NIEMINEN, P, (1978). Soil stabilisation with gypsum lime. *Proceedings International Conference on By-products and Waste in Civil Engineering, Paris*, **1**, 229-235.

QUEIROZ DE CARVALHO, J.B. (1981). Amorphous materials and lime stabilised soils. *Proceedings 10th International Conference on Soil Mechanics and Foundation Engineering, Stockholm*, **3**, 761-762.

ROGERS, C.D.F. & BRUCE, C.J. (1991). *Slope stabilisation using lime slope stability engineering*, Thomas Telford, London, 367-374.

TSYTOVICH, N.A., ABELEV M.Y. & TAKHIROV, I.G. (1971). Compacting saturated loess soils by means of lime piles. *Proceedings 4th International Conference on Soil Mechanics and Foundation Engineering, Budapest*, 837-847.

VAN IMPE, W.F. (1989). *Soil improvement techniques and their evaluation.* Balkema, Rotterdam.

WOOD, D.M. (1985). Some fall-cone tests. *Geotechnique*, **35**, 64-68.

4-7 LABORATORY STUDY OF SOIL-FLUID INTERACTION BEHAVIOUR

R.K. SRIVASTAVA, A.V. JALOTA & B. SINGH,
Department of Civil Engineering, M.N.R. Engineering College Allahabad, India.

ABSTRACT

Indiscriminate disposal of industrial waste water is likely to affect the strength, deformation and seepage behaviour of soils. In addition, if it comes into contact with fresh water, contamination of water used for human consumption could pose a health hazard. The present study discusses the interaction behaviour of a typical alluvial soil and waste water discharged from a fertilizer plant (collected from two locations). It has been found that whilst the effluent meets the standards set by pollution control authorities, it may nevertheless modify the geotechnical engineering properties of the soil.

INTRODUCTION

Although industrial activity is necessary for the socio-economic progress of a country, it generates large amounts of solid and liquid wastes which may become a source of ground contamination. Pollutants allowed to permeate into the ground will affect the soil/rock in various ways and there has been a growing concern amongst geotechnical and environmental engineers over the detrimental effects both on the strength, deformation and seepage behaviour of soils as well as on the quality of fresh water in nearby areas which may be used for human consumption.

Most of the studies on soil-fluid interactions and particularly those concerned with waste water pollutants have been carried out during the last decade. Some of the important and useful work to date includes that reported by Pavilsonsky (1985), Anderson and Brown (1981), Peterson and Gee (1985), Gibson (1985), Fang (1985), Mitchel and Madsen (1987), Kirov (1989) and Pakhemov and Yegorov (1990). Emphasis has been on the reduction of the subsurface movement of fluids, especially hazardous waste. Recently held conferences, however, have drawn attention to the complex chemistry of the fluid and soil-rock interaction and the need for both development and evaluation of testing techniques as well as a standardization of procedures for design of barriers used in the disposal of wastes.

The present study has been carried out on the interaction between fertilizer plant effluent and a typical alluvial soil found in the plains of Northern India. The soil was first tested with distilled water and then with the waste water, using a triaxial cell as a permeameter (Figure 1). The advantage of this arrangement was that the soil sample could be encased in a rubber membrane and saturation ensured by the use of back pressure. The flexible membrane reduces problems of leakage between the specimen and the sides of the permeameter which would give spuriously high values of permeability.

This paper presents the engineering and chemical properties of the soil and the chemical analyses of the fertilizer plant effluent both before and after their interaction.

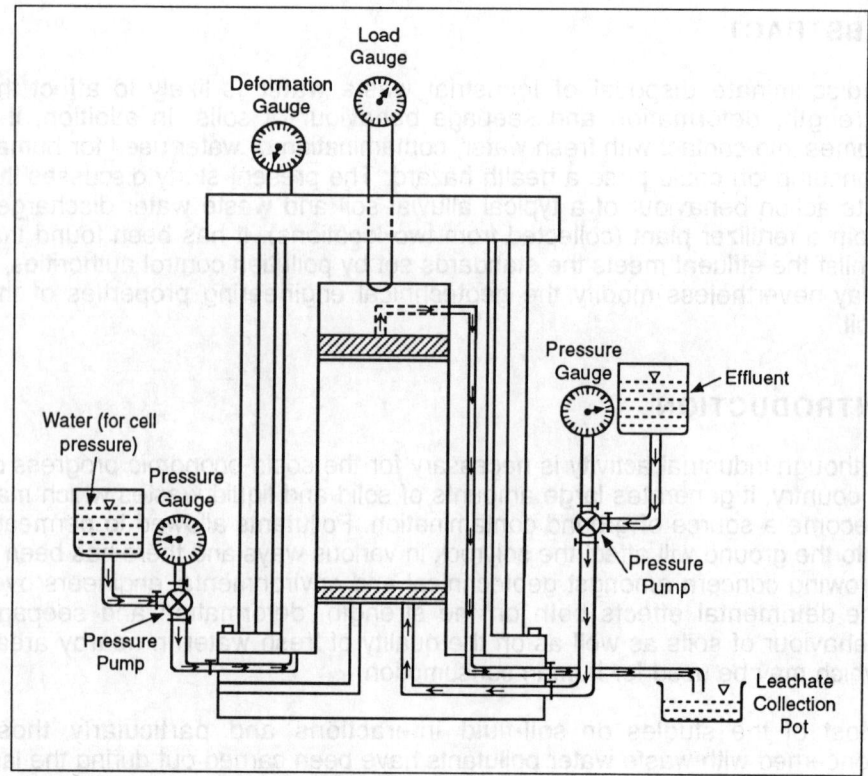

Figure 1: Schematic diagram of triaxial cell for permeability.

PROPERTIES OF THE SOIL

The initial engineering, chemical and plant effluent properties of the soil are presented in Table I and the results of the chemical analyses of the fertilizer plant effluent in Table II. The effluent samples were collected from two locations about 2 km and 3 km from the point of discharge.

Table III presents the changed properties of the soil following the interaction with the fertilizer plant effluent in the permeability tests and Table IV the properties of the effluent.

Table I: Properties of the soil.

Property	Value
Liquid limit	32%
Plastic limit	22%
Clay content	15%
Silt content	65%
Sand content	20%
Optimum moisture content	15%
Maximum dry density	1.74 Mg/m^3
Permeability	1.01 x 10^{-8} m/s
Strength parameters on soil compacted to MDD and saturated	
(a) Cohesion (c_u)	117.7 kPa
(b) Angle of Internal friction ($ø_u$)	0°
Coefficient of compression (Cc)	0.197
pH	8.10
Ca^{2+} Total	150 ppm
Exchangeable	30 ppm
Na^+ Total	20 ppm
Exchangeable	14 ppm
K^+ Total	450 ppm
Exchangeable	12 ppm
Mg^{2+} Total	17 ppm
Exchangeable	8 ppm
Fe^{2+} Total	1.4 ppm
Exchangeable	0.7 ppm
Cu^{2+} Total	3.8 ppm
Exchangeable	0.26 ppm

Table II: Properties of effluent (industrial waste water).

Property	Location I (3 km)	Location II (2 km)
pH	7.25	7.65
Conductivity	1550 μmho	1308 μmho
K^+	12 ppm	16 ppm
Ca^{2+}	51 ppm	49 ppm
Na^+	126 ppm	102 ppm
PO_4^{2-}	1251 ppm	1026 ppm
Cl^-	304 ppm	199 ppm
Total hardness (as $CaCO_3$)	396 ppm	402 ppm
Calcium hardness	208 ppm	229 ppm
Magnesium hardness	188 ppm	173 ppm

DISCUSSION

Comparison of engineering properties of the soil before and after its interaction with fertilizer plant effluent (Tables I and III) indicates that in general there was an increase in consistency limits (16.7% in liquid limit), permeability (138.6%) and coefficient of compression (24.3%) and decrease in shear strength (25%). In this particular case of soil fluid

Table III: Modified properties of soil after interaction with effluent from the two locations.

Property	Interaction with effluent from Location I	Interaction with effluent from Location II
Liquid limit	37%	37%
Plastic limit	25%	26%
Permeability	2.41×10^{-8} m/s	2.32×10^{-8} m/s
Strength parameters on soil compacted to MDD and saturated with effluents		
(a) Cohesion (c_u)	87.3 kPa	90.2 kPa
(b) Angle of internal friction (ϕ_u)	0°	0°
Coefficient of compression Cc	0.243	0.245

Table IV: Properties of effluent after passing through soil.

Property	Location I	Location III
pH	7.40	7.20
Conductivity	1456 µmho	1316 µmho
K^+	6.3 ppm	6.0 ppm
Ca^{2+}	46 ppm	28 ppm
Na^+	60 ppm	82 ppm
PO_4^{2-}	0.55 ppm	0.50 ppm
Cl^-	224 ppm	182 ppm
Total hardness (as $CaCO_3$)	823 ppm	680 ppm
Calcium hardness	563 ppm	382 ppm
Magnesium hardness	260 ppm	298 ppm

interaction, therefore, the resultant soil has deteriorated from the geotechnical engineering point of view. The decrease in shear strength indicates a reduced bearing capacity while the increase in liquid limit and coefficient of compression reflect its increased susceptibility to settlement indicating difficulties and thus an overall problem for any future reclamation of the disposal site or the adjacent areas. The large increase in permeability also suggests a serious environmental pollution problem as the higher permeability soil would allow a greater seepage of effluent to nearby fresh water bodies (ground water, wells, ponds, lake, river etc.) and it is likely this would increase with increasing permeability of the soil even over a short period. This has been verified from visual observations at areas near the disposal channel where the taste of water in the wells has also changed. Indeed, villagers are not able to use the water from this well for cooking and drinking purposes and an alternate source of water supply has had to be provided.

As can be seen in Tables II and IV, various parameters, e.g. hardness, pH, conductivity, amount of total and exchangeable cations and anions, have been modified because of the soil-effluent interaction. Some of the chemicals have been absorbed from the effluent into the soil while chemicals from the soil moved into the effluent leaching out from the soil. In this study the cation content of the effluents decreased and the total hardness increased after the interaction.

It must be appreciated that the results reported here were obtained from a short duration study of soil in a fertilizer plant effluent environment; the long term effects may well be more drastic.

CONCLUSIONS

The following conclusions may be drawn from the present study:

1. The strength of soil decreases as a result of the interaction with the effluent, causing an increase in the cost of constructing on or using this material.

2. The increase in coefficient of compressive strength indicates the soil has become prone to settlement and more appropriate measures would have to be taken in the design and construction of foundations in this type of soil.

3. Following the interaction effect, the soil was found to have a greater permeability hence there would be an increased likelihood of waste water contaminating fresh water bodies within a short length of time.

4. The hardness of the water was also found to have increased, which would adversely affect the use of the water, especially in rural areas.

REFERENCES

ANDERSON, D. & BROWN, L. (1981). Organic leachate effects on the permeability of clay liners. In: *7th Annual Research Symposium on Land Disposal of Hazardous Wastes*, D.W. Schultz (ed.), EPA-600/9-81-0026, U.S. Environmental Protection Agency, Cincinnati, OH, 45268.

GIBSON, A.H. (1985). Permeability testing on clayey soil and silty sand-bentonite mixture using acid liquor. *Hydraulic Barriers in Soil and Rock, ASTM, STP* 874, 140-154.

FANG, H.Y. (1985). Soil-pollutant interaction effects on the soil behaviour and the stability of foundation structures. *Symposium on Environmental Geotechnics and Problematic Soil and Rock, Asian Institute of Technology*, 127-146.

KIROVE, B. (1989). Influence of waste water on soil deformations. *Proceedings of the XIIth International Conference Soil Mechanics and Foundation Engineering, Rio de Janeiro, Brazil*, 1881-1882.

MITCHELL, J.K. & MADSEN, F.T. (1987). Chemical effects on clay hydraulic conductivity. In: R.D. Woods (ed.) *Geotechnical Practice for Waste Disposal, Geotechnical Special Publication*, **13**, 87-116.

PAKHOMOV, S.I. & YEGOROV, YU. K. (1990). Clay and liquid wastes interaction in environmental geology. *Proceedings of the 6th International Congress, International Association of Engineering Geology, Amsterdam, Netherlands.*

PAVILONSKY, V.M. (1985). Varying permeability of clayey soil linings. *XIth International Conference Soil Mechanics and Foundation Engineering, San Francisco, USA.*

PETERSON, S.R. & GEE, G.W. (1985). *Interaction between acidic solution and clay liners: Permeability and neutralization. Hydraulic Barriers in Soil and Rock, ASTM, STP 874, 229-243.*

Session 5 - Environmental

5-1	Performance of inactive barrier clays in contact with municipal solid waste leachate R.M. Quigley	439
5-2	Assessment of chemical buffering capability of a micaceous soil R.N. Yong, B.K. Tan & A.M.O. Mohamed	451
5-3	A European perspective on landfill engineering A. Street	463
5-4	Classification of excavated materials for disposal purposes - a case history on the analysis of ground chemistry C. Mirza	473
5-5	Physical and chemical reactions between seepage ground water and fault gouge within a dam foundation rockmass Z. Huyuan & H. Wenfeng	481
5-6	Factors controlling the solubility of substances in natural waters A. Sahinci & A. Turkman	489
5-7	Microbiological aspects of the generation of impermeability in high sand content greens infested with black plug layering D.R. Cullimore & S. Nilson	499
5-8	Biogenic sulphuric acid attack on concrete in sewer environments T. van Mechelen & R. Polder	511
5-9	Determination of the chemical resistance of cements using small scale accelerated tests V. Paul & S. Uberoi	525
5-10	Calcium aluminate mortars and concretes: an application to sewer pipes in harsh environments T. Dumas	541
5-11	The influence of geoenvironmental dissolved sulphates on the swelling process in concrete W. Prince & R. Perami	553
5-12	Damage to concrete caused by alkaline metal hydroxides W. Prince & R. Perami	559

5-13 Standard hydrogeochemical models and their possible application to engineering construction problems involving ground water 569
J.H. Tellam & J.W. Lloyd

5-14 Development of a computer assisted program to project the well plugging risk index (wpri) for existing or planned water wells 591
D.R. Cullimore & G. Alford

5-15 Chemical and microbiological effects on construction dewatering systems 607
W. Powrie, T.O.L. Roberts & S.A. Jefferis

5-16 Chemical erosion in lateritic soil: a case study 617
M.R. de Ruijter

5-17 Ultrasonic monitoring of the acid rain impact on building stones 623
J. Pininska & P. Lukaszewski

5-18 Environmental biochemical decay control on Jeronimos Monastery cloister, Lisbon, Portugal 633
L. Aires-Barros & A.M. Mauricio

5-1 PERFORMANCE OF INACTIVE BARRIER CLAYS IN CONTACT WITH MUNICIPAL SOLID WASTE LEACHATE

R.M. QUIGLEY
Geotechnical Research Centre, Faculty of Engineering Science,
The University of Western Ontario, London, Ontario, Canada N6A 5B9.

ABSTRACT

Clayey soil barriers comprise the primary liner system for containment of municipal solid wastes and their resultant leachates throughout the world. This paper presents a short review of the diffusive and advective processes whereby such liners become contaminated with leachate chemicals, followed by an engineering assessment of two such barriers in southern Ontario, Canada.

It will be demonstrated that the barriers at both sites have undergone a long-term decrease in hydraulic conductivity concurrent with extensive chemical migration, primarily by diffusion.

INTRODUCTION

Modern landfill sites are constructed with a relatively impervious barrier system designed to limit the escape of pollutants to environmentally acceptable levels. The barrier itself may consist of a clay liner only, a flexible membrane in contact with a clay liner or a complex multi-liner system (Cope, 1987; Daniel, 1987). Leachate collection layers and pipes are normally present on top of each liner system so that large hydraulic head drops do not occur across the liners and concentrated leachate is continuously removed from the system. With required hydraulic conductivity values as low as 10^{-8} cm/s and very small advective velocities, diffusion becomes the major process for pollutant migration at many sites. At a typical landfill with a low k clayey barrier about 1 m thick, the advective breakthrough time might be close to 100 years compared to diffusion related breakthrough of chemicals in 10 years or less. The diffusion process obviously requires detailed consideration at all landfill sites.

In addition to the problem of early arrival of chemicals to underlying aquifers, it is equally important to know whether this non-advective transport of chemicals is likely to adversely affect the in situ hydraulic conductivity of the barrier. Consideration of this aspect to the problem forms the second part of this paper.

THE DIFFUSION PROCESS

Diffusion is the process whereby dissolved chemicals present at "high" concentration migrate towards areas where they are present in lower concentration, in the absence of other types of gradient. The process, while simple in concept, is very complex in reality since a concentration gradient generates osmotic flow, electro-osmotic flow and often hydraulic flow, all coupled with the diffusion part of the migration. In dilute "free" solution at $t = 8°C$, the diffusion coefficient, D_o, typically varies from ~5×10^{-6} cm^2/s for a divalent cation to ~15×10^{-6} cm^2/s for a monovalent species such as K^+ or Cl^-.

In a soil water system as illustrated on Figure 1, increased flow path tortuosity, liquid viscosity, back flow by osmosis, etc. greatly slow the rate of migration and the diffusion coefficient, D_s, for conservative Cl^- might decrease to a value of ~6.5×10^{-6} cm^2/s. If adsorption affects the behaviour of a species (such as K^+ or the heavy metals), D_s, drops to a very low values as the effect of the distribution coefficient kicks in. A fairly comprehensive review of the process for civil engineering use was presented by Quigley, Yanful and Fernandez (1987a).

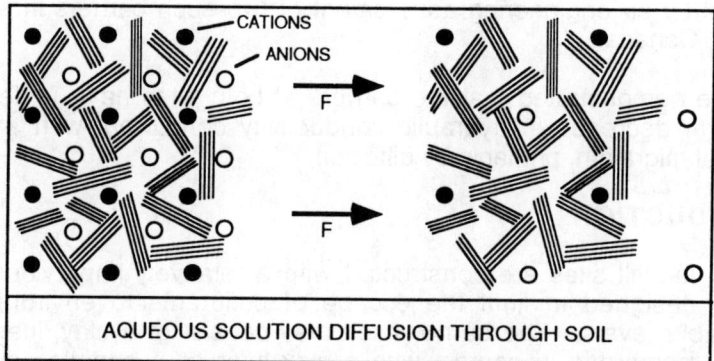

Figure 1: Schematic of diffusion by cations and anions from high to low concentration through a saturated clayey soil.

When MSW leachate comes into contact with a freshwater clayey barrier, the time rate migration into the clay follows Fick's second law which has the form:

$$\frac{dC}{dt} = D_s \frac{d^2C}{dx^2}$$

where C = species concentration; t = time; D_s = effective diffusion coefficient in soil = real diffusion coefficient for a conservative species; and x = distance from the high salt interface.

A plot of the error function solution to the above equation is presented on Figure 2 for the conservative or non-reactive species, chloride. As chloride is an abundant constituent in most MSW leachates it is an excellent migration tracer for barriers and plumes. The D_s of 6.5×10^{-6} cm^2/s employed is typical of values measured at many landfill sites and in laboratory tests on clays which have water contents of between 17 and 25%.

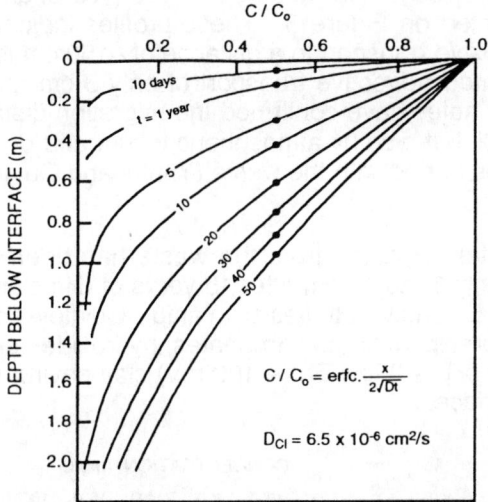

Figure 2: Relative concentration-depth-time plots for chloride diffusive migration in a soil with w = 15 to 25%

The curves clearly show that in only 5 years or so measurable chloride should be exiting from the base of a 1 m thick clayey barrier in the complete absence of field advection.

There are other important aspects of the curves on Figure 2. One is that diffusion slows up dramatically with time as the length of the path increases and dc/dx decreases. Diffusion studies should be done within 10 years of leachate contact if one wishes to monitor significant advance through clay from one year to the next. Secondly, in the early stages of diffusion the rate of advance of the chemical front is extremely rapid and a thin bentonitic barrier would be transmitting full diffusion fluxes within a few days in spite of k-values possibly as low as 10^{-10} cm/s.

EXAMPLES OF FIELD DIFFUSION

Diffusion profiles of different ages are presented in the next section for the well published Confederation Road landfill at Sarnia, Canada and the Keele Valley Landfill at Toronto, Canada.

(A) THE CONFEDERATION ROAD LANDFILL

The small Confederation Road landfill is sited on about 30 m of lacustrine clay having a hydraulic conductivity of ~2×10^{-8} cm/s. A very slight downward gradient of 0.25 is responsible for an average linearized downward pore water velocity of approximately 0.24cm/annum (Goodall and Quigley, 1977).

Diffusion profiles for Na+ at t = 6 and 12a (years) and for Cl- at t = 12a are presented on Figure 3. These profiles indicate clearly that the actual diffusive transport to a distance of ~150 cm is far in advance of the estimated advective transport of only 3 cm at t = 12a. Profiles from other holes have confirmed the migration distances not only by Na+ and Cl- but also by atmospheric tritium and del oxygen-eighteen from leaves buried with the waste (Yanful and Quigley, 1990).

Heavy metal migration from the waste has been greatly retarded, reaching only 10 to 15 cm after 15 years of diffusion as illustrated on Figure 4. Extensive studies by Yanful, Quigley and Nesbitt (1988) indicated precipitation as carbonates, hydroxides and even organics in the high pH (~8) low Eh (~-100 mV) clay environment close to the waste interface.

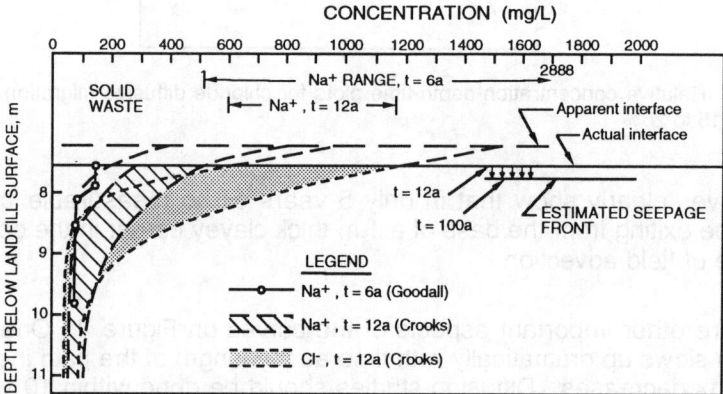

Figure 3: Pore fluid concentration of Na+ and Cl- (mg/l) in clay below waste after 6 and 12 years of diffusion (t = 6a and 12a) [adapted from Goodall and Quigley, 1977; Crooks and Quigley, 1984; Quigley and Rowe, 1987].

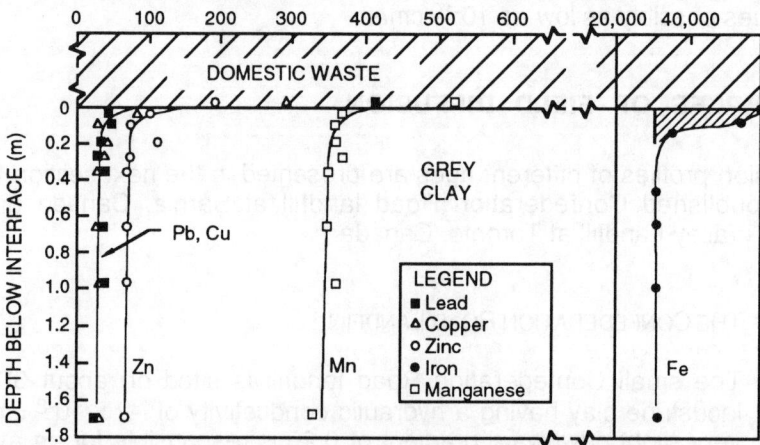

Figure 4: Total metal concentrations in the bulk soil below the waste at t = 16a (after Quigley et al, 1987).

This same 50 cm of clay directly below the waste was shown to support much higher bacterial growth (~10^6 colony forming units/g of soil) compared to 10^3 at depth.

(B) THE KEELE VALLEY COMPACTED LINER

The Keele Valley landfill north of Metropolitan Toronto is separated from a previously contaminated sand layer by an engineered, 1.2 m thick, compacted clayey till liner having a specified hydraulic conductivity of < 1 x 10^{-8} cm/s. A continuous series of papers has flowed from this huge project with respect to liner performance with time. The latest of these papers by King, Quigley, Fernandez, Reades and Bacopoulos (1992) describes compatibility testing, field diffusion by exhumation studies and electrical conductivity sensor monitoring, and hydraulic conductivity by field lysimeters compared to extensive laboratory k testing.

Previously published but modified diffusion profiles for this site are presented on Figures 5 and 6 for an exhumed section of clayey liner and its overlying sand cushion/drainage layer. Waste had been in contact with this section of liner for an estimated 4.25 years at the time of exhumation.

The chloride profile is shown extending from the waste/sand interface for a total distance of ~90 cm of which 60 cm represents penetration into the clay. Linearized seepage velocities, v_s, of ~2.4 cm/a are estimated for most of the clay barrier at this site. If the sand was saturated with fresh water when waste disposal was started, the seepage front would have travelled only ~10 cm into the sand in 4.25a. Leachate chemicals would therefore have only reached the one-third point in the 31 cm thick sand cushion by advective transport. The overall profile shown is therefore attributed primarily to molecular diffusion starting at the top of the sand.

The chloride profile and the cation profiles shown on Figure 6 have now been confirmed by several exhumations and study programs at other landfills. Na^+ is normally slightly retarded behind Cl^-, K^+ is heavily retarded by preferential adsorption and fixation, and Ca^{2+}, the dominant adsorbed cation, is normally desorbed to form a "visible" concentration hump or hardness halo.

The calculated sodium adsorption ratios (SAR) vary from close to 20 at the top of the sand to ~8 at the clay interface. They then decrease steadily with depth to values of ~1 at a distance of 40 cm below the sand/clay interface.

Further confirmation of migration is illustrated by the electrical conductivity sensor data presented on Figure 5b. The data for 1988 are estimated to be roughly 4 years following waste disposal. At this time, the conductivity ratio (actual to original background conductivity) had increased in those sensors located above a depth

5-1

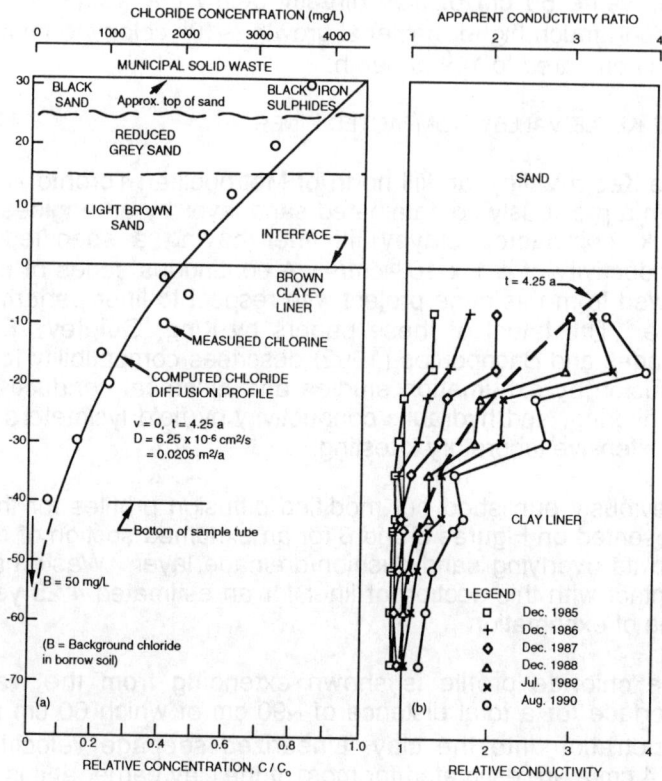

Figure 5: Pore water chloride and field conductivity at t ≈ 4.25a: a) actual and calculated chloride profile from exhumation: b) apparent conductivity ratio from field sensors (adapted from Quigley et al, 1990; King et al, 1992).

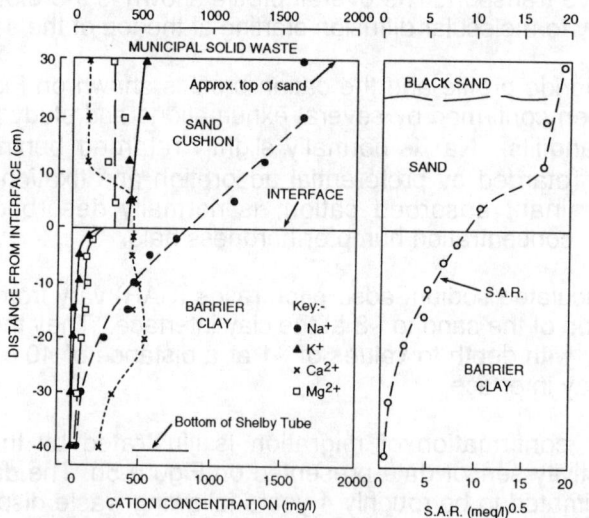

Figure 6: Pore water cation profiles in the Keele liner at t = 4.25a (adapted from Quigley et al, 1990).

of ~60 cm into the clay. This depth corresponds closely to the ~60 cm of chloride ion penetration into the clay shown on Figure 5a at another section of the liner.

HYDRAULIC CONDUCTIVITY

(A) CONFEDERATION ROAD LANDFILL

At this site, the zone contaminated by diffusion was about 1.5 m thick at the time hydraulic conductivity was checked in 1983/84. Two procedures were used: (1) oedometer testing and calculation of k; and (2) direct measurement of k using squeezed pore fluid as the permeant (Quigley, Fernandez, Yanful, Helgason, Margaritis and Whitby, 1987b).

The oedometer test results obtained on samples from a 1983 borehole are presented on Figure 7 adapted from Quigley et al (1987b). The k-values calculated for stress levels just above and just below the preconsolidated stress of the clay ($\sigma_p' \approx 170 kPa$) yielded two curves of k vs depth. Both curves show a slight decrease in k in the 20 cm zone just below interface. As the corresponding average water contents were both constant with depth, the changes in k may be attributed to chemical factors. A plot of sodium adsorption ratio, SAR, included with the profiles in Figure 7, suggests that expanded double layers associated with 8 to 10% Na^+ adsorption may have influenced the decrease in the k-values somewhat.

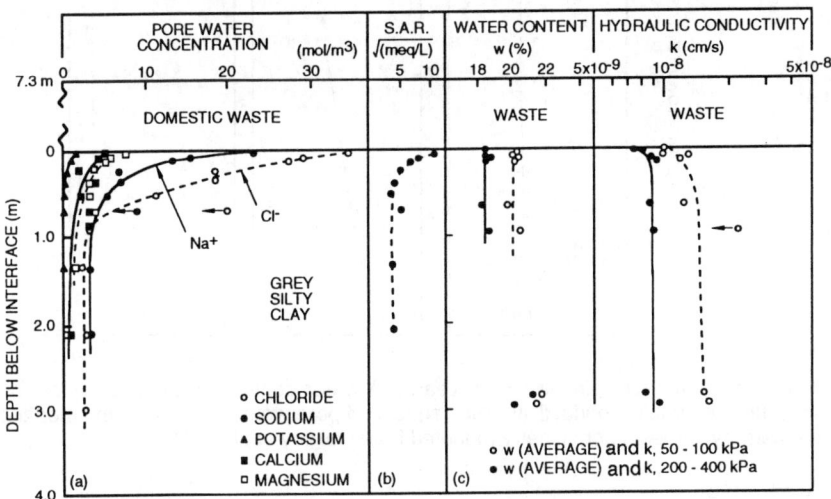

Figure 7: Summary plot at t = 15 years: a) pore water chemistry; b) sodium adsorbtion ratio in $(meq/l)^{0.5}$; c) hydraulic conductivity from oedometer tests, all versus depth at BH #83-2 (adapted from Quigley et al, 1987).

The oedometer k-values calculated for the stress levels below σ_p' were about 1.7×10^{-8} cm/s in the unaltered clay 1 m below the interface and correspond closely to field values reported by Goodall and Quigley (1977).

A second set of directly measured k-tests on samples from a 1984 borehole are presented on Figure 8. The k vs depth profile shows a similar decrease in k towards the interface in the top 50 cm of the clay profile. In all tests the combined static and seepage stresses were kept below the σ_p' of 170 kPa to prevent unwanted consolidation. Nevertheless, the water content profiles at this location suggest a correlation between final water content and k which was not apparent for the oedometer k-test results. The porosimetry data shown on Figure 8c for a nearby borehole indicate decreases in mean and modal diameters towards the interface. These probably result from heavy metal precipitation and possibly drying at the interface at the time of excavation. Regardless of their explanation, a decrease in pore size would correlate with a decrease in k.

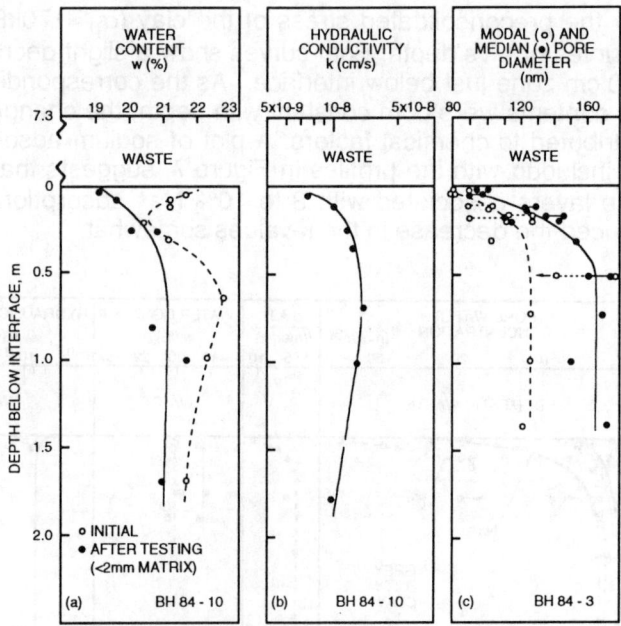

Figure 8: Summary plot at $t \approx 16$ years: a) water content of k-test samples; b) directly measured hydraulic conductivity using squeezed pore water; c) median and modal pore size by mercury intrusion porosimetry (adapted from Quigley et al, 1987).

(B) KEELE VALLEY COMPACTED LINER

In addition to large numbers of triaxial tests performed on both as-compacted and exhumed samples, special lysimeter k-tests are reported by King et al (1992) for the compacted Keele liner dated from the initial time of waste loading. The results illustrated on Figure 9 show a long slow decrease in k from as-constructed values of ~1 x 10^{-7} cm/s to mean "final" values of ~4 x 10^{-9} cm/s. The latter values correspond to about 5 years of waste disposal, increases in vertical effective stresses from zero to at least 160 kPa, and corresponding liner compression.

Chemical analyses of the lysimeter water showed that Na^+ was just starting to arrive after 5 years of combined advection and diffusion; however, large amounts of desorbed Ca^{2+} were arriving as a hardness front. The elution chemistry was very similar to that produced during clay/leachate compatibility testing in the laboratory. The latter testing produced significant chemical changes in the adsorption regime of the clay but did not alter k at the stress levels of 160 kPa employed for the laboratory testing.

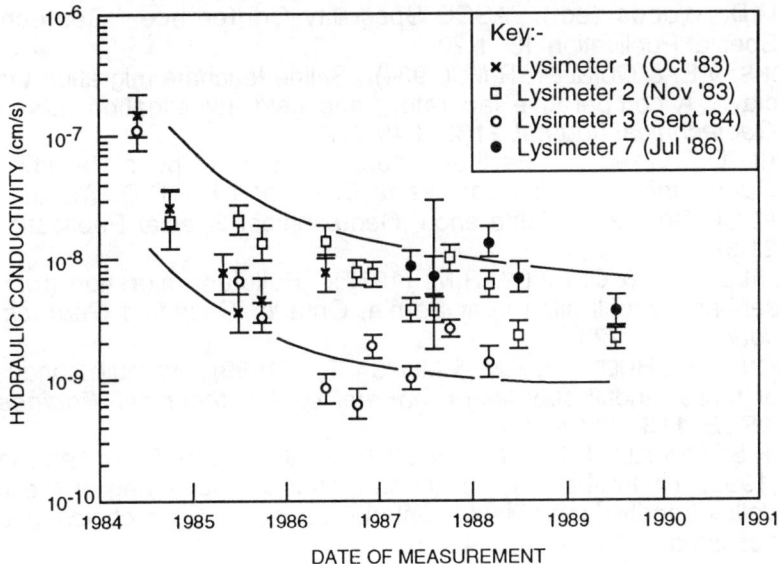

Figure 9: Estimated hydraulic conductivity based on shallow lysimeter effluent flow rates (from King et al, 1992).

CONCLUDING REMARKS

Chemical profiles generated at two landfill sites have been presented to illustrate the extremely important role that diffusion plays in the absence of advection. At both sites clayey barriers were in contact with MSW leachate, both high in Na^+ and Cl^-.

In situ and laboratory studies of the hydraulic conductivity of the two barriers have shown no adverse reaction caused by contact with leachate. In fact, the compacted Keele liner has improved with respect to k, probably mostly as a result of stress application (waste loading) and consolidation. The natural, "unloaded" clay at the Confederation Road landfill has retained its low k of 2×10^{-8} cm/s and even improved somewhat with 20 or 30 cm on the interface. No adverse effects from MSW leachate are apparent. Concurrent barrier studies by Gordon, Huebner and Miazga (1989) confirmed the highly satisfactory performance of three Wisconsin barriers beneath similar MSW landfills.

Both clayey soils discussed are inactive, the clay minerals consisting primarily of illite and chlorite. It should not be assumed that the successful behaviour of these two barriers can be extrapolated to the use of c-axis contractible smectites (bentonites) used as a barrier material.

REFERENCES

COPE, F.W. (1987). Design of waste containment structures (synthetic liner based systems). In: *Geotechnical Practice for Waste Disposal 1987*, R.D. Woods (ed.), ASCE Speciality Conference, Geotechnical Special Publication 13, 1-20.

CROOKS, V.E. & QUIGLEY, R.M. (1984). Saline leachate migration through clay: A comparative laboratory and field investigation. *Canadian Geotechnical Journal*, **21(2)**, 349-362.

DANIEL, D.E. (1987). Earthen liners for land disposal facilities. In: *Geotechnical Practice for Waste Disposal 1987*, R.D. Woods (ed.), ASCE Speciality Conference, Geotechnical Special Publication 13, 21-39.

GOODALL, D.C. & QUIGLEY, R.M. (1977). Pollutant migration from two sanitary landfill sites near Sarnia, Ontario. *Canadian Geotechnical Journal*, **14**, 223-236.

GORDON, M.E., HUEBNER, P.M. & MIAZGA, T.J. (1989). Hydraulic conductivity of three landfill clay liners. *Journal of Geotechnical Engineering, ASCE*, **115**, 1148-1160.

KING, K.S., QUIGLEY, R.M., FERNANDEZ, F., READES, D.W. & BACOPOULOS, A. (1992). Hydraulic conductivity and diffusion monitoring of the Keele Valley Landfill Liner, Maple, Ontario. *Canadian Geotechnical Journal*, Accepted for Vol. 29, No. 4 or 5.

QUIGLEY, R.M. & ROWE, R.K. (1986). Leachate migration through clay below a domestic waste landfill, Sarnia, Ontario, Canada: Chemical interpretation and modelling philosophies. In: *Hazardous and Industrial Solid Waste Testing and Disposal:*, D. Lorenzen, R.A. Conway, L.P. Jackson, A. Hamza, C.L. Perket and W.J. Lacy, (eds.), Sixth Volume, ASTM STP 933, American Society for Testing and Materials, Philadelphia, 93-103.

QUIGLEY, R.M., FERNANDEZ, F., YANFUL, E., HELGASON, T., MARGARITIS, A. & WHITBY, J.L. (1987b). Hydraulic conductivity of contaminated natural clay directly below a domestic landfill. *Canadian Geotechnical Journal*, **24(3)**, 377-383.

QUIGLEY, R.M., YANFUL, E.K. & FERNANDEZ, F. (1987a). Ion transfer by diffusion through clayey barriers. In: *Geotechnical Practice for Waste*

Disposal 1987, R.D. Woods (ed.), ASCE Speciality Conference, Geotechnical Special Publication No. 13, 137-158.

YANFUL, E.K., QUIGLEY, R.M. & NESBITT, H.W. (1988). Heavy metal migration at a landfill site, Sarnia, Ontario, Canada - 2: Metal partitioning and geotechnical implications *Applied Geochemistry*, **3**, 623-629.

YANFUL, E.K., QUIGLEY, R.M. (1990). Tritium, oxygen-18 and deuterium diffusion at the Confederation Road landfill site, Sarnia, Ontario, Canada. *Canadian Geotechnical Journal*, **27**, 271-275.

5-2 ASSESSMENT OF CHEMICAL BUFFERING CAPABILITY OF A MICACEOUS SOIL

R.N. YONG
Geotechnical Research Centre (GRC), McGill University, Montreal, Canada

B.K. TAN
Department of Geology, University of Kebangsaan Malaysia, Bangi, Malaysia

A.M.O. MOHAMED
Geotechnical Research Centre (GRC), McGill University, Montreal, Canada

ABSTRACT

The chemical buffering capability of a micaceous soil and its admixtures is studied using leaching column tests, to determine its suitability for use as a natural soil engineered barrier for containment of municipal solid waste. Analysis showed the natural soil alone possesses satisfactory characteristics such as a pH of 7 to 8 and a permeability of 10^{-9} m/sec. In addition it had a high adsorption or retention capability for cations especially the heavy metals Pb^{2+} and Zn^{2+}. However, tests conducted on the natural micaceous soil and soil admixtures showed the addition of a natural clay from Eastern Canada containing carbonates significantly improves the properties of the natural soil, whereas the addition of kaolinite as a clay material reduces its adsorption capabilities. The attenuation of heavy metals by all three soil types is nonetheless very high. Retention of heavy metals is mainly through precipitation (high pHs) and cation exchange.

INTRODUCTION

Land disposal of municipal and industrial solid wastes continues to be the most favoured method of waste management, in spite of the many arguments raised with respect to the need for conservation of precious land resources and the high potential for contamination of groundwater from leachates emanating from the waste piles. The common "engineering" answer for safe containment of waste piles via landfilling practices lies in the use of "engineered barriers" designed to isolate the waste in landfills through the use of "impermeable containers". The rationale for such an approach lies in the expectation that the impermeable boundaries created by the "engineered barriers" will completely prevent escape of leachates, thereby accomplishing the requirement for "waste isolation".

Reliance on the capability of double-membrane synthetic liners to function for many years (100+?) as completely impermeable barrier membranes is often misplaced, inasmuch as there is no evidence to date to support an ad infinitum capability of membranes. In many countries, the use of double-membrane synthetic liners as the only barrier is specifically prohibited, i.e. the membrane must be supported by an engineered barrier constructed of natural clay material, of sufficient thickness to provide proper contaminant buffering, if the worst-case scenario of leachate escape is encountered. This requirement is not unreasonable and constitutes the operating set of conditions for many countries - with and without the added "bonus" of a double-membrane synthetic liner.

When engineered clay barriers are to be used as clay membrane liners for landfills, a necessary prerequisite is the assessment of the compatibility of the clay material to the influent leachate. Both theory and experience with contaminant-clay interaction mechanisms suggest that contaminant transport processes in and through the clay barrier depend on the properties and characteristics of the clay soil constituents and the contaminants. Interaction processes could produce detrimental effects, especially in regard to increases in the transport coefficients eg fractures, cracks, shrinkage of barrier material.

The study reported in this paper was designed to examine the performance of a candidate micaceous soil and its mixtures vis-a-vis leaching column tests. The particular items of interest included a study of the adsorption characteristics of the soils with respect to the retention and migration of heavy metals through the soil columns. The results of the study are then used to evaluate the suitability of the soil as a potential landfill liner and adsorbent of pollutants.

MATERIALS AND METHODS

Three types of soil materials were used in the leaching column tests, namely:

i. Natural Micaceous Soil (hydrobiotite-vermiculite); designated V,
ii. Natural Soil + Kaolinite (20% by dry weight), designated K,
iii. Natural Soil + Eastern Canada Clay/Quebec Clay (25% by dry weight), designated Q.

The dominant minerals of the natural micaceous soil were illite, phlogopite, hydrobiotite and vermiculite. The properties of the soil are as summarized in Table I. The soils were compacted at their optimum moisture contents using static compaction, into plexiglass cylinders. These were later reassembled as leaching cells, identical to those used in the study reported by Yong, Warith and Boonsinsuk. (1986). For each soil type (V, K, Q), three leaching cells were used, for example V1, V3, V5, representing leaching with 1 pore volume (pv), 3 pore volumes and 5 pore volumes of leachate.

Leaching was performed using a municipal solid waste (MSW) leachate spiked with heavy metals (Pb^{2+}, Zn^{2+}) and cations (Na^+, K^+, Mg^{2+}, Ca^{2+}) in the form of chlorides. The pH of the reconstituted (spiked) MSW leachate was also lowered by adding concentrated HCl. The chemical composition of the reconstituted leachate is shown in Table II. A constant air pressure of 12.0 or 15.0 p.s.i. was applied on the influent, i.e. an equivalent water head

Table I: Summary of Soil Properties.

Property	Values
Specific surface area (m^2/g)	90-206
Cation exchange capacity (C.E.C.) (meq/100g)	7-19
Standard Proctor Compaction:	
Maximum dry density (Mg/m^3)	1.83-1.84
Optimum moisture content (%)	14.4-16.1
Permeability (falling head test) (m/sec)	$1.2\text{-}2.3 \times 10^{-9}$

of 8.4 or 10.6 m depending on the sample height. This created hydraulic gradients ranging from 87.2 to 89.4 for almost all the samples, except for sample Q5 (hydraulic gradient = 97.0).

During the leaching process, effluents were collected at every 0.5 pv and analysed. At the end of the 1 pv, 3 pv and 5 pv series, the soil samples were extruded and cut into 1 cm thick slices. These sliced samples were analysed for pore fluid contents and concentrations of soluble and exchangeable ions.

The analytical procedures were performed in accordance with the procedures specified in the Geotechnical Research Centre (GRC) Laboratory Manual. Pore fluid soluble ions and the soil exchangeable cations were determined using batch equilibrium testing (ASTM, 1984). Determination of species and concentrations of cations and heavy metals was accomplished using atomic absorption spectrophotometry (AAS) after various appropriate dilutions. Chloride was determined using titration with $AgNO_3$. The pH and specific conductivity were determined using a pH meter and the electrophoretic mass-transport analyzer respectively.

Table II: Chemical compositions of the reconstituted leachate (all ionic concentrations are in ppm).

pH	1.3
Specific Conductivity ($\mu S/cm$)	16.8
Na^+	346.0
K^+	164.8
Mg^{2+}	43.8
Ca^{2+}	95.4
Pb^{2+}	1372.2
Zn^{2+}	1141.6
Cl^-	5258.4

RESULTS AND DISCUSSIONS

PERMEABILITY/HYDRAULIC CONDUCTIVITY

The variation in the permeabilities (k values) of the samples for the soils was not significant. In general, the k values of all the samples decreased slightly, with initial values between 0.5×10^{-9} m/sec for the K samples and

2.4×10^{-9} m/sec for the V soil, with increasing number of influent pore volumes during the leaching process. At the end of 5 pore volumes of influent leaching they were about 10% less than the original k values for the three different types of soil samples, i.e. within the range of experimental error. It can be concluded that leaching only caused a very slight decrease in the permeabilities of the three types of soils examined at the initial stages of leaching. Ranking of the permeabilities of the soils showed kV > kQ > kK, and that they remained within the same ranking and order of magnitude (10^{-9}) during the leaching process.

PH

The pH of the original MSW leachate was measured as 7.06. However, for the leaching column tests the pH of the spiked MSW leachate was buffered to 1.33 by the addition of concentrated HCl, to facilitate the study of the migration of the heavy metals (Pb^{2+}, Zn^{2+}) through the soil columns. This procedure was designed to avoid precipitation of the heavy metals (e.g. Pb^{2+} precipitates at pH values > 5 etc). Within the time period of the leaching tests, dissolution degradation of the clay minerals would not be expected to be measurable or significant; this was confirmed by the pH-effluent curves shown in Figure 1. The equilibrium pH attained in the lower portion of the leached soil columns is shown in Figure 2. The initial pH of the soil-water system, shown for the case where pore volume (pv) is zero, in Figure 1, is in general close to the pH of the effluent after one pv application, demonstrating the capability of the system to function as a chemical buffer against the entry of the acidic spiked MSW leachate, as demonstrated by the ability of the effluent pH to remain near the original pH value.

The results given in Figure 1 show the effluent pH to be greatest for sample Q and least for sample K. This is attributed to the differences in the mineralogical compositions. The Eastern Canada Clay used for the Q soil

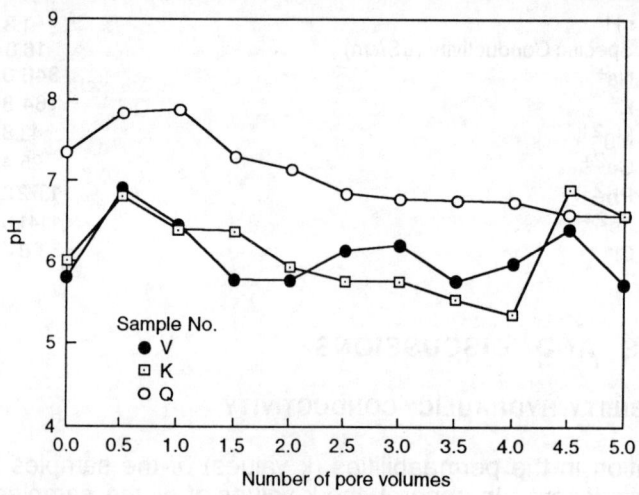

Figure 1: Effluent pH, comparing samples V, K and Q.

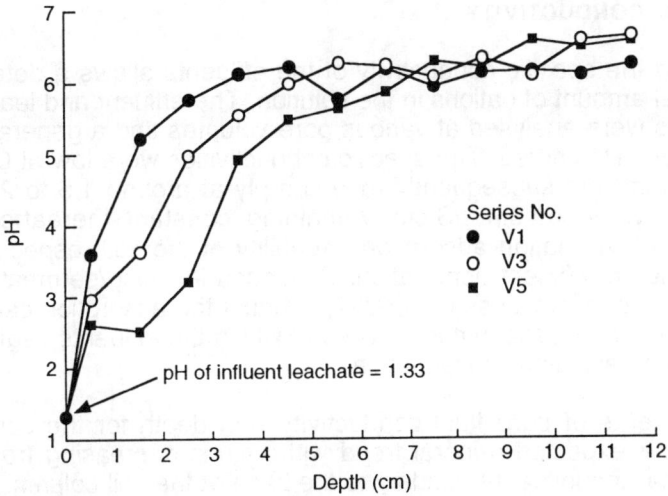

Figure 2: pH variation with depth, sample V.

mixtures contained abundant carbonates, up to 30-40% (calcite, $CaCO_3$ and dolomite, $CaMg(CO_3)_2$) (Wang, 1984; Quigley et al., 1982), which would encourage a higher pH while the kaolinite in the K soil would contribute to a low pH value.

PH VARIATIONS WITH SOIL COLUMN THICKNESS (DEPTH)

The ability of the soil to chemically buffer against the acidic leachate can also be seen clearly in Figure 2, which shows the pH variation with depth (sample thickness) after 1, 3, and 5 pvs. At the completion of each test, the "leached" soil was sectioned into 10 mm thick slices and the soil and pore fluid for each section analysed. As seen from Figure 2, the influent acidic pH (1.33) is quickly buffered and remains close to the natural soil sample pH throughout the length of the sample, i.e. during the course of transport of the contaminants through the sample. The decrease in the equilibrium pH as a result of increasing the number of influent pvs is at the influent end of the sample and provides an indication as to the capability of the system to continue to buffer against the acidic leachate. With the fifth pv, the equilibrium pH condition is reached, approaching 60 mm beyond the influent boundary. If simple linear extrapolation is made, a very conservative estimate would suggest that the 120 mm thick sample could be capable of buffering the spiked MSW leachate for up to 10 pvs of influent leachate. The results shown in Figure 2 indicate that beyond the 10 pv extrapolation, the pH value will decrease gradually, probably falling below an acceptable value after about 15 pv of influent leachate. Simple calculations relating pore volume leachate to total leachate produced in a contained waste landfill will provide information as to the capability of the particular buffer material thickness to function well. The thickness of the buffer material must then be increased to ensure full buffer functioning capability.

SPECIFIC CONDUCTIVITY

Measuring the specific conductivity of the effluents allows a determination of the total amount of cations in the solution. The effluent and leached soils pore fluids were analysed at various pore volumes and a general trend for all soil types was noted. The specific conductivities were low at 0.5 pv (0.2 to 0.7 µS/cm) but subsequently rose sharply at around 1.5 to 2.5 pv to a maximum value of 3.4 µS/cm, remaining constant thereafter. This is indicative of the cation adsorption capability of the soil, especially at the initial stages of effluent penetration. Continued leaching/permeation by the leachate, at pvs in excess of about 2, exhaust the adsorption capacities of the soils and an equilibrium is reached at which the effluents begin to attain a constant total cation concentration.

Measurements of pore fluid conductivity with depth for the soil columns showed the expected general trend with values decreasing from the top portion near the influent boundary to the base of the soil column, indicating the greater amounts of cations present in the pore fluid at the higher levels. As noted in the buffering performance of the soil (Figures 1 and 2) the conductivity values were also greatest for the carbonate rich Q soils and least for the kaolinite mixture.

BREAKTHROUGH CURVES

Breakthrough curves were obtained for the various cations (Na^+, K^+, Mg^{2+}, Ca^{2+}, Pb^{2+} and Zn^{2+}) and the anion Cl^- for the three types of soils (V, K, Q). These curves for chloride, Cl^-, are shown in Figure 3. The chloride ion is non-specific and conservative, hence attenuation of the Cl^- is low. For the V and K soils, breakthrough (ratio of effluent concentration Ce to influent concentration Co, Ce/Co = 0.5) of Cl^- occurred at pv = 1.2. The Ce/Co values increase sharply from 0.5 pv to peak at Ce/Co of about 1 at 2 pv, thereafter remaining constant at Ce/Co about 1. For the Q soils, breakthrough occurred at a pv of about 1.7. Comparing soil types V, K and Q it can be seen that V and K soils show very similar behaviour with respect to attenuation of Cl^-. The difference in the Cl^- breakthrough curve between the Q soils and the V/K soils is similar to the difference in the cation (Na^+, K^+, Mg^{2+} and Ca^{2+}) breakthrough curves observed between Q soils and the V/K soils.

For the Na^+ tests, breakthrough occurred at 2.7 pv for K, 3.0 pv for V and 3.7 pv for Q. In the case of K^+, none of the soils showed breakthrough occurring within the pv range studied. The highest relative concentration attained at 5.0 pv was less than 0.5, i.e. Ce/Co about 0.3. The failure to attain breakthrough for K^+ over the pv range studied is due to the fact that K^+ is often adsorbed and incorporated into the interlayer lattice of micaceous soils (illite, micas, etc) since all three soils studied (V, K, Q) are basically micaceous containing illite, phlogopite, hydrobiotite, vermiculite etc.

For Mg^{2+}, all three soils also showed very high relative concentrations with Ce/Co of up to about 30, indicating excessive elution of Mg^{2+} from the soil

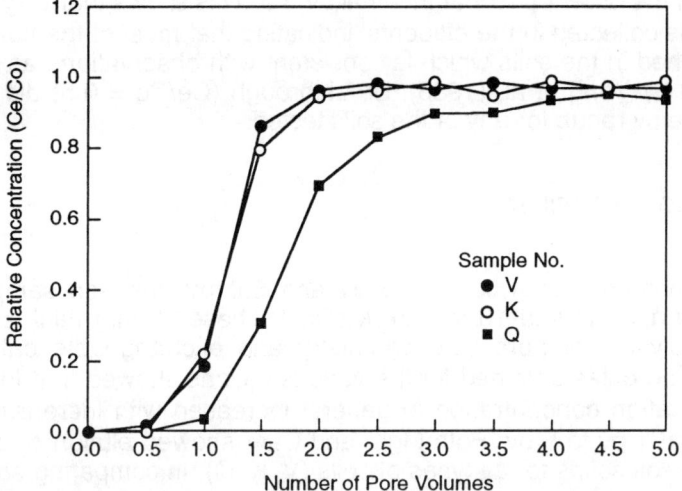

Figure 3: Cl⁻ breakthrough curves for V, K and Q.

solids during the leaching process. For all soil types, breakthrough (Ce/Co = 0.5) occurred at around 0.5 pv. Again a similarity in behaviour of the V and K soils was evident, with V soils attaining slightly higher Ce/Co than K soils. The Q soils showed lower Ce/Co values for pv < 2.5, implying the slower elution or better adsorption capacities of the Q soils as compared to the other two types. However, for pv > 2.5, the Q soils showed the highest Ce/Co values. This would indicate that with excessive leaching (pv > 2.5), the carbonates in the Q soils begin to release more Mg^{2+} into the pore fluids, exceeding that eluted from the micaceous soils alone.

The elution phenomenon was further demonstrated by the high relative concentrations of up to Ce/Co = 10 to12 measured for the case of Ca^{2+}, indicating high elution of Ca^{2+} from the soil solids during the leaching process as noted for Mg^{2+}. For the V and K soils, breakthrough occurred at pv about 0.6 for the V and K soils and pv = 0.5 for the Q samples. A similarity in behaviour between the V and K soils was again observed, with K soils showing slightly higher maximum Ce/Co values at 11 to 12 for K compared to 0 to 10 for V soils while slower elution of Ca^{2+} for Q soils (pv < 2.5) once again indicated the better adsorption capacities of the Q soils. For pv > 2.5, however, the behaviour of the Q soils was similar to that of the other two, with Q soils showing an intermediate maximum Ce/Co value between 10 and 11. The high relative concentrations of Ca^{2+} and Mg^{2+} in the effluents collected can also be attributed to cation exchange or replacement by Pb^{2+} and Zn^{2+} as the heavy metals were introduced in high concentrations in the leachate (> 1000 ppm). It would appear from all the results that the Q soils showed higher attenuation of cations (Na^+, K^+, Mg^{2+} and Ca^{2+}) as well as the chloride anion.

An examination of the results of the breakthrough curves for Pb^{2+} and Zn^{2+} showed that the relative concentration values were all extremely low, i.e.

Ce/Co in the order of 10^{-4} at the final pv = 5. This shows that very little Pb^{2+} or Zn^{2+} is collected in the effluents, indicating that most of the heavy metals are retained in the soils which is consistent with observations as previously reported by Yong et al, (1986). Breakthrough (Ce/Co = 0.5) did not occur within the pv range for any of the soils tested.

MIGRATION PROFILES

After leaching with 1.0 pv, 3.0 pv and 5.0 pv, the soil samples were extruded and cut into 10 mm thick slices. These 10 mm thick slices were then analysed for pore fluid chemistry and exchangeable cations. The migration profiles obtained for the various cations showed that for Na^+ and K^+, the cation concentration in general increased with increasing permeation from 1 pv to 5 pv. Both Mg^{2+} and Ca^{2+} showed elution or desorption from the soil solids for all types of soils (V, K, Q). In comparing soil types V, K, Q, it was seen that the Q soils retained higher concentrations of Mg^{2+} and Ca^{2+}, as expected, because of the release of these cations from carbonates reacting with the leachate.

The migration profiles for Pb^{2+} shown in Figure 4 indicate drastic reductions in the Pb^{2+} values in all the soil types studied from 1372 ppm (initial Pb^{2+} concentration in leachate) to zero ppm. The migration profiles showed a slight increase in Pb^{2+} concentrations from 1 pv to 5 pv as seen in the upper half of the soil columns for V and K soils. The lower half of the soil columns indicated total attenuation of the Pb^{2+} ion. Results for V and K soils are similar with V somewhat in excess of K, while the Q soils showed almost complete attenuation from 10 mm downwards. The order of retention capability is Q > V > K.

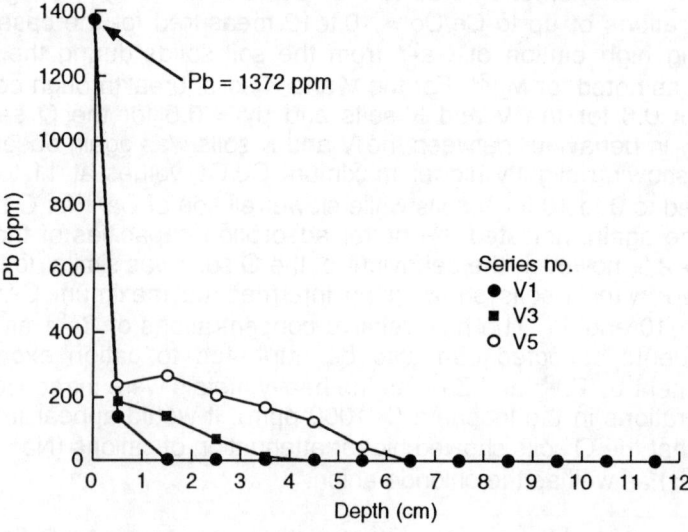

Figure 4: Pb migration profile, sample V.

In the case of Pb^{2+}, the migration profiles for Zn^{2+} also showed considerable reduction or attenuation, from 1142 ppm (initial Zn^{2+} concentration in leachate) to zero ppm. The Zn^{2+} concentrations increased with increasing number of pore volumes of leachate permeated and the depth of migration also increased accordingly. However, for the V and K soils, total attenuation of Zn^{2+} occurred at depths > 80 mm; for the Q soils, total attenuation occurred at depths > 20 mm. The order of retention capability for Zn^{2+} was the same as for Pb^{2+}. Comparing Pb^{2+} and Zn^{2+}, Zn^{2+} was seen to be slightly more mobile, with a slightly greater depth of migration of up to 80 mm for Zn^{2+} compared to about 60 mm for Pb^{2+} and with slightly higher concentrations of Zn^{2+} despite the lower intitial concentration of 1142 ppm compared to 1372 ppm Pb^{2+} in leachate.

ADSORPTION ISOTHERMS

The relation between absorbed and soluble cations in soils can be shown in the form of adsorption isotherms which depict the equilibrium distribution between the amount of cation adsorbed and the cation concentration remaining in the pore fluid at equilibrium conditions. For soil suspensions, adsorption isotherms can normally be obtained through batch equilibrium tests and the resulting plots are usually satisfactory. This may not necessarily be the case with leaching column tests. The adsorption isotherms for the soils studied do not show exceptional behaviour and can be approximated by either the Langmuir or Freundlich models.

EXCHANGEABLE CATIONS

To obtain a better insight into the adsorption or retention of cations by the soil columns, the variation of exchangeable (adsorbed) cations in relation to soil sample depth was examined. The exchangeable cations were determined using NH_4OAc through batch equilibrium tests. The results show that exchangeable Na^+ was higher in the upper half of the soil column at the initial stages of leaching. With increasing permeation by leachate, the exchangeable Na^+ in the lower half of the soil column increased. Similar trends were observed for K^+.

An examination of the variation of exchangeable Pb^{2+} with depth (Figure 5) showed that the V and K soils demonstrate similar behaviour, i.e. high absorbed Pb^{2+} at the upper levels of the soil column and greater adsorption with increasing permeation (e.g. pv 1 to pv 5). The Pb^{2+} also migrates deeper down into the soil column with increasing permeation. The similarity between the V and K soils is striking. For the Q soils, adsorption of Pb^{2+} is confined to the top 20 to 30 mm below which no exchangeable Pb^{2+} is detected (total attenuation at upper 30 mm). For the Q soils, adsorption of Pb^{2+} also increases with increasing permeation. The adsorption capacity of Q soils being greatest and K soils least. The similarity between V and K soils is again striking. Zn^{2+} showed similar results to Pb^{2+}.

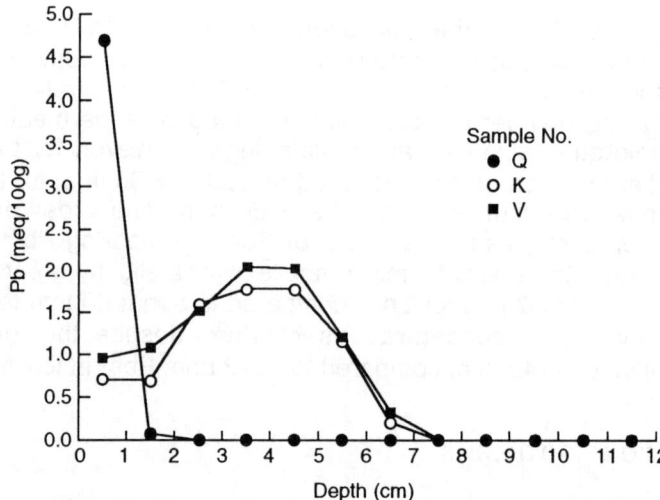

Figure 5: Exchangeable cation, sample V, 5 pv, Pb.

From the results of the exchangeable cations studied, (Na^+, K^+, Mg^{2+}, Ca^{2+}, Pb^{2+} and Zn^{2+}), an indication of the cation exchange capacities of the soils and their variations with depth can be obtained. As can be seen in Figure 6, the Sum of Exchangeable Cations shows that Q > V > K, i.e. with Q about 40-50 meq/100 g, V about 10 to 30 meq/100 g and K about 10 to 20 meq/100g. This is to be expected since Q soils contain Quebec Clay which is known to have higher cation exchange capacity (C.E.C.) of about 60 meq/100 g (Warith, 1987) while K soils contain kaolinite which has a lower C.E.C. (about 5-15 meq/100g, Yong and Warkentin, 1975). The natural micaceous soils have C.E.C. values ranging from 10 to 20 meq/100 g (see Table I).

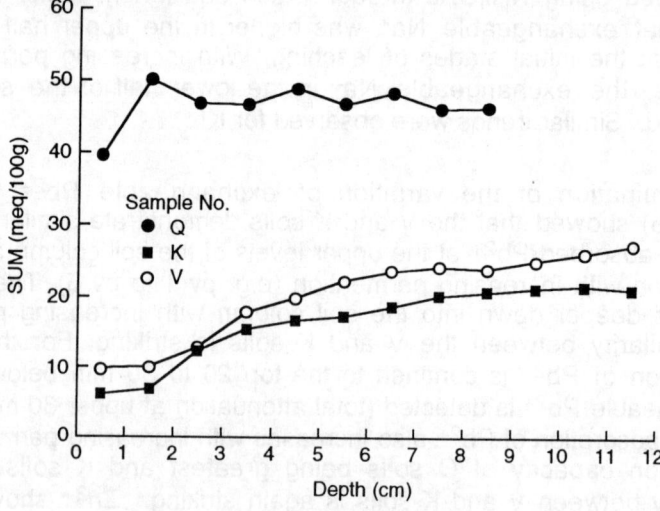

Figure 6: Sum of exchangeable cations.

CONCLUDING REMARKS

The tests designed in this study were intended to provide information concerning the capability of the natural soil (and its mixtures) to function as a proper clay barrier against a specific type of MSW leachate. The use of "spiking" to increase the heavy metals concentration in the natural MSW leachate obtained from a field site provided the opportunity to examine the more "drastic" conditions of contaminant assault - a procedure which seeks to maximise contaminant-soil interaction effects. The performance of the material in response to attack from the spiked MSW leachate should be assessed on the basis of chemical buffering capability and permeability alterations. The buffering performance and capability of the materials tested appear to be consistent with the predictions and analyses reported previously by Yong, Warkentin, Phadungchewith and Galvez (1990).

The implications arising from an evaluation of the performance of the natural micaceous soil, and also from the use of the Eastern Canadian clay containing carbonates, are encouraging; suggesting that a proper soil engineered barrier can be constructed - provided care is taken with compaction control and placement. From the test results, the following concluding remarks can be made.

i. The permeabilities of the three soil types (V,K,Q) tested ranged from $0.4 - 1.9 \times 10^{-9}$ m/sec. Leaching with the spiked MSW leachate resulted in only a very slight decrease in the permeabilities of the soils indicating the k compatibility between the soils tested and the leachate.

ii. The effluent pH's ranged from 5.4 to 7.9 despite the very low influent leachate pH (1.3). This indicates the high buffering capacities of the soils, in particular the Q soils, where effluent pH's show $Q > V > K$. The ability of the soils to maintain buffer capability after 5 pv of influent leachate suggests that the material is likely to be suitable for engineered barrier construction.

iii. The pH values of the pore fluids of the soil columns increase from the top (influent boundary) rising from the low leachate pH value to the natural soil pH value after permeation of a few centimetres into the sample. This supports the observations made above concerning the buffering capability of the soils and further testifies to the capability of the soil to function as a competent engineered material. Simple extrapolation procedures can determine the optimum thickness of the barrier required to buffer against the total leachate load anticipated from the waste pile.

iv. The results of the information concerning effluent and pore fluid analyses together with the breakthrough performance of the various cations all support the observations made above.

v. The differences in the rentention/adsorption and buffering capacities of the Q, V and K soils are attributed to differences in the composition of the soils, notably the presence of carbonates in the Q soils. Whilst only the V soils (natural micaceous soils) perform satisfactorily in regard to having a low permeability (10^{-9} m/sec), high pH and high adsorption/retention of cations, especially Pb^{2+} and Zn^{2+}, the addition of Quebec Clay/ Champlain Sea Clay (Q soils) enhanced its performance. The addition of kaolinite (K soils) to the V soils, however, reduced its performance vis-a-vis adsorption or retention of cations. The C.E.C. of Q was about 40 to 50 meq/100g, V about 10 to 30 meq/100g and of K about 10 to 20 meq/100g.

vi. As with previous studies on heavy metal retention for clay soils of similar mineralogy, retention of heavy metals for the soils was studied mainly through precipitation (high pHs of >5) and cation exchange. The results show the latter being confined to the upper portions of the soil columns and are consistent with observations reported by Yong et al. (1986).

ACKNOWLEDGEMENTS

The authors acknowledge the technical assistance provided by the staff of the Geotechnical Research Centre.

REFERENCES

AMERICAN SOCIETY FOR TESTING AND MATERIALS, (1984). Standard Test Method for Distribution Ratios by the Short Term Batch Method. *ASTM Standards*, **4-08**, 766-733.

BOWDERS, J.J., DANIEL, D.E., BRODERICK, G.P. & LILJESTRNA, D.H. (1986). Methods for testing the compatibility of clay liners with landfill leachate. In: *Hazardous and Industrial Solid Waste Testing: Fourth Symposium, ASTM STP-886*, J.K. Petros, Jr., W.J. Lacy, R.A. Conway, (eds.) American Society for Testing and Materials, Philadelphia, 1986, 233-250.

QUIGLEY, R.M., SETKI, A.J., BOONSINSUK, P., SHEERAN, D.E. & YONG, R.N. (1982). Geologic Control on Soil Composition and Properties, Lake Ojibway Clay Plain, Matagami, Quebec, *Proceedings of the 25th Canadian Geotechnical Conference*, 460-479.

WANG, B.W. (1984). *Compositional Effects on Soil Suction*. Unpublished MEng thesis, McGill University, Montreal, Canada.

WARITH, M.A. (1987). *Migration of leachate solution through clay soil*. Unpublished PhD thesis, McGill University, Montreal, Canada.

YONG, R.N. & WARKENTIN, B.P. (1975). *Soil Properties and Behaviour*. Elsevier, New York, 499 pp

YONG, R.N., WARITH, M.A. & BOONSINSUK, P. (1986). Migration of leachate solution through clay liner and substrate. *ASTM STP-933*, 208-225.

YONG, R.N., WARKENTIN, B.P., PHADUNGCHEWITH, Y. & GALVEZ, R. (1990). Buffer capacity and lead retention in some clay materials. *Water, Air and Soil Pollution Journal*, **53**, 53-67.

5-3 A EUROPEAN PERSPECTIVE ON LANDFILL ENGINEERING

A. STREET
MRM Partnership, Queen Victoria House, Redland, Bristol BS6 6USA

ABSTRACT

The importance of understanding the amount and nature of municipal waste and problems related to its disposal are highlighted. The situation in a number of European countries is discussed and comparison made with that for the U.K. The paper includes the main points in the EC Landfill Directive and discusses the implications of these.

INTRODUCTION

Significant changes are taking place within the waste management industry, both within the U.K. and Europe as a whole, largely resulting from the introduction of stricter environmental controls at national and international (European Community) level.

It is estimated that about 70% of the municipal waste and 35% of the industrial waste produced throughout the European Community is landfilled. Within the U.K. over 90% of "controlled" wastes are landfilled. The adoption of an integrated approach to waste management, including clean technology and recycling, will ultimately reduce the overall amounts of waste for final disposal to landfill, although it is very difficult to predict the extent to which this may occur.

One thing is certain, however, landfill has to remain a vital part of waste management policy for the U.K. and other European countries. No matter how effective recycling, waste minimisation, incineration and pre-treatment strategies may prove to be, landfill is the final disposal route for residues and those wastes which cannot be recycled or pre-treated.

Given this dependency on landfill it is interesting to note the different approaches adopted across Europe.

FRANCE

The complexity of laws and regulatory controls governing waste management policy in France has led to differing levels of achievement. Within the existing regulatory framework three classes of landfill are defined, namely:

- CLASS I   -   authorized industrial wastes
- CLASS II   -   municipal and equivalent (commercial) wastes
- CLASS III -   inert and construction industry type wastes (wood, demolition products, concrete etc.)

Depending on the nature of waste produced the Ministry of the Environment issue orders and/or circulars laying down technical recommendations to which reference should be made by the Prefects (Regional Authorities) in issuing authorisations. These technical instructions ("instruction techniques") describe waste type, site selection criteria, operational controls and aftercare procedures.

For Class I landfills, technical instructions have been produced dealing with existing and new industrial waste sites. Engineering and site selection criteria include:

- a sub-base of at least 5m of material with a hydraulic conductivity (k) of $\leq 1 \times 10^{-9}$ m/s
- a basal geomembrane to contain leachate within each cell
- a basal leachate drainage system ($k \geq 1 \times 10^{-4}$ m/s) to maintain a maximum of 30cm of head of leachate above the base.

For Class II landfills engineering and site selection criteria are less stringent and include:

- a sub-base of at least 5m of material with a hydraulic conductivity (k) $\leq 1 \times 10$-6 m/s
- a low permeability final cover layer.

No specific engineering criteria are set for Class III, inert waste sites. Co-disposal is not practised in France. There is a legal requirement for consulting engineers to design and supervise landfill engineering works. Operational and aftercare insurance funds are currently being set up by the two major private waste operators in France.

GERMANY

The multi-barrier concept (Steif, 1990) forms the basis for planning, operation and aftercare of landfills in Germany. Waste prevention and stabilisation are fundamental precursors to existing, and more especially future, landfilling policy. Long term strategy (ie, greater than 15 years) is to incinerate all domestic and industrial waste producing a more stable, inert residue for final landfill disposal.

The German system of landfill classification is similar to that outlined in the Draft EC Landfill Directive (Council Directive, 1991), based on the following categories of waste:

- contaminated soil
- construction industry waste
- domestic waste
- industrial waste
- toxic or hazardous waste

The latest draft of the German Federal Regulations (TA Siedlungsabfall) stipulates that domestic waste landfills (Deponieklasse II) should have a composite liner system comprising a synthetic membrane (HDPE) and 0.7m of compacted clay having a maximum permeability of 1×10^{-10} m/s. For inert waste landfills (Deponieklasse I) the basal liner may be reduced to 0.5m of compacted clay having a permeability of 1×10^{-9} m/s. For toxic or hazardous waste sites the clay layer must be increased to a minimum thickness of 1.5m. A summary of the German technical requirements for toxic and hazardous waste landfills is presented in Table I.

Table I: German technical requirements: toxic and hazardous waste landfills.

1. SUBSOIL

    A) GEOLOGICAL BARRIER
    High absorption potential
    For soft rock $d = \geq 10m$; $k \leq 1 \times 10^{-7}$ m/s
    For hard rock $d = \geq 30m$; $k \leq 1 \times 10^{-8}$ m/s

    B) SUBGRADE

    sufficient bearing capacity
    inclination $\geq 3\%$
    if necessary additional measures

    C) GROUNDWATER LEVEL
    not higher than 1m below
    subgrade top level
    if necessary additional measures

2. BASAL SEALING SYSTEM

    COMBINED SEALING SYSTEM
    (i) MINERAL SEALING
    (low permeability clay layer)
    $d \geq 1.5m$; clay mineral content $\geq 10\%$
    $k \leq 1 \times 10^{-10}$ m/s
    inclination $\geq 3\%$

    (ii) SYNTHETIC FLEXIBLE MEMBRANE
    $d \geq 2.5mm$; approved material

3. SURFACE SEALING SYSTEM

    COMBINED SEALING SYSTEM
    (i) MINERAL SEALING
    $d \geq 0.5m$; clay mineral content $\geq 10\%$
    $k \leq 1 \times 10^{-10}$ m/s
    inclination $> 5\%$

    (ii) SYNTHETIC FLEXIBLE MEMBRANE
    $d \geq 2.5mm$; specified material

4. PERVIOUS BLANKET WITH DRAINAGE PIPES

    pervious blanket, grain size 16/32 mm
    $d \geq 0.3$
    $k \geq 1 \times 10^{-3}$ m/s
    Drainage pipes dn 300 (above basal liner only)

The German Authorities require a high standard of engineering, with independent consultants being responsible/liable for ensuring that the works are designed and constructed to the criteria stipulated. Most landfills are constructed above natural ground level using gravity leachate drainage systems to facilitate leachate removal.

Licensing procedures in Germany are similar for all types of landfill. The issuing authority is always at Laender (State) level, which through a process of public consultation may impose stricter recommendations than those set at national level.

The Germans do not operate landfill sites on a co-disposal basis. The most toxic wastes are normally landfilled in secure, high technology, "mono-disposal" sites.

ITALY

The Italian system of landfill classification is quite complex as illustrated by Figure 1. Waste quality assessment criteria have been adopted which dictate the standards of landfill or treatment required.

In Italy a set of national regulations exists for each class of landfill, although these regulations are currently being reviewed. The regulations cover the following aspects of landfill design:

- location of the site in relation to water resources and inhabited areas;
- geological and geotechnical features;
- groundwater protection and engineering of the basal liner system;
- leachate collection systems;
- surface water drainage; and
- site restoration, including engineering of the landfill capping system.

At present, minimum geotechnical requirements are given for all landfill classes except Class IIa (inert wastes) and Class III (special encapsulated waste). It should be noted that each landfill project may be approved by the appropriate Italian Regional Authority which can impose additional barriers or other amendments to the lining system. With the exception of Class IIa and III landfills, the emphasis is again on the adoption of a multi-barrier approach comprising double synthetic basal liners or composite basal lining systems similar to those used in the United States. Composite sealing layers are not however specified for final cover systems over any class of landfill.

HOLLAND

In Holland any waste can be landfilled subject to the provision of appropriate measures to prevent water pollution and other environmental hazards. Regulatory controls on landfill engineering do not appear to be as

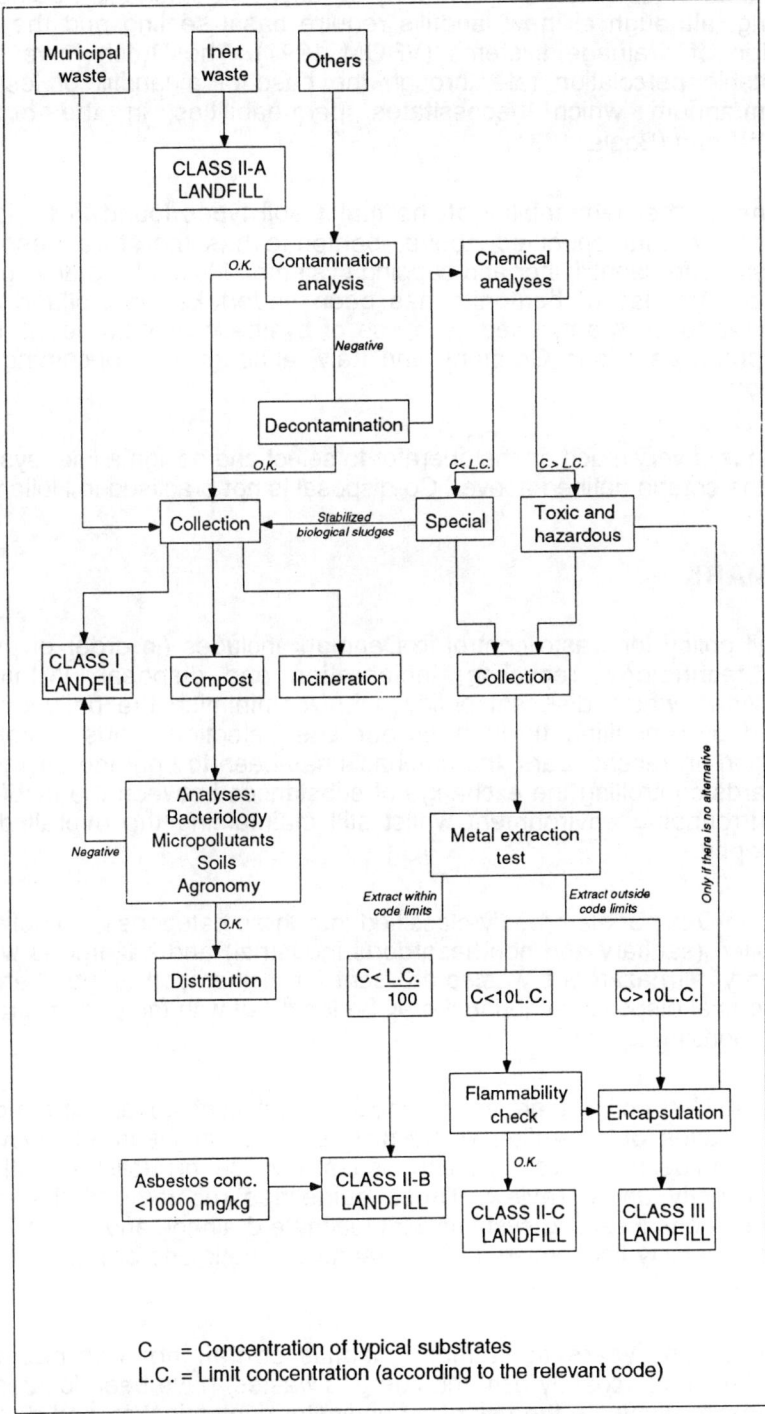

Figure 1: Italian classification system for wastes and landfills.

specific as those in force in Germany or Italy. The emphasis has been on capping, although all new landfills require basal sealing and the incorporation of drainage systems (VROM 1991). The Dutch have set an acceptable percolation rate through the base of a landfill of less than 20 mm/annum which necessitates permeabilities in the order of 1×10^{-10} m/s (Boels, 1991).

In general, the permeability of the major soil types found in Holland is higher than that specified above; bentonite has therefore been used extensively for landfill liner and capping systems. Much of the development work on the use of bentonite has been undertaken in Holland and it continues to be widely used. The use of synthetic membranes is not so widespread as it is in Germany and Italy, although it is becoming more common.

The onus is very much on the operator to select and design a liner system to meet the criteria defined above. Co-disposal is not practised in Holland.

DENMARK

Overall policy for waste control in Denmark includes (in order of priority) clean technology, recycling, incineration and disposal. Within this framework, waste disposal policy aims to minimise the pollution risks caused by landfilling through proper site selection, construction and operation. In recent years, the emphasis has been to upgrade engineering standards controlling the exchange of substances between the landfill and the surrounding environment whilst still maintaining the overall design philosophy.

Waste in Denmark is broadly classified into three categories, namely inert, controlled (sanitary and non-hazardous industrial) and hazardous wastes. In theory, all hazardous wastes are sent for incineration and/or treatment prior to final disposal, although it may be landfilled with the prior approval of the responsible authority.

Similar to U.K. policy, there are no specific regulations governing the design and operation of controlled waste disposal sites in Denmark. However, national guidelines covering these issues have been published (DIF) and are currently under review. These guidelines recommend the use of composite basal sealing systems with leachate drainage and collection and low permeability final covers (incorporating synthetic and/or natural sealing systems).

During recent years most mono-landfills containing non-hazardous industrial waste (coal fly ash and flue gas wastes) have been located near the coast to eliminate the risk of groundwater contamination and designed according to a "controlled leaching" strategy. There are no specific geotechnical recommendations for inert disposal sites.

Since 1983, a remediation programme has been established to clean up old contaminated landfill sites at a cost of approximately 53 million pounds to the Danish Government.

BELGIUM

National policy within Belgium is administered by a number of governmental bodies which outline general environmental policy and implement EC Directives. Three regions exist within Belgium - Wallonie, Flemish and Brussels - who (since 1981) are responsible for formulation of waste disposal policy.

Broadly speaking waste is categorised into inert (Class III), municipal (Class II) and industrial (Class I). Other waste such as pesticides, toxic waste and radioactive waste are subject to national regulations and are generally not acceptable for landfilling.

Waste disposal policy varies considerably between regions as a result of differing demands; for example, in the Brussels region waste is either incinerated or exported to other regions for final disposal. Current policy aims to achieve total containment of Class I and II wastes, but stabilisation through selection, treatment, solidification and combustion prior to deposition is now becoming an integral part of waste disposal policy. Funding through the Belgium government for the clean up of old landfill sites has also now been initiated.

HOW DOES THE U.K. COMPARE?

Landfill waste disposal in the U.K. currently operates within a well defined framework comprising four distinct elements:

- Planning controls
- Site licensing
- Enforcement
- Defined standards

The Control of Pollution Act (COPA - 1974) broadly defines operational requirements for landfills. This has now been superseded by the Environmental Protection Act (1990) which incorporates the requirements set out in COPA and strengthens these with more stringent controls over the surrender of licences and introduces a duty for authorities and operators to monitor and remedy closed landfill sites. National policy classifies waste into three broad categories, namely household, commercial and industrial wastes. Other terms in common usage include:

- Special Wastes: those containing, or contaminated with materials that are dangerous to human health, flammable waste and certain medical products.
- Difficult Wastes: those which could be harmful to the environment, either in the short or long term.

Whilst most other European countries have, in general, introduced a formal landfill classification system, no such system exists in the U.K. However, by considering the waste types received at landfills three broad categories of site can be identified:

- Multi-disposal sites: where more than one controlled waste type are deposited together (e.g. household and commercial).

- Co-disposal sites: where limited amounts of difficult wastes (solids and liquids) are landfilled with household and commercial waste to utilise the microbiological and chemical processes within the landfill to immobilise, neutralise or breakdown the difficult waste.

- Mono-disposal sites: where only one type of waste is landfilled. This usually applies to inert waste only sites where no specific site preparation would normally be required.

One of the principal changes that has taken place in the U.K. over the last year or two is the move away from "dilute and attenuate" landfill sites towards development on a containment basis. The principal driving forces behind this move are the National Rivers Authority (NRA), who are seeking to protect groundwater resources and no longer view dilution by groundwater as an acceptable means of leachate treatment/disposal, Waste Regulation Authorities who are imposing stricter controls on landfill gas migration, and the anticipation of further EC legislation.

While certain sites may benefit from a degree of containment by virtue of geological conditions, most would require varying degrees of engineering to ensure protection against leachate and landfill gas migration.

In the U.K. the most commonly used landfill sealing material is natural clay. Whilst there are no statutory U.K. guidelines or design criteria for clay liners, or indeed any other aspect of landfill engineering, several guidance documents have been produced. These include Waste Management Paper No. 26 and "Earthworks for Landfill Sites" prepared by the NRA North West Region. The emphasis is on the method of emplacement and the need for careful control of the material type and its moisture content rather than purely adopting an "end result" permeability specification.

The U.K. approach is based upon each site being considered on its own merits related to the type of waste, local geological and hydrogeological conditions and the nature and extent of any pollution risk. A lining system is normally required to offer a primary and secondary defence. Where a "secondary" defence exists by virtue of the in-situ ground conditions the minimum requirement for a liner would normally be for a 1.0m thick clay liner compacted to achieve a permeability of no greater than 1×10^{-9} m/s. An approach which is becoming more common is to adopt a composite system utilising compacted clay and synthetic (HDPE) liner since this combination is considered to offer the most effective protection against mechanical and chemical failure.

For any containment landfill the liner must be backed up by an effective leachate drainage and collecting system. In many cases a maximum allowance for leachate head above the liner will be defined in the site licence.

IMPLICATIONS OF THE EC LANDFILL DIRECTIVE

The most significant piece of EC legislation concerning the landfilling of waste is the proposed Landfill Directive (Landfill, 1991). The Commission's formal proposal follows six preliminary drafts indicating the level of concern expressed by national governments and the chemical and waste management industries.

The main features of the Directive include:

- strict categorisation of landfill sites (hazardous, municipal/non-hazardous, inert);
- banning of certain wastes;
- defining site selection criteria;
- effective natural or synthetic barriers to protect soil and groundwater;
- comprehensive surface water, leachate and gas management/control procedures;
- comprehensive procedures for detailing permits and reporting to the waste authority;
- effective procedures for accepting, monitoring and controlling landfilled waste; and
- defined closure and aftercare procedures.

The Commission's acceptance of the principle of co-disposal has been welcomed by those involved in the U.K. waste and chemical industries. However concerns have been expressed, both in the U.K. and other European Community countries, over the basis for leachate testing, the possible ambiguity in the definition of "dilution" and the lack of availability and nature of insurance facilities for landfills.

Introduction of a formal waste and landfill classification, together with specific requirements on the permeability and thickness of the substratum to the landfill base, appears on the face of it to bring the U.K. more into line with the approach adopted in other European countries where classification systems and engineering specifications have already been introduced. This should not, however, necessarily be seen as implying that the U.K. has fallen behind the rest of Europe in terms of landfill engineering standards. The effect, however, has been to reinforce the movement towards well engineered containment sites with an associated increase in disposal costs.

CONCLUDING REMARKS

It can be seen that by comparison with other European countries the U.K. currently appears to lack specific statutory guidance on landfill classification and related engineering design criteria. The onus is very much on each site being considered on its own merits with the engineering elements of the scheme being developed according to site specific requirements related to risk. Given the planning and licensing framework adopted in the U.K. this site specific approach appears to work well since the regulatory authorities are fully involved in the development process. This should not, however, be seen as an excuse for the landfill design engineer - consultant or contractor - to avoid facing up to their responsibilities and potential liabilities. We must avoid the temptation to see the NRA or Waste Regulation Authorities as a free source of advice on engineering design matters.

The situation is of course going to change as a result of implementation of the EC Landfill Directive. On the face of it, the result will be to bring the U.K. more into line with the approach adopted in other European countries where classification systems and engineering specifications have already been introduced. It would be wrong to view this as an indication that the U.K. has necessarily fallen behind the rest of Europe in terms of landfill engineering standards. U.K. landfill design practice compares very favourably with the rest of Europe and we in the U.K. should strongly oppose any ill informed views which may be expressed to the contrary.

ACKNOWLEDGEMENT

The Author is grateful to MRM Partnership for permission to prepare and present this paper.

REFERENCES

BOELS, D. (1991). *Aspects of Landfill Design in the Netherlands.*
COUNCIL DIRECTIVE ON THE LANDFILL OF WASTE (91/C 190/01), (1991).
GDA. *Recommendations. Geotechnics of Waste Disposal Sites and Abandoned Sites.* (English Translation from German)
HMSO, (1990). *Environmental Protection Act.*
SANITARY LANDFILL LINERS. *DIF recommendations DS/R 466 (In English).*
STIEF, K. (1990). Multi barrier concept in West Germany. In: *Sanitary Landfilling,* Stegmann, R. and Cossu, R. (eds).
VROM, (1991). *Guidelines for the Construction of impervious final covers for waste product storing sites and landfills (634/EA91/0006/16895) (English).*

5-4 CLASSIFICATION OF EXCAVATED MATERIALS FOR DISPOSAL PURPOSES - A CASE HISTORY ON THE ANALYSIS OF GROUND CHEMISTRY

C. MIRZA
Strata Engineering Corporation, Don Mills, Ontario, Canada

ABSTRACT

Guidelines for soil and debris removal from construction sites within the greater Metropolitan Toronto area in the Province of Ontario, Canada, have been in use for some years. They are based on provincial environmental legislation and regulations. The soil or debris intended for removal from the construction site is tested for acceptability as lake or open water fill material as lakefill disposal is generally the most economical as the city borders Lake Ontario. If lakefill soil phytotoxicity criteria are not met, the other options range from ordinary terrestrial disposal to disposal as a registered waste at a licensed landfill or waste handling facility, all of which are considerably more expensive alternatives than lake filling.

This paper describes how ground chemistry and its interpretation play a significant role in the cost of construction and emphasises the importance of documentation and communication between engineering consultants, contractors, architects and owners.

INTRODUCTION

The de-commissioning and disposal of contaminated land is a key environmental consideration whenever the sale, transfer or re-development of a property is involved. In the province of Ontario, Canada, legislation places the responsibility for site clean-up on current and past owners. In Metropolitan Toronto, a city of over 2.2 million inhabitants on the north shore of Lake Ontario, guidelines are in place for site clean-up by excavation and removal of the material to an off-site disposal facility. These guidelines require that any soil leaving a construction site must meet certain ground chemistry standards for the type of disposal option selected. The most economical means of soil disposal in Toronto is in open or confined water at two or three major land reclamation sites bordering the city's south shoreline. Hence, every contractor attempts to dispose of soil from a construction site within the city as "lakefill". If the soil phytotoxicity standards for either open or confined lake disposal areas are not met, the soil is disposed of as terrestrial landfill at a slightly higher cost. Should the soil be found to be unacceptable by terrestrial landfill standards, it is taken to a licensed landfill site at a considerably higher cost.

A case history is presented to illustrate current regulations and guidelines for ground chemistry in construction. Lessons learned in interpretation and project documentation are described.

ENVIRONMENTAL REGULATIONS AND GUIDELINES

Current environmental legislation in the Province of Ontario, Canada, has recently been summarised by Phyper and Ibbotson (1991). The decommissioning and disposal of contaminated lands is covered by the Environmental Protection Act and its various regulations, which are being revised as experience is gained with a variety and complexity of site clean-up issues (MOE, 1989).

Within Metropolitan Toronto the responsibility for ensuring quality of fill materials used in the reclamation of land from Lake Ontario has been delegated to the Metropolitan Toronto and Region Conservation Authority (MTRCA). The lakefill quality guidelines were designed as an overall management system (Figure 1) to protect the present and future aquatic environment of lake fill locations. The improved Lakefill Quality Control Program (MTRCA, 1988) or ILQCP, enforces environmental controls on materials intended for lakefill disposal. The ILQCP management system is essentially a decision-making framework which provides for a rational approach to the disposal of soil from construction sites within Metropolitan Toronto at two or more established land reclamation sites on the north shore of Lake Ontario.

Most contractors operating within City limits find the Lake Ontario lakefill locations to be convenient and economical for soil disposal from excavated sites. On the other hand, general terrestrial disposal, such as at farms and golf courses outside the city limits, are both less productive and more expensive, involving haulage out of the City, traffic delays and much higher tipping fees.

The cost of disposal at a designated landfill site is about twenty times greater than that of disposal at a general terrestrial site, which in turn is about four to five times greater than that for lakefill disposal.

For soil considered sufficiently toxic to be classified as a hazardous waste, the costs of removal escalate even further as disposal is allowed only at designated hazardous waste receiving sites, of which there is currently only one in the entire Province.

IMPROVED LAKEFILL QUALITY CONTROL PROGRAM

The ILQCP classifies construction sites as either small or large. Small sites are those which generate less than 200 m^3. They are exempt from certain testing and auditing requirements of the guidelines while larger sites require a pre-disposal environmental inspection, site audit and environmental report. Large sites must be registered with the MTRCA before construction commences and on approval, Bills of Lading are issued by the MTRCA which allow the carrier to enter the lakefill disposal gates subject to further visual verification of the incoming load by the gate operator. Sites which contain 'complex' materials, ie those unsuitable for lakefill disposal, require on-site supervision to ensure that only acceptable materials are actually sent to the lakefill site.

CLASSIFICATION OF EXCAVATED MATERIALS FOR DISPOSAL PURPOSES 475

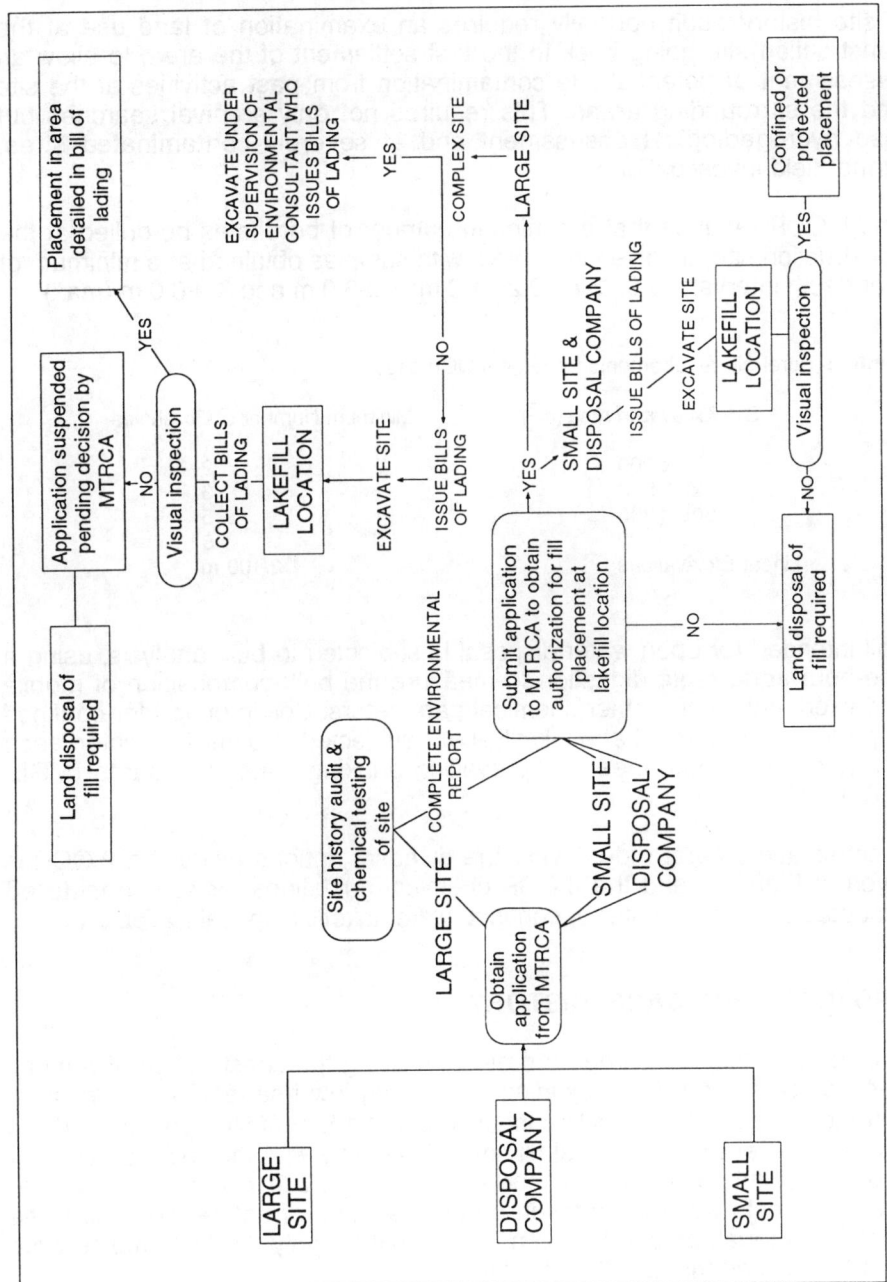

Figure 1: ILQCP management guidelines for lakefill disposal (MTRCA, 1988).

A site history audit normally requires an examination of land use at the construction site going back to the first settlement of the area, to allow an assessment of potential site contamination from past activities at the site and the surrounding areas. This requires not only archival searches but also hydrogeological assessment and at severely contaminated sites, further field investigations.

The ILQCP requires that a minimum number of boreholes be drilled at the construction site, as given in Table I, with samples obtained at a minimum of four depth intervals: 0-0.25 m; 0.25-1.0 m; 1.0-3.0 m and 3.0-6.0 m (max).

Table I: Borehole requirements for large ILQCP sites

Site Excavation Area (m^2)	Minimum Number of Boreholes
< 200	2
200-500	3
500-1000	5
> 1000	8
Linear Excavations	2/100 m

Soil intended for open water disposal is subjected to bulk analysis, using a one-hour aqua regia digestion to measure the bulk composition of mobile and fixed metals and other chemical parameters. Soil intended for confined or protected placement at a lakefill site is subjected to a distilled water leach analysis which must meet the prevailing drinking water standards (MOE, 1984).

A schematic diagram identifying the disposal options using the ILQCP is given in Figure 2 and the ILQCP chemical guidelines for soil considered acceptable for open water or confined water lakefill disposal in Table II.

PROJECT AND CASE HISTORY

This case history concerned a project involving the construction of a multi-level residential building adjoining an existing low rise facility in the city of Toronto. The > 1000 m^2 site contained about 1.5 m of fill material overlying fairly competent native glacial till soil. To comply with the requirements for large sites, inspections were carried out and a site history audit made. Chemical tests were undertaken on two samples of natural soil and three from fill material; as can be seen from Table III, only the fill samples were found to exceed the ILQCP guidelines.

On the basis of these results and especially in view of the exceedance of parameters under strict control, the fill material at the site could not be classified as suitable for lakefill disposal under the ILQCP options, although the natural soils could be disposed of in open waters as lakefill. Whilst it may have been possible to re-test the fill material, in view of the expense involved the contractor decided to send it to a designated landfill site, but before it could be accepted a distilled water leach analysis was required. A

CLASSIFICATION OF EXCAVATED MATERIALS FOR DISPOSAL PURPOSES

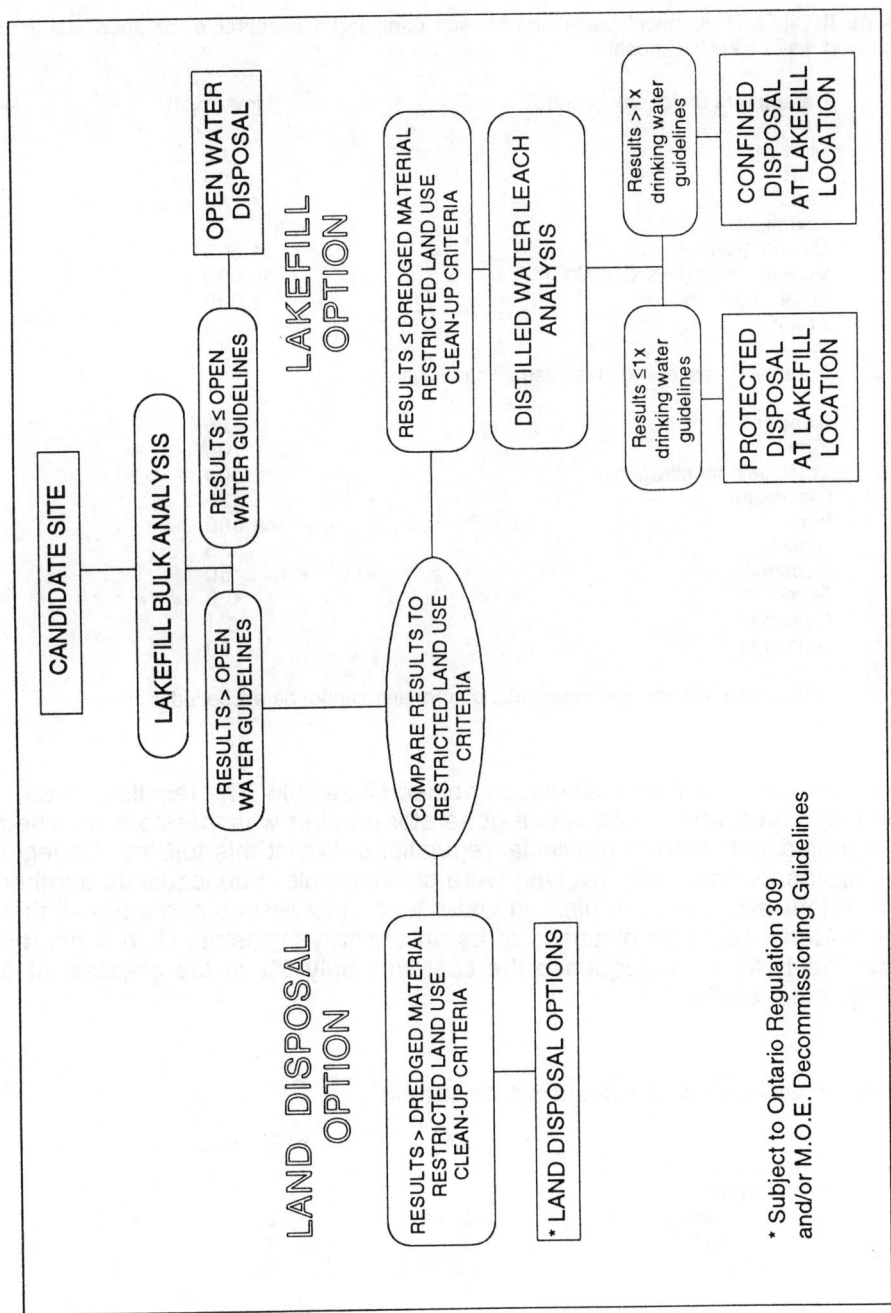

Figure 2: ILQCP decision framework for soil disposal from construction sites (MTRCA, 1988).

Table II: ILQCP chemical guidelines for soil considered acceptable for open water or confined water lakefill disposal.

Parameters under strict control	Limit (mg/l)
Mercury*	0.3
PCB*	0.05
Lead*	50
Cadmium*	1
Oil and grease	1,500
Volatile solids (loss @ 600°C)	60,000
Total phosphorous	1,000
Arsenic	8
Parameters assessed on a case by case basis	
Copper	25
Zinc	100
Total kjeldahl nitrogen	2,000
Chromium	25
Iron	40,000
Nickel	25
Cobalt	50
Silver	0.5
Cyanide	0.1
Ammonia	100

*These are parameters under strict control and cannot be exceeded.

test was carried out on a single, combined fill sample. The results proved a high lead level and a solid waste generator number was therefore obtained in keeping with current provincial regulations. Whilst this fulfilled the legal requirements, the costs involved were considerable. Subsequently another consultant examined the distilled water leach test results and advised that the material could be disposed of as an ordinary terrestrial fill in a nearby open field. As a consequence the cost was only 5% of the disposal at a designated landfill site.

Table III: Exceedances of tested fill material samples

	mg/l	No of Samples
Parameters under strict control		
Mercury	0.3-0.6	3
PCB	0.1	1
Lead	303-1300	3
For case by case basis		
Copper	30-49	3
Zinc	108-535	3
Iron	16000-17000	3
Ammonia	160-240	2
For restricted land use		
Lead (375 mg/L)	1300	1
Mercury (0.5 mg/L)	0.65	1

CONCLUSIONS

This case history highlights several important considerations regarding the classification of fill material for disposal purposes. In view of the widely disparate costs involved as well as the environmental implications, the way in which samples are collected and tested should be clarified. In particular, where exceedances are small or do not apply to all the samples, additional testing should be carried out and procedures for obtaining a second expert opinion should be clearly defined.

REFERENCES

MOE, (1984). *Water Management Goals, Policies, Objectives and Implementation Procedures.* Ontario Ministry of the Environment.

MOE, (1984). *Guidelines for the de-commissioning and clean-up of sites in Ontario.* Ontario Ministry of the Environment.

MTRCA, (1988). *The Manual for an Improved Lakefill Quality Control Program.* The Metropolitan Toronto and Region Conservation Authority, 26pp.

PHYPER, J-D. & IBBOTSON, B. (1991). *The Handbook of Environmental Compliance in Ontario.* McGraw-Hill Ryerson Ltd, Scarborough, Ontario, 346 pp.

5-5 PHYSICAL AND CHEMICAL REACTIONS BETWEEN SEEPAGE GROUND WATER AND FAULT GOUGE WITHIN A DAM FOUNDATION ROCK MASS

Z. HUYUAN & H. WENFENG
Department of Geology, Zanzhou University, China

ABSTRACT

As the water level within a reservoir rises, the local ground water within the rock mass of the dam foundation rock mass is diluted and gradually replaced by the seepage flow from the reservoir. The chemical composition of the ground water may result in physical and chemical reactions between fault gouge and seepage flow along the fault zone. Studies show that with the development of the seepage the reactions occur into two stages. In the first, the dissolution of soluble salts in fault gouge is predominant while the second stage is dominated by cation exchange. The effects of these reactions on the shear strength of fault gouge is discussed, taking into account the seepage characteristics within the fault zone.

INTRODUCTION

The Longyang Gorge Power Station, located in the Qinghai Province of the People's Republic of China, is the first stairstep power station along the Yellow River. The 320,000 kw 178 m high gravity arch dam is constructed of concrete and with a reservoir capacity of 2.47 x 10 10 m^3, is the largest in China. Construction of the power station began in 1978 and is currently nearing completion.

The geological conditions at the dam site are very complicated and hence a number of eminent experts in the geotechnical and engineering geological fields were invited to participate in the field investigations and act as consultants on the project.

The right and left abutment rock masses are both cut by faults and liable to shear sliding. The infilled faults controlling the stability of the abutment rock masses were therefore excavated and filled with concrete while the faulted rocks themselves were grouted with cement.

As can be seen in Figures 1 and 2, the right abutment rock mass is cut by fracture planes F18 and F120 and hence could slide along fault F18. Clearly, if the shear strength of F18 is affected by the reaction due to seepage flow, this will have implications for the stability of the right abutment.

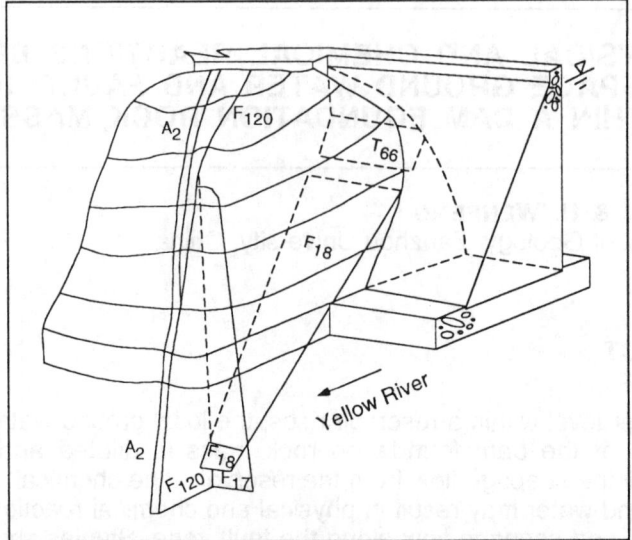

Figure 1: Stereoscopic scheme of the right abutment of the Longyang Dam.

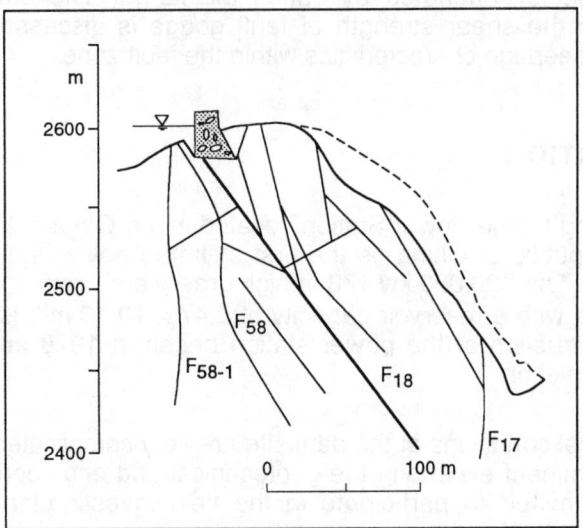

Figure 2: Cross section of the right abutment of the Longyang Dam.

ENGINEERING GEOLOGICAL PROPERTIES OF FAULT F18

The geology of F18 is cataclastic rock, breccia, mylonite and fault gouge; the thickness varying from 1 to 4 m. As the engineering geological properties of the fault gouge are the poorest and this material occurs throughout the fault zone, it is this which controls the shear strength of F18 and to some extent therefore, the stability of the right abutment rock mass.

The fault gouge of F18 was formed by intensive grinding of the parent granite. Its red coloration implies the presence of Fe^{3+} as a consequence of weathering. The chemical composition, physical properties and grain size are given in Tables I, II and III respectively.

Table I: Chemical and clay mineral composition of the fault gouge.

Chemical Composition (%)

SiO_2	Al_2O_3	CaO	K_2O	Fe_2O_3	MgO	FeO	Na_2O	TiO_2	P_2O_5	MnO
73.2	12.37	2.23	2.09	1.43	1.14	0.67	0.50	0.27	0.11	0.03

Clay Mineral Composition (%)

Illite	Montmorillonite	Kaolinite
53.3	28.0	18.7

Table II: Range and mean of the grain size composition of the fault gouge material.

Grain Size (%)

	10-4 mm	4-2 mm	2-0.5 mm	0.5-0.25 mm	0.25-1.10 mm	0.10-0.05 mm	0.05-0.005 mm	<0.005 mm
max	37.0	-	28.0	12.0	12.0	14.0	62.0	30.0
min	5.4	-	9.9	6.0	6.0	6.0	6.0	6.0
mean	17.3	-	17.3	8.7	9.0	10.4	17.3	20.0

Table III: Range and mean of the physical properties of the fault gouge material.

	Specific gravity	Natural density (g/cm^3)	Dry density (g/cm^3)	Moisture content (%)	Liquid limit (%)	Plastic limit (%)	Plasticity index
max	2.72	2.36	2.24	24.8	43.4	20.9	22.5
min	2.67	1.95	1.56	5.2	23.8	11.3	12.5
mean	2.70	2.23	2.03	10.9	31.0	15.4	15.6

As can be seen from Table I, the fault gouge characteristically has high concentrations of iron and potassium and lower levels of magnesium. Table II indicates 20% of the material is less than 0.005 mm, hence is clay or very fine mineral fragments. From Table I it can be seen that the clay minerals are most probably illitic although montmorillonite is common. The index properties are low for illite and montmorillonite-rich material. It is assumed therefore that the fault gouge contains much fine rock flour. In view of the high percentage of fine material and the nature of the clay minerals it is anticipated the fault gouge has a very low permeability.

VARIATION OF CHEMICAL COMPOSITION OF THE GOUGE PORE WATER

The original pore water within the fault gouge material was natural ground water. With the rise in reservoir level, however, seepage from the reservoir replaced this natural groundwater, resulting in an increase in pH from 7.99 to 8.33, a decrease in salt content from 2,550 to 310 mg/l and a change in chemical composition from Na-Cl to Ca-HCO_3 (Table IV).

Table IV: Chemical composition of the natural and reservoir waters (mg/l).

	Natural ground water	Reservoir water
pH	7.99	8.33
Salt content	2,550	310
Na^+, K^+	546.68	44.60
Ca^{2+}	145.89	52.91
Mg^{2+}	65.58	12.52
Cl^-	976.02	18.44
SO_4^{2-}	557.16	74.93
HCO_3^-	231.87	211.73
CO_3^{2-}	9.00	11.10
OH^-	0	0
Chemistry	Na-Cl	Ca-HCO_3

PERMEABILITY TESTS

METHOD

The samples used in the permeability tests were undisturbed fault gouge taken from the F18 fault with water from the Longyang Gorge Power Station reservoir water as the permeating liquid. A flexible walled triaxial cell was chosen as the latex membrane ensures a tight contact with the sample under confining pressure, minimising leakage; the relationship between hydraulic conductivity and effective stress can be resolved and the head pressure can be controlled.

At intervals throughout the permeability test the effluent liquid was tested for cation content.

RESULTS

From the results given in Table V it can be seen that the undisturbed fault gouge material has a very low permeability, with hydraulic conductivity (k) as low as 3×10^{-10} m/s. Although this decreased with increased effective stress, the change was so small that hydraulic conductivity can be considered constant, ie the fault gouge was effectively compacted under natural stress conditions.

Table V: Hydraulic conductivity of fault gouge for different effective stresses.

Effective stress (kg/cm^2)	1.0	1.5	2.0	2.5	3.0
Hydraulic conductivity (x10^{-10} m/s)	3.67	3.54	3.27	3.26	3.10

The analyses of cation concentration were carried out using atomic absorbtion; see Table VI and Figures 3 and 4. The variations in Na$^+$ and Ca^{2+} concentration occurred in two stages. Initially the concentrations of Na$^+$ and Ca^{2+} were greater in the effluent than the influent, indicating some soluble salts were dissolved into the permeating liquid. This is supported by the analysis of soluble salts (Table VII). After about 50 hours the concentration of Ca^{2+} in the effluent also decreases and the Na$^+$ is greater than that in the influent. It is likely that by this second stage the dissolution of soluble salts has almost ceased and the main reaction is that of cation exchange.

Table VI: Concentration of cations in influent and effluent.

	pH	K$^+$ (mg/l)	Na$^+$ (mg/l)	Ca^{2+} (mg/l)	Mg^{2+} (mg/l)
Influent	8.84	16.9	10.8	46.0	5.25
Effluent					
Sampling time (hrs)					
5	8.48	8.3	335	120	9.65
11	8.26	6.0	164	103	3.85
17	8.44	2.5	83.5	90	23.4
23	8.48	3.5	73.5	79	6.9
38	8.80	1.5	55	49	5.9
53	8.72	6.5	58	46	6.8
77	8.80	5.5	59	29	3.6
98	8.68	2.5	52	32	5.7

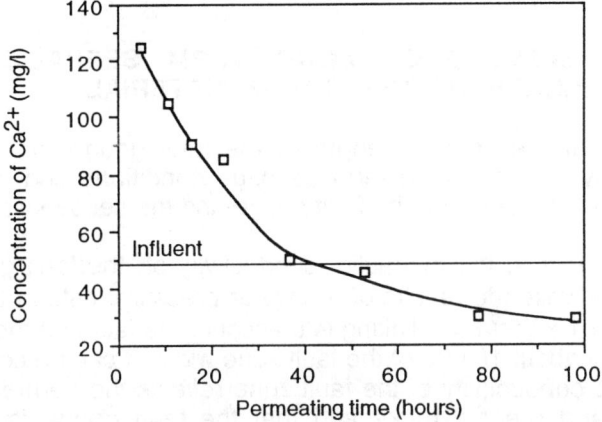

Figure 3: Concentration of Ca^{2+} in effluent versus permeating time.

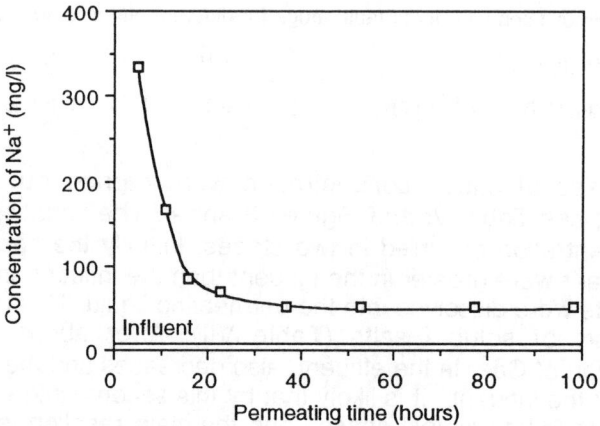

Figure 4: Concentration of Na^+ in effluent versus permeating time.

Table VII: Soluble salts content in the fault gouge before and after the permeameter test.

Sample	K^+	Na^+	Ca^{2+}	Mg^{2+}
		(mg/100 g soil)		
Before permeameter	3.90	5.06	47.54	8.71
After permeameter	3.65	3.75	18.84	4..35

Because the fault gouge is immersed in the natural ground water containing NaCl for a long time, Na^+ is the main cation in the diffuse layer of the gouge particle. However, the permeating liquid contains a high percentage of Ca^{2+} cation, which may be absorbed upon the gouge particle surface due to its smaller cation diameter and high electrical charge compared with Na^+. Thus the Na^+ cation in the original diffuse layer of the gouge particle surface moves into the permeating liquid (by de-absorption). The cation exchange reaction results in decreased concentrations of Ca^{2+} and increased Na^+ concentrations in the effluent.

POSSIBLE EFFECTS OF LONG TERM SEEPAGE ON THE SHEAR STRENGTH OF THE GOUGE MATERIAL

It is likely that the shear strength of the fault gouge material will be influenced by both the long term seepage conditions and the physical-chemical reactions between the fault gouge and the seeped waters.

As seen in Table V, the hydraulic conductivity of the fault gouge is only 3×10^{-10} m/s while the results of the water pressure tests in the F18 fault zone indicate 1×10^{-7} m/s. Taking into account the fact that the fault gouge occupies only about 1/1000 of the fault zone width, it can be concluded that the hydraulic conductivity of the fault zone reflects the permeability of the cataclastic and breccia rocks and that the fault gouge is a relatively impermeable subzone within the fault. As a result of this relative impermeability, it is difficult for the seepage water to replace the original ground water within the fault gouge hence it is unlikely to have a significant effect.

The reaction between the fault gouge and seepage flows occurs in two forms; the dissolution of soluble salts and cation exchange. The former will result in a decreased shear strength while the effect of the latter is to enhance it as the diffuse layer of the gouge particle surface will thin when the Ca^{2+} cation is absorbed, i.e. contrary effects. It is believed these effects are very slight and that the seepage flow in the future will not significantly affect the shear strength of the gouge.

CONCLUSIONS

The study showed that if the seepage flow from the dam encountered the fault gouge material:

a) the chemistry of the gouge pore water would change from Na-Cl to $Ca-HCO_3$

b) two reactions might occur; the dissolution of soluble salts in the fault gouge material and an exchange of cations

However, as a result of the impermeable nature of the fault gouge material, it is unlikely that significant mixing/replacement of the natural ground water would take place. In the event of this occurring, the effects of the reactions noted above would not only be small but would also counterbalance each other, thus no noticeable variation in shear strength would result.

REFERENCES

BEAR, F.E. (1964). *Chemistry of the Soil.*
NORTHWEST CHINA HYDROELECTRIC INVESTIGATION AND DESIGN INSTITUTE, (1986). *Engineering Geology, Report of the Technical Design of the Longyang Gorge Power Station, Volume 3.*
XIAN-GONG, Z. et al, (1986). Engineering Geological Classification of fault rocks. *Proceedings of the Fifth Congress of the International Association of Engineering Geology*, **2**, 479-486.
ZARUBA, Q. (1976). Engineering Geology.

5-6 FACTORS CONTROLLING THE SOLUBILITY OF SUBSTANCES IN NATURAL WATERS

A. SAHINCI
Akdeniz University, Faculty of Engineering, Isparta, Turkey
A. TURKMAN
Dokuz Eylul University, Faculty of Engineering, Izmir, Turkey

ABSTRACT

As water dissolves many salts and some types of organic matter, natural waters always contain some impurities. The proportion and characteristics of such impurities determine its suitability for different purposes. This paper discusses the composition of some waters in terms of the physical, chemical and biological factors.

PHYSICAL FACTORS

THE ABUNDANCE OF ELEMENTS IN THE EARTH'S CRUST

The ultimate source of all dissolved ions in natural waters is the mineral assemblage in rocks in the earth's crust. Dissolved inorganic constituents in natural waters are classified according to their concentration as major, minor and trace constituents.

The major ions including Ca, Mg, Na and K are abundant in the earth's crust while such elements as Rb, Cs and Be, being rare in the earth's crust, are generally minor constituents in natural waters. However, the relation between the abundance of elements in the earth's crust and their concentration in natural waters is not always simple, as the different salts have varying characteristics as regards their potential solubility.

Solubility, which is defined as "the total amount of solute species which can be retained permanently in solution under a given set of conditions" differs from one substance to another, hence such elements as Si, Al, Fe and Mn, whilst abundant in the earth's crust, are not found in natural waters to any great extent (Figure 1a).

The ratio of the concentration of a given element in sea water and fresh waters (in ppm) to its concentration in the earth's crust are symbolised as CC_s and CC_f respectively. These values are directly proportional to the solubilities of elements. As seen in Figure 1b, the elements with CC_s ratios greater than one are Cl, H, Br, S and O, while only O and H have CC_f ratios greater than one (Figure 1b).

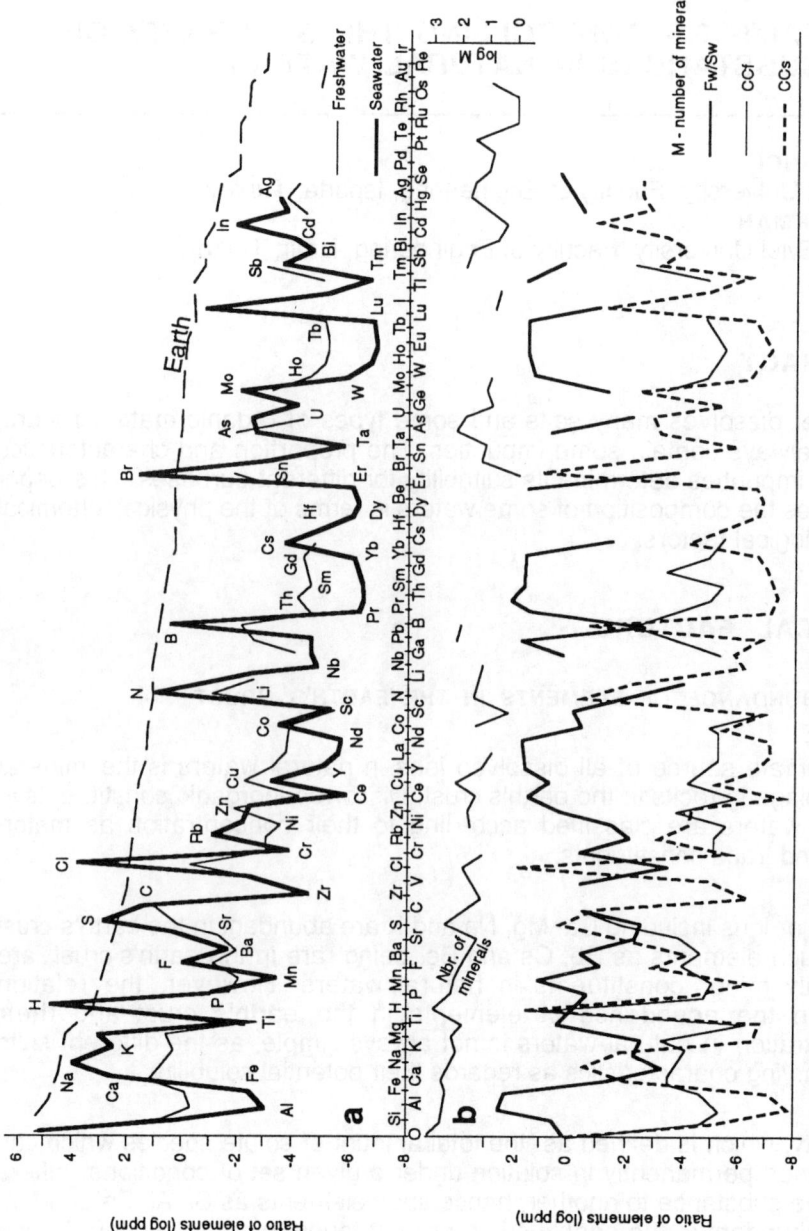

Figure 1: a) The concentrations of elements in seawater and fresh waters; b) The ratios of concentration of elements in sea water and fresh waters to their concentration in the Earth's crust and the number of free minerals of elements.

TEMPERATURE AND PRESSURE

In general the solubility of salts in natural waters increases with increasing temperature but is not greatly affected by pressure. On the other hand, the solubility of gases in natural waters decreases with increasing temperature and decreasing partial pressure.

In deep aquifers, where high temperature and pressure conditions pertain, the equilibrium constants for the geochemical reactions differ from those at the surface. Thus in deep aquifers $CaCO_3$ precipitation is not observed. As soon as the pressure decreases in the wells, however, precipitation takes place at the inner surfaces of the wells.

Other physical factors which affect the solubility of substances include time in contact with water, size of minerals and rocks, viscosity of water, porosity and permeability of the medium and flow rate of water. In general, compounds at the corners of the crystal formations dissociate most easily. The surface contact per unit volume of water (specific surface) is greater for small crystals. For water molecules to be able to penetrate into the crystals, rock must be fractured and porous. If the movement of water in the rock is slow, the contact time between the rock and water will be longer and consequently more dissolution will take place. Fossil and petroleum waters are highly mineralised due to their confinement in rocks over a long period.

CHEMICAL FACTORS

INTERNAL CHEMICAL FACTORS

Among the factors controlling solubility, chemical properties are the most important. These include the energy of ionisation, the ionic radius, the ionic valence and the ionic potential. From the periodic table of elements it can be seen that:

> Within the groups, atomic or ionic radius increases from top to bottom;

> Within the periods atomic or ionic radius decreases from left to right.; thus it may be inferred that the two ions which are diagonally one beneath the other in the periodic system have approximately the same atomic or ionic radius. This property is important for the similarity of geochemical properties of elements.

> The polarizing effect increases with increasing positive charge and decreasing radius of the ion.

The polarisability of elements increases with ionic radius and in the case of anions, with the number of electrons gained. Due to the structure of water, polar elements are dissolved more easily in water than non-polar ones.

The solubility of a compound decreases nearer the ionic radius of the elements in the compound. For example, the ionic radius of Ag^+ is 1.26 Å; of

Li$^+$, 0.68 Å and of Cl$^-$, 1.81 Å. Thus the solubility of AgCl is smaller than that of LiCl. However this rule does not apply in the presence of other effects such as ionisation energy, ionisation potential, pH and Eh.

Ionisation Energy

Ionisation energy increases from left to right in the periodic table of elements, thus alkali elements have less ionisation energy and dissolve more readily in water than the other cations.

Ionic potential

Ionic potential may be defined as the ratio of the valence (Z) of the ion to the ionic radius (r). Goldschmidt (1937) divides the elements into three main groups on the basis of ionic potential values. The first group, with an ionic potential of less than 3, consists of elements which form real ionic compounds such as Na^+, K^+, Ca^{2+}, Mg^{2+}, Sr^{2+}, Fe^{2+}, Mn^{2+}. The ionic potential of the second group lies between 3 and 12 and includes the elements which form amphoteric oxides, eg Be^{2+}, V^{3+}, Fe^{3+}, Ti^{3+}, Cr^{3+}, Ca^{2+}, etc. These elements show a tendency to hydrolyse in water and form limited soluble hydroxides hence are named the "hydrolysate group". The third group includes the elements having ionic potentials greater than 12. The elements in this group are easily dissolved in an oxidising medium; examples include N^{3+}, C^{4+}, S^{6+}, P^{5+}, Se^{6+}, Br^{7+}, B^{3+} (Figure 2).

From the above it is possible to conclude that, other conditions being equal, the attraction of the cations by water molecules is proportional to the ionic potential.

Alkali metal salts (ionic potential < 3) formed by halogen, carbonates and sulphates, are readily soluble in water. They are followed by halogen and bicarbonate salts of alkaline earth elements (Ca^{2+} and Mg^{2+}) and halogen, bicarbonate and sulphate salts of Mn^{2+}, Fe^{2+}, Co^{2+}, Ni^{2+}, Cu^{2+}, Zn^{2+} etc. On the other hand, $SrSO_4$, $BaCO_3$, $ZnHCO_3$, TiO_2 (rutile), $ZrSiO_4$ (zircon), ThO_2 (thoriante), SnO_2 (cassiterite), $HFeO_2 \cdot nH_2O$ (limonite), $n(Co,Ni)O \cdot MnO_2 \cdot H_2O$ (asbolite), $Fe(OH)_3$, $Co(OH)_3$ and AgCl salts are only slightly soluble in natural waters.

The solubility of elements in natural waters is determined by their capacity to form complex ions, which in turn depends on the structure of the electron configuration of the atom.

Ionic potential gives a general picture of the solubility of elements in natural waters. The dependence of the solubility of elements on their internal structure has a very complex nature and is determined by the structure of the electron configuration of the atom. In general the most soluble elements in natural waters have an 8 and 18 electron structure of the atomic shell, as in Na, K, Pb, Cs, Mg, Ca, Cu, Zn, Cd, Hg, Ge, Br and Pb. The elements less soluble in water have a transitional character with an unfilled inner electron shell (d shell), eg Fe, Co, Ni, Cr, V, Ti, Zr, Mn, Pt and others. The elements having configurations with 8 and 18 filled electron shells easily give away

FACTORS CONTROLLING SOLUBILITY OF SUBSTANCES IN NATURAL WATERS

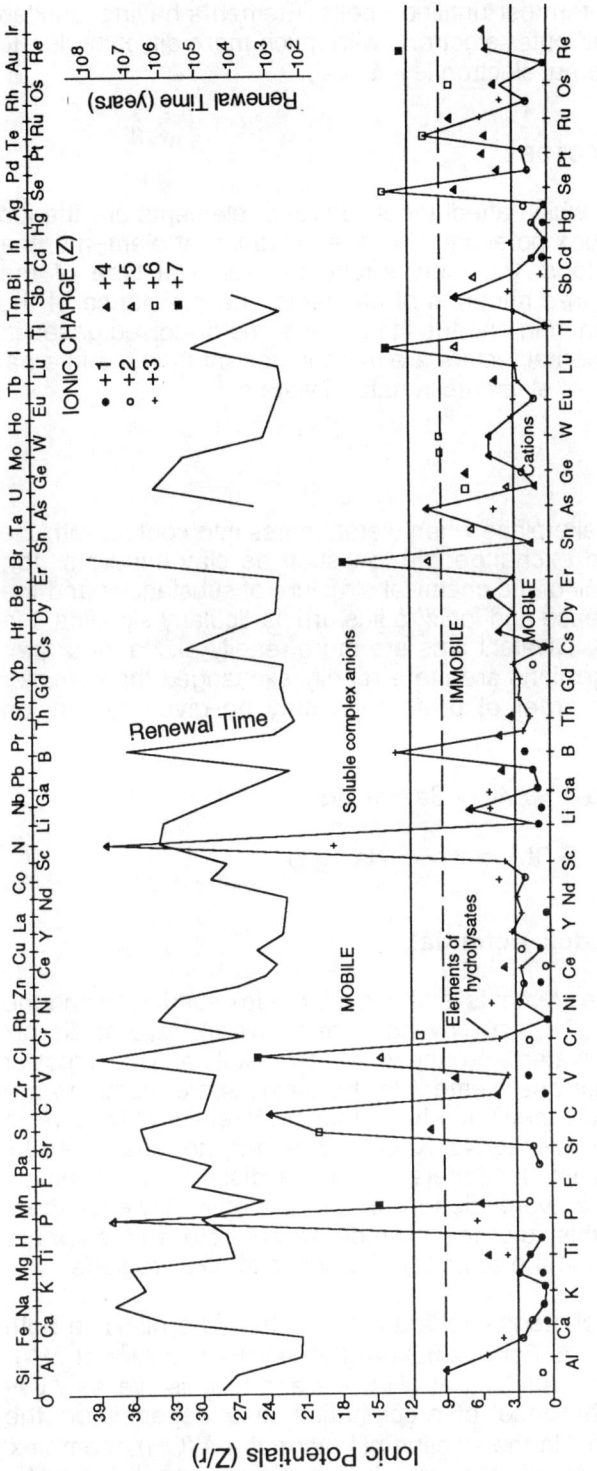

Figure 2: The ionic potentials of elements.

the electrons from the outermost unfilled shells. Elements having unfilled inner shells give away their outer electrons with much more difficulty due to this tendency to fill the internal electron levels first.

EXTERNAL CHEMICAL FACTORS

The main external factors which affect the solubility of elements are the ion exchange, the pH, the redox potential (Eh), the migration of elements, the influence of the similar ions, the ionic effect, the renewal time of the elements, the number of free minerals of elements, the resistance of the minerals to chemical weathering and the influence of the dissolved gases in water. In general these external factors are more important than the internal influences on the solubility of elements in natural waters.

Ion exchange

Ion exchange processes take place when water comes into contact with the substances containing an exchangeable ion such as clay minerals. Ion exchange occurs as a result of the chemical structure of substances and the polarity of water. Ionic valence and ionic radius are particularly significant in the ion exchange process. Divalent ions are more readily exchanged than monovalent ions and larger ions are more readily exchanged than smaller ones. Thus the following order of preference may be given for an ion exchange process:

$$Ba^{2+} > Sr^{2+} > Ca^{2+} > Mg^{2+}$$

$$Cs^+ > Rb^+ > K^+ > Na^+ > Li^+$$

Influence of pH and redox potential

The solubility of most of the elements is affected by redox conditions and the pH of the medium. Generally, fast flowing waters (rivers) have acidic pH values while slow flowing waters (groundwater) are alkaline. The effect of pH on solubility differs from one element to the other; some elements are better dissolved in an acidic medium while other preferentially dissolve in an alkaline medium and others (eg Na, K, Cl, I, Br and B) are unaffected by the pH value of the medium. In general elements dissolve in an acidic medium and precipitate as hydroxides in an alkaline one. However there are many exceptions to this rule; for example, carbonate and sulphate complexes of uranium extend the solubility of uranium to alkali regions.

Some substances are amphoteric in character and therefore dissolve both in acidic and basic conditions. For example Al_2O_3 dissolves as Al^{3+} at pH = 0, it starts to precipitate as $Al(OH)_3$ at pH = 4.1 and to dissolve as AlO_2 complex at pH = 8.9. The onset of precipitation also depends on the concentration of free Al (III). In the alkaline pH range the $Al(OH)_4^-$ complex controls the solubility of Al. Its structure has been confirmed by NMR measurements.

The pH also affects the adsorption and ion exchange processes in water. For example the montmorillonite adsorbs less metallic ions in acidic medium than alkaline due to the preferential adsorption of H+ ions in acidic medium.

The solubility of elements also depends on the redox potential of the water medium. While some elements are easily dissolved in an oxidising medium, others may settle. Generally surface waters contain more oxygen than groundwaters due to the turbulence and air-water contact. Elements whose solubility is not affected by pH-Eh conditions include Ca, Na, K, Mg, Cl, Br, I, F, Sr, P while those which dissolved under oxidizing conditions include S, B, Mo, U, Se, Re, Cu, Co, Ni, Hg, Ag, Au etc. Fe, Mn and S reach their highest oxidising state (Fe^{3+}, Mn^{4+}, S^{6+}) under oxidising conditions.

In an acidic medium H+ becomes dominant and the elements such as S, B, Mo, V, U, Se, Zn, Cu, Co, Ni, Ht, Ag, Au, As, Cd, dissolve. The elements which dissolve in neutral and alkali medium are S, B, Mo, V, U, Se, Re, As, Cd, Ph, Li, Rb, Be, Bi, Sb, Ge, Cs, Tl. The elements which dissolve under reducing conditions vary according to the presence or absence of H_2S. When H_2S is absent in a reducing medium, Fe^{2+} an Mn^{2+} may be abundant. Elements such as V and Cu form insoluble precipitates under reducing conditions. In a reduced medium where H_2S gas is present methane and other gases may also exist. Iron and some other elements may form insoluble sulphide precipitates.

Migration of Elements

Migration may be defined as the movement of elements from one medium to another. There is a close relationship between migration and solubility. The elements which migrate without being affected by the pH-Eh condition of the water are most soluble in natural waters, eg Cl, Br, I, Ca, Na, Mg etc. On the other hand, some elements migrate slowly and hence more extensive dissolution occurs, eg K. Potassium migrates slowly as the chemical weathering of the feldspars which contain it, such as orthoclase, microline ($KAlO_8$) and micas is very slow. In addition, the sorption of potassium by clay minerals and vegetation also reduces the migration. The migration and solubility characteristics of such alkali elements as Li, Rb and Cs are similar to those of potassium. The salts of nitrogen, one of the basic constituents of organic matter, also dissolve readily and migrate slowly.

The Mutual Interaction of Elements

The presence of specific elements may increase or decrease the solubility of others. For example, when lead or iron ions are added to water containing molybdenum, they precipitate as vulphenite ($PbMoO_4$) or ferrimolybdate ($Fe_2(MoO_4)_3 \cdot 8H_2O$). The precipitation of $PbMoO_4$ depends not only on the presence of lead and molybdate but also on their concentrations. Similarly, silver ions are precipitated as AgCl in presence of chloride ions. The solubility equilibrium and its kinetics are affected by the conditions of the medium, such as pH, temperature, ligands etc. The common ion effect is to decrease the solubility of elements in water.

The solubility of salts increases in the presence of other salts without any common ion. For example, the solubility of $CaCO_3$ increases in the presence of the dissolved NaCl. This is known as the ionic strength effect.

The Renewal Time of Elements

The renewal time may be defined as the ratio of the total content of a given element in the ocean to its annual transport by rivers. The elements having a short renewal time (100-400 years) include Al, Th, Fe, Pb etc. The renewal time of the silicon is less than 15,000 years but a number of elements have much greater renewal times, eg Mg (1.29×10^7 years), Na (7.25×10^7 years), Cl (1.02×10^8 years) and Br (1.38×10^8 years) - Figure 2. The elements with high solubility also have long renewal times.

Free Minerals of Elements

Generally there is a direct relationship between solubility and the number of free minerals of the elements. Free minerals of elements such as Si, Al, Fe and Mn are numerous in nature although they are only slightly soluble in water. This may be explained by their concentration in the earth's crust. The formation of free minerals depends on the chemical characteristics of elements as well as their abundance in the earth's crust. The elements forming few free minerals are unlikely to reach saturation, hence they rarely precipitate from water. The elements which are not plentiful in the earth's crust (eg Sn, Pr, Ga, Nd, Rb, Gd, Yb, Hf, Dy, Er, Eu, Tb, Lu, Tm, Re) do not have free minerals; on the other hand the elements with abundant free minerals precipitate easily, eg O(1364), Si(432), Al(320), Fe(300), Ca(385), H(908), P(122) - Figure 1b.

Resistance of Minerals to Chemical Weathering

In general the minerals most susceptible to chemical weathering ionise more rapidly and the minerals formed at high pressure and temperature are more easily weathered than those formed at low pressure and temperature. Thus the resistance to chemical weathering may be written in the following order:

$$\text{oxides} > \text{silicates} > \text{carbonates and sulphates}$$

Some elements which are highly soluble do not dissolve in water due to the resistance of the mineral to chemical weathering. As an example, the high resistance of feldspars and micas to chemical weathering prevents the passage of potassium to water.

Dissolved Gases in Water

The gases which influence the solubility of elements are CO_2, O_2 and H_2S. Carbon dioxide is important for the chemical equilibrium between carbonates, bicarbonates and CO_2 while oxygen is significant for the redox conditions of the water and H_2S influences the precipitation of some elements. The latter also provokes the solid formation of elements such as Cu, Pb, Zn, Ag, Cd, U, Mo and some sulphide compounds.

Biological Factors

Biological factors mainly affect the oxygen and carbon dioxide content of water, which in turn affects the solubility of elements. Aquatic vegetation and chlorophyls in water consume the dissolved CO_2, thus influencing $CaCO_3$ precipitation. During the consumption of CO_2 oxygen is released and the redox condition of water is affected. In the presence of oxygen, aerobic bacteria act as a catalyst for the biogeochemical reactions.

CONCLUSIONS

In order to appreciate the factors controlling the solubility of substances in natural waters it is very important to understand the chemical nature of the water. The quality of water constrains its usage hence it is important to understand what influences the composition of natural waters. It appears that a number of of different factors are important and it is their direct and interactive effects which make prediction difficult.

REFERENCES

GARRELS, R.M. & VE CHRIST, C.L. (1965). *Solution, minerals, and equilibria.* Harper's Geoscience Series.

GOLDSCHMID, V.M. (1937). The principles of distribution of chemical elements in minerals and rocks. *Journal of the Chemistry Society*, 655-673.

HEM, J.D. (1971). Study and interpretation of the chemical characteristics of natural water. *Geological Survey Water Supply*, Paper 1472.

KRAUSKOPF, K.B. (1979). *Introduction to geochemistry.* McGraw-Hill.

LEVINSON, A.A. (1980). *Introduction to exploration geochemistry.* Applied Publications Ltd.

MASON, B. (1966). *Principles of geochemistry.* John Wiley.

PAGENKOPF, G.D. (1978). *Introduction to natural water chemistry.* Vol 3, Marcel Dekker.

PEREL'MAN, A.I. (1977). *Geochemistry of elements in the supergene zone.* U.S. Department of Commerce, National Technical Information Service, Springfield, Va 22151.

ROSLER, H.J. & VE LANGE, H. (1972). *Geochemical table.* Elsevier.

SCHOELLER, H. (1962). *Les eaux souterraines.* Masson et Cie.

SHVARTSEV, S.L., UDODOV, P.A. & RASSKAZOV, N.M. (1975). Some features of the migration of microcomponents in natural waters of the supergene zone. *Journal of Geochemical Exploration*, 433-439.

STUMM, W. VE & MORGAN, J.J. (1970). *Aquatic chemistry. An introduction emphasizing chemical equilibria in natural waters.* Wiley-Interscience.

SAHINCI, A. (1986). *Geochemistry of natural waters.* Dokuz Eylül Univ. Faculty of Engineering, Izmir (Turkey).

5-7 MICROBIOLOGICAL ASPECTS OF THE GENERATION OF IMPERMEABILITY IN HIGH SAND CONTENT GREENS INFESTED WITH BLACK PLUG LAYERING

D.R. CULLIMORE & S. NILSON
Regina Water Research Institute, University of Regina, Saskatchewan, Canada

ABSTRACT

High sand content golfing greens have become subjected to severe infestations of black plug layering over the last twenty years. Characteristically, this involves a loss in permeability of the greens; severe stresses in the turf grass leading to chlorosis, secondary infections and in some cases die back; the generation of odours and erosion of the playing surface. The consequential loss of playability has severe economic implications for the operation of such infested golf courses.

Investigations have revealed that the black plug layer is one part in the maturation cycle of biofouling by a microbial consortium. This consortium includes iron related bacteria, sulphate reducing bacteria and methanogens. Nutritional regimes found to stimulate this consortium within low exchange capacity porous media included urea, ammonium and phosphorus (as phosphate) while the generation of the black pigments (sulphides) was stimulated by the presence of iron, manganese and sulphur (as sulphate or elemental sulphur). These layering events were duplicated in the field as well as in laboratory sand cores.

Permeability was investigated using an ascending head permeometer in which the rate of entrance transmissivity (infiltration) was measured as the inverse of the rate of ascendance of the water column head. This was achieved using a computer controlled infiltrometer (INFILTREX 7, Droycon Bioconcepts Inc., Canada). Bioimpaired transmissivity was recorded where the transmissivity was reduced below 2 mm/sec, with severe impairment occurring at less than 0.5 mm/sec. Control strategies have been developed based on the routine evaluation of the entrance transmissivity of greens and the appropriate implementation of penetrants, aerification and shifts in the irrigation patterns has allowed the infestations to be controlled in the municipal golf courses operated by the City of Regina.

INTRODUCTION

Both as a recreational pastime and a professional sport, golf requires the highest possible quality of the playing surface to maximize the controllability of the movement of the golf ball. Such surfaces are commonly achieved by the monoculturing of specific types of turf grass in a manner which ensures a reliable substrate. This is of particular importance on the putting greens where the rolling of the golf ball in a prescribed direction becomes critical. To achieve this quality of surface, it is common to employ advanced agronomic practices in the culture of the turf grass and provide a suitable underlying porous medium to encourage good drainage, a stable surface contour and a deep penetration of the turf grass roots into the soil. The practice of using single source easily assimilable fertilizers (e.g., urea, inorganic ammonium based salts) has supported the management and health of the turf grass but has also generated a response within the microflora inhabiting the underlying porous medium. This has taken the form of a "monoculturing", in which the number of component species become reduced and the nature of the growth more aggressive in competing with the turf grass for water, oxygen, nutrients and living space.

One effect of these practices is the generation of a microbially induced fouling (MIF) of the soils (porous media) which impacts and directly competes with the turf grass. One such MIF phenomenon is known as the black plug layer (BPL) which has been reported as occurring in golf courses around the world. The major impact of BPL is through a biological occlusion (plugging) of the soil pore (interstitial) spaces. Such BPL occlusions are particularly common within the high sand content golf course turf grass greens throughout North America.

A BPL infestation goes through a number of stages of maturation. Its occurrence is generally first recognized by the appearance of intensely black globular particles, usually at 200 to 300 mm depth in the soil or sand profile. It is generally believed that the black colour is due to two events during the initial growth of the BPL: the bioaccumulation of iron and manganese and the generation of hydrogen sulphide (H_2S) within the occlusion. The net effect of these two events is the production of the black metal sulphides which remain entrapped within the occlusion and, where there is a surplus of H_2S, produce a smell of "bad eggs", which may appear to emanate from the surface of the infested green. In the next stage, the BPL assumes a dark, "slimy" form and occurs closer to surface of the green (Shank, 1987) and/or in depressions and natural drainage slopes (Berndt and Vargas, 1987). At this stage of maturation, the BPL extends laterally, commonly at depths of between 5 and 100 mm and competes directly with the turf grass by the creation of a range of structures including:

i) microglobular processes which attach directly to the turf grass roots;

ii) interconnective lateral and peninsulate plates which extend the infestation laterally; and

iii) vertical minor and major columnar processes which abut close to (3 to 8 mm distance) the surviving roots.

A successful maturation of the BPL will lead to the loss of turf grass cover and the lateral and vertical structures of the BPL may fuse and elevate within the green to lie immediately beneath the surface (0 to 20 mm depth). Under these circumstances, black globular particles up to 4 mm in diameter may extend above the surface of the green and there is a higher risk of H_2S venting from the green. A final stage sometimes noted in a BPL event is an increase in the "fragility" of the structures, a tendency for the BPL layers to descend and dissipate and the colour to change from an intense black (sulphide dominated) to brown as some of the metal sulphides become oxidized.

Microbiological factors are responsible for the generation of a BPL event. Dominant components within a BPL include aerobic bacteria capable of producing copious amounts of extracellular polymeric substances (ECPS), sulphate reducing bacteria (SRB) and micro-algae, which are also capable of producing mucilaginous ECPS. It is the ECPS which predominantly fill the void spaces in the porous medium with a high proportion of "bound" water present. Such occlusions then prevent the movement of water down through the infested lateral zones and restrict the passage of oxygen. As a result of this and the fact that the BPL is an anaerobic entity, "pools" of water may accumulate on it within the porous medium. Micro-algae flourish within these "pools" of water, causing additional competitive problems for the stressed turf grass.

Following a severe die back due to BPL, the surface of the green may become less playable and begin to erode as water is unable to penetrate into the bio-occluded porous media. Furthermore, the surface may become uneven as the BPL zones expand in volume through the retention of gases (e.g., methane, nitrogen, carbon dioxide, hydrogen) as a result of biological catabolic activities. While an initial response may be the establishment of temporary alternative greens, the long term solution has to be the management of the BPL infestation so that it does not have such a significant impact on the playability of the green.

Early detection and management of a BPL infestation is necessary to eliminate the detrimental effects on golf course greens. Methods for the detection of the BPL, either directly or through some impact effect, may be simplistic or sophisticated. Direct visual observation of black stratifications within soil profiles (obtained by coring), the recognition of sulphurous odours present within the soil and the recording of turf grass root length are three simple approaches employed. However these methods are vulnerable to misinterpretation as other forms of bio-occlusion not associated with an obvious BPL event (i.e. void of black coloration) may not be recorded. Measurement of the length of turf grass roots can prove to be somewhat more effective in the monitoring of a BPL infestation but the severity of the infestation would have heightened by this time.

Sophisticated methods of BPL observation are also being implemented including computer assisted tomography (CAT) scanning for the location of BPL within a soil core. The high iron content (80 to 160 ppm, w/v) allows the BPL structures within the core to be easily observed throughout the soil profile (Cullimore and Lindenbach, 1989). Monitoring of the entrance transmissivity of water into the green at a depth of 10 mm may also prove valuable. Where severe bio-occlusion (plugging) is occurring, the rate of flow of water into the soil will be severely impaired. The INFILTREX 8 system (Droycon Bioconcepts Inc., Regina, Canada) involves an ascending head computer-assisted infiltrometer. This unit measures the water impedance characteristics of the test porous medium from a constant head (800 mm). Non-entrancing water surplus to demand forms a recordable ascending head which inversely relates to the volume of infiltrated water under standard conditions; consequently the less permeable the soil, the greater the probability of a BPL event occurring within the soil profile. The application of this technology in the monitoring of a BPL infestation in a municipal golf course in Regina, Canada, is described below.

Bacterial components within a BPL event may also be identified through an analysis of the methylated ester linked fatty acid (MEFA) components which are inevitably present in the viable cells and which have unique relationships in each microbial strain. Through the application of a standard methodology, involving a quantitative and qualitative GC analysis for the MEFA, isolated pure cultures can be successfully identified. This method has proved to be quick (Moss, Dees and Guerrant, 1980), reliable and reproducible (Roy, 1988; Mallory and Sayler, 1984). A modification of this technology which allows soil core samples to be analyzed using a culture enrichment/MEFA analysis technique is described below. This methodology allows the recovery of three MEFA which through both their presence and ratio, indicate that a BPL event is occurring in the sample under test.

MEFA ANALYSIS FOR THE CONFIRMATION OF BPL

METHODOLOGY

This technique is used for soil samples which have been taken from a zone where there is a confirmed or suspected BPL event occurring. Where there are no obvious signals of an infestation, soil should be taken from sites where BPL is most likely to be generating. Studies using synthetically MIF induced BPL events suggest such sampling sites would probably be from a depth of 10 to 20 mm (shallow) or 40 to 100 mm (deep) from the zone furthest away from any turf grass roots. Intense competition can occur between the roots of the turf grass and the BPL bacterial consortia infesting the soil. The likelihood of turf grass survival and recovery would appear to be linked to the proximity of the BPL to the roots (ie. the closer the BPL is to the roots, the lower the probability of survival).

Four grams (± 0.2g) of soil samples should be removed from the core and placed into aluminium foil capped sterile 250 ml Erlenmeyer flasks containing 50 ml Trypticase Soy Broth (TSB) and with 50% of the base being covered by a layer of 2 mm (diameter) glass beads. The 8% soil dilution in the TSB medium is immediately placed on an Orbital Shaker (#3590, Lab Line Instruments, Inc., Melrose Park, Illinois) for 30 minutes at 400 rpm at room temperature (22°C ±1°C). This causes the soil to disperse into suspension. After agitation, the soil suspension is allowed to settle for 10 minutes. A 9 ml aliquot of this suspension (taken at the midpoint down the profile) is next pipetted aseptically into sterile 15 ml screw cap test tubes. Where the porous medium being examined is high in organic content (e.g. a peat enriched sand), a 10^{-1} dilution of the aliquot into TSB is used. Each test tube containing the 9 ml aliquot (or dilution) is then incubated at 28°C for 7 days to allow the intrinsic microflora a period of time to adjust and grow under the cultural conditions provided. Following the incubation period, each tube is centrifuged using an IEC Centra-8R centrifuge at 4,000 rpm for 20 minutes at room temperature. This process causes most of the microbial cells which have grown in the TSB medium to form into a "pellet" in the bottom of the centrifuge tube and consequently most of the biomass would also be concentrated there. The supernatant is pipetted off to obtain the pellet which is subjected to standard saponification and preparation procedures prior to GC analysis for MEFA incorporating the MIDI software (Microbial I.D., Delaware).

For each sample a print out is obtained which includes an XY plot (fingerprint) of the MEFA present scaled by retention time prior to elution (X axis) and the areas for each individual MEFA peak (Y axis). Additionally, there is a comprehensive print out of all the MEFA recovered including relative abundance (% of total MEFA) and concentrations relative to a standard. These data are interpreted by a selection of aerobic and anaerobic software identification packages which attempt to ascribe a genus and species to each sample.

In this methodology, original soil samples are used involving a wide range of microbial strains and so the MEFA recovered would represent the combined MEFA developed by the strains able to survive and grow under the cultural conditions presented in the TSB culture medium. It was therefore not possible to obtain any identification of individual bacterial strains by this technique but it was possible to examine the range and ratios of the MEFA recovered by this analytical technique.

INTERPRETATION OF THE MEFA ANALYSIS FOR BPL PRESENCE

Examination of MEFA indicated that soils characterized as being infested with the BPL fouling (e.g., by cultural characteristics, visible evidence and impaired permeability) incorporated three specific MEFA entities which occurred in a particular abundance and with specific relationships. These biomarkers for a BPL event confirmation were designated Alpha, Beta and Gamma and were determined to be present in BPL infested soils and absent in control (uninfested) soils. Alpha and Gamma biochemical markers were observed in all BPL samples while the Beta biochemical marker was commonly observed in BPL soil samples.

The three biomarkers are defined by their position and relative quantities within the output MEFA data, i.e. by the retention time (mins) prior to peak signal reception (RT) and equivalent chain length (ECL) which projects the theoretical number of carbon atoms which would normally be present in the averaged straight chain molecule. These criteria are used to define the various biomarkers (standard deviation given in brackets): (1) Alpha, RT 1.977 (±0.005), ECL 7.431 (±0.013); (2) Beta, RT 2.627 (± 0.004), ECL 8.841 (±0.013); and (3) Gamma, RT 3.056 (±0.005), ECL 9.773 (±0.015). It was found that these three signals occurred routinely in BPL fouled soil or sand green core samples but did not occur in "healthy" samples where no such evidence was present or contrived. One further marker routinely present both in the control and the BPL fouled samples, designated as Delta, has an RT of 10.040 (±0.012) and ECL of 15.996 (±0.008). This value was used in a routine manner as a baseline for determining the likelihood of a significant BPL event being in progress.

Comparisons between the Alpha, Beta, Gamma and Delta were performed to determine the relevant ratios between each of the three biochemical markers for BPL and also with Delta. Comparison of the biochemical marker data was performed using the peak areas (PA) obtained for the four biochemical markers where these were present. The PA for the three BPL related biochemical markers (i.e., Alpha, Beta and Gamma) were each compared to the PA obtained for Delta. This was calculated using the formula:

$$PPA_x = \frac{PA_x}{PA_d} \cdot 100$$

where PPA is the percentage of the peak areas of either Alpha, Beta or Gamma when compared to the PA for Delta (PA_d). Subscript "x" represents the specific biochemical marker (eg., PA_d is the PA value for Delta).

In occurrences of BPL events, the Alpha and Gamma biochemical markers appear to be linked in a different manner than the Beta biochemical marker. When using the PPA_x values, the following observations may be made with respect to the PPA Alpha and Gamma values:

1) where PPA is > 10 there is a probability that a BPL is present;

2) where the PPA is > 40 there is a high probability that a BPL event is occurring;

3) where the PPA is > 80 there is an extremely significant potential for an active BPL to have generated;

4) where the PPA < 1.0 there is a low possibility of significant BPL presence;

5) both Alpha and Gamma should normally be present for a BPL infestation to be confirmed.

For the PPA Beta values, the interpretation is more related to the maturation of the BPL than to its presence/absence. Interpretation of the PPA values for Beta reflect this:

1) the absence or very low PPA value (< 1.0) for Beta does not indicate that BPL is absent - it could be in a very immature state;

2) where the PPA for Beta exceeds 10 then there is a reasonable probability that the BPL is maturing; and

3) if the PPA for Beta exceeds 50 it is probable that a very severe and mature BPL infestation is present.

Where the Alpha/Gamma and Beta diagnoses coincide, then a greater weighting may be placed on the interpretation.

Detection of these MEFA biochemical markers, which appear to be uniquely related to a BPL fouling of golf greens, suggests that this analytical system could be used to provide the earliest possible recognition of a potential problem. This would form an extension of the methodology that has already been proved to yield reliable and reproducible results for MEFA profile based identification of many species of bacteria (Mallory and Saylor, 1984; Roy, 1988). Earlier detection of the BPL activity in a green may allow a rapid response to be initiated. Remedial measures may include aerification (Kershasky, 1987), alteration of irrigation strategies (Cullimore and Nelson, 1989; Berndt and Vargas, 1989), fertilizer selection (Cullimore, 1991; Berndt and Vargas, 1987) and penetrant application (Templeton, 1988).

Confirmation of the presence of a BPL in the samples under examination was by the rate of occurrence of specific reaction patterns in the BART^{Tm} biodetectors (Droycon Bioconcepts Inc., Regina, Canada). Bacterial consortia including the BPL microorganisms would trigger specific reaction patterns (Rx) within a certain time period (days of delay, dd). To conduct these confirmatory tests, the BART^{TM} biodetectors were precharged with 15 ml of sterile isotonic saline solution immediately before 0.1 g (±0.02 g) of the soil sample under test was added. The biodetectors were not shaken but left at room temperature and inspected on a twice daily basis. When a reaction occurred, the Rx and the day of delay were recorded. There was a high probability of BPL being present where the following data were obtained:

IRB-BART^{TM}, primary Rx = 10, dd < 3;

SRB-BART^{TM}, Rx = 1, 2 or 3, dd < 3;

SLYM-BART^{TM}, primary or secondary Rx = 6, dd < 5.

MONITORING GOLF GREEN PERMEABILITIES BY INFILTROMETRY

BPL infested golf turf grass greens may exhibit symptoms ranging from a loss in permeability in the green, the occurrence of a visible lateral layer of intense black material within the green (usually within the top ten centimetres) and unpleasant odours being given off from the green. As the BPL appears inevitably to lead to radical reductions in the permeability of the green, a technique for the routine monitoring of the entrance transmissivity of water into the green would prove of value.

To monitor the plugging, the rate of infiltration of water from a constant head to an ascending column was used. A series of computer-controlled infiltrometers have been constructed over the period from 1988, terminating in the INFILTREX 8 unit in May of 1991. These infiltrometers measure the rate of water flow into the green as an inverse of the ascending head within the water column being charged from just above sand/soil level. Three photo-optic detectors (POD) are set in the ascending column at heights above the soil surface of 100 mm (POD1), 200 mm (POD2) and 300 mm (POD3). The discharge point into the soil is set at a depth of 10 mm below surface level to reduce the risk of severe lateral and vertical leakage from the column. The 26 mm diameter column is of copper construction, with a free discharge area of 5.1 cm^2. The active volume of water in the ascending column head from the surface level to POD1 is 58.3 ml while the volume between POD1 and POD2 is 50.7 ml, with the same volume between the upper PODs. The monitoring time to fill the ascending column head with a sealed discharge port (control reading) is set by controlling downflow rate from the static water head set at 800 mm. A satisfactory fill time to POD3 (acceptable by the software) is 20 (± 2) secs.

Operation of the INFILTREX 8 involves the initial establishment of baseline data by sealing the discharge port and recording the times in which the three POD responses occur. These then become the standard readings for a series of site readings. Normally five infiltrations are performed on each green using the pre-obtained standard. Once the POD3 has responded, each infiltration value (IV) is obtained for the POD1, POD2 and POD3 and a mean IV obtained.

From field and laboratory experiences it would appear that the mean IV for the entrance transmissivity of water into the soils using this technique varies from 0.01×10^{-6} m/sec (close to totally plugged) to as high as 7.0×10^{-6} m/sec (unbiofouled high sand content green). The critical point for the onset of a major biofouling by BPL appears to occur as the mean IV obtained falls within the range from 2.0×10^{-6} m/sec down to 0.5×10^{-6} m/sec, with radical losses of transmissivity (i.e., permeability) at IV values of less than 0.1×10^{-6} m/sec.

The position and structural integrity of the BPL event may also be observed by the manner in which the water from the infiltrometer moves down into the fouled soil. In an unfouled (highly permeable) porous medium there would be ongoing downward movement restricted only by the occurrence of saturation. However in a biofouled medium, such water flows would be impeded. The infiltrometer can detect some MIF events from anomalies in the times taken for the POD sensors to be triggered by the rising water column in the ascending head. Three major lateral plug formations which can be determined by these time shifts are:

i) a deep (> 25 mm) lateral confluent (DLC) plug;
ii) a shallow (< 10 mm) lateral confluent (SLC) plug and;
iii) a fragile disintegrating lateral (FDL) plug formation.

The time intervals between the triggering of each of the three POD units can allow the prediction of the form of lateral plugging which is occurring. As the water ascends the sensing column it will trigger, in turn, POD1, POD2 and then POD3. The time intervals (T) may be computed using the following equations:

$$T1 = T_{POD1}$$
$$T2 = T_{POD2} - T_{POD1}$$
$$T3 = T_{POD3} - T_{POD2}$$

where the Tx values are timed from the initiation of the infiltration reading to the triggering of the PODx. Where a DLC event is occurring, water will initially flow rapidly into the green but then become impeded. It would be expected that T1 > T2 > T3. For an FDL event, water will be initially entrapped closer to the surface and then, as the plug begins to lose integrity (collapse), the water flows faster. Here, it may be expected that the relationship would be T1 < T2 < T3. For an SLC event, the occlusion (plugging) would be close to the surface of the green. This would impede water infiltration almost completely so that all T values would be similar (ie., T1 = T2 = T3) and close to those obtained for the control values using a sealed discharge port. It becomes possible therefore, by comparison of the T values, to predict the depth and integrity of any lateral fouling which is impeding water infiltration.

Another factor which may be detected with the INFILTREX 8 is an additional MIF termed Localized Dry Spot (LDS). In an LDS, microbial growth occurs very high in the soil profile and also on the turf grass itself causing complete occlusion of the porous media. As a result, the soil at depths of 30 mm may be completely dry and any precipitation simply runs off over the surface. When infiltrometry is performed on an LDS, two responses can be observed. One effect is that the IV generated have a negative value. This is because the water does not even penetrate through the fouled turf grass but is simply diverted up the ascending column and rapidly triggers the PODs. The second response is for the water to penetrate into the fouled turf grass, exit down the discharge port but return to the surface via fractures in the occluded soil, resulting in very high and erratic IV values.

The software program controlling the INFILTREX 8 can perform the necessary steps to determine when there is a probability that a DLC, SLC, FDL or a LDS is occurring. The facility is built into the software to print out both the IV values for the number of replicate determinations performed (commonly five) and also, through the "expert" system, to extrapolate the nature of any MIF occlusive event.

In practice, the INFILTREX system has been used to monitor the permeabilities of high sand content golf greens as an indicator of the potential development of an MIF event. In Regina there have been major infestations of both BPL and LDS for the last five years. By the selective monitoring of the permeabilities of the greens, sufficient warning has been obtained to implement remedial action before the playability was affected to a significant extent. This experience suggests the following monitoring program for green permeability could be established for new and existing golf courses:

1. Monitor each green in the spring, summer and fall. Where all greens are $> 2 \times 10^{-6}$ m/sec, continue 1; if specific greens are below 2×10^{-6} m/sec then go to 2.

2. Monitor greens monthly during the growing season. Where the IV of the greens falls to $<1.0 \times 10^{-6}$ m/sec, institute remedial measures; if IV falls to $<0.2 \times 10^{-6}$ m/sec, go to 3; if IV $>2.0 \times 10^{-6}$ m/sec, return to 1.

3. Monitor greens weekly with particular attention being paid to stressed areas of the green (additional readings); undertake restorative measures and assess short- and long- term improvements which may have resulted. Where the IV of the greens falls to $< 0.1 \times 10^{-6}$ m/sec, go to 4; if IV improves to $>1.0 \times 10^{-6}$ m/sec, go to 2.

4. Remove green from play, conduct intense restoration programme including aerification, use of penetrants, shifts in irrigation practices. Monitor reaction to treatments and observe stresses. Where the IV rises to $> 0.5 \times 10^{-6}$ m/sec, return to 3.

SUMMARY

Early experiments revealed the BPL was a complex biofouling event involving a mixture of iron related bacteria, *Bacillus* species, pseudomonads and both sulphate and sulphur reducing bacteria. These bacteria grew in a consortium in complex structures involving lateral plates, vertical columns and spherical forms. The intensely black pigmentation was due primarily to the presence of iron sulphides, deposited within the structures as a result of a chemical reaction between the hydrogen sulphide generated by the consortium of bacteria and the metal ions (principally iron) bioconcentrated within the layers. This layering developed under conditions of oxygen starvation (anaerobic) and gradually filled the spaces between the sand (or soil) particles. Water could no longer pass freely down the soil profile but tended to accumulate above the BPL. Competition for oxygen, water and living space between these consortial bacteria in the BPL and the grass roots caused stresses to develop in the turf grass leading to the early symptoms. At the same time, water would be "ponding" both in and upon the surface of the green, which would encourage the growth of a variety of micro-algae which have been linked by several research teams to the BPL phenomenon. Concurrently, the additional stress on the turf grass and accumulating organic material (due to decomposition of roots and growth of microalgae) causes a parallel increase in the fungal (mould) population.

Not surprisingly, there has been considerable controversy over the cause of the BPL events as various researchers have focused on the rampant and obvious populations of either the micro-algae or the fungi overlying the bacterially dominated BPL. Management of the fouled greens has been therefore focussed on modifying the environment within the green to one less conducive to the growth of the bacterially dominated BPL. Control factors considered include:

>Increasing the oxygen entry into the BPL zone;

>Applying penetrants to disperse, or at least traumatize, the incumbents in the BPL; and

>Shift to fertilizer practices less supportive of the BPL.

Green #14 at a Regina Municipal Golf Course was selected for the research, development and application of these techniques. Changes in the management practices included the use of more intense but less frequent irrigation (condition 1); aerification (condition 2); application of penetrants (condition 3); and reduction in the level of fertilizer application (condition 4) as well as the use of foliar fertilizers (condition 5) and the abandonment of the use of urea-based fertilizers (condition 6). These fertilizers had been shown in laboratory trials to stimulate the radical growth of BPL dominated conditions. By September of 1987, green #14 was returned to play with the turf grass recovering from a 50% die back event.

ACKNOWLEDGEMENTS

The authors wish to acknowledge the National Research Council of Canada for financial support through the Natural Sciences and Engineering Research Council grant-in-aid (DRC #GP0005073) and the Industrial Research Assistance Programme (-H, -L and -M), the City of Regina for both financial support (data interpretation and consulting; SN, scholarship) and use of the golf course facilities, the Canadian Turf Grass Research Foundation, and the Aquatrol Corporation of America (SN, scholarship). The following are also thanked for their personal commitment to the project: Art Salverson, Ron Graham, (City of Regina); Ken Nelson (Sutton Creek Golf Course, Ontario); Dr A.T. Abiola (University of Regina); Jeff Reihl, Evan Ostryzniuk, Steve Reihl and Julie Driver (Droycon Bioconcepts Inc., Regina) and Natalie Ostryzniuk (Regina Water Research Institute) for the preparation of the manuscript.

REFERENCES

BERNDT, W.L. & VARGA, J.M. (1987). Sulphate Reduction in Turf Grass. *Divot News*, **25(4)**, 34-40.

BERNDT, W.L., VARGAS, J. & MELVIN, B. (1989). Sulfur, Organic Matter and the Black Layer. *Golf Course Management*, **57(3)**, 44-45, 48, 50.

CULLIMORE, D.R., LINDENBACH, S., NELSON, K., JACK, T.R. & MARENTETTE, D. (1989). Preliminary Observation on the Use of Computer Assisted Topography for the Diagnosis of Black Plug Layering in Sand Greens of a Golf Course in Saskatchewan. *Biofouling*, **1**, 279-285.

CULLIMORE, D.R. & NELSON, K. (1989). Investigating the causes and formation of the black plug layers in turf. *Super News of Manitoba, Newsletter of the Golf Course Superintendents' Association of Manitoba*, 3-6.

CULLIMORE, D.R. (1991). *Black plug layer research: a compendium of technical and scientific papers*. Publ Regina Water Research Institute, University of Regina, ISSN 0843-848X ISBN 0-7731-0198-5 125 pp.

KERSHASKY, J. (1987). The Black Layer. *The Grass Roots*, **14(2)**, 6-7.

MALLORY, L.M. & SAYLER, G.S. (1984). Application of FAME (Fatty Acid Methyl Ester) analysis in the numerical taxonomic determination of bacterial guild structure. *Microb. Ecol.*, **10**, 283-296.

MOSS, C.W., DEES, S.B. & GUERRANT, G.O. (1980). Gas-liquid chromatography of bacterial fatty acids with a fused-silica capillary column. *J. Clinical Microbiol.*, **12(1)**, 127-130.

NILSON, S. (1991). A Literature Survey on the Black Plug Layer Phenommenon in Turf Grass Greens on Golf Courses. In: *Black Plug Layer Research* (ed. D.R. Cullimore) publ. Regina Water Research Institute, University of Regina ISSN 0843-848X ISBN 0-7731-09198-5. 66-122.

ROY, M. (1988). Use of fatty acids for the identification of phytopathogenic bacteria. *Plant Disease*, **72(5)**, 460.

SHANK, B. (1987). The Black Layer: Nature's Revenge?. *Sports Turf* **3(6)**, 11.

TEMPLETON, A.R. (1988). Managing vegetation by managing water: the role of soil wetting agents, Paper presented at the Chemical Vegetation Management Symposium, PGRSA Annual Meeting, San Antonio, TX, 1-8.

5-8 BIOGENIC SULPHURIC ACID ATTACK ON CONCRETE IN SEWER ENVIRONMENTS

T. VAN MECHELEN & R. POLDER
TNO-Building and Construction Research, PO Box 49, 2600 AA Delft,
The Netherlands

ABSTRACT

This investigation deals with the determination of the rate of attack of concrete in sewer manholes and subsections by biogenic sulphuric acid attack (BSA), microstructural analysis of the deteriorated concrete by scanning electron microscopy (SEM) and the development of a simple method for measuring the aggressivity of a particular sewer section. The chemistry of the production of sulphuric acid by *Thiobacilli* is briefly described. The impact of this acid on concrete specimens can be seen in SEM-micrographs; the rate of attack was established during non-accelerated exposure in sewer manholes over a three year period. The SEM shows details of the attacked concrete layer which consists predominantly of gypsum and a gel-like substance. However, penetration of sulphate into the sound concrete proceeds beyond this gypsum-rich outer layer. Examples of concrete specimens exposed in sewers and concrete cylinders taken from pipes in seriously deteriorated sewer sections are presented. The highest rate of attack of concrete was 3 mm/year. The causes of this attack are discussed.

INTRODUCTION

Biogenic sulphuric acid attack (BSA) is a mechanism that takes place in sewers, causing degradation of the concrete. Until a few years ago the problem was only recognised in hot climates, eg California and Australia. However in the seventies the process caused severe damage in Hamburg's large diameter sewers and examples of severe BSA have also been found in The Netherlands with its moderate marine-influenced climate. As details of the BSA mechanisms can be found, for example, in Thistlewaite (1972); Bielecki and Schremmer (1987); Schremmer (1990) and Hillman (1990), the process is described only briefly here.

Bacteria in the anaerobic slime layer on the underwater pipe surface reduce sulphates and other sulphur compounds in the sewage to sulphide; the sulphide is liberated into the sewage and consequently oxidised to sulphate if the sewage contains sufficient oxygen. If there is insufficient oxygen available, however, as may be the case if the sewage residence time is long or if there is no contact with air (as in highly filled sewers), the sulphide remains in the reduced form and may be liberated to the sewer atmosphere as hydrogen sulphide, the well known odorous gas. Here it can be dissolved in the moist layer on the concrete surface and subsequently oxidised to sulphuric acid by the sulphur oxydizing bacteria *Thiobacilli*.

Different types of *Thiobacilli* are responsible for the subsequent lowering of the sewer wall pH (Milde, Sand, Wolff and Bock, 1983). At the beginning *Thiobacillus intermedius* and *Thiobacillus novellus* grow on the alkaline or neutral concrete wall. As the pH drops to 6 the number of *Thiobacillus neapolitanus* increases. At pH 5 *Thiobacillus thiooxidans* start to grow. This acid resistant species, also known as *Thiobacillus concretivorus*, can produce sulphuric acid of about 10% w/w having a pH below 1. All these bacteria appear to be commonly present in ground and ground water, albeit in relatively low numbers (Bielecki and Schremmer, 1987). In favourable conditions, such as in sewers with a high level of sulphur, they multiply rapidly (Milde et al, 1983).

Concrete is a material consisting usually of inert aggregate particles bound together by an alkaline cement matrix which consists mainly of calcium silicate hydrates. As the acid solution dissolves the matrix, gypsum and other degradation products are formed which may be visible on the concrete surface or washed away to expose the gravel in the surface. Under favourable conditions the degradation rate may be up to 6 mm per year. This process, called biogenic sulphuric acid attack because of the essential role played by micro-organisms, is the most important internal degradation mechanism in concrete sewers.

In the literature consideration is given not only to the processes and mechanistic aspects of BSA but also to the materials involved, such as the type of cement and aggregate. Some investigations (Sand, Bock and White, 1987) seem to support the view that the more alkaline the cement, the less degradation by BSA is observed. This would favour ordinary Portland cement types over blended cements. In this instance the use of limestone aggregate (an alkaline material) is preferred over the use of inert siliceous aggregate. However, as regards BSA, high alumina cement, which has no free lime at all, appears to have superior properties regarding BSA over ordinary Portland cement and blast furnace slag cement (Bock, Sand, Kirstein and Rammelsberg 1989; Dumas, 1990) despite being less alkaline than the other two cement types. Blast furnace slag cement with at least 65% slag has been used in sewers and other structures in The Netherlands for decades. Although specific data about resistance to BSA are not available, BFSC concrete has better resistance against chlorides and sulphates because of its denser mix. The addition of fly ash and silica fume to the concrete also improves the resistance of concrete against the permeation of aggressive substances.

In 1987 TNO Building and Construction Research and the Public Works Department of the Municipality of Rotterdam began experiments on BSA. These were carried out in real sewer systems in order to obtain results with practical relevance especially with regard to measurement and assessment of the level of aggressivity, the degradation rate of the concrete and the influence of the concrete composition (ie cement type).

Two hundred concrete specimens were exposed to sewer environments in ten manholes (Polder and van Mechelen, 1989). Each time samples were placed or removed chemical tests were undertaken in order to characterise the sewers. After exposure the mass loss and the penetration of acid into the concrete was determined.

EXPOSURE OF CONCRETE IN A SEWER ENVIRONMENT

The results of experiments after two years of exposure have been described in an earlier paper (van Mechelen and Polder, 1990). Three concrete types were exposed to the sewer environment. The concrete was made as similar as possible to production sewer pipes using each of the three cement types as marketed in The Netherlands. The cements used were ordinary Portland cement (OPC), blast furnace slag cement (BFSC, containing about 70% blast furnace slag) and Portland fly ash cement (PFAC, containing 25% fly ash) and the concrete mixes were made with sand and gravel (d_{max} = 8 mm). As a reference, specimens were included from a normal production sewer pipe made with BFSC ((d_{max} = 12 mm). Concrete quality of both laboratory and production concrete was high; the water/cement ratio was about 0.3 and consequently the concrete had low porosities (12.5 to 15.9 volume %) and high strength (flexural strength between 7.6 and 9.6 N/mm^2). The specimens were about 150 x 45 x 45 mm in size; all surfaces of the exposed specimens were sawn to avoid the influence of cement-rich surface layers. Five specimens of each mix were mounted in steel frames, each containing 20 specimens. These were suspended in each of ten manholes at the level of the crown of the pipes, as shown schematically in Figure 1. The specimens were washed only when the water level was high, in the same way as the crown of the pipes in field situations.

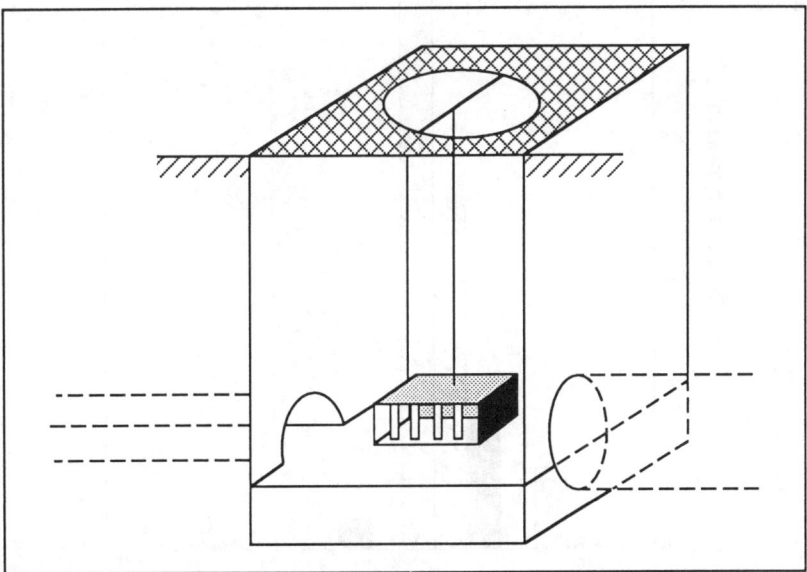

Figure 1: Specimens in a steel frame suspended in a manhole.

All selected exposure locations were gravity sewers, transporting mainly domestic sewage by a mixed system. The pipes ranged from circular 400 mm diameter to egg shaped 800/1200 mm. A special aspect of the Rotterdam sewer system is that many sewer sections and pipes are suffering from uneven settlement, which increases sewage residence times.

Specimens have been on test since September 1987. The first series of samples were taken in January 1988, after 3.5 months of exposure; the second series in October 1988 after 13 months and the third in August 1989 after two years' exposure. In September 1990 a fourth and final series of samples were taken out, representing three years' exposure.

MASS LOSS AND DEGRADATION OF CONCRETE

Mass loss of concrete specimens proceeds gradually with time. In the first months only superficial attack of the specimens was visually observed. This was either discoloration or loss of cement matrix around the gravel. After one, two and three years of exposure severe damage could be observed in some manholes and clear differences in mass loss between the manholes were noted (Figure 2).

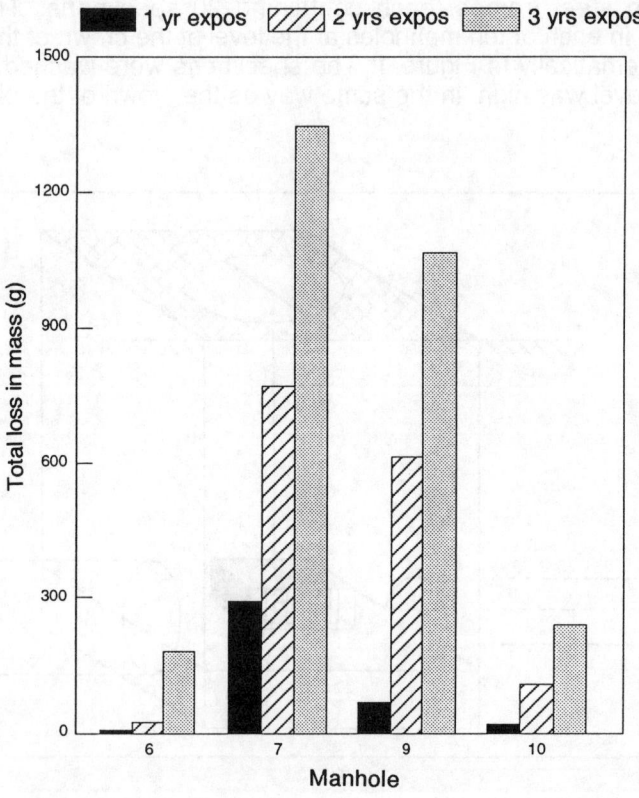

Figure 2: Mass loss after 1, 2 and 3 years of exposure in manholes 6, 7, 9 and 10.

It was appreciated that mass loss is not the only form of degradation hence the penetration of the acid was also tested. This showed a gypsum layer had been produced around the specimen which does not contribute to the strength of the concrete and can therefore be seen as loss of material. The acid penetration was measured by spraying an indicator liquid on a freshly cleaved concrete surface. Generally a surface layer with a pH < 5 was observed attached to the still alkaline concrete. The boundary seemed to be sharp, extending less than 1 mm. The mass loss and the acid penetration (or neutralisation) are accounted for in the degradation rate. After two years of exposure differences in degradation rates between the concrete types were noted (van Mechelen and Polder, 1990) but these were no longer significant after three years of exposure.

The most severe degradation rate was found in manholes 7 and 9 with a mean degradation rate (mass loss + acid penetration) of 3.3 and 2.2 mm/year.

ALTERATIONS OF THE CONCRETE DURING BSA

The degradation of the concrete specimens was studied by Scanning Electron Microscopy (SEM) and Energy Dispersive X-ray Analysis (EDAX).

Samples of the different types of concrete were investigated. The samples were large enough to include both the degraded surface layer and the non-degraded bulk concrete. For all concrete types the following were observed:

The surface consisted of gypsum, sometimes in the presence of another degradation product, mainly silica gel with or without alumina (Plate 1). Large amounts of gypsum were observed immediately below the surface. Deeper into the material, between 1 and 3 mm below the degraded surface,

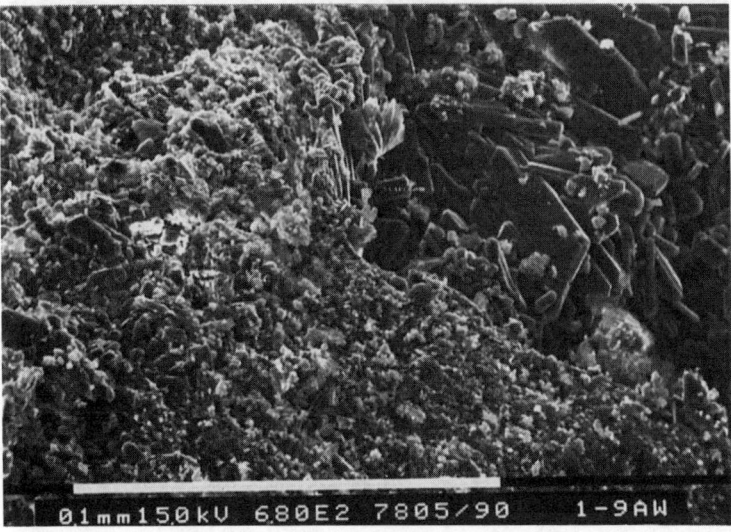

Plate 1: Degraded concrete surface showing gypsum and silica gel.

gypsum and hardened cement paste were present together. Between 3 and 5 mm from the degraded surface there was a layer of increased sulphate, supporting the idea that the boundary between non-degraded and degraded concrete is not sharp and that there is a transition zone from non-degraded material via increased sulphate levels to the degraded gypsum and silica-gel containing zone. Accelerated testing of concrete specimens on immersion in sulphuric acid solutions (Attiogbe and Rizkalla, 1988) has also indicated the presence of decreasing sulphate levels from the degraded surface zone to the internal non-degraded material.

The transition zone indicates sulphate penetrates into the non-degraded concrete without completely converting it. The SEM analysis showed penetration of this sulphate beyond the zone of neutralisation (pH < 5) thus the sulphate penetration has not yet had an influence on the alkalinity of the concrete. In most cases the sulphate penetration is about 2 mm deeper than the neutralisation layer identified by spraying indicator liquid.

In some specimens ettringite (tricalciumaluminosulphatehydrate) was found directly below the degraded concrete layer. Plate 2 shows ettringite in a void and Plate 3 ettringite in the presence of gypsum 4 mm under the surface. Ettringite is a typical reaction product of sulphate and the cement mineral tricalciumaluminate. The source of the sulphate may have been the calcium sulphate originally present in the cement as a retarder. However, during the first stages of cement hydration this sulphate reacts with tricalciumaluminate to form ettringite which is subsequently converted to monosulphate. In most cases therefore, no ettringite is to be expected. Consequently the source of sulphate is probably not from the internally present calcium sulphate but from elsewhere. It is possible ettringite formation is due to sulphate penetration either from (sub)surface gypsum or

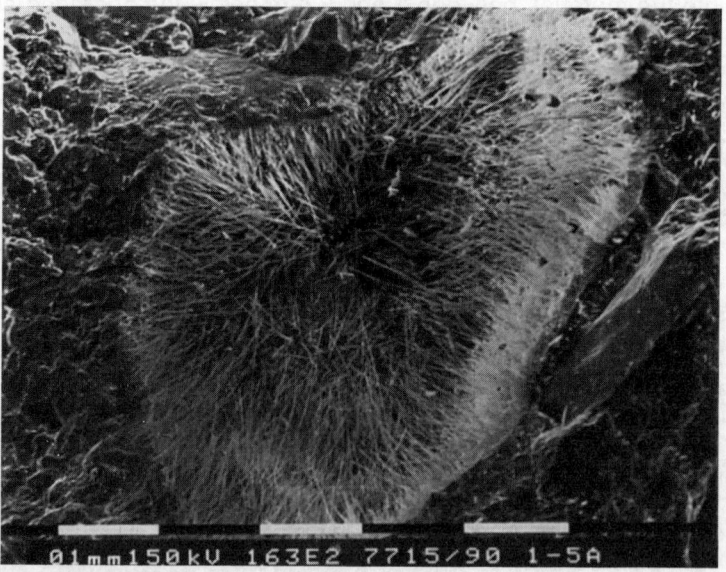

Plate 2: Ettringite crystals in a void in the concrete.

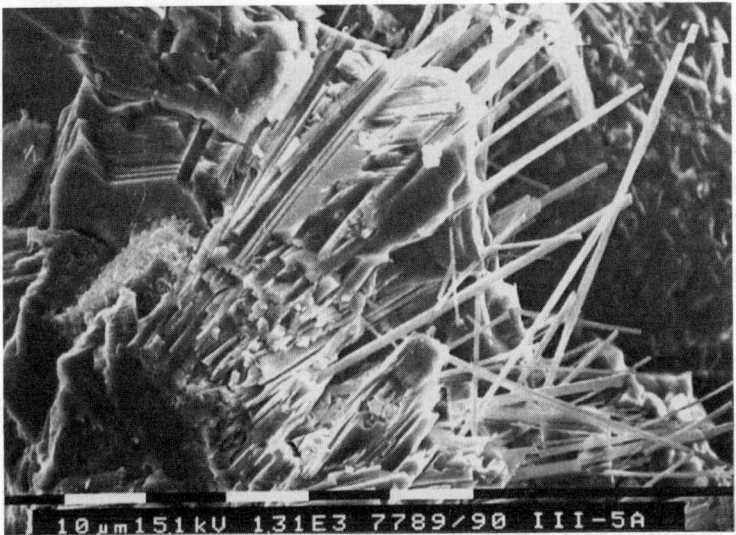

Plate 3: Ettringite and gypsum crystals 4 mm below the surface.

directly from sulphuric acid. SEM micrographs and element mapping appear to support this view. This would mean that by the BSA mechanism, in addition to the direct sulphuric acid attack at the surface, deeper in the concrete some sulphate transport takes place leading to ettringite formation. The importance of this in the overall BSA process is not yet clear. It is possible the sulphate attacks the existing structure and thus facilitates the sulphuric acid attack when the zone with active BSA approaches the transition and non-degraded zones.

The work also identified a magnesium enriched layer in BFSC specimens at about 2 mm below the degraded concrete. This probably reflects the variation in solubility of the different oxides present in the degraded concrete. At low pH the magnesium will be more soluble than the other oxides; in dissolved form it will penetrate the concrete but it may be precipitated as soon as it reaches a zone where the pH is higher than in the degraded layer. As OPC contains less magnesium, it was not detected in the OPC concrete specimens.

CHEMICAL CHARACTERISATION OF SEWER ENVIRONMENT

In addition to the collection of concrete specimens from the sewer manholes, tests were carried out to determine the characteristics of a sewer environment and to investigate the relation between the sewer environment and the degradation rate observed on the concrete specimens. Oxygen and sulphide content were measured together with the pH and temperature of the sewage and the pH of the concrete surface above the water line. The presence of heavy turbulences, eg falling water in the main stream, outlet of pressure pipes, the sewage flow and the level of the sewage in every pipe section were noted.

These investigations showed an association between a low oxygen content of the sewage as measured using an oxygen specific electrode and the tendency to produce sulphide. Sulphide itself is the source of hydrogen sulphide gas and the fuel for BSA. Sulphide was measured by a sulphide specific electrode in sewage samples buffered at high pH. Sewage pH and temperature influence the sulphide build-up and H_2S liberation, hence the overall BSA activity. The sewage level may provide an indication of uneven settlement and consequent stagnation in certain sections. Stagnation increases the residence time of the sewage, which strongly promotes sulphide build-up.

The pH of the concrete surface is a direct measure of the actual BSA activity; the lower the pH of the concrete surface the more active the BSA. The surface pH was measured by spraying the concrete with two different pH indicator solutions and observing the resulting colour by a video camera suspended from a tripod in the manhole. The indicator solutions were chosen to distinguish the most important pH regions: pH 1 and lower; pH between 1 and 3; pH between 3 and 5; pH above 5. These pH levels were expected to correspond approximately to the very aggressive, moderately aggressive, weakly and non-aggressive conditions respectively.

The results of sewer environment characterisation tests are given in Table I. Sewage pH ranged from 7.2 to 7.7 in August 1991 and is fairly constant over the years. As can be seen from Table I there is considerable variation over the seasons. In summer sewage temperatures are high, oxygen contents low and sulphide contents high. This leads to a low pH of the concrete surface, indicating a high BSA activity. In winter the situation is reversed. With low sewage temperatures there is more oxygen and less sulphide so that the concrete surface pH is higher, indicating a lower BSA activity. All the measurements except those in August 1990 were carried out after a period without any substantial rainfall hence the undiluted and most aggressive state of the sewer in that period was measured. However, in August 1990 there was an enormous rainfall two days before the amounts were recorded which may account for the variation in the August 1989 and August 1990 measurements. The unusually high rainfall led to dilution of the sulphide, lowering of the temperature, the production of more oxygen and presumably flushing of the concrete surface by the rise in water level during the flow through of the rain which in some cases increased the concrete surface pH. Further conclusions about the characteristics of the sewer environment are given in van Mechelen and Polder (1990).

It is of note that manholes 7 and 9 have the largest degradation rate. In these manholes water from side-streams falls turbulently into the main stream. The side stream which enters manhole 7 is from a side sewer which suffers from uneven settlement and therefore has very high sulphide levels. This anaerobic sewage falls into manhole 7 and results in an increased transfer of sulphide to the air, exacerbating BSA on the concrete sewer wall (Figure 3).

Table I: Results of sewage analysis for each manhole at the time of the final analysis.

						Manhole					
	Date	1	2	3	4	5	6	7	8	9	10
Temp (°C)	Sep 1987	18	18	19	19	18	18	18	18	19	18
	Jan 1988	9	11	13	12	12	11	12	12	12	11
	Oct 1988	16	16	17	17	17	17	17	17	18	16
	Aug 1989	22	22	23	22	22	23	23	22	24	23
	Aug 1990	20	20	21	21	20	20	20	20	21	20
	Aug 1991	20		21		22	22	22	22	24	22
O_2 (mg/l)	Sep 1987										
	Jan 1988	7.3	3.4	4.9	4.2	2.4	2.3	2.8	5.9	2.2	1.6
	Oct 1988	4.4	4.7	4.8	3.2	2.7	3.2	4.0	6.1	3.9	3.1
	Aug 1989	1.3	0.0	1.2	0.0	0.0	0.0	0.0	0.0	0.1	0.0
	Aug 1990	2.0	1.2	0.6	2.1	0.0	0.3	0.9	2.2	0.0	0.0
	Aug 1991	1.8		0.8		0.0	0.0	0.1	0.3	0.0	0.0
S^{2-} (mg/l)	Sep 1987	3.5	2.3	1.8	1.4	4.1	4.7	5.1	2.4	4.7	8.3
	Jan 1988	1.0	1.4	1.4	1.0	2.6	3.1	3.4	1.2	2.4	5.6
	Oct 1988	2.2	1.4	1.5	1.1	3.8	4.1	3.9	1.9	5.0	7.0
	Aug 1989	9.0	6.0	6.8	3.5	10.7	10.3	11.3	5.6	18.5	31.0
	Aug 1990	11.5	1.2	3.2	0.8	2.7	3.0	6.0	1.4	3.4	7.5
	Aug 1991	3.0		6.2		6.6	9.4	8.1	4.9	8.4	13.0
pH	Sep 1987	1	5	1	5	3	1	1	5	1	3
	Jan 1988	3	5	5	5	5	5	1	5	3	3
	Oct 1988	1	5	1	5	1	3	1	5	1	1
	Aug 1989	1	3-5	1	3	1	1-3	1	1-3	1	1
	Aug 1990	1	5	1	5	3	1	1	1	1	1
	Aug 1991	1		1		1	1	1	1	1	1
Turbulence 0 = no; 1 = yes		1	0	1	0	0	1	1	1	1	0

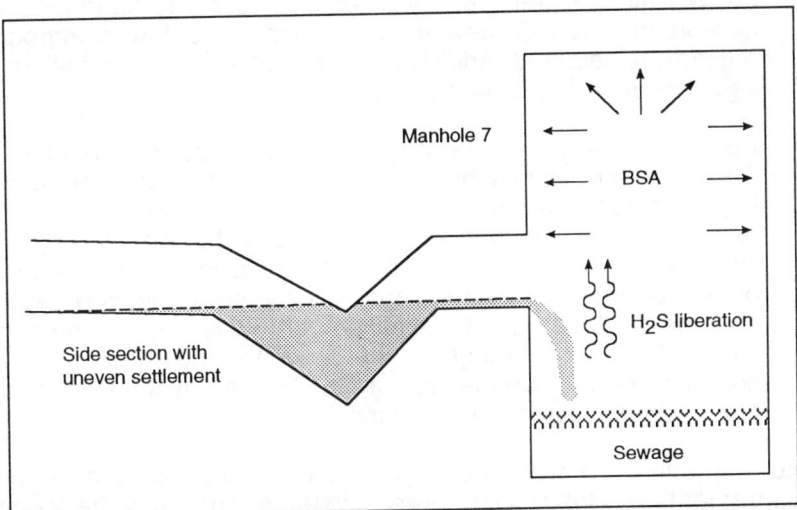

Figure 3: Situation in manhole 7; anaerobic sewage falls in the manhole resulting in an increased transfer of hydrogen sulphide gas to the air.

ASSESSMENT OF THE AGGRESSIVITY IN A PARTICULAR SEWER

Using the data obtained from the sewer systems in Rotterdam a practical method for aggressivity assessment has been derived. An advantage of this method is that relatively simple field and laboratory measurements are the basis for establishing the aggressivity in a particular sewer section.

The most important parameters are:

- a) total sulphide content of the sewage
- b) surface pH of the manhole concrete
- c) turbulence of the sewage flow into the manhole and
- d) sewage temperature.

Of these, sulphide provides the fuel for BSA; sewage turbulence promotes the transfer of H_2S from the sewage to the atmosphere and concrete surface pH shows the level of the actual acid production process. The temperature is important as it was noted that only the most aggressive spot maintained a high level of BSA activity at lower temperatures.

A modification of the previously described aggressivity assessment method of van Mechelen and Polder (1990) is proposed (Table II). This weighs the input data to assess the level of BSA activity and provides three possible courses of action:

A: BSA is not present or activity is only in a very weak form. Degradation rates are up to about 0.5 mm concrete/year. No further measures are necessary.

B: BSA is present in a mildly active form. Degradation rates are up to about 1 mm concrete/year. If sewage temperature during inspection is more than 20°C, new measurements in a lower temperature regime are required. Additional investigation of the condition of the sewer system may be necessary.

C: BSA is present in a strongly or very strongly active form. Degradation rates may be up to about 3 mm concrete/year and even higher rates may occur. If measurements were taken at a temperature higher than 20°C they should be recorded again in a lower temperature regime. Additional investigation of the condition of the sewer system and the quality of the sewer pipes is necessary. This may include the determination of the thickness of the concrete pipe wall and the strength of the concrete pipes, locating sulphide sources and/or turbulences in the sewer system. Appropriate protection measures must be considered.

Table II provides a framework in which to interpret the results of measurements in inspected sewer systems and may be helpful in assessing priorities in the management and maintenance of attacked sewer systems; action 'C' sewers require high priority attention; 'B' sewers may be attended to later.

Table II: Decision table for the aggressivity to concrete of a particular sewer section based on a single inspection.

sulphide (mg/l)	<1			1-5			5-15			>15		
pH-concrete (±)	5	3	1	5	3	1	5	3	1	5	3	1
Turbulence, (S = strong, W = weak)	S,W	S,W	S,W	S,W	S,W	S,W	S,W	S,W	S,W	S,W	S,W	S,W
Temperature												
<15°	A,A	B,A	C,B	B,A	C,B	C,C	C,B	C,C	C,C	C,C	C,C	C,C
15-20°	A,A	A,A	B,A	A,A	B,A	C,B	B,A	C,B	C,C	C,B	C,C	C,C
>20°	A,A	A,A	A,A	A,A	A,A	B,A	A,A	B,A	C,B	B,A	C,B	C,C

A, B, C denote urgency of action (see text)

Table III demonstrates the use of this method in assessing the aggressivity of the ten Rotterdam manholes. It can be seen that the manholes with the highest degradation rates for the concrete prisms, numbers 1 7, 9 and 10 almost always resulted in a 'C' rating. Manhole 3 is an exception and although a 'C' rating is obtained it shows a low degradation rate of the concrete prisms.

It is clear that manholes containing sewage at higher temperatures (eg the measurements in August 1989) are often rated 'C'; the higher temperatures and sulphide formation causing a strong BSA. During the short warm period this appears to be a relatively common phenomena in the inspected manholes. It is therefore recommended that measurements are repeated at lower sewage temperatures. The measurements in January show the best relation with the degradation rates. At that time only the manholes with a high degradation rate were attributed a 'C' rating. It is therefore preferable to inspect sewer manholes during the cold season as if conditions are found to be aggressive then it is likely they will be so throughout the year. Degradation will be most severe in those cases.

Table III: Comparison of measured degradation rates in inspected manholes and assessment provided by methods given in Table II.

Manhole	Degradation rate mm/year	Assessment by decision table				
		Sept 1987	Jan 1988	Oct 1988	Aug 1989	Aug 1990
1	1.0	C	C	C	C	C
2	0.5	A	A	A	A	A
3	0.7	C	B	C	C	B
4	0.5	A	A	A	A	A
5	0.6	A	A	B	B	A
6	0.7	C	B	B	C	C
7	3.3	C	C	C	C	C
8	0.6	A	B	A	C	B
9	2.2	C	C	C	C	B
10	0.9	B	C	C	C	B

The assessment also shows that measurements should be carried out after a certain period without rain. This will give results of the undiluted state of the sewage and be more representative of the severity of the BSA to which the concrete is exposed.

APPLICATION OF BSA ASSESSMENT

Practical tests of aggressivity assessment in sewer environments were carried out in Rotterdam in a 28 year old sewer subsection in which severe concrete degradation had been identified using television cameras. This section was inspected in both winter and summer periods and showed very high sulphide levels, low oxygen and low pH of the concrete surface. Assessment by the decision table gave mainly 'B' and 'C' ratings, consistent with the visually observed attack.

The BSA was caused by uneven settlement of the sewer pipes. Recent alluvium in the Rotterdam area consolidated both unevenly and at a relatively high annual rate. This has resulted in stagnation in the sewage flow, long residence times and subsequent depletion of oxygen. In this way sulphides built up and the sewer section was suffering degradation. In addition, at the beginning of the section where it was connected to blocks of flats with only a few drains into the sewer, at some hours of the day the amount of sewage induced a high degree of turbulence.

The combination of high sulphide levels and concentrated, intermittent turbulence determined BSA activity and damage to the concrete. Effective service life was fifteen to twenty-eight years. When the section was replaced it was possible to test some of the concrete sewer pipes after removal. The degradation rate was found to be between 1.0 and 2.2 mm/year over the total sewer section lifetime. This was consistent with the assessment produced using the decision table and the exposure experiments in the ten manholes. Concrete quality of the non-attacked parts proved to be good.

CONCLUSIONS

Exposure experiments on concrete specimens in sewer manholes during three years have yielded much information regarding the process and parameters of biogenic sulphuric acid attack (BSA), the influence on the concrete material, BSA aggressivity measurement and assessment of BSA aggressivity.

Degradation of the concrete specimens occurred in most manholes but the degradation rate varied considerably, ranging from between 0.5 and 3 mm/year. This degradation could be related to the sewer environment. High sulphide contents of the sewage combined with high turbulences gave rise to more degradation. High sulphide contents were associated with long residence times in areas with uneven settlement of the soil and consequently of the pipes.

Microstructural analysis revealed details of changes caused by BSA in the concrete. Gypsum and other degradation products were formed and penetration of sulphate into the material was noted. A transition zone between the sound concrete and the degraded (sub)surface zone was present, which contained hardened cement paste with a higher than average sulphate content. The penetration of sulphate was deeper than the neutralisation layer of the concrete.

Differences in degradation rate between concrete types were not found to be significant and SEM analysis did not identify any important differences in attack.

It has been shown that by means of a few simple measurements the aggressivity of the sewer environment due to BSA in a particular section can be characterised. Input of the key parameters into a decision allows an assessment of the aggressivity of the sewer environment for concrete and what action should be taken. The best assessments are made when measurements are carried out in the cold season and after a period without appreciable rainfall. Although this method is only semi-qualitative in form, it can be used successfully in discriminating between conditions of no aggressivity, mild aggressivity and strong aggressivity. With this information priorities can be made in the maintenance policy of the sewer authorities.

REFERENCES

ATTIOGBE, E.K. & RIZKALLA, S.H. (1988). Response of concrete to sulfuric acid attack. *ACI Materials Journal, November-December*, 481-488.

BIELECKI, R. & SCHREMMER, H. (1987). Biogenic sulphuric acid corrosion in gravity sewers. *Heft der Mitteillungen des Leichtweiss-Institut fuer Wasserbau der Technische Universitaet Braunschweig.*, **94**.

BOCK, E., SAND, W., KIRSTEIN, K. & RAMMELSBERG, J. (1989). Untersuchungen zur Bestandigkeit von Zementmortel-auskleidungen duktiler Gussrorhe gegenuber biogener Schwefelsaure-Korrosion. *Korrespondenz-Abwasser*, **36(12)**, 1376-1387.

DUMAS, TH. (1990). Calcium aluminates cementitious binders: an answer to bacterial corrosion. *Proceedings CSTB-CEFRACOR Symposium Corrosion-Deterioration in Building, November 1990, Paris.*

HILLMAN, P.F. (1990). Microbiological effects on materials in sewerage systems. *Proceedings of the Federation of European Microbiological Societies Symposium, Microbiology in Civil Engineering, Cranfield, U.K.* Howsam, P. (ed.), E. & F.N. Spon, London.

MILDE, K., SAND, W., WOLFF, W. & BOCK, E. (1983). Thiobacilli of the corroded concrete walls of the Hamburg sewer system. *Journal of General Microbiology*, **129**, 1327-1333.

POLDER, R.B. & VAN MECHELEN, A.C.A. (1989). Assessment of Biogenic Sulphuric Acid Agressivity of Sewer Environments. *Proceedings 2nd International Conference on Pipeline Construction, Hamburg.*

SAND, W., BOCK, E. & WHITE, D.C. (1987). Biotest system for rapid evaluation of concrete resistance to sulphur-oxidizing bacteria. *Materials Performance*, **26**, 14-17.

SCHREMMER, H. (1990). Abschaetzung der Biogenen Schwefelsaurekorrosion in Abwasserkanale. *Korrespondenz-Abwasser*, **11**, 1332-1339.

THISTLEWAYTE, D.K.B. (1972). *Control of Sulphides in Sewage Systems.* Butterworths, Sydney.

VAN MECHELEN, A.C.A. & POLDER, R.B. (1990). Degradation of Concrete in Sewer Environments by Biogenic Sulphuric Acid Attack. *Proceedings of the Federation of European Microbiological Societies Symposium Microbiology in Civil Engineering, Cranfield, U.K.*, Howsam P. (ed.), E & F.N. Spon, London.

5-9 DETERMINATION OF THE CHEMICAL RESISTANCE OF CEMENTS USING SMALL SCALE ACCELERATED TESTS

V. PAUL & S. UBEROI
Building Research Establishment, Garston, Watford, WD2 7JR

ABSTRACT

Small scale accelerated tests using 10 mm x 10 mm x 60 mm mortar prisms have been carried out to determine the chemical resistance of cements. The effect of sulphates, acids and phenol solutions and sulphate/acid, sulphate/acid/phenol mixtures on ordinary Portland cement (OPC) and OPC/blastfurnace slag, OPC/pulverised fuel ash admixtures was investigated.

Only sulphate solutions containing 50,000 mg SO_4^{2-}/l (>10 times Class 4, BRE Digest 363) caused any degradation to the mortar prisms. No degradation was observed due to attack by phenol, acid or weaker sulphate, 2000 mg SO_4^{2-}/l (Class 2-3, Digest 363) solutions. Solutions containing more than one chemical were found to be no more aggressive than solutions containing individual compounds. The most consistent increase in durability was shown by mortar containing a lower water/cement ratio; increase in durability was also observed with the use of blastfurnace slag.

INTRODUCTION

Research has shown that concrete is susceptible to degradation in a number of naturally occurring environments (Bartholomew, 1979; Harrison, 1989; Eglinton, 1975; Gutt and Harrison, 1977). Laboratory experiments have also shown that compounds such as sulphates (Lawrence, 1990), acids (Fattuhi and Hughes, 1983) and phenols (Smith, 1985) are deleterious towards concrete. The majority of this work has focused on the effects of single substances; the effect of combinations of chemicals has only occasionally been studied. Combinations of chemicals have generally been found to produce a more aggressive environment for the target material. A report by the Water Research Centre (Crathorne and James, 1990) found that chemicals that do not generally permeate polyethylene pipes, in the presence of a second chemical can rapidly contaminate water flowing in them. Land which has been contaminated during a previous industrial history may contain a cocktail of chemicals (contaminants) which may enhance the deterioration of the concrete, therefore the effects of mixtures of chemicals on the durability of concrete needs to be investigated.

Laboratory studies on the effect of chemicals on concrete have shown that permeability, cement type and C_3A content can all affect durability (Lawrence, 1990; Fattuhi and Hughes, 1983; Smith, 1985). The rate of ingress of the contaminant solution into the concrete determines the rate of the chemical degradation process in the concrete (Mehta, 1987); thus, the permeability of concrete is the principal factor influencing durability. Permeability can be decreased by either increasing the cement content or reducing the water/cement ratio. Selection of an appropriate cement type is also important in increasing concrete durability. Partial replacement of ordinary Portland cement (OPC) with either ground granulated blastfurnace slag or pulverised fuel ash has been shown to improve the sulphate resistance of concretes. BRE Digest 363 (BRE, 1991) recommends that for concrete exposed to sulphate attack, such cements can now be used in soils of Class 4 and 5 (soils containing 3.0 to 6.0 and >6.0 %SO_4 respectively). However, whether such partial replacement leads to improved performance when subjected to other aggressive agents is unclear (Harrison, 1987).

The Building Research Establishment (BRE) has begun a four point test programme in order to investigate the durability of concrete in contaminated land. The programme consists of

a) accelerated tests involving the immersion of small mortar prisms in both single contaminant solutions and solutions of mixed contaminants;

b) longer term immersion tests using concrete cubes in solutions of contaminants;

c) exposure of concrete cylinders to slurries or suspensions to which a mixture of contaminants has been added; and

d) a field trial involving the burial of concrete cylinders in a contaminated land site.

The programme will investigate the effects of reducing permeability and of partial replacement of OPC with ground granulated blastfurnace slag or pulverised fuel ash on concrete durability. An integral part of the test programme is the use of solutions of contaminants that accurately reflect the chemical conditions concrete may normally be subjected to in contaminated land. These solutions are being formulated from contamination profiles that have been compiled for gasworks, chemical works and scrapyards from a number of site investigation reports obtained by BRE. The profiles also provide a guide to the nature and concentration of the major contaminants likely to be found on contaminated land associated with particular industries. The use of solutions of contaminants corresponding to the chemical conditions associated with particular contaminated land sites should allow for a greater understanding of the long term performance of concrete structures in such sites. The preliminary results of the accelerated tests are presented here.

ACCELERATED TESTS

A number of small scale accelerated tests (Mehta and Gjorv, 1974; Koch and Steinegger, 1960; ASTM, 1987) have been used to determine the performance of cements in aggressive environments. Such tests can give an indication of a cement's ability to resist chemical attack within two or so months. Three such accelerated tests were reported in a recent study by Osborne (1989): the Mehta test (Mehta and Gjorv, 1974); the Koch-Steinegger square mortar prism test (Koch and Steinegger, 1960) and a flat mortar prism test. These were assessed to see whether any of them could be recommended as an acceptance test for determining the sulphate resistance of cements. Reasonable agreement between the results of the three small-scale accelerated tests was found. Comparison with longer term concrete cube test data also showed that the data obtained from the small scale accelerated tests gave a reasonable correspondence (Osborne, 1991). The flat prism test extended to a measurement period of six months was considered to provide the best rapid indication of cement behaviour in sulphate conditions. The test also employed a similar storage routine to that for the assessment of concrete cubes. The sulphate resistance of cements was measured by the change in dimensions of the prism. As not all chemical attack results in expansion, the flat mortar prism test was not considered suitable for the present programme. It was therefore decided to use the square mortar prism test which shares some of the advantages of the former test, notably in employing a similar storage routine to that used in concrete cube immersion testing. The square prism test measures sulphate resistance by noting the bending strength of the prism, a technique which could also be employed to measure chemical resistance in other solutions.

EXPERIMENTAL PROCEDURE

The square mortar prisms (10 mm x 10 mm x 60 mm) were prepared in moulds according to German specification DIN 1164 (DIN, 1978) using mortar composed of one part by weight of cement (ordinary Portland cement [OPC] or OPC and partial replacement material), one part by weight of standard sand I (fine), two parts by weight of standard sand II (coarse) and 0.6 parts by weight of de-ionised water. The water was added to a mixing bowl and mixed with the cement for 30 seconds at low speed using a high speed food mixer. Hand mixed sand was added over the next 30 seconds before mixing for a further 60 seconds at high speed. The moulds were half filled with mortar and compacted on a vibrating table for 15 seconds. They were then filled over the next 30 seconds and vibrated for a further 75 seconds. The mortar was levelled off and the moulds placed in a moist air curing cabinet at 100 % RH for one day before being demoulded and cured in water for 20 days at 20°C.

Four prisms were placed in de-ionised water and four in the contaminant solution. The solutions were changed every two weeks and after eight weeks the bending strengths of the samples stored in water and in solution were measured. The cement was classed as having good chemical resistance (C_r) if the ratio of the bending strength of the solution cured prisms b_s and the water cured prisms b_w was greater than 0.8.

$$C_r = (b_s/b_w) \qquad [1]$$

MATERIALS

The chemical composition and batch numbers of the cement, ground granulated blastfurnace slag (ggbs) and pulverised fuel ash (pfa) are shown in Table I.

The Portland cement was chosen because it contained a medium C_3A content (C_3A = 8.18). The ggbs and pfa were both commercially available materials. The ggbs was donated by the Frodingham Cement Company with an alumina (Al_2O_3) content of 12.73%: BRE Digest 363 (1991) sets a maximum level of 15% Al_2O_3 in the slag to be used in soils of Classes 2 and 3.

Four mortar compositions were studied.

(a) Control mix: One part by weight OPC, one part by weight standard sand I, two parts by weight standard sand II and 0.6 parts by weight de-ionised water;

(b) As (a) except water 0.55 parts by weight;

(c) As (a) except OPC replaced by OPC/ggbs 30:70;

(d) As (a) except OPC replaced by OPC/pfa 75:25.

Table I: Chemical analysis of Portland cement, blastfurnace slag and pulverised fuel ash.

Oxide	Portland Cement OPC 856	Blast Furnace Slag FCC[1] 59	Pulverised Fuel Ash Fiddlers Ferry 897
SiO_2	20.55	35.77	50.60
Al_2O_3	5.07	12.73	26.90
Fe_2O_3	3.10	0.58	8.84
CaO	64.51	38.91	2.39
MgO	1.53	9.27	1.45
K_2O	0.73	0.54	3.40
Na_2O	0.15	0.32	0.97
TiO_2	0.26	0.67	0.99
P_2O_5	0.19		0.28
MnO		0.45	
Mn_2O_3	<0.01		0.06
BaO	0.01		0.15
SrO	0.18		0.05
ZnO			0.08
SO_3	2.53	0.15	0.71
S^{2-}		0.89	
Cl		0.02	
LOI at 1000°C	1.58	0.54	3.40
Total	100.39	100.84	100.27
Free CaO	0.96		
Fineness m^2/kg		421.0	
Calc. C_3A	8.18		

[1] Fordingham Cement Company

SOLUTIONS OF CONTAMINANTS

The susceptibility of mortars and structural concretes to attack by sulphates (Lawrence, 1990), acids (Fattuhi and Hughes,1983) and phenols (Smith, 1985) has been known for many years. An extensive body of data exists on the performance of mortars, cements and concretes in solutions of varying concentrations of the above chemicals. In order not to duplicate this work only concentrations with direct relevance to building on contaminated land were investigated.

The concentrations of sulphate, acid and phenol used in the first set of immersions are shown in Table II. The values studied correspond to the Interdepartmental Committee on the Redevelopment of Contaminated Land trigger concentrations (DOE, 1987).

The action of hydrocarbons on concrete was also investigated. The solution used was a sample of a free phase organic fraction obtained from a site where slow leakage of petroleum products and solvent had occurred over a number of years. The major components of this fraction were toluene, xylene (ortho, meta and para), benzene, 1,2,4 trimethyl benzene, chlorobenzene and dichloromethane.

Table II: Concentration of sulphate, acid and phenol used in primary immersion tests.

a)	2.9% Na_2SO_4 anhydrous solution (2000 mg SO_4^{2-}/l)
b)	7.4% Na_2SO_4 anhydrous solution (50000 mg SO_4^{2-}/l)
c)	5.1% $MgSO_4.7H_2O$ solution (2000 mg SO_4^{2-}/l)
d)	12.8% $MgSO_4.7H_2O$ solution (50000 mg SO_4^{2-}/l)
e)	HCl pH 3
f)	HCl pH 5
g)	0.1% Phenol (1000 mg/l), HCl pH5 solution
h)	2.9% Na_2SO_4 anhydrous, HCl pH 5 solution
i)	5.1% $MgSO_4.7H_2O$, HCl pH 5 solution
j)	2.9% Na_2SO_4, HCl pH 5, 0.1% phenol solution
k)	Free phase organic fraction containing: toluene, xylene, benzene, 1,2,4 trimethyl benzene chlorobenzene and dichloromethane as major components

Solutions which more accurately reflect the chemical conditions found on contaminated land were formulated using data from the contamination profiles (Appendix 1). Median rather than mean values were used to formulate solutions because it was felt that this would give a more accurate reflection of the concentration of contaminants associated with such sites. Mean values were thought to be unreliable as an indication of the extent of contamination associated with an "average" contaminated land site because of bias which can be introduced by a single extremely high value. More aggressive conditions were simulated by using the 75th percentile and 90th percentile values to formulate solutions. Table III shows the mixtures of contaminants used.

Table III: Solutions of contaminants used in second immersion experiment.

Concentration of Contaminants in Solution

	Gas Works			Chemical Works			Scrap Yards		
	Median	75th Percentile	90th Percentile	Median	75th Percentile	90th Percentile	Median	75th Percentile	90th Percentile
pH	7.8	7.3[1]	6.2[2]	8.0	7.7[1]	7.5[2]	7.8	7.5[1]	7.0[2]
Sulphate	835	3400	28660	11160	24930	48000	987	2384	5200
Sulphide	34	150	320	0	2	16	0	22	40
Cyanide	6	32	290	0	2	20	2	3	6
Phenol	1	5	27	3.5	17	690	0	2	3
Chloride				218	489	960	24	40	80

All concentrations except for pH expressed as mg/l
[1] Value corresponds to 25th percentile
[2] Value corresponds to 10th percentile

RESULTS

The results of the accelerated immersion tests for the chemical resistance of cements are shown in Table IV.

SULPHATES

No degradation of the mortar prisms was detected on exposure to a solution containing 2000 mg SO_4^{2-}/l after eight weeks. In the majority of cases a small increase in the flexural strength of the prisms was detected. In contrast after exposure to solutions containing 50000 mg SO_4^{2-}/l severe spalling of the prisms occurred in all but three cases. In only two cases, OPC (w/c 0.55) in strong $MgSO_4$ solution and OPC/ggbs 30:70 (w/c 0.6) in strong Na_2SO_4 was C_r above 0.8 indicating that the mortars were sulphate resistant.

ACIDS

All four mortars had good to excellent resistance when exposed to the two acid solutions. The storage routine employed in the Koch-Steinegger test (1960) involves changing the solution every two weeks to simulate an environment where there is occasional movement of ground water. Concrete reacts with the solution to form reaction products which may temporarily halt the degradation before they are washed away by ground water movement. The cycle then starts again. In the experiment it was found that the acidic solution was neutralised within 2 days for the pH 5 solution and 4 days for the pH 3 solution. Thus it seems likely that the alkalinity of the cement in the mortar prisms neutralises the acidic solution before any

DETERMINATION OF THE CHEMICAL RESISTANCE OF CEMENTS

Table IV: Results from accelerated immersion for chemical resistance of cements.

Mortar composition		Solution of Contaminants									
		a	b	c	d	e	f	g	h	i	j	k
Mix (a) OPC w/c 0.6												
Cr		0.99	0.44	1.15	0.61	0.92	0.89	0.89	1.05	0.96	0.94	1.00
Chemical Resistance		Good	Poor	Excellent	Poor	Good	Good	Good	V. Good	Good	Good	Good
Mix (b) OPC w/c 0.55												
Cr		1.17	0.79	1.17	0.92	1.06	1.04	0.95	1.10	1.00	1.09	0.91
Chemical Resistance		Excellent	Moderate/Good	Excellent	Good	V. Good	V. Good	Good	V. Good	Good	V. Good	Good
Mix (c) OPC/ggbs 30:70 w/c 0.6												
Cr		1.01	0.99	1.06	0.24	0.82	0.91	0.94	0.96	0.98	0.94	1.03
Chemical Resistance		Good	Good	V. Good	Poor	Good	Good	Good	Good	Good	Good	V. Good
Mix (d) OPC/pfa 75:25 w/c 0.6												
Cr		1.00	0.45	1.05	0.51	1.01	1.00	0.92	1.10	1.05	0.84	0.92
Chemical Resistance		Good	Poor	V. Good	Poor	Good	Good	Good	V. Good	V. Good	Good	Good

degradation occurs. It is worth noting that in practice failures of concrete due to acid attack have only been reported when there is a constant flow of acidic water or where there has been an uncontrolled discharge of mineral acid waste (Harrison, 1987). This suggests that whilst cement is potentially vulnerable to attack by acids, the rate of attack is negligible unless there is a constant flow of acidic water to wash away any reaction products, as implied by the results of the experiment. In order to test the effect of an environment where there is constant replenishment of acid, further tests are being undertaken with the solution kept to a pH of 3 by daily addition of concentrated hydrochloric acid (37% HCl content).

PHENOL

Phenol has been found to be aggressive to concrete (Smith, 1985; Biczok, 1972) although the exact mechanism of attack is unclear. Biczok (1972) attributed the adverse effect of phenol to a physical expansion mechanism whilst Smith (1985) considered it is due to an inhibition and modification of the hydration process. Smith (1985) found that concrete exhibited a loss of strength after two years' exposure to solutions containing 0.2% and 0.5% by volume of phenol. He concluded that there may be an adverse effect on concrete at concentrations as low as 0.1% phenol (1000 mg/l).

Exposure of the mortar prisms to a solution containing 1000 mg/l of phenol produced no visible signs of degradation after eight weeks. For all four mixes C_r was found to be greater than 0.8, indicating that their chemical resistance to phenol was good.

MIXTURES OF CONTAMINANTS

Neither the sulphate/acid nor the sulphate/acid/phenol mixtures showed any sign of being more aggressive than the single contaminant solutions. The C_r values for the samples exposed to these solutions did not vary significantly from those exposed to single contaminants.

In the case of the sulphate/acid mixture it may be that it was only the effect of sulphate on the mortar prisms that was recorded as the acid was neutralised within two days. The similarities between the C_r values for the prisms exposed to the sulphate/acid mixture and those exposed to sulphate would tend to support this. For prisms exposed to the sulphate/acid/phenol mixture a small decrease in the C_r value compared to similar prisms exposed to sulphate was detected.

ORGANIC COMPOUNDS

Immersion of the prisms in the free phase organic fraction obtained from a contaminated land site produced no visible signs of degradation. This was confirmed by the C_r values found for the four mixes. In each case the prisms were found to have good chemical resistance to attack by organic compounds.

DISCUSSION

The performance of concrete in aggressive conditions is dictated by concrete mix design and binder type. The rate determining step in the degradation process is the ingress of the chemical into the concrete (Mehta, 1987). Thus the use of dense impermeable concrete will to a large extent prevent its degradation. A reduction in permeability and consequent increase in durability can most easily be achieved by a reduction in the water/cement ratio. The results of the experiments show that only the mortar prisms with a water/cement ratio of 0.55 can be classed as having good sulphate resistance in both the $MgSO_4$ and Na_2SO_4 solutions containing 50000 mg SO_4^{2-} /litre

In contrast, the partial replacement of OPC by pfa and ggbs was found to improve only the sulphate resistance of mortars exposed to Na_2SO_4 solutions. Samples containing pfa and ggbs were found to have a lower chemical resistance rating compared to the control mix in the strong $MgSO_4$ solution. The exact mechanism by which ggbs and pfa increase sulphate resistance is not completely understood and a number of theories have been suggested to explain the enhanced durability of concrete with pfa and ggbs (Biczok, 1972; Mathews, 1987). These include:

a) dilution of the C_3A content of the cement;

b) removal of the free $Ca(OH)_2$ by reaction with pfa or ggbs;
The conversion of the free $Ca(OH)_2$ to silica and alumina gels provides both a chemical and physical base for improved chemical resistance. The gel is thought to seal the capillary pores making the concrete less permeable as well as forming a protective coating over the vulnerable aluminate compounds. The reduction in the amount of $Ca(OH)_2$ available for reaction permits a higher concentration of sulphate before enough gypsum is formed to cause an expansive reaction. The rate of ingress of SO_4^{2-} into concrete is also reduced due to the reduction in the concentration of OH^- ions available for counterdiffusion.

c) a reduction in the stability of hydrated calcium aluminates;
The solubility of ettringite, while low in a saturated lime solution, increases as the lime in the solution decreases. Removal of the free $Ca(OH)_2$ establishes a condition in which ettringite remains in solution and expansion does not occur.

It is likely that a combination of all three mechanisms is responsible for the increased sulphate resistance of concrete on partial replacement of OPC with pfa or ggbs. However, the reduction in the permeability of the concrete is likely to be the most important factor in imparting increased durability against other aggressive chemicals.

The lower chemical resistance of mortars containing ggbs and pfa compared to the control mix in the strong $MgSO_4$ solution is thought to be due to the reduction in the $Ca(OH)_2$ content which occurs when OPC is partially replaced by ggbs and pfa. $MgSO_4$ reacts with the $Ca(OH)_2$ in the hardened cement paste to form the highly insoluble $Mg(OH)_2$ (Biczok, 1972). This accumulates on the surface of the concrete to form a protective layer. The extent to which this protective layer forms is dependent on the concentration of the $MgSO_4$ and the amount of $Ca(OH)_2$ in the hardened cement paste. Thus whilst cements containing high levels of $Ca(OH)_2$ may have a high chemical resistance in $MgSO_4$ solutions, cements with low levels of $Ca(OH)_2$ will be particularly vulnerable to high concentrations of $MgSO_4$. This has recently been recognised by BRE and in Digest 363 (1991) the use of ggbs and pfa cements in soils of Class 4 and 5 has been limited to those containing <1.0 g/l of magnesium in the ground water.

The small scale accelerated tests form the initial part of a four point test programme as described in the introduction. The results obtained from these tests give an indication of durability of the cement type. Later tests on concrete cubes in solution and cylinders in doped slurries should give a more accurate description of the effects of chemicals on structural concrete. Finally, a field trial involving the burial of concrete cylinders of different mix designs will determine the validity of the laboratory tests results. The information obtained from these tests will be used as part of a predictive model. It is hoped that such a model will allow the long term performance of structural concrete in contaminated land to be determined without resorting to long term durability testing.

CONCLUSIONS

Cement is only vulnerable in extremely aggressive solutions. Only sulphate solutions containing 50000 mg SO_4^{2-}/l (>10 times Class 4, Digest 363) caused any degradation of the mortar prisms. No degradation was observed due to attack by phenol, acid and sulphate, 2000 mg SO_4^{2-}/l (Class 2-3, Digest 363) solutions.

No synergistic effect was observed in the presence of multiple contaminants. Solutions of contaminants containing sulphate/acid and sulphate/phenol/acid were found to be no more aggressive than solutions containing the individual compounds.

Permeability of cement based materials is the principal factor influencing their durability. Reduction of the water/cement ratio had the greatest effect on the durability of the mortar prisms. Mortar prisms containing a water/cement ratio of 0.55 had good chemical resistance in all the solutions studied.

Improved sulphate resistance is obtained on partial replacement of OPC by ggbs on exposure to Na_2SO_4.

The use of cements containing ggbs and pfa in conditions where high concentrations of Mg^{2+} ions are present should be carefully considered since the experiments showed that the use of ggbs and pfa produced no increase in chemical resistance in $MgSO_4$ solution containing 50000 mg SO_4^{2-}/l.

ACKNOWLEDGEMENTS

The work described has been carried out as part of a research programme funded by the Department of the Environment. BRE would like to thank the Frodingham Cement Company Ltd. for providing the equipment and materials and Groundwater Technology Inc. Ltd for providing the free phase organic fraction.

REFERENCES

AMERICAN SOCIETY FOR TESTING AND MATERIALS, (1987). *Test method for length change of hydraulic cement mortar exposed to sulphate solution.* ASTM, Philadelphia, ASTM C1012.

BARTHOLOMEW, R.F. (1979). The protection of concrete piles in aggressive ground conditions: an international appreciation. *Proceedings of the Conference on Recent Developments in Design and Construction of Piles, Institution of Civil Engineers, London.*

BICZOK, I. (1972). *Concrete corrosion and concrete protection.* Hungarian Academy of Science, Budapest, English edition.

BUILDING RESEARCH ESTABLISHMENT, (1991). Concrete in sulphate-bearing soils and groundwater, *Digest 363.*

CRATHORNE, B. & JAMES, C. (1990). *Effects of soil contaminants on materials used for distribution of water.* WRC report for DOE, 2222-M1.

DEPARTMENT OF THE ENVIRONMENT, (1987). *Guidance on the assessment and redevelopment of contaminated land.* ICRCL 59/83 2nd Edition.

DIN, 1978. *Portland, blastfurnace and pozzolanic cements.* DIN, Berlin, DIN 1164 Part 1.

EGLINTON, M.S. (1975). Review of concrete behaviour in acidic soils and groundwaters. *CIRIA Technical Note*, **69**.

FATTUHI, N I. & HUGES, B.P. (1983). Effect of acid attack on concrete with different admixtures or protective coatings. *Cement and Concrete Research*, **13**, 655-665.

GUTT, W.H. & HARRISON, W.H. (1977). The chemical resistance of concrete. *Concrete*, **11(5)**, 35-37.

HARRISON, W.H. (1987). Durability of concrete in acid soils and waters. *Concrete, February*, 18-24.

HARRISON, W.H. (1992). *Long-term tests on sulphate resistance of concrete at Northwick Park: Third Report.* Building Research Establishment, BR 164.

KOCH, A. & STEINEGGER, H. (1960). An accelerated method for testing the sulphate resistance of cements. *Zement-Kalk-Gips*, (Portland Cement Association Foreign Literature Studies No. 306), **13**, 317-324.78

LAWRENCE, C.D. (1990). Sulphate attack on concrete. *Magazine of Concrete Research*, **42(153)**, 249-264.

MATHEWS, J.D. (1987). *Pulverised fuel ash - its use in concrete: Part 2 influences on durability.* Building Research Establishment, IP 12/87.

MEHTA, P.K. (1987). *Concrete: structure, properties and materials.* Prentice-Hall Inc., New Jersey.

MEHTA, P.K. & GJORV, O.E. (1974). A new test for sulphate resistance of cement. *J. Test Evaluation*, **2(6)**, 510-515.

OSBORNE, G.J. (1986). The durability of concrete made with gasifier-slag cement. *Durability of Building Materials*, **4**, 151-177.

OSBORNE, G.J. (1989). Determination of the sulphate resistance of blastfurnace slag cements using small-scale accelerated methods of test. *Advances in Cement Research*, **2(5)**, 21-27.

OSBORNE, G.J. (1991). The sulphate resistance of Portland and blastfurnace slag cement concretes. *2nd CANMET/ACI International Conference on "The durability of concrete"*, Montreal, Canada.

SMITH, M.A. (1985). The effect of phenol on concrete. *Magazine of Concrete Research*, **37(133)**, 234-237.

APPENDIX I — GASWORKS CONTAMINATION PROFILE

Contaminant	No of Samples	Range [1,2] Observed	Mean[1]	Median	75th Percentile	90th Percentile
pH	900	1.2-12.4	7.5	7.8	7.3[3]	6.2[4]
Acid soluble sulphate	618	0-373000	9216.4	835.0	3400	28660
Water soluble sulphate	383	0-16	1.1	1.0	1.5	2.0
Acid soluble sulphide	837	0-5300	159.2	34.0	150.0	320.0
Elemental sulphur	63	0-25000	3061.9	1300	3900	8300
Total cyanide	867	0-38000	336.6	6.0	32.0	290.0
Free cyanide	550	0-930	20.6	0.8	5.0	27.0
Complex cyanide	205	0-38000	789.9	0.1	3.1	400.0
Total Arsenic	556	0-1270	15.2	5.0	13.5	30.0
Total Cadmium	636	0-503	2.8	0.0	1.0	3.0
Total Chronium	588	0-410	26.3	23.0	31.0	42.0
Total Copper	622	0-580	47.8	26.5	57.0	110.0
Total Lead	664	0-166000	532.5	91.0	193.5	520.0
Total Manganese	490	0-6500	543.6	355.0	620.0	1300
Total Nickel	622	0-456	32.9	23.5	36.0	72.0
Total Zinc	650	0-10500	146.3	65.0	130.0	295.0
Available Copper	123	0-71	6.2	0.0	10.0	20.0
Available Lead	91	0-48000	564.3	5.0	15.0	35.0
Available Nickel	123	0-20	1.3	0.0	2.0	5.0
Available Zinc	123	0-320	21.4	10.0	25.0	41.1
Zinc equivalent	32	0-311	82.5	62.6		
Phenol	907	0-6573	27.7	1.0	5.0	27.0
Toluene extract	944	0-505490	5781.6	1175	3685	12000
Coal tar	503	0-175000	3590.5	600.0	2900	7700
Naphthalene	125	0-47000	478.1	0.0	4.0	56.0
Acenaphthylene	125	0-1710	28.2	0.0	1.6	30.0
Acenaphthene	16	0-89	7.6	0.0	0.0	21.0
Fluorene	125	0-3200	67.0	0.0	2.2	53.0
Phenanthrene	125	0-7400	185.5	4.2	22.0	120.0
Anthracene	99	0-104	6.3	0.0	5.1	12.7
Fluoranthene	125	0-11100	284.4	11.4	44.1	300.0
Pyrene	125	0-6500	149.5	9.5	32.2	190.0
Benzo(a)anthracene	42	0-140	10.3	0.0	10.0	24.0
Chrysene	125	0-137	10.7	4.6	8.1	26.8
Benzo(b)fluoranthene	16	0-99	11.9	0.0	6.8	56.0
Benzo(k)fluoranthene	16	0-32	7.5	0.0	14.5	22.0
Benz(a)pyrene	125	0-113	9.7	2.0	8.4	28.1
Dibenzo(a,h)anthracene	16	0-63	5.7	0.0	4.2	15.0
Benzo(a,h,i)perylene	16	0-84	10.9	0.0	11.5	34.0
Indeno(1,2,3,cd)pyrene	16	0-19	5.2	1.3	10.4	15.0
Biphenyl	83	0-32	0.5	0.0	0.0	0.0
Acridine	83	0-13	0.3	0.0	0.0	0.0
Carbazole	83	0-13	3.5	0.0	2.8	6.2
Benzofluorene	83	0-48	3.1	0.8	2.9	7.1
4,5 Benzopyrene	83	0-66	7.3	2.7	9.1	19.7
Perylene	83	0-29	4.0	2.7	3.9	7.0

Notes:
[1] All concentrations expressed as mg/kg except water soluble sulphate as g/l.
[2] Values below the detection limit taken as zero.
[3] Corresponds to 25th percentile.
[4] Corresponds to the 10th percentile.

APPENDIX I (CONT'D): SCRAP YARD CONTAMINATION PROFILE

Contaminant	No of Samples	Range[1,2] Observed	Mean[1]	Median	75th Percentile	90th Percentile
pH	402	2.6-11.6	7.8	7.8	7.5[3]	7.0[4]
EC	402	52-9000	731.0	515.8	882.0	1619
Acid soluble sulphate	331	0-24000	2104.1	987.0	2384	5200
Water soluble sulphate	19	0-2.1	1.0	1.2	1.7	1.9
Acid soluble sulphide	285	0-226	12.3	0.0	22.0	40.0
Total cyanide	306	0-236	3.7	2.0	3.0	6.0
Chloride	126	0-248	34.3	24.0	40.0	80.0
Total Antimony	285	0-139	4.7	0.0	4.0	10.0
Total Arsenic	307	0-141	16.3	11.0	20.0	33.0
Total Barium	285	0-3400	208.6	0.0	250.0	580.0
Total Cadmium	297	0-150	4.4	0.8	2.3	11.0
Total Chromium	331	0-2467	10.5	41.0	71.0	174.0
Total Cobalt	285	0-81	11.5	11.0	18.0	26.0
Total Copper	379	0-10667	364.7	103.0	260.0	800.0
Total Lead	379	0-13520	458.2	130.0	310.0	990.0
Total Manganese	285	80-19000	1032.8	730.0	1174	2100
Total Mercury	306	0-40	0.8	0.1	0.4	1.4
Total Molybdenum	285	0-210	5.6	0.0	0.0	18.0
Total Nickel	379	0-845	77.9	43.0	77.0	172.0
Total Vandium	285	0-193	51.0	47.0	65.0	83.0
Total Zinc	379	0-62666	915.0	175.0	423.0	1600
Phenol	308	0-213	2.1	0.0	2.0	3.0
Toluene extract	291	0-25910	1218.4	170.0	1010	3470
Naphthalene	9	0-0.3	0.08			
Acenaphthylene	8	0-0.09	0.02	0.00	0.02	0.09
Acenaphthene	6	0.00	0.00	0.00	0.00	0.00
Fluorene	9	0-0.18	0.04	0.00	0.04	0.18
Phenanthrene	10	0-0.84	0.14	0.02	0.20	0.55
Anthracene	10	0-1.20	0.27	0.04	0.30	1.02
Fluoranthene	9	0-153	0.23	0.00	0.23	1.53
Pyrene	10	0-3.40	0.59	0.04	0.30	2.65
Benzo(a)anthracene	8	0-0.80	0.18	0.01	0.33	0.80
Chrysene	6	0-1.00	0.20	0.00	0.20	1.0
Benzo(b)fluoranthene	6	0-0.05	0.83	0.00	0.00	0.5
Benzo(k)fluoranthene	6	0-0.90	0.23	0.00	0.50	0.9
Benz(a)pyrene	6	0-0.60	0.10	0.00	0.00	0.6
Dibenzo(a,h)anthracene	6	0-0.70	0.12	0.00	0.00	0.7
Benzo(a,h,i)perylene	6	0-0.40	0.07	0.00	0.00	0.4
Indeno(1,2,3,cd)pyrene	6	0-1.60	0.33	0.00	0.40	1.6

Notes:
1. All concentrations expressed as mg/kg except water soluble sulphate as g/l.
2. Values below the detection limit taken as zero.
3. Corresponds to 25th percentile value.
4. Corresponds to 10th percentile value.

APPENDIX I (CONT'D): CHEMICAL WORKS CONTAMINATION PROFILE

Contaminant	No of Samples	Range [1,2] Observed	Mean[1]	Median	75th Percentile	90th Percentile
pH	223	1.8-12.4	8.0	8.0	7.7[3]	7.5[4]
Acid soluble sulphate	223	940-277670	22586.1	11160	24930	48000
Water soluble sulphate	216	0.4-2.7	1.0	0.9	1.3	2.0
Acid soluble sulphide	224	0-368	7.6	0.0	2.0	16.0
Total cyanide	196	0-2250	59.3	0.0	2.0	20.0
Free cyanide	24	2-1200	150.2	22.0	205.0	400.0
Chloride	224	13-43240	651.7	218.0	489.0	960.0
Total Arsenic	200	3-3850	259.3	65.0	148.0	575.0
Total Cadmium	199	0-840	7.2	0.0	2.0	5.0
Total Chromium	199	2-1347	79.8	39.0	59.0	120.0
Total Copper	199	9-703000	4257.4	180.0	347.0	1380.0
Total Lead	199	14-40000	1533.6	530.0	1040	458.0
Total Mercury	198	0-63	4.3	2.0	3.1	10.6
Total Nickel	199	2-137	34.4	28.0	41.0	66.0
Total Zinc	199	33-106000	2836.4	351.0	568.0	1410.0
Phenol	224	0-211000	12725	3.5	17.0	690.0
Toluene extract	223	0-369400	20561.9	1300	12800	59300
Coal tar	154	0-369400	20818.5	5600	8755	90800
Naphthalene	16	0.0	0.0	0.0	0.0	0.0
Acenaphthylene	16	0-38	4.8	0.0	8..0	15.0
Acenaphthene	16	0-100	22.7	8.0	31.0	100.0
Fluorene	16	0-100	24.6	8.0	46.0	77.0
Phenanthrene	16	0-715	121.1	27.0	184.5	331.0
Anthracene	16	0-546	44.8	0.0	4.0	108.0
Fluoranthene	16	0-238	66.8	23.0	100.0	238.0
Pyrene	16	0-131	24.1	11.5	23.0	85.0
Benzo(a)anthracene	16	0-254	23.6	0.0	11.5	54.0
Chrysene	16	0-277	33.2	8.0	42.0	54.0
Benzo(b)fluoranthene	16	0-0	0.0	0.0	0.0	0.0
Benzo(k)fluoranthene	16	0-0	0.0	0.0	0.0	0.0
Benz(a)pyrene	16	0-31	1.9	0.0	0.0	0.0
Dibenzo(a,h)anthracene	16	0-0	0.0	0.0	0.0	0.0
Benzo(a,h,i)perylene	16	0-0	0.0	0.0	0.0	0.0
Indeno(1,2,3,cd)pyrene	16	0-0	0.0	0.0	0.0	0.0

Notes:
1. All concentrations expressed as mg/kg except water soluble sulphate as g/l.
2. Values below the detection limit taken as zero.
3. Corresponds to 25th percentile value.
4. Corresponds to 10th percentile value.

5-10 CALCIUM ALUMINATE MORTARS AND CONCRETES: AN APPLICATION TO SEWER PIPES IN HARSH ENVIRONMENTS

T. DUMAS
Lafarge Fondu International, 157 av Charles de Gaulle,
92521 Neuilly sur Seine, France.

ABSTRACT

Solving the problems encountered in sewerage networks is a matter of growing importance in almost all parts of the world. Special emphasis is currently being placed on the corrosion phenomena caused by both chemical and bacterial activity. Laboratory studies and investigations on particularly aggressive sites are highlighting new aspects of this subject and suggest that calcium aluminate based binders can be appropriate in such environments.

INTRODUCTION

Durability is a complex concept, including the performance of a given product under various constraints in a dynamic system. Among these constraints are chemical stress, mechanical stress, thermal stress and biological factors. Both time and the nature of the material are important factors. Concretes and mortars are heterogeneous materials which will react differently in a given stress environment depending on the specific chemical and structural characteristics of the binder and of the aggregates.

This paper presents a case study discussing the performance obtained with calcium aluminate based mortars and concretes in the very specific conditions encountered in sewerage systems exposed to bacterial corrosion.

ANALYSIS OF THE PROBLEM

Bacterial corrosion in sewage environments has been appreciated for a long time, particularly in countries such as Australia and southern USA. Indeed for many years it was thought that this sort of trouble was largely restricted to warm climates. However, it is now known that it may occur in any country and an increasing number of surveys and research studies have been initiated around the world in the last few years. Concern has increased, particularly in developed countries where environmental questions such as the quality of water have become of greater importance.

In general, bacterial corrosion in these cases can be described as the production of sulphuric acid on the walls of sanitary systems by transformation of the sulphur compounds contained in the effluent. Sulphur is effectively present in most effluents under various forms, such as sulphate ions and amino acids with thiol groups in proteins. Under anaerobic conditions, these sulphur species transform into hydrogen sulphide (H_2S) as a gaseous compound dissolved in the effluent. This is typically the case in flatter, larger sewers when there may be inadequate flow rates and consequently long retention times particularly when higher temperatures exist. Dormitory cities, resorts and new townships are therefore particularly sensitive and warm climates are of special concern.

According to the classic laws of diffusion, the H_2S passes continuously into the air of the sewer. This phenomenon is enhanced by turbulence conditions so that pressure drops; elbows, irregular beddings and local disturbances are therefore likely to be sites of high H_2S concentration, up to 200 ppm (ASCE Manual 69, 1989; Colin, 1987; Thistlethwayte, 1972).

The H_2S collects and concentrates in the upper parts of the system (pipe crown and manholes) where it is oxidised into S^o (sulphur) and deposited on the wall. Bacterial activity (*thiobacilli*) converts sulphur into sulphuric acid, H_2SO_4; see schematic diagram, Figure 1.

The active growth of a population of such bacteria in a favourable environment is likely to induce pH values as low as 2 to 5; even lower pH values have been encountered on some sites.

In ordinary concrete structures the acid attacks the cementitious material producing a pastey mass of gypsum. Sand aggregate then drops into the waste water, exposing fresh cementitious material to further acid attack, reactivating and accelerating the degradation process (Cochet, 1990; Polder, 1989).

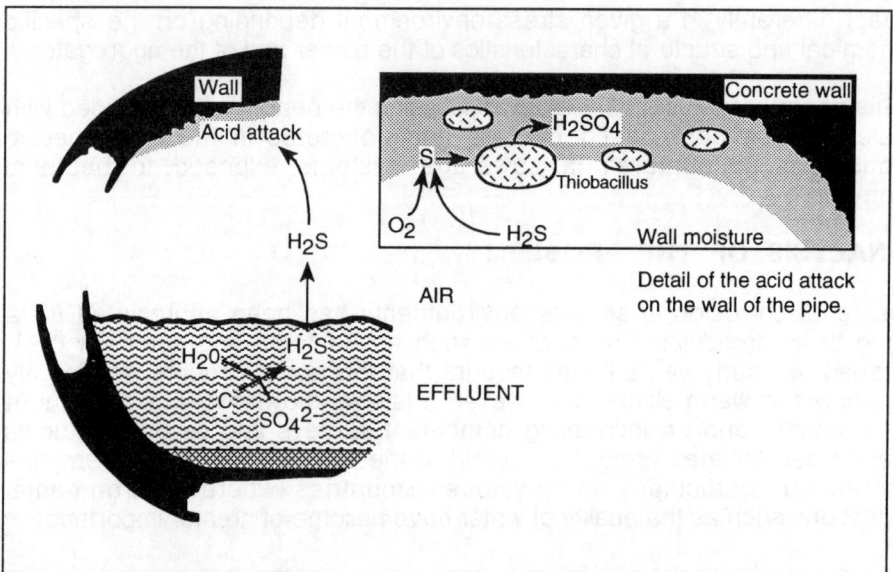

Figure 1: Biogenic corrosion.

To date very few surveys have been undertaken which are sufficiently precise to allow a reasonable estimation of the extent of bacterial corrosion to be made but in 1989 NIPO reported that 16% of the damage affecting the Dutch network was linked to this phenomenon (NIPO, 1988). The alternative strategies available to tackle this problem include:

1. Preventing the formation of H_2S in the effluent:
   Shorter sewer lines and retention times reduce the amount of H_2S produced. Oxidants such as oxygen, chlorine and hydrogen peroxide oxidize the sulphides in the effluent. Metals such as iron, chromium, copper, zinc, nickel or cadmium precipitate the sulphides in the effluent but may be toxic to the environment.

2. Limiting the release of H_2S in the air:
   Smoother and more regular profiles reduce the turbulence.

3. Preventing H_2S from concentrating in some parts of the system:
   Ventilation drives the gas away while periodic flooding dilutes the sulphuric acid and washes it from the pipe wall.

4. Resisting the acid attack:
   Using materials able to withstand corrosion with regard to the specific environment and to the expected lifetime of the structures.

The following sections report on investigations conducted with calcium aluminate base cementitious materials.

THE CALCIUM ALUMINATE SYSTEM

Ordinary concretes and mortars are known to perform poorly in acidic environments. The explanation is well known and depends on the high basicity of Portland cement and hydrates. In 1847 Vicat suggested that a cement with a high $(SiO_2 + Al_2O_3 + Fe_2O_3) / (CaO + MgO)$ ratio would exhibit better performance than Portland cement. As a result the manufacture of Ciment Fondu Lafarge (C.F.L.) was developed in 1908.

This cement is made by simultaneous fusion of bauxite and limestone in a reverberatory furnace at about 1450°C. The molten product decants continuously from the furnace and solidifies on a pan conveyor. The newly solidified "clinker" is then ground in ball mills to cement fineness.

The Rankin diagram, shown in Figure 2, illustrates the mineralogical composition of CFL. The principal phase here is the calcium monoaluminate CA (about 40%), with various ferrites comprising about 35% of the material and minor phases such as C_2S about 6%, $C_{12}A_7$ about 2%, C_2AS about 2% and glass about 10%.

The hydration of the aluminates follows the Le Chatelier law. The anhydrous material dissolves congruently and very rapidly in water until the saturation concentration is reached. After a germination period, mass

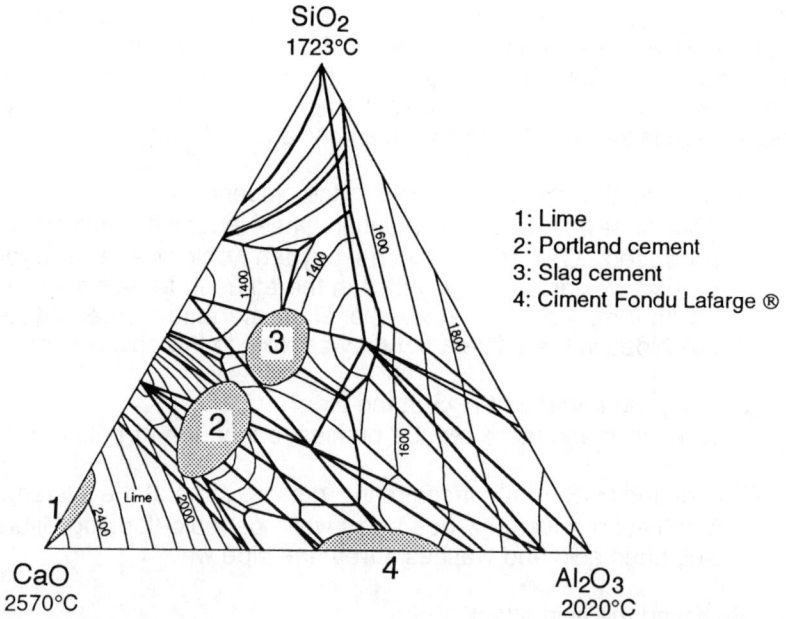

Figure 2: Rankin diagram.

precipitation of the hydrates occurs. The principal final hydrate phase is $Ca_3Al_2O_6.6H_2O$, associated with crystalline gibbsite $Al(OH)_3$. The iron combines in hydrates as a partial substitution for Al^{3+} ions.

It is notable that CFL does not release free lime when hydrating which gives it a major advantage over traditional cements as regards corrosion resistance. For this reason CFL based concretes rapidly gained a prominent position in applications in sewage-rich environments. The first work using CFL based concrete recorded in France was the construction of an in situ grouted collector in the area of Marseilles in 1926. Thereafter it rapidly became used worldwide in a variety of applications from new works to the rehabilitation of pre-cast pipes, pipelines etc (Mathieu 1987; Mathieu 1989).

DURABILITY INVESTIGATIONS ON SITE

Investigations have recently been reported from a number of countries where calcium aluminate base mortars and concretes have been emplaced since the 1930s. This paper discusses two case studies, in Malaysia and Egypt. In each case the pH on the wall was measured, the temperature recorded and observations made of the condition of the concrete and mortar surfaces which were tested for consistency using the Schmidt hammer. Photographs and samples were taken whenever possible.

ON SITE OBSERVATIONS IN KUALA LUMPUR, MALAYSIA

The municipality of Kuala Lumpur has been specifying CFL for sewage applications since the 1950s. With the help of an independent laboratory and the technical services of the municipality, thirty-two of the sixty-four manholes were inspected along the thirty-year old Pantai trunk sewer line. The temperature was 34°C. Corrosion was noticed particularly in the end section of the network. The Portland cement concrete (OPC) constituting the manhole chambers was corroded, especially at the ceiling, exposing the aggregates. The pH values measured were around 4, justifying the use of CFL rather than other alternatives. As seen in Plates 1a and b, the pipes made with CFL were in good condition; this visual observation was confirmed by the testing.

Plate 1a: CFL lined sewer pipes after 30 years - Kuala Lumpur, Pantai, Malaysia.

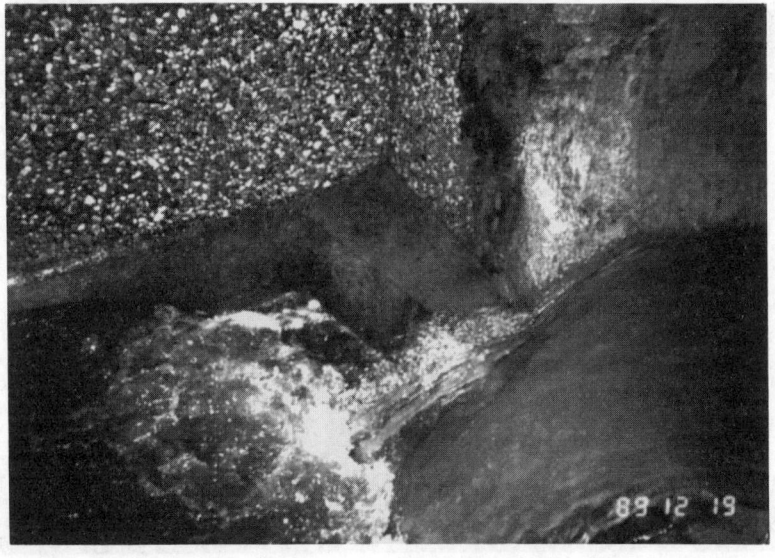

Plate 1b: The OPC manhole chamber is corroded in the same environment.

ON SITE OBSERVATIONS IN GREATER CAIRO, EGYPT

The chosen site was El Berka, Greater Cairo. The domestic waste is discharged into CFL lined sewer pipes after being transported through an approximately 5 km long open air sewer to the treatment farm. This line was installed in 1981. The flow level was about 50% with very high flow rates and in addition the line was curved, creating turbulences favourable to H_2S off-gassing.

The manhole inspected was made of Portland cement concrete (OPC). The 1500 mm diameter pipe was protected internally with a centrifuged CFL mortar lining. The pH was between 1 and 2, both on the walls of the manhole chamber and at the crown of the pipe.

The Portland concrete structure was badly damaged and reinforcement had clearly been required (Plate 2a). The CFL mortar lining was intact however and responded hard to the Schmidt rebound hammer (Plate 2b).

LABORATORY ANALYSES OF SAMPLES

The CFL samples were analysed by X ray diffraction and by scanning electron microscopy. Calcium sulphate (gypsum) was observed as a major exogenic mineral phase in each case (Figure 3), in addition to gibbsite from the CFL and materials from the aggregates. Further observations showed the presence of other salt compounds based on iron and sulphur ions although the exact composition of these salts could not be determined.

Plate 2a: OPC precast manhole chamber after 10 years - Greater Cairo, Egypt.

Plate 2b: CFL pipe lining in the same location after 10 years - Greater Cairo, Egypt.

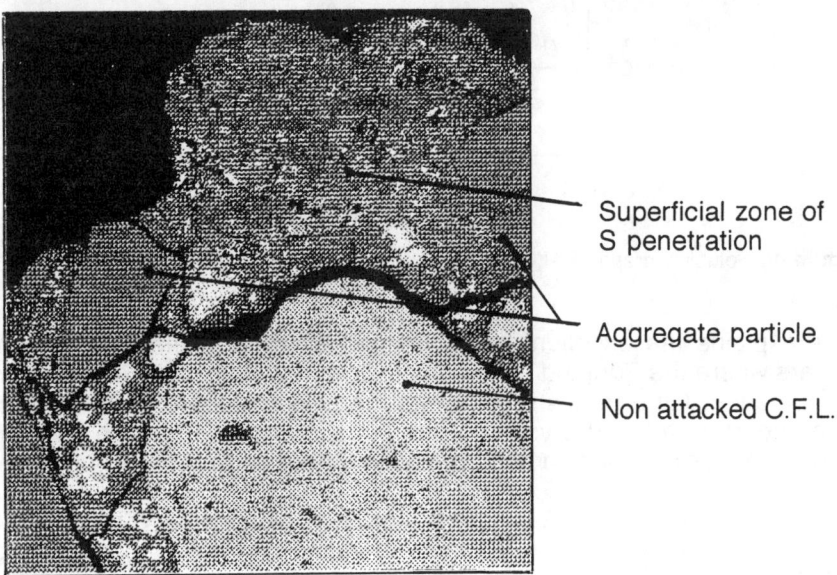

Figure 3: Surface of CFL mortar after exposure to H_2S corrosion in a sewer site (scanning electron microscopy).

DISCUSSION:
UNDERSTANDING OF THE CORROSION PHENOMENA

The above observations suggest CFL performs well even over long periods in aggressive sewage environments. The following section investigates to what extent these results confirm the theory as understood from the laboratory studies.

Investigations were conducted in the Lafarge laboratories with neat paste prisms immersed in acidic solutions. The prism samples, 2 x 2 x 5 cm in size, were cured for three days at 50°C. Each was immersed in a 50 cc acidic solution at various pHs. The solution was renewed twice daily (Bayoux et al, 1990).

Unlike Portland cement based mortars and concretes, CFL does not contain free lime hence it withstands acid corrosion more easily than traditional cements as the initial acid reaction can only occur through difficult hydrolysis of (a) the stable hydrates and (b) the crystalline gibbsite $Al(OH)_3$. Due to its stability, the latter never occurs at pH values higher than 3.5, as illustrated in Figure 4.

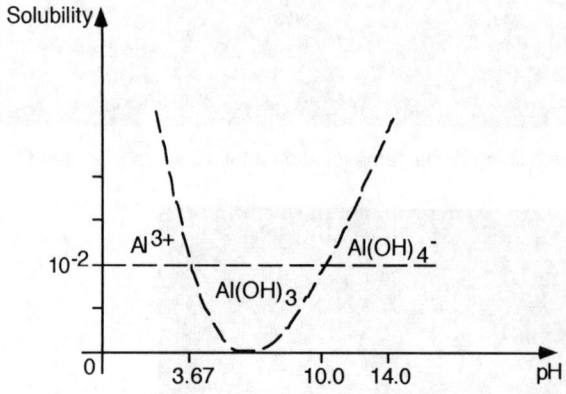

Figure 4: Solubility graph of $Al(OH)_3$.

This explains the excellent performance observed in the Kuala Lumpur sewers where the Portland base cementitious materials have suffered much deterioration. Lime reacts readily with the acid forming expansive calcium sulphate and ettringite crystals at the surface and deep inside the matrix, leading to loss of cohesion and promoting further corrosion.

At pH values lower than 3.5 both the calcium aluminate hydrates and the aluminium hydroxyde start to react, liberating calcium ions according to reactions [1] and [2]:

$$(CaO)_3.(Al_2O_3).6(H_2O) + 6H^+ => 3Ca^{2+} + 2Al(OH)_3 + 6H_2O \qquad [1]$$

$$Al(OH)_3 + 3H^+ ==> Al^{3+} + H_2O \qquad [2]$$

Corrosion here is a progressive frontal phenomenon, evident from both inspection and sample analysis. Scanning electron microscopy analysis and element mapping of corroded samples clearly showed that sulphur compounds do not penetrate inside the mortar but remain at the surface (Plate 3). Increasing acidity leads to formation of a protective calcium

Plate 3: Surface of samples having been exposed to H_2S corrosion (scanning electron microscopy).

sulphate layer according to reaction [3] which is in perfect agreement with on site observations:

$$Ca^{2+} + SO_4^{2-} \iff CaSO_4 \qquad [3]$$

Similarly, equation [2] becomes theoretically possible, which would result in the formation of aluminium salts according to equation [4], with additional formation of complex salts and gels of calcium, aluminium and iron coming from the cement. To some extent this explains the good performance of CFL in very harsh environments such as the sewers in Cairo.

$$2Al^{3+} + 3SO_4^{2-} \iff Al_2(SO_4)_3 \qquad [4]$$

However, it does not explain the poor results of Portland cement in the same conditions: Portland related cements also form a considerable quantity of calcium sulphate salts and some laboratory studies conducted in the past could lead to the conclusion that OPC performs well under very acidic conditions (pH 1, ASCE Manual, 1989). However, these studies took no account of the degradation pattern of the concrete: with CFL there is a linear degradation of the surface of the concrete due to the presence of the acid. On the other hand, with Portland related cements the attack progresses erratically, such that by the time sufficient salts are formed durability is already impaired.

Other arguments concerning the reactions are related to the nature of the bacterial attack.

EXPERIMENTS IN A BACTERIAL BREEDING CHAMBER

As mentioned previously, the environment where sulphur oxidizing bacteria (*thiobacilli*) grow and produce acid is a very complex system and difficult to simulate in the laboratory by traditional methods. It was therefore decided to design a carefully controlled bacterial breeding chamber which would take account of the interactions between the bacteria and the building material (substratum) and which could be used as a reference system for comparative testing as well as basic investigations.

The chamber was built by the Department of Microbiology at the University of Hamburg, working in close collaboration with the municipality. Concrete specimens (60 x 11 x 7 cm in size and scored with 1.8 cm squares) were placed on end with the base standing in 10 cm of water as shown in Plate 4. The chamber was maintained at 30°C and 100% humidity with 10 ± 1 ppm hydrogen sulphide. The samples were periodically sprayed with *thiobacillus* cultures which had been isolated from the sewer system. The pH at the surface of the concrete was measured by pH indicator sticks. The surface cubes were broken off at different times and washed in a sterile solution before analysis of the bacterial population. The weight losses were measured on each cube sample (Sand, 1987; Bock, Sand, Kirstein and Rammelsberg, 1989). The initial investigations were conducted with Portland cement samples and the results showed two major conclusions:

1) The rate of corrosion (loss of weight) was proportional to the quantity of bacteria cells on the sample.

2) After one year in the chamber the corrosion observed was comparable to that in samples taken after eight years in the sewers of Hamburg.

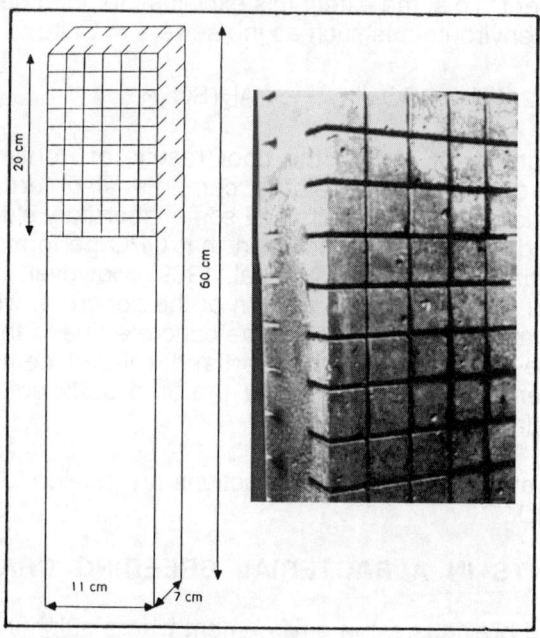

Plate 4: Concrete samples as tested in the bacterial breeding chamber.

Comparative studies were conducted on different mortars; the results are summarised in Figure 5.

Three conclusions may be drawn:

1) The calcium aluminate based products perform better than those based on traditional cements (Bock, Sand, Kirstein and Rammelsberg, 1989) even at very low pH. This confirms the site inspections.

2) The chemistry of bacteria induced corrosion differs from that seen in standard acid corrosion tests.

3) Although the bacterial population grows on both calcium aluminate compounds and other materials the associated corrosion differs. This involves the bacterial activity in some way.

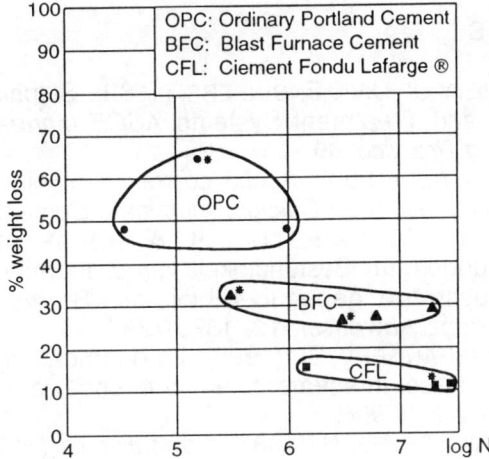

Figure 5: Graph of the loss of weight observed with Ordinary Portland Cement, Blast Furnace Cement and CFL after 344 days exposure in a bacterial breeding chamber.

CONCLUSIONS

1. The theory predicts that CFL concrete will perform well when the pH is higher than 3.5.

2. Results observed in the field are better than can be predicted and show good performance even in harsh environments (pH 1.5).

3. These results have been confirmed by research carried out in an artificial sewer taking into account the growth and activity of the bacteria.

4. The apparent difference in results is related to the following considerations:

a) Biological: the bacteria do not have the same activity on all substrata;

b) Chemical: bacterial corrosion is related to high humidity environments rather than standard aqueous solutions;

c) Degradation pattern of the concrete: corrosion is a frontal progressive phenomenon with CFL whereas it progresses erratically with other cements. This is not taken into account in standard tests where the pH is suddenly imposed on the system and maintained constant at a given value.

Further laboratory investigations should take these points into consideration.

REFERENCES

AMERICAN SOCIETY OF CIVIL ENGINEERS, (1989). Sulphide in Waste Water Collection and Treatment Systems. *ASCE Manuals and Reports on Engineering Practice*, **69**.

BAYOUX, J.P. ET AL. (1990). Acidic corrosion of high alumina cement. *Midgley Symposium on Calcium Aluminate Cements*.

BOCK, E., SAND, W., KKIRSTEIN, K. & RAMMELSBERG, J. (1989). Untersuchungen zur Beständigkeit von Zementmörtel-auskleidungen duktiler Gussrohre gegenüber biogener Schwefelsäure-Korrosion. *Korrespondenz Abwasser*, **12**, 1376-1387.

COCHET, C. & DERANGERE, D. (1990). La dégradation du béton dans les ouvrages d'assainissement en présence d'hydrogène sulfuré. *Cahiers du CSTB*, 906.

COLIN, F. & MUNK-KOEFED, N. (1987). *Formation de l'H2S dans les réseaux d'assainissement - Conséquences et remèdes*. ed. Agence de Bassin Rhône Méditerranée Corse.

LAFARGE COPPEE RECHERCHE (France). Rapports internes, 1987-1991.

MATHIEU, A. SOUKATCHOFF, P. 1987. Revêtement intérieur de mortier de ciment alumineux pour canalisations destinées au transport d'eaux usées. *L'eau, l'industrie, les nuisances*; 111.

MATHIEU, M., LANGENFELD, M & RAMMELSBERG, J. (1989). Auskleidung von duktilen Gussrohren mit Tonerdeschmelzzementmörtel für den Abwassertransport. *Korrespondenz Abwasser*, **10**, 1027-1036.

NIPO (1988). Kwantitatief vrij-verval riolering onderzoek in Nederland.

POLDER, R. & VAN MECHELEN, T. (1989). Assessment of biogenic sulfuric acid agressivity of sewer environment. *2nd International Conference on Pipeline Construction, Hamburg*, 23-26, 601-615.

SAND, W. (1987). Importance of hydrogen sulphide, thiosulphate, and methyl-mercaptan for growth of thiobacilli during simulation of concrete corrosion. *Applied and Environmental Microbiology*; 1645-1648.

THISTLETHWAYTE, K.B. (1972). *Control of sulphides in sewerage systems*. Butterworths Pty Ltd, Chastwood, New South Wales, Australia.

5-11 THE INFLUENCE OF GEOENVIRONMENTAL DISSOLVED SULPHATES ON THE SWELLING PROCESS IN CONCRETE

W. PRINCE & R. PERAMI
Laboratoire de Minéralogie et Matériaux, 39, Allée Jules Guesde
31400 Toulouse, France

ABSTRACT

Concrete in constructions near saliferous rocks can be damaged by the swelling of ettringite if they are in contact with sulphates. The present study shows that the reaction depends on the nature of the sulphates.

If the sulphate is of an alkali metal (eg Na_2SO_4) the hydrolysis of the aluminates is countered and ettringite cannot form easily. If the sulphate is of an alkaline earth, however (eg $CaSO_4$) aluminates hydrolyse well and ettringite can form. The paper provides an explanation of these results.

INTRODUCTION

Suitably activated clay minerals react more or less quickly with lime and sulphates to give ettringite, $Al_2O_3 3CaO 3CaSO_4.31H_2O$, the compound which is mainly responsible for the setting of over-sulphated cements. This can cause considerable problems when it crystallizes late in mortar that has already hardened, leading to expansion and cracking in the concrete, a reduction in its mechanical properties and possibly causing serious damage to the structures (Nakamura, Sudoh and Akaiwa, 1968; Schwiete, Ludwig and Jager, 1973; Mehta, 1982).

Late ettringite crystallization in concrete implies the presence of:

1) activated aluminium or silicon-aluminium-containing materials and

2) a source of sulphates.

The natural activation of silico-aluminates may result from a variety of factors such as intense grinding of mylonitized granite rocks or lamination of schist-gneiss metamorphic rocks. Laboratory experiments showed that when samples of healthy granite are subjected to grinding they become reactive in the presence of lime and sulphates and produce ettringite.

The activation may also result from gradual changes in the rocks, leading to the formation of more or less crystallized clay minerals. Thus, depending on the hydrothermal and climatic conditions, the structure of feldspar changes, giving sequences such as vermiculite - montmorillonite - kaolinite. These amorphous or pseudo-amorphous neoformed minerals readily release their silica and alumina and are capable of reacting in the same way as activated clays (Prince, 1982; Prince and Perami, 1990).

Sources of sulphates are plentiful and may be provided by saline rocks of the evaporite family, the most abundant crystallized members of which are anhydrite and gypsum. In France these are to be found in large massifs in the Pyrenees and in certain geological layers dating from the end of the Primary, Permo-Trias and Keuper. The source may also be ordinary rocks into which gypsum has been injected during tectonic movements because of its ductility, forming diapir folds. In addition, the oxidation of pyritic rocks, sulphate-rich water, sea water or waste water containing various pollutants all provide sources of sulphate.

It is of note however, that sulphates are almost always accompanied by cations such as Na^+, K^+, Ca^{2+}, Mg^{2+}, etc. and this paper considers the influence of the nature of the compensating cations on ettringite crystallisation.

Two extreme cases are discussed, corresponding to the presence of very large quantities of either alkaline ions, or alkaline earth ions. The respective effects of alkaline and alkaline earth sulphates on the formation of ettringite are examined.

ACTION OF ALKALI METAL SULPHATES

In order to demonstrate the mechanisms involved, chemical products of "rectapur" quality and industrial kaolin of well-defined composition were used in the study. The kaolin, activated by thermal treatment, was mixed with lime and sulphates in the presence of water. The paste obtained was kept in a saturated atmosphere for increasing intervals of time. At each stage, the products of the reaction were analysed by X-ray diffractometry.

Examples of the diffractograms recorded after 2 hours, 24 hours and 72 hours of reaction between the kaolin, lime and sodium sulphate are shown in Figure 1. The only identifiable neoformed phases are:

- tetracalcium aluminate with 19 water molecules, $Al_2O_3 4CaO.19H_2O$

- tricalcium aluminate with 6 water molecules, $Al_2O_3 3CaO.6H_2O$

- hydrated monocalcium silicate, $SiO_2 CaO.H_2O$ or CSH.

Even after being kept for 28 days, the tetracalcium aluminate remained stable in the reaction medium and ettringite peaks could not be identified on the diagrams.

Clearly, ettringite does not crystallize easily in the presence of alkali-metal sulphates. This could be explained by the mechanism discussed below.

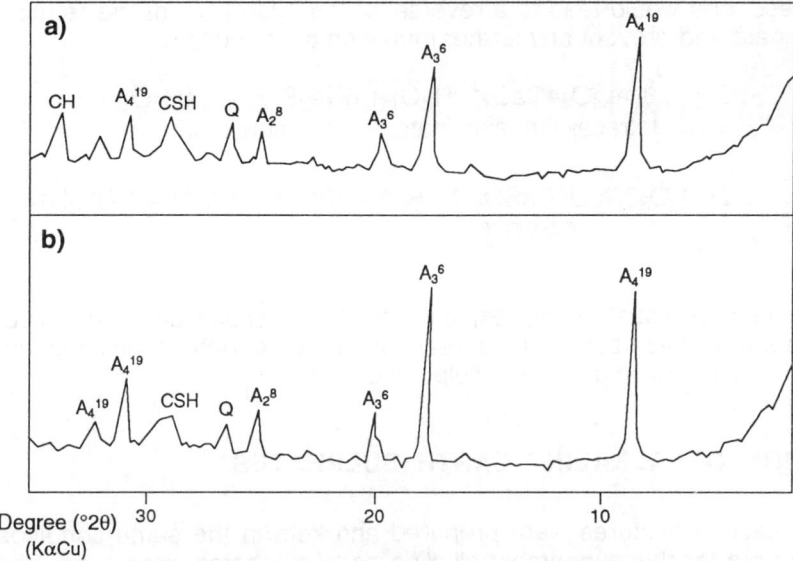

Figure 1: XRD of the reaction products obtained with activated kaolin, lime and sodium sulphate. a) After 24 hours; b) After 72 hours. A_4^{19} - tetracalcic aluminate hydrated with $19H_2O$, A_2^8 - dicalcic aluminate hydrated with $8H_2O$, CH - lime, CSH - monocalcic hydrated silicate, Q - quartz.

REACTION MECHANISM IN THE PRESENCE OF ALKALI METAL SULPHATES

Initially the kaolin reacts with the lime to give tetracalcium aluminate as the principal phase. This only slightly soluble aluminate is partially hydrolyzed through reaction [1]. The activities of the ions in solution are then determined by the solubility product Ks of equation [2]. This solubility product is constant for a given temperature and any additional quantity of one of the OH^-, $Al(OH)_4^-$ or Ca^{2+} ions must lead to a reversal of the hydrolysis and stabilization of the tetracalcium aluminate.

$$Al_2O_3 4CaO 19H_2O \Leftrightarrow 4Ca^{2+} + 2Al(OH)_4^- + 6OH^- + 12H_2O$$
(tetracalcium aluminate) [1]

$$Ks = [Ca^{2+}]^4 [Al(OH)_4^-]^2 [OH^-]^6 \qquad [2]$$

In the presence of alkali-metal sulphates (e.g. Na_2SO_4), the reaction which would give ettringite requires that the tetracalcium aluminate should already be dissolved and that the ions liberated (Ca^{2+}, $Al(OH)_4^-$ and OH^-) combine with the sulphate ions.

Equation [3] gives this reaction, showing that the crystallization of ettringite is accompanied by the release of a large number of OH^- and $Al(OH)_4^-$ ions. Given that the solubility product Ks must remain constant, the accumulation of these ions would lead to a reversal of the hydrolysis of the tetracalcium aluminate and prevent any further formation of ettringite.

$$3(Al_2O_3 4CaO.19H_2O) + 6Na_2SO_4 + 14H_2O$$
$$\text{tetracalcium aluminate} \quad\quad \text{anhydrite}$$
$$\Downarrow$$
$$2(Al_2O_3 3CaO 3CaSO_4.31H_2O) + 12Na^+ + 10OH^- + 2Al(OH)_4^- \quad\quad [3]$$
$$\text{ettringite}$$

This process satisfactorily explains both the results and the difficulty of ettringite crystallization in a reaction medium rich in alkaline ions or principally containing alkaline sulphates.

ACTION OF ALKALINE EARTH SULPHATES

The reaction mixtures were prepared and kept in the same conditions as before but for this experiment alkaline earth sulphates were used. Figure 2 shows the diffractograms obtained after 2 hours, 6 hours and 24 hours of reaction using anhydrite as the source of sulphates.

The evolution of the $A^4{}_{19}$ and E peaks shows that tetracalcium aluminate forms in the first few hours after mixing but then gradually disappears while ettringite develops.

These experiments demonstrate that the tetracalcium aluminate formed in the presence of anhydrite is subsequently transformed into ettringite, whereas that which crystallizes in the presence of alkaline sulphates remains stable in the reacting medium. An explanation for this difference can be given by the following process, during which the tetracalcium aluminate formed initially is first hydrolyzed, freeing Ca^{2+}, $Al(OH)_4^-$ and OH^- ions. These then react with the alkaline earth sulphate to give ettringite.

REACTION MECHANISMS IN THE PRESENCE OF ALKALINE EARTH SULPHATES

Equation [4] expresses the process described. All the compounds formed in this reaction give precipitates and there is therefore no accumulation of OH^- ions as is the case for the alkaline sulphates. Under these conditions, $Al(OH)_3$ constitutes the stable form of the alumina and there is nothing to prevent the hydrolysis of the tetracalcium aluminate and the formation of ettringite from continuing.

Similar reactions can be written by replacing the anhydrite by magnesium sulphate. Brucite is then precipitated instead of lime and the ettringite can

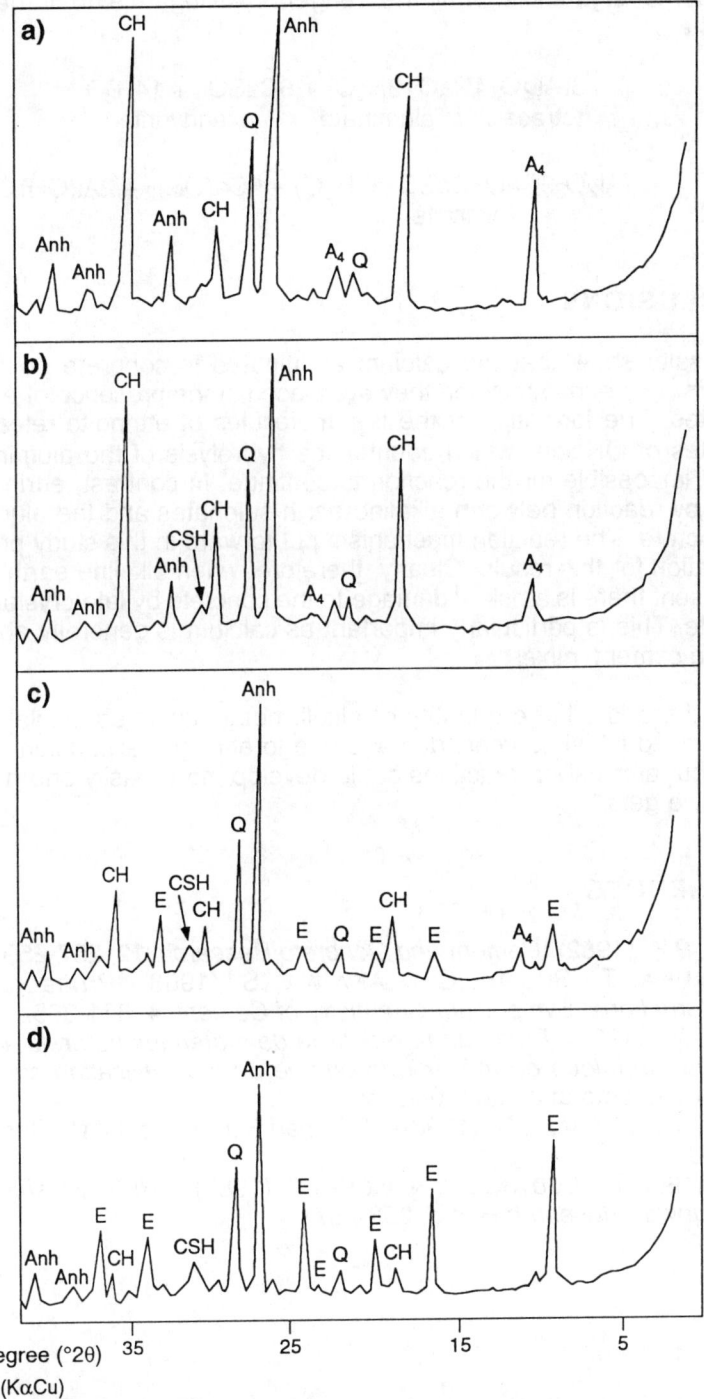

Figure 2: XRD of the reaction products obtained with activated kaolin, lime and anhydrite. a) After 2 hours; b) After 4 hours; c) After 6 hours; d) After 24 hours. Q - quartz, Anh - anhydrite, E - ettringite, A_4 - tetracalcic hydrated aluminate, CSH - hydrated calcic silicate.

continue to crystallize until the reagents giving rise to it have been exhausted.

$$3(Al_2O_3 4CaO 19H_2O) + 6CaSO_4 + 14H_2O$$
tetracalcium aluminate   anhydrite
$$\Downarrow$$
$$2(Al_2O_3 3CaO 3CaSO_4 31H_2O) + 6Ca(OH)_2 + 2Al(OH)_3 \qquad [4]$$
ettringite

CONCLUSIONS

The results show that the calcium aluminates in concrete do not easily transform into ettringite when they are placed in the presence of alkali metal sulphates. The formation of the first molecules of ettringite releases large quanitites of OH^- ions which counter the hydrolysis of the aluminates and make it impossible for the reaction to continue. In contrast, ettringite forms rapidly by reaction between alkaline earth sulphates and the aluminates of the concrete. The reaction mechanism put forward in this study provides an explanation for this result. Clearly, therefore, when alkaline earth sulphates are present there is a risk of damage to the concrete by late crystallization of ettringite. This is particularly important as calcium is generally abundant in Portland cement mixes.

Where there is a large quantity of alkali metal sulphates available locally, e.g. from industrial sources, damage due to ettringite expansion should not occur, but alkali-silica reactions could develop more easily and give rise to expansive gels.

REFERENCES

MEHTA, P.K. (1982). *Cement and Concrete Research*, **12**, 257-259.
NAKAMURA, T., SUDOH, G. & AKAIWA, S. (1968). *Proceedings 5th International Symposium Chemistry of Cement*, **4**, 351-365.
PRINCE, W. (1987). *Etude de la réactivité de matériaux naturels activés par voie thermique ou mécanique en vue de leur utilisation comme liant*, Thèse Doctorat d'Etat, Toulouse.
PRINCE, W. & PERAMI, R. (1990). *C.R.Acad.Sci., Paris*, **t.310**, Série II, 535-540.
SCHWIETE, H.E., LUDWIG, W. & JAGER, P. (1983). *SR90, Washington-DC, Highway Research Board*, 353-367.

5-12 DAMAGE TO CONCRETE CAUSED BY ALKALINE METAL HYDROXIDES

W. Prince & R. Perami
Laboratoire de Minéralogie et Matériaux, 39 Allee Jules Guesde, 31400, Toulouse, France

ABSTRACT

An investigation has been carried out into the mechanisms of the reaction which occurs in concrete between siliceous aggregate and alkali, causing damage to structures. It is shown that the the most important factor is not the alkaline-metal-ions content but rather the availability of OH⁻ ions. In order to prevent deterioration due to alkali-aggregate reactions, therefore, the medium must be depleted of OH⁻ ions. The paper discusses how silica fume may fulfill this role.

INTRODUCTION

Since their creation, rocks have been subjected to changes due to environmental influences. The original minerals pass through different stages of transformation and give rise to other minerals in equilibrium with new conditions. Natural conditions are generally neutral or slightly acid while the pH of cement paste is very basic. When rocks come into contact with cement and particularly when they are mixed with cement as an aggregate, therefore, their equilibrium is disrupted and the minerals begin to dissolve progressively leading to specific reactions. Among these are the well-known alkali-silica and alkali-silicate reactions where silanol and siloxane groups of siliceous materials are attacked by the alkali. The result is a gel which expands and can cause cracking and degradation of the structure.

To avoid the risk of concrete degradation by alkali-reaction, one of the precautions taken in recent years is the limitation of the alkali content in the cement; ASTM recommends an available upper limit of 0.6% of Na_2O equivalent (ASTM, 1987; AFNOR, 1981). However, it has been observed that while the use of some Portland cements with a very low alkali content does not preclude an alkali-aggregate reaction, this effect was not found in other cements with higher Na_2O contents. A high quantity of alkaline metal ions is probably not a deciding criterion for the triggering of alkali-aggregate reactions therefore.

In order to obtain a better understanding of the mechanisms involved, a study has been made of the products which form from some simple mixtures containing siliceous materials and with Na^+, OH^- and Ca^{2+} ions. Several papers discuss silico-aluminate materials, eg Prince (1987) and Prince and Perami (1991).

The siliceous materials used in the present study were samples of cristobalite and silica fume, each of which contained about 96% (± 2%) of silica. The other reagents employed were high grade chemical products supplied by the Prolabo company.

In each case the constituents were thoroughly mixed and then wetted in order to obtain a homogeneous paste. This paste was poured into cylindrical moulds and kept for 24 hours in a saturated atmosphere. At the end of this period a specimen was removed from the mould, dried and then crushed for X-ray diffractometric analysis. All the diffractograms were recorded under the following experimental conditions: copper anticathode, $K\alpha$ Cu radiation, sensitivity 400 cps, scanning speed 1°(2θ) per minute.

ALKALI-SILICA REACTION

The respective parts played by the Na^+ ions and OH^- ions in the alkali-silica reaction were studied using cristobalite as the siliceous material. Two types of mixture supplied Na^+, OH^- and silica: cristobalite and alkali and cristobalite, sodium chloride and lime.

ALKALI-SILICA REACTIONS OBTAINED FROM THE CRISTOBALITE AND ALKALI MIXTURE

The mixture contained 70% cristobalite and 30% alkali (NaOH or KOH), the alkali having been dissolved in demineralised water before being mixed with the cristobalite. Figure 1 shows the diffractograms of the initial cristobalite and those recorded after the reaction with NaOH and KOH.

Diffractogram 1a was obtained after the potassium hydroxide attack. It can be seen that almost all the cristobalite has reacted with the alkali and produced a new compound, $KHSiO_5$. Diffractogram 1b, recorded after the sodium hydroxide attack, shows a complete disappearance of the cristobalite peaks but some new peaks due to the crystallisation of Na_2SiO_3. In each case the alkali-silica reaction has caused the dissolution of the initial material and has yielded new phases, more or less crystallised.

ALKALI-SILICA REACTIONS OBTAINED FROM THE CRISTOBALITE, SODIUM CHLORIDE AND LIME MIXTURE

In the previous reactions Na^+, K^+ and OH^- ions are furnished by sodium hydroxide and potassium hydroxide. In the internal liquid of the cristobalite, sodium chloride and lime mixture the high solubility of the sodium chloride gave a high concentration of Na^+ while the OH^- ion concentration was limited and fixed by the lime solubility. In this way, the relative importance of these two types of ions could be demonstrated.

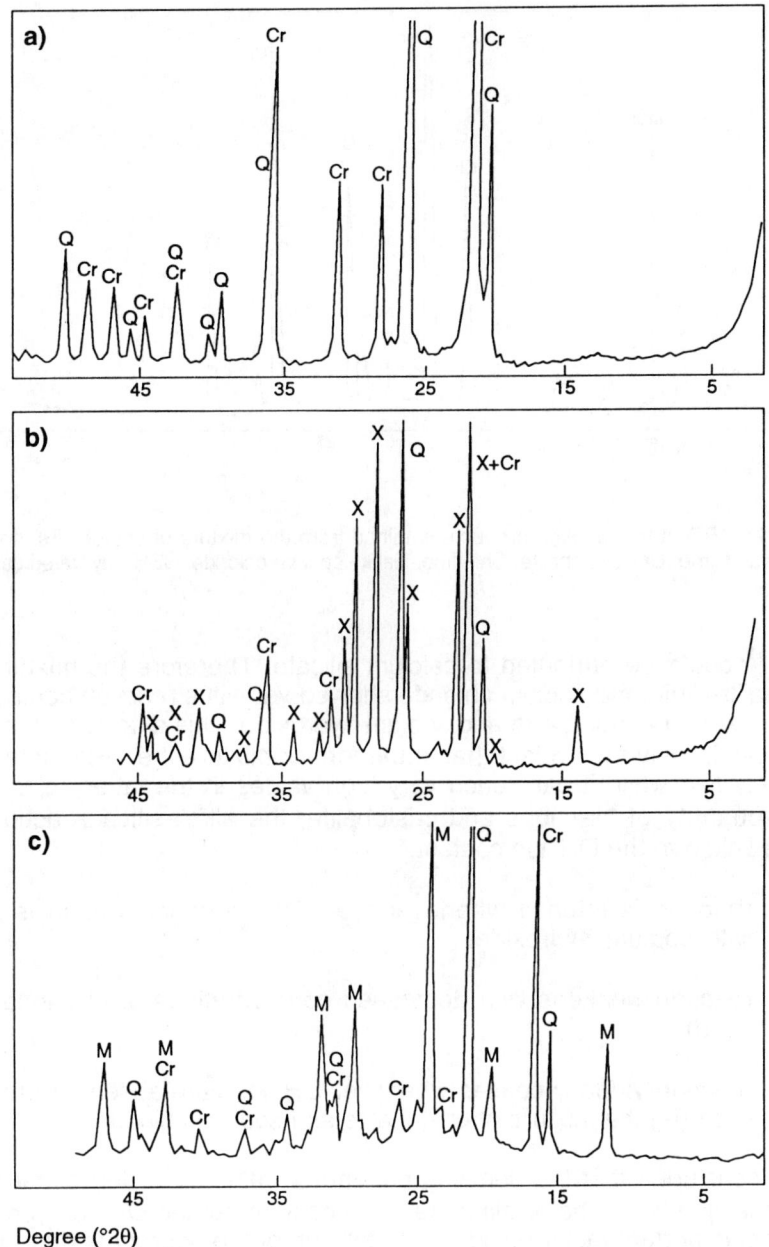

Figure 1: XRD of the products obtained by the reaction between cristobalite and alkalis. a) Cristobalite; b) After reaction with KOH; c) After reaction with NaOH. Cr - cristobalite, Q - quartz, X - $KHSiO_5$, M - Na_2SiO_3.

The mixture analysed contained 50% cristobalite, 25% sodium chloride and 25% lime. The XRD of the reaction products are shown in Figure 2. It can be seen that the peaks of cristobalite remain dense in spite of the basicity of the medium and the availability of Na^+ ions. Only a weak peak

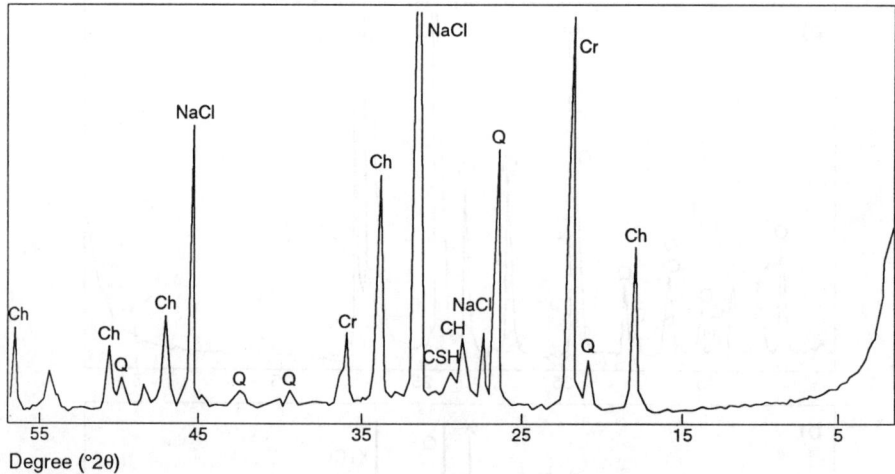

Figure 2: XRD of the reaction products obtained from the mixture of cristobalite, sodium chloride and lime. Cr - cristobalite, Ch - lime, NaCl - sodium chloride, CSH - hydrated calcium silicate.

at 0.3 Å could be attributed to calcium silicate. Therefore the mixture of cristobalite, lime and sodium chloride showed very little reaction compared to the mixture of cristobalite and sodium hydroxide. This difference may be explained largely by the fact that in the former mixture the OH^- content is relatively low while it can reach very high values in the latter; ie for the same quantity of Na^+ ions and cristobalite, the alkali-silica reaction is closely linked to the OH^- ion content:

> the reaction is intense whenever the OH^- ion concentration is high (eg with sodium hydroxide)

> the reaction weakens with decreased concentrations of OH^- ions (eg with lime)

> the reaction virtually ceases whenever OH^- ions are absent or are not renewed (eg the mixture of cristobalite and sodium chloride)

It was concluded that the deciding parameter influencing degradation by alkali-reaction is not the alkaline metal ion content but the OH^- ion content; the most important factor being the dissolution of the aggregate due to the effect of OH^- ions.

THE ROLE OF CALCIUM IONS IN THE ALKALI-SILICA REACTION

Applying these experimental results to concretes would suggest alkali-aggregate reactions would generally be slow as OH^- ions come mainly from lime. Lime brings not only OH^- ions but also Ca^{2+} ions to the medium however, hence the impact of the latter was also investigated. Two cases were considered, corresponding to two levels of Ca^{2+} ion concentrations in the internal liquid.

ALKALI REACTION WITH LOW CA^{2+} ION CONTENT

An examination was made of the products resulting from the reaction of sodium hydroxide dissolved in water and mixed with lime and cristobalite. In the resulting paste the lime solubility and therefore the Ca^{2+} ion content was very low as the alkali hydroxide brought out a large quantity of OH^- ions. Thus this mixture simulated the conditions of a concrete whose internal liquid contained few Ca^{2+} ions but a high level of hydroxide.

Figure 3 shows the diffractogram recorded from this mixture. Again it can be seen that the cristobalite peaks disappear due to the effect of the OH^- ions. A silicate containing both sodium and calcium was created instead of the sodium silicate produced when the mixture was composed only of cristobalite and sodium hydroxide. It appeared that the Ca^{2+} ions partly invaded the space previously occupied by the Na^+ ions, the substitution being partial due to the low Ca^{2+} ion content of the internal liquid. In this case, therefore, the alkaline metal ions reacted mainly to stabilise and neutralise the siliceous colloids released during the dissolution but could be replaced by the Ca^{2+} cations.

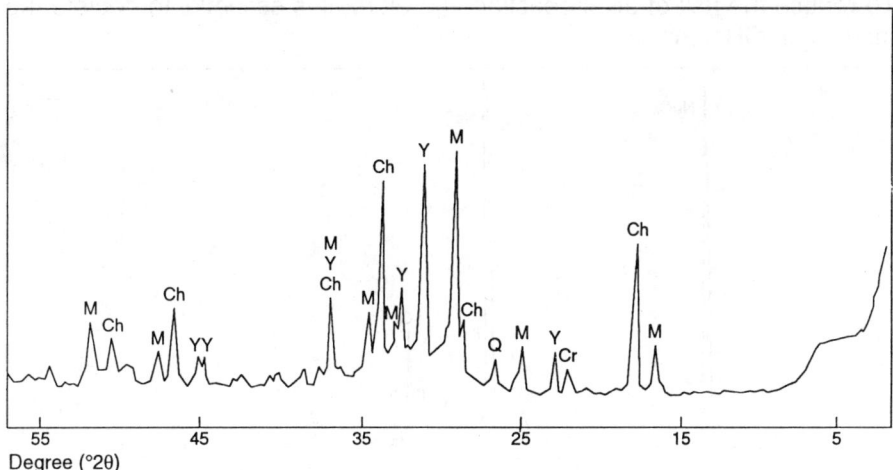

Figure 3: XRD of the reaction products obtained from the mixture of cristobalite, sodium hydroxide and lime. Ch - lime, Q - quartz, M - Na_2SiO_3, Y - $Na_2O_2CaO_2SiO_2H_2O$.

ALKALI REACTION WITH HIGH CA^{2+} ION CONTENT

A pastey mixture made of cristobalite, sodium hydroxide and calcium chloride was prepared. This had the same calcium ion content as the mixture used in the previous experiment but was richer in calcium as calcium chloride is highly soluble.

The diffractogram of the resulting products is shown in Figure 4. The cristobalite peaks have disappeared and some new peaks appeared related to the sodium chloride and hydrated calcium silicate. The main feature is that the silicate obtained in this case contains only calcium and not both sodium and calcium as previously. The cristobalite dissolved completely giving a hydrated calcium silicate rather than the double sodium and calcium silicate. The high calcium ion content in the medium explains this result which suggests the Ca^{2+} ions replace all the Na^+ ions.

It is concluded that:

1. the availability of OH^- ions is the principal factor affecting the dissolution of the aggregate;

2. the siliceous colloids created during this dissolution are neutralised by alkaline metal ions and give rise to a more or less crystallised alkali silicate;

3. the Ca^{2+} ions replace the alkaline metal ions and gradually transform the alkali silicate into a hydrated calcium silicate. This replacement restores the alkaline metal ions which can again intervene. In these conditions the reaction does not require a very high proportion of alkaline metal ions.

To reduce the risk of alkali-reaction, therefore, it is essential to deplete the medium in OH^- ions.

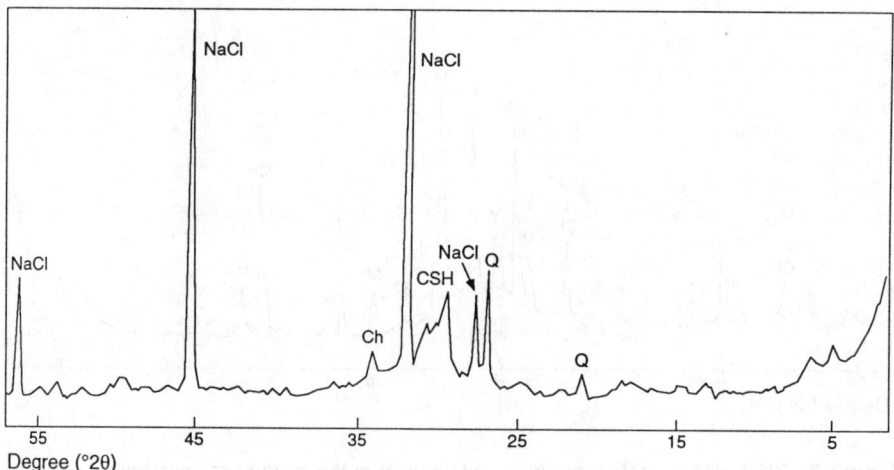

Figure 4: XRD of the reaction products obtained from the mixture of cristobalite, sodium hydroxide and calcium chloride. NaCl - sodium chloride, Q - quartz, Ch - lime, CSH - hydrated calcium silicate.

ROLE OF SILICA FUME IN REDUCING THE RISK OF ALKALI-REACTION

The role which silica fume can play in the depletion of OH^- ions was examined (Grattan-Bellow, 1981; Regourd, Hornain, Montureux, Poitevin &

Peuportier, 1982). Two cases were considered involving the presence or absence of calcium

ALKALI-SILICA FUME REACTION WHEN CALCIUM IONS ARE NOT AVAILABLE

The products formed when silica fume reacted with an alkaline solution was studied. The diffractograms recorded using sodium hydroxide and potassium hydroxide are presented in Figures 5a and 5b respectively. It can be seen that the large silica fume peak which initially centred around 4.04Å disappeared but another peak located around 3.03Å can be identified associated with a gel of alkali-silica. The resulting products differ according to the siliceous materials used. Although cristobalite and silica have the same silica content, the former reacts with alkali and gives two crystallised phases ($KHSiO_5$ or Na_2SiO_3) while the latter reacts to give a gel due to a quick mass setting.

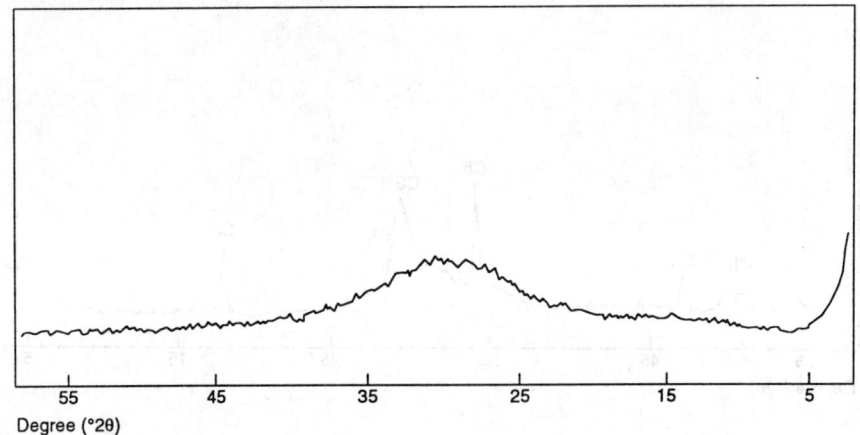

Figure 5a: XRD of the reaction products obtained from the mixture of silica fume and sodium hydroxide.

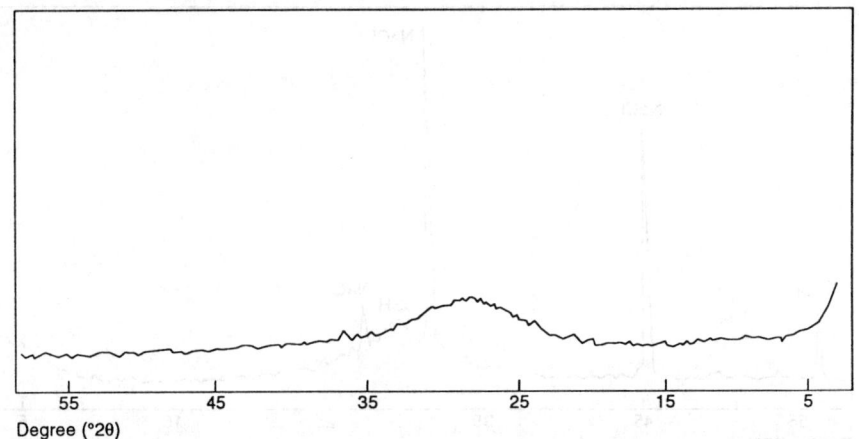

Figure 5b: XRD of the reaction products obtained from the mixture of silica fume and potassium hydroxide.

It is clear that when calcium ions are not available the reaction kinetics are particularly important. The aggregates which can give a siliceous gel are those which quickly release large quantities of silica in the medium. When silica fume is introduced into concrete, however, it reacts quickly and depletes the medium in OH⁻ ions. The risks of further degradation by alkali-aggregate reaction are then reduced.

ALKALI-SILICA FUME REACTION IN THE PRESENCE OF CALCIUM IONS

In this study, two mixtures were considered, one containing silica fume, sodium hydroxide and lime and the other silica fume, potassium hydroxide and calcium chloride. Figure 6 presents the diffractograms obtained from each.

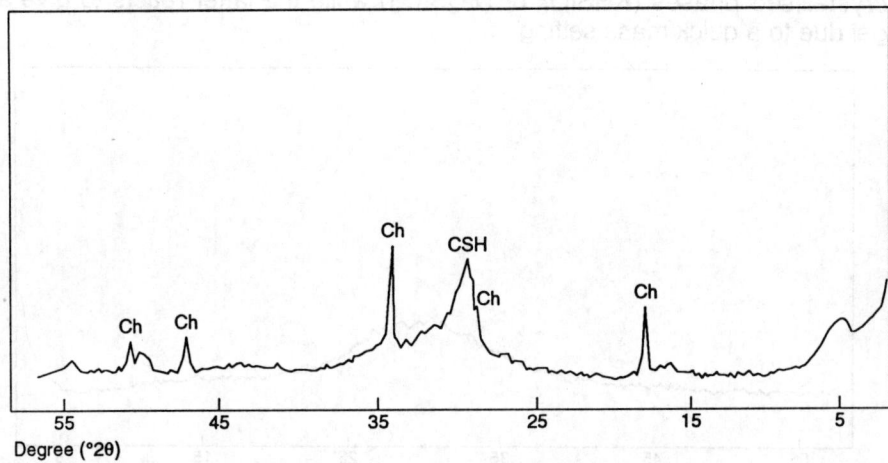

Figure 6a: XRD of the reaction products obtained from the mixture of silica fume, sodium hydroxide and lime. Ch - lime, CSH - hydrated calcium silicate.

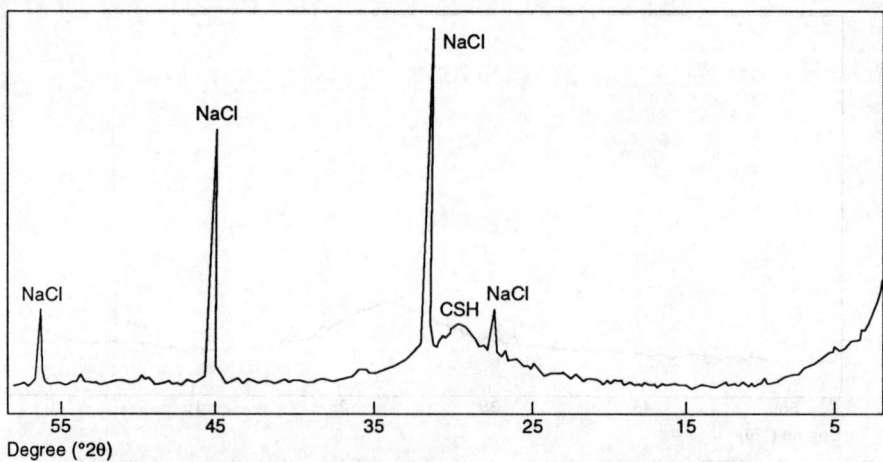

Figure 6b: XRD of the reaction products obtained from the mixture of silica fume, sodium hydroxide and calcium chloride. NaCl - sodium chloride, CSH - hydrated calcium silicate.

In both cases the reaction product was mainly a hydrated calcium silicate. This more crystallised phase is different from the gel of alkali-silicate which forms when calcium is lacking. As can be seen in Figure 6 the alkaline metal ions crystallise massively as sodium chloride or potassium chloride, confirming the hydrated calcium silicate is very poor in Na^+ or K^+ ions. It would appear that the alkaline metal ions of the gel were replaced progressively by calcium ions and that the gel was finally converted to a hydrated calcium silicate.

CONCLUSIONS

The present study shows that the alkaline metal ion content in concrete is not the deciding factor in silica reactions but that the availability of the OH^- ion is crucial. The role of alkaline metal ions is to stabilise and neutralise the siliceous colloid released during the dissolution of the aggregate.

The alkali-aggregate reaction leads to a crystallised phase or a gel depending on the kinetics of the reaction. The aggregates producing gels are those which can release large quantities of silica in the medium.

Depending on the Ca^{2+} ion content, the reaction product is either an alkaline metal silicate or a silicate containing both alkali metal and calcium or a hydrated calcium silicate. The study showed the alkali-aggregate reaction is severe whenever the concentration of OH^- ions is high, slow when the OH^- ion concentration is lower and almost zero when these ions are absent or are not restored.

To prevent the risks of degradation by alkali-aggregate reaction, the role of silica fume is to consume the OH^- ions in the medium.

REFERENCES

AFNOR, P. (1981). *Essai de stabilité dimensionnelle en milieu alcalin*, 18-585.
ASTM, (1987). *Annual book of American Society for Testing and Materials Standards*.
GRATTAN-BELLOW, P.E. (1981). A review of test methods for alkali-aggregate. *Expansivity of concrete aggregates, Cape Town, S252/9*.
PRINCE, W. (1987). *Etude de la réactivité de matériaux naturels activés par voie thermique ou mécanique en vue de leur utilisation comme liant*. Unpublished PhD, Toulouse.
PRINCE, W. & PERAMI, R. (1991). Influence de la soude dans les réactions entre alcalis et granulats de type silico-aluminates. *Cement and Concrete Research*, **25**, 575-580.
REGOURD, M., HORNAIN, H., MONTUREUX, B., POITEVIN, P. & PEUPORTIER, H. (1982). Altération de la liaison patede ciment-granulats dans le béton - La réaction alcalis-granulats. *Coloque International du R.I.L.E.M., Toulouse*.

5-13 STANDARD HYDROGEOCHEMICAL MODELS AND THEIR POSSIBLE APPLICATION TO ENGINEERING CONSTRUCTION PROBLEMS INVOLVING GROUND WATER

J.H. TELLAM & J.W. LLOYD
School of Earth Sciences, University of Birmingham, Birmingham B15 2TT

ABSTRACT

There exists a wide range of hydrogeochemical computer codes which are used extensively by geologists but much less widely by engineers. This paper reviews the types of model available, describes various example codes, illustrates some simple applications and comments on the limitations of such models. It is suggested that hydrogeochemical models could be used much more widely in routine engineering work than they are at present.

INTRODUCTION

Over the last few decades many hydrogeochemical models have been developed for investigations involving groundwater and sea water chemistry. The simplest of these are designed to predict aqueous speciation given a major component chemical analysis whilst the most powerful can deal with kinetic reactions or take into account many tens of components. Relatively user-friendly PC codes of a wide range of these models are now commercially available but although their use is becoming increasingly widespread in geological circles, with some important exceptions, such models do not yet seem to be as common in the engineering world.

This article provides an introduction to the available model codes and illustrates their use in the context of some simple groundwater problems related to engineering construction.

A BRIEF REVIEW OF SOME OF THE MORE WIDELY AVAILABLE HYDROGEOCHEMICAL CODES

TYPES OF MODEL

Hydrogeochemical models can be split into three categories :

(a) aqueous speciation models ("speciation models");
(b) models which can predict the outcome of reactions ("reaction models"); and
(c) models which combine groundwater flow and chemical reaction ("flow/reaction models").

General reviews of available hydrogeochemical models include Nordstrom and Ball (1984), Sposito (1985), Bonazountas (1986), Runnells (1987), Rice (1988) and Domenico and Schwartz (1990, p. 515-525). As well over fifty codes have been published in the North American literature alone (Runnells, 1987), the following review is necessarily far from complete. Table I summarizes the attributes of some of the more commonly available codes.

SPECIATION MODELS

Speciation models are essential for all types of chemical calculations and hence are contained within reaction and flow/reaction models as well as being available separately. Codes whose purpose is only to calculate speciation include WATEQF (Plummer, Jones and Truesdell, 1976) and WATSPEC (Wigley, 1978). Such models differ not only in presentation, computer language and equation-solving method but also more importantly in choice of components and their associated thermodynamic data values. Most of the programs will calculate the concentrations and activities of all the species considered (including ion pairs) and present a list of saturation indices for various minerals:

$$\text{Saturation index (SI)} = \log \left(\frac{\text{Ion Activity Product}}{\text{Solubility Product (K)}} \right)$$

It is thus possible to determine whether a water is under- (SI<0) or over-saturated (SI>0) with respect to, for example, calcite (SIC = log [$(Ca^{2+})(CO_3^{2-})$/K]).

An important use of such models is in processing a large amount of data to provide a general picture of the aggressiveness or incrustation potential of a set of groundwaters. The relationship of saturation indices and other indices of water chemical aggressiveness is discussed below. Most codes will deal with the effects of temperature change and some deal with saturation states of redox-sensitive species (e.g. pyrite).

REACTION MODELS

There are a range of available reaction codes (Table I). Several enable an equilibrium system between a groundwater and a set of solids to be calculated whilst others allow kinetic reactions to be followed in time. Examples of the former "heterogeneous" equilibrium models include MINTEQA2 (Felmy, Girvin and Jenne, 1984; Allison, Brown and Novo-Gradac, 1990) and EQ3NR (Wolery, 1983).These models enable predictions to be made of the chemical composition of a water at equilibrium with, for example, calcite or portlandite. Other specific conditions can be enforced, such as fixing the partial pressure of CO_2. MINTEQA2 also allows equilibrium sorption reactions to be modelled.

More powerful codes, the reaction path codes, enable the progress of a reaction to be mapped in time. At the simplest end of the spectrum, MIX2

Table I: Summary of some common hydrogeochemical models.

Model	Comments*	Availability**	Reference
Speciation			
WATSPEC	M, T.	A	Wigley (1978)
WATEQF	M, T.	EA	Plummer et al (1976)
Reaction			
SOLMNEQ	M, T. Heterogeneous equilibrium model	A	Kharaka & Barnes (1973)
EQ3NR	M, T. Heterogeneous equilibrium model	EA	Wolery (1983)
EQ6	M, T. Reaction path. Real or arbitrary time kinetics. Flow or non-flow conditions ***	EA	Wolery (1989)
MIX2	M. Reaction path model	A	Plummer et al (1975)
PHREEQE	M, T. Reaction path model. Electron balance. Exchange	EA	Parkhurst et al (1980)
MINTEQA2	M, T. Heterogeneous equilibrium model. Several sorption options including organic substrates and complexers	EA	Allison et al (1990)
PATHI	Reaction path model	A	Helgeson (1970)
GEOCHEM	M, T. Organics. Heterogeneous equilibrium. Sorption	S	Sposito & Mattigod (1980)
Reaction/Flow			
CHEMTRN)	Combined chemical and flow research models	NA	Miller & Benson (1983)
TRANQL)		NA	Cederberg et al (1985)
CHEQMATE	Chemical model based on PHREEQE	NA	Haworth et al (1988a, b)

*   M   = major species including their ion pairs (Na, K, Ca, Mg, Cl, SO_4 HCO_3)
     T   = trace species (ie any non-major species and their ion pairs)

**   EA   = easily available (ie commercially or from authors)
     A   = available (eg codes are published)
     NA   = not really available (eg research codes)

***   Flow conditions allow reaction products to be left behind

(Plummer, Parkhurst and Kosiur, 1975) allows the user to add reactants to the water step-by-step without specific reference to an actual time frame (Table I). A more sophisticated version of this model (PHREEQE) was later developed by Parkhurst, Thorstenson and Plummer (1985). EQ6 (Wolery, 1989) can also be operated on a reaction-path principle where small increments of the reactants are added and the equilibrium at every stage is calculated: however, it can also incorporate reaction rate laws using experimental kinetic data (e.g. Lasaga, 1984) to predict the real-time changes in chemistry. The main problems with the latter approach are the considerable uncertainty associated with the reaction site constraints and the lack of good data.

FLOW / REACTION MODELS

In flow/reaction models both groundwater flow and chemical reaction are taken into account. Ideally this requires the incorporation into the model of reaction rate data and equations, but sometimes the assumption of local equilibrium has to be made in the interests of tractability. In the latter case the water is assumed to be in chemical equilibrium at every point in its flow path (Valocchi, 1985).

Most codes in this category are research tools and are not commercially available. Examples include CHEQMATE (Haworth, Sharland, Tasker and Tweed, 1988a and b) which was developed for radioactive waste application and based on a mixing cell idea with PHREEQE (Parkhurst et al., 1985) as the basic chemical code, CHEMTRN (Miller and Benson, 1983), TRANQL (Cederberg et al., 1985) and ion exchange programs by many authors (e.g. Valocchi, Street and Roberts, 1981; Appelo and Willemsen, 1987). Radstake, Attia and Lennaerts (1988) used a mixing cell model combined with PHREEQE to model irrigation problems. Because of the heavy computer requirements, the need for kinetic data and the theoretical problems associated with kinetic calculations, it is likely to be some time before such programs are widely available commercially.

The most widely available codes for dealing with combined flow and reaction are those which deal with linear sorption isotherms or decay according to some empirical exponential law. EQ6 can deal with situations where precipitated minerals are left behind by flow.

EXAMPLE APPLICATIONS

Table II gives some problems which could be examined using hydrogeochemical models. In the following discussion four codes (WATEQF, MIX2, MINTEQA2, and EQ3NR/EQ6) are used to illustrate some of the problems listed. Chemical analyses of the groundwaters used in the illustrations are given in Table III. In each case the calculations presented are rather simple and designed only to give an idea of the scope of calculation possible. In a real application more complex calculations might be carried out and a wide-ranging sensitivity analysis performed to check on the reliability of the conclusions.

CO_2 DEGASSING

Groundwater is usually enriched in CO_2 by a factor of 10-100 compared with surface waters. The predicted partial pressures of CO_2 given in Table III are calculated by WATEQF, and represent the CO_2 pressure of a

Table II: Example applications of hydrogeochemical models.

Problems	Examples
CO_2 degassing	Incrustation of wells, pipelines. Water treatment.
Rock dissolution	Limestone, dolomite, evaporation strength losses. Attack of facing materials.
Weathering	Exposed rock. Fill/water interaction. Tunnel drainage waters.
Concrete/cement/mortar	Attack by groundwaters : $SO_4/Cl/CO_2$. Foundations, piles, pipelines. Walls, tunnels and buildings.
Salt precipitation	Damage to roads, structures, following evaporation.
Steel corrosion	Reinforcing. Steel structures. Wells.

hypothetical gas phase were it to be in contact with the water: the atmosphere's partial pressure is around $10^{-3.5}$ atmospheres (ie pP_{CO_2} = 3.5). The high CO_2 contents in groundwater can give rise to problems associated with corrosion (e.g. of steels) if the CO_2 is not allowed to degas, or with incrustation if it is. Degassing can be modelled easily using MIX2 (Table I); Figure 1(a) shows the effect of the degassing of waters 1 and 3 (Table III) to the point of equilibration with atmospheric CO_2 content. Although the two groundwaters are from sites within a few kilometres of each other, the effect of degassing is very different. No carbonate precipitation occurs in groundwater 1 but in the case of groundwater 3 considerable precipitation is possible giving rise to potential problems with well screen, pipeline or drain clogging. Figure 1(b) shows the predicted effect if a rise in temperature accompanies the degassing, which in this case is minimal.

CO_2 degassing increases the pH of the water and may give rise to other, unwanted, precipitation reactions, for example involving iron. The results of Figure 1(c) were calculated using EQ3NR assuming equilibrium with atmospheric O_2.

ROCK DISSOLUTION : CARBONATES

Dissolution of rock sequences by groundwater may have a considerable effect on strength, eg dissolution of a limestone bed in a mudstone sequence below a dam. Sometimes carbonate aggregate can be prone to dissolution, especially by CO_2-rich waters. Different waters have very different abilities to dissolve carbonate. Table III includes calcite saturation indices and shows that several of the groundwaters can dissolve calcite (SIC<0). MIX2 can again be used to find out how much calcite can be dissolved by some of the calcite-undersaturated waters shown in Table III. The results are presented in Figure 2. Two extreme cases have been modelled; open system conditions where the CO_2 contents of the waters have been held constant at their initial values and closed system conditions where no access to an infinite CO_2 source is allowed. Modelling stopped at

Table III: Analyses of waters used in calculations.

Sample	Ca	Mg	Na	K mg/l	Cl	SO4	HCO3	NO3	pH	pP_{CO2}*	SIC*@	SID*@	SIG*@
1 Haydock, Lancashire (Triassic sandstone)	29	27	34	5.2	78	144	5	4.5	4.8	1.04	-4.7	-9.3	-1.8
2 Haydock, Lanchashire (Triassic sandstone)	51	17.5	21	3.4	70	45	127	-	6.5	1.48	-1.4	-3.1	-2.0
3 Croft, near Warrington (Triassic sandstone)	108	40	24	3.3	16	6	558	0	7.4	1.76	0.4	0.57	-2.7
4 Well 9, N.Madras aquifer (Quaternary sands)	19	4.5	20	2.0	28	6	81	-	7.60	2.66	-0.6	-1.4	-
5 Glazebrook, near Warrington (surface water)	64	27	105	11.6	125	235	103	19	7.26	2.30	-0.63	-1.33	-1.8
6 Near Warrington (Triassic sandstone)	102	61	105	3.8	57	229	481	0	7.50	1.92	0.36	0.73	-1.3
7 Agden, near Warrington (Triassic evaporite)	1224	411	122500	105	192000	4690	35	-	5.75	For saturation indices see Table VIII pPco2 calculation uncertain			
8 Croft (Triassic sandstone)	162	110	14470	45	22150	940	150	-	7.5		=	=	=
9 Gatewarth, near Warrington (Triassic sandstone)	2700	800	1400	92	28200	1160	58	-	6.5		=	=	=
10 Croft (Permian sandstone)	1900	200	70000	140	105000	3600	24	-	7.93		=	=	=
11 Sea water	40	1350	10500	380	19000	2700	140	-	8.0				

* Calculated using WATEQF (Plummer at I, 1976)
@ Saturation indices: SIC = calcite; SID = dolomtie; SIG = gypsum

APPLICATION OF HYDROGEOCHEMICAL MODELS TO GROUND WATER PROBLEMS

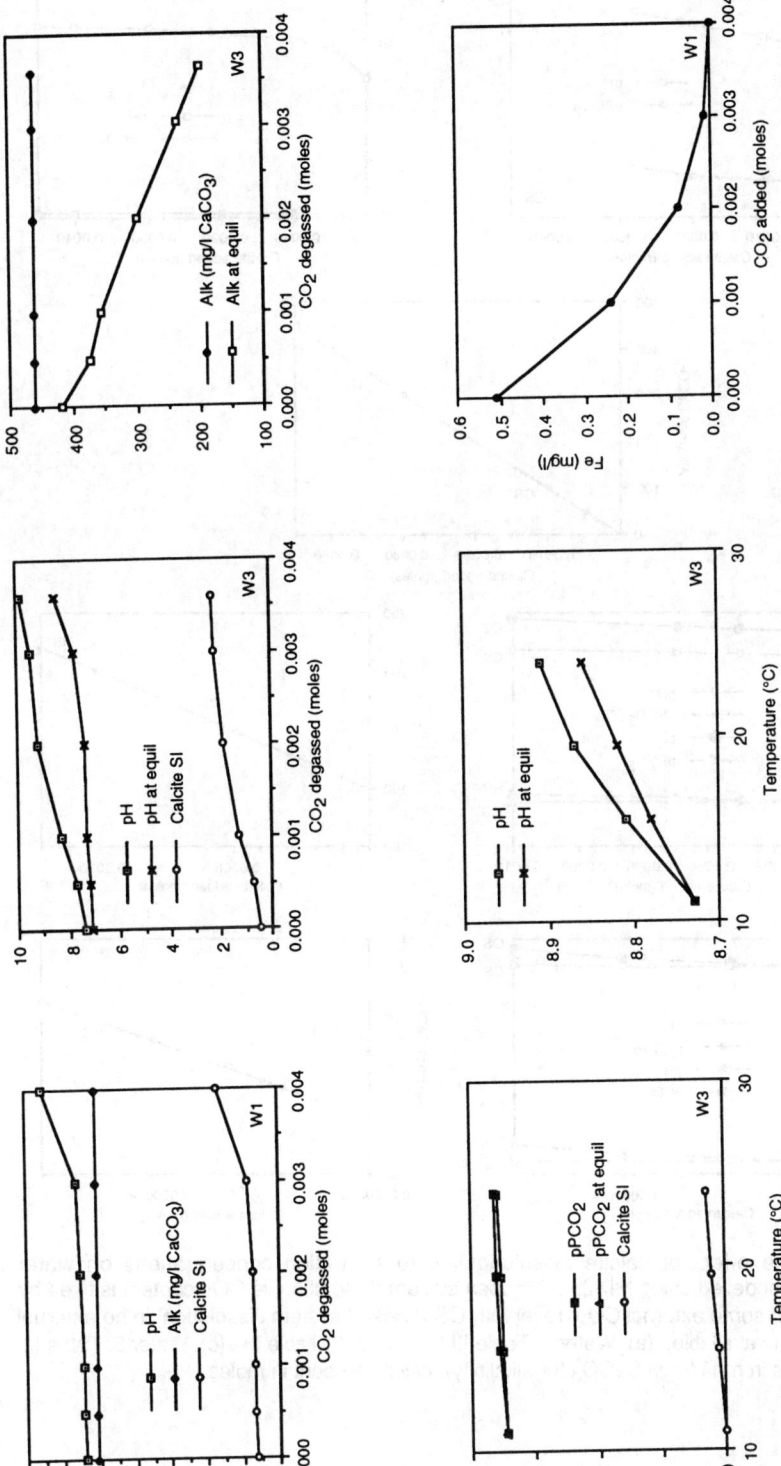

Figure 1: The effect of degassing of CO_2 to atmospheric concentrations ($P_{CO_2} = 10^{-3.5}$) modelled using MIX2 ((a) and (b)) and MINTEQA2 ((c)). (a) Waters 1 and 3 (W1, W3), Table III. (b) The effect of temperature change on Water 3, Table III following degassing. (c) The effect on Fe concentrations of CO_2 degassing; Water 1, Table III. Concentrations in mg/l ($CaCO_3$ for alkalinity); CO_2 concentrations in moles. Results for before and after equilibration with calcite ("in equil") are shown where degassing results in oversaturation. SI = saturation index.

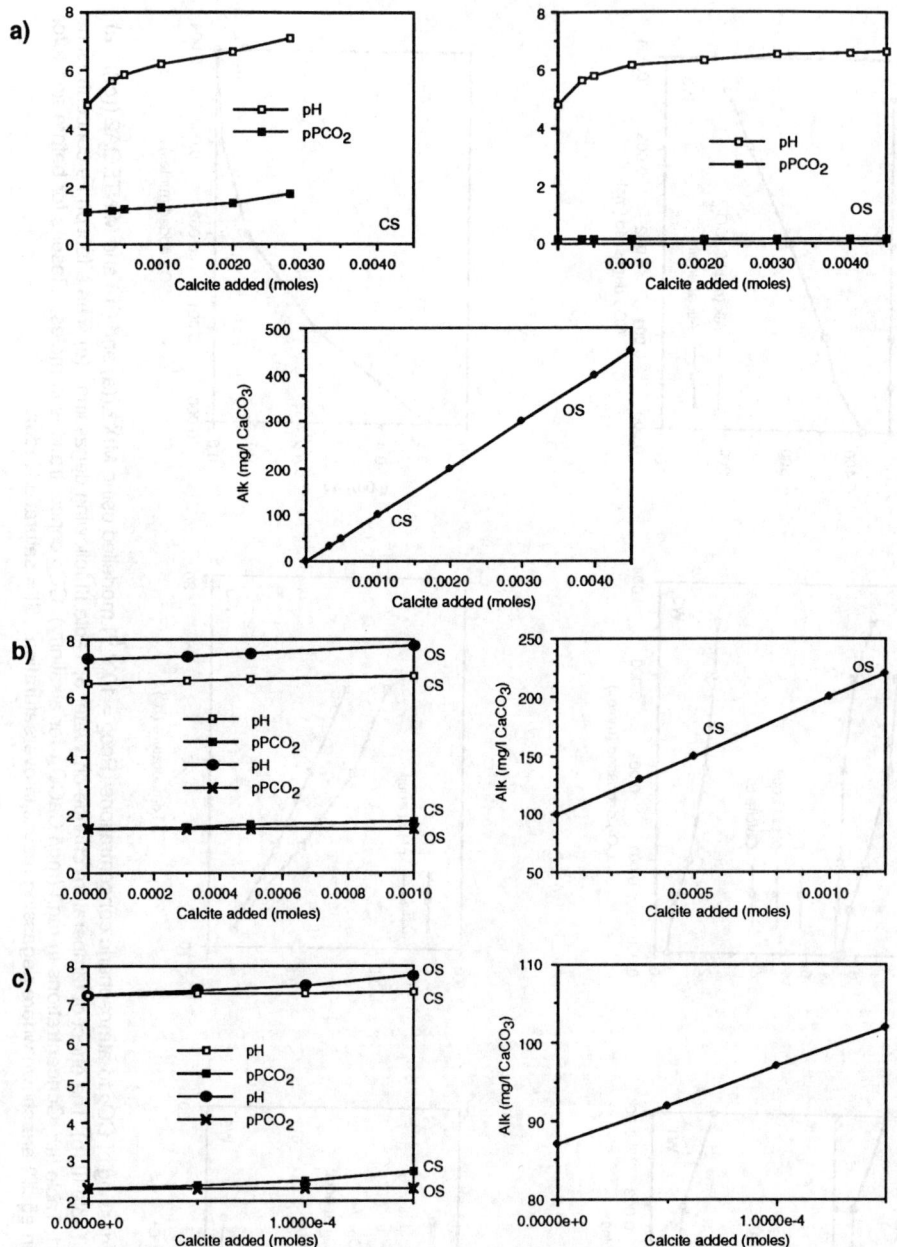

Figure 2: The effect of calcite dissolution up to saturation concentrations on water chemistry as modelled using MIX2. OS = open system dissolution, ie CO_2 content is fixed by interaction with some external CO_2 reservoir: CS = closed system dissolution, ie no external CO_2 reservoir is available. (a) Water 1, Table III (b) Water 2, Table III. (c) Water 5, Table III. Concentrations in mg/l (mg/l $CaCO_3$ for alkalinity): calcite amount in moles.

the point of saturation with respect to calcite. Although the two groundwaters differ in their calcite dissolution potential, both dissolve much more calcite than the surface water example in Table III despite the latter being unusually rich in CO_2 for a surface water.

If waters mix together, the saturation state of the mixture is not linearly related to the mixing properties (Bogli, 1964; Wigley and Plummer, 1976). The non-linearity can be predicted by MIX2 for example (Figure 3).

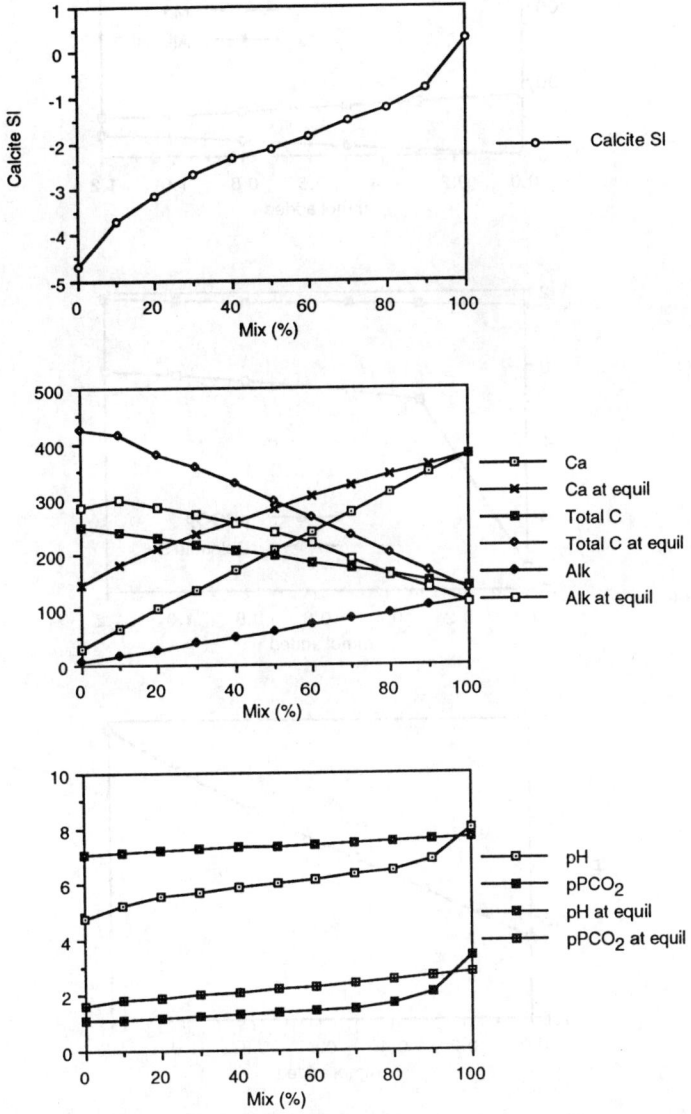

Figure 3: The effect of mixing seawater (Water 10, Table III) in various amounts (0-100%) and a groundwater (Water 1, Table III) as modelled using MIX2. The chemistry of the mixed water after equilibration with calcite is also shown. Concentrations in mg/l (mg/l $CaCO_3$ for alkalinity). SI = saturation index.

If a water is oversaturated with respect to calcite but undersaturated with respect to dolomite, dolomite dissolution and calcite preciptation can occur together. Figure 4 illustrates this effect as modelled using MIX2.

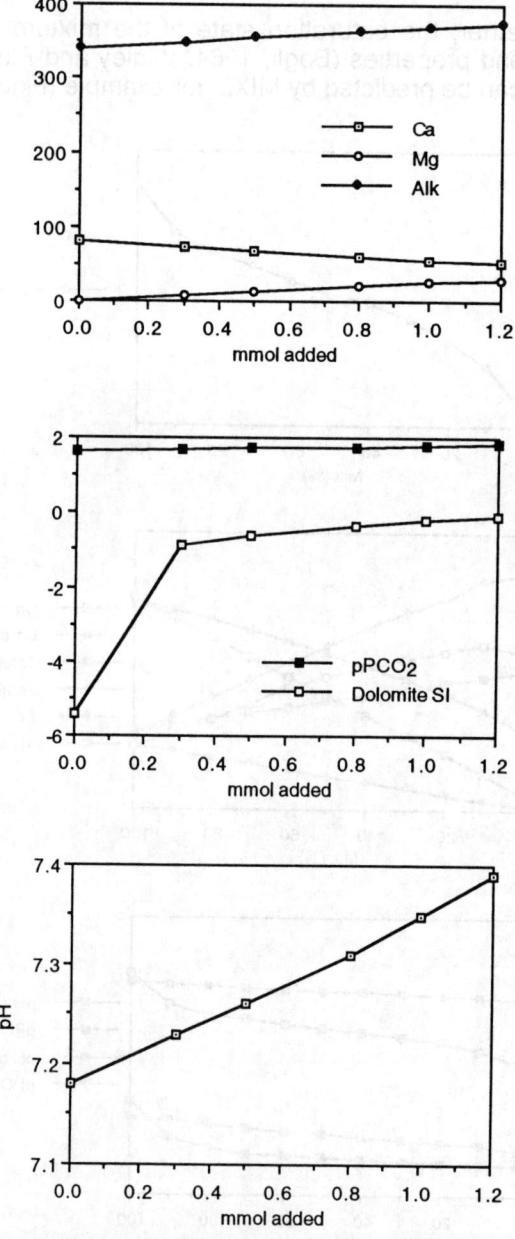

Figure 4: The effect of incongruent dissolution of dolomite (ordinate in mmoles added) on water chemistry as modelled using MIX2. Water 6, Table III, with Mg initially set at zero and SO_4 and Cl reduced appropriately to avoid an electrical imbalance. Concentrations in mg/l (mg/l $CaCO_3$ for alkalinity). SI = saturation index.

INDICES OF THE AGGRESSIVENESS AND PRECIPITATION-POTENTIAL OF WATERS WITH RESPECT TO THE CARBONATE SYSTEM

The previous two sections have dealt with the potential of waters either to attack engineering materials because of the presence of dissolved CO_2 or to deposit carbonate on the surfaces of engineering structures. The calcite saturation index has been used to indicate which effect is likely, but many other more widely used indices are available. Some of these are listed and compared with the calcite saturation index in Table IV. As can be seen from Table IV, the indices are closely related to each other and the choice as to which to use is a matter of convenience (Rossum and Merrill, 1983).

Under most conditions the usual methods of calculating the indices in Table IV are quite adequate. However, using a hydrogeochemical model to calculate the indices has the advantage that extreme cases (e.g. high or low pH, or where unusual ionic proportions are present) are automatically taken into account with reasonable accuracy. In addition, the amount of carbonate which must be dissolved or precipitated to bring the water into equilibrium can easily be assessed without recourse to further approximations, nomographs, or specially written programs (e.g. Rossum and Merrill, 1983), even if other conditions (such as specific CO_2 concentrations) need to be enforced.

WEATHERING

Weathering reactions are usually considered to involve silicates and sulphides, although in principle they could include any reaction involving the thermodynamic adjustment of rocks and soils to prevailing environmental conditions. Silicate reactions, especially sulphide, are particularly important when considering the degradation of new rock cuttings and the interaction of newly placed fill material with groundwater and acid tunnel-drainage waters for example.

Weathering reactions are extremely complex, often involving poorly understood networks of bacterially-catalysed redox reactions. Nevertheless, some idea of the scale of a chemical problem can be obtained by modelling. Table V shows the results of a simple model of pyrite oxidation in equilibrium with the atmosphere as might occur in a tunnel (a) and a case where the resultant water comes into contact with a carbonate source (b). The calculations were made using EQ3NR/6.

CEMENT AND CONCRETE INTERACTIONS WITH GROUNDWATERS

The reactions occurring within a cement/aggregate/water system are extremely complex. Codes such as PHREEQE and EQ3NR/6 have been used to investigate such systems in the context of radioactive waste disposal (e.g. Cross et al., 1987) and for example, EQ3NR/6 contains within it the ability to deal with a wide range of relevant solid species (Table VI).

Table IV: Common indices of the carbonate-system aggressiveness/precipitation potential of water and their relationship to the saturation index for calcite.

	Name	Definition	Relationship to Saturation Index **	Reference
1.	Saturation Index (SI)	$SIC = \log\left(\frac{(Ca^{2+})(CO_3^{2-})}{K}\right)$	SIC = SIC	
2.	Langelier Index (LI)	$LI = pH - pH_s^*$	LI = SIC	Langelier (1936)
3.	Ryznar Index (RI)	$RI = 2pH_s - pH^*$	$RI = pH_s - SIC$	Ryznar (1944)
4.	Aggressiveness Index (AI)	$AI = pH + \log_{10}([Ca^{2+}][Alk])$ (concs. in mg/l $CaCO_3$)	$AI = SIC + \log_{10}(5 \times 10^9 K/K_2)$@ $\approx SIC + 12$	Rossum and Merrill (1983)
5.	Driving Force Index (DFI)	$DFI = \frac{[Ca^{2+}][CO_3^{2-}]}{K}$	$DFI = 10^{SIC}$	Rossum and Merrill (1983)
6.	Momentary Excess (ME)	$ME^2ME[(Ca^{2+}) + (CO_3^{2-})]$ $-K = 0$	$ME^2ME[(Ca^{2+}) + (CO_3^{2-})]$ $+[(Ca^{2+}][CO_3^{2-}](1-10^{-SIC})=0$	Rossum and Merrill (1983)
7.	Calcium Carbonate Precipitation Potential (CCPP)	$CCPP = 50000 (Alk - Alk_{eq})$ where Alk_{eq} = Alk after precipitation of calcite	$CCPP^2 10^{-10} - CCPP 10^{-5}([Ca^{2+}] + [CO_3^{2-}])$ $+ [Ca^{2+}][CO_3^{2-}](1-10SIC) = 0$	Merrill (1978)

* pH_s = pH at which the water would be at saturation (not the pH after dissolving calcite). Alk = alkalinity in mg/l $CaCO_3$ except in the case of CCPP where alkalinity is in equivalents. [ ] = concentration.

** The relationships ignore the fact that SIC is usually calculated by taking into account both activity coefficients and ion pairs.

@ K_2 = dissociation constant of HCO_3^- (i.e. $K_2 = (H^+)(CO_3^{2-})/(HCO_3^-)$)

Table V: Pyrite oxidation by Water 1 (Table III), assuming it to be initially saturated with respect to an atmospheric P_{O2} of 0.2 atmospheres.

	pH	E_H mV	P_{O2} atmos	Calcite SI	Haematite SI	Pyrite SI	Fe* mg/l	SO_4^- mg/l	HS^- mg/l
(a) Pyrite oxidation									
Initial water	4.78	935	0.2	-6.2	0.0	<-10	0	144	0
After equilibration with pyrite	4.08	43	0.0	-8.1	-6.0	0.0	4.16	158	0
(b) Pyrite oxidation in the presence of calcite									
Initial water	4.78	936	0.2	-6.2	0.0	<-10	0	144	0
After equilibration with pyrite and calcite	6.99	-167	0.0	0.0	0.0	0.0	0.005	157	0

* Total Fe, total SO_4

Table VI: Solid phase species of particular relevance to cement calculations and present in the standard version of EQ3NR/6.

Portlandite	$Ca_3 Si_2 O_7.3H_2O$	$Ca_5 Si_6 O_{17}.21/2H_2O$
$Ca_2 SiO_4$	$Ca_4 Si_3 O_{10}.3/2H_2O$	$Ca Si_2O_5 . 2H_2O$
Larnite ($Ca_2 Si O_4$)	$Ca_6 Si_6 O_{18}.H_2O$	$Ca_2 Si_3 O_8 . 5/2 H_2O$
$Ca_3 Si O_5$	$Ca_5 Si_6 O_{17}.3H_2O$	Ferrosilite
$Ca_2 SiO_4.7/6 H_2O$	$Ca_5 Si_6 O_{17}.11/2 H_2O$	Gehlenite ($Ca_2 Al_2 SiO_7$)
Ferrite-2-Ca	Ferrite-Mg	Lime
Ferrite-Ca		

SALT DAMAGE

Salt damage, to roads for example, is common in some parts of the world (Obika, Freer-Hewish and Foskes, 1989). It may be initiated by evaporation of groundwater near ground level with various salts being precipitated as the water evaporates. At low to moderate concentrations (up to around sea water levels) the standard activity coefficient models are adequate, but beyond this level it is advisable to use the Pitzer model (Pitzer, 1973). The only readily available package to include the Pitzer activity coefficient calculations is EQ3NR/6.

To avoid straying too far into this field, two simplified problems have been examined: the interaction of two rather different groundwaters firstly with pure portlandite ($Ca(OH)_2$) and secondly with portlandite and an example calcium silicate hydrate ($Ca_3 Si_2 O_7. 3H_2O$). The results, given in Figure 5 (MIX2) and Table VII (MINTEQA2), show the effect of the hydroxides in raising the pH, converting available carbon into CO_3 and precipitating calcite. The equilibrium pH is in excess of 12 and it is also clear that the final chemistry and the degree of dissolution and precipitation is very dependant upon the original water chemistry.

Figure 6 shows calculations of evaporation of a groundwater carried out using MIX2 : in this case calcite is the only mineral to precipitate. Table VIII

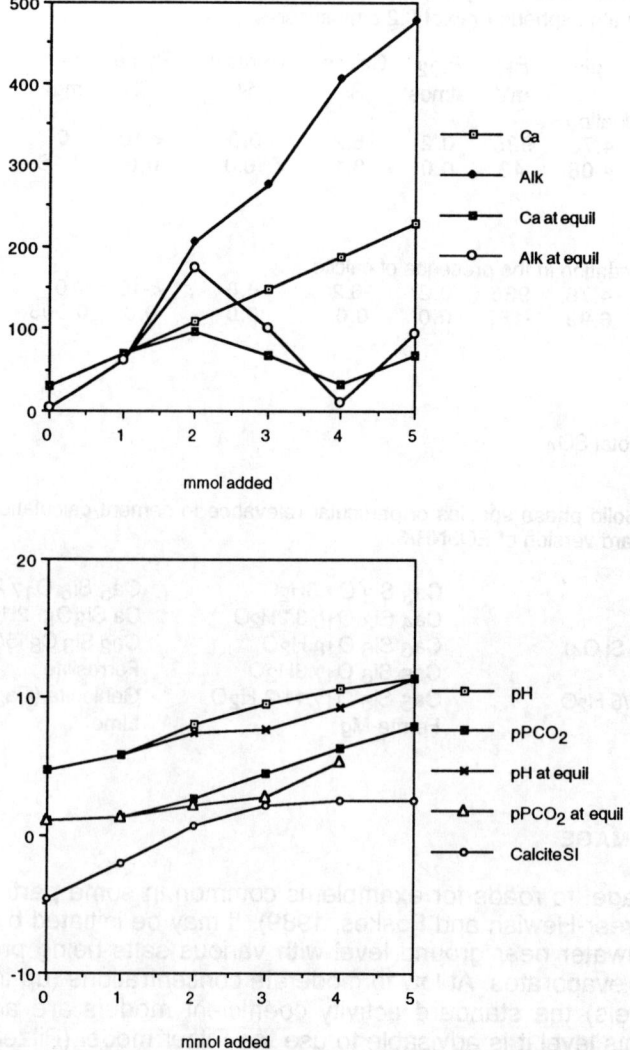

Figure 5: The effect of dissolving portlandite (ordinate in mmoles added) on water chemistry as modelled using MIX2. Water 1, Table III. Both the chemistry of the water immediately after the dissolution and subsequently after equilibration with calcite ("in equil") are shown. Concentrations in mg/l (mg/l $CaCO_3$ for alkalinity).

provides examples of the saturation states of some very saline groundwaters as predicted by EQ3NR/6.

METAL CORROSION

As is the case with cement reactions, metal corrosion is a highly complex phenomenon. Despite this some progress can be made towards modelling a metal/water system. For example, following a previous published conceptual model for crevice corrosion of stainless steel, Haworth et al.

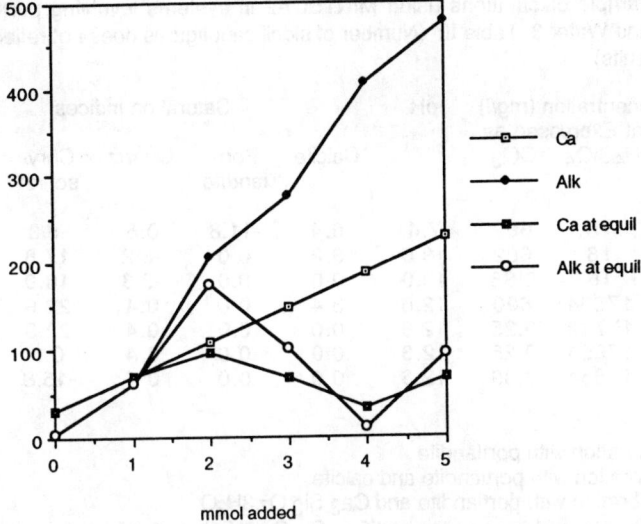

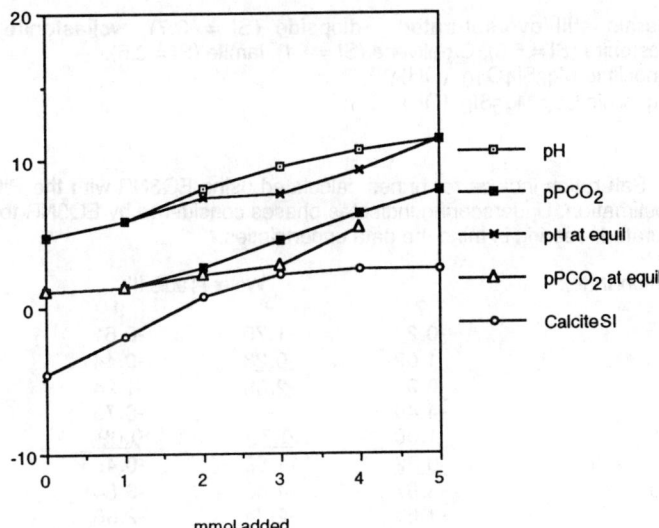

Figure 6: The effect of water chemistry as modelled using MIX2. Water 4, Table 3. The amount of evaporation is indicated by a factor: a factor of 3 means that the H_2O component has been decreased to a third of its original value. Concentrations in mg/l (mg/l $CaCO_3$ for alkalinity). OS = open system with respect to CO_2, CO_2 fixed at atmospheric concentrations. CS = closed system allowing CO_2 to vary.

(1988(a)) used CHEQMATE, which incorporates PHREEQE, to quantify the process. Even without the sophistication of Haworth et al.'s (1988(a)) model it is possible to obtain a general appreciation of corrosion potential in particular circumstances, if only by using models to predict CO_2 content or the total amount of iron solution assuming no reaction products accumulate.

Table VII: Example calculations using MINTEQA2 in systems involving portlandite, $Ca_3Si_2O_7 \cdot 3H_2O$ and Water 3, Table III (Number of significant figures does not reflect estimated accuracy of results).

	Total Concentration (mg/l) of Element Expressed as			pH	Saturation Indices				
	Ca	H_4SiO_4	CO_3		Calcite	Portlandite	Quartz	Chrysotile*	Tremolite**
(a)	108	16	602	7.4	0.4	-11.8	0.5	-6.3	-7.3
(b)	1068	16	602	13.0	3.2	0.0	-3.2	17.8	39.1
(c)	693	16	0.36	13.0	0.0	0.0	-3.3	18.0	39.3
(d)	7348	17664	600	12.6	3.4	0.0	0.4	22.9	64.6
(e)	6968	17712	0.25	12.6	0.0	0.0	0.4	22.9	64.6
(f)	6996	17654	0.25	12.6	0.0	0.0	0.4	0.0	26.3
(g)	6996	17654	0.39	12.6	0.0	0.0	0.4	-15.8	0.0

(a) Initial Water
(b) After equilibration with portlandite
(c) After equilibration with portlandite and calcite
(d) After equilibration with portlandite and $Ca_3Si_2O_7 \cdot 3H_2O$
(e) After equilibration with portlandite and $Ca_3Si_2O_7 \cdot 3H_2O$, and calcite
(f) After equilibration with portlandite and $Ca_3Si_2O_7 \cdot 3H_2O$ calcite and chrysotile
(g) After equilibration with portlandite and $Ca_3Si_2O_7 \cdot 3H_2O$ calcite and tremolite

Other minerals still oversaturated: diopside (SI = 3.7), wollastonite (SI = 6.2), pseudowollastonite (SI = 5.3), Ca-olivene (SI = 4.1), larnite (SI = 2.5).
* Monoclinic $Mg_6Si_4O_{10}(OH)_8$
** Monoclinic $Ca_2Mg_5Si_8(OH)_2$

Table VIII: Saturation indices for brines calculated using EQ3NR with the Pitzer approach to activity estimation. Underscoring indicates phases considered by EQ3NR to be saturated or over-saturated, bearing in mind the data uncertainties.

Saturation Index	Water (Table III)			
	7	8	9	10
Anhydrite	<u>-0.28</u>	-1.75	-0.61	-0.47
Aragonite	-1.09	<u>-0.28</u>	-0.44	<u>0.52</u>
Basanite	-0.96	-2.38	-1.24	-1.12
Bloedite	-4.40	-	-6.75	-5.42
Dolomite	-1.00	<u>0.76</u>	<u>0.09</u>	<u>1.80</u>
Gypsum	<u>-0.32</u>	-1.59	-0.45	-0.40
Hexahydrite	-3.67	-4.42	-3.64	-3.95
Mirabilite	-1.61	-2.74	-2.86	-1.70
Sodium Carbonate	-7.30	-7.11	-	-6.13
Pentahydrite	-3.90	-	-3.97	-4.24
Sylvite	-3.13	-4.49	-4.09	-3.36
Thenardite	-1.40	-3.48	-3.59	-
Bischofite	-5.36	-	-7.28	-6.46
Brucite	-6.28	-4.41	-5.33	-2.75
Calcite	-0.92	<u>-0.12</u>	<u>-0.27</u>	<u>0.69</u>
Hemihydrite	-1.13	-2.55	-1.41	-1.29
Epsomite	-3.54	-4.19	-	-3.77
Glauberite	-0.81	-4.36	-3.33	-1.03
Halite	<u>-0.20</u>	-2.35	-2.27	-0.93
Kainite	-6.99	-	-	-
Kieserite	-4.60	-5.83	-5.04	-5.16
Magnesite	-1.71	-0.76	-1.27	-0.51
Nesquehonite	-4.72	-3.49	-4.00	-3.37

LIMITATIONS

Despite the existence of some very sophisticated codes there are a number of shortcomings with hydrogeochemical models, some of which are described below.

Thermodynamic Data

Thermodynamic data do not exist for all the species which may be of interest and those data which are available are of variable quality. However there are continuing programmes of data collation and testing which allow for updating of code data bases and many codes allow data to be modified either for a specific run or permanently, so that the user can either carry out sensitivity analyses or update his version of the code when new data are published. As a result of differences in thermodynamic data bases, different models give slightly different predictions; a detailed comparison of model predictions is presented by Nordstrom et al. (1979) and Read et al. (1990).

Perhaps a more fundamental data problem is associated with the fact that many minerals are not pure. Some codes take into account minerals in solid-solution series, but the effect of a particular lattice substitution in kaolinite for example, is not easily determinable.

Kinetic data, especially precipitation data, are sparse and often depend on factors beyond the scope of the model, eg flow turbulence or pore architecture. The degree to which a water is in equilibrium with particular mineral phases can often be assessed approximately by general chemical knowledge: olivene may be the predicted thermodynamically stable phase, but in the context of surface processes it may be unlikely to form within the time period of an engineering structure. Similarly, several codes allow an empirical approach to kinetics to be followed where at each step in the reaction progression only a given amount of dissolution is allowed to occur.

Bacterially-mediated Reactions

In the present context it is not feasible to attempt to model deterministically the processes of bacterial mediation. Again, successful application of the codes depends on a knowledge of what chemical reactions are likely to become important through bacterial mediation.

Redox Reactions

Waters are rarely in complete redox equilibrium, but at the present time there is little theoretical guidance on predicting the states of disequilibrium. Many models assume complete equilibrium; a few (e.g. EQ3NR/6) allow a degree of disequilibrium to be modelled. Again, chemical judgement must be applied.

Organic Species

Although most models deal only with inorganic species, MINTEQA2 for example does include some organic species and the very active research in this area suggests that incorporation of organics is likely in the future.

Sorption

Few codes deal explicitly with sorption (MINTEQA2 is again one exception). However, some degree of sorption or exchange modelling can be done by appropriate "manual" manipulation outside the code: for example, a reaction might be defined where Ca is added and Mg removed, thus modelling ion exchange. Again care is needed, especially in major ion exchange since the effects of flow are so important where competitive exchange is occurring.

Flow

Through dispersion, the presence of immobile water of different chemistry, the removal of reaction products and the supply of new reactants, flow can have a considerable effect on chemical reactions. An acidic water ponding on concrete may well be of little consequence, but if the water flows past the concrete at a rate of 1000 m^3/d, the extent of chemical attack is likely to be considerably more significant. To deal with such systems a model should ideally be able to take account of the chemistry of both the solid surface (including weathering product coatings) and the water. As discussed earlier, computationally this is a formidable task. The nearest readily available codes have come to dealing with the problem is the ability of EQ3NR/6 to allow water to flow over a solid and leave behind reaction products.

PREDICTION UNCERTAINTIES

Unlike the use of groundwater flow codes, for example, where the final model emerges as a result of careful comparison with field data, hydrogeochemical models often cannot be tested against existing data: they are used to extrapolate, given theoretical constraints. Unless a very critical attitude is adopted, this can lead to problems. The theoretical framework is far more complex than that of groundwater hydraulics and very often the simple intuition which is such a good guide for flow modellers is lacking when complex reactions are being considered. It is thus sometimes difficult to determine whether an unexpected result is due to incorrect intuition, incorrect presentation of the problem to the code or a bug in the code.

INPUT CHEMICAL ANALYSIS DATA

Finally, it is very important that good input data are obtained and analyses are of the correct type. For example, although an analysis of total Fe may be very accurate, unless the code can deal with colloids, the prediction of the model may have little relevance. Another example is the need for field-determined pH values.

CONCLUSIONS

There are many widely available hydrogeochemical codes which can deal with reactions in quite complex solid/aqueous systems.

Speciation models calculate chemical speciation from a given water analysis and indicate which mineral phases are likely to precipitate or are able to dissolve. Such models could predict, for example, how aggressive water was to a limestone aggregate.

Reaction models allow prediction of water and precipitate chemistry either at equilibrium or with some regard to kinetics, thus the effect water coming into contact with a given engineering material may be predicted.

Flow/reaction models allow prediction of chemical reactions in flowing systems. Because of the computational effort and the inherent complexity, such models are not yet widely available.

Several limitations of the existing codes have been discussed. These relate both to imperfect experimental data sets and to imperfectly understood chemistry. Despite these shortcomings, however, hydrogeochemical models may provide a very useful method for investigating many fairly routine chemical questions in construction engineering - not just the questions posed in research or for radioactive waste engineering.

ACKNOWLEDGEMENTS

We would like to thank Dr. M. Ivanovich, AEA Technology, Harwell Laboratories for supplying the version of MINTEQA2 used in this study. The MIX2 version was modified by Dr. J.A. Heathcote and subsequently by Mr. P Eggerton : the EQ3NR/6 version was supplied by Dr. D. Savage and his group at the British Geological Survey.

REFERENCES

ALLISON, J.D., BROWN, D.D. & NOVO-GRADAC, K.J. (1990). *MINTEQA2/PRODEFA2, A geochemical assessment model for environmental systems* : Version 3.0 user's manual. Environmental Research Laboratories, Offices of Research and Development, U.S. Environmental Protection Agency, Athens, Georgia, 30613, 106pp.

APPELO, C.A.J. & WILLEMSEN, A. (1987). Geochemical calculations and observations on salt water intrusion I. A combined geochemical/ mixing cell model. *Journal of Hydrology*, **94**, 313-330.

BOGLI, A. (1964). Mischungkorrosion, ein Beitrag zum Verkartongsproblem. *Erkunde, Arch. Wisse. Geogr.*, **13**, H1/2.

BONAZOUNTAS, M. (1986). Chemical fate modelling in soil systems: a state-of-the-art review. In: *Scientific basis for soil protection in the European Community*, Barth, H. and L'Hermite, P. (eds.), Elsevier Applied Science, London, 487-566.

CEDERBERG, G.A., STREET, R.L. & LECKIE, J.O. (1985). A groundwater mass-transport and equilibrium chemistry model for multi component systems. *Water Resources Research*, **21**, 1095-1104.

CROSS, J.E., EWART, F.T. & TWEED, C.J. (1987). *Thermochemical modelling with application to nuclear waste processing and disposal.* Harwell Laboratory Report AERE R12324.

DOMENICO, P.A. & SCHWARTZ, F.W. (1990). *Physical and chemical hydrogeology.* John Wiley and Sons, New York, 824 pp.

DREVER, J.I. (1982). *The geochemistry of natural waters.* Prentice-Hall, Englewood Cliffs, USA.

FELMY, A.R., GIRVIN, D.C. & JENNE, E.A. (1984). *MINTEQ - a computer program for calculating aqueous geochemical equilibria.* U.S. Environmental Protection Agency, Athens, Georgia, EPA-600/3-84-032.

HAWORTH, A., SHARLAND, S.M., TASKER, P.W. & TWEED, C.J. (1988 a). *A guide to the coupled chemical equilibrium and migration code CHEQMATE.* Harwell UKAEA Safety Studies (Nirex Radioactive Waste Disposal) NSS/R113, 31 pp.

HAWORTH, A., SHARLAND, S.M., TASKER, P.W. & TWEED, C.J. (1988 b). *Extensions to the coupled chemical equilibrium and migration code CHEQMATE.* Harwell UKAEA Safety Studies (Nirex Radioactive Waste Disposal) NSS/R115, 21 pp.

HELGESON, H.C., BROWN, T.H., NIGRINI, A. & JONES, T.A. (1970). Calculation of mass transfer in geochemical processes involving aqueous solutions. *Geochim. Cosmochim. Acta*, **34**, 569-592.

KHARAKA, Y.K. & BARNES, I. (1973). SOLMNEQ: solution mineral equilibrium computations. *National Technology Information Service Technical Report.* PB 214-899, 82 pp.

LANGELIER, W.F. (1936). The analytical control of anti-corrosion water treatment. *Journal of the American Water Works Association*, **28**, 1500-1521.

LASAGA, A.C. (1984). Chemical kinetics of rock-water interactions. *J. Geophysical Research*, **89**, 4009-4025.

MERRILL, D.T. (1976). Chemical conditioning for water softening and corrosion control. In: *Water treatment plant design for the practising engineer*, Sanks, R.L. (ed.), Ann Arbor Science Pub., Michigan, 497-565.

MILLER, D. & BENSON, L. (1983). Simulation of solute transport in a chemically reactive heterogeneous system: model development and application. *Water Resources Research*, **19**, 381-391.

NORDSTROM, D.K. & BALL, J.W. (1984). Chemical models, computer programs and complexation in natural waters. In: *Complexation of trace metals in natural waters*, Kramer, C.J.M. and Duinker, J.C. (eds.), Nijhoff-Junk Publ., The Hague, The Netherlands, 149-164.

NORDSTROM, D.K., PLUMMER, L.N., WIGLEY, T.M.L., WOLERY, T.J., BALL, J.W., JENNE, E.A., BASSETT, R.L., CREROR, D.A., FLORENCE, T.M., FRITZ, B., HOFFMAN, M., HOLDREN, G.R., LAFON, G.M., MATTIGOD, S.V., MCDUFF, R.E., MOREL, F., REDDY, M.M., SPOSITO, G. & THRAILKILL, J. (1979). A comparison of computerized models for equilibrium calculations in aqueous systems. In: *Chemical modelling in aqueous systems*, Jenne, E.A. (ed.), *Chemical Society Symposium Series*, **93**, 857-892.

OBIKA, B., FREER-HEWISH, R.J. & FOOKES, P.G. (1989). Soluble salt damage to thin bituminous road and runway surfaces. *Quarterly Journal of Engineering Geology*, **22**, 59-73.

PARKHURST, D.L., THORSTENSON, D.C. & PLUMMER, L.N. (1980). PHREEQE - a computer program for geochemical calculations. *U.S. Geological Survey Water Resources Investigations*, **80-96**, 210 pp.

PITZER, K.S. (1973). Thermodynamics of electrolytes. I. Theoretical basis and general equations. *Journal of Physical Chemistry*, **77**, 268-277.

PLUMMER, L.M., JONES, B.F. & TRUESDELL, A.H. (1976). WATEQF- a FORTRAN IV version of WATEQ, a computer program for calculating chemical equilibrium of natural waters. *U.S. Geological Survey Water Resources Investigation Report*, **76-13**, 61 pp.

PLUMMER, L.N., PARKHURST, D. & KOSIUR, D.R. (1975). MIX2, a computer program for modelling chemical reactions in natural waters. *U.S. Geological Survey Water Resources Investigation Report*, **61-75**.

RADSTAKE, F., ATTIA, F.A.R. & LENNAERTS, A.B.M. (1988). Forecasting groundwater suitability for irrigation - a case study in the Nile Valley, Egypt. *Journal of Hydrology*, **98**, 103-119.

READ, D., BROYD, T.W. & COME, B. (1990). The CHEMVAL project - an international study aimed at the verification and validation of equilibrium speciation and chemical transport models. *GEOVAL 90 Symposium on Validation of Geosphere Flow and Transport Models, Stockholm.*

RICE, R.E. (1986). The fundamentals of geochemical equilibrium models with a listing of hydrochemical models that are documented and available. *Document GWMI 86-04 of the International Ground Water Modelling Center, TNO-DGV Inst. Applied Geoscience, POB 285, 2600 AG, Delft, The Netherlands.*

ROSSUM, J.R., & MERRILL, D.T. (1983). An evaluation of the calcium carbonate saturation indexes. *Journal of the American Water Works Association*, **75**, 95-100.

RUNNELLS, D.D. (1987). American research in the geochemical modelling of groundwater. *Journal of the Geological Society of India*, **29**, 135-144.

RYZNAR, J.W. (1944). A new index for determining the amount of $CaCO_3$ scale formed by water. *Journal of the American Water Works Association*, **36**, 473-486.

SPOSITO, G. (1985). Chemical models of inorganic pollutants in soils. *CRC Critical Reviews in Environmental Control*, **15**, No. 1.

SPOSITO, G. & MATTIGOD, S.V. (1980). GEOCHEM : a computer program for the calculation of chemical equilibrium in soil solutions and other natural water systems. *Department of Soil and Environmental Science Report, University of California, Riverside, California*, 92 pp.

VALOCCHI, A.J., STREET, R.L. & ROBERTS, P.V. (1981). Transport of ion-exchanging solutes in groundwater: chromatographic theory and field simulation. *Water Resources Research*, **17**, 1517-1527.

VALOCCHI, A.J. (1985). Validity of the local equilibrium assumption for modelling sorbing solute transport through homogeneous soils. *Water Resources Research*, **21**, 808-820.

WIGLEY, T.M.L. (1978). WATSPEC : a computer program for determining the equilibrium speciation of aqueous solutions. *British Geomorphological Research Technical Bulletin*, **20**, 16 pp.

WIGLEY, T.M.L. & PLUMMER, L.N. (1976). Mixing of carbonate waters. *Geochim. Cosmochim. Acta*, **40**, 989-995.

WOLERY, T.J. (1983). *EQ3NR. A computer program for geochemical aqueous speciation-solubility calculations : user's guide and documentation.* Lawrence Livermore Laboratory, University of California, Livermore, California 94550, UCRL-53414, 191 pp.

WOLERY, T.J. (1989). *EQ6. A computer program for reaction path modelling of aqueous geochemical systems: user's guide and documentation.* Lawrence Livermore Laboratory, University of California, Livermore, California 94550, 253 pp.

5-14 DEVELOPMENT OF A COMPUTER ASSISTED PROGRAM TO PROJECT THE WELL PLUGGING RISK INDEX (WPRI) FOR EXISTING OR PLANNED WATER WELLS

D.R. CULLIMORE
Regina Water Research Institute, University of Regina, Regina, Saskatchewan, Canada, S4S OA2
G. ALFORD
ARCC Inc., Daytona Beach, Florida, U.S.A.

ABSTRACT

Water wells are sometimes subjected to a significant level of biofouling in the porous media associated with the active hydraulic movement of ground water towards the zone of pumping within the well. Such fouling commonly involves the formation of biofilms which change the flow characteristics as well as the properties of the water being pumped from the well. These effects may be observed in a variety of ways such as an increase in drawdown associated with the loss in transmissivity, radical elevations in metals content due to either corrosivity, sloughing or inactivation of the biofilms, generation of odours and colours and rampant increases in complaints from the users.

To predict if these events will occur, a factorial weighting has been generated for the various physical, chemical, microbiological and geohydrological factors considered to be potentially significant to these nuisance occurrences, e.g., loss in production by plugging. These weighted factors have been compartmentalized within the major categories to generate a (Water) Well Plugging Risk Index (WPRI). Each of the factors where data are presented are weighted both directly and in relation to other known synergistic factors to generate a single digit index where zero would reflect a zero plugging risk and nine would reflect a certainty of bioimpairment occurring within months. Confidence in the WPRI value calculated would relate to the size of the data base used to calculate the index. A total of one hundred and ten parameters have been incorporated into the computation of the WPRI and the confidence level in the calculated index increases with the number of parameters introduced.

INTRODUCTION

Traditionally, it has been considered that the natural microbiological intrusion into porous media and installations associated with a developed water well were minimal and largely confined to the water column, well screen and the packing material immediately around the well. However, recent findings indicate the ubiquity of microorganisms within the various porous media (Fliermans and Hazen, 1990) surrounding a water well (Cullimore, 1990), with a focused activity frequently occurring at the reduction-oxidation interface. The effect of this zone of magnified activity is a concentration of microbially induced fouling (MIF) leading to changes in the characteristics of the water passing through this biofouled interface (Cullimore, 1987).

Product (postdiluvial) water which has passed through an MIF event around a water well may cause changes in the production capacity of the well and also the water quality characterization. This is likely to occur over a long period and in a manner related to the state of maturation of the various biological structures within the MIF zone. Symptoms of the formation associated with a focused MIF include increasing drawdowns during pumping, lower production capacities and sporadic degenerations in water quality (Smith, 1992). Eventually the bioimpedance of the water flow will become sufficiently intense that the porous medium becomes occluded (plugged).

Water quality in the postdiluvial water may pass through a number of phases during the maturation of the MIF. Initially, the various biofilms forming within the focused MIF zone will entrap and concentrate various organic and inorganic chemicals suspended or dissolved in the passaging water. These bioaccumulates may be actively assimilated by the incumbent cells in the biofilm (Cullimore, 1992). It may be anticipated that within the biofilm, the concentration of the actively assimilated chemicals will be influenced by the rate of biodegradation, as some of these compounds are utilized both catabolically and as a source for nutrients. Passive accumulates within the biofilm may continue to increase in concentration throughout the active lifespan of the MIF event (Gehrels and Alford, 1990). These bioaccumulates commonly include heavy metals and, in particular, iron and manganese. In water wells which are sufficiently biofouled to have become plugged, subsequent recoveries using the patented blended chemical heat treatment (BCHT) process have revealed iron burdens within the dispersed biofilms ranging from 10,000 to 40,000 mg Fe/Kg in biozones close to the well. Other metals may also be accumulated to varying extents.

Active accumulates participate in the catabolic and nutrient uptake activities of the incumbent microflora in the composite biofilms. The rate at which these active accumulates are assimilated and utilised by the microflora will influence the rate at which the biomass will grow (Cullimore, 1987) and eventually reach sufficient dimensions to cause a plugging (i.e. 60 to 80% occupancy of the interstitial spaces).

In the management of a producing water well it is important to project the likelihood that a well will lose an attainable capacity as a result of an MIF plugging. For successful management of the water well, ongoing maintenance should recognize the level of risk that may be present as a result of biofouling. Such activities should take account of not only the bulk of nutrients being transmitted to the well with hydraulic flow but also the prevalent environmental conditions. It is proposed to develop a well plugging risk index (WPRI) based on a weighted evaluation of the parameters which are known to influence an MIF event.

DATA ENTRY MECHANISM

Data entry may include as few data points as five or six characteristics or as many as thirty or more. These would be entered upon request under five major categories: chemical, physical, biological, construction and operation. The user would be requested to enter whatever data may be available within each of these major domains.

The initial screen advisory would generate a file name for the well and give location and relevant calendar information. Following this, the user would be presented automatically with each of the five major categories as single on-screen editable files. These would be formatted as shown below:

CHEMICAL PARAMETERS

File #:   screen: 1
Please enter all chemical parameters as parts per million (ppm) or milligrams per litre (mg/L). Where there is more than one entry for each parameter, enter each in turn with a comma in between each entry (maximum of five entries).

A1	Iron (total):	A12	Phosphorus (phosphate, soluble):
A2	Iron (soluble):	A13	Phosphorus (total filterable):
A3	Manganese (total):	A14	Phosphorus (total):
A4	Sodium:	A15	Carbon, Total Organic (filtered):
A5	Potassium:	A16	Carbon, Total Organic (not filtered):
A6	Calcium:	A17	Carbon, Dissolved Organic:
A7	Magnesium:	A18	Carbon, Inorganic:
A8	Nitrogen (total):	A19	Total Dissolved Salts:
A9	Nitrogen (nitrate-N):	A20	Oxygen, Dissolved:
A10	Nitrogen (ammonium - N):	A21	Calcium hardness:
A11	Nitrogen (Kjeldahl):		

PHYSICAL PARAMETERS

File #:   Screen: 2
Please enter the data in the standard format to match the criteria given in brackets following each parameter. Up to five data points may be entered per parameter, place a comma between each data entry.

B1	pH:	B4	Conductivity (ohms/cms);
B2	Temperature (°C):	B5	Turbidity (FTUC):
B3	Temperature (°F):	B6	Turbidity (NTU):

BIOLOGICAL PARAMETERS

File #:   Screen: 3
Enter the data in the format given in brackets after each parameter is listed. Note that up to five data points may be entered separated by commas.

5-14

C1	Coliforms (Total, TC/100ml):
C2	Coliforms (Fecal, FC/100ml):
C3	Enterococci (Fecal Streptococci, FS/100ml):
C4	Standard Plate Count (Bacteria, cfu/ml):
C5	Heterotrophic Plate Count (Bacteria, cfu/ml):
C6	Pseudomonads (cfu/ml):
C7	dd Iron Bacteria (IRB-BART™ days of delay):
C8	dd Sulphate Reducing Bacteria (SRB-BART™ days of delay):
C9	dd Slime Forming Bacteria (SLYM-BART™ days of delay):
C10	RM Pattern Iron Bacteria (IRB-BART™ Rx):
C11	Gallionella reported (0-absent, 1-present):
C12	*Sphaerotilus-Leptothrix* reported (0-absent, 1-present):
C13	Bad Egg Odour noted (0-never, 1-occasionally, 2-always):
C14	Slimes noted (0-never, 1-occasionally, 2-always):

CONSTRUCTION DETAILS

File #:   Screen: 4

Please indicate whether the data is going to be in metric or U.S. Imperial measurement prior to answering subsequent questions; the parameter format will be shifted between the two formats depending on the answer to question D1.

D1	Metric or U.S. Imperial (0-metric, 1-U.S. Imperial):
D2	Month and Year of commissioning (month by number followed by a comma and the last two digits of the year):
D3	Total Depth of Well (feet-metres):
D4	Internal Diameter of Well (inches-cms):
D5	Length of Screen (feet-metres):
D6	Width (radius) of Gravel Pack (feet-metres):
D7	Porosity of the Pack Material Ground Well (%):
D8	Porosity of Media in the Aquifer (%):
D9	Screen Material (0-stainless, 1-plastic, 2-fibreglass):
D10	Casing (0-steel, 1-plastic, 2-fibreglass):
D11	Screen Slot Width (1/1,000"-mm):
D12	Pump Type (0-submersible, 1-airlift, 2-other):

MANAGEMENT OPERATION

File #:   Screen: 5

The manner in which a water well is operated will also influence the rate of biofouling and hence the WPRI.

E1	Metric or U.S. Imperial measurements (0-metric, 1-US Imperial):
E2	Specific capacity when developed (USgp-m^3/hr):
E3	Original drawdown (feet-metres):
E4	Current drawdown (feet-metres):
E5	Original pumping rate (USgpm-m^3/hr):
E6	Current pumping rate (USgpm-m^3/hr):
E7	Percentage time pumping (%):
E8	Pumping routine (0-on line continuously, 1-back up role):

E9   Average length of down time (hours):
E10   Year of installation
E11   Year of data report

INTERPRETATION STRATEGY, WPRI

In generating a WPRI, the data base may not contain entries for all of the questions nor will the individual entries for the same questions be identical in all cases. The logical interpretation should begin with a basic status report on the well and its operating performance level using screens 4 and 5 followed by a determination of any "obvious" evidences of plugging. This would be followed by a theoretical evaluation of the plugging risk using data obtained through screens 1, 2 and 3. Each parameter would be subjected to three levels of interpretation during the WPRI prediction risk process:

(a) a simple arithmetic risk assessment based on the size of the parameter;
(b) automatic interactive relationships between the parameter under study and other parameters which may further serve to interpret the likelihood of a plugging event occurring; and
(c) a weighting factor which would be applicable to each individual parameter in the composite evaluation.

Each parameter would therefore be presented in a list which would include consideration of all potentially significant values and their direct and indirect evaluation. These will be presented in a close coded format for each parameter where AN would be the parameter; $conc_l$ the lowest concentration ever associable with plugging and $conc_h$, critical concentration at which plugging is considered to be an inevitable event. For example,

$$C4, 10, 500$$

would indicate that the data base is relevant to the standard plate count (C4) and that the $conc_l$ would be where only 10 or less cfu/ml were recorded while the $conc_h$ would occur where there were 500 or more cfu/ml.

Secondary importance is given to interactive parameters which may have a synergistic effect on the parameter under evaluation with respect to the risk of plugging. These interactive factors are shown as separate entities within brackets following the principal parameter. The relationship between the data for the two parameters (i.e. principal parameter, PP and interactive parameter IP) is given factorially (Fact):

$$Fact. = IP / PP$$

This Fact. now presents a factorial relationship (as $Fact_l$ and $Fact_h$) between the interactive parameter and the principal parameter (i.e. 0.001 to 0.4 for the lowest and highest risk concentrations relatable to plugging respectively). For example, if the principal parameter had a value of 10 and the interactive parameter a value of 100, the Fact. value would be (100)/(10) which would equal 10.

For some of the principal parameters in screen 1 and 2 (physical and chemical), there is an optimal range within which plugging may be possible whereas higher or lower values may not be so supportive. These parameters use a different formatting. Here the optimal range is shown from the lowest to highest value within the range separated by a double dash (--). To the left and right of this range, and separated by commas, are respectively the lower (<) and upper (>) concentration ranges which may actually severely reduce the risk of a microbially induced fouling (MIF) event being involved in any plugging.

Two parameters not used in the calculation of the WPRI are required for the projection of the dominant microbial components likely to be involved in an MIF. These are considered where a WPRI indicates that a risk of plugging does exist.

Parameter weighting by key parameter (a and b)

A1, 0.01, 5.0 (A2, 0.8, 0.1) (A3, 10.0, 0.1) (C7, 1,000, 10,000).
A2, 0.005, 1.0
A3, 0.001, 0.5
A4, -, - (A5, 0.1, 1.0)
A5, -,
A6, -, -(A7, 1.0, 0.1)
A7, -, -.
A8, 0.1, 2.0 (A9, 0.1, 0.8) A10, 0.1, 0.7) (A11, 0.2, 0.8) (A15, 0.001, 0.2) (A16, 0.005, 0.15) (A17, 0.01, 0.2) (A14, 0.01, 0.25) (A12, 0.01, 0.25) (A13, 0.005, 0.15) (C4, 100, 1,000) (C5, 100, 1,000)
A9, 0.1, 20 (A10, 0.1, 5.0) (A11, 0.2, 10.0) (A20, 10, 0.1) (C1, 0.1, 100) (C2, 0.01, 10) (C3, 0.5, 500) (C6, 10, 1,000) (C9, 5, 500) (C13, 0, 2.0).
A10, 0.01, 0.75 (A11, 1.0, 20.0) (C1, 1.0, 100).
A11, 0.05, 1.5 (A15, 200, 4) (A16, 400, 10) (A17, 100, 8) (C1, 0.1, 20) (C4, 10, 1,000).
A12, 0.01, 0.5 (A13, 1.0, 0.05) (A14, 1.0, 0.15) (A15, 0.0001, 0.01)
A13, 0.02, 0.75 (A14, 1, 0.1) (A15, 0.0001, 0.03) (A17, 0.0001, 0.02)
A14, 0.01, 0.75 (A15, 0.0001, 0.02)
A15, 0.05, 2.0 (A16, 0.1, 5) (A17, 0.5, 2) (A20, 2, 0.1)
A16, 0.1, 5 (A17, 1.0, 0.1) (#A20, 2.0, 0.1) (C1, 0.001, 10) (C2, 0.0001, 1.0) (C3, 0.01, 10) (C4, 10, 1,000) (C5, 10, 1,000) (C6, 5, 500) (C7, 5, 50) (C8, 0.01, 1) (C9, 10, 1,000)
A17, 0.05, 0.1
A18<0.5, 10--200, >1,000
A19<5, 50-60,000, >150,000
A20, 10, 1.5 (C1, 0.01, 10) (C2, 0.001, 1) (C3, 0.01, 1) (C4, 0.1, 100) (C5, 0.1, 100) (0.1, 10) (C7, 0.01, 10)
B1<3.5, 6.5--8.8, >10.5
B2<1.5, 8.5--38.5, >75
B3<35, 50--100, >180
B4<1, 100-4,000, >10,000

B5, 0, 0.1
B6, 0, 0.1
C1, 0, 10 (C2, 0.01, 1) sewage (C3, 1, 0.01) manure (C4, 10, 100) (C5, 10, 100) (C6, 10, 1,000)
C2, 0, 10
C3, 0, 100
C4, 10, 500
C5, 10, 500
C6, 5, 1,000
C7, 10, 100
C8, 1, 10
C9, 10, 1,000
C10*1, 2, 7, 3, 4, 6, 8, 5, 9, 10*
C11, 0, 1
C12, 0, 1
C13, 0,2
C14, *0, 3, 2, 1*

Note: where there is a superscripted asterisk (*), data are not quantitative but qualitative. The data reflect a specific reaction pattern commonly seen from particular groups of microorganisms where reacting in the biodetector.

These factors (A1 to C13) are the primary factors to be used in the calculation of the WPRI. As each factor or pair of interacting factors (i.e. PP and IP) would not necessarily have the same level of importance in the computation of the WPRI, a series of weighting factors (WF) have to be introduced. Such WF values would bias the significance of the data relative to the other attained data.

CALCULATION OF THE WPRI

All the principal and interactive parameters considered significant are listed above. In this simple format, the initial AN (letter, number) code gives the principal or interactive parameter which is then normally followed by two numbers separated by a comma. The first number represents a condition where there is generally considered to be a zero plugging risk (i.e. zero value) for that parameter. The second number (after the comma) represents a condition under which plugging would be most likely to occur (i.e. value equalling one). Data distributed between these two values (i.e. zero and one) would be considered as posing a linear factorial risk where the risk is arithmetically relatable to the position of the data between the zero and certain risk of plugging. For example, if the theoretical parameters were to be established at A24, 1, 11 a rare zero risk would exist with a reading of 1.0 and a maximal risk would be established at a reading of 11. If a data value of 6.0 was obtained then the risk factor (RF) could be calculated as being 0.5 using a linear factorial relationship. The equation would be:

Risk Factor risk (RF) = $(D_x - D_l)/(D_h - D_l)$ [1]

where D_x is the data obtained, D_l is the zero risk reading and D_h is the maximal risk reading. RF values computed as less than 0.0 are considered to be zero and factorial risks of > 1.0 are considered to be 1.0. RF values therefore are initially computed between values of zero (0.0) and one (1.0) only.

Each RF computed for the various principal and interactive factors are not considered as being of equal merit. For example, factor A15 (total organic carbon) is more likely to generate plugging than A4 (sodium). It is proposed therefore to provide a weighting modifier (WM). This would be achieved by selecting a multiplier value from x0.5 to x10 depending upon the particular values for each of the factors generatable. The factors used in the evaluation of the WPRI are listed below. Each item has a parameter code followed by a comma and then the WM value.

A1,x10	A1A2,x3	A1A3,x2	A1C7,x4
A2,x5	A3,x7	A4A5,x1	A6A7,x2
A8,x8	A8A9,x5	A8A10,x2	A8A11,x4
A8A2,x5	A8A13,x2	A8A14,x7	A8A15,x1
A8A16,x3	A8A17,x1	A8C4,x5	A8C5,x5
A9,x4	A9A10,x1	A9A11,x3	A9A20,x4
A9C1,x1	A9C2,x0.5	A9C3,x2	A9C6,x6
A9C9,x6	A9C13,x2	A10,x2	A10A11,x0.5
A10C1,x2	A11,x4	A11A15,x1	A11A16,x2
A11A17,x1	A11C1,x1	A11C4,x3	A12,x4
A12A13,x2	A12A14,x5	A12A15,x4	A13,x6
A13A14,x3	A13A17,x4	A14,x8	A14A15,x4
A15,x6	A15A16,x8	A15A17,x5	A15A20,x5
A16,x5	A16C3,x4	A16C5,x4	A16C4,x4
A16C2,x2	A16C3,x4	A16C5,x4	A16C5,x4
A16C6,x6	A16C7,x5	A16C8,x5	A16C9,x6
A17,x10	A18,x2	A19,x1	A20,x3
A20C1,x1	A20C2,x1	A20C3,x2	A20C4,x2
A20C5,x2	A20C6,x4	A20C7,x4	B1,x1
B2,x1	B3,x1	B4,x1	B5,x0.5
B6,x0.5	C1,x6	C1C2,x6	C1C3,x7
C3,x6	C4,x4	C5,x4	C6,x8
C7,x8	C8,x6	C9,x10	C10,x9
C11,x8	C12,x8	C13,x10	C14,x10

Each evaluation would be unique including a particular combination of the above key and interactive parameters. The validity of the final WPRI can be calculated using the weighted factors (where data are available). The sequence for calculating the WPRI would involve a number of stages.

Firstly, each principal and relevant interactive parameter would have the RF calculated (equation 1). Where the parameter involved an optimal range (--) and a less than (<) and greater than (>) level, the optimal range would be considered to equal 1.0. In the ranges outside the optimate range but not outside the extreme levels (i.e. < and >) then the RF would be calculated using the one of the following formulae:

$$\text{Modified } RF_I = (D_x - D_<) / (D_{mno} - D_<) \quad \text{or} \quad [2]$$

$$\text{Modified } RF_h = (D_> - D_x) / (D_> - D_{mxo}) \quad [3]$$

where RFl and RFh are the RF factors for data (Dx) values outside of the optimal range (D_{mno}--D_{mxo}) at the low (<) and high (>) data ranges respectively.

Once the RF values have been calculated, they are modified to allow for the relative weighting of the data. This is done using the appropriate WM value for the principal or principal - interactive parameter. The modified RF (MRF) is calculated using the following formula:

$$MRF = RF * WM \quad [4]$$

A series of MRF values are created equivalent to the number of Dx entered into the program for screens 1, 2 and 3. The WPRI may be calculated as the relationship between the sum of the MRF values obtained and the maximum possible value where all of the RF values had equalled 1.0 (i.e. maximal plugging risk) and each MRF calculated equalled the WM in value.

$$MRF_{max} - 1.0 \times WM \quad [5]$$

where each RF value computed equals 1.0.

The WPRI is computed in a series of steps in which each possible RF is calculated (using either equations 1, 2 or 3, whichever is appropriate to the Dx value and the parameter/s). Once all of the possible RF values have been calculated, the next stage is to calculate both the MRF (equation 4) and the MRF_{max} (equation 5). It is now necessary to generate the sum for all of the MRF values (SMRF) and the sum for all of the MRF_{max} values (XMRF):

$$SMRF = Sum (MRF_x .. MRF_z) \quad [6]$$

$$XMRF = Sum (MRF_{max.x} .. MRF_{max.z}) \quad [7]$$

where x to z represents each MRF value in the calculable range. Once these values are calculated then the WPRI may be obtained as a factorial relationship of the SMRF to the XMRF:

$$WPRI = SMRF / XMRF \quad [8]$$

The WPRI calculated would have a value between 0.0 (low plugging risk) and 1.0 (very high plugging risk). This value, however, may have been obtained from a vast number or simply a single data point. To generate a confidence in the WPRI, two approaches are used. Firstly, the number of principal parameters utilised (NNP_u) is compared to the total number of principal parameters (NPP_t) available (maximum, 38). Secondly, the total number of replicate data (D_x) points for all of the principal parameters utilised (Sum D_x) in relation to the maximum possible number (i.e., $NNP_u \times 5$). Confidence (C_{wpri}) is based equally upon the C_{app} (see equation 9) and the C_{dxt} (see equation 10).

$$C_{app} = NPP_u / NPP_t \qquad [9]$$

$$C_{dxt} = \text{Sum } D_x / (NPP_u \times 5) \qquad [10]$$

$$C_{wpri} = (C_{app} + C_{dxt}) \times 100 \qquad [11]$$

In calculating the C_{wpri} (equation 11) a percentile confidence value is obtained. Here the critical range from a low level of confidence to a high level is over the C_{wpri} values from 30 to 80%. Values of greater than 80% C_{wpri} would be considered as giving a high level of confidence in the WPRI calculated. Where C_{wpri} values are less than 30% then relatively little confidence may be placed in the calculated WPRI.

WATER WELL PRODUCTIVITY (WWP)

This may be presented as a simple interpretation of screen 5 statistics starting with the pump rate shifts:

original capacity	= E5
present pump rate	= E6
original drawdown	= E3
present drawdown	= E4

From this the shift towards increased drawdown and loss in the rate of production can be calculated, both of which may be the result of an MIF plugging event. It could be that the management practice is to control one of these parameters and allow the other to vary. Consequently either the loss in pumping rate and/or any increase in drawdown may be considered as evidence that some form of plugging event may be occurring. The calculation of the loss in pumping rates over time would be displayed as the percentage of productivity lost which could be attributable to a plugging event. This would be calculated as:

$$L = \% \text{ Loss in pumping} = ((E5-E6)/(E5)) \times 100$$

Increases in drawdown resulting from impairment of the well for any reason may be calculated as the percentage increase in the drawdown which has occurred based on the formula:

$$D = \% \text{ Increase in Drawdown} = ((E4 - E3)/(E3)) \times 100$$

Plugging may be suspected where the D or L values exceed 10%. The following scales relate the L value and the risk:

L VALUE	RISK
10 to 25	Plugging may be occurring
26 to 40	Plugging impeding well, control
41 to 60	Severe plugging, use control strategies
>60	Serious risk of failure due to plugging

For the D value, the percentage increase in drawdown is not set within finite bounds and therefore is not so easily interpreted as the L value. Verbal interpretation should therefore be limited to a recording of the D value and where it exceeds 10%, the statement that "plugging appears to be a potential cause of the increased drawdown".

The rate of plugging may be calculated assuming a linear loss in flow since the well was installed (E10) to the year of the data entry (E11). The percentage loss in pumped flow may be estimated on a per annum basis as:

$$\text{Annual \% pumped flow loss} = (L)/(E11-E10)$$

The management of the well itself will clearly affect the rate at which plugging can occur. The following scenario can be used to evaluate the practices as given on screen 5.

Ideally, a well being pumped continuously is likely to biofoul and plug less quickly than a well in discontinuous use or "sidelined" into a backup role. Optimized performance (OP) may be calculable using E7, E8 and E9: where

$$A = 100 - E7$$
$$B = E8 \times 100$$
$$C = E9 \times 4$$
$$OP = (100 - ((A + B + C)/3))$$

Here, an OP of 100 would be a continuously operating well with no down time. An OP of 0 would be a well which was not being pumped at all. Improvements in the OP can be achieved by the following action:

1. where E7 is < 70, increase to > 80
2. where E8 is 1 bring well back on line
3. where E9 > 8 reduce by at least 50%.

FORM OF THE BIOFOULING

In control management strategies there should be some reference to the nature of the MIF causing the plugging projected by the WPRI. This may be based primarily on the screen 3 relationships. Five major MIF scenarios can be addressed:

BFi	=	Iron related bacterial biofouling
BFp	=	Pseudomonad dominated biofouling
BFa	=	Anaerobic (corrosive) biofouling
BFs	=	Sewerage pollution
BFm	=	Animal manure pollution
BFza	=	Hazardous waste site (aerobic)
BFzn	=	Hazardous waste site (anaerobic)

5-14

The differentiation of each form of biofouling would be by the evaluation of the entries.

BFi

    C7 > 0.8 x C9
    C7 > 2.0 x C8
    C4 or C5 > 100
    C10 - 1, 2, 3, 4, 6 or 7
    C14 - 1 or 2
    C11 or C12 positive would be confirmatory where A15 or 16 < 2.0 and A1 or A2 is > 0.5, while A20 is > 0.5.

Where the first four parameters (or those available) support the BFi but C11, C12, A15 and A16 are not confirmatory, the iron related bacteria present may be considered to be heterotrophic iron related bacteria (hirb) and possibly related to a BFp event: the BFp parameters should be checked for confirmation. If not confirmed as a BFp event, the dominant biofouling organisms are hirb-types. Where confirmation is obtained, the biofouling may be considered to be caused by iron related bacteria.

BFp

Pseudomonads tend to be dominant in waters where the following conditions apply:

    A15 or A16 > 5
    A8 > 0.2
    A14 > 0.01
    B2 < 25 or B3 < 75
    A1 or A2 < 1

Confirmation of the dominance of pseudomonads in the MIF event would involve the following parameters:

    C10 - 8, 9 or 10
    C6 > 0.5 x C5 or C4
    C7 < 0.5 x C9
    C14 - 2 or 3

Biofouling is often associated with pollution events involving high levels of specific organics (e.g. gasoline).

BFs

The occurrence of human domestic sewerage contaminating a water well and stimulating a biofouling event may involve the following parameters:

    C2 > 1.0
    C1 > 10
    C1/C3 > 1.0
    C8 < 6.0

DEVELOPMENT OF PROGRAM TO PROJECT THE WELL PLUGGING RISK INDEX

      C10 either 5 or 10
      C13 - 1 or 2
      C4 or C5 > 100
      C6/C4 < 0.3
      C6/C5 < 0.3
      C11 and C12 - 0.00
      C14 - 2 or 3
      A9 > 1.0
      A12 > 0.2
      A14 > 0.5

It should be noted that the R_x value for a SLYM-BART™ would commonly be 6 while a C10 would frequently be 5 as the R_x1 reaction. Where these parameters are confirmed, this would support the hypothesis that the MIF event would include human sewerage contamination.

BFa

A predominantly anaerobic biofouling would include some unique features which differentiate this form of MIF. These include:

      A20 < 1.5
      C14 - 2 or 3
      C8 < 7
      C10 - 5 (frequent initial) or 10
      C13 - 2 or 3
      C6 < (C4 or C5) x 0.1
      C9 - frequent R_x, 6
      C1 > 10
      C8 > C9 x 0.5

Where these are confirmatory, an anaerobic form of biofouling may be suspected.

BFm

Where the water well is being influenced by discharges from animal husbandry operations, the parameters will be somewhat different to those for the BFs, notably:
      C1/C3 < 1.0
      C2 < C3 x 0.3

Other parameters are the same as for BFs.

BFza

In waters from a hazardous waste site which has been under aerobic or nitrate supplemented management, aerobic bacterial activity will be very aggressive, hence the following events are likely:

C9 < 4.0 (R_x commonly 5 or 6)
C10 - 8, 9, 10 or 5
FLOR-BART™ - dd < 4.0

BFzn

Waters within a hazardous waste site may also be heavily populated with anaerobic bacteria and the water samples are likely to prove:

C7 < 5.0
C8 < 8.0 (R_x may shift rapidly to 3)
C10 - initial R_x commonly 5

It should be noted that where a BFz is being monitored, some of the components may affect the stability of the BART™ construction materials (e.g. various plastics). It is recommended that a trial test be conducted first to determine the durability of the biodetector under the conditions generated by the water sample.

INTERPRETATION OF THE BF PROJECTIONS

In most cases, not all of the factors will have been addressed in the initial data loading. Confidence in the prediction of a BF will have a direct percentile relationship to the degree of information available. An indication of the percentage confidence with which this prediction is made should always be given. In addition, only the highest probable BF will be indicated unless there are two calculated BF values with a similar (± 5%) value. The confidence of the interpretation will be based on the following formula:

$$\% \text{ confidence} = \frac{\# \text{ of parameters supporting hypothesis}}{\text{total number of parameters listed}} \times 100$$

SUMMARY

A method has been described which would allow the determination of the risk of a MIF driven plugging event occurring in a water well and the degree of confidence which may be placed on the prediction. Where an MIF plugging event is already occurring, the data base routinely collected during the operation of the water well may be examined to determine the current productivity and future deterioration that may be expected. Additionally, where screen 3 parameters have been monitored, it becomes possible to project the nature of microbial consortia driving the fouling. This can be either aerobic (dominated by pseudomonad bacteria), iron rich with radical plugging (formed by iron related bacteria), anaerobic with a high risk of corrosivity (incorporating sulphate reducing bacteria and iron related bacteria), human sewerage pollution (including a high residual population of enteric bacteria with *Escherichia coli* present), or animal wastes (including a predominance of anaerobic bacteria and the enterococcal bacteria).

The stages described form the basis for the development of a management strategy for the water in question. This may involve:

(1)   restriction or relocation of the microbial activity to reduce the impact level;

(2)   restoration of productivity in a recognized MIF plugged water well through rehabilitation practices (such as "shock" chlorination, acidization and blended chemical heat treatment); and

(3)   future management designed to contain the MIF event at an economical cost with a satisfactory product quality and delivery capability.

ACKNOWLEDGEMENTS

The authors wish to acknowledge the National Research Council of Canada Natural Sciences and Engineering Research Council for a grant-in-aid (DRC, # GP0005073) and also Karim Nagvi (aspects of computer software development), Dr Abimbola T. Abiola, Neil Mansuy and Jeff Reihl for conducting field experiments during various phases in the development of this project, and Natalie Ostryzniuk for the preparation of the manuscript.

REFERENCES

CULLIMORE, D.R. (1987). Physico-chemical factors influencing the biofouling of groundwater. *Proceedings of the International Symposium on Biofouled Aquifers: Prevention and Restoration* (Cullimore, D.R. (ed.)) Publ. Am. Water Resources Assoc., Techn. Publ. TPS-87-1, 23-26.

CULLIMORE, D.R. (1990). Microbes in civil engineering: biofilms and biofouling. *FEMS Symposium No. 59 Microbiology in Civil Engineering* (Howsam, P., (ed.)) Publ. E. & F.N. Spon, London, 15-23.

CULLIMORE, D.R. (1992). *Practical manual for groundwater microbiology.* Publ. Lewis Publishing, Michigan, U.S.A. in press.

FLIERMANS, C.B. & HAZEN, T.C. (eds.) (1990). *Proceedings of the First International Symposium on Microbiology of the Deep Subsurface.* Publ. Westinghouse Savannah River Company, Aitken, SC, U.S.A., 711 pp.

GEHRELS, J. & ALFORD, G. (1990). Application of physio-chemical treatment techniques to a severely biofouled community well in Ontario, Canada. In: *Water Wells: Monitoring, Maintenance, Rehabilitation*, Howsam, P. (ed) Publ. E. & F.N. Spon, London, 219-235.

SMITH, S.A. (1992). Monitoring methods for iron/manganese biofouling. *Publ. Amer. Water Works Assoc. Res. Found.*, Denver, Co, U.S.A., 149 pp (in press).

5-15 CHEMICAL AND MICROBIOLOGICAL EFFECTS ON CONSTRUCTION DEWATERING SYSTEMS

W. POWRIE
Queen Mary and Westfield College, London
T.O.L. ROBERTS
WJ Engineering Resources Ltd, Watford
S.A. JEFFERIS
Golder Associates, Maidenhead

ABSTRACT

This paper describes some of the ways in which chemical and microbiological phenomena can affect the operation and even the viability of construction dewatering systems. The problems addressed are clogging due to biofouling; microbially induced corrosion; calcium carbonate precipitation and the disposal of groundwater abstracted from contaminated sites. The circumstances in which these problems may arise are discussed in relation to actual case histories. The implications for construction dewatering systems are considered and some conclusions are drawn concerning the situations in which these problems are most likely to occur.

INTRODUCTION

A construction dewatering system consists essentially of an array of pumped wells. Such systems are used to lower temporarily the groundwater level in the vicinity of an excavation below the natural water table, in order that the sides and base of the excavation remain dry and stable. The method of groundwater extraction depends on the required well depth and capacity, as indicated in Table 1.

In wellpoint systems, the well risers are joined to a common vacuum main at ground level with up to about 100 wellpoints operated by one suction pump. As the extraction of groundwater from the wellpoints depends entirely on the vacuum generated in the header main, the operational depth of these systems is limited to about 6 m. With submersible pumps ("deepwell") and ejector systems, a pumping device is installed in each individual well. An ejector is a water-driven jet pump which will pump both air and water. Ejectors are used primarily in fine soils where flow rates are low and a vacuum may be generated within the well provided that the well is sealed. The characteristics of all three systems are described in more detail by Powers (1981).

Construction dewatering systems are designed primarily on the basis of the permeability of the soil and the proposed excavation geometry. Chemical and microbiological effects can also be important and may even affect the viability of a system, particularly where it is required to remain in operation for months or a year rather than weeks or days. This paper is concerned with some of the ways in which chemical and microbiological effects can influence the operation and performance of construction dewatering systems. It is not intended to be an exhaustive catalogue of problems.

Table I: Typical operating ranges for three common forms of construction dewatering system.

	Well Depth (m)	Wellscreen diameter (mm)	Well Yield litre/sec	Soil Permeability m/sec
Wellpoints	<6	25-50	<1	10^{-3} to 10^{-5}
Submersible pumps	10-40*	150-250	1-20*	10^{-3} to 10^{-5}
Ejectors	<35	50-75	<0.5	10^{-5} to 10^{-7}

* These figures represent typical ranges in practice, rather than performance limits which will depend on the power and size of the pumps installed.

CLOGGING DUE TO BIOFOULING

Biofouling is the term applied to the growth of a film or mass of live bacteria, dead cells and inorganic detritus, which can clog the wellscreen, filter pack, pump and pipework leading to a reduction in both the capacity and effectiveness of a dewatering system. The mechanism and processes of biofouling are described in detail by Howsam (1988). Essentially the bacteria are sessile; they need a supply of oxygen, iron and other nutrients in order to survive and often thrive at an aerobic/anaerobic interface.

In groundwater engineering, the most common form of biofouling is associated with the iron-related bacteria *Gallionella*, although other forms of bacteria may be present in consortium (Howsam, 1988). The susceptibility to *Gallionella* biofouling of the various forms of dewatering system outlined above was investigated by Powrie, Roberts and Jefferis (1990) on the basis of a number of case histories. Table II gives tentative trigger levels for *Gallionella* biofouling according to the type of dewatering system, the concentration of iron in the groundwater and the pumped flow rate. All of these factors were found to have a significant influence on the extent and rate of development of the biofilm.

In the case studies reported previously, the biofouling did not appear to affect the wellscreen and filter pack significantly. The deterioration in the performance of the dewatering system (which was severe in at least one case) was due to an accumulation of biomass on the pumps and pipework. On cleaning the affected components, the original yield of the well was re-established.

Table II: Tentative trigger levels for susceptibility to *Gallionella* biofouling

a) Submersible pumps

Iron concentration mg/litre	Frequency of cleaning
<5	6 to 12 months
5-10	0.5 to 1 month
>10	weekly

b) Ejectors

Iron concentration mg/litre		Frequency of cleaning
High flow (>20 litres/min/ejector)	Low flow (<10 litres/min/ejector)	
<2	<5	6 to 12 months
2-5	5-10	monthly
5-10	10-15	weekly

More severe biofouling was encountered during a recent project which involved an array of nine pumped deepwells of 28m depth, installed in a ring at a spacing of approximately 25m. The wells were screened through the superficial gravel deposit and also through the deeper Bunter Sandstone aquifer, which was confined by a 5m thick layer of boulder clay. Pumping tests carried out at the site had indicated only a very limited connection between the two more permeable strata prior to the installation of the dewatering system.

After the dewatering system had been in operation for about four months, it became apparent that performance was deteriorating due to biofouling. The biomass was clearly iron-related and laboratory analysis confirmed that *Gallionella* was the main species of bacteria present. A video camera survey of the inside of the well indicated that the biomass was growing on the wellscreen, the pump and the riser pipe. Average well yields fell by 20 to 30% over a two to three month period. Pump removal/cleaning and well development by airlift restored yields to their original values.

Unfortunately it was not possible to obtain separate samples of groundwater from the two aquifers after biofouling had been identified as a problem. However previous tests had indicated that the groundwater in the Bunter Sandstone was slightly brackish, with an iron content of about 5 ppm. No data on the chemistry of the groundwater from the unconfined gravel aquifer were available. It was apparent during drilling that there either was or had been some leakage into the upper aquifer from the adjacent sewage treatment works.

Not all of the wells exhibited the same level of biofouling. Seven of the nine wells showed moderately severe fouling within two months of cleaning, with biofilm thicknesses of 15 to 25mm inside and outside the pump and well riser. In most of these wells the iron deposit was very soft and easily removed. However in three of the seven wells the deposit was hard and more difficult to remove. This suggests that it may have been chemical rather than biological in nature. The eighth well suffered exceptionally severe fouling, with the 76mm diameter riser pipe becoming totally blocked after three months. Remarkably, in the ninth well there was no visual evidence of biofouling at all.

In addition to the nine pumped wells, two further wells which were not pumped showed signs of extensive fouling. This tends to confirm the view that the fouling was largely triggered by the mixing within the well of groundwater from the upper and lower aquifers. Presumably the water from the upper aquifer provided the bacteria with oxygen and nutrients while the water from the lower aquifer was the source of iron. It is probable that the biomass observed inside the well using the video camera had developed primarily on the inside of the wellscreen where the two groundwaters mixed. The ease with which the well yields were restored suggests that significant biofouling of the external filter pack adjacent to each aquifer had not occurred. The observed variations in the extent of the fouling in individual wells may have been the result of local contamination of the groundwater in the upper aquifer by leakage from the sewage treatment works.

Although van Beek (1989) describes a case of biofouling of wellpoint filters which occurred as a result of the mixing of very shallow oxygen-containing groundwater with deeper iron-containing groundwater, it seems that biofouling of wellpoint systems is comparatively rare. However, iron biofouling has been encountered in a more recent wellpoint dewatering scheme in silty estuarine sand of permeability 5×10^{-5} m/sec. The system comprised about 100 wellpoints, giving a drawdown of approximately 4.5m below the initial water table 1.3m below ground level.

The initial flow rate was high (about 30-40 litres/sec), but after a few days the system was trimmed back and about half of the wellpoints were switched off to prevent their drawing air. The total extraction flow rate was steady at approximately 5 litres/sec for the remaining twenty-two weeks of operation. On dismantling the wellpoint system it was found that the plastic wellpoints and the flexible connectors to the vacuum main (swings) were almost blocked by an accumulation of biomass. Slight biofouling was also apparent in the steel header main. No deterioration in drawdown was apparent, but this was probably because the flow rate was only 0.05 to 0.1 litre/sec per wellpoint, or approximately 5-10% of the normal maximum capacity. It may have been significant that the wellpoint system was pumping brackish water which could also have been a contributory factor in the two cases of biofouling of submersible pumps reported previously as well as the sewage treatment works described above.

In a second case of biofouling of a wellpoint system, the biomass was present in the plastic header main and flexible swing connectors (which were also made of plastic) while the galvanized steel wellpoints remained comparatively clean. This may occur because the biofilm adheres more readily to plastic but it is perhaps more likely that the zinc galvanizing is toxic to the bacteria forming the biofilm. Experience with over 100 wellpoint systems suggests that biofouling of any significance is comparatively rare: although this could be due to the use of galvanized steel rather than plastic pipework in many of these cases.

MICROBIALLY INDUCED CORROSION

Two mechanisms of corrosion associated with the presence of aerobic bacteria such as *Gallionella* are described by Stott (1988). The first involves the establishment of *Gallionella* and sulphate reducing bacteria in consortium. The *Gallionella* can create an environment which is sufficiently anaerobic for the sulphate reducing bacteria to grow and can use iron sulphides (a product of sulphate reduction) as a source of reduced iron. The second mechanism is the localized pitting of stainless steel by chloride ions. Chloride ions can penetrate the passive oxide layer which protects stainless steels from corrosion. Under normal conditions repassivation takes place by the formation of fresh oxide using dissolved oxygen. However beneath a biofilm of *Gallionella* conditions are anaerobic and repassivation cannot occur. Anodic corrosion sites are established with continued corrosion leading quite rapidly to the formation of flask-shaped cavities in the steel.

Submersible pumps, which are generally manufactured from stainless steel and submerged in the borehole during operation, can be susceptible to these forms of attack. Out of more than thirty deepwell projects, chloride corrosion of stainless steel submersible pumps underneath a *Gallionella* biofilm was identified in only one. Although the corrosion did not result in an in-service failure, all of the pumps in an eight-well dewatering system were found to be badly pitted after twelve months in operation. In this case, the dewatering system was adjacent to an estuary, pumping brackish water with a significant chloride content.

It seems possible that submersible pumps could be vulnerable in conditions where a consortium of *Gallionella* and sulphate reducing bacteria could flourish. For this reason, groundwater containing significant concentrations of both iron and sulphates should probably be treated with caution.

CALCIUM CARBONATE PRECIPITATION

The precipitation of calcium carbonate is a well-known phenomenon in chalk groundwaters (Driscoll, 1986). The problem arises because calcium bicarbonate (which may be produced by the action of carbonic acid in rainwater on chalk) is soluble but unstable except in the presence of excess carbon dioxide. If the excess carbon dioxide is removed (for example as a result of degassing following a pressure drop as groundwater enters a well), then the following reaction occurs:

$$Ca(HCO_3)_2 \rightarrow CaCO_3 + H_2O + CO_2 \uparrow \qquad [1]$$

Calcium carbonate ($CaCO_3$) is relatively insoluble and may precipitate as a white deposit. The situation is further complicated by the fact that calcium carbonate can remain in supersaturated solution without precipitation.

However, precipitation of calcium carbonate from a supersaturated solution can be triggered by an event which causes nucleation.

The calcium/carbon dioxide system is very complex, but some insight can be obtained using geochemical modelling programs such as PHREEQE (Parkhurst, Thorstenson and Plummer, 1980). These programs can give an indication of whether a system is undersaturated, saturated or supersaturated with respect to calcium carbonate, but cannot be used to predict rates of precipitation.

One instance of calcium carbonate precipitation has been observed in a wellpoint system which was being used to dewater a chemically contaminated site. In one region of the site a spillage of acid had occurred. The acid reacted with the calcium carbonate which was present as a minor constituent of the soil to produce a solution of calcium bicarbonate. Deposition of calcium carbonate occurred in the header main while the wellpoints drawing the contaminated groundwater and their swing connectors remained unaffected. Analysis using PHREEQE confirmed that deposition could have been the result of the removal of carbon dioxide from solution due to the vacuum in the header main. However if this were the sole cause, deposition would have been expected in the swing connector as well as in the header main, although possibly not in the well riser where the hydrostatic pressure of the water column would inhibit degassing. That no deposition occurred in the swing may have been due to the failure of the calcium carbonate to nucleate on the wall of the flexible plastic connector. Equally, the deposition in the header main could have been exacerbated by the mixing therein of water extracted from different parts of the site. The addition of a small amount of alkali can lead to precipitation by equation 1 due to disturbance of the carbon dioxide balance.

The contamination of groundwater with alkali can occur as a result of the careless placement of concrete. If wet concrete is allowed to fall through water - for example, through failure to use a tremie pipe while backfilling redundant wells or boreholes - the alkali from the cement may be dissolved and dispersed into the groundwater, causing a substantial rise in pH and the potential nucleation of calcium carbonate. This effect has been observed to devastating effect in one deepwell system in chalk. Under normal circumstances, however, calcium carbonate clogging of dewatering systems in chalk does not appear to be a problem (e.g. Roberts and Preene, 1989).

TREATMENT AND DISPOSAL OF CONTAMINATED GROUNDWATER

The continuing pressures on building land in the U.K. have led to a substantial increase in the redevelopment of second-hand sites, some of which may be contaminated as a result of their previous use. If construction dewatering is required in connection with the redevelopment of a site, it is possible that the extracted groundwater may be contaminated to some degree and special steps must be taken to ensure that it is disposed of safely. Several options are available.

1. Discharge to lake, stream or river, for which consent from the National Rivers Authority (NRA) would be required. Such consent would be subject to stringent controls on water quality. Treatment and monitoring of the groundwater prior to discharge will almost certainly be essential in all but the most mild and harmless cases of contamination.

2. Recharge of the groundwater on site, for which NRA and/or water authority consent would be required. Such consent is unlikely to be forthcoming if there is any risk of spreading or otherwise exacerbating the extent of groundwater contamination.

3. Discharge to foul sewer. A consent from the water authority would be required, which would include flow and quality limits. The precise limits would depend on the nature of the contamination and the spare capacity at the local sewage treatment works. The water authority normally charge for disposal on the basis of both quality and volume. These charges can represent a very significant proportion of the total operating costs of a dewatering system.

4. Transport the extracted groundwater away from the site using tankers to a licensed disposal facility. Unless flows are very low indeed, the cost of this would generally be prohibitive. Furthermore, the number of disposal facilities which can accept liquid waste is quite small.

The most appropriate procedure will be dictated by discharge quality and flow rate and the availability of surface water/foul sewer discharge points. In any event, the cost of on-site treatment or disposal may well be significant. To some extent, these costs may be limited by careful consideration of the design of the dewatering system. Flow rates may be controlled by dewatering only certain sections of the site at any one time. Water at depth may be less contaminated than water nearer the surface, in which case wells with deeper screens can improve the quality of the discharge substantially. Low density immiscible contaminants can be separated more easily if they have not been emulsified by passing the extracted groundwater through a centrifugal pump. Positive displacement piston pumps may be used as an alternative.

CONCLUDING REMARKS

Dewatering systems, particularly deepwells, are often considered to be similar to water supply wells. Chemical and microbiological effects in water supply wells are fairly well documented (e.g. Howsam, 1990). However this paper has shown that with dewatering systems, the chemical and microbiological aspects differ in a number of respects.

Water supply wells are very deep with comparatively shallow drawdowns and microbiological effects will often build up near the water surface where aerobic conditions prevail. Dewatering wells are almost always over-pumped, with the water in the well drawn down to the pump intake so that the biomass develops around the pump and inside the riser pipe. Dewatering wells may draw water from more than one aquifer, in which case the mixture of aerated water from near the ground surface with water from a deeper aquifer may generate favourable conditions for the growth of a biofilm. Aerated water from near the surface is often excluded from water supply wells, which draw water from a deeper level.

While the tentative trigger levels shown in Table II for biofouling of submersible pump and ejector systems do not seem unreasonable, there is some evidence that the susceptibility of dewatering systems to biofouling may also be affected by other factors. In particular, it seems that galvanized screens and pipework may be more resistant to biofouling than plastic, possibly due to the toxicity of the zinc. It may also be that biofouling is more likely to occur in a dewatering system which pumps brackish rather than fresh water. These matters are currently the subject of continuing investigation and research.

Carbonate clogging of water supply wells does occur in certain circumstances. Dewatering systems can be very severely affected by carbonate clogging when groundwaters of different chemical composition are allowed to mix or when contamination with a strong alkali such as wet cement occurs.

The environmental and economic problems associated with the extraction of contaminated groundwater which requires treatment prior to disposal are becoming increasingly important. The level of sophistication required in the site investigation for and the design, installation and operation of dewatering systems at contaminated sites seems set to increase as the true costs of the treatment and disposal of the effluent groundwater become apparent to the construction industry.

REFERENCES

DRISCOLL, F.G. (1986). *Groundwater and Wells*. 2nd edition, Johnson Division: Minnesota.
HOWSAM, P. (1988). Biofouling in wells and aquifers. *Journal of the Institution of Water and Environmental Management*, 2, 209-215.
HOWSAM, P. (ed.) (1990). *Water wells: monitoring, maintenance, rehabilitation*. E and F.N. Spon, London.
PARKHURST, D.L., THORSTENSON, D.G. & PLUMMER, L.N. (1980). *PHREEQE: a program for geochemical modelling*. U.S. Geological Survey, Water Resources Division.
POWERS, J.P. (1981). *Construction dewatering: a guide to theory and practice*. Wiley, New York.
POWRIE, W., ROBERTS, T.O.L. & JEFFERIS, S.A. (1990). Biofouling of site dewatering systems. *Proceedings of the Conference on Microbiology in Civil Engineering, Cranfield Institute of Technology*, 341-352.
ROBERTS, T.O.L. & PREENE, M. (1989). Case studies of construction dewatering in chalk. *Proceedings of the International Chalk Symposium, Brighton*.

STOTT, J.F.D. (1988). Assessment and control of microbially-induced corrosion. *Journal of the Institute of Metals*, **4(4)**, 224-229.

VAN BEEK, C.G.E. (1989). Rehabilitation of clogged discharge wells in The Netherlands. *Quarterly Journal of Engineering Geology*, **22**, 75-80.

5-16 CHEMICAL EROSION IN LATERITIC SOIL: A CASE STUDY

M.R. DE RUIJTER
Fugro-McClelland Engineers B.V., Veurse Achterweg 10, Leidschendam, The Netherlands.

ABSTRACT

At a Nigerian Oil Refinery, soil contamination with caustic soda has resulted in considerable foundation problems. A failure mechanism is discussed. It is concluded that the main cause of the apparent soil instability is increased dispersive behaviour, but climate, foundation and discharge system design and low groundwater levels may have exacerbated the problem.

INTRODUCTION

In this case study, a situation is presented where a shift in the chemical balance of the soil due to soil contamination has resulted in considerable problems. The aim of this paper is to increase awareness of the possible effects of chemical changes on the mechanical properties of soils and to provide those involved in design and installation of structures with the information necessary to prevent similar problems.

Fugro-McClelland Engineers B.V. (FME) was asked to identify the principal factors leading to the apparent disappearance of a large volume of soil causing collapse of a shallow foundation at the processing area of a petrochemical plant. The collapse revealed a subsoil cavity of over 50 m^3 which appeared to originate at a leaking joint in the underground discharge system (Figure 1). When a small tunnel leading from the cavity was excavated, a second cavity was found and several other tunnels which seemed to be leading further.

The following approach was adopted for the investigation:

1. Assessment of in-situ soil conditions. Special attention was given to the possible occurrence of discontinuities such as tunnels, animal burrows or tree roots.

2. Delineation of the affected area.

3. Assessment of the changes that had occurred to the soil causing the problems.

GEOLOGICAL, GEOMORPHOLOGICAL AND CLIMATOLOGICAL CONDITIONS

The plant is located in Nigeria, close to the discharge mouth of a large river in an essentially flat topography. The general geology comprises

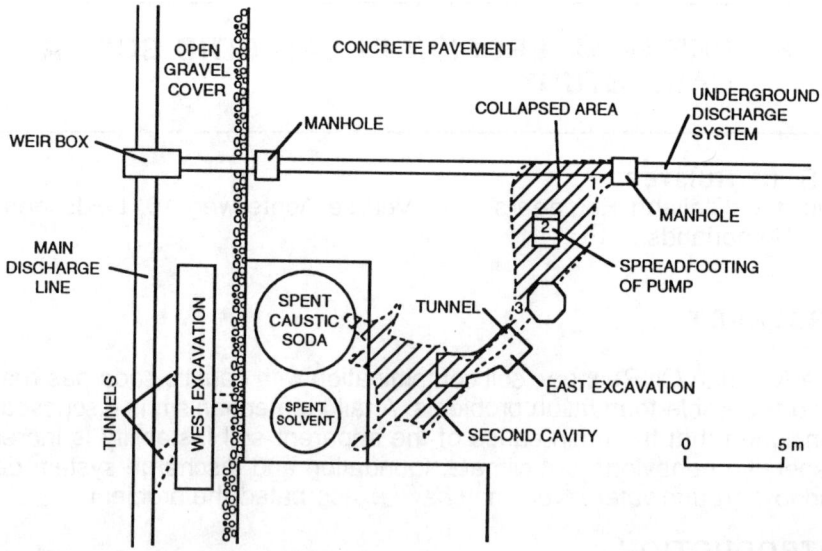

Figure 1: Detail of process area showing extent of cavity formation and tunnel development. Sample locations are indicated: 1 - upstream, 2 - midstream, 3 - downstream.

Pleistocene sands overlying thick sandy and clayey deltaic deposits (Henkel, 1982). The high temperature, combined with high seasonal rainfall and possibly the production of organic acids, has caused an accelerated rate of breakdown of the topsoil, resulting in a lateritic profile. The groundwater table is approximately 10 m below ground level, with a variation of up to 5 m (Akpokodje, 1987).

SOIL CONDITIONS

The surface soil layer under consideration consists of a mixture of sand and clay. The clay fraction varies between 20% and 30% and is considered to be mostly kaolinite (Akpokodje, 1987). This clay holds together the relatively unweathered quartz grains of sand grade. The brown-yellow colour is related to the abundance of ferric oxide, which may cause some cementation.

The dry unit weight of the soil is approximately 16 kN/m^3, which corresponds with a void ratio of 0.68. This ratio is relatively high for sand (indicating loose to medium dense conditions), and relatively low for typical clays which have void ratios of 0.7 to 1.0. The natural pH of the soil is 4 to 5.

OBSERVATIONS

Soil samples from the first cavity showed a markedly raised pH. Field tests using indicator paper showed a pH of 9 to 10; further tests in FME's laboratory according to BS 1377: 1990 showed a pH closer to 7.

The increased pH indicates contamination of the soil with caustic soda through the leaking discharge system.

The laboratory testing program centred on the dispersability of the soil and the effects of caustic soda on the soil properties (Table 1). The samples marked "upstream", "midstream" and "downstream" have been obtained from the bottom of the first cavity at the locations indicated in Figure 1. The sample marked "natural" was obtained from a nearby location, and represents uncontaminated soil.

Table I: Properties of soil samples.

Sample	Depth (m)	Atterberg Limits w_l (%)	w_p (%)	I_p (%)	Fraction <2μ (%)	<63μ (%)	Emerson Class	Pinhole Class	Sp Gr (g/cc)	Org Cont (%)	Carb Cont (%)	pH
Natural	1.0	39	15	24	22	33	6(6)	ND1	2.67	1.2	7	4.0
					27	36		(PD1)		1.4	8	4.2
Up stream	2.5	22	11	12	22	28	5(5)	PD1	2.65		6	6.5
					15	26			2.65		3	6.9
Down stream	2.5	44	12	32	34	45	6(5)		2.69		6	6.0
									2.68		7	6.8
Mid stream	2.5				12	32		D2				

Brackets indicate results from tests with a 10% NaOH solution as solvent.

The Pinhole Test (Sherard, Dunnigan, Decker and Steele, 1976) showed that under normal conditions the soil is erosion resistant (ND 1), but soil contaminated with caustic soda showed dispersion (PD 1). It was observed that while for uncontaminated soil the flow remained linear with time, in contaminated soil the flow was reduced as a result of collapse of the central part of the sample following the removal of clay and ferric oxide. This is shown in Figure 2 where the total amount of water flushed through the sample is plotted against time.

In a modified test, a 10% caustic soda solution was dripped on an uncontaminated sample around the pinhole. The effluent was hardly colloidal. After cutting the sample, it became apparent that dispersion had taken place, but the clay was washed in just below the surface and along the sides of the pinhole.

Determination of the Emerson Class number by the Crumb Test (Standards Association of Australia, 1980) confirmed the findings of the Pinhole Test. In this test, the behaviour of air dried or remoulded crumbs of soil on immersion in water is observed.

The Atterberg Limits tests showed a marked reduction in plastic limit, liquid limit and plasticity index for the contaminated soil, which is to be expected when sodium becomes the dominant ion adsorbed (e.g. Grim, 1962).

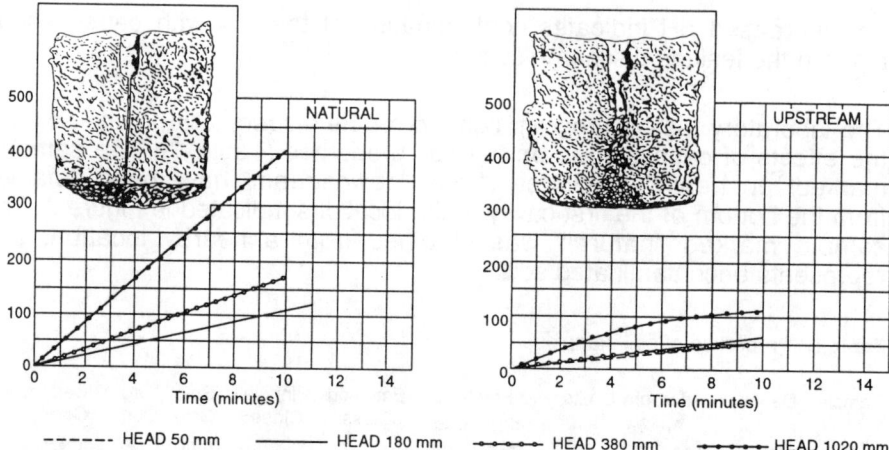

Figure 2: During the pinhole tests, water flow remains nearly constant in uncontaminated samples and increases with increasing head; in contaminated samples (here represented by the upstream sample) water flow decreases with time. The insets show drawings of cut samples after testing. In the uncontaminated sample (left) the pin hole has remained intact, while in the upstream sample clay and silt were washed out and the centre of the sample has collapsed (right).

POSTULATED FAILURE MECHANISM

This section describes a tentative scenario leading to failure as observed at the location. The scenario is based on the results of the field and laboratory investigations and includes details of the underground discharge system design.

The primary cause of the development of leaks in the system is believed to be thermal expansion and contraction of the pipes. Apart from surface water, caustic soda, sulphuric acid and other waste products, steam was drained through this system, increasing the temperature significantly. The development of cracks may have been advanced by chemical attack.

The final section of the pipeline was raised 0.25 m for design purposes. At the connection with the main system, an overflow with a height of 0.75 m was placed. Therefore a water pressure of 1 m was constantly exerted at the location of the developing crack. This water pressure may have resulted in a water jet through the developing crack, causing soil collapse by combined chemical and mechanical action.

The cross-sectional area of the crack at the time of investigation has been estimated at 0.01 m^2. As an indication of the capacity of discharge a calculation was made using Bernouilli's equation for orifices in a reservoir with a constant hydraulic head. Under the conditions described, discharge through the crack could be as high as 80 to 160 m^3/h.

Obviously the calculated flow through the crack will initially be reduced by the permeability of the soil, as illustrated by the permeability coefficient k_v of approximately 2×10^{-8} m/s. But the pinhole tests have indicated that contaminated soil is dispersive under a head of 1 m, resulting in soil collapse. Cationic exchange reactions take place nearly instantly in kaolinite (Krauskopf, 1985) and consequently soil collapse and cavity development will be rapid.

Cavity formation could have been the result of soil collapse and soil transport. Assuming clay and silt are washed out of the affected soil by percolating water and the remaining sand densifies to a void ratio of 0.40, the final soil volume may be less than 10% of the original volume. However this type of failure cannot progress far in the absence of considerable water pressure. The pinhole test using caustic soda has indicated that at low water pressures deposition of contaminated clay will take place, reducing the overall permeability and limiting flow to weaker (more permeable) spots.

If fines are flushed through the soil as indicated, variations in the percentage fines of the different samples would be expected, with small fine fractions where fines are washed out and larger fractions where fines are washed in. The particle size analyses clearly show differences in the fractions passing the 63 μm sieve, with the highest percentage passing from the downstream sample (Table I). These data should however be used with caution since comparison with data from Henkel (1982) shows that the observed differences may be due to natural variation.

It was concluded that cavity development may originally have been the result of soil collapse, but soil transport has played a significant role during later development. Such transport could have initially taken place along small interstices, which are enlarged in the process. At times of high discharge, increased flow and water pressure may induce the development of secondary cavities.

CONCLUSIONS

Fugro-McClelland Engineers studied the causes contributing to cavity development and foundation collapse in a lateritic soil. Contamination with caustic soda was identified as the main cause, but climate, foundation and discharge system design and the low groundwater level may all have added to the extent of the problem.

Similar foundation collapses have been observed earlier (Henkel, 1962). In this case, the problem could well have been prevented had the literature been interpreted correctly.

Proposed remedial actions included repair of the discharge system, backfilling of cavities and tunnels, and chemical grouting of contaminated soil. It will be clear that any of these measures will be difficult and expensive in an area already occupied by an operational plant. Chemical grouting, aimed at filling undetected cavities and stabilisation of the soil, may prove to be problematic in view of the low permeability of the soil and the susceptibility to piping.

REFERENCES

AKPOKODJE, E.G. (1987). The engineering geological characteristics and classification of the major superficial soils of the Niger Delta. *Engineering Geology*, **23**, 193-211.

GRIM, R.E. (1962). *Applied clay mineralogy*, McGraw-Hill Book Company, Inc., London.

HENKEL, D.J. (1982). Geology, geomorphology and geotechnics. *Geotechnique*, **32(3)**, 175-192.

KRAUSKOPF, K.B. (1985). *Introduction to Geochemistry*, 2nd ed., McGraw-Hill Book Company, ISBN 0-07-Y66382-3.

SHERARD, J.L., DUNNIGAN, L.P., DECKER, R.S. & STEELE, E.F. (1976). Pinhole test for identifying soils. *Journal of the Geotechnical Engineering Division*, **102(GT1)**, 69-85.

STANDARDS ASSOCIATION OF AUSTRALIA, (1980). A.S. 1289.C3.1: Determination of Emerson Class Number of a Soil. *Methods of Testing Soils for Engineering Purposes Part C - Soil Classification Tests*.

5-17 ULTRASONIC MONITORING OF THE ACID RAIN IMPACT ON BUILDING STONES

J. PININSKA & P. LUKASZEWSKI
Warsaw University, Faculty of Geology, Zwirki i Wigury 93, 02-086 Warszawa, Poland

ABSTRACT

Acid rain resulting from emissions of oxides of carbon, sulphur and nitrogen into the atmosphere, affects areas far beyond the source of the pollution. The contamination of soil, water and vegetation has received much attention in recent years but the effects on constructions in general and building stones in particular are less frequently considered, despite the fact that, being passively subjected to acid rain, they provide a good indication of the cumulative effects of acid rain in the locality.

This paper discusses the results of controlled corrosion on samples of building stones commonly used in Poland. In addition to providing information on the extent of deterioration caused by acid rain, these tests also assist in the assessment of measures required for both preventative and remedial works.

INTRODUCTION

Air pollution with oxides of carbon, sulphur and nitrogen produces acid rain. Precipitation waters with pH below 7 prevail in industrialised regions and in the vicinities of power stations with coal furnaces. Due to atmospheric circulation, however, acid rain extends well beyond regional and international boundaries. Countries emitting oxides retain only part of their own pollution, the rest migrates abroad. As a result, distribution of acidogenic atmospheric pollutants is uneven and the chemistry of acid rains varies regionally (Kulig, 1989; Malecka, 1991).

In Poland the pH of rain waters averages between 3.9 and 5.9 while extreme values vary from 3.0 in Sudetes where thousands of acres of National Park forests have been destroyed (Plate 1) to 8.3 in north eastern Poland. In the capital, Warsaw, pHs of between 4.0 and 5.5 have been recorded.

The effects of acid rain on soil, water and vegetation is well known but its impact on construction in general (Plate 2) and on building stones in particular has received much less attention. Building stones passively subjected to acid rain corrosion become cumulative indicators of acid rain impact in their vicinity. Once polished stones on building facades become matt, their surfaces quickly show micro-relief features due to removal of the less resistant matrix (Plate 3) and bands of seepage sagging (Plates 4 and 5).

Plate 1: National Park Forest in Sudetes, destroyed by acid rain.

Plate 2: Impact of acid rain on construction (lamp post).

Plate 5: Impact of acid rain on sandy limestone: note the difference in appearance of surface protected by sign-board (just removed) in the upper right corner.

Plate 4: Bands of sagging

Plate 3: Microrelief developed on originally polished surface (Devonian limestone).

Where street traffic is heavy, exhaust fumes penetrate the pores of stone and, in the presence of water, exacerbate the corrosion.

A knowledge of the depth and extent of corrosion or destruction is important when assessing renovation and preventative measures.

LABORATORY TESTS

The extent of corrosion was controlled on standard samples of sedimentary and metamorphic building stones commonly used in Poland:

A) METAMORPHIC ROCKS:

Gneisses and schists from the Lower Silesia region frequently found in historic buildings and bridges;

B) CARBONATE - SEDIMENTARY ROCKS:

   i) Greyish-red Upper Devonian limestones from the Holy-Cross Mountains containing many fossils which produce a highly decorative effect on polished surfaces and hence are often used as facing slabs or tiles.

   ii) The Upper Permian (Zechstein) conglomerates, which are dominantly composed of grey, brown, blackish and reddish boulders, cobbles and pebbles derived from carbonate Devonian and Carboniferous rocks with cherry-red and brown cementation and multi-layered white and coloured calcite veins. These highly decorative rocks, known as the King Sigmunt's Conglomerate, are still in use in monumental buildings for such architectural features as walls, stairs, balustrades, door frames, portals etc.

   iii) The whitish Upper Cretaceous marly gauzes from the Lublin Upland area are commonly used in rural houses, being easily shaped, light and porous.

For comparison, standard clay red-brick samples were also tested.

Undisturbed rock samples were subjected to standard testing including acoustic determination of the longitudinal wave velocity (V_L) by the passing through method (Pininska, 1982; Pininska & Lukaszewski, 1991). High frequency (1 MHz) was used in the acoustic testing because of the inhomogeneity of the tested material. The natural properties of the samples, including their standard longitudinal wave velocities, are presented in Table I.

Samples were then subjected to capillary ascent of pH 4 acid solution over several months, to simulate the action of acid rain on exposed stones. Changes in the longitudinal wave velocity in the tested samples indicated

Table I: Natural properties of the samples.

Rock	Utilisation	Technical Data				
		Compression Strength (MPa)	Soaking (%)	Volume Density (T/m³)	Abrasion Test Boehme (%)	Abrasion Test Deval (%)	Initial Longitudinal Wave Velocity (m/s)
1	2	3	4	5	6	7
Lower Silesia Gneiss	Historical Buildings and Bridges	20-65	0.5-1.4	2.68-2.80		<1	5200
Holy-Cross Mountains Limestone	Facing Slabs and Floor Tiles	100-150	0.06-0.15	2.69-2.70	0.64-0.66		6400
Carbonate Conglomerate	Finishings Architectural Details	90-150	0.13-0.20	2.69-2.70	0.81-0.88		6300
Lublin Upland Marly Gauze	Country Buildings	15-30	12.7-36.4	1.40-1.92		8-28	1900
Red Brick	Common Brick Masonry	10					2200

progressive saturation and the development of acid-corroded zones. After drying the samples were again tested for final longitudinal velocity (V_{LF}). By substraction of standard longitudinal velocity (V_L) from final longitudinal velocity (V_{LF})) a corrosion indicator (DC) was determined.

RESULTS

The strongest impact of the acid solution was noted for bricks and gauzes with a moderate effect measured on limestones and carbonate conglomerates. Schists and gneisses were least affected, confirming previous observations (Hutchinson, 1980; Tombers, 1991). Visual observations of the surfaces of the samples indicated changes in colour and increasing roughness. In the limestones and conglomerates increased roughness is due to the exposure of the more resistant aggregates of the rock fabric.

Changes in the longitudinal wave velocities recorded during the action of the acid solution are given in Figure 1, the reference levels being the initial velocities (V_I) previously determined on the undisturbed rock samples.

The most susceptible samples (brick and gauze) showed three phases of change in longitudinal wave velocities:

Phase I - reduction in velocity caused by initial corrosion of the surficial zone of the rock;

Phase II - increase in velocity above the reference level caused by pore saturation with acid solution

Phase III - final decrease in velocity despite pore saturation caused by development of intra- and intergranular defects (microcracks)

In gauzes (Figure 1a) decrease in velocity begins immediately after the acid solution is applied and reaches 95% to 85% of standard velocity in 20 days. Saturation increases the velocity by up to 6% above the standard value before it starts to decrease again. After drying the gauze samples retain approximately 88% of their initial longitudinal wave velocity as determined on the undisturbed rock.

Three phases of change were also observed for the bricks (Figure 1b) but the final decrease in velocity measured on the dried samples did not exceed 5%, reaching 94% of the initial velocity.

In the limestones of Devonian age, the decrease in velocity appeared after a longer period than for the gauzes or bricks and reached 3% at Phase I (Figure 1c). Due to less penetration of the pores by the acid solution the increase in velocity at Phase II was small, reaching only 2% above the

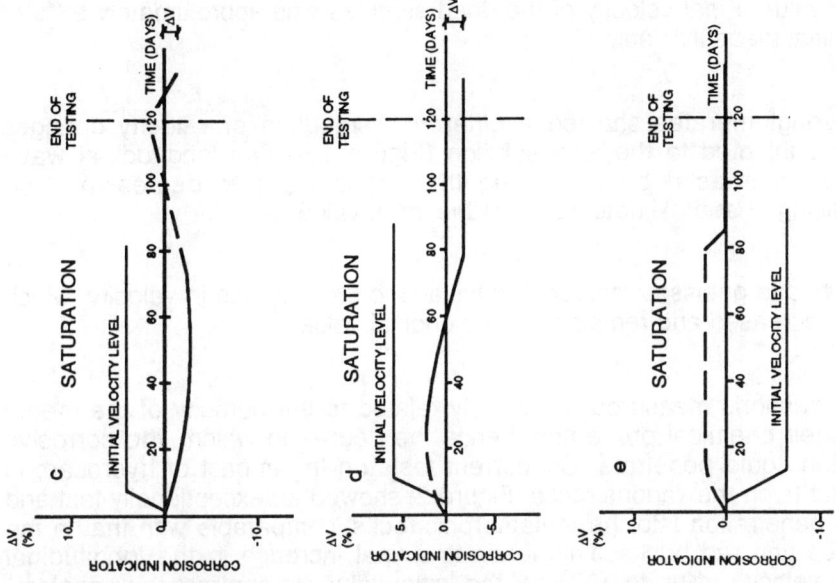

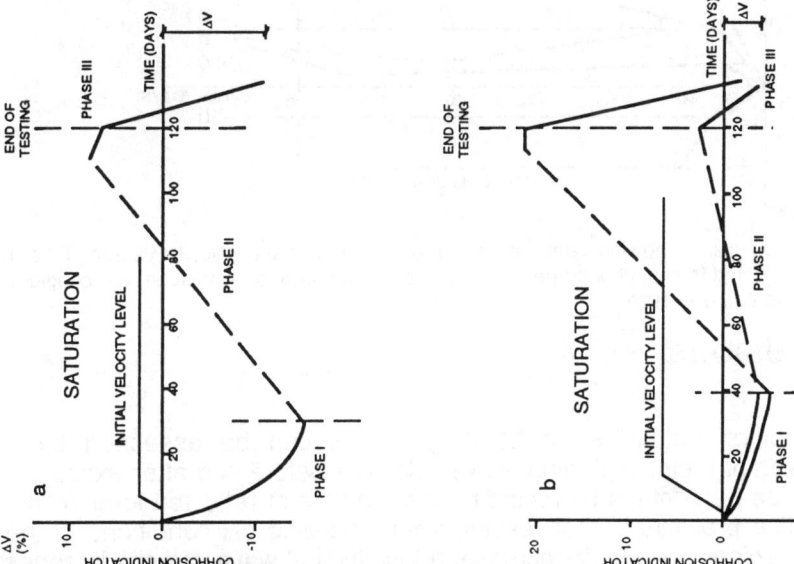

Figure 1: Diagram showing changes of the longitudinal wave velocity versus time in samples subjected to acid solution: a - gauze, b - red brick, c - limestone, d - conglomerate (carbonate), e - gneiss.

initial value. Final velocity of the dried samples was approximately 98% of the initial measurement.

The conglomerates showed a different distribution of velocity changes when subjected to the acid solution (Figure 1d). The longitudinal wave velocity increased by 3% during the Phase I before decreasing and stabilising at approximately 96% of the initial value.

Schists and gneisses recorded an initial 3 to 5% increase in velocity, which then decreased and remained at the original value.

The variations measured are closely related to the porosity of the media and their chemical properties, hence the degree to which the corrosive solution could penetrate. Concurrent tests on the impact of hydrocarbon pollutants on the various rocks (Figure 2) showed an exceptionally fast and deep penetration into the metamorphic rocks, comparable with that in the gauzes and red bricks. This leads to a final increase in the longitudinal wave velocity of up to 107% of the initial value. In contrast the conglomerates subjected to hydrocarbon pollutants reacted with a final decrease of 2.5% from the initial velocity.

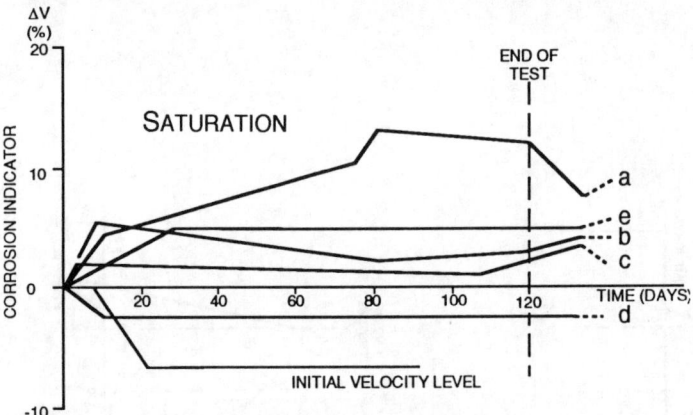

Figure 2: Diagram showing changes of the longitudinal wave velocity versus time in samples subjected to impact of crude oil: a - gauze, b - red brick, c - limestone, d - conglomerate (carbonate), e - gauze.

CONCLUSIONS

The effect of pollutants on building stones can be assessed by a comparison of their longitudinal wave velocities before and after exposure. This can be determined by recording the run-time of reflected longitudinal waves. The preliminary test results, from modelling pH conditions in the Warsaw region, showed a decrease in longitudinal wave velocity in zones affected by acid rain waters of by up to 12% in the case of porous carbonate rocks.

REFERENCES

HUTCHINSON, T.C. (1980). *Effects of acid precipitation on terrestrial ecosystems*. Pergamon. Press. New York.

KULIG, A. (1989). Monitoring wplywu zakwaszonych wod na srodowisko. *Mat. Symp. Chemizm opadow atmosferycznych. Wyd*, Uniwersytetu Lodzkiego. Lodz.

MALECKA, D. (1991). Opady atmosferyczne jako wazny czynnik ksztaltujacy chemizm wod podziemnych. *Przeglad Geologiczny*, **1**, Warszawa.

PININSKA , J. (1982). Comparison of some dynamic non-destructive test results for different geological conditions. *Proceedings of the European Symposium on Penetration Testing. ESOPT*. Amsterdam.

PININSKA, J. & LUKASZEWSKI, P. (1991). The relationships between post-failure state and compression strength of Sudetic fractured rocks. *Bulletin of the International Association of Engineering Geology*, **43**, Paris.

TOMBERS, J. (1991). *Untersuchungen zur Salzverteilung in verboten Naturstein*. Doctors Diss. Universität des Saarlandes. Saarbrücken.

5-18   ENVIRONMENTAL BIOCHEMICAL DECAY CONTROL ON JERONIMOS MONASTERY CLOISTER, LISBON, PORTUGAL

L. AIRES-BARROS & A.M. MAURICIO
Laboratorio de Mineralogia et Petrologia, Instituto Superior Tecnico, Av Rovisco Pais, 1096 Lisboa Codex

ABSTRACT

The sixteenth century Jeronimos monastery is approximately 250 m from the River Tagus. The stone used in its construction is a Cretaceous limestone, mainly from the Cenomanian and Turonian. No lichens are found from the base of the first floor on the south wall of the Cloister Gallery to a height of about 1.2 m. From here to approximately 3 m, however, soft white lichens are abundant. At higher levels a hard greyish fungi is observed which gradually disappears towards the top of the structure. The lichen-limestone interface permits the genesis of oxalate minerals such as wewellitte and weddellite.

The monastery wall has been monitored for air temperature, relative humidity, dew point and surface air temperature and the data input to a specially created database. A clear relationship can be seen between the spatial relationship of the lichen and nanoclimate variables. Graphical representations of climatic data gathered on the wall during a twelve-month period are presented, emphasising these relationships.

INTRODUCTION

Jeronimos monastery is located approximately 250 m to the north of the Tagus river about 8 km east of its mouth. It was built by order of King D Manuel I both for religious services and as a royal mausoleum. It is one of the most important examples of the "Manuelino" architectural style and in 1982 was classified as a world heritage site.

The main building stone is Lios Limestone which was obtained from quarries in Lisbon which are no longer worked. Lios Limestone is a Cretaceous (mainly Cenomanian and Turonian) conchiferous limestone which has been extensively used for revetting and building stone. It has an average porosity of 0.31% and a density of 2.703 Mg/m^3. Its colour ranges from white to yellowish white and reddish white.

From a visual examination of the monastery wall it can be seen that a very thin brownish patine of dirt more or less uniformly covers the cloister surface. This is very difficult to remove. Lichens are aggregated in small, approximately circular colonies averaging about 20 mm diameter. These occur in a few very disperse patches over the eastern wall but are practically non-existent over the northern and western walls. On the southern wall, however, approximately two-thirds of the surface is covered in an elongated elipse pattern, more or less centred relative to the wall geometry (Figure 1).

The contamination pattern has three levels. The first reaches about 1.2 m high relative to the floor and there are practically no lichens. The second level develops to about 3 m high and has extensive lichen growth for about 30 m of the wall length. At the lower extremes of this level the patches are mainly whitish and thin, becoming greyish-green and thicker with increasing height. Further up the wall there is a sharp increase in the density such that they gradually join to become large, greyish-green, irregularly shaped continuous patches each covering several square centimetres of the wall surface. At about this 3 m level the patches are smaller, gradually decreasing in density with increasing height. Aires-Barros et al (1989) classified and analysed these lichen and the lichen-limestone interface and referred to the existence of the oxalate minerals wewellite and weddellite.

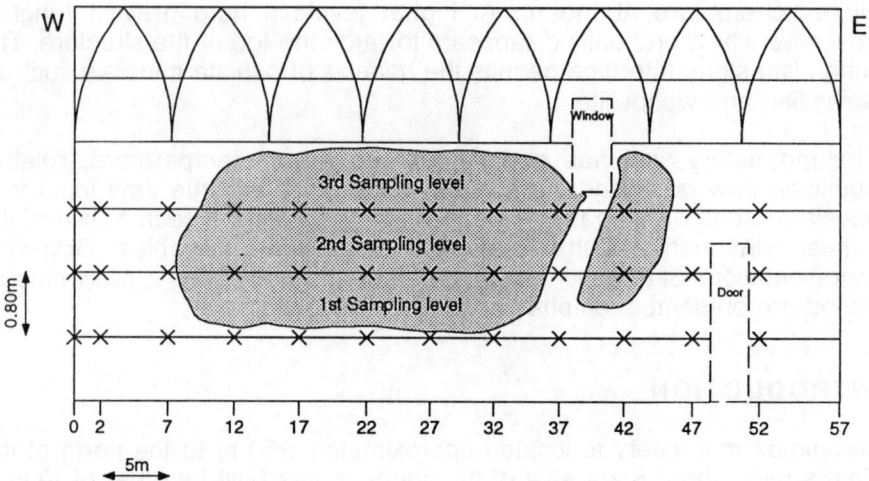

Figure 1: Main geometric characteristics, sampling and contamination pattern on the south wall of the monastery.

THE STUDY

Lichenic abundance and distribution patterns and their connection with the cloister nanoclimate were studied in order to allow not only a better understanding of the phenomena but also to formulate a strategy to prevent this kind of destruction, perhaps by identifying a product toxic to this species of lichen but which would not have a detrimental effect on the rock itself. The procedure followed was:

1) macroclimate characterisation, global and local
2) characterisation of the monument microclimate

3) wall nanoclimate instrumentation
4) data analysis
5) conclusions

MACROCLIMATE CHARACTERISATION

The climate of Portugal is mainly influenced by humid oceanic air masses arriving dominantly from the northwest and southwest quadrants, hence with relatively little pollution. Lisbon has a Mediterranean climate, with an average temperature of about 17°C. The precipitation reaches its maximum during the winter, with intermediate values during the spring and lesser ones during the autumn but almost no rainfall during the summer. Thus while the yearly average precipitation reaches 0.76 m there is a long period of dry weather coinciding with the months with the highest temperatures. Both relative humidity and temperature vary seasonally, the maximum average temperature in August being 18°C and the minimum average relative humidity of 40% occurring in September. The average lowest temperature of 5.5°C and the highest average relative humidity of 70% are both in February (Aires-Barros et al, 1990).

MONUMENT MICROCLIMATE

Microclimate data on Jeronimos monastery has been obtained from monitoring devices remaining in situ for three years. The apparatus provided data on rain water (pH, conductivity and ion concentrations), sun radiation, air temperature and relative humidity, dew point, air particulate content and gaseous pollution (SO_2, NO_x, HCl, O_3, CO). Figures 2 to 8 illustrate some of the results for 1990.

As can be seen, the rainwaters are slightly acid to neutral, the pH variation ranging between 5.5 and 7.2. The precipitation reaches a maximum in November with intermediate values being measured in December, January, March and April and smaller levels in February, May, June and September. No rainfall was recorded in July and August.

The predominant anions content of the rainwaters are $SO_4 > Cl > NO_3$ (Figure 6). NO_3^- and SO_4^{2-} appear to be strongly correlated, at least between January and September. Higher Cl^- contents seem to be related to maritime winds.

Figure 8 shows the average monthly gaseous air pollution in the upper cloister. It can be seen that the mean values of NO_x are higher and more erratic than the SO_2 content which appears to be related to the amount of SO_4^{2-} (Figure 6). There also appears to be a negative correlation between air SO_2 and water SO_4^{2-} concentrations during the rainy months. The correlation between the amount of rainfall nitrate and air NO_2 with water is not so good.

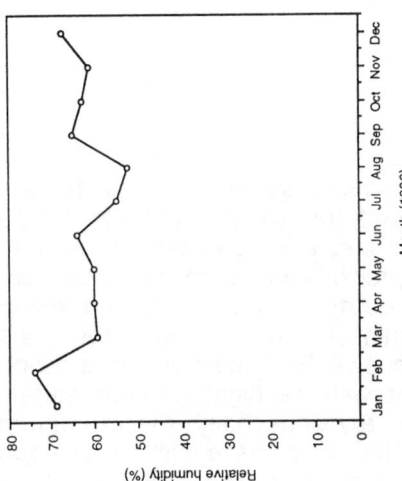

Figure 2: Upper cloister gallery temperature plot.

Figure 3: Upper cloister gallery relative humidity.

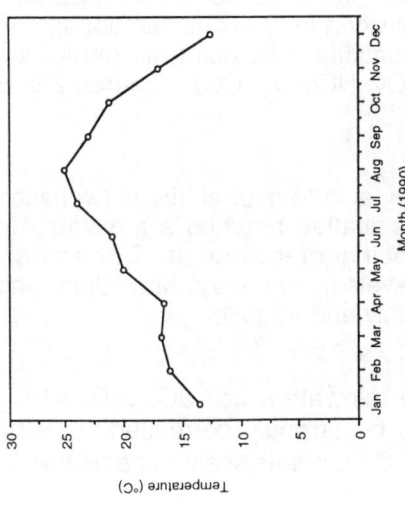

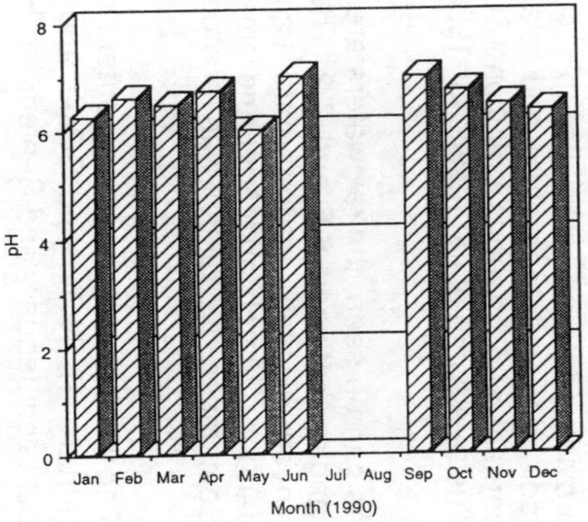

Figure 4: Cloister terrace rain water pH plot.

Figure 5: Cloister terrace average precipitation.

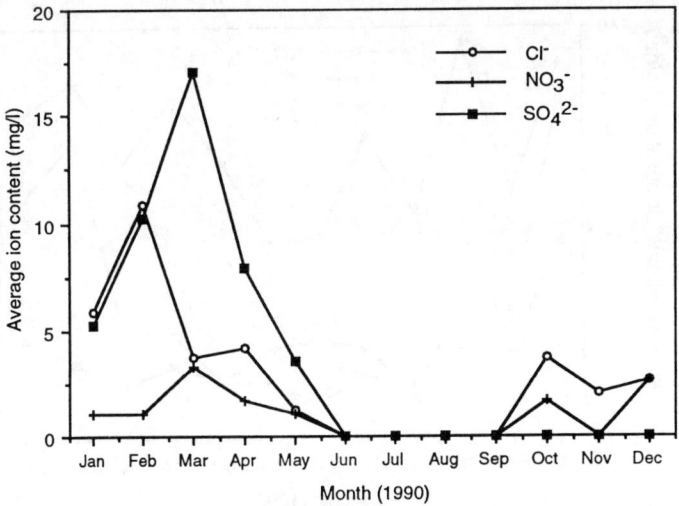

Figure 6: Monthly average rain water ion content plot.

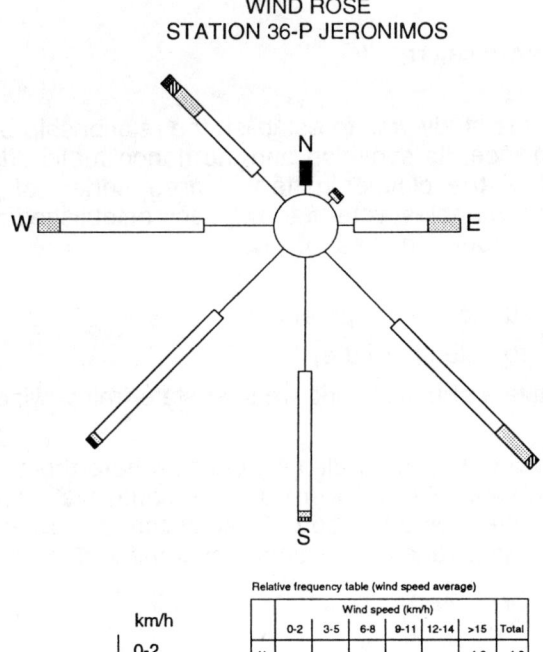

Figure 7: Average horizontal wind speed and direction.

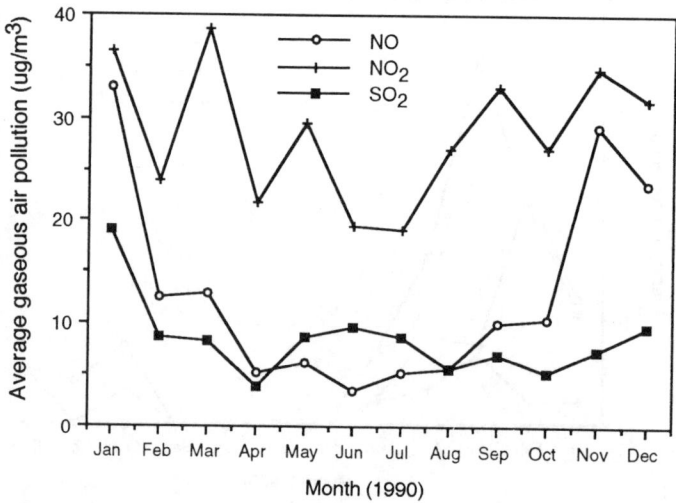

Figure 8: Monthly average gaseous air pollution plot.

WALL NANOCLIMATE

The aim of the study was to establish the relationship between the lichenic flora occurrence, its evolution and the nanoclimatic characteristics on the south wall of the cloister gallery. Three series of measurements of nanoclimatic variables were made at each established network point, three times a week during the year of study:

a) rock surface air temperature;
b) dew point temperature;
c) relative air humidity and weather state (rainy, windy etc).

Three control points were also measured where there are no lichens over the eastern wall, about 15 m from the south wall. The need to access gradient variation as a function of wall distance allowed the inclusion of air temperature as a variation measured over the surface and at 0.5 m from the surface.

DATA ANALYSIS

The results of the monitoring are presented in Figures 9 to 23. Taking average values at each sample point, five types of rock surface temperature, dew point temperature and relative air humidity maps were created: dry season maps, wet season maps, morning maps, afternoon maps and overall maps using Krigging as the interpolating method. This is the best linear unbiased interpolation method and cartography is one of the applications.

ENVIRONMENTAL BIOCHEMICAL DECAY CONTROL ON MONASTERY CLOISTER   639

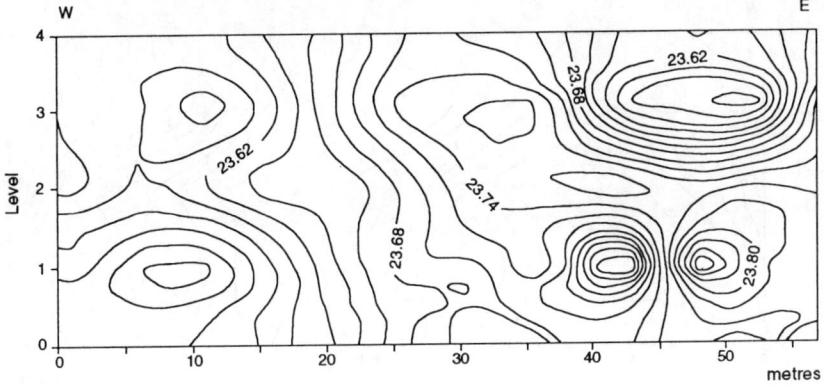

Figure 9: Dry season air temperature.

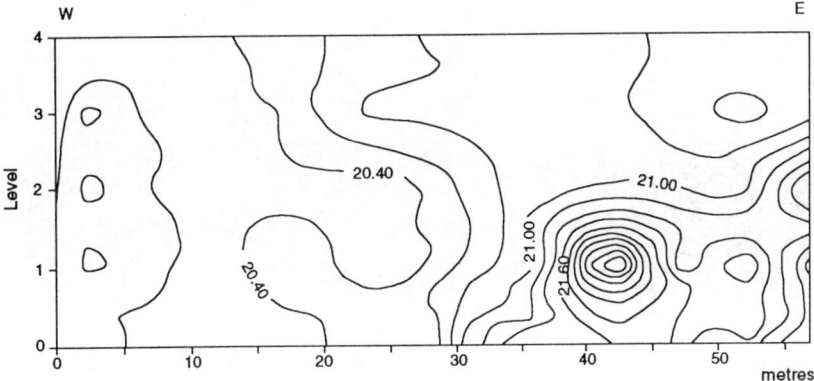

Figure 10: Dry season dew point temperature.

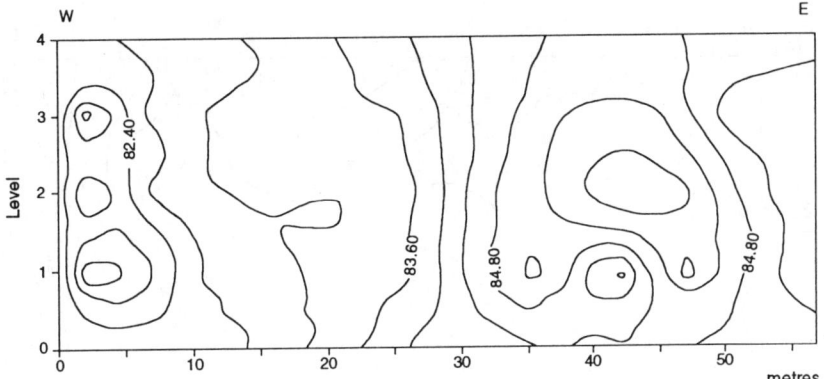

Figure 11: Dry season relative humidity.

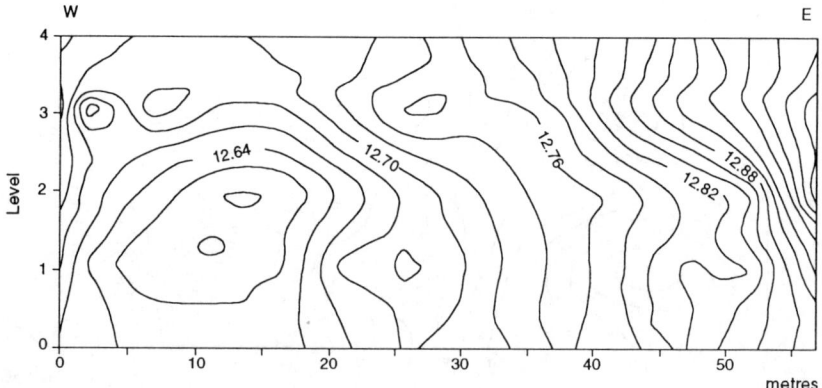

Figure 12: Wet season air temperature.

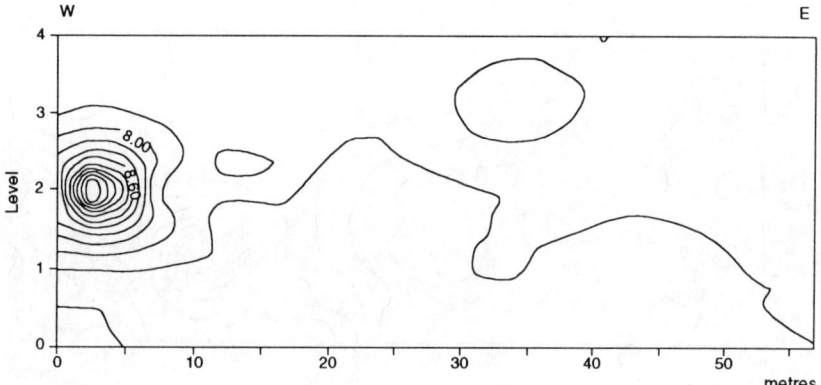

Figure 13: Wet season dew point temperature.

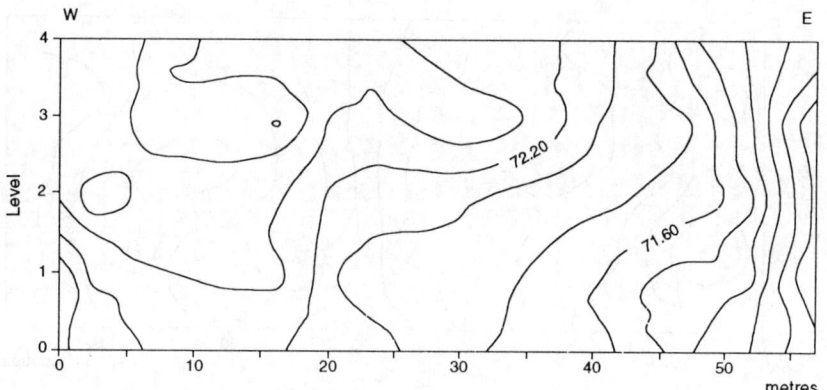

Figure 14: Wet season relative humidity.

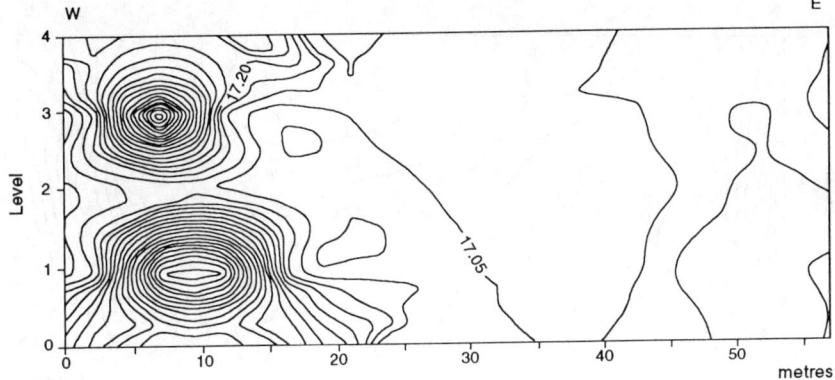

Figure 15: Morning air temperature.

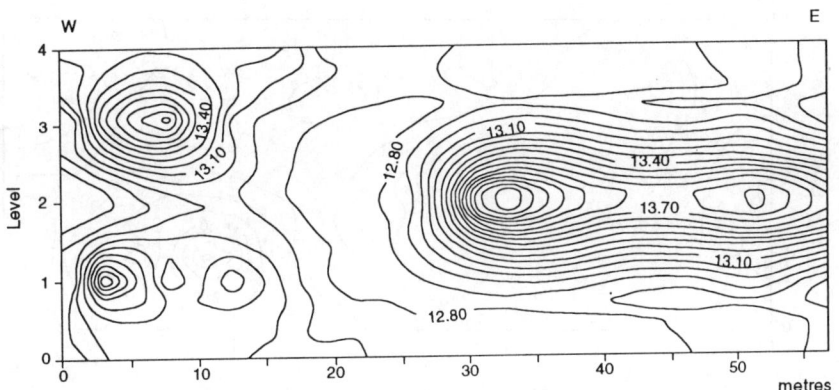

Figure 16: Morning dew point temperature.

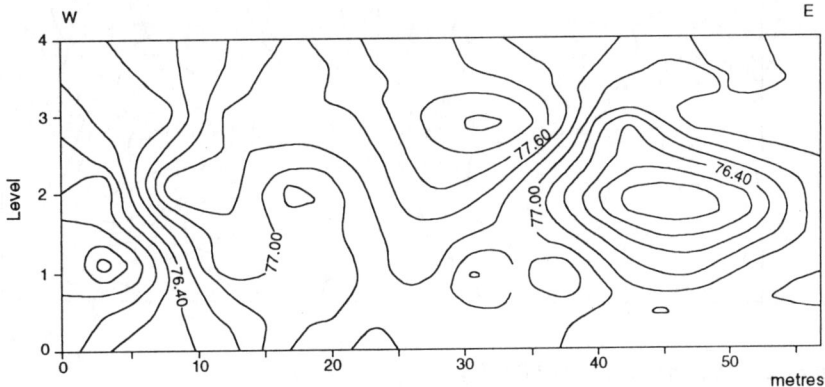

Figure 17: Morning relative humidity.

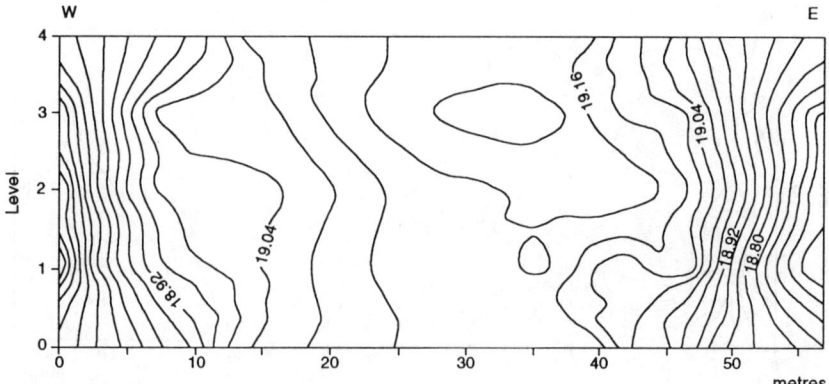

Figure 18: Afternoon air temperature.

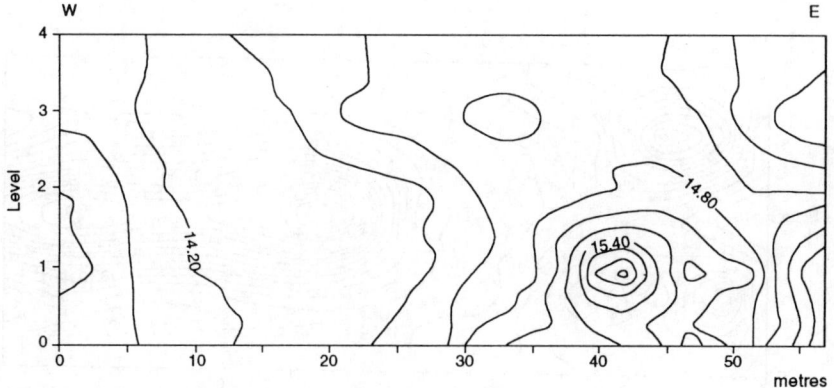

Figure 19: Afternoon dew point temperature.

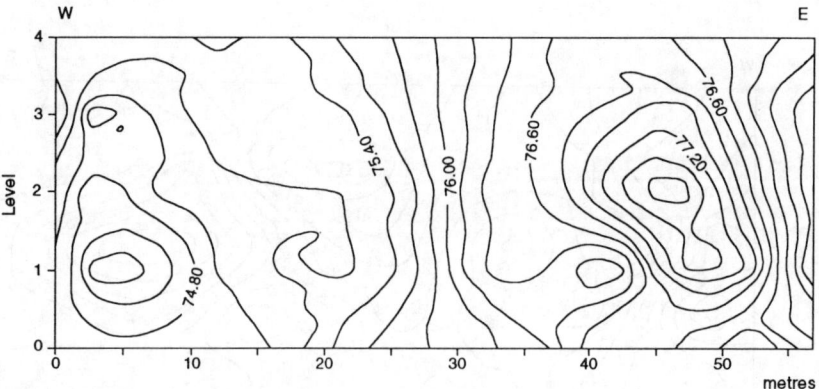

Figure 20: Afternoon relative humidity.

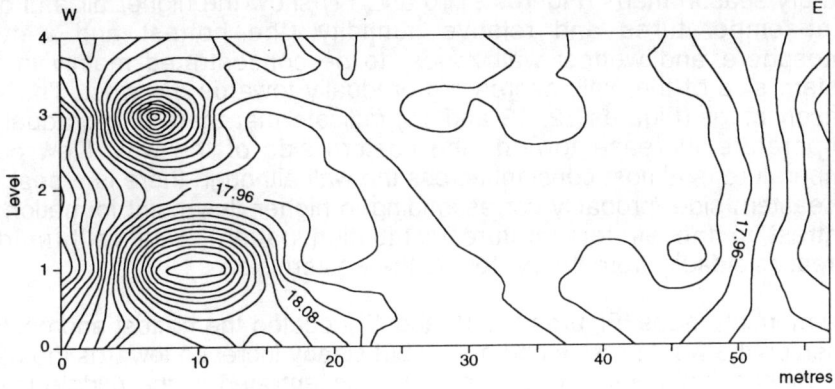

Figure 21: Yearly air temperature.

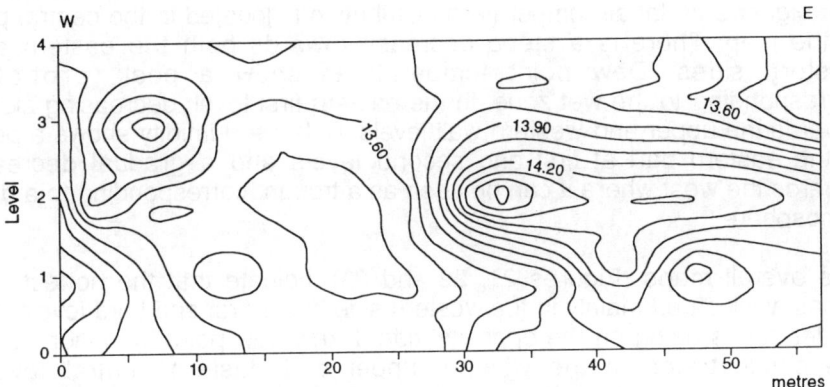

Figure 22: Yearly dew point temperature.

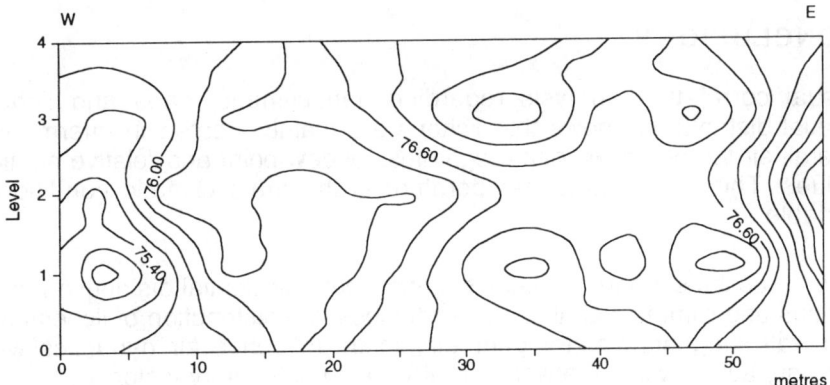

Figure 23: Yearly relative humidity.

The dry season maps (Figures 9, 10 and 11) show the higher air and dew point temperatures and relative humidity (the hottest and wettest atmosphere and wettest wall zones) to be concentrated mainly in the eastern side of the wall, decreasing gradually towards the west. The wet season maps (Figures 12, 13 and 14) indicate that there is a gradual air temperature increase towards the eastern side of the wall. Dew point temperature is almost constant across the wall although there is a peak on the eastern side, probably corresponding to higher dew point formation. In contrast to the air temperature distribution pattern, relative humidity increases steadily from the eastern to the western side.

The morning maps (Figures 15, 16 and 17) located the hottest and coldest zones on the western side and a slow but steady increase towards the east. The higher dew point temperatures are concentrated in the middle levels on the eastern part of the map with a steeper decrease towards the upper and lower levels. The relative humidity map shows higher values in the upper and central parts. The afternoon maps (Figures 18, 19 and 20) show that higher and flat air temperature distribution is located in the central part of the map. There is a steep decrease towards both the eastern and western sides. Dew point temperatures show a peak, probably corresponding to the wet zone, in the eastern first level, decreasing slowly towards the upper and western wall levels. Relative humidity shows a peak in the eastern part at first and second levels and a gradual decrease towards the west where it can be seen as a trough, corresponding to a drier atmosphere.

The overall maps (Figures 21, 22 and 23) indicate that the hottest wall zones are located mainly in the western side at the first and third levels, the colder zones being on the opposite side. Likely dew point formation zones are concentrated in the western upper and eastern central levels, decreasing more or less steeply towards upper and lower levels. Wetter air humidity is recorded in the first and third levels in the central and eastern zones with the drier areas concentrated mainly at first level in the west.

CONCLUSIONS

Visual correlation analysis regarding nanoclimatic maps and lichenic spatial distribution shows that lichens are mainly located in intermediate overall air temperature zones regardless of dew point and relative humidity values. This suggests air temperature is the main climatic conditioning variable.

These results support a relationship between the spatial distribution of the lichens and climatic variables. Nevertheless, the distribution of lichens over the wall suggests cloister geometry might influence air dynamics while capillary action within the wall may also be a contributory factor.

It is hoped to modify and extend the sampling programme in order to obtain more information on the above micrometeorological variables as well as on the other suggested variables, at least on an hourly basis. This will allow more accurate maps to be compiled as well as further characterisation of the nanoclimate and quantitative correlations to be established between nanoclimate and biological decay.

ACKNOWLEGEMENTS

This research work was supported by JNICT (Contract No PMCT/C/CEN/96/90).

REFERENCES

AIRES-BARROS, L., et al, (1989). The decay of limestone caused by saxicolous lichens: the case of the Monastery of Jeronimos (Lisboa-Portugal). *Proceedings 1st International Symposium on The Conservation of Monuments in the Mediterranean Basin, Bari*, 225-229.

AIRES-BARROS, L., et al, (1990). Weathering and air pollution action on limestones of Jeronimos Monastery, Lisbon, Portugal. *Mem e Not*, **109**, 117-130, Coimbra.

CONCLUDING REMARKS

D.A. GREENWOOD
Cementation Piling & Foundations, Denham Way, Maple Cross, Rickmansworth, Herts WD3 2SW

In opening the Conference, Professor Severn referred to the subject matter crossing the boundaries of disciplines, involving scientists and engineers. In that I have the perspective of a geotechnical engineer, I might be regarded as a non-scientist in this context.

The Conference subjects may be considered esoteric by some. It is gratifying that they attracted a mixture of both experts and novices, scientists and engineers, to debate the subjects which are increasingly inter-disciplinary. They involve complex sciences and engineering technology. The development of the subjects is no longer at a state where ignorance is bliss. This is now unacceptable.

The scene setting and theme lectures were almost frightening in highlighting the range of contaminants, physical situations, biological growths and engineering design oversights which can combine in almost unexpected ways to yield intractable problems with unexpected ramifications and magnitude: the subject matter of the Conference was perhaps more about the irregular performance of construction.

The issues relate to inherent variability of chemicals, of soils, of gas emissions, microbes and physical circumstances. Everything is evolving and changing continuously so that the problems to be solved are difficult to grapple with. Solutions can generate problems by changing environments or circumstances. It is usually inappropriate to consider a static situation as in a structural engineering problem: the goal-posts move and solutions must recognise evolving possibilities.

Often difficulties with structures arise because of some highly localised environment. The problem frequently becomes one of identifying exactly where the precise difficulty lies, both in completed structures which are not performing but, much more difficult, with foresight to recognise where a detail in the concept may be potentially defective.

New technology is producing new materials with unknown durability. How long should we expect such materials to last? Population pressures create new problems of waste which may have consequences for a very long time. Acid rain and decaying anchor chains have been cited as unexpected examples.

The ramifications of the problems of individual structures and of new technology, population growth and resource depletion are not easily predictable. Detection of effects is much easier. Part of our developing technology should be better identification of processes, not previously considered. By knowing about them, we can address them.

Even if knowledge is incomplete, awareness of issues is important. Such awareness directs attention to aspects of problems which might otherwise be missed. Awareness and knowledge can then be used in deriving engineering solutions. Knowledge and awareness of processes in action is a start, but for engineering structures the effects need to be quantified in order to assess their significance in a specific context.

We have some historical evidence, for example durability of gas pipes or of sheet piles buried in the ground which can help in judging whether in the practical sense the effects of these physical processes are significant. There is concern at any time with specific problems not generalities. All engineering issues become site specific and it is often the detail which fails rather than the concept.

One of the outstanding problems currently recurring seems to be how and where to tests; how many tests, what size sample? Does testing contaminate or disturb the environment we are examining? What are the relevant tests for the problem in hand?

It is of some economic importance and commercial significance that standards for testing are established, so that results of tests are known to be comparable and relevant to the context of the enquiry. It is already important to consider training of technicians undertaking large numbers of tests to ensure consistent and appropriate techniques so that analysis of results is reliable. We should try to work towards setting standards for tests.

Political exploitation and emotional reaction must not be allowed to exaggerate the practical significance of pollutants. There is a danger of over-reacting. Bugs and pollution are with us and always have been. We need to learn how to manage them and live with them. This is a large problem for regulators. Blanket rules are not site specific, but the practical problems are. Of course the public must be protected, but to what magnitude of cost in the context of overall economic development? The drafting of directives for a new community such as that evolving in Europe for widely differing circumstances is a formidable if not impractical task. Perhaps it is much better to consider much smaller environments and allow them to formulate appropriate rules for their circumstances. In this context it was encouraging to learn of the flexibility of the Toronto rules for wastes based on reasonable use of sites with some negotiable limits for pollutants. Regulations should take account of feasibility and cost/benefit. Professional engineers should be expected to perform professionally and sometimes to judge the unjudgeable. They are in the best situation to assess and deal with each site.

Another question arising is whether regulations should be advisory or mandatory in a field of activity where knowledge and technology are rapidly changing and advancing. It seems to me that there should be a balance according to the circumstances.

Engineering risk analysis seems to be a way forward. The suggestion for developing a scoring system for rating to sort out complexity and classify risks should be encouraged. This is analogous to the beneficial attempts at Rock Mass ratings in a likewise intractable field. Also encouraging was a move towards engineering pragmatism in the conception of pollution control projects. We have to engineer for changing conditions in which construction activity is a major change. It was encouraging to hear of designs which kept the extent of control measures in perspective. Some of these were relatively simple like passive venting and proper detailing of building foundations; consolidation of fill materials to maintain low permeability and designs to circumvent problems altogether.

As the commercial world moves increasingly towards litigation to resolve what are perceived as engineering mistakes, we need to define more accurately how to investigate sites. It would be helpful if some Codes of Practice could be developed. Since most of construction spending is on housing and small buildings, there is little cash to invest in technical developments, especially as it is not at this stage easy to perceive what would be most beneficial. In this situation pragmatic regulations are essential and professional practitioners should help to develop them.

Thus the Conference highlighted the need for awareness of potential physical and biological processes at work, so that they can be circumvented or managed at design stage. Conformity and standardisation of testing is a pre-requisite of assessment of practical risk for which a system of analysis of complexities is potentially within reach. Meantime, regulations and simple pragmatism in concepts are essential if costs of controlling the natural environment are to be contained.

This Conference made a good start in identifying these issues. We shall probably need another one in a few years' time to see to what extent these possibilities have been realised and to carry us further forward.

REFERENCE INDEX

ADAMS, R.S. & ELLIS, R. (1960), **2-9**, 184
ADAMSE, A.D. et al.(1972), **2-9**, 184
AFNOR, P. (1981), **5-12**, 567
AIRES-BARROS, L. et al. (1989), **5-18**, 645
AIRES-BARROS, L. et al. (1990), **5-18**, 645
AKPOKODJE, E.G. (1987), **5-16**, 622
AL-KHAFAJI, A.W.N. & ANDERSLAND, O.B. (1981), **4-5**, 411
ALLISON, J.D. et al. (1990), **5-13**, 587
ALLSOPP, D. & SEAL, K.J. (1986), **1-3**, 46; **3-4**, 248
ALY SABRY, M.M. & PARCHER, J.V. (1979), **4-6**, 428
AMINABHAVI, T.M. & BALUNDGI, R.H. (1990), **3-4**, 248
AMOS, C.L. (1989), **4-4**, 396
AMOS, C.L. et al. (1988), **4-4**, 396
AMOS, C.L. et al. (1992), **4-4**, 396
ANDERSON, D. & BROWN, L. (1981), **4-7**, 435
ANDERSON, W.F. & CRIPPS., J.C. (1992), **2-2**, 86
ANGELOVA, R. (1987), **2-7**, 158
ANGELOVA, R. & EVSTATIEV, D. (1990), **2-7**, 158
ANON. (1960), **2-2**, 86
APPELO, C.A.J. & WILLEMSEN, A. (1987), **5-13**, 587
ASCE, (1989), **5-10**, 552
ASTM, (1984), **5-2**, 462
ASTM, (1987), **5-9**, 535; **5-12**, 567
ASTM, (1992), **2-6**, 145
ATLAS, R. M. (1984), **1-3**, 46
ATLAS, R.M. & BARTHA, R. (1987), **1-3**, 46
ATTIOGBE, E.K. & RIZKALLA, S.H. (1988), **5-8**, 523
AUDLEY-CHARLES, M.G. (1970), **3-8**, 308

BAGONZA, S.E. et al. (1987), **3-1**, 208
BARCLAY, L.M. & OTTEWILL, R.H. (1970), **1-1**, 15
BARNES, I. (1964), **4-3**, 381
BARNES, I. & CLARKE, F.E. (1969), **4-3**, 381
BARONIO, B. & BERRA, M. (1986), **3-1**, 208
BARRON, J.LAC. & LEUKING, D.R. (1990), **2-3**, 99
BARTHOLOMEW, R.F. (1979), **3-1**, 208; **5-9**, 535
BASTIDA, R. (1971), **4-2**, 366

BAYOUX, J.P. et al. (1990), **5-10**, 552
BEAR, F.E. (1964), **5-5**, 487
BECKER, G.F. & DAY, A.L. (1905), **2-1**, 74
BEGHEIJN, L.TH. (1979), **2-3**, 99
BELL, F.G. (1988), **4-6**, 428
BELL, F.G. & COULTHARD, J.M. (1990), **4-6**, 428
BELL, F.G. & TYRER, M.J. (1987), **4-6**, 428
BERKELEY, R.C.W. et al. (1980), **1-3**, 46
BERNDT, W.L. & VARGA, J.M. (1987), **5-7**, 509
BERNDT, W.L. et al. (1989), **5-7**, 509
BERNER, R.A. (1968), **3-1**, 208
BERNER, R.A. (1970), **2-1**, 74
BERNER, R.A. (1971), **3-1**, 208
BESSEMS, E. (1979), **3-4**, 248
BEZRUK, V. (1958), **2-7**, 158
BICZOK, I. (1972), **5-9**, 535
BIELECKI, R. & SCHREMMER, H. (1987), **5-8**, 523
BIRD, M.J. et al. (1989), **3-3**, 232
BLACK, K. (1989), **4-4**, 397
BLATT, H. et al. (1972), **1-2**, 33
BOCK, E. et al. (1989), **5-8**, 523; **5-10**, 552
BOELS, D. (1991), **5-3**, 472
BOGLI, A. (1964), **5-13**, 587
BOKEN, E. (1958), **2-5**, 130
BONAZOUNTAS, M. (1986), **5-13**, 587
BOOTH, G.H., COOPER, A. W. &.TILLER, A.K. (1967), **4-1**, 352
BOSSERT, I. AND BARTHA, R. (1985), **1-3**, 46
BOWDERS, J.J. et al. (1986), **5-2**, 462
BRANKEVICH, G.J. et al. (1990), **4-2**, 366
BRIERLEY, C.L. et al. (1989), **1-3**, 46
BRINDLEY, G.W. (1951), **2-6**, 145
BRINDLEY,G.W. & ROBINSON, K. (1946), **2-6**, 145
BROCK, T.M. & MADIGAN, M.T. (1991), **1-3**, 46
BROMS, B.B. & BOMAN, P. (1979), **4-6**, 428
BROWN, A.E. (1946), **3-4**, 248
BROWNE, M.A.E. et al. (1984), **4-5**, 412
BRYHN, O.R. et al. (1983), **4-6**, 428
BRE, (1990), **2-9**, 185
BRE, (1991), **2-1**, 74; **2-4**, 109; **2-9**, 185; **5-9**, 535
BRE, (1991), **2-6**, 145
BSI, (1975 & 1990), **2-1**, 74; **2-2**, 86; **2-6**, 145
BSI, (1981), **2-4**, 109
BURCHILL, S. et al. (1981), **4-5**, 412
BUTLER, N.J. & EGGINS, H.O.W. (1966), **3-4**, 248

CADMUS, E.L. (1977), **3-4**, 249

CALLAGHAN, I.C. & OTTEWILL, R.H. (1974), **1-1**, 15
CAMPBELL, R. (1983), **1-3**, 46
CAMPBELL, R. (1990), **1-3**, 46
CAMPBELL, R. & MARTIN, M.H. (1990), **1-3**, 46
CEBULA, D.J. & OTTEWILL, R.H. (1981), **1-1**, 15
CEBULA, D.J. et al. (1979), **1-1**, 15
CEDERBERG, G.A. et al. (1985), **5-13**, 588
CSMRS, (1986), **3-7**, 291
CHANDLER, R.J. & DAVIS, A.G. (1973), **3-8**, 308
CHANDLER, R.J. et al. (1992), **2-6**, 145
CHANDRA, S. & CHEN, G.L. (1985), **3-7**, 291
CHANDRA, S. & GARCIA, E.B. (1984), **3-7**, 291
CHARACKLIS, W.G. (1981), **4-2**, 366
CHARACKLIS, W.G. & MARSHALL, K.C. (1990), **4-2**, 366
CHENEY, J.E. (1988), **2-6**, 145
CHENG, B.T. (1972), **2-5**, 131
CHIGIRA, M. (1990), **3-9**, 320
CHIGIRA, M. & SONE, K. (1991), **3-9**, 320
CHOW, D.C. (1988), **3-10**, 326
CHRISTIAN, H.A. et al. (1990), **4-4**, 397
CLARK, C.J. & CRAWSHAW, D.H. (1979), **2-5**, 131
COCHET, C. & DERANGERE, D. (1990), **5-10**, 552
COLE, B.A. et al. (1977), **3-7**, 291
COLE, D.C.H. & LEWIS, J.G. (1960), **3-2**, 223
COLES, S.M. (1979), **4-4**, 397
COLIN, F. & MUNK-KOEFED, N. (1987), **5-10**, 552
COLIN, G. et al. (1986), **3-4**, 249
COLMER, A.R. et al. (1950), **2-1**, 74; **2-3**, 99
CONNOLLY, R.A. (1971), **3-4**, 249
CONNOLLY, R.A. (1972), **3-4**, 249
COOKE, R.U. (1981), **3-1**, 208
COOKE, R.U. & SMALLEY, I.J. (1968), **3-2**, 223
COOLING, L.F. & WARD, W.H. (1948), **2-6**, 145
COPE, F.W. (1987), **5-1**, 448
CORK, D.J. & KRUEGER, J.P. (1991), **1-3**, 46
CORRENS, C.W. (1949), **2-1**, 74; **3-2**, 223
COUMOULOS, D.G. (1977), **3-7**, 291
COUNCIL DIRECTIVE ON THE LANDFILL OF WASTE, (1991), **5-3**, 472
COVENEY, R.M. & PARIZEK, E.J. (1977), **2-1**, 74
COWARD, M.F. & JONES, G.W. (1931), **2-9**, 185

CRAGNOLINO G. & TUOVINEN, O.H. (1984), **4-2**, 366
CRATHORNE, B. & JAMES, C. (1990), **5-9**, 535
CREEDY, D.P. (1989), **2-9**, 185
CROFT, J.B. (1964), **4-6**, 428
CROOKS, V.E. & QUIGLEY, R.M. (1984), **5-1**, 448
CROSS, J.E. et al. (1987), **5-13**, 588
CULLIMORE, D.R. (1987), **5-14**, 605
CULLIMORE, D.R. (1990), **5-14**, 605
CULLIMORE, D.R. (1991), **5-7**, 510
CULLIMORE, D.R. (1992), **5-14**, 605
CULLIMORE, D.R. & NELSON, K. (1989), **5-7**, 510
CULLIMORE, D.R. et al. (1989), **5-7**, 509

DABORN, G.R. (1989), **4-4**, 397
DABORN, G.R. et al. (1992), **4-4**, 397
DADE, B.W. et al. (1990), **4-4**, 397
DANIEL, D.E. (1987), **5-1**, 448
DAVIS, G.W. (1988), **3-4**, 249
DE BOER, P.L. (1981), **4-4**, 397
DE COSTE, J.B. (1972), **3-4**, 249
DE SEGONZAC, G.D. (1970), **1-2**, 33
DECHO, A.W. (1990), **4-4**, 397
DECKER, R.S. & DUNNIGAN, L.P. (1977), **3-7**, 291
DEGENS, E.T. & MOPPER, K. (1975), **4-5**, 412
DOE, (1987), **5-9**, 535
DOE, (1991), **2-9**, 185
DOE, (1992), **2-9**, 185
DEXTER, S.C. (1976), **4-2**, 366
DEXTER, S.C. & GAO, G.Y. (1987), **4-2**, 367
DEXTER, S.C. et al. (1986), **4-2**, 367
DIAMOND, S. (1976), **2-7**, 158
DIN, (1978), **5-9**, 535
DIXON, J.B. et al. (1982), **3-9**, 320
DOMENICO, P.A. & SCHWARTZ, F.W. (1990), **5-13**, 588
DOORNKAMP, J.C. et al. (1980), **3-1**, 208
DREVER, J.I. (1982), **5-13**, 588
DRISCOLL, F.G. (1986), **5-15**, 614
DUMAS, TH. (1990), **5-8**, 523
DUMBLETON, M.J. (1962), **4-6**, 429

EDGAR, L.A. & PICKETT-HEAPS, J.D. (1984), **4-4**, 397
EDMUNDS, W.M. (1973), **4-3**, 381
EDMUNDS, W.M. & BATH, A.N. (1976), **4-3**, 381
EDMUNDS, W.M. & COOK, J.M. (1987), **3-3**, 232
EDWARDS, C. (1990), **1-3**, 46
EDYVEAN, R.G.J. (1984), **4-2**, 367
EDYVEAN, R.G.J. (1987), **4-2**, 367
EDYVEAN, R.G.J. & VIDELA, H.A. (1992), **4-1**, 352

EFIRD, K.D. & LEE, T.S. (1979), **4-2**, 367
EGGESTAD, Å. & SEM, H. (1976), **4-6**, 429
EGGINS, H.O.W. & OXLEY, T.A. (1980), **3-4**, 249
EGLINTON, M.S. (1975), **5-9**, 535
ELIAS, V. (1989), **3-4**, 249
ERDO, T. & TSURATA, T. (1969), **3-10**, 327
EVANS, J.D. & WILSON, A.C. (1992), **2-5**, 131
EVANS, L.S. (1970), **3-2**, 223
EVSTATIEV, D. (1965), **2-7**, 158

FANG, H.Y. (1985), **4-7**, 435
FASISKA, E. *et al.* (1974), **2-2**, 86
FATTUHI, N I. & HUGES, B.P. (1983), **5-9**, 535
FELMY, A.R. *et al.* (1984), **5-13**, 588
FIELDS, R.D. & RODRIGUEZ, F. (1975), **3-4**, 249
FLANAGAN, C.P. & OLMGREN, G.G.S. (1977), **3-7**, 291
FLIERMANS C.N. & BROCK, T.D. (1972), **2-1**, 74
FLIERMANS, C.B. & HAZEN, T.C. (1990), **5-14**, 605
FOLLENFANT, H.G. (1975), **3-3**, 232
FONSECA, M.S. (1989), **4-4**, 397
FOOKES, P.G. (1978), **2-1**, 74
FOOKES, P.G. & COLLIS, L. (1975), **2-1**, 74
FOOKES, P.G. & FRENCH, W.J. (1977), **3-1**, 209
FOOKES, P.G. *et al.* (1985), **3-1**, 209
FORD, J.E. (1986), **3-5**, 274
FORSYTHE, P. (1977), **3-7**, 291
FOSTER, R.H. & GAZZARD, I.J. (1975), **4-6**, 429
FRENCH, W.J. (1976), **3-1**, 209
FRENCH, W.J. (1978), **3-1**, 209
FRENCH, W.J. *et al.* (1982), **3-1**, 209
FRITZ, J.S. & YAMAMURA, S.S. (1955), **4-3**, 381
FROBISHER, M. (1968), **3-4**, 249
FROSTICK, L.E. & McCAVE, I.N. (1979), **4-4**, 397

GABRIELE, P.D. & IANNUCCI, R.M. (1984), **3-4**, 249
GARRELS, R.M. & CHRIST, C.L. (1965), **3-1**, 209; **4-3**, 381; **5-6**, 497
GAUDY, A. & GAUDY, E. (1980), **2-3**, 99
GAYLARDE, C.C. & VIDELA, H.A. (1987), **4-2**, 367
GEESEY, G.C. (1982), **4-2**, 367
GEHRELS, J. & ALFORD, G. (1990), **5-14**, 605
GHIORSE, W.T. & WILSON, J.T. (1988), **1-3**, 46
GIBBONS, A. (1989), **3-4**, 249

GIBBS, R.J. (1977), **1-2**, 33
GIBSON, A.H. (1985), **4-7**, 435
GILKES, P.W. *et al.* (1991), **2-5**, 131
GILLOTT, J.E. (1987), **4-6**, 429
GILLOT, J.E. *et al.* (1974), **2-2**, 86
GOLDSCHMID, V.M. (1937), **5-6**, 497
GOODALL, D.C. & QUIGLEY, R.M. (1977), **5-1**, 448
GOODMAN, G.T. & ROBERTS T.M. (1971), **2-5**, 131
GORDON, M.E. *et al.* (1989), **5-1**, 448
GRAINGER, P. (1984), **1-2**, 33
GRANT, J. & GUST, G. (1987), **4-4**, 397
GRATTAN-BELLOW, P.E. (1981), **5-12**, 567
GRAY, D.H. & OHASHI, H. (1983), **3-10**, 327
GREENWAY, D.F. (1987), **3-10**, 327
GRIFFIN, G.J.L. & URBIE, M. (1984), **3-4**, 249
GRIFFIN, J.J. *et al.* (1965), **1-2**, 33
GRIGOROVA, R. & NORRIS, J.R. (1990), **1-3**, 46
GRIM, R.E. (1962), **2-6**, 145; **5-16**, 622
GRIM, R.E. (1968), **4-6**, 429
GRIM, R.E. *et al.* (1937), **2-6**, 145
GRIMALDI, F.S. *et al.* (1966), **2-1**, 74
GUTT, W.H. & HARRISON, W.H. (1977), **5-9**, 535

HAMILTON, W.A. & SANDERS, P.F. (1984), **4-2**, 367
HAMILTON, W.A. *et al.* (1988), **4-2**, 367
HANDY, R. *et al.* (1959), **2-7**, 159
HANSBO, S. (1957), **4-4**, 397; **4-6**, 429
HARDY, J.A. & BOWN, J.L. (1984), **4-2**, 367
HARRISON, W.H. & TEYCHENNE, (1981), **3-1**, 209
HARRISON, W.H. (1977), **3-1**, 209
HARRISON, W.H. (1987), **3-1**, 209; **5-9**, 535
HARRISON, W.H. (1992), **5-9**, 535
HAWKINS, A.B. (1984), **4-6**, 429
HAWKINS, A.B. & HIGGINS, M. (1992), **2-1**, 74
HAWKINS, A.B. & McDONALD, C. (1992), **3-8**, 308
HAWKINS, A.B. & McDONALD, C. (1992), **3-8**, 308
HAWKINS, A.B. & PINCHES, G.M. (1986), **2-1**, 74
HAWKINS, A.B. & PINCHES, G.M. (1987a), **2-1**, 74; **2-2**, 86; **2-3**, 99; **2-4**, 109
HAWKINS, A.B. & PINCHES, G.M. (1987b), **2-1**, 74
HAWKINS, A.B. & PINCHES, G.M. (1988), **2-1**, 74
HAWKINS, A.B. & PINCHES, G.M. (1992), **1-2**, 33; **2-1**, 74

HAWKINS, A.B. & PINCHES, G.M. (1992), **2-4**, 109
HAWKINS, A.B. & WILSON, S.L.S. (1990), **2-2**, 86
HAWKINS, A.B. et al. (1986), **3-8**, 308
HAWKINS, A.B. et al. (1988), **3-8**, 308
HAWKINS, A.B. et al. (1988), **3-9**, 320
HAWKINS, A.B. et al. (1989), **4-5**, 402
HAWKINS, A.B. et al. (1991), **4-5**, 412
HAWORTH, A. et al. (1988 a), **5-13**, 588
HAWORTH, A. et al. (1988 b), **5-13**, 588
HEAF N.J. (1981), **4-2**, 367
HEALTH AND SAFETY EXECUTIVE, **2-9**, 185
HEAP, W.M. & MORRELL, S.H. (1968), **3-4**, 249
HEERTEN, G. (1986), **3-6**, 282
HEINZELMANN, C. & WALLISCH, S. (1991), **4-4**, 397
HELGESON, H.C. et al. (1970), **5-13**, 588
HEM, J.D. (1961), **4-3**, 381
HEM, J.D. (1971), **5-6**, 497
HENKEL, D.J. (1982), **5-16**, 622
HICKS, R.J. et al. (1990), **1-3**, 46
HILLMAN, P.F. (1990), **5-8**, 523
HILT, G.H. & DAVIDSON, D.T. (1960), **4-6**, 429
HMSO, (1990), **5-3**, 472
HO, C.S. (1975), **3-10**, 327
HO, C.S. (1986), **3-11**, 334
HOBBS, N.B. (1986), **4-5**, 412
HOECHST-CELANESE CORPORATION (1988), **3-4**, 249
HOFFMAN, R. (1992), **4-1**, 352
HOFMANN, V. et al. (1933), **2-6**, 145
HOLLAND, A.F. et al. (1974), **4-4**, 397
HOLM, G. et al. (1983), **4-6**, 429
HORTA, J.C. & DE, O.S. (1985), **3-2**, 223
HOWER, J. et al. (1976), **1-2**, 33
HOWSAM, P. (1988), **5-15**, 614
HOWSAM, P. (1990), **5-15**, 614
HOWSAM, P. (1991), **1-3**, 46
HUANG, J.-C., SHETTY, A.S. & WANG, M-S. (1990), **3-4**, 249
HUECK, H.J. (1974), **3-4**, 249
HUTCHINS, S.R. et al. (1986), **2-3**, 99
HUTCHINSON, T.C. (1980), **5-17**, 631

INGLES, O.G. & METCALF, J.B. (1972), **3-11**, 334
ISRAELACHVILI, J.N. & ADAMS, G.E. (1978), **1-1**, 16
IU, K.L. et al. (1981), **2-5**, 131

JACKSON, S.D. & CRIPPS, J.C. (1992), **2-2**, 86
JAILLOUX, J-M. & VERDU, J. (1990), **3-4**, 249
JAKUBOWSKI, J.A. et al. (1983), **3-4**, 249

JAMES, A.N. & LUPTON, A.R.R. (1978), **3-1**, 209
JANUSZKE, R.M. & BOOTH, E.H.S. (1984), **3-2**, 223
JEANS, C.V. (1978), **3-8**, 308
JENKYNS, H.C. (1980), **2-1**, 74
JENSEN, V. et al. (1986), **1-3**, 46
JOHNSEN, R. & BARDAL, E. (1986), **4-2**, 367
JOHNSON, S.W. & BLAKE, J.M. (1867), **2-6**, 145
JOHNSTON, T.A. & EVANS, J.D., (1985), **2-5**, 131
JONES, H.A. & NEDWELL, D.B. (1990), **2-9**, 185
JOSHI, R.C. et al. (1981), **4-6**, 429
JUKES-BROWN, A.J. (1902), **3-8**, 308
JURAN, R. (ed.) (1989), **3-4**, 249

KAPLAN, A.M. (1977), **3-4**, 250
KAPLAN, A.M. et al. (1970), **3-4**, 250
KELLY, D.P. et al. (1979), **2-1**, 74
KENNARD, M.F. et al. (1967), **2-5**, 131
KENNEDY, V.C. & ZELLWEGER, G.W. (1974), **4-3**, 381
KERSHASKY, J. (1987), **5-7**, 510
KÉZDI, A. (1979), **4-6**, 429
KHARAKA, Y.K. & BARNES, I. (1973), **5-13**, 588
KING, K.S. et al. (1992), **5-1**, 448
KING, R.A. (1981), **4-1**, 353
KING, R.A. & MILLER, J.D.A. (1971), **4-2**, 367
KING, R.A. et al. (1973), **4-2**, 367
KING, R.A. et al. (1976), **4-2**, 367
KINSMAN, D.J.J. (1926), **3-2**, 223
KIROVE, B. (1989), **4-7**, 435
KLAUSMEIER, R.E. (1972), **3-4**, 250
KLAUSMEIER, R.E. & JONES, W.A. (1961), **3-4**, 250
KOCH, A. & STEINEGGER, H. (1960), **5-9**, 535
KOERNER, R.M. (1986), **3-5**, 275
KOERNER, R.M. (1990), **3-4**, 250
KRAUSKOPF, K.B. (1967), **2-5**, 131
KRAUSKOPF, K.B. (1979), **5-6**, 497
KRAUSKOPF, K.B. (1985), **5-16**, 622
KRAVITZ, J.M. (1970), **4-4**, 397
KRUMBEIN, W.E. (1978), **1-3**, 47
KUENEN, J.G. & TUOVINEN, O.H. (1981), **2-1**, 75
KUJALA, K. & NIEMINEN, P. (1983), **4-6**, 429
KUJALA, K. (1983), **4-6**, 429
KULIG, A. (1989), **5-17**, 631
KUSHNER, D.J. (1978), **1-3**, 47
KUSTER, E. (1979), **3-4**, 250
KUZNETSOVA, T. et al. (1989), **2-7**, 159

LAGARO, R.C. & MOH, Z.C. (1970), **3-11**, 334

LAGUROS, J. & DAVIDSON, D. (1963), **2-7**, 159
LAKE, R.D. (1987), **3-3**, 232
LAMBE, T. et al. (1959), **2-7**, 159
LAMBE, T.W. & WHITMAN. R.V. (1970), **4-4**, 398
LAMING, D.J.C. (1982), **3-8**, 308
LANE, K.S. & WASHBURN, D.E. (1946), **2-2**, 87
LANGELIER, W.F. (1936), **5-13**, 588
LANGMUIR, D. (1965), **4-3**, 381
LASAGA, A.C. (1984), **5-13**, 588
LAWRENCE, C.D. (1990), **5-9**, 535
LAZAR, V. (1975), **3-4**, 250
LEA, F.M. (1970), **3-1**, 209
LEE, D.H. & TEN, Y.M. (1987), **3-11**, 334
LEONARDS, G.A. et al. (1991), **3-7**, 291
LEVINSON, A.A. (1980), **5-6**, 497
LICINA, G.J. (1989), **4-2**, 367
LITTLE, C. et al. (1992), **4-4**, 398
LOCHER, F.W. & SPRUNG, S. (1975), **3-1**, 209
LOGANI, K.L. & HECTOR, M. (1979), **3-7**, 291
LOVE, L.G. et al. (1984), **2-1**, 75
LOW, P.F. (1987), **1-1**, 16
LUBETKIN, S.D.et al. (1984), **1-1**, 16
LUNDGREN, D.G. & SILVER, M. (1980), **2-3**, 99
LYBIMOVA, T. & IAGODOVSKAYA, T. (1961), **2-7**, 159
LYNCH, J.L. & EDYVEAN, R.G.E. (1988), **4-2**, 367
LYONS, J.S. et al. (1981), **1-1**, 16

MACDONALD, A. & REID, J.M. (1991), **2-5**, 131
MACLACHLAN, J. et al. (1966), **3-4**, 250
MACLEAN, R.D. (1966), **3-3**, 232
MALECKA, D. (1991), **5-17**, 631
MALLORY, L.M. & SAYLER, G.S. (1984), **5-7**, 510
MANCINELLI, R.L. et al. (1981), **2-9**, 185
MANSFIELD, F. et al. (1990), **4-2**, 367
MANZENRIEDER, H. (1983), **4-4**, 398
MARSH, T.J. & DAVIES, P.A. (1983), **3-3**, 232
MARSHALL, R.A. (1990), **3-4**, 250
MARTIN, C. (1988), **2-5**, 131
MASON, B. (1966), **5-6**, 497
MASON, E. (1989), **3-1**, 209
MATHEWS, J.D. (1987), **5-9**, 535
MATHIEU, A. & SOUKATCHOFF, P. (1987), **5-10**, 552
MATHIEU, M. et al. (1989), **5-10**, 552
MATHUR, S.P. & ROUATT, J.W. (1975), **3-4**, 250
MCDANIEL, T.N. & DECKER, R.S. (1979), **3-7**, 291
MEADOWS, P. & TAIT, J. (1989), **4-4**, 398
MEADOWS, P. et al. (1990), **4-4**, 398

MEHTA, A.J. & PARTHENIADES, E. (1982), **4-4**, 398
MEHTA, P.K. (1982), **5-11**, 558
MEHTA, P.K. (1987), **5-9**, 535
MEHTA, P.K. & GJORV, O.E. (1974), **5-9**, 535
MEIGH, A.C. (1976), **3-8**, 308
MELVILL, A.L. & MACKELLAR, D.C.R. (1980), **3-7**, 291
MENZEL, B. et al. (1990), **3-4**, 250
MERRILL, D.T. (1976), **5-13**, 588
MILDE, K. et al. (1983), **5-8**, 523
MILLER, D. & BENSON, L. (1983), **5-13**, 588
MILLER, J. (1988), **3-9**, 320
MILLER, J.D.A. (1971), **1-3**, 47
MILLER, L.P. (1949), **4-2**, 367
MILLER, M.C. et al. (1977), **4-4**, 398
MILLS, J. & EGGINS, H.O.W. (1974), **3-4**, 250
MINER, R.J. (1972), **3-4**, 250
MISH, F.C. (1986), **3-4**, 250
MITCHELL, J.K. (1981), **4-6**, 429
MITCHELL, J.K. & MADSEN, F.T. (1987), **4-7**, 435
MITSUNASHI (1976), **3-9**, 320
MOE, (1984a), **5-4**, 479
MOE, (1984b), **5-4**, 479
MOH, Z.C. (1962), **2-7**, 159
MOLLICA, A. et al. (1984), **4-2**, 368
MONTAGUE, C.L. (1986), **4-4**, 398
MOORE, W.L. & MASCH, F.D. (1962), **4-4**, 398
MORGAN, D.J. (1978), **2-6**, 145
MORTLAND, M.M. (1970), **4-5**, 412
MOSEBACH, R. (1951), **3-2**, 223
MOSS, C.W. et al. (1980), **5-7**, 510
MOUM, J. & ROSENQVIST, I. (1959), **2-1**, 75
MOUM, M.J. et al. (1968), **4-6**, 429
MTRCA, (1988), **5-4**, 479

NABARRO, F.R.N. & JACKSON, P.J. (1958), **3-2**, 223
NADEAU, P.H. et al. (1984), **1-2**, 33
NAKAMURA, T. et al. (1968), **5-11**, 558
NARULA, P.L. & SHARDA, Y.P. (1990), **3-7**, 291
NETTERBERG, F. (1970), **3-1**, 209
NETTERBERG, F. (1979), **3-1**, 209
NETTERBERG, F. (1984), **3-1**, 209
NETTERBERG, F. et al. (1974), **3-1**, 209; **3-2**, 224
NETTERBERG, F. et al. (1987), **3-1**, 209
NEUMANN, A.C. et al. (1970), **4-4**, 398
NICHOLLS, R. (1952), **2-7**, 159
NIEMINEN, P, (1978), **4-6**, 429
NILSON, S. (1991), **5-7**, 510
NINHAM, B.W. (1985), **1-1**, 16
NIPO (1988), **5-10**, 552

NIXON, P.J. (1978), **2-1**, 75; **2-2**, 87; **2-4**, 109
NORDSTROM, D.K. & BALL, J.W. (1984), **5-13**, 588
NORDSTROM, D.K. et al. (1979), **5-13**, 588
NW CHINA HIDI, (1986), **5-5**, 487
NOWELL, A.R.M. et al. (1981), **4-4**, 398
NOWELL, A.R.M. & JUMARS, P.A. (1987), **4-4**, 398
NOWELL, A.R.M. et al. (1989), **4-4**, 398

O'BRIEN, N.R. & SLATT, R.M. (1990), **1-2**, 33
OBIKA, B. & FREER-HEWISH, R.J. (1988), **3-2**, 224
OBIKA, B. et al. (1985), **3-1**, 210
OBIKA, B. et al. (1989), **3-2**, 224; **5-13**, 589
OPPENHEIM, D.R. & PATERSON, D.M. (1990), **4-4**, 398
OSBORNE, G.J. (1986), **5-9**, 536
OSBORNE, G.J. (1989), **5-9**, 536
OSBORNE, G.J. (1991), **5-9**, 536
OSMAN, J.L. et al. (1971), **3-4**, 250
OSMAN, J.L. et al. (1972), **3-4**, 250
OTTEWILL, R.H. (1991), **1-1**, 16

PAGENKOPF, G.D. (1978), **5-6**, 497
PAKHOMOV, S.I. & YEGOROV, YU. K. (1990), **4-7**, 436
PANKHURST, E.S. et al. (1971), **3-4**, 250
PARCHURE, T.M. (1984), **4-4**, 398
PARKHURST, D.L. et al. (1980), **5-13**, 589; **5-15**, 614
PATERSON, D.M. (1989), **4-4**, 398
PATERSON, D.M. (1990), **4-4**, 398
PATERSON, D.M. et al. (1986), **4-4**, 398
PATERSON, D.M., et al. (1990), **4-4**, 399
PAUL, M.A. et al. (1992), **4-5**, 412
PAULIC, M. et al. (1986), **4-4**, 399
PAULING, L. (1930), **2-6**, 145
PAVILONSKY, V.M. (1985), **4-7**, 436
PENNER, E. et al. (1973), **2-1**, 75; **2-2**, 87; **2-4**, 109
PERELMAN, A.I. (1977), **5-6**, 497
PETERSON, S.R. & GEE, G.W. (1985), **4-7**, 436
PHYPER, J-D. & IBBOTSON, B. (1991), **5-4**, 479
PININSKA , J. (1982), **5-17**, 631
PININSKA, J. & LUKASZEWSKI, P. (1991), **5-17**, 631
PITZER, K.S. (1973), **5-13**, 589
PLUMMER, L.N. et al. (1975), **5-13**, 589
PLUMMER, L.M. et al. (1976), **5-13**, 589
POLDER, R. & VAN MECHELEN, T. (1989), **5-8**, 523; **5-10**, 552
POPE, D.H. (1986), **4-2**, 368
POTTER, P.E. et al. (1980), **1-2**, 33; **2-2**, 87

POTTS, J.E. et al. (1972), **3-4**, 250
POWERS, J.P. (1981), **5-15**, 614
POWRIE, W. et al. (1990), **5-15**, 614
PRIMMER, T.J. & SHAW, H.F. (1987), **1-2**, 33
PRINCE, W. (1987), **5-11**, 558; **5-12**, 567
PRINCE, W. & PERAMI, R. (1990), **5-11**, 558
PRINCE, W. & PERAMI, R. (1991), **5-12**, 567
PYE, K. & MILLER, J.A. (1990), **2-2**, 87; **2-3**, 99; **3-9**, 320

QUEIROZ DE CARVALHO, J.B. (1981), **4-6**, 429
QUIGLEY, R.M. & ROWE, R.K. (1986), **5-1**, 448
QUIGLEY, R.M. & VOGAN, R.W. (1970), **2-1**, 75; **2-2**, 87; **2-4**, 109
QUIGLEY, R.M. et al. (1973), **2-1**, 75
QUIGLEY, R.M. et al. (1982), **5-2**, 462
QUIGLEY, R.M. et al. (1987a), **5-1**, 448
QUIGLEY, R.M. et al. (1987b), **5-1**, 448

RACKLEY, F.A. & TILLER, A.K. (1988), **4-1**, 353
RADSTAKE, F. et al. (1988), **5-13**, 589
RAINEY, T.P. & ROSENBAUM, M.S. (1989), **3-3**, 232
RAINWATER, F.H. & THATCHER, L.L. (1960), **4-3**, 381
RAISWELL, R. & BERNER, R.A. (1986), **2-2**, 87
RAMACHANDRAN, V.S. et al. (1981), **3-11**, 334
RANKILOR, P.R. (1981), **3-4**, 250; **3-5**, 275; **3-6**, 282
RANKILOR, P.R. (1986), **3-5**, 275; **3-6**, 282
RANKILOR, P.R. (1988), **3-5**, 275
RANKILOR, P.R. (1989), **3-5**, 275; **3-6**, 282
RANKILOR, P.R. (1990), **3-5**, 275; **3-6**, 282
READ, D. et al. (1990), **5-13**, 589
REGOURD, M. et al. (1982), **5-12**, 567
RICE, R.E. (1986), **5-13**, 589
RILEY, J.P. (1958), **2-1**, 75
RISSEEUW, P. & SCHMIDT, H.M. (1990), **3-4**, 251
ROBERTS, T.O.L. & PREENE, M. (1989), **5-15**, 614
ROBINS, N.S. & MILODOWSKI, A.E. (1986), **3-3**, 232
ROGERS, C.D.F. & BRUCE, C.J. (1991), **4-6**, 429
RONBEN, C. (1979), **3-1**, 210
ROSE, A.H. (1981), **1-3**, 47
ROSLER, H.J. & VE LANGE, H. (1972), **5-6**, 497

ROSSUM, J.R. & MERRILL, D.T. (1983), **3-1**, 210; **5-13**, 589
ROY, M. (1988), **5-7**, 510
RUNNELLS, D.D. (1987), **5-13**, 589
RUSSEL, D.J. & PARKER, A. (1979), **2-2**, 87; **2-3**, 99; **3-9**, 320
RYZNAR, J.W. (1944), **5-13**, 589

SAHINCI, A. (1986), **5-6**, 497
SAMPSON, L.R. et al. (1987), **3-1**, 210
SAND, W. (1987), **5-10**, 552
SAND, W. et al. (1987), **5-8**, 523
SAX, N.I. & LEWIS, R.J. (1987), **3-4**, 251
SCHNEIDER, H. (1987), **3-5**, 275
SCHNEIDER, H. & GROTH, M. (1987), **3-4**, 251
SCHOELLER, H. (1962), **5-6**, 497
SCHREMMER, H. (1990), **5-8**, 523
SCHWIETE, H.E. et al. (1983), **5-11**, 558
SCOFFIN, T.P. (1970), **4-4**, 399
SCOTTO, V. et al. (1985), **4-2**, 368
SCULLION, J. & EDWARDS, R.W. (1980), **2-5**, 131
SEAL, K.J. & EGGINS, H.O.W. (1981), **3-4**, 251
SEAL, K.J. & PANTKE, M. (1988), **3-4**, 251
SEAMAN, R.N. & VENKATARAMAN, B. (1976), **3-4**, 251
SHANK, B. (1987), **5-7**, 510
SHAPIRO, O. & MAGIER, L. (1991), **3-4**, 251
SHAW, H.F. & PRIMMER, T.J. (1991), **1-2**, 33
SHAW, H.F. (1981), **1-2**, 33
SHEARMAN, D.J. et al. (1972), **3-8**, 308
SHERARD, J.L. et al. (1972), **3-7**, 292
SHERARD, J.L. et al. (1976a), **3-7**, 292
SHERARD, J.L. et al. (1976b), **3-7**, 292; **5-16**, 622
SHERRELL, F.W. (1979), **2-2**, 87
SHVARTSEV, S.L. et al. (1975), **5-6**, 497
SILVERMAN, M.P. & LUNGREN, D.G. (1959), **2-3**, 99
SINGER, A. (1980), **1-2**, 33
SINGH, N. & OJHA, P. (1981), **2-7**, 159
SKEMPTON, A.W. (1953), **4-5**, 412
SKEMPTON, A.W. & PETLEY, D.J. (1970), **4-5**, 412
SMITH, K.A. & ARAH, J.R.M. (1985), **1-3**, 47
SMITH, M.A. (1984), **3-1**, 210
SMITH, M.A. (1985), **5-9**, 536
SMITH, S.A. (1992), **5-14**, 605
SOTTON, M. et al. (1982), **3-4**, 251
SPEARS, D.A. & TAYLOR, R.K. (1972), **3-9**, 320
SPEARS, D.A. et al. (1971), **2-5**, 131
SPERLING J. & COOKE, R.U. (1985), **3-1**, 210

SPERLING, C.H.B. & COOKE, R.U. (1985), **3-2**, 224
SPOSITO, G. & MATTIGOD, S.V. (1980), **5-13**, 589
SPOSITO, G. (1985), **5-13**, 589
SRODON, J. et al. (1986), **2-6**, 145
SAA, (1980), **5-16**, 622
STEWARD, H.E. & CRIPPS, J.C. (1983), **2-3**, 99
STIEF, K. (1990), **5-3**, 472
STOTT, J.F.D. (1988), **5-15**, 615
STUMM, W. & LEE, G.F. (1961), **2-5**, 131
STUMM, W. VE & MORGAN, J.J. (1970), **5-6**, 497
SZKLARSKA-SMIALOWSKA, Z. (1986), **4-2**, 368

TABER, J. (1916), **3-2**, 224
TABOR, D. & WINTERTON, R.H.S. (1969), **1-1**, 16
TAYLOR, R.K. & CRIPPS, J.C. (1984), **2-1**, 75; **2-2**, 87
TAYLOR, R.K. (1984), **2-5**, 131
TEME, S.C. (1990a), **4-3**, 382
TEME, S.C. (1988), **4-3**, 381
TEME, S.C. (1990b), **4-3**, 382
TEMPLE, K.L. & COLMER, A.R. (1951), **2-3**, 99
TEMPLETON, A.R. (1988), **5-7**, 510
TERRY L.A. & EDYVEAN, R.G.J. (1981), **4-2**, 368
TERRY L.A. & EDYVEAN, R.G.J. (1986), **4-2**, 368
THENG, B.K.G. (1974), **4-5**, 412
THISTLETHWAYTE, K.B. (1972), **5-8**, 524; **5-10**, 552
THOMAS, C.J. et al. (1988), **4-2**, 368
TILLER, A.K. (1988a), **4-1**, 353
TILLER, A.K. (1988b), **4-2**, 368
TILLER, A.K. (1990), **4-1**, 353
TIRPAK, G. (1970), **3-4**, 251
TOLANSKY, S. (1948), **1-1**, 16
TOMBERS, J. (1991), **5-17**, 631
TOMLINSON, M.J. (1980), **4-3**, 382
TRRL, (1952), **2-6**, 145
TSUCHII, A., et al. (1990), **3-4**, 251
TSYTOVICH, N.A. et al. (1971), **4-6**, 430

UNDERWOOD, J.C. et al. (1992), **4-4**, 399
UPSHER, F.J. & ROSEBLADE, R.J. (1984), **3-4**, 251

VALOCCHI, A.J. (1985), **5-13**, 589
VALOCCHI, A.J. et al. (1981), **5-13**, 589
VAN BEEK, C.G.E. (1989), **5-15**, 615
VAN BREEMON, N. (1988), **2-1**, 75
VAN IMPE, W.F. (1989), **4-6**, 430
VAN MECHELEN, A.C.A. & POLDER, R.B. (1990), **5-8**, 524
VAN OLPHEN, H. (1963), **1-1**, 16

VAUX, R. (1988), **4-2**, 368
VERWEY, E.J.W. & OVERBEEK, J.TH G. (1948), **1-1**, 16
VIDELA, H.A. (1989a), **4-2**, 368
VIDELA, H.A. (1989b), **4-2**, 368
VIDELA, H.A. *et al.* (1988), **4-2**, 368
VIDELA, H.A. *et al.* (1989), **4-2**, 368
VISHNIAC, W.V. (1974), **2-1**, 75; **2-4**, 109
VON PETTENKOFER, M. (1871), **2-9**, 185
VOS, P.C. *et al.* (1988), **4-4**, 399
VROM, (1991), **5-3**, 472

WALDRON, L.J. (1977), **3-10**, 327
WANG, B.W. (1984), **5-2**, 462
WANG, Y. (1970), **3-11**, 334
WARD, R.C. (1975), **3-2**, 224
WARITH, M.A. (1987), **5-2**, 462
WARRINGTON, G. & SCRIVENER, R.C. (1980), **3-8**, 308
WEINERT, H.H. & CLAUSS, M.A. (1967), **3-2**, 224
WELLMAN, H.W. & WILSON, A.T. (1965), **2-2**, 87; **3-2**, 224
WEYL, P.K. (1959), **3-2**, 224
WHITFIELD, M. (1971), **4-3**, 382
WHITTAKER, A. (1972), **3-8**, 308
WIGLEY, T.M.L. & PLUMMER, L.N. (1976), **5-13**, 590
WIGLEY, T.M.L. (1978), **5-13**, 589
WILKINSON, T.G. (1983), **4-2**, 368
WILLIAMS, G.R. & DALE, R. (1983), **3-4**, 251

WILLIAMS, P.L. *et al.* (1969), **3-4**, 251
WILSON, A. & EVANS, J.D. (1991), **2-5**, 131
WINKLER, E.M. & SINGER, P.C. (1972), **2-1**, 75; **2-2**, 87
WINKLER, E.M. (1975), **3-2**, 224
WOLERY, T.J. (1983), **5-13**, 590
WOLERY, T.J. (1989), **5-13**, 590
WOOD, D.M. (1985), **4-6**, 430
WOODWARD, H.B. & USSHER, W.A.E. (1911), **3-8**, 308
WOOLMINGTON, A.D. & DAVENPORT, J. (1983), **4-2**, 368
WRIGLEY, N. (1987), **3-6**, 282
WU, C.M. (1967), **3-10**, 327
WU, T.H. *et al.* (1988a), **3-10**, 327
WU, T.H. *et al.* (1988b), **3-10**, 327

XIAN-GONG, Z. et al, (1986), **5-5**, 487

YANFUL, E.K. & QUIGLEY, R.M. (1990), **5-1**, 449
YANFUL, E.K. *et al.* (1988), **5-1**, 449
YEN, F.S. & TSAI, Y.H. (1985), **3-10**, 327
YONG, R.N. & WARKENTIN, B.P. (1975), **5-2**, 462
YONG, R.N. *et al.* (1986), **5-2**, 462
YONG, R.N. *et al.* (1990), **5-2**, 462
YOUNG, R.N. & SOUTHARD, J.B. (1978), **4-4**, 399

ZARUBA, Q. (1976), **5-5**, 487